COMPUTATIONAL METHODS FOR HEAT AND MASS TRANSFER

Series in Computational and Physical Processes in Mechanics and Thermal Sciences
(Formerly the Series in Computational Methods in Mechanics and Thermal Sciences)

W. J. Minkowycz and E. M. Sparrow, Editors

Anderson, Tannehill, and Pletcher, Computational Fluid Mechanics and Heat Transfer

Aziz and Na, Perturbation Methods in Heat Transfer

Baker, Finite Element Computational Fluid Mechanics

Baker, Finite Element Computational Fluid Mechanics, Second Edition

Beck, Cole, Haji-Sheikh, and Litkouhi, Heat Conduction Using Green's Functions

Carey, Computational Grids

Comini, del Giudice, and Nonino, Finite Element Analysis in Heat Transfer

Heinrich and Pepper, Intermediate Finite Element Method: Fluid Flow and Heat Transfer Application

Jaluria, Computer Methods for Engineering

Koenig, Modern Computational Methods

Majumdar, Computational Methods for Heat and Mass Transfer

Patankar, Numerical Heat Transfer and Fluid Flow

Pepper and Heinrich, The Finite Element Method

Pepper and Heinrich, The Finite Element Method, Second Edition

Shih, Numerical Heat Transfer

Shyy, Udaykumar, Rao, and Smith, Computational Fluid Dynamics with Moving Boundaries

Tannehill, Anderson, and Pletcher, Computational Fluid Mechanics and Heat Transfer, Second Edition

PROCEEDINGS

Chung, *Editor*, Finite Elements in Fluids

Chung, *Editor*, Numerical Modeling in Combustion

Haji-Sheikh, *Editor*, Integral Methods in Science and Engineering

Shih, *Editor*, Numerical Properties and Methodologies in Heat Transfer

COMPUTATIONAL METHODS FOR HEAT AND MASS TRANSFER

PRADIP MAJUMDAR

Taylor & Francis
Taylor & Francis Group
New York London

Reprint 2017

CRC Press
Taylor & Francis Group
6000 Broken Sound Parkway NW, Suite 300
Boca Raton, FL 33487-2742

ISBN 978-1-1380-4486-9

Library of Congress Cataloging-in-Publication Data

Computational methods for heat and mass transfer / by Pradip Majumdar.
p. cm. -- (Series in computational and physical processes in mechanics and thermal sciences)
Includes bibliographical references and index.
ISBN 1-56032-994-7 (acid-free paper)
1. Fluid mechanics--Mathematical models. 2. Heat--Transmission--Mathematical models. 3. Mass transfer--Mathematical models. I. Title. II. Series.

QA901.M346 2005
532'.001'5118--dc22 2005005302

Visit the Taylor & Francis Web site at
http://www.taylorandfrancis.com

and the CRC Press Web site at
http://www.crcpress.com

Printed in India by Saurabh Printers Pvt. Ltd.

For sale in India, Pakistan, Nepal, Bhutan, Bangladesh and Sri Lanka only.

CONTENTS

To my
parents, Snehalata and Rati Ranjan
wife, Srabani
children, Diya and Ishan

PREFACE

With the availability of faster computers, there is a growing demand to find graduating engineers with the basic skills of computational methods for heat transfer, mass transfer, and fluid dynamics. Computational fluid dynamics and heat transfer and finite element codes are becoming standard tools in the computer-aided design and analysis of processes and products involving coupled transport and multi-physic phenomena. This book introduces the fundamentals of the *finite difference–control volume* and *finite element* methods, two important computational techniques for solving heat and mass transfer and fluid flow problems. Theory and procedures are presented in a simple and straightforward manner with detailed derivations and illustrations. Numerous classic problems in heat transfer, mass transfer, and fluid flow are included throughout the book to help students understand the concepts and procedures thoroughly, express computational methodology in computer code, and solve more complex problems on their own.

The book is organized in three parts. Part I, Basic Equations and Numerical Analysis, presents reviews of basic equations of heat transfer, mass transfer, and fluid dynamics in chapter 1; basic concepts of numerical approximations and errors in chapter 2; numerical solution techniques for systems of linear algebraic equations in chapter 3; and numerical integrations and quadrature formulas in chapter 4. Part II, Finite Difference–Control Volume Method, discusses the finite difference–control volume method in chapters 5 through 8. Two different approaches for discretization procedures are presented, one based on finite difference formulas derived using truncated Taylor series expansions and the other based on integrating a control volume. Part III, Finite Element Method, is comprised of chapters 9 through 14. In these chapters, both the Galerkin and variational methods of finite element formulation are explained. Basic concepts of the calculus of variations and variational method of approximations are also presented for clear understanding of the finite element method. The topics of the finite difference–control volume and finite element methods cover basic theory and examples dealing with one-dimensional steady state diffusion problems, multi-dimensional problems, transient or unsteady state problems, and convection problems. Solution procedures are worked out for a number of problems in hand calculations to demonstrate the solution algorithm. Since it is expected that students may use different programming languages, no programs are included; rather, the algorithms are presented in the form of pseudo-code for students to express them using their choice of programming language. End-of-chapter problems are included to help students gain further understanding of the techniques.

This book is primarily written as an introductory text for senior undergraduate and first-year graduate level courses, but it is comprehensive enough to be used as a reference book for independent study by students. However, practicing engineers and scientists involved in developing code and using commercial code for the analysis and design of the processes and products involving heat transfer, mass transfer, and fluid dynamics will also find the text useful.

It is not the intention of this book for all fourteen chapters to be covered in a first course at either the senior undergraduate or first-year graduate level. chapter 1, a review of basic governing equations and boundary conditions, may be assigned to students as a self-study reading assignment. chapters 2, 3, and 4 are included primarily for students who have had no prior course in numerical analysis. The material presented in chapter 3 is quite comprehensive, and is presented as a reference for advanced computing. It is not expected that all topics presented in chapter 3 be covered in the first course.

I would like to express my sincere gratitude and thanks to Prof. E. M. Sparrow and Prof. W. J. Minkowycz for their extremely valuable comments, advice, and encouragement. I would also like to thank other reviewers for their constructive comments, and Summers Scholl for her unflagging editorial support, patience, and understanding.

The book evolved from several years of teaching a course on computational heat transfer and fluid dynamics. I would like to thank the students who used the notes that led to the present manuscript, and students who helped in preparing the solutions to many example problems. Thanks are also due to Joy Hadley for her typing skills.

Finally, I wish to express my appreciation to my wife, Srabani, and children, Diya and Ishan, for their unlimited support, understanding, and patience during the long hours of preparation of the manuscript. They were a constant source of motivation for me to continue and complete the manuscript.

NOMENCLATURE

A	area; coefficient matrix; area of a finite element
$[A]$	coefficient matrix
$\vec{A}$	area vector
a	elements of a matrix; absorption coefficient
a's	coefficients in discretization equations given by the control volume method
a, b, c, d	vectors in tridiagonal matrix algorithm
$\vec{a}$	acceleration vector
C	capacitance in unsteady diffusion equation 7.1
c	mass concentration of species; known vector
$\lfloor c^e \rfloor$	local capacitance matrix
$\{c\}$	known vector
c_p	specific heat at constant pressure
c_v	specific heat at constant volume
D	diffusivity or diffusion coefficient
$[D]$	matrix with elements defined by the derivatives of shape function
E	total energy; emissive power in radiation heat transfer; error
e	specific energy; residual
F	force; shape factor in radiation heat transfer
$\vec{F}$	force vector
$\{F\}$, $\mathbf{F}$	force vector in global system of equations in finite element method
f^e	element force vector in finite element method
f	function; force vector
f''	surface of boundary flux
f_0	Fourier number; stability parameter in heat transfer
g	acceleration due to gravity

H	total enthalpy; Householder matrix; height
h	enthalpy; step size; convection transfer coefficient or film coefficient, film thickness
h_c	convection heat transfer coefficient
h_D, h_m	convection mass transfer coefficient
I	integrals; functional in variational method; power intensity
I_0	power intensity at $x=0$
$\{I\}$	vector for inter-element contributions
$\hat{i}, \hat{j}, \hat{k}$	unit vectors in x, y, and z directions
i	specific internal energy
i, j, k, l	indices for nodal points representing x, y, z, and t positions; represents nodes of a finite element
$\mathbf{J}$	Jacobian matrix
j	diffusive mass flux
$[K]$, $\mathbf{K}$	global matrix in finite element method
k	thermal conductivity
$\lfloor k^e \rfloor$, k^e	element characteristic matrix in finite element method
$[k_\Gamma]$, k_D	element characteristic matrix due to diffusion term
$[k_I]$	element characteristic matrix due to inertia term
$[k_p]$	element characteristic matrix due to pressure term
$[k_s]$	element characteristic matrix due to linearized source term
$[k_b]$	element characteristic matrix due to boundary condition
L	length; lower triangular matrix
L_e	element length
Le	Lewis number
L_i, L_j, L_k	line coordinates in two-dimensional finite element method
l	elements of lower triangular matrix
l_i, l_j, l_k	line coordinates in one-dimensional finite element method
l_e, h_e	length and height of two-dimensional finite element
M	molecular weight
$\dot{m}$	mass flow rate; mass generation or consumption
$\dot{m}'''$	volumetric mass generation or consumption
$\dot{m}''$	mass generation or consumption per unit area
m_i	mass fraction of species i
N	shape function in finite element method
$[N]$	row vector of shape function
Nu	Nusselt number
NX, NY, NZ	number of nodes in x, y, and z directions
n_p	number of nodes in an element
$\hat{n}$	unit vector in normal direction
P	perimeter; Peclet number, constants in discretization equations given by control volume method
P_i, Q_i	constants TDMA method
p	pressure

Q	orthogonal matrix in QR factorization method
$\dot{Q}$	heat transfer rate across system boundary
$\dot{Q}'''$	volumetric heat generation
q	heat transfer rate
q_R	radiation heat transfer
q''	heat flux or heat transfer rate per unit area
R	resistance; reminder term in Taylor series; residual
r, θ, z	cylindrical coordinate system
r, θ, ϕ	spherical coordinate system
Re	Reynolds number
S	source
$\bar{S}$	average source term
S_0, S_1	constants in linearized source term
Sh	Sherwood number
Sc	Schmidt number
T	temperature
t	time
U	upper triangular matrix
$\{U^e\}$	element unknown vector
$\{U\}$	global unknown vector
u, v, w	velocity components in x, y, and z directions
u_r, u_θ, u_z	velocity components in r, θ, and z direction
u_r, u_θ, u_ϕ	velocity components in r, θ, and ϕ direction
u	elements of upper triangular matrix
V	volume, volume of three-dimensional finite element
$\vec{V}$	velocity vector
$\dot{\forall}$	volume flow rate
W	width
w_i	weighting factors in Gauss–Legendre quadrature
X	extensive property in equation 1.29
x, y, z	Cartesian coordinate system
x	extensive property per unit mass in equation 1.29; mole fraction, unknown vector
$\{x\}$	unknown vector
x_i	mole fraction of species
x'_i	function points in Gauss–Legendre quadrature
y	vector in LU decomposition method

Greek Letters

Γ	transport property
$[\Gamma]$	transport property matrix
Λ	relaxation parameter
Δx	step size; grid size; mesh size
α	thermal diffusivity

β	eigenvalue
λ	weighting factor in time approximation scheme
γ_0	stability parameter
ε	extensional strain; emissivity factor; constant in equation 1.155
ε_t	true relative error; true percent relative error
ε_a	approximate error; approximate percent relative error
ε_s	specified tolerance limit
θ	angular rotation; dimensionless temperature
ϕ	dependent variable
ψ	stream function; approximation function in variational method
ρ	density
σ	normal stress; Stefan–Boltzman constant
τ	tangential or shear stress
δ	operator in variational method
$\vec{\omega}$	rotation vector
μ	molecular viscosity
ν	kinematic viscosity; atomic diffusion volume in equation 1.154
$\vec{\zeta}$	vorticity vector
Φ	viscous dissipation
ξ, η, ζ	natural coordinates in finite element method

Subscript

ad	adsorption
b	boundary; bottom
C	cold
D	diffusion
H	hot
l	left
nb	neighbor nodes
r	right
s	surface; source
t	top
w	wall
∞	ambient

Superscript

*	estimated or approximate value
k	index for iteration
/	corrected value
l or m	index for time increment

Part I
BASIC EQUATIONS AND NUMERICAL ANALYSIS

Chapter

1

REVIEW OF BASIC LAWS AND EQUATIONS

The numerical solutions of heat and mass transfer start from the basic rate and conservation equations for these processes. In this chapter we will give a brief review of the fundamental concepts, principles and basic equations in fluid dynamics, heat and mass transfer. Knowledge and understanding of the basic concepts and principles are essential to analyze these processes using computational techniques.

1.1 BASIC EQUATIONS

Analysis of fluid flow and heat transfer starts with the statement of the basic conservation laws. This includes: (a) conservation of mass, (b) Newton's second law of motion or conservation of momentum, and (c) first law of thermodynamics. Also, we need additional relations such as the equation of state or constitutive relations. These basic laws are the same as those in mechanics and thermodynamics. However, these laws are reformulated in a manner suitable for fluid flows, heat and mass transfer.

We can formulate the basic laws in terms of infinitesimal or finite systems and control volumes. Formulation based on infinitesimal system and control volumes leads to ***differential formulation***. Solution of the differential formulation gives detailed knowledge or point-to-point behavior of fluid flows and heat transfer, whereas the ***integral formulation*** that results from the finite system and control volume gives gross or average behavior. In many fluid flow and heat transfer problems, we may additionally apply the integral formulation to check the accuracy of the computational solution of the differential equations by observing the overall balance of mass, momentum, and energy.

1.2 FLUID FLOW

1.2.1 Fluid Properties

Fluid properties are classified into four basic types: (1) *Thermodynamic properties* such as temperature, pressure, density, enthalpy, and entropy; (2) *Transport properties* such as viscosity thermal conductivity, and mass diffusivity; (3) *Kinematics properties* such as velocity, acceleration, angular velocity, vorticity, and deformation or strain rate; and (4) other miscellaneous properties such as surface tension. Among these, *kinematics properties* are truly the properties of the fluid motion itself rather than that of the fluid.

Kinematics of fluid and kinematic properties

Velocity Field The velocity at a point in a flow field is defined as the velocity of a fluid particle that passes through that point. The velocity of a fluid particle is defined in the same way as in dynamics, i.e. as the rate of change of the particle's position vector with respect to time. So, velocity at any point of a flow field is written as

$$\vec{V} = \frac{d\tilde{x}}{dt} \tag{1.1}$$

In Cartesian co-ordinate system, where we define the position vector as $\tilde{x} = \hat{i}x + \hat{j}y + \hat{k}z$, the velocity vector can be written as

$$\vec{V}(x, y, z, t) = \hat{i}\frac{dx}{dt} + \hat{j}\frac{dy}{dt} + \hat{k}\frac{dz}{dt} \tag{1.2a}$$

or

$$\vec{V}(x, y, z, t) = \hat{i}u + \hat{j}v + \hat{k}w \tag{1.2b}$$

where $u = dx/dt$, $v = dy/dt$, and $w = dz/dt$ are the x, y, and z component velocities respectively.

Similarly, the velocity vectors in cylindrical and spherical co-ordinate systems are defined in Figure 1.1.

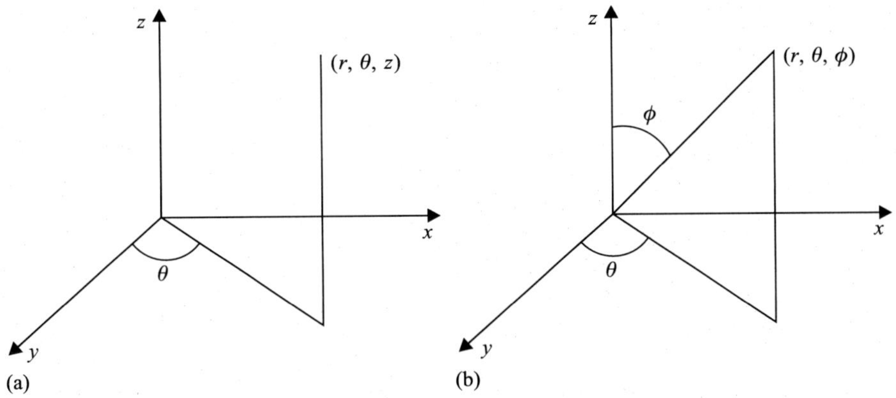

Figure 1.1 Co-ordinate systems: (a) cylindrical and (b) spherical

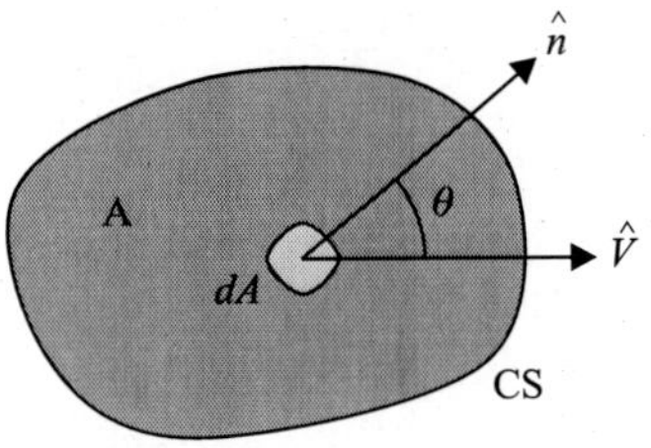

Figure 1.2 Arbitrary control volume surface

Cylindrical:

$$\vec{V}(r, \theta, z, t) = \hat{i}u_r + \hat{j}u_\theta + \hat{k}u_z$$

Spherical:

$$\vec{V}(r, \theta, \phi, t) = \hat{i}u_r + \hat{j}u_\theta + \hat{k}u_\phi$$

Volume and Mass Flow Rates Once the velocity field is known, we can calculate the volume flow rate, mass flow rate and average velocity for flow through any fixed or arbitrary control surface (CS) as shown in Figure 1.2.

Volume Flow Rate Volume flow rate through the differential area dA is

$$d\dot{\forall} = \left(\vec{V} \cdot \hat{n}\right) dA = \vec{V} \cdot d\vec{A} \tag{1.3}$$

Integrating over the whole area, we get the total flow rate

$$\dot{\forall} = \int_{\text{CS}} \left(\vec{V} \cdot \hat{n}\right) dA = \int \vec{V} \cdot d\vec{A} \tag{1.4}$$

Mass flow rate

$$\dot{m} = \int_{CS} \rho\left(\vec{V} \cdot \hat{n}\right) dA \tag{1.5}$$

Average velocity

$$\bar{V} = \frac{\dot{\forall}}{A} = \frac{\int_{CS} \vec{V} \cdot \vec{dA}}{A} \tag{1.6}$$

Acceleration The acceleration of a fluid particle at any point on the flow field is defined as the rate of change of velocity of the particle with time. For a velocity field, the ***acceleration*** can be derived as

$$\vec{a} = \frac{D\vec{V}}{Dt} = \frac{\partial \vec{V}}{\partial t} + u\frac{\partial \vec{V}}{\partial x} + v\frac{\partial \vec{V}}{\partial y} + w\frac{\partial \vec{V}}{\partial z} \tag{1.7}$$

The term $D\vec{V}/Dt$ on the left-hand side is called total derivative or material derivative. It represents the total acceleration of the fluid particles. It is evident from the right-hand side of the expression that the total acceleration has two major components: the first term $\partial\vec{V}/\partial t$ is the local acceleration, and the remainder is convective acceleration given by

$$u\frac{\partial \vec{V}}{\partial x} + v\frac{\partial \vec{V}}{\partial y} + w\frac{\partial \vec{V}}{\partial z}$$

It can be mentioned here that these nonlinear convective terms pose considerable mathematical and numerical difficulties. The expression for acceleration can be written in vector notation as

$$\vec{a} = \frac{D\vec{V}}{Dt} = \frac{\partial \vec{V}}{\partial t} + \left(\vec{V} \cdot \nabla\right)\vec{V} \tag{1.8}$$

The scalar component equations are written as

***Cartesian coordinate* (*x*, *y*, *z*)**

$$a_x = \frac{Du}{Dt} = \frac{\partial u}{\partial t} + u\frac{\partial u}{\partial x} + v\frac{\partial u}{\partial y} + w\frac{\partial u}{\partial z} \tag{1.9a}$$

$$a_y = \frac{Dv}{Dt} = \frac{\partial v}{\partial t} + u\frac{\partial v}{\partial x} + v\frac{\partial v}{\partial y} + w\frac{\partial v}{\partial z} \tag{1.9b}$$

$$a_z = \frac{Dw}{Dt} = \frac{\partial w}{\partial t} + u\frac{\partial w}{\partial x} + v\frac{\partial w}{\partial y} + w\frac{\partial w}{\partial z} \tag{1.9c}$$

***Cylindrical coordinate* (r, θ, z)**

$$a_r = \frac{Du_r}{Dt} = \frac{\partial u_r}{\partial t} + u_r\frac{\partial u_r}{\partial r} + \frac{u_\theta}{r}\frac{\partial u_r}{\partial \theta} - \frac{u_\theta^2}{2} + u_z\frac{\partial u_r}{\partial z} \tag{1.10a}$$

$$a_\vartheta = \frac{Du_\theta}{Dt} = \frac{\partial u_\theta}{\partial t} + u_r\frac{\partial u_\theta}{\partial r} + \frac{u_\theta}{r}\frac{\partial u_\theta}{\partial \theta} + \frac{u_r u_\theta}{r} + u_z\frac{\partial u_\theta}{\partial z} \tag{1.10b}$$

$$a_z = \frac{Du_z}{Dt} = \frac{\partial u_z}{\partial t} + u_r\frac{\partial u_z}{\partial r} + \frac{u_\theta}{r}\frac{\partial u_z}{\partial \theta} + u_z\frac{\partial u_z}{\partial z} \tag{1.10c}$$

***Spherical coordinates* (r, θ, ϕ)**

$$a_r = \frac{Du_r}{Dt} = \frac{\partial u_r}{\partial t} + u_r\frac{\partial u_r}{\partial r} + \frac{u_\phi}{r}\frac{\partial u_r}{\partial \phi} + \frac{u_\theta}{r\sin\phi}\frac{\partial u_r}{\partial \theta} - \frac{u_\theta^2 + u_\phi^2}{r} \tag{1.11a}$$

$$a_\theta = \frac{Du_\theta}{Dt} = \frac{\partial u_\theta}{\partial t} + u_r\frac{\partial u_\theta}{\partial r} + \frac{u_\phi}{r}\frac{\partial u_\theta}{\partial \phi} + \frac{u_\theta}{r\sin\phi}\frac{\partial u_\theta}{\partial \theta} + \frac{u_r u_\theta}{r} - \frac{u_\theta u_\phi \cot\phi}{r} \tag{1.11b}$$

$$a_\phi = \frac{Du_\phi}{Dt} = \frac{\partial u_\phi}{\partial t} + u_r\frac{\partial u_\phi}{\partial r} + \frac{u_\phi}{r}\frac{\partial u_\phi}{\partial \phi} + \frac{u_\theta}{r\sin\phi}\frac{\partial u_\phi}{\partial \phi} + \frac{u_r u_\phi}{r} - \frac{u_\theta^2 \cot\phi}{r} \tag{1.11c}$$

Motion or Kinematics of Fluid Element In order to describe the motion or kinematics of fluid, let us consider a differential control volume of a fluid element of volume dv in Cartesian co-ordinate as shown in Figure 1.3.

Let us assume that a fluid element initially at a point p at time t then moves to a new position at time $t+dt$. As the fluid element moves in the flow field, it may experience several things. These include: (1) translation, (2) rotation, (3) linear deformation or extensional strain, and (4) angular deformation or

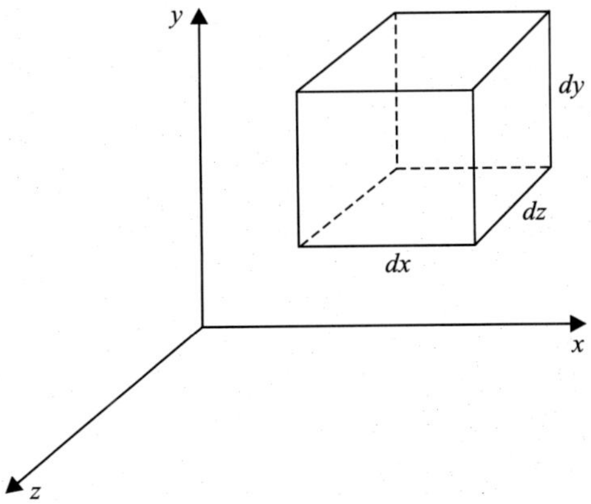

Figure 1.3 Differential control volume in Cartesian co-ordinate

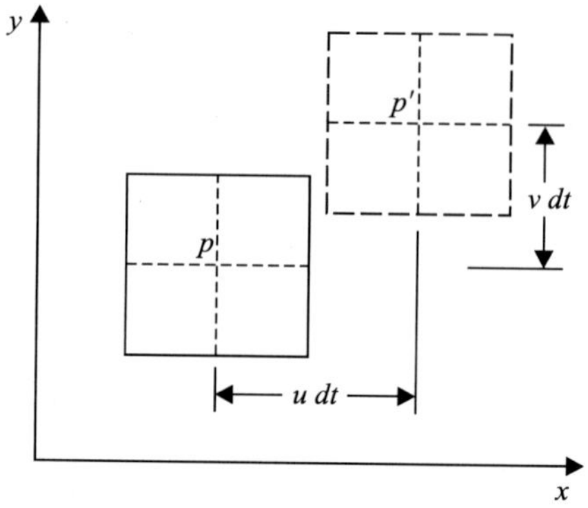

Figure 1.4 Translation in the x–y plane

shear strain. These are discussed briefly by considering a motion on an x–y plane. Similar motions can also be illustrated for y–z and z–x planes.

Translation The ***translation*** is defined by the displacements $u\,dt$ and $v\,dt$ of the point p to p' as shown in Figure 1.4.

The rates of translations are u and v. For three-dimensional fluid flows, the translations are defined by velocity components u, v, and w.

Rotation The fluid ***rotation***, $\vec{\omega}$, is a vector quantity with rotation about all three coordinate axes, and it is written as

$$\vec{\omega} = \hat{i}\omega_x + \hat{j}\omega_y + \hat{k}\omega_z \tag{1.12}$$

where ω_x, ω_y, and ω_z are rotations about x, y, and z axes respectively.

Each of these rotational components is defined as the average angular velocity of any two mutually perpendicular lines in a differential element. Figure 1.5 shows the rotation of such an element in the x–y plane.

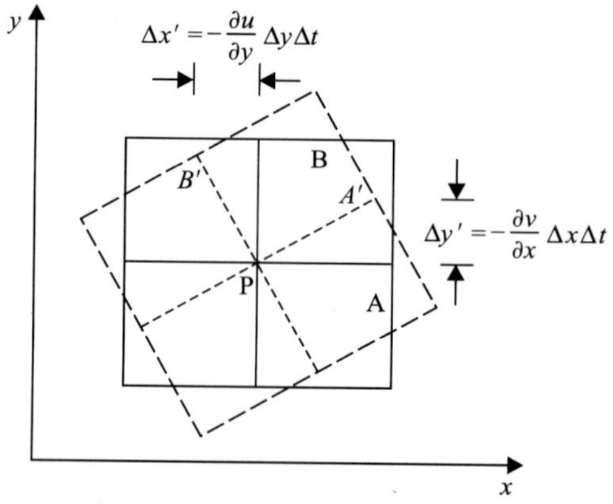

Figure 1.5 Rotation of a fluid element in the x–y plane

The rotation ω_z of this element is defined as the average angular velocity of two mutually perpendicular lines PA and PB.

The angular velocity of line PA is given as

$$\omega_{\text{PA}} = \lim_{\Delta t \to 0} \frac{\Delta\theta}{\Delta t} = \lim_{\Delta t \to 0} \frac{\Delta y'/\Delta x}{\Delta t} = \frac{\partial v}{\partial x} \tag{1.13}$$

and the angular velocity of line PB is given as

$$\omega_{\text{PB}} = \lim_{\Delta t \to 0} \frac{\Delta\phi}{\Delta t} = \lim_{\Delta t \to 0} \frac{\Delta x'/\Delta y}{\Delta t} = -\frac{\partial u}{\partial y} \tag{1.14}$$

The rotation of the fluid element about the z-axis is expressed as

$$\omega_z = \frac{1}{2}\left(\frac{\partial v}{\partial x} - \frac{\partial u}{\partial z}\right) \tag{1.15a}$$

Similarly, the rotation of the differential fluid element about the x-axis and the y-axis is given as

$$\omega_x = \frac{1}{2}\left(\frac{\partial w}{\partial y} - \frac{\partial v}{\partial z}\right) \tag{1.15b}$$

and

$$\omega_y = \frac{1}{2}\left(\frac{\partial u}{\partial z} - \frac{\partial w}{\partial x}\right) \tag{1.15c}$$

Substituting, the complete expression of rotation is written as

$$\vec{\omega} = \frac{1}{2}\left[\hat{i}\left(\frac{\partial w}{\partial y} - \frac{\partial v}{\partial z}\right) + \hat{j}\left(\frac{\partial u}{\partial z} - \frac{\partial w}{\partial x}\right) + \hat{k}\left(\frac{\partial v}{\partial x} - \frac{\partial u}{\partial y}\right)\right] \tag{1.15d}$$

We can express this in a compact form using vector notation as

$$\vec{\omega} = \frac{1}{2}\left(\nabla \times \vec{V}\right) \tag{1.15e}$$

Another important quantity of great interest in fluid mechanics is the ***vorticity***, which is defined as

$$\vec{\zeta} = 2\vec{\omega} = \nabla \times \vec{V} \tag{1.16}$$

It can be mentioned here that a flow is irrotational if vorticity $\vec{\zeta} = 0$ or angular velocity $\vec{\omega} = 0$. Also, an initial irrotational flow field can only become

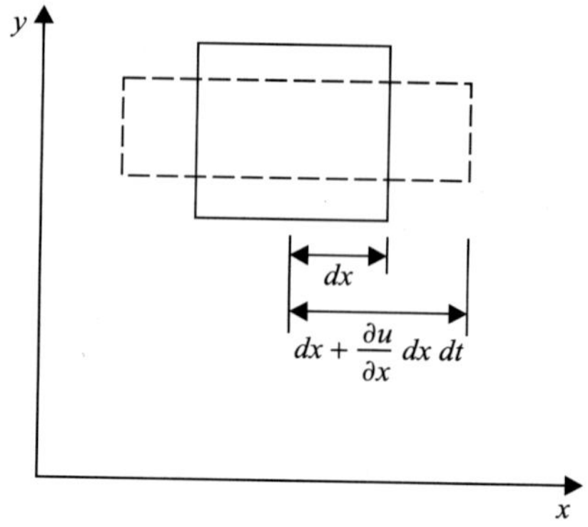

Figure 1.6 Linear deformation of the fluid element in the x–y plane

rotational with a simultaneous angular deformation caused by viscous or shear stresses.

Linear Deformation or Extensional Strain The ***linear deformation*** or ***extensional strain*** is defined as the fractional increase in length of one side of the element (see Figure 1.6).

We can express the extensional strain or linear deformation of side dx as

$$\varepsilon_{xx}\,dt = \frac{[dx + (\partial u/\partial x)dxdt] - dx}{dx}$$

Thus, the extensional strain rate in the x-direction is

$$\varepsilon_{xx} = \frac{\partial u}{\partial x} \tag{1.17a}$$

Similarly, the y and z components of extensional strain rates are

$$\varepsilon_{yy} = \frac{\partial v}{\partial y} \tag{1.17b}$$

and

$$\varepsilon_{zz} = \frac{\partial w}{\partial z} \tag{1.17c}$$

Angular Deformation or Shear Strain The two-dimensional ***angular deformation*** or ***shear strain*** is defined as the average decrease of the angle between two lines PA and PB as shown in Figure 1.7.

Thus the shear strain rate for the x–y plane is

$$\varepsilon_{xy} = \frac{1}{2}\left(\frac{d\theta}{dt} + \frac{d\phi}{dt}\right) \tag{1.18}$$

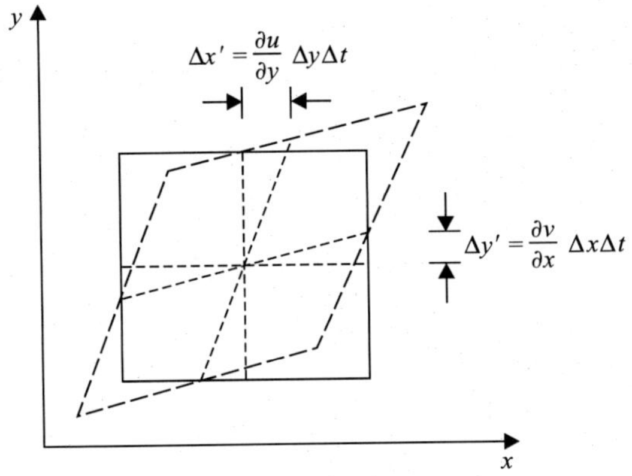

Figure 1.7 Angular deformation of fluid element in the x–y plane

Now, noting that $d\theta/dt = \partial v/\partial x$ and $d\phi/dt = \partial u/\partial y$, we get

$$\varepsilon_{xy} = \frac{1}{2}\left(\frac{\partial v}{\partial x} + \frac{\partial u}{\partial y}\right) \tag{1.19a}$$

Similarly, the shear strain rates for the y–z and z–x planes are

$$\varepsilon_{yz} = \frac{1}{2}\left(\frac{\partial w}{\partial y} + \frac{\partial v}{\partial z}\right) \tag{1.19b}$$

and

$$\varepsilon_{zx} = \frac{1}{2}\left(\frac{\partial u}{\partial z} + \frac{\partial w}{\partial x}\right) \tag{1.19c}$$

Also, the shear strain rates are symmetric; that is $\varepsilon_{xy} = \varepsilon_{yx}$, $\varepsilon_{yz} = \varepsilon_{zy}$ and $\varepsilon_{zx} = \varepsilon_{xz}$.

We can write all the components of extensional and shear strains as a symmetric second-order strain tensor

$$\varepsilon_{ij} = \begin{pmatrix} \varepsilon_{xx} & \varepsilon_{xy} & \varepsilon_{xz} \\ \varepsilon_{yx} & \varepsilon_{yy} & \varepsilon_{yz} \\ \varepsilon_{zx} & \varepsilon_{zy} & \varepsilon_{zz} \end{pmatrix} \tag{1.20}$$

Type of Forces in a Fluid. Forces in a fluid can be categorized into ***surface forces*** and ***body forces***. The ***surface forces*** are exerted by direct contact at normal as well as tangential to the surfaces. These forces include pressure force and viscous force. ***Body forces*** on the other hand are external forces developed without any body contact and are distributed over the entire volume of the fluid. These forces include gravitational forces, magnetic forces, etc.

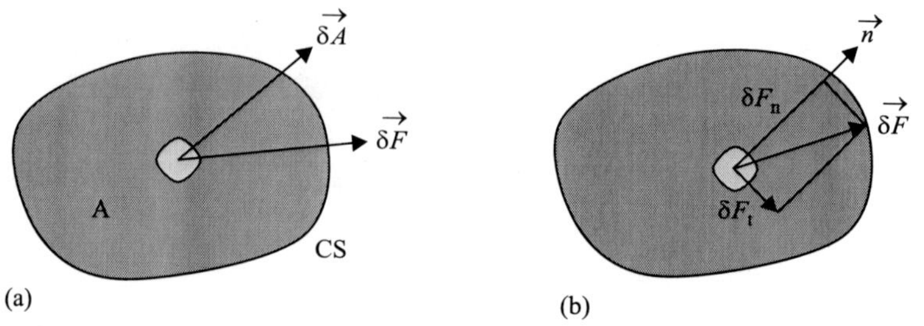

Figure 1.8 Forces on an arbitrary area

These forces are a direct function of mass. For example, the gravitational force that acts on a fluid element of volume $d\forall$ is

$$d\vec{F}_\mathrm{B} = \rho\vec{g}\, d\forall \tag{1.21}$$

where

ρ = density of the fluid
g = acceleration due to gravity

Stress Field Stresses in a medium result from surface forces acting on some portion of the medium. The stress is defined as a force per unit area. Both force and area are vector quantities, and stress is a tensor quantity of second order containing nine components. Let us consider an area dA through a point A as shown in Figure 1.8.

The unit vector, $\hat{n}$, gives the orientation of the area. The force acting on the area can be resolved into two components: a normal component, δF_n, and a tangential component, δF_t. Normal and shear stresses now can be defined as

$$\text{Normal stress} = \lim_{\delta A_n \to 0} \frac{\delta F_\mathrm{n}}{\delta A_\mathrm{n}} \tag{1.22}$$

$$\text{Shear stress} = \lim_{\delta A_\mathrm{n} \to 0} \frac{\delta F_\mathrm{t}}{\delta A_\mathrm{n}} \tag{1.23}$$

Stress on a Plane Let us consider a differential force $\delta\vec{F}$ that acts on a positive x-plane of area δA_x as shown in Figure 1.9.

If we resolve the force δF into three components δF_x, δF_y, and δF_z in x, y, and z directions, respectively, then the three stress components on the x-plane can be defined as

$$\sigma_{xx} = \lim_{\delta A_x \to 0} \frac{\delta F_x}{\delta A_x} = \text{normal stress on the } x \text{ plane in the } x \text{ direction} \tag{1.24a}$$

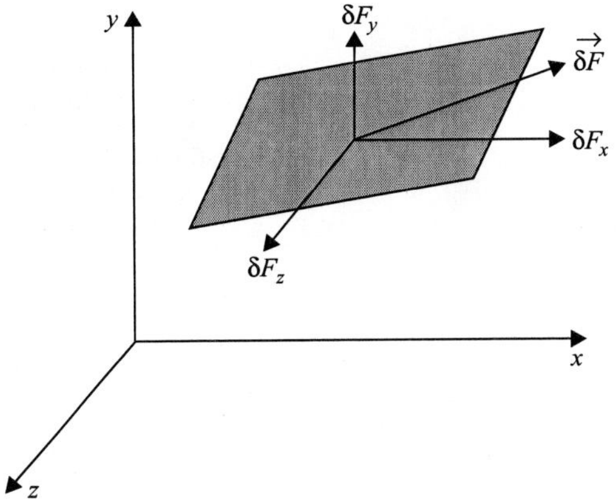

Figure 1.9 Forces on a positive x-plane

$$\tau_{xy} = \lim_{\delta A_x \to 0} \frac{\delta F_y}{\delta A_x} = \text{tangential or shear stress on the } x \text{ plane in the } y \text{ direction} \tag{1.24b}$$

$$\tau_{xz} = \lim_{\delta A_x \to 0} \frac{\delta F_z}{\delta A_x} = \text{tangential or shear stress on the } x \text{ plane in the } z \text{ direction} \tag{1.24c}$$

Similarly, we may have three component stresses σ_{yy}, τ_{yx}, and τ_{yz} on the y-plane of area δA_y, and σ_{zz}, τ_{zx} and τ_{zy}, on the z-plane of area δA_z.

Stress at a Point Stress at a point is given by nine stress components on three mutually perpendicular planes and can be expressed as a nine-component stress tensor

$$T_{ij} = \lim_{\delta A_i \to 0} \frac{\delta F_j}{\delta A_i}, \quad \text{where, } i = x, y \text{ and } z; j = x, y \text{ and } z \tag{1.25a}$$

or

$$T_{ij} = \begin{bmatrix} \sigma_{xx} & \tau_{xy} & \tau_{xz} \\ \tau_{yx} & \sigma_{yy} & \sigma_{yz} \\ \tau_{zx} & \tau_{zy} & \sigma_{zz} \end{bmatrix} \tag{1.25b}$$

Stresses on a Differential Element For a differential element, there are six surfaces (two x-planes, two y-planes, and two z-planes) on which stresses act, as shown in Figure 1.10

All stresses are shown as positive. A stress component is positive when the direction of the stress component and the plane on which it acts are

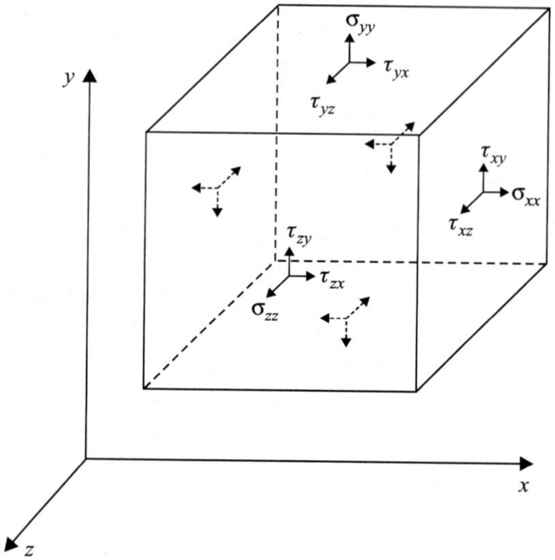

Figure 1.10 Stresses on a differential element

either both positive or both negative. For example, τ_{yx} represents shear stress on a positive y plane acting in a positive x-direction or shear stress on a negative y plane acting in a negative x direction.

1.2.2 Basic Equations in Integral Form

There are two basic approaches for analyzing fluid flow and heat transfer problems: ***system*** and ***control volume***. In the following sections we will present basic equations for a system and for a control volume.

Basic Equations for a System

Conservation of Mass The conservation of mass for a system states that mass M of the system is constant, so we have

$$\left.\frac{dM}{dt}\right|_{\text{System}} = 0 \tag{1.26a}$$

where

$$M_{\text{System}} = \int_m dm = \int_{\forall} \rho \, d\forall \tag{1.26b}$$

Conservation of Momentum Newton's second law states that the time rate of change of momentum of a system is equal to the sum of all externally applied forces, and is expressed as

$$\left.\frac{d\vec{L}}{dt}\right|_{\text{System}} = \vec{F} \tag{1.27a}$$

where the linear momentum $\vec{L}\big|_{\text{System}}$ is given by

$$\vec{L}\Big|_{\text{System}} = \int_m \vec{V}\, dm = \int_{\forall} \vec{V}\rho\, d\forall \tag{1.27b}$$

and $\vec{F}$ includes all surface and body forces that act on the system and expressed as

$$\vec{F} = \vec{F}_S + \vec{F}_B \tag{1.27c}$$

The First Law of Thermodynamics – Conservation of Energy The first law of thermodynamics for a system undergoing a process is stated as

$$\left.\frac{dE}{dt}\right|_{\text{System}} = \pm\dot{Q} \pm \dot{W} + \dot{Q}_{\text{gen}} \tag{1.28a}$$

where $\dot{Q}$ is the heat transfer rate across the system boundary, $\dot{W}$ is the work done by the system on the surrounding or vice-versa, E is the total energy content of the system, and $\dot{Q}_{\text{gen}}$ is the energy generation in the system. The total energy is expressed as

$$E_{\text{System}} = \int_m e\, dm = \int_{\forall} e\rho\, d\forall \tag{1.28b}$$

and the specific energy e is a sum of specific internal energy (i), kinetic energy, and potential energy, and written as

$$e = i + \frac{V^2}{2} + gZ \tag{1.28c}$$

It can be noted that all conservation statements, given by Equation 1.26 to Equation 1.28, involve a time derivative of an extensive property of the system (such as mass M, linear momentum $\vec{L}$ and energy E). These extensive properties can be represented by a single symbol, X, which can be expressed in a general form

$$X = \int_m x\,dm = \int_{\forall} x\rho\,d\forall \tag{1.29}$$

where x is the extensive property per unit mass, and equals 1 for the mass balance Equation 1.26, equals $\vec{V}$ for the momentum balance Equation 1.27, and equals e for the energy Equation 1.18.

The basic equation for the system analysis is converted into a control volume analysis by simply using the ***Reynolds' transport equation*** (Bennett and Myers, 1982), which is stated as

$$\left.\frac{dX}{dt}\right|_{\text{System}} = \left.\frac{\partial X}{\partial t}\right|_{\text{CV}} + \int_{\text{CS}} x\rho\vec{V}\cdot d\vec{A} \tag{1.30}$$

where the first term on the right-hand side of the equation represents the rate of change of the extensive property X within the system. The second term represents the net rate of flow of the extensive property through the control volume surface. Using ***Reynold's transport equation***, we can get all the basic conservation equations for a control volume as summarized below.

Basic Equations for a Control Volume

Conservation of Mass Using $X = M$ and $x = 1$, we get the conservation of mass equation for a control volume as

$$\frac{\partial}{\partial t}\left(\int_{\text{CV}} \rho\,d\forall\right) + \int_{\text{CS}} \rho\vec{V}\cdot d\vec{A} = 0 \tag{1.31a}$$

For steady flow

$$\int_{\text{CS}} \rho\vec{V}\cdot d\vec{A} = 0 \tag{1.31b}$$

For steady flows, the net mass flow rates into a control volume must be equal to the net mass flow out of the control volume.

For incompressible flow in a nondeformable control volume

$$\int_{\text{CS}} \vec{V}\cdot d\vec{A} = 0 \tag{1.31c}$$

In incompressible flows, the net volume flow rate into a control volume must be equal to the net volume flow rate out of the control volume.

Conservation of Momentum Substituting $X = \vec{L}$ and $x = \vec{V}$ into Equation 1.29, we obtain the conservation of momentum for a control volume from Equation 1.27 and Equation 1.30 as

$$\frac{\partial}{\partial t}\left(\int_{\forall} \vec{V}\rho \, d\forall\right) + \int_{CS} \vec{V}\rho\, \vec{V} \cdot d\vec{A} = \sum \vec{F}_S + \sum \vec{F}_B \tag{1.32}$$

where

$$\sum \vec{F}_S = \text{sum of all surface forces}$$

$$\sum \vec{F}_B = \text{sum of all surface forces}$$

The momentum equation can be written in three component scalar equations in different coordinate systems. For example, in Cartesian coordinate system, we can write

$$x\text{-component}: \quad \frac{\partial}{\partial t}\left(\int_{\forall} u\rho \, d\forall\right) + \int_{CS} u\rho\vec{V} \cdot d\vec{A} = \sum \vec{F}_{Sx} + \sum \vec{F}_{Bx} \tag{1.33a}$$

$$y\text{-component}: \quad \frac{\partial}{\partial t}\left(\int_{\forall} v\rho \, d\forall\right) + \int_{CS} v\rho\vec{V} \cdot d\vec{A} = \sum \vec{F}_{Sy} + \sum \vec{F}_{By} \tag{1.33b}$$

$$z\text{-component}: \quad \frac{\partial}{\partial t}\left(\int_{\forall} w\rho \, d\forall\right) + \int_{CS} w\rho\vec{V} \cdot d\vec{A} = \sum \vec{F}_{Sz} + \sum \vec{F}_{Bz} \tag{1.33c}$$

The first term on the left-hand side represents the rate of change of momentum inside the control volume. The second term is the net rate of momentum flux through the control volume. The sum of these two terms is equal to the sum of all surface and body forces acting on the control volume.

The First Law of Thermodynamics Substituting $X = E$ and $x = e$, we obtain the conservation of energy for a control volume from Equation 1.28 and Equation 1.30 as

$$\frac{\partial}{\partial t}\left(\int_{\forall} e\rho \, d\forall\right) + \int_{CS} e\rho\vec{V} \cdot d\vec{A} = \pm\dot{Q} \pm \dot{W} + \dot{Q}_{\text{gen}} \tag{1.34}$$

1.2.3 Differential Analysis of Fluid Motion

To obtain detailed point-to-point knowledge, we need to have the basic equations in differential form. These basic differential equations can be obtained by considering infinitesimal systems and control volume, and applying conservation principles. A good understanding of these equations is important because they give many insights or explain physics clearly. Detailed discussions on differential analysis of fluid motions are given in the fluid mechanics books by Fox and McDonald (1998), White (1988), and Schlichting (2000). A brief summary of these equations is given in the following section.

Conservation of Mass

Cartesian (x, y, z) Coordinate System The differential equation for conservation of mass, or the continuity equation, is derived by considering a cubic infinitesimal control volume ΔV surrounding the point p, as shown in the Figure 1.11. At point p, the density is assumed to be ρ and the velocity is given by $\vec{V} = \hat{i}u + \hat{j}v + \hat{k}w$.

Considering the mass balance over a control volume of infinitesimal size, the conservation of mass or continuity equation is given as

$$\frac{\partial \rho}{\partial t} + \frac{\partial(\rho u)}{\partial x} + \frac{\partial(\rho v)}{\partial y} + \frac{\partial(\rho w)}{\partial z} = 0 \tag{1.35a}$$

or

$$\frac{\partial \rho}{\partial t} + \nabla.\rho\vec{V} = 0 \tag{1.35b}$$

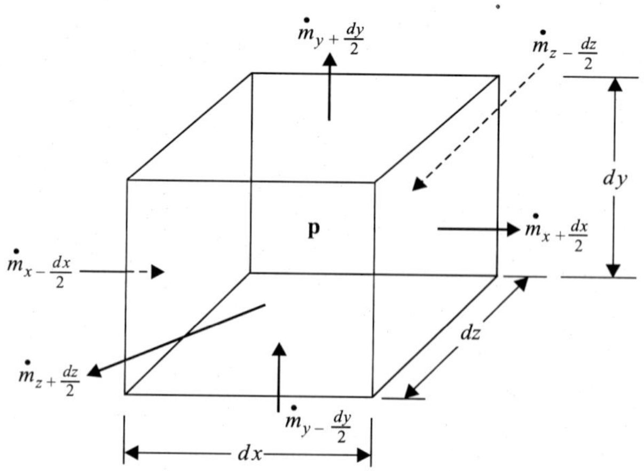

Figure 1.11 Differential control volume in Cartesian co-ordinate

Equation 1.35 is simplified for special cases as follows.

Incompressible Flow For incompressible flow, density is constant, i.e. $\rho =$ constant and Equation 1.35 reduces to

$$\frac{\partial u}{\partial x} + \frac{\partial v}{\partial y} + \frac{\partial w}{\partial z} = 0 \tag{1.36a}$$

or

$$\nabla \cdot \vec{V} = 0 \tag{1.36b}$$

We have indicated before that the three derivative terms of Equation 1.36a represent the extensional strain in the x, y and z directions. So the conservation of mass equation simply represents the net dilation of the fluid element as zero.

Steady Flow For steady flow the density is independent of time, and so conservation of mass equation transforms into

$$\frac{\partial \rho u}{\partial x} + \frac{\partial (\rho u)}{\partial y} + \frac{\partial (\rho w)}{\partial z} = 0 \tag{1.37a}$$

or

$$\nabla \cdot \rho\vec{V} = 0 \tag{1.37b}$$

Cylindrical Coordinate System In a cylindrical coordinate system, we consider a cylindrical infinitesimal control volume where the mass conservation equation is given as

$$\frac{\partial \rho}{\partial t} + \frac{\rho u_r}{r} + \frac{\partial (\rho u_r)}{\partial r} + \frac{1}{r}\frac{\partial (\rho u_\theta)}{\partial \theta} + \frac{\partial (\rho u_z)}{\partial z} = 0 \tag{1.38a}$$

or

$$\frac{\partial \rho}{\partial t} + \frac{1}{r}\frac{\partial (\rho r u_r)}{\partial r} + \frac{1}{r}\frac{\partial (\rho u_\theta)}{\partial \theta} + \frac{\partial (\rho u_z)}{\partial z} = 0 \tag{1.38b}$$

Incompressible flow

$$\frac{1}{r}\frac{\partial (r u_r)}{\partial r} + \frac{1}{r}\frac{\partial (u_\theta)}{\partial \theta} + \frac{\partial (u_z)}{\partial z} = 0 \tag{1.39}$$

Steady flow

$$\frac{1}{r}\frac{\partial (r\rho u_r)}{\partial r} + \frac{1}{r}\frac{\partial (\rho u_\theta)}{\partial \theta} + \frac{\partial (\rho u_z)}{\partial z} = 0 \tag{1.40}$$

Spherical Coordinate System In a spherical coordinate system, we consider an infinitesimal control volume where the mass conservation equation is given as

$$\frac{\partial \rho}{\partial t}+\frac{1}{r^2}\frac{\partial\left(\rho r^2 u_r\right)}{\partial r}+\frac{1}{r\sin\phi}\frac{\partial\left(\rho u_\phi \sin\phi\right)}{\partial \phi}+\frac{1}{r\sin\phi}\frac{\partial(\rho u_\theta)}{\partial \theta}=0 \tag{1.41}$$

Conservation of Momentum. The conservation of momentum equation is derived from Newton's second law of motion, which expresses proportionality between applied force and resulting acceleration of a particle. So, the conservation of momentum for an infinitesimal fluid mass is

$$\Delta\vec{F}=\Delta m\vec{a}=\Delta m\frac{D\vec{V}}{Dt} \tag{1.42}$$

or

$$\Delta\vec{F}_{\mathrm{B}}+\Delta\vec{F}_{\mathrm{S}}=\rho\frac{D\vec{V}}{Dt}\Delta\forall \tag{1.43}$$

where $\Delta\vec{F}$ is the net applied force on the fluid element composed of two types of force, net body force $\Delta\vec{F}_{\mathrm{B}}$ and net surface force $\Delta\vec{F}_{\mathrm{S}}$. $\vec{a}$ is the acceleration of the fluid element and is derived in the Eulerian system as

$$\vec{a}=\frac{D\vec{v}}{Dt}=\frac{\partial\vec{V}}{\partial t}+u\frac{\partial\vec{V}}{\partial x}+v\frac{\partial\vec{V}}{\partial y}+w\frac{\partial\vec{V}}{\partial z} \tag{1.44}$$

The body force has three components written as

$$\Delta\vec{F}_{\mathrm{B}}=\text{body force}=\hat{i}\Delta F_{\mathrm{B}x}+\hat{j}\Delta F_{\mathrm{B}y}+\hat{j}\Delta F_{\mathrm{B}z} \tag{1.45}$$

Assuming that the body force is due to gravity only, we can express as

$$\Delta\vec{F}_{\mathrm{B}}=\left(\hat{i}\rho g_x+\hat{j}\rho g_y+\hat{k}\rho g_z\right)\Delta\forall=\rho\vec{g}\Delta\forall \tag{1.46}$$

The surface force $\Delta\vec{F}_{\mathrm{s}}$ is computed from the symmetric stress tensor given by Equation 1.25. For example, the estimation of net surface force in the x direction is obtained by considering all x-component stresses on six surfaces of the fluid element as shown in Figure 1.12. If the x-component stresses at the center of the differential element are taken to be σ_{xx}, τ_{yx}, and τ_{zx}, then stresses acting in the x direction on each face of the element are derived from a truncated Taylor series expansion as demonstrated in Figure 1.12.

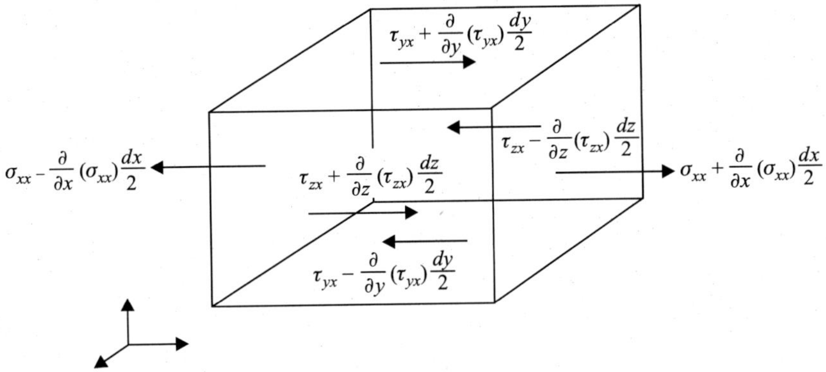

Figure 1.12 x-component stresses on a fluid element

Adding all forces in the x direction, we get the net x-component force

$$dF_{sx} = \left(\frac{\partial \sigma_{xx}}{\partial x} + \frac{\partial \tau_{yx}}{\partial y} + \frac{\partial \tau_{zx}}{\partial z}\right) dx\, dy\, dz \qquad (1.47a)$$

Similarly, the net y-component and z-component surface forces are given as

$$dF_{sy} = \left(\frac{\partial \tau_{yx}}{\partial x} + \frac{\partial \sigma_{yy}}{\partial y} + \frac{\partial \tau_{yz}}{\partial z}\right) dx\, dy\, dz \qquad (1.47b)$$

$$dF_{sz} = \left(\frac{\partial \tau_{zx}}{\partial x} + \frac{\partial \tau_{zy}}{\partial y} + \frac{\partial \sigma_{zz}}{\partial z}\right) dx\, dy\, dz \qquad (1.47c)$$

Substituting Equation 1.45 to Equation 1.47 into Equation 1.44 and dividing through by the volume $d\forall = dx\, dy\, dz$, we obtain the equation of motion in component form as

x-momentum

$$\rho\left(\frac{\partial u}{\partial t} + u\frac{\partial u}{\partial x} + u\frac{\partial u}{\partial y} + w\frac{\partial u}{\partial z}\right) = \rho g_x + \left(\frac{\partial \sigma_{xx}}{\partial x} + \frac{\partial \tau_{yx}}{\partial y} + \frac{\partial \tau_{zx}}{\partial z}\right) \qquad (1.48a)$$

y-momentum

$$\rho\left(\frac{\partial v}{\partial t} + u\frac{\partial v}{\partial x} + v\frac{\partial v}{\partial y} + w\frac{\partial v}{\partial z}\right) = \rho g_y + \left(\frac{\partial \tau_{yx}}{\partial x} + \frac{\partial \sigma_{yy}}{\partial y} + \frac{\partial \tau_{yz}}{\partial t}\right) \qquad (1.48b)$$

z-momentum

$$\rho\left(\frac{\partial w}{\partial t} + u\frac{\partial w}{\partial x} + v\frac{\partial w}{\partial y} + w\frac{\partial w}{\partial z}\right) = \rho g_z + \left(\frac{\partial \tau_{zx}}{\partial x} + \frac{\partial \tau_{zy}}{\partial z} + \frac{\partial \sigma_{zz}}{\partial z}\right) \qquad (1.48c)$$

Equation 1.48 is applicable to the motion of any fluid that satisfies the continuum assumption. In order to solve this equation for the flow field or flow properties, it is necessary to express the stresses in terms of measurable quantities such as velocity vector $\vec{V}$ and its derivatives. This is done through an established relationship between stress (T_{ij}) and deformation rate or strain rate (ε_{ij}), which is known as ***Stokes viscous deformation law*** (Schlichting and Gersten, 2000; White, 1988).

Stokes Viscous Deformation Law The viscous deformation law for Newtonian fluids was first postulated by Stokes in 1845. Basic assumptions made in deriving this relationship are: (1) fluid is isotropic and continuous, (2) the shear stress–strain rate relation is ***linear***, and (3) the stress equals the hydrostatic pressure P for cases with zero strain rate. Stokes postulates lead to the following deformation law.

Normal stresses

$$\sigma_{xx} = -p + 2\mu\varepsilon_{xx} - \tfrac{2}{3}\mu\nabla\cdot\vec{V} \tag{1.49a}$$

$$\sigma_{yy} = -p + 2\mu\varepsilon_{yy} - \tfrac{2}{3}\mu\nabla\cdot\vec{V} \tag{1.49b}$$

$$\sigma_{zz} = -p + 2\mu\varepsilon_{zz} - \tfrac{2}{3}\mu\nabla\cdot\vec{V} \tag{1.49c}$$

Shear stresses

$$\tau_{yx} = \tau_{xy} = 2\mu\varepsilon_{xy} \tag{1.50a}$$

$$\tau_{zx} = \tau_{xz} = 2\mu\varepsilon_{xz} \tag{1.50b}$$

$$\tau_{yz} = \tau_{zy} = 2\mu\varepsilon_{yz} \tag{1.50c}$$

Using Equation 1.18 and Equation 1.19 for strain rates, we can write the deformation law as

$$\tau_{xy} = \tau_{yx} = \mu\left(\frac{\partial v}{\partial x} + \frac{\partial u}{\partial y}\right) \tag{1.51a}$$

$$\tau_{yz} = \tau_{zy} = \mu\left(\frac{\partial w}{\partial y} + \frac{\partial v}{\partial z}\right) \tag{1.51b}$$

$$\tau_{zx} = \tau_{xz} = \mu\left(\frac{\partial u}{\partial z} + \frac{\partial w}{\partial x}\right) \tag{1.51c}$$

and

$$\sigma_{xx} = -p - \tfrac{2}{3}\mu\nabla\cdot\vec{V} + 2\mu\frac{\partial u}{\partial x} \tag{1.51d}$$

$$\sigma_{yy} = -p - \tfrac{2}{3}\mu\nabla\cdot\vec{V} + 2\mu\frac{\partial v}{\partial y} \tag{1.51e}$$

$$\sigma_{zz} = -p - \tfrac{2}{3}\mu\nabla\cdot\vec{V} + 2\mu\frac{\partial w}{\partial z} \tag{1.51f}$$

The Navier–Stokes Equations Substituting deformation law (1.51) into equation of motion (1.48), we get the ***Navier–Stokes equations***, which represent the momentum or equation of motion for Newtonian viscous fluid

$$\rho\frac{Du}{Dt} = \rho g_x - \frac{\partial p}{\partial x} + \frac{\partial}{\partial x}\left[\mu\left(2\frac{\partial u}{\partial x} - \frac{2}{3}\nabla\cdot\vec{V}\right)\right]$$
$$+\frac{\partial}{\partial y}\left[\mu\left(\frac{\partial u}{\partial y} + \frac{\partial v}{\partial x}\right)\right] + \frac{\partial}{\partial z}\left[\mu\left(\frac{\partial w}{\partial x} + \frac{\partial u}{\partial z}\right)\right] \tag{1.52a}$$

$$\rho\frac{Dv}{Dt} = \rho g_y - \frac{\partial p}{\partial y} + \frac{\partial}{\partial x}\left[\mu\left(\frac{\partial u}{\partial y} + \frac{\partial v}{\partial x}\right)\right]$$
$$+\frac{\partial}{\partial y}\left[\mu\left(2\frac{\partial v}{\partial y} - \frac{2}{3}\nabla\cdot\vec{V}\right)\right] + \frac{\partial}{\partial z}\left[\mu\left(\frac{\partial v}{\partial z} + \frac{\partial w}{\partial y}\right)\right] \tag{1.52b}$$

$$\rho\frac{Dw}{Dt} = \rho g_z - \frac{\partial p}{\partial z} + \frac{\partial}{\partial x}\left[\mu\left(\frac{\partial w}{\partial x} + \frac{\partial u}{\partial z}\right)\right]$$
$$+\frac{\partial}{\partial y}\left[\mu\left(\frac{\partial v}{\partial z} + \frac{\partial w}{\partial y}\right)\right] + \frac{\partial}{\partial z}\left[\mu\left(2\frac{\partial w}{\partial z} - \frac{2}{3}\nabla\cdot\vec{V}\right)\right] \tag{1.52c}$$

Incompressible Fluids and Constant Viscosity For incompressible fluids, density is constant, and conservation of mass equation simplifies to

$$\frac{\partial u}{\partial x} + \frac{\partial v}{\partial y} + \frac{\partial w}{\partial z} = 0 \tag{1.53a}$$

or

$$\nabla\cdot\vec{V} = 0 \tag{1.53b}$$

Substituting $\nabla \cdot \vec{V} = 0$, and further assuming a fluid with constant viscosity (μ), Equation 1.52 reduces to

$$\rho\left(\frac{\partial u}{\partial t} + u\frac{\partial u}{\partial x} + v\frac{\partial u}{\partial y} + w\frac{\partial u}{\partial z}\right) = \rho g_x - \frac{\partial p}{\partial x} + \mu\left(\frac{\partial^2 u}{\partial x^2} + \frac{\partial^2 u}{\partial y^2} + \frac{\partial^2 u}{\partial z^2}\right) \quad (1.54a)$$

$$\rho\left(\frac{\partial v}{\partial t} + u\frac{\partial v}{\partial x} + v\frac{\partial v}{\partial y} + w\frac{\partial v}{\partial z}\right) = \rho g_y - \frac{\partial p}{\partial y} + \mu\left(\frac{\partial^2 v}{\partial x^2} + \frac{\partial^2 v}{\partial y^2} + \frac{\partial^2 v}{\partial z^2}\right) \quad (1.54b)$$

$$\rho\left(\frac{\partial w}{\partial t} + u\frac{\partial w}{\partial x} + v\frac{\partial w}{\partial y} + w\frac{\partial w}{\partial z}\right) = \rho g_z - \frac{\partial p}{\partial z} + \mu\left(\frac{\partial^2 w}{\partial x^2} + \frac{\partial^2 w}{\partial y^2} + \frac{\partial^2 w}{\partial z^2}\right) \quad (1.54c)$$

and in vector notation

$$\rho\frac{D\vec{V}}{dt} = \rho\vec{g} - \nabla p + \mu\nabla^2\vec{V} \quad (1.54d)$$

It can be mentioned here that Equation 1.54 is not suitable for viscous-incompressible non-isothermal flow, particularly for liquids for which viscosity is a strong function temperature, and hence a function of space co-ordinates.

Frictionless Flow or Inviscid Flow For frictionless flow, viscosity $\mu = 0$ and flow is defined as inviscid flow. Substituting $\mu = 0$, Equation 1.54 gives the momentum equation for inviscid flow as

$$\rho\frac{D\vec{V}}{Dt} = \rho\vec{g} - \nabla p \quad (1.55)$$

Equation 1.55 is also known as the ***Euler equation.***

1.2.4 Boundary Conditions for Flow Field

Solution of conservation of mass and momentum equations require mathematically tenable and physically realistic boundary conditions, which represent the physical conditions at extreme values of the independent variables. A detailed description of hydrodynamic boundary conditions is given in Goldstein (1965), White (1988), Panton (1984), and Schlichting and Gersten (2000). The following are the most commonly used boundary conditions.

Solid–Fluid Interface

Tangential Component Velocity We normally assume that at the solid–fluid interface there is no relative speed difference or finite discontinuities of velocities between the solid surface and the fluid in contact.

So, the fluid velocity equals the velocity of the solid surface with which it moves. Such a condition is known as ***no-slip condition***. This is expressed mathematically as

$$u = u_w \tag{1.56}$$

where u_w is the velocity of the solid wall.

For more discussions on the validity of no-slip conditions and use of slip conditions, refer to White (1988).

Normal Component Velocity The velocity component v normal to the wall becomes zero for an impermeable wall. However, it becomes nonzero for a permeable or porous wall through which fluid is sucked in or blown out. For example, for a wall with flow suction this condition is stated mathematically as

$$v = v_0 \tag{1.57}$$

where v_0 is the normal velocity at the surface.

Fluid–Fluid Interface

Liquid–Liquid Interface Consider two liquids in contact as shown in Figure 1.13.

The velocity as well the shear stress is continuous across the interface and these are stated as follows.

No-slip condition for axial component:

$$u_1 = u_2 \tag{1.58}$$

Continuity in normal component velocity:

$$v_1 = v_2 \tag{1.59}$$

Continuity in shear stress:

$$\tau_1 = \tau_2 \tag{1.60a}$$

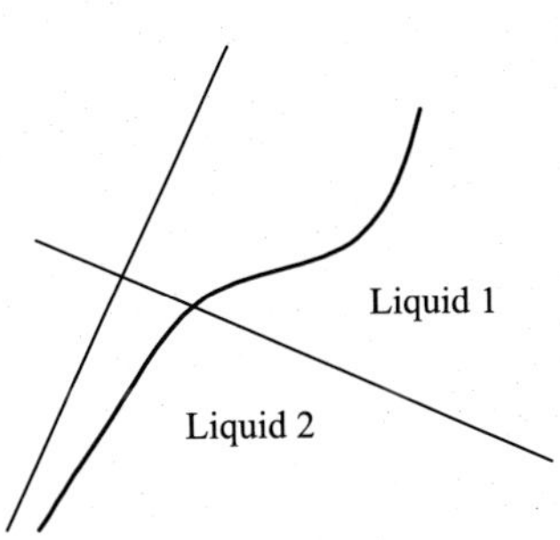

Figure 1.13 Liquid–liquid interface

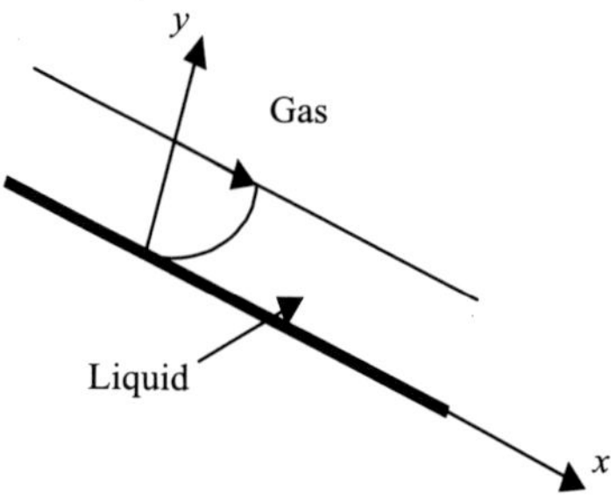

Figure 1.14 Liquid–gas interfaces

or

$$\mu_1 \frac{\partial u_1}{\partial n} = \mu_2 \frac{\partial u_2}{\partial n} \tag{1.60b}$$

It can be noted that, for dissimilar viscosity, there exists a discontinuity in the velocity gradients.

Liquid–Gas Interface At liquid–gas interfaces the momentum flux and hence the velocity gradient in the liquid is negligibly small compared to that in the gas phase since the viscosity of the liquid is much greater than that of the gas (Figure 1.14).

So, the interface boundary condition for the liquid phase is stated as

$$\frac{\partial u_l}{\partial n} = 0 \tag{1.61}$$

Deformable Solid–Fluid Interface. There are many problems in which the solid is coupled to the fluid through deformation and dynamic interaction. In such cases the equation of motion for solving the fluid and solid to be solved simultaneously, which satisfy continuity of velocity and stress at the interface, is

$$\tau_f = \tau_s \tag{1.62}$$

where τ_f is the fluid stress given, for example, by Newtonian fluid behavior and τ_s is the solid stress given, for example, by Hook's law.

1.3 HEAT TRANSFER

Heat transfer is a science that determines how and at what rate heat energy is transferred as a result of a temperature gradient or difference. There are three basic modes of heat transfer: conduction, convection, and radiation.

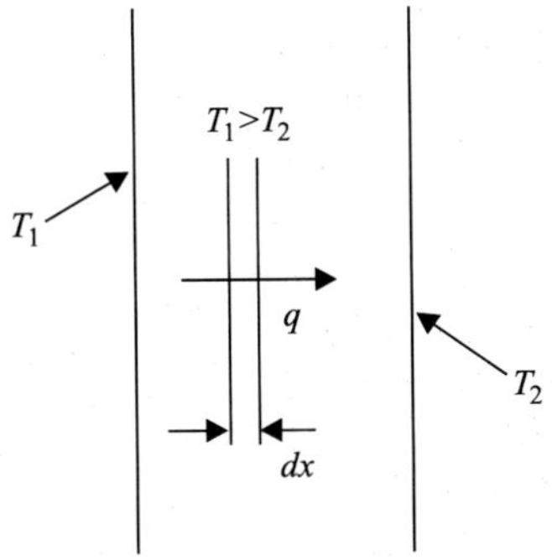

Figure 1.15 Heat conduction through a plane wall

1.3.1 Basic Modes and Transport Rate Equation

Conduction Heat Transfer. This mode is important primarily for heat transfer in solid and stationary fluid.

Conduction Rate Equation. The conduction rate equation is governed by Fourier's law, which states that, in a homogeneous substance, the local heat flux is proportional to the local temperature gradient. For one-dimensional heat flow in x-direction, such as in a plane slab (Figure 1.15), this law is written as

$$\frac{q}{A} = -\frac{dT}{dx} \tag{1.63}$$

Introducing the constant of proportionality

$$q'' = -k\frac{dT}{dx} \tag{1.64}$$

where $q'' = q/A =$ heat flow per unit area per unit time or heat flux, and k is the thermal conductivity of the material. The negative sign is included to ensure that heat flows in the direction of decrease in temperature.

Similarly, we can write the heat conduction rate equation in y and z directions. In general, the heat flux is a vector quantity and expressed as

$$\vec{q} = -k\nabla T \tag{1.65}$$

where Equation 1.65 can be expressed in different coordinate systems as follows.

Cartesian Coordinate

$$\vec{q} = -k\nabla T = \hat{i}q''_x + \hat{j}q''_y + \hat{k}q''_z \tag{1.66}$$

where

$$q''_x = -k\frac{\partial T}{\partial x}, \quad q''_y = -k\frac{\partial T}{\partial y} \quad \text{and} \quad q''_z = -k\frac{\partial T}{\partial z}$$

Cylindrical Coordinate

$$\vec{q} = -k\nabla T = \hat{i}q''_r + \hat{j}q''_\theta + \hat{k}q''_z \tag{1.67}$$

where

$$q''_r = -k\frac{\partial T}{\partial r}, \quad q''_\theta = -k\frac{1}{r}\frac{\partial T}{\partial \theta} \quad \text{and} \quad q''_z = -k\frac{\partial T}{\partial z}$$

Spherical Coordinate

$$\vec{q} = -k\nabla T = \hat{i}_r q''_r + \hat{i}_\phi q''_\phi + \hat{i}_\theta q''_\theta \tag{1.68}$$

where

$$q''_r = -k\frac{\partial T}{\partial r}, \quad q''_\phi = -k\frac{1}{r}\frac{\partial T}{\partial \phi} \quad \text{and} \quad q''_\theta = -k\frac{1}{r\sin\phi}\frac{\partial T}{\partial \theta}$$

We can also express the heat flux normal to any arbitrary surface (Figure 1.16) as

$$q''_n = \vec{q}'' \cdot \hat{n} = -k\frac{\partial T}{\partial n} \tag{1.69}$$

Also, the total heat flow through a surface area A_n is estimated by integrating the local heat transfer rate over the entire area as

$$q_n = \int_{A_n} q''_n dA_n = -\int_{A_n} k\frac{\partial T}{\partial n} dA_n \tag{1.70}$$

For uniform heat flux over the area, Equation 1.70 reduces to

$$q_n = -kA_n\frac{\partial T}{\partial n} \tag{1.71}$$

The ***Fourier*** law given by Equation 1.71 is valid for an ***isotropic*** material for which the thermal conductivity is directionally independent. In an ***anisotropic*** medium, thermal conductivity depends on coordinate directions. One special category of this is the orthotropic medium such as in laminated composites, and the heat flux vector given by Equation 1.71 for Cartesian co-ordinate system is written as

$$\vec{q} = -\left(\hat{i}k_x\frac{\partial T}{\partial x} + \hat{j}k_y\frac{\partial T}{\partial y} + \hat{k}k_z\frac{\partial T}{\partial z}\right) \tag{1.72}$$

where the heat flux components are

$$q''_x = -k_x\frac{\partial T}{\partial x}, \quad q''_y = -k_y\frac{\partial T}{\partial y} \quad \text{and} \quad q''_z = -k\frac{\partial T}{\partial z}$$

Convection Heat Transfer. The convection heat transfer occurs between a moving fluid and an exposed solid surface. Let us consider the fluid flow over a solid surface at a temperature T_S as shown in Figure 1.16. The fluid upstream temperature and velocity are T_∞ and u_∞ respectively.

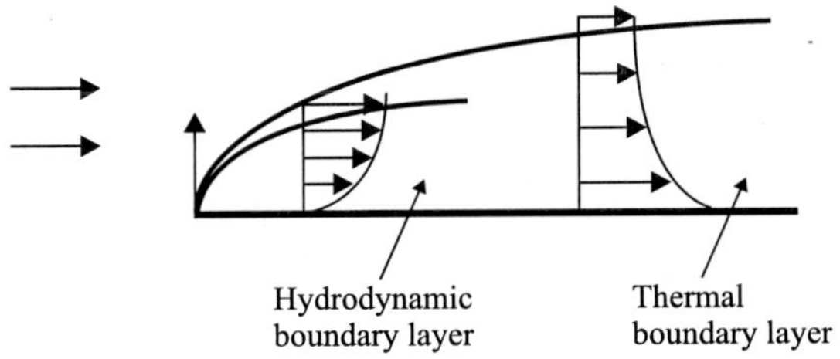

Figure 1.16 Hydrodynamic and thermal boundary layers for flow over a solid surface

Due to the effect of the viscosity or no-slip condition, a thin fluid region develops, known as the hydrodynamic boundary layer, inside which velocity varies from the solid surface velocity to the outer stream velocity, u_∞. Similarly, there is a thermal boundary layer inside which fluid temperature changes from surface temperature T_S to outer fluid temperature T_∞. Since the fluid is stationary at the solid surface, the heat is transferred by conduction normal to the surface, and is expressed by the conduction rate equation 1.64 as

$$q''_S = -k_f\left.\frac{\partial T}{\partial y}\right|_{y=0} \tag{1.73}$$

where k_f is the thermal conductivity of the fluid.

In order to estimate the heat flux by convection, the temperature gradient or the temperature distribution in the thermal boundary layer needs to be known. This temperature distribution depends on the surface geometry, transport and thermo-physical properties, and the nature of the fluid motion or the velocity field, and is determined by solving the energy equation along with the mass and momentum equations. Based on the nature of the flow field, the convection heat transfer is classified as ***forced convection***, ***free or natural convection*** or ***phase change heat transfer***, such as in condensation and boiling. In ***forced convection*** some external forces generated by pumps, fans or winds induce the flow field. On the other hand, for ***free or natural convection***, the flow is induced by natural forces such as buoyancy or marangoni forces. The buoyancy force arises from density variations, which is caused by temperature variations. In both forced and free convections, energy being transferred is in the form of sensible heat of the fluid. On the other hand, in ***phase change heat transfer***, the energy transfer is in the form of latent heat of the fluid, and the flow field is created due to the formation of vapor bubbles, as in boiling heat transfer, or due to the condensation of vapor on a solid surface, as in condensation heat transfer.

Irrespective of this classification of convection heat transfer, the overall effect is given by a ***convection rate equation***, governed by ***Newton's law of cooling***, expressed as

$$q''_c \propto (T_S - T_\infty) \tag{1.74}$$

We introduce a proportionality constant

$$q''_c = h_c(T_S - T_\infty) \tag{1.75}$$

where h_c, the constant of proportionality, is called the ***convection heat transfer coefficient*** or ***film coefficient***. The defining equation for the convection heat transfer coefficients is obtained by recasting Equation 1.75 as

$$h_c = \frac{q''_c}{(T_S - T_\infty)} \tag{1.76}$$

Combining Equation 1.73 and Equation 1.75 we obtain the alternate defining equation for the convection heat transfer coefficient

$$h_c = \frac{-k_f(\partial T/\partial y)\big|_{y=0}}{(T_S - T_\infty)} \tag{1.77}$$

It is important to note that the temperature gradient inside the thermal boundary layer varies along the surface, and hence the convection heat transfer coefficient varies locally from point to point. The local convection coefficient is defined based on local heat transfer rate, as

$$dq = h_c dA(T_S - T_\infty) \tag{1.78}$$

The total heat transfer rate can either be calculated by integrating the local heat transfer rate over the area or using Equation 1.75 with an average heat transfer coefficient $\bar{h}_c$, which is defined as

$$\bar{h}_c = \frac{1}{A}\iint_A h_c dA \tag{1.79}$$

Convection heat transfer coefficients are derived for many flow conditions in the form of a correlation. For forced convections the correlations are of the form

$$\text{Nu} = f(\text{Re}, \text{Pr})$$

where

$$\text{Nu} = \text{Nusselt number} = \frac{h_c L_c}{k}$$

$$\text{Re} = \text{Reynolds number} = \frac{\rho U_c L_c}{\mu}$$

$$\text{Pr} = \text{Prandtl number} = \frac{\mu c_p}{k} = \frac{\nu}{\alpha}$$

L_c and U_c represent the characteristics length and velocity in the problem. A list of many correlations for different flow geometries and conditions can be found in Incropera and DeWitt (2002), Holman (1997), and Mills (1999).

Thermal Radiation Thermal radiation is the intermediate portion (0.1–100 μm) of the electromagnetic radiation emitted by a substance as a result of its temperature. Thermal radiation heat transfer involves transmission and exchange of electromagnetic waves or photon particles as a result of temperature difference.

The thermal radiation emitted by a substance encompasses a range of wavelength (λ), and it is referred to as spectral distribution. The total black-body emissive power is obtained by integrating the spectral emissive power over the entire range of wavelengths and is derived as

$$E_b = \sigma T^4 \tag{1.80}$$

where σ is the Stefan–Boltzman constant ($5.6697 \times 10^{-8}\ \text{W/m}^2 \cdot \text{K}^4$) and T is the temperature in absolute scale.

Equation 1.80 is referred to as the ***Stefan–Boltzman law***, which gives total emissive power of the ***black*** or ***ideal body*** at a given temperature. It is called black-body radiation because a body which behaves according to this law appears black, since there is no reflection of any incident radiation from it. For a nonblack body, the emitted power is less than the black body emissive power, and it is given as

$$E = \varepsilon E_b \tag{1.81a}$$

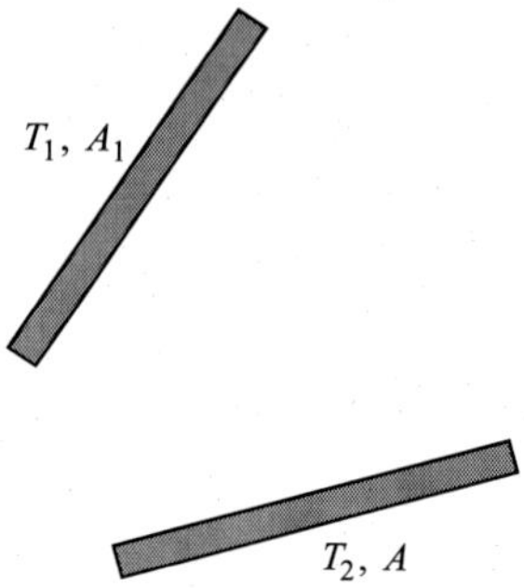

Figure 1.17 Radiation exchange between two black bodies

or

$$E_b = \varepsilon\sigma T^4 \tag{1.81b}$$

where ε is the emissivity factor of a ***nonblack surface*** and is defined as the ratio of its emissive power to that of a black body, i.e. $\varepsilon = E/E_b$. In addition to the emissivity factor, other material optical properties such as absorptivity, transmissivity and reflectivity, are also essential for heat transfer analysis involving radiation.

Radiation Heat Exchange. Based on the Stefan–Boltzmann law, ***the radiation heat transfer between two black surfaces*** through a nonparticipating medium is expressed as

$$q_R = \sigma A_1 F_{ij}\left(T_i^4 - T_j^4\right) \tag{1.82a}$$

Using a reciprocity relation $A_i F_{ij} = A_j F_{ji}$, Equation 1.82 can also be written as

$$q_R = \sigma A_j F_{ji}\left(T_i^4 - T_j^4\right) \tag{1.82b}$$

where A_i, A_j are areas of two surfaces, T_i, T_j are temperatures of two surfaces, and F_{ij}, F_{ji} are the shape factors or view factors.

The shape factor F_{ij} is defined as the fraction of energy leaving surface A_i that is intercepted by surface A_j. The shape factor depends on the size, shape, and orientations of the two surfaces i and j.

Shape factors are computed for many common geometries and configurations and presented in the form of formulas and graphs (Sparrow and Cess, 1978; Siegel and Howell, 1992).

Equation 1.82 can also be written in an electrical network analogy as

$$q = \frac{E_{bi} - E_{bj}}{1/A_i F_{ij}} = \frac{E_{bi} - E_{bj}}{R_{ij}} \tag{1.83}$$

where $R_{ij} = 1/A_iF_{ij}$ is termed as the ***radiation space resistance*** between the two surfaces i and j.

The radiation heat exchange between two diffuse gray surfaces is derived as

$$q = \frac{E_{bi} - E_{bj}}{\frac{1-\varepsilon_i}{\varepsilon_i A_i} + \frac{1}{A_iF_{ij}} + \frac{1-\varepsilon_j}{\varepsilon_j A_j}} = \frac{\sigma T_i^4 - T_j^4}{\frac{1-\varepsilon_i}{\varepsilon_i A_i} + \frac{1}{A_iF_{ij}} + \frac{1-\varepsilon_j}{\varepsilon_j A_j}} \tag{1.84}$$

Equation 1.83 can also be written in an electrical network analogy as

$$q = \frac{E_{bi} - E_{bj}}{R_{ij}} \tag{1.85}$$

where

$$R_{ij} = \frac{1-\varepsilon_i}{\varepsilon_i A_i} + \frac{1}{A_iF_{ij}} + \frac{1-\varepsilon_j}{\varepsilon_j A_j}$$

is the total radiation resistance between two gray surfaces i and j. The term $(1-\varepsilon)/\varepsilon A$ is defined as the ***radiation surface resistance***.

For many applications, the expression for radiation heat transfer is given in a linearized form as

$$q = h_r A(T_i - T_j) \tag{1.86}$$

where

$$h_r = \sigma(T_i + T_j)(T_i^2 + T_j^2)/R_{ij} \tag{1.87}$$

1.3.2 The First Law of Thermodynamics and Heat Equation

In order to estimate the heat transfer rates in a medium, we need to determine the temperature distribution or temperature field resulting from the physical conditions or thermal boundary conditions that exist at the boundary. The temperature field is determined by solving the heat equation, which is a statement of conservation of energy or the first law of thermodynamics. Let us consider a differential control volume in a Cartesian coordinate, and various forms of energy transfer across the control volume surfaces as shown in Figure 1.18.

The ***first law of thermodynamics*** for a differential control volume can be derived from Equation 1.34 stated for a control volume as

$$\pm\dot{Q} \pm \dot{W} + \dot{Q}_{gen} = \frac{\partial}{\partial t}\int_{\forall} e\rho d\forall + \int_{CS} e\rho\vec{V}\cdot d\vec{A} \tag{1.88}$$

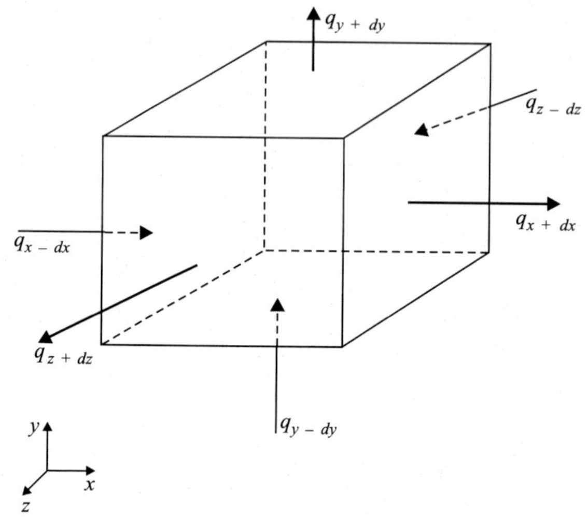

Figure 1.18 Differential control volume for energy balance

The first term $\dot{Q}$ represents heat transfer across the control volume surfaces by conduction ($\dot{Q}_C$) and by radiation heat transfer ($\dot{Q}_R$). The net conduction heat transfer per unit area per unit time across the differential control volume surfaces is derived as

$$\dot{Q}_C = -\left(\frac{\partial q_x}{\partial x} + \frac{\partial q_y}{\partial y} + \frac{\partial q_z}{\partial z}\right) dx\, dy\, dz = -\left(\frac{\partial q_x}{\partial x} + \frac{\partial q_y}{\partial y} + \frac{\partial q_z}{\partial z}\right) d\forall \tag{1.89}$$

or

$$\dot{Q}_C = -(\nabla \cdot \vec{q})\, d\forall \tag{1.90}$$

The net radiation heat transfer across the control volume surface is given as

$$\dot{Q}_R = (\nabla \cdot \vec{q}_R)\, d\forall \tag{1.91}$$

Combining Equation 1.90 and Equation 1.91, we get total heat transfer rate through the control volume surface as

$$\dot{Q} = -(\nabla \cdot \vec{q})\, d\forall + (\nabla \cdot \vec{q}_R)\, d\forall \tag{1.92}$$

The second term $\dot{W}$ in Equation 1.88 represents the energy transfer across the control volume surfaces due to the net work done by the surface forces on the differential elements. The three components of this work are given as

$$\dot{W}_x = \left[\frac{\partial}{\partial x}\{u(\sigma_{xx} - p)\} + \frac{\partial}{\partial y}(u\tau_{yx}) + \frac{\partial}{\partial z}(u\tau_{zx})\right] d\forall \tag{1.93a}$$

$$\dot{W}_y = \left[\frac{\partial}{\partial x}(v\tau_{xy}) + \frac{\partial}{\partial y}\{v(\sigma_{yy} - p)\} + \frac{\partial}{\partial z}(v\tau_{zy})\right] d\forall \tag{1.93b}$$

$$\dot{W}_z = \left[\frac{\partial}{\partial x}(w\tau_{xz}) + \frac{\partial}{\partial y}(w\tau_{yz}) + \frac{\partial}{\partial x}\{w(\sigma_{zz} - p)\}\right] d\forall \tag{1.93c}$$

The total work done by all surface stresses is given as

$$\dot{W} = \left[\frac{\partial}{\partial x}\{u\sigma_{xx} + v\tau_{xy} + w\tau_{xz}\} + \frac{\partial}{\partial y}\{u\tau_{yx} + v\sigma_{yy} + w\tau_{yz}\}\right.$$
$$\left. + \frac{\partial}{\partial z}\{u\tau_{zx} + v\tau_{zy} + w\sigma_{zz}\}\right] d\forall - \left[\frac{\partial}{\partial x}(up) + \frac{\partial}{\partial y}(vp) + \frac{\partial}{\partial z}(wp)\right] d\forall \tag{1.94}$$

The third term $\dot{Q}_{\text{gen}}$ in Equation 1.88 represents the total heat generation within the differential element and is expressed as

$$\dot{Q}_{\text{gen}} = \dot{Q}''' \, d\forall \tag{1.95}$$

Where $\dot{Q}'''$ is the volumetric heat generation rate. Energy generation may occur as a result of chemical reaction, absorption of nuclear radiation, electrical heat generation, or absorption of solar or laser energy in a semitransparent medium.

The fourth term in Equation 1.88 represents the rate of change of energy within the volume, which for the case of negligible potential energy is the rate of change of internal energy and kinetic energy given as

$$\frac{\partial}{\partial t}\int_{\forall} e\rho \, d\forall = \frac{\partial}{\partial t}\left(\rho i + \tfrac{1}{2}\rho V^2\right) d\forall \tag{1.96}$$

where i is the internal energy per unit mass of the fluid in the differential element and V is the magnitude of the local velocity defined at the center of the element.

The fifth term in Equation 1.88 represents the net energy flux across the control volume surfaces due to fluid motion, and expressed as

$$\int_{\text{cs}} e\rho\vec{V}\cdot d\vec{A} = \left[\frac{\partial}{\partial x}\left(\left(i + \frac{V^2}{2}\right)\rho u\right) + \frac{\partial}{\partial y}\left(\left(i + \frac{V^2}{2}\right)\rho v\right)\right.$$
$$\left. + \frac{\partial}{\partial z}\left(\left(i + \frac{V^2}{2}\right)\rho w\right)\right] dx\, dy\, dz$$

or

$$\int_{CS} e\rho\vec{V}\cdot d\vec{A} = \left[\nabla\left(i+\frac{V^2}{2}\right)\rho\vec{V}\right]d\forall \tag{1.97}$$

Substituting Equation 1.92 and Equation 1.94 to Equation 1.97 into Equation 1.88 we have

$$\begin{aligned}\frac{\partial}{\partial t}\left(\rho\left(i+\frac{V^2}{2}\right)\right) &= -\nabla\cdot\vec{q}+(\nabla\cdot\vec{q}_{\mathrm{R}})-\nabla\left[\rho\left(i+\frac{V^2}{2}\right)\vec{V}\right]\\ &\quad -\nabla\cdot\left(p\vec{V}\right)+\dot{Q}'''+\nabla\cdot\left(\tau\vec{V}\right)\end{aligned} \tag{1.98}$$

Equation 1.98 is further simplified and regrouped as

$$\rho\frac{D}{Dt}\left(i+\frac{V^2}{2}\right) = -\nabla\cdot\vec{q}+\nabla\cdot\vec{q}_{\mathrm{R}}-\nabla\left(p\vec{V}\right)+\dot{Q}'''+\nabla\cdot\left(\tau\vec{V}\right) \tag{1.99}$$

Equation 1.99 is referred to as the total energy balance equation in a strictly conservative form. A more popular form in heat transfer analysis, known as the thermal or internal energy balance, is obtained by subtracting the mechanical energy equation

$$\rho\frac{D}{Dt}\left(\frac{V^2}{2}\right) = -\vec{V}\nabla p+\vec{V}\nabla\cdot\tau \tag{1.100}$$

from Equation 1.99, and written as

$$\rho\frac{Di}{Dt} = -\nabla\cdot\vec{q}+\nabla\cdot\vec{q}_{\mathrm{R}}-p\nabla\cdot\vec{V}+\dot{Q}'''+\tau\nabla\cdot\vec{V} \tag{1.101}$$

The thermal energy equation can also be written in terms of enthalpy by substituting $h = i+(p/\rho)$ into Equation 1.101 as

$$\rho\frac{Dh}{Dt} = -\nabla\cdot\vec{q}+\nabla\cdot\vec{q}_{\mathrm{R}}+\frac{dp}{dt}+\dot{Q}'''+\tau\nabla\cdot\vec{V} \tag{1.102}$$

With the substitution of Equations 1.51a–f for normal stresses and shear stresses, and Equation 1.65 for conduction transport rate equation in Equation 1.101 and Equation 1.102, we get

$$\rho\frac{Di}{Dt} = \nabla\cdot(k\nabla T)+\nabla\cdot\vec{q}_{\mathrm{R}}-p\nabla\cdot\vec{V}+\dot{Q}'''+\Phi \tag{1.103}$$

and

$$\rho \frac{Dh}{Dt} = \nabla \cdot (k\nabla T) + \nabla \cdot \vec{q}_{\mathrm{R}} + \frac{Dp}{Dt} + \dot{Q}''' + \Phi \tag{1.104}$$

where Φ is called the viscous dissipation term given as

$$\Phi = 2\left[\left(\frac{\partial u}{\partial x}\right)^2 + \left(\frac{\partial v}{\partial y}\right)^2 + \left(\frac{\partial w}{\partial x}\right)^2\right] + \left[\left(\frac{\partial u}{\partial y} + \frac{\partial v}{\partial x}\right)^2 + \left(\frac{\partial v}{\partial z} + \frac{\partial w}{\partial y}\right)^2 + \left(\frac{\partial w}{\partial x} + \frac{\partial u}{\partial z}\right)^2\right]$$
$$- \frac{2}{3}\left[\frac{\partial u}{\partial x} + \frac{\partial v}{\partial y} + \frac{\partial w}{\partial z}\right]^2 \tag{1.105}$$

In most heat transfer problems, thermal energy equations (1.103) and (1.104) are written solely in terms of primary variables such as temperature by expressing internal energy and enthalpy in terms of primary variables using thermodynamic relations (Sonntag et al., 1998).

The temperature-based energy equation is given in two different forms

$$\rho c_v \frac{DT}{Dt} = \nabla \cdot (k\nabla T) + \nabla \cdot \vec{q}_{\mathrm{R}} - T\left(\frac{\partial p}{\partial T}\right)_v \left(\nabla \cdot \vec{V}\right) + \dot{Q}''' + \Phi \tag{1.106}$$

and

$$\rho c_p \frac{DT}{Dt} = \nabla \cdot (k\nabla T) + \nabla \cdot \vec{q}_{\mathrm{R}} + \beta T \frac{DP}{Dt} + \dot{Q}''' + \Phi \tag{1.107}$$

The energy equation is simplified for some commonly used cases as follows

1. ***Incompressible liquid***. For incompressible fluid $\nabla \cdot \vec{V} = 0$, $\beta = 0$, and Equation 1.106 and Equation 1.107 reduce to

$$\rho c \frac{DT}{Dt} = \nabla \cdot (k\nabla T) + \nabla \cdot \vec{q}_R + \dot{Q}''' + \Phi \tag{1.108}$$

2. ***Ideal gas***. For an ideal gas $Pv = RT$ it gives $(\partial p/\partial T)_v = R/v$, and this reduces the internal energy-based energy equation (1.103) to

$$\rho c_v \frac{DT}{Dt} = \nabla \cdot (k\nabla T) + \nabla \cdot \vec{q}_{\mathrm{R}} - p\left(\nabla \cdot \vec{V}\right) + \dot{Q}''' + \Phi \tag{1.109}$$

Also, for the ideal gas it can be shown that $\beta = 1/T$ and the enthalpy-based energy equation (1.107) reduces to

$$\rho c_p \frac{DT}{Dt} = \nabla \cdot (k\nabla T) + \nabla \cdot \vec{q}_R + \frac{DP}{Dt} + \dot{Q}''' + \Phi \tag{1.110}$$

3. ***A simplified heat equation***. A simplified model applicable to many convection problems is derived based on the following assumptions: (a) constant thermal conductivity, k; (b) zero internal heat generation, $\dot{Q}'''$; (c) negligible viscous dissipation Φ; and (d) negligible compressibility effect and (e) negligible radiation heat transfer rate, i.e., $\vec{q}_R = 0$. The energy equation for such a model is deduced from Equation 1.107 as

$$\rho c_p \frac{DT}{Dt} = k\nabla^2 T \tag{1.111}$$

In terms of specific coordinate systems, we can write

***Cartesian* (*x*, *y*, *z*)**

$$\rho c_p \left(\frac{\partial T}{\partial t} + u\frac{\partial T}{\partial x} + v\frac{\partial T}{\partial y} + w\frac{\partial T}{\partial z} \right) = k\left(\frac{\partial^2 T}{\partial x^2} + \frac{\partial^2 T}{\partial y^2} + \frac{\partial^2 T}{\partial z^2} \right) \tag{1.112a}$$

***Cylindrical* (*r*, *θ*, *z*)**

$$\rho c_p \left(\frac{\partial T}{\partial t} + u_r\frac{\partial T}{\partial r} + \frac{u_\theta}{r}\frac{\partial T}{\partial \theta} + u_z\frac{\partial T}{\partial z} \right) = k\left[\frac{1}{r}\frac{\partial}{\partial r}\left(r\frac{\partial T}{\partial r} \right) + \frac{1}{r^2}\frac{\partial^2 T}{\partial \theta^2} + \frac{\partial^2 T}{\partial z^2} \right] \tag{1.112b}$$

***Spherical* (*r*, *θ*, *ϕ*)**

$$\begin{aligned} &\rho c_p \left(\frac{\partial T}{\partial t} + u_r\frac{\partial T}{\partial r} + \frac{u_\phi \partial T}{r\partial \phi} + \frac{u_\theta}{r\sin\phi}\frac{\partial T}{\partial \theta} \right) \\ &= k\left[\frac{1}{r^2}\frac{\partial}{\partial r}\left(r^2\frac{\partial T}{\partial r} \right) + \frac{1}{r^2\sin\phi}\frac{\partial}{\partial \phi}\left(\sin\phi\frac{\partial T}{\partial \phi} \right) + \frac{1}{r^2\sin^2\phi}\frac{\partial^2 T}{\partial \theta^2} \right] \end{aligned} \tag{1.112c}$$

While dealing with high viscous flows with large velocity gradients, the model above is modified by adding the viscous dissipation term Φ, which takes into account the internal heating due to viscous dissipation

$$\rho c_p \frac{DT}{Dt} = k\nabla^2 T + \Phi \tag{1.113}$$

Such problems are encountered in lubrication, thermo-mechanical analysis of metal forming processes, and convective heat transfer involving high viscous fluids and/or high velocities.

4. ***Heat conduction in solids – heat diffusion equation***. For solids, $\vec{V} = 0$ and the energy equation (1.113), along with the inclusion of the volumetric heat generation term, transforms into a form known as the ***heat diffusion equation*** for solids

$$\rho c \frac{\partial T}{\partial t} = \nabla \cdot (k\nabla T) + Q''' \tag{1.114}$$

and for constant thermal conductivity it reduces to

$$\frac{1}{\alpha}\frac{\partial T}{\partial t} = \nabla^2 T + Q''' \tag{1.115}$$

where $\alpha = k/\rho c$ is called the thermal diffusivity of the material.

Equation 1.114 and Equation 1.115 are written in a specific co-ordinate system as given in Table 1.1.

Table 1.1 Heat Diffusion Equation in Specific Coordinate Systems

Cartesian coordinate

$$\rho c\frac{\partial T}{\partial t} = \frac{\partial}{\partial x}\left(k\frac{\partial T}{\partial x}\right) + \frac{\partial}{\partial y}\left(k\frac{\partial T}{\partial y}\right) + \frac{\partial}{\partial z}\left(k\frac{\partial T}{\partial z}\right) + \dot{Q}''' \tag{1.116a}$$

$$\frac{1}{\alpha}\frac{\partial T}{\partial t} = \frac{\partial^2 T}{\partial x^2} + \frac{\partial^2 T}{\partial y^2} + \frac{\partial^2 T}{\partial y^2} + \frac{\dot{Q}'''}{k} \tag{1.116b}$$

Cylindrical coordinate

$$\rho c\frac{\partial T}{\partial t} = \frac{1}{r}\frac{\partial}{\partial r}\left(kr\frac{\partial T}{\partial r}\right) + \frac{1}{r^2}\frac{\partial}{\partial \theta}\left(k\frac{\partial T}{\partial \theta}\right) + \frac{\partial}{\partial z}\left(k\frac{\partial T}{\partial z}\right) + \dot{Q}''' \tag{1.116c}$$

$$\frac{1}{\alpha}\frac{\partial T}{\partial t} = \frac{1}{r}\frac{\partial}{\partial r}\left(r\frac{\partial T}{\partial r}\right) + \frac{1}{r^2}\frac{\partial^2 T}{\partial \theta^2} + \frac{\partial^2 T}{\partial z^2} + \frac{\dot{Q}'''}{k} \tag{1.116d}$$

Spherical coordinate

$$\rho c\frac{\partial T}{\partial t} = \frac{1}{r^2}\frac{\partial}{\partial r}\left(kr^2\frac{\partial T}{\partial r}\right) + \frac{1}{r^2\sin\phi}\frac{\partial}{\partial \phi}\left(k\sin\phi\frac{\partial T}{\partial \phi}\right) + \frac{1}{r^2\sin^2\phi}\frac{\partial}{\partial \theta}\left(k\frac{\partial T}{\partial \theta}\right) + \dot{Q}''' \tag{1.116e}$$

$$\frac{1}{\alpha}\frac{\partial T}{\partial t} = \frac{1}{r^2}\frac{\partial}{\partial r}\left(r^2\frac{\partial T}{\partial r}\right) + \frac{1}{r^2\sin\phi}\frac{\partial}{\partial \phi}\left(\sin\phi\frac{\partial T}{\partial \phi}\right) + \frac{1}{r^2\sin^2\phi}\frac{\partial^2 T}{\partial \theta^2} + \frac{\dot{Q}'''}{k} \tag{1.116f}$$

Equation 1.114 and Equation 1.115 can further be simplified in some special cases as follows.

1. *Unsteady heat conduction without heat source*

$$\rho c\frac{\partial T}{\partial t} = \nabla \cdot (k\nabla T) \tag{1.117a}$$

$$\frac{1}{\alpha}\frac{\partial T}{\partial t} = \nabla^2 T \tag{1.117b}$$

2. *Steady state heat conduction with heat source*

$$\nabla \cdot (k\nabla T) + \dot{Q}''' = 0 \tag{1.118a}$$

$$\nabla^2 T + \frac{\dot{Q}'''}{k} = 0 \tag{1.118b}$$

3. *Steady state heat conduction without heat source*

$$\nabla \cdot (k \nabla T) = 0 \tag{1.119a}$$

$$\nabla^2 T = 0 \tag{1.119b}$$

1.3.3 Initial and Boundary Conditions for Heat Transfer

In order to determine the temperature distribution and heat transfer rate in the medium, we need to solve the appropriate form of the differential heat diffusion equation using the physical condition in time, known as the ***initial condition***, and physical conditions in space, known as the ***boundary conditions***.

Initial Condition. The initial condition is needed for problems that are transient or time dependent. Since the heat diffusion equation is first-order in time, only one initial condition is needed. The initial condition specifies the known temperature distribution in the medium at an instant of time, usually at the origin of time, and it is stated as

$$\text{At time } t = 0, \quad T(\vec{x}, 0) = f(\vec{x}) \quad \text{or} \quad T_1 \tag{1.120}$$

Boundary Conditions. Temperature boundary conditions are derived from the physical condition that exists at the boundary. Some of the most common boundary conditions are described as follows.

1. Boundary condition of the first kind or *Dirichlet boundary condition* (Figure 1.19). The surface temperature is specified as a constant temperature or a function of space and time, i.e.

$$T = T_S(\vec{x}, t) \quad \text{on the surface } S \tag{1.121}$$

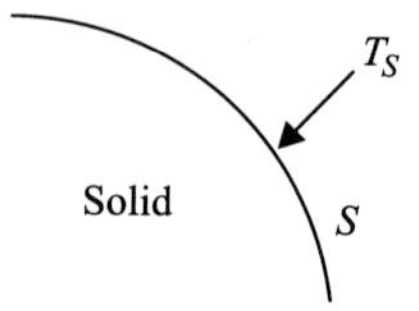

Figure 1.19 Boundary condition of the first kind

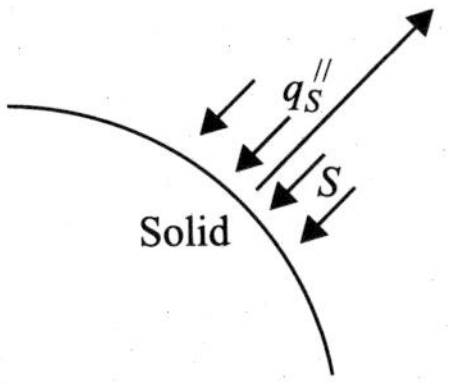

Figure 1.20 Boundary condition of the second kind

where $T_S(\vec{x}, t))$ is a function of time and position or it could be a constant surface temperature, T_S. For the special case of $T_S = 0$, the boundary condition is referred to as the *homogeneous boundary condition of the first kind.*

2. The boundary condition of the second kind or *Newmann condition* (Figure 1.20). The heat flux at the surface is specified to be constant or a function of space and time, i.e.

$$q_n'' = \pm q_S''(\vec{r}, t) \quad \text{or} \quad \pm q_S'' \tag{1.122a}$$

or

$$-k\frac{\partial T}{\partial n} = \pm q_S''(\vec{r}, t) \quad \text{or} \quad \pm q_S'' \tag{1.122b}$$

where $q_S(\vec{x}, t)$ is a function of time and position or it could be a constant surface heat flux q_S'', and q_n'' is the conduction heat transfer rate per unit area as given by the ***Fourier's*** law of heat conduction equation (1.69), and shown as positive in the outward normal direction to the surface. For a heat transfer rate in an inward normal direction to the surface, the heat transfer rate q_n'' will be considered as negative, and the boundary condition will be stated as

$$-q_n'' = \pm q_S''(\vec{r}, t) \quad \text{or} \quad \pm q_S'' \tag{1.123a}$$

or

$$k\frac{\partial T}{\partial n} = \pm q_S''(\vec{r}, t) \quad \text{or} \quad \pm q_S'' \tag{1.123b}$$

For the special case of

$$q_S'' = 0 \quad \text{or} \quad \left.\frac{\partial T}{\partial n}\right|_S = 0 \tag{1.124}$$

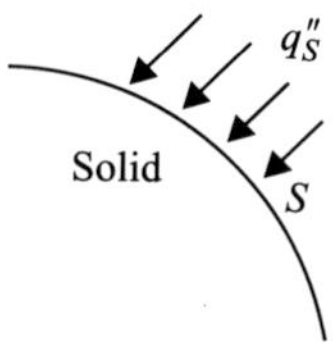

Figure 1.21 Homogeneous boundary condition of the second kind

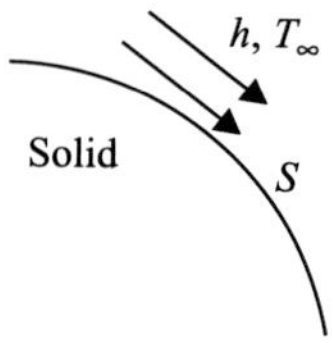

Figure 1.22 Boundary condition of the third kind

the boundary condition is referred to as the *homogeneous boundary condition of the second kind* (Figure 1.21). This is also referred as *adiabatic condition*, or *perfectly insulated condition* or *symmetric condition*.

3. The boundary condition of the third kind or *convective surface condition*

$$\pm q_n'' = h_c(T_s - T_\infty) \tag{1.125a}$$

or

$$\pm k \frac{\partial T}{\partial x}\bigg|_s = h_c(T_s - T_\infty) \tag{1.125b}$$

For combined heat losses by convection and radiation from the surface, an energy balance leads a boundary condition as

$$\pm k \frac{\partial T}{\partial x}\bigg|_s = h_c(T_s - T_\infty) + h_r(T_s - T_{sur}) \tag{1.126}$$

where h_r is the radiation film coefficient given by Equation 1.87 as

$$h_r = \sigma(T_s + T_{sur})(T_s^2 + T_{sur}^2)/R_{s,\,sur} \tag{1.127}$$

1.4 MASS TRANSFER

Mass transfer is a science that describes the relative motion of species in a mixture of chemical species as a result of mass concentration gradient or difference. There are two basic modes of mass transfer: ***mass transfer by***

diffusion and ***mass transfer by convection.*** There is a close similarity between heat transfer and mass transfer in terms of the transport rate equation and the transport conservation equation. Let us briefly define some of the basic mixture properties before describing mass transport rate and conservation equations in the following section. For more comprehensive discussion, refer to Sherwood et al. (1975), Bird et al. (2002), Incropera and DeWitt (2002), Kays and Crawford (1993), Mills (1999), Mills (2001), and Treybal (1980).

The amount of any species i in a mixture of chemical species is given in terms of mass density or concentration ρ_i (kg/m^3) on mass basis. This is also defined on a molar basis as molar density or concentration C_i (kmol/m^3). The mass density and concentration are related through molecular weight M_i as

$$C_i = \frac{\rho_i}{M_i} \tag{1.128}$$

The total mass density ρ (kg/m^3) and total molar density C (mole/m^3) are defined as

$$\rho = \sum_i \rho_i \tag{1.129}$$

and

$$C = \sum_i C_i \tag{1.130}$$

The quantity of a species can also be expressed in terms of mass fraction m_i or mole fraction x_i, which are defined as

$$m_i = \frac{\rho_i}{\rho} \tag{1.131}$$

and

$$x_i = \frac{C_i}{C} \tag{1.132}$$

1.4.1 Basic Modes and Transport Rate Equation

Diffusion Mass Transfer. This mode is primarily important for mass transfer in a stationary medium such as in a solid and stationary fluid. The ***diffusion rate equation*** is given by ***Fick's law***, which expresses the transfer of a species 1 in a binary mixture of 1 and 2 as

$$\vec{j}_1 = -\rho D_{12} \nabla m_1 \tag{1.133a}$$

or

$$\vec{\bar{j}}_1 = -cD_{12}\nabla x_1 \tag{1.133b}$$

where

$\vec{j}_1$ = diffusive mass flux of species 1 (kg/m^2 · s)
$\vec{\bar{j}}_1$ = diffusive molar mass flux of species 1 (kmol/m^2 · s)
m_1 = mass fraction of species 1
x_1 = mole fraction of species 1
ρ = local mixture density (kg/m^3)
D_{12} = binary diffusion coefficient or mass diffusivity (m^2/s)

In scalar form, the equation can be written in three components and in a Cartesian coordinate system as

$$j_{x_1} = -\rho D_{12}\frac{\partial m_1}{\partial x} \tag{1.134a}$$

$$j_{x_1} = -\rho D_{12}\frac{\partial m_1}{\partial y} \tag{1.134b}$$

$$j_{x_1} = -\rho D_{12}\frac{\partial m_1}{\partial z} \tag{1.134c}$$

Mass Diffusion Coefficient As we can see ***Fick's law*** is analogous to ***Fourier's law*** for heat conduction and introduces in an analogous way the ***mass density*** or ***binary diffusion coefficient*** as a material transport property. Empirical correlations for a binary diffusion coefficient are derived based on kinetic theory and using experimental data. A widely used such correlation is given as

$$D_{12} = \frac{0.00100T^{1.75}\left(\frac{1}{M_1} + \frac{1}{M_2}\right)^{1/2}}{P\left[\left(\sum v\right)_1^{1/3} + \left(\sum v\right)_2^{1/3}\right]^2} \tag{1.135}$$

where M_1 and M_2 are the molecular weights of species 1 and 2, respectively, and the quantity $\left(\sum v\right)$ represents the summation of atomic-diffusion volumes for each species of the binary. A list is given in Sherwood et al. (1975).

A commonly used theoretical equation for the binary diffusion coefficient at low pressure is given as

$$D_{12} = \frac{0.001858T^{1.5}((1/M_1) + (1/M_2))^{1/2}}{P\sigma_{12}\Omega_D} \tag{1.136}$$

where

T = temperature in K
P = pressure in atm
$\Omega_D = f(kT/\varepsilon_{12})$
$\sigma_{12}, \varepsilon_{12}$ = Lennard–Jones force constants
k = Boltzman constant

The binary values of σ_{12} and ε_{12} are computed from the values of component species based on the following rules

$$\sigma_{12} = \tfrac{1}{2}(\sigma_1 + \sigma_2) \tag{1.137a}$$

and

$$\frac{\varepsilon_{12}}{k} = \left(\frac{\varepsilon_1}{k}\frac{\varepsilon_2}{k}\right)^{1/2} \tag{1.137b}$$

Values of $\Omega_D = f(kT/\varepsilon_{12})$, ε/k and σ for many common pure substances are given in Sherwood et al. (1975).

In a porous media the diffusion mechanism can be of three different types: ordinary diffusion, Knudsen diffusion, and surface diffusion. If the pores are much larger than the mean free path length, then the molecules collide with each other more frequently than with the pore walls, and ordinary diffusion is assumed to be the dominant diffusion mechanism. Knudsen diffusion is encountered in smaller pores and/or at lower pressure or density. In this case, molecules collide more frequently with the walls than with each other. The Knudsen diffusion coefficient is given, based on kinetic theory, as

$$D_{1k} = 2/3a\bar{v}_1 \tag{1.138}$$

where a is the effective pore radius in meters and $\bar{v}_1$ is the average molecular speed of species, and given as

$$\bar{v}_1 = \left(\frac{8RT}{\pi M_1}\right) \tag{1.139}$$

Substituting the value for the gas constant and combining Equation 1.138 and Equation 1.139, the expression for Knudsen diffusion co-efficient is given as

$$D_{1k} = 97a\left(\frac{T}{M_1}\right)^{1/2}, \quad \text{m}^2/\text{s} \tag{1.140}$$

where T is temperature in Kelvin and M_1 is the molecular weight.

In an intermediate range of pore sizes, both ordinary and Knudsen diffusion contributes to the transfer of the species in the media. In this range, the combined ordinary and Knudsen diffusion can be represented by assuming parallel resistances expressed as

$$D_1 = \left(\frac{1}{D_{12}} + \frac{1}{D_{1k}} \right)^{-1} \tag{1.141}$$

For even smaller pores and for species adsorbed on pore surfaces, the dominant mechanism is surface diffusion in which molecules diffuse along the surface in the direction of decrease in concentration.

Other important factors that have to be considered for porous media are the presence of the tortuous path and changes and reduction in the cross-sectional area of pore channels. The effective diffusivity is expressed as

$$D_{1,e} = \frac{\varepsilon}{\tau} D_1 \tag{1.142}$$

where ε is the ***porosity*** of the media that accounts for the reduction of the free area for diffusion due to the presence of the solid phase and τ is the ***tortuosity factor*** that accounts for the increase in the diffusional path due to the tortuous path of the pores.

Convection Mass Transfer. The convection mass transfer is analogous to convection heat transfer and occurs between a moving mixture fluid species and an exposed solid surface. Like hydrodynamic and thermal boundary layers, a concentration boundary layer forms over the surface if the free stream concentration of a species 1, $C_{1\infty}$, differs from species concentration at the surface, C_{1S}.

Since the fluid is stationary at the solid surface, the mass of species 1 is transferred by diffusion normal to the surface, and is expressed by the diffusion rate equation 1.33 as

$$m''_{1S} = -D_{12} \left. \frac{\partial C_1}{\partial y} \right|_{y=0} \tag{1.143}$$

Like convection heat transfer, the mass flux of the species 1 is given by a ***convection rate equation*** as

$$m''_{1S} = h_m (C_{S1} - C_{\infty 1}) \tag{1.144}$$

where h_m is called the ***convection mass transfer coefficient*** or ***mass film coefficient***. The defining equation for the convection mass transfer coefficients is obtained as

$$h_m = \frac{m''_{1S}}{(C_{1S} - C_{1\infty})} \tag{1.145}$$

and

$$h_{\mathrm{m}} = \frac{-D_{12}(\partial C_1/\partial y)\big|_{y=0}}{(C_{1\mathrm{S}} - C_{1\infty})} \tag{1.146}$$

The correlations for convective mass transfer coefficient are given in a similar manner as in convections heat transfer, but in terms of mass transfer parameters as

$$\mathrm{Sh} = f(\mathrm{Re}, \mathrm{Sc}) \tag{1.147}$$

where

$$\mathrm{Sh} = \text{Sherwood number} = \frac{h_{\mathrm{m}} L_{\mathrm{c}}}{D_{12}} \tag{1.148}$$

$$\mathrm{Sc} = \text{Schmidt number} = \frac{\nu}{D_{12}} \tag{1.149}$$

Relation Between Heat and Mass Transfer. Examination of the governing equations and boundary conditions shows that mass transfer phenomena is closely related to heat transfer. So, the correlations for heat and mass transfer can be used interchangeably. The relation between heat and mass transfer convection coefficients is given by

$$\frac{h}{h_{\mathrm{m}}} = \rho c_{\mathrm{P}} \mathrm{Le}^{1/2} \tag{1.150}$$

where $\mathrm{Le} = \alpha/D_{12}$ is defined as the Lewis number.

1.4.2 Conservation of Mass Species and Mass Concentration Equation

Like heat transfer rates, the species mass flux can be determined from the mass concentration field by solving the species mass concentration equation, which is a statement of conservation of mass species. The species mass concentration equation is derived following the procedure outlined for conservation of energy as

$$\frac{DC_1}{Dt} = \nabla \cdot (D_{12} \nabla C_1) + \dot{n}_1 \quad \text{in molar form} \tag{1.151a}$$

and

$$\frac{D\rho_1}{Dt} = \nabla \cdot (D_{12} \nabla \rho_1) + \dot{m}_1 \quad \text{in mass basis} \tag{1.151b}$$

where $\dot{n}_1$ and $\dot{m}_1$ are the volumetric generation or consumption of the species 1 per unit volume in molar basis and mass basis respectively.

In Cartesian coordinate systems, we can write the equation as

***Cartesian* (*x*, *y*, *z*)**

$$\left(\frac{\partial C_1}{\partial t} + u\frac{\partial C_1}{\partial x} + v\frac{\partial C_1}{\partial y} + w\frac{\partial C_1}{\partial z}\right)$$

$$= \frac{\partial}{\partial x}\left(D_{12}\frac{\partial C_1}{\partial x}\right) + \frac{\partial}{\partial y}\left(D_{12}\frac{\partial C_1}{\partial y}\right) + \frac{\partial}{\partial z}\left(D_{12}\frac{\partial C_1}{\partial z}\right) + \dot{n}_1 \quad (1.152a)$$

or

$$\left(\frac{\partial \rho_1}{\partial t} + u\frac{\partial \rho_1}{\partial x} + v\frac{\partial \rho_1}{\partial y} + w\frac{\partial \rho_1}{\partial z}\right)$$

$$= \frac{\partial}{\partial x}\left(D_{12}\frac{\partial \rho_1}{\partial x}\right) + \frac{\partial}{\partial y}\left(D_{12}\frac{\partial \rho_1}{\partial y}\right) + \frac{\partial}{\partial z}\left(D_{12}\frac{\partial \rho_1}{\partial z}\right) + \dot{m}_1 \quad (1.152b)$$

The second, third and fourth terms on the left-hand side of Equation 1.152 represent the net transport of the species 1 by convective motion in x, y and z directions, respectively. The first three terms on the right-hand side of the equation represent the net tranport by diffusion. The last term on the right-hand side represents the generation or consumption of the species 1 by processes such as absorption, adsorption/desorption, and other chemical or electro-chemical reactions.

Mass Diffusion Equation in Solids. For solids, $\vec{V} = 0$ and the mass equation (1.151a) transforms into a form known as the ***mass diffusion equation*** for solids

$$\frac{\partial C_1}{\partial t} = \nabla \cdot (D_{12}\nabla C_1) + \dot{n}_1 \quad (1.153)$$

and for constant mass diffusivity, it reduces to

$$\frac{1}{D_{12}}\frac{\partial C_1}{\partial t} = \nabla^2 C_1 + \frac{\dot{n}_1}{D_{12}} \quad (1.154)$$

In Cartesian coordinates, this takes the form

$$\frac{1}{D_{12}}\frac{\partial C_1}{\partial t} = \frac{\partial^2 C_1}{\partial x^2} + \frac{\partial^2 C_1}{\partial y^2} + \frac{\partial^2 C_1}{\partial z^2} + \frac{\dot{n}_1}{D_{12}} \quad (1.155)$$

1.4.3 Initial and Boundary Conditions for Mass Transfer

Initial Condition. The initial condition is specified in terms of known concentration distribution in the medium at an instant of time. For example, it can be stated as

$$\text{at time } t = 0, \quad C(\vec{x}, 0) = f(\vec{x}) \quad \text{or} \quad C_t \tag{1.156}$$

Boundary Conditions. Concentration boundary conditions are specified in the same manner as in temperature based on the physical condition that exists at the boundary. Some of the most common boundary conditions are described as follows:

1. *Specified surface concentration.* The surface concentration is specified as a constant concentration or a function of space and time, i.e.

$$C = C_S(\vec{x}, t) \quad \text{on the surface } S \tag{1.157}$$

where $C_S(\vec{x}, t)$ is a function of time and position or it could be a constant surface concentration, C_S.

2. *Surface mass generation and consumption.* The mass flux at the surface is specified to be constant or a function of space and time, i.e.

$$-D_{12}\left.\frac{\partial C}{\partial x}\right|_S = \pm m''_S(\vec{x}, t) \tag{1.158}$$

where $m_S(\vec{x}, t)$ is a function of time and position or it could be a constant surface heat flux, m''_S.

For an ***impermeable surface*** $m''_S = 0$ and the boundary condition reduces to

$$\left.\frac{\partial C}{\partial x}\right|_S = 0 \tag{1.159}$$

3. *The convective surface condition*

$$\pm D_{12}\left.\frac{\partial C}{\partial x}\right|_s = h_m(C_s - C_\infty) \tag{1.160}$$

1.5 MATHEMATICAL CLASSIFICATION OF GOVERNING EQUATIONS

The governing differential equations for fluid flow and heat transfer can be transformed into a standard form depending on the physical nature under consideration. We will give a brief discussion on the mathematical classification of the different forms of these governing differential equations. This will help in selecting appropriate discretization and solution procedures. Let us consider a general two-dimensional second-order partial differential equation of the form

$$a(x,y)\frac{\partial^2\phi}{\partial x^2}+b(x,y)\frac{\partial\phi}{\partial x\partial y}+c(x,y)\frac{\partial^2\phi}{\partial y^2}+F(x,y,\phi,\frac{\partial\phi}{\partial x},\frac{\partial\phi}{\partial y})=0 \tag{1.161}$$

The equation is linear for

$$F(x,y,\phi,\frac{\partial\phi}{\partial x},\frac{\partial\phi}{\partial y})=d\frac{\partial\phi}{\partial x}+e\frac{\partial\phi}{\partial y}+f\phi+g \tag{1.162}$$

where coefficients a, b, c, d, e, and f are arbitrary functions of independent variables x and y. The dependent variable ϕ and its normal derivatives are known on a smooth curve Γ in the x–y plane. A combination of these coefficients leads to different standard types of partial differential equations such as ***elliptic***, ***parabolic***, and ***hyperbolic***. Results arise from the existence of a unique solution for all second derivatives at a point P on the curve. The necessary and sufficient conditions for such a solution state that if $a(\mathrm{P})\neq 0$ the curve will be characteristic at the point P if and only if the slope of the curve at the point satisfies

$$\frac{dy}{dx}=\frac{b\pm\sqrt{b^2-4ac}}{a} \tag{1.163}$$

Three special cases can be highlighted.

1. If $b^2-4ac=0$ at point P, there is only one characteristic direction at P and Equation 1.161 is parabolic at P. For example, the diffusion equation $\partial\phi/\partial t=\alpha\partial^2\phi/\partial x^2$ is parabolic in the x–t plane.
2. If $b^2-4ac<0$, no real curve can satisfy Equation 1.163 and is not characteristic at point P, and Equation 1.161 is said to be ***elliptic*** at P. For the two-dimensional Laplace equation, $\partial^2\phi/\partial x^2+\partial^2\phi/\partial x^2=0$, we get $b^2-4ac=-1$, so this equation is elliptic in the x–y plane.
3. If $b^2-4ac>0$, there are two characteristic directions at P, and Equation 1.161 is said to be ***hyperbolic***. For the wave equation of the form $\partial^2\phi/\partial t^2=\beta\partial^2\phi/\partial x^2$, we get $b^2-4ac=\beta$, i.e., $b^2-4ac>0$ for a positive value of β. So, the wave equation is hyperbolic in the x–t plane. More discussions about this classification are given by Anderson et al. (1984), Courant and Hilbert (1962), and Stakgold (1979).

This mathematical classification is often tied with the concepts of one-way and two-way coordinates (Patankar, 1980). A one-way coordinate is defined as the one in which the properties at a given point are influenced by changes in properties on only one side of that point, i.e. either upstream or downstream side of that point. In two-way coordinates, the properties at a point are influenced by changes in properties on both sides, i.e., both upstream and downstream of that point. While the term parabolic represents a one-way coordinate, the elliptic represents a two-way concept. Time is always one-way, based on the fact that the flow or temperature variable at a given time is only influenced by changes at a previous time. Space coordinates, on the other hand, are normally two-way co-ordinates. However, in some fluid flow and convection heat transfer problems, it is assumed to be one-directional or can be treated as parabolic. This situation arrives when there is a strong unidirectional flow in the coordinate direction so that the significant influences travel only from upstream points to downstream points. It should be remembered that diffusion processes have two-way characteristics. However, for strong unidirectional flow, convection overpowers diffusion and makes the space coordinate one-way.

PROBLEMS

1.1 Start from the Navier–Stokes equations and derive the governing equations in Cartesian coordinates for low Reynolds number flow, referred to as ***creep flow***.

1.2 Make appropriate assumptions and drive the two-dimensional boundary-layer equations for flow over a flat plate in Cartesian coordinates.

1.3 Consider the flow of liquid film in steady and laminar motion, falling down an inclined surface with slope θ from the vertical direction of the gravity. The thickness, h, of the film is constant along the length. Assume the flow as fully-developed and at zero-pressure gradient, i.e., $dP/dx = 0$.

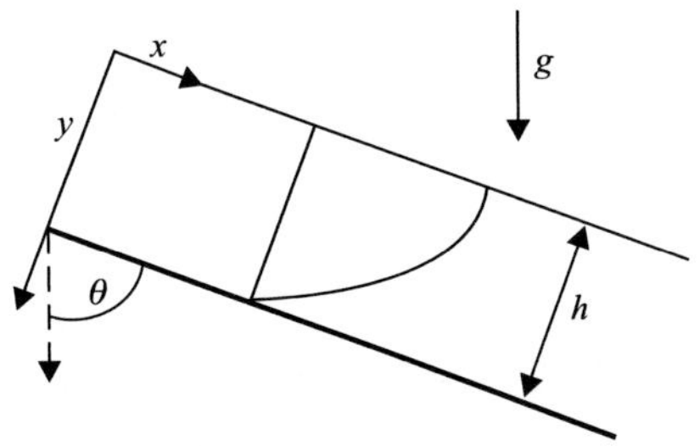

Derive the mathematical statement of the problem from mass and momentum equations of fluid flow for determining velocity distribution and mass flow rate of the liquid film.

1.4 Consider the flow of a viscous incompressible fluid in a square cavity shown in the figure below. The flow is induced by the motion of the top wall at a constant velocity U_w. Derive the governing equation and boundary conditions.

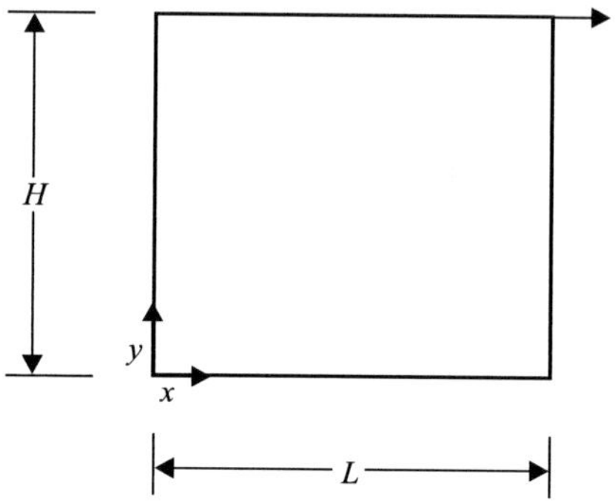

1.5 Consider the fully-developed laminar steady flow of Newtonian viscous fluid through a two-dimensional channel formed by two infinite parallel plates as shown below. The lower plate is stationary and the upper plate is moving in the x direction with a constant speed, U_w, in the positive x-direction.

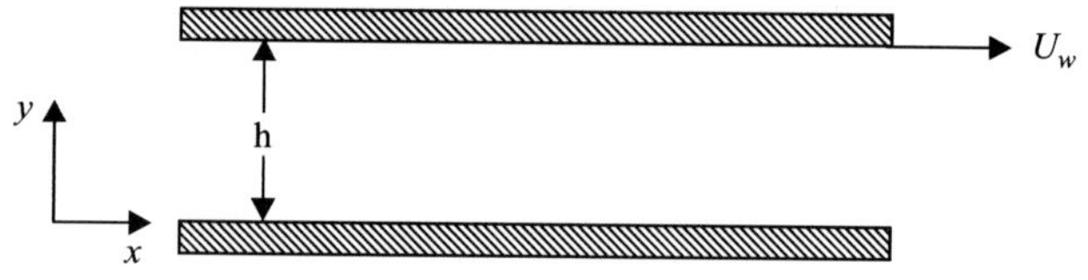

Derive the mathematical statement of the problem for determining from mass and momentum equations of fluid flow, referred to as Couette flow.

1.6 Consider the problem of developing laminar steady flow of Newtonian viscous fluid flow in a two-dimensional channel formed by two infinite parallel stationary plates as shown below. The upper plate is maintained at a constant temperature T_w and the lower plate is adiabatic.

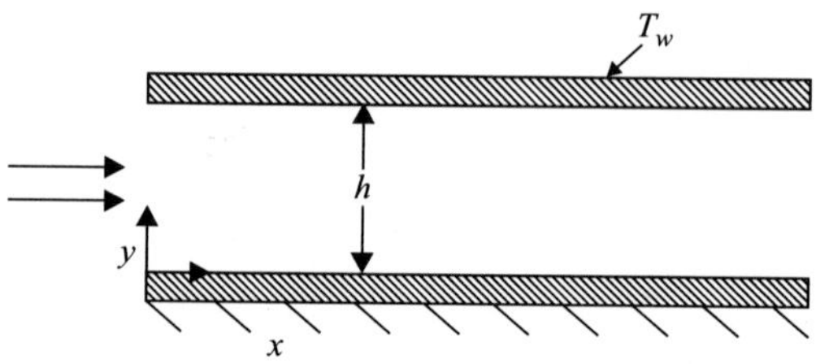

What are the continuity and x and y momentum equations for this pressure-driven ($dp/dx < 0$) channel flow? What is the appropriate energy equation with significant viscous heat dissipation?

1.7 Consider one-dimensional unsteady state heat conduction in a plane slab of thickness L. Initially the slab is at a uniform temperature of T_I. At time $t > 0$, the left side of the slab is subjected to uniform constant surface heat flux q''_s and the right surface is maintained at a constant temperature, T_R. State the mathematical statement (governing equation, initial conditions and boundary conditions) for determining the transient temperature distribution in the slab.

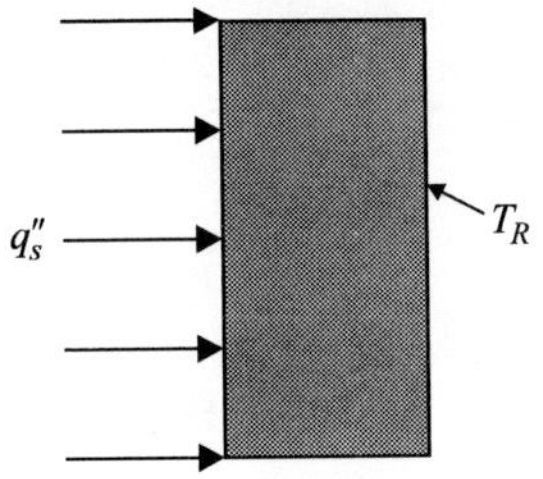

1.8 Consider two-dimensional steady-state conduction in a rectangular slab with the top surface maintained at a high temperature. All other surfaces are subjected to convection condition with h_c and T_∞.

State the mathematical statement (governing equation and boundary conditions) for determining the steady-state temperature distribution in the slab.

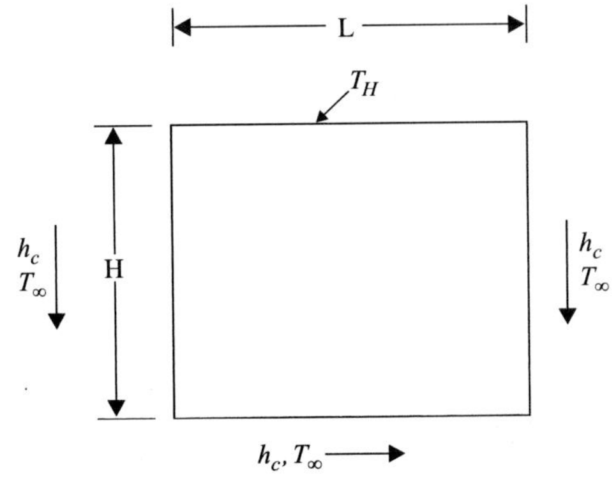

1.9 Consider the diffusion of moisture from an air stream in a two-dimensional channel to the porous adsorbing material felt lined on the bottom of the air flow channel as shown. The felt is supported at the bottom by an aluminum plate, thus making it impermeable to mass diffusion. Air enters the channel at uniform velocity u_{in} and uniform concentration C_{in}. Initial concentration distribution in the felt is uniform at C_0. Assume constant rate of moisture adsorption m'''_{ad} and state the governing equations and boundary conditions for determining the transient moisture concentration distribution in the channel and in the solid felt.

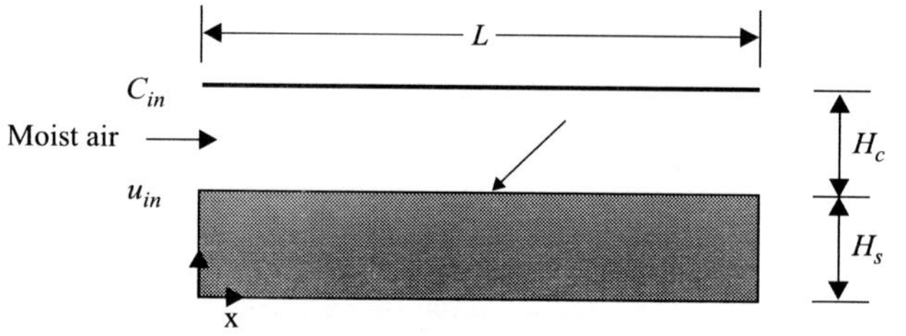

1.10 The figure below shows the schematic diagram of laser heating of a material. The incident laser beam is absorbed by the material and rapidly heats up the thin layer of the material surface and forms a thin layer of molten pool.

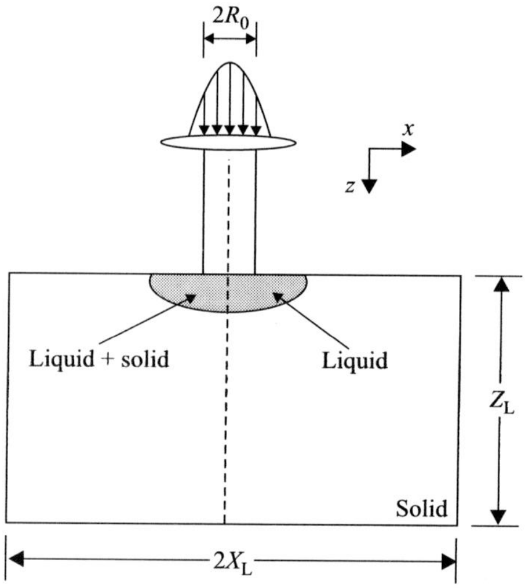

What are the governing equations in terms of continuity and x and z component momentum equations and boundary condition for the flow field in the liquid pool as developed by the surface tension force? What are the heat equations in the solid and liquid region and the associated boundary conditions?

CHAPTER

2

APPROXIMATIONS AND ERRORS

Understanding the concept of error is important, not only for the effective use of numerical methods, but also to get an insight into improving the mathematical model and taking into account other variables and parameters. Numerical solutions always deviate from exact solutions, giving rise to errors since inaccuracies are introduced while approximating mathematical formulas and representing a quantity or a data by a computer numbering system. Inaccuracies are also introduced during arithmetic operations using data. Exact solutions are also subject to errors as computations are performed retaining only a finite number of digits. One of the major challenges of numerical methods is to retain the error within certain acceptable limits. There are two basic categories of error, ***round off error*** and ***truncation error***, which are related directly to numerical methods and termed as computational errors. In addition, there are other sources of error that are not directly related to the computational methods themselves such as ***gross blunders***, ***formulation or modeling errors*** and ***data uncertainty***. In this chapter we will discuss primarily the computational errors.

2.1 TRUNCATION ERROR

Truncation errors are those that result from using an approximation in place of exact mathematical operations and quantities. The mathematical statement of fluid flow and heat transfer problems usually involves derivatives or integrals of different orders. Approximations to such quantities are obtained by retaining a certain number of terms in a power series instead of using an infinite number of terms to obtain exact results. Let us consider the problem in which a quantity is approximated by a truncated power series. If a function $\phi(x)$ and its $n+1$ derivatives are continuous on an interval $[x_{i+1}, x_i]$, then the value of the function at x_{i+1} is given in terms of the function value and its derivatives at another close point x_i by the Taylor series

$$\phi(x_{i+1}) = \phi(x_i) + \phi'(x_i)h + \frac{\phi''(x_i)}{2!}h^2 + \frac{\phi'''(x_i)}{3!}h^3 + \cdots + \frac{\phi^{(n)}(x_i)}{n!}h^n + R_n \quad (2.1)$$

where h is the step size defined by $h = \Delta x = x_{i+1} - x_i$. R_n is called the ***reminder term*** that represents the error in approximating the function by the sum of the first $n+1$ terms.

Accuracy of the approximation increases with increase in number of terms retained in the truncated series. The rest of the terms are grouped as a reminder term, which basically represents the ***truncation error***, and it is expressed in integral form as

$$R_n = \int_{x_i}^{x_{i+1}} \frac{(x_{i+1} - x_i)^n}{n!} \phi^{(n+1)}(x')\, dx' \quad (2.2)$$

An alternative form of the reminder term is obtained by using the second mean value theorem for the integral (Equation 2.2) as

$$R_n = \frac{\phi^{n+1}(\xi)}{(n+1)!} h^{(n+1)} \quad (2.3)$$

where ξ is a value of x that lies between x_{i+1} and x_i.

The major difficulty in evaluating the truncation error exactly by using Equation 2.2 or Equation 2.3 is that the function $\phi(x)$ and its $(n+1)$ derivatives are not known for problems for which a numerical solution is sought. Additionally, the value of ξ is not known exactly. However, these expressions for truncation error give a relative estimate of the truncation error. In order to express the relative error, the reminder term given by Equation 2.2 or Equation 2.3 is also expressed as $R_n = O(h^{n+1})$, and this means that the truncation error is of h^{n+1} order.

It can be seen that the ***truncation error*** is directly proportional to the step size, h, and so gives an idea about comparative errors of computational methods based on the Taylor series. For example, if the error is $O(h)$, halving the step size will halve the error. If the error is of $O(h^2)$, then halving the step size will quarter the error. The important conclusion that we can make here is that retaining additional terms in the truncated Taylor series decreases the truncation error. In many cases, if h is sufficiently small, the first- and second-order terms will account for a disproportionately high percentage of the error. In such cases only a few terms are retained in order to get an adequate estimate, and the desired accuracy of the solution is obtained by reducing the step size progressively. However, in many fluid flow and heat transfer problems we may reach a limit to decrease the step size any further. With decrease in step size, number calculation steps increase and, hence, it will increase the round-off error and will require increased computer memory. In such cases we will have to use higher-order approximation. In general, higher-order approximations converge faster, but require additional computations.

The physical interpretation of the order of such approximations can further be demonstrated in Figure 2.1, and with the following cases.

In the case of a ***zero-order approximation***, only the first term is retained in the estimation of the function $\phi(x_{i+1})$ as

$$\phi(x_{i+1}) \approx \phi(x_i) \tag{2.4a}$$

with error of the order $O(h)$. Such an approximation is good when the unknown function being approximated has a very small variation or is close to constant over the interval x_{i+1} and x_i, as shown in Figure 2.1.

In the case of a ***first-order approximation***, the first two terms are retained in the estimation of the function $\phi(x_{i+1})$ as

$$\phi(x_{i+1}) \approx \phi(x_i) + \phi'(x_i)h \tag{2.4b}$$

with error of the order $O(h^2)$. In such an approximation, the unknown function is assumed to vary in a linear straight-line manner over the interval x_{i+1} and x_i as shown in the figure.

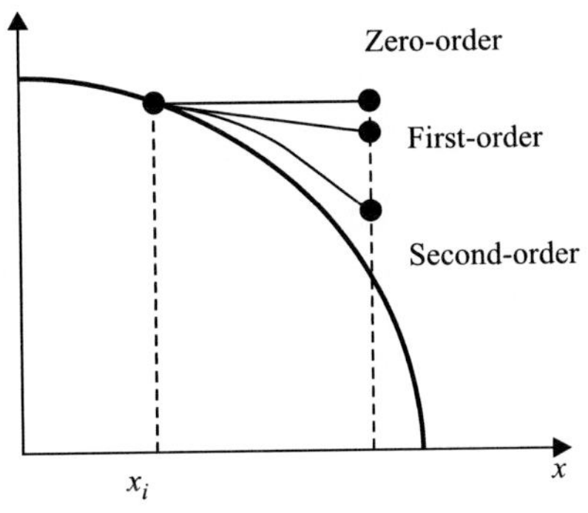

Figure 2.1 Order of approximation

In the case of a ***second-order approximation***, the first three terms are retained in the estimation of the function $\phi(x_{i+1})$ as

$$\phi(x_{i+1}) \approx \phi(x_i) + \phi'(x_i)h + \frac{\phi''(x_i)}{2!}h^2 \tag{2.4c}$$

with error of the order $O(h^3)$. In such an approximation, the unknown function is assumed to vary in a quadratic manner over the interval x_{i+1} and x_i as shown in the figure.

It can be noted that, with increased order of approximations, the function is more closely approximated.

Example 2.1: Truncation Error: Estimate the value of the function $\phi(x) = \sin x$ at $x = \pi/6$, based on the values of the function $\phi(x)$ and its derivatives at $x = \pi/5$ by considering one to six terms in the Taylor series.

Solution: In this problem, the step size is

$$h = \frac{\pi}{6} - \frac{\pi}{5} = -\frac{\pi}{30}$$

The zero-order approximation is computed as

$$\phi\left(\frac{\pi}{6}\right) \approx \phi\left(\frac{\pi}{5}\right)$$

or

$$\sin\left(\frac{\pi}{6}\right) \approx \sin\left(\frac{\pi}{5}\right) = 0.587\,785\,252\,292\,47$$

The first-order approximation is computed as

$$\phi\left(\frac{\pi}{6}\right) \approx \phi\left(\frac{\pi}{5}\right) + \phi'\left(\frac{\pi}{5}\right)h$$

or

$$\sin\left(\frac{\pi}{6}\right) \approx \sin\left(\frac{\pi}{5}\right) + \cos\left(\frac{\pi}{5}\right) \times \left(-\frac{\pi}{30}\right) = 0.503\,065\,190\,753\,89$$

The second-order approximation is computed as

$$\phi\left(\frac{\pi}{6}\right) \approx \phi\left(\frac{\pi}{5}\right) + \phi'\left(\frac{\pi}{5}\right)h + \frac{\phi''(\pi/5)}{2!}h^2$$

or

$$\sin\left(\frac{\pi}{6}\right) \approx \sin\left(\frac{\pi}{5}\right) + \cos\left(\frac{\pi}{5}\right) \times \left(-\frac{\pi}{30}\right) + \frac{-\sin(\pi/5)}{2!}\left(-\frac{\pi}{30}\right)^2 = 0.499\,842\,297\,468\,93$$

The procedure is repeated for the third- to fifth-order approximations and the results are summarized in the table below.

Order (n)	$\phi(x) = \sin(\pi/6)$
0	0.587 785 252
1	0.503 065 190
2	0.499 842 297
3	0.499 997 140
4	0.500 000 086
5	0.500 000 011
∞	0.5

We note that, with the addition of a greater number of terms in the Taylor series, the truncation error decreases and the approximate value approaches the exact solution.

2.2 ROUND-OFF ERROR

Round-off errors originate in all numerical calculations because actual numbers are approximated while storing and performing arithmetic on a computer. This is due to the fact that computers can only represent quantities with a finite number of digits, i.e., retain only a fixed number of significant figures during calculations. Numbers such as π, e, $\sqrt{15}$, etc., cannot be expressed by a finite number of significant figures, and hence cannot be represented exactly. This discrepancy introduced by the omission of ***significant figures*** is called the ***round-off error***, and it depends primarily on the ***computer number system***. In order to obtain a better understanding of the round-off error, a clear understanding of the concept of significant figures and computer number systems is essential.

2.2.1 Significant Figures or Digits

The ***significant digits*** or ***figures*** of a number are those that can be used with confidence. They correspond to some certain digits plus one estimated digit. We must also remember that zeros are not always used as significant figures.

For example, heat flux data 26 400 W/m^2 may have ***three***, ***four*** or ***five*** significant digits depending on whether the zeros are known with confidence. This uncertainty can be resolved by using scientific notation such as 2.64×10^4, 2.640×10^4 and 2.6400×10^4. Significant figures are important to particularly express an *approximate computational solution* with confidence. We should decide that our approximate solution is correct to some significant digits.

Example 2.2: Significant Figure: Consider a temperature reading from a temperature gauge or a thermometer with a resolution of 1°C. A temperature reading of 62.25°C has three certain digits 62.2 and one estimated digit 0.05, which is one half of the smallest scale division. So, the significant figure for temperature is 62.25°C.

2.2.2 Computer Number System

The information that is stored in a computer may be numerical data such as numbers or non-numeric data that consist of alphabets, digits, and symbols. Digital computers store all information in binary form defined by different fundamental units such as ***bit***, ***byte***, and ***word***. A bit is a binary digit, i.e., a zero (0) or one (1). A ***byte*** is a larger unit, which consists of 8 bits, and a *word* is an even bigger unit, which consists of a number of bytes or a string of bits. The word lengths could be 16 bits, 32 bits or 64 bits, depending on the type of computer. For example, a 32-bit word consists of 4 bytes or 32 bits. Numbers are stored in one or more words.

A number system is a way to represent a number. Some of the number systems that are commonly being used are the decimal or the 10-base system, octal or 8-base system, and the binary or 2-base system. Each computer uses a specific number system to express numbers. For example, integer numbers are whole numbers with no fractional part, and they are represented by a fixed-point numbering system such as

$$(I_n)_b = \pm(d_{n-1}d_{n-2}\cdots d_2 d_1 d_0)_b \tag{2.5}$$

where b is the base of the number system, n is the number of digits, and the digits d_j represent any integer values from $j = 0, 1, 2, \ldots (b-1)$.

For example, in a decimal system the number is evaluated as

$$\begin{aligned}(I_n)_{10} &= (45\,682)_{10} = 10^4(4) + 10^3(5) + 10^2(6) + 10^1(8) + 10^0(2)\\ &= 40\,000 + 5000 + 600 + 80 + 2\\ &= 45\,682\end{aligned}$$

and the decimal value of a binary number is evaluated as

$$\begin{aligned}(I_n)_2 &= (11\,010\,1)_2 = 2^5(1) + 2^4(1) + 2^3(0) + 2^2(1) + 2^1(0) + 2^0(1)\\ &= 32 + 16 + 0 + 4 + 0 + 1 = (53)_{10}\\ &= 32 + 16 + 0 + 4 + 0 + 1 = (53)_{10}\\ &= (53)_{10}\end{aligned}$$

A real number is expressed using a ***floating point system*** of the form

$$(R_n)_b = \pm\, d_1 d_2 d_3 \cdots d_n \times b^e \tag{2.6}$$

where b is the base of the computer number system, digits $d_1d_2d_3 \cdots d_n$ are defined as the ***mantissa*** (m) of the number, e is the exponent, and n is the finite length of the mantissa. Each d is an integer between 0 and $b-1$, and $d_1 \neq 0$. The mantissa is usually normalized in order to make sure that $d_1 \neq 0$ and this gives $1/b \leq m \leq 1$.

For example, a number such as 0.000 465 789 1 is stored as $0.000\,465\,7 \times 10^0$ in a floating point number system that allows only seven significant digits to be stored and without any normalization. However, with the use of normalization, the floating-point number system will store the number as $0.465\,789\,1 \times 10^{-3}$.

Another important item to be remembered is that the size of a ***word*** in terms of number of binary digits or bits limits the number of significant digits that can be retained in storing and representing a number. For example, in a typical 32-bit computer with 1 bit for the sign, 1 bit for the sign of the exponent, 7 bits for exponent (e), and 23 bits for the mantissa (n), the computer will represent a number with about seven decimal places or seven significant decimal digits because the least significant bit in the mantissa represents 2^{-23} or approximately 10^{-7}. The accuracy can be improved by using an additional ***word*** in representing the floating-point number. In a ***double-precision*** or ***extended-precision*** computation, floating point numbers are represented by two ***words*** with approximately twice as many bits in the mantissa, and this leads to double the number of significant digits or significant decimal places. However, double-precision computations require additional computer storage and is slower in computation than that in a single-precision computation.

Round-off error is introduced since a computer approximates a number consisting of more digits than the number of significant digits. The actual approximation is achieved in one of two ways: ***chopping*** or ***rounding***. ***Chopping*** means that any quantity falling within an interval of length Δx will be retained as a quantity with the lower bound of the interval, and this leads to an upper bound of the error as Δx. On the other hand, in ***rounding*** any quantity falling within an interval Δx will be retained as the nearest allowable number, i.e., a number given either by the lower bound or the upper bound of the interval, whichever is the nearest. This leads to an upper bound of the error as $\Delta x/2$.

2.2.3 Machine Epsilon

Often, error bounds are expressed in normalized form in order to take into account the fact that the interval Δx increases as the magnitude of the number itself increases. For ***chopping***, the error bound is expressed as

$$\frac{|\Delta x|}{|x|} \leq \varepsilon \tag{2.7}$$

and for rounding the error it is expressed as

$$\frac{|\Delta x|}{|x|} \leq \frac{\varepsilon}{2} \tag{2.8}$$

where ε is defined as the ***machine epsilon*** that is computed for a specific computer as

$$\varepsilon = b^{1-n} \tag{2.9}$$

and where b is the number base and n is the number of significant digits in the mantissa.

The value of the machine epsilon (ε) varies from computer to computer depending on the *word* length, number of words used (i.e., one or two), *base* (b), and type of number-approximation (*rounding* or *chopping*) used.

Example 2.3: Machine Epsilon: In a computer the value of $\pi = 3.141\,592\,654$ is approximated as 3.141 59 by chopping. What is the number of significant digits retained in the computer?

Solution: The error bound is expressed as

$$\left|\frac{3.141\,592\,654 - 3.141\,59}{3.141\,592\,654}\right| \leq b^{1-n}$$

or

$$0.845 \times 10^{-6} < 10^{1-7}$$

Hence, the number of significance digits for the approximation of π is 7.

Aside from the ***computer number system in representing and storing a number***, there could be ***round-off*** error due to mathematical operations such as

Addition, Subtraction, Multiplication, Division, etc. The loss of significant numbers during *subtraction* is the greatest source of ***round-off error***. Quite often we can use double precision to reduce the effect of round-off errors. However, extra price is paid in terms of more memory and execution time.

The inequality in Equation 2.7 and Equation 2.8 specifies the worst possible cases. In the following section we will see the significance of the absolute magnitude of the ***round-off*** error as well as the normalized form of the error, and the *machine epsilon* in the convergence and stopping criteria in an iterative numerical scheme.

2.3 ERROR DEFINITIONS

As mentioned before, numerical errors arise from the use of an approximation to represent exact mathematical operations and quantities. The relationship between true or exact value and the approximate value is

$$\text{true value} = \text{approximate value} + \text{error} \tag{2.10}$$

Note that in the above equation the error term represents the true error, and this is expressed as

$$\text{true error } (E_\text{t}) = \text{true value, } \phi_\text{tv} - \text{approximate value, } \phi_\text{av} \tag{2.11}$$

Normalizing the error with respect to the true value, we express the ***true relative error*** as

$$\varepsilon_\text{t} = \frac{E_\text{t}}{\phi_\text{tv}} \tag{2.12}$$

and ***true percent relative error*** as

$$\varepsilon_\text{t} = \frac{E_\text{t}}{\phi_\text{tv}} \times 100\% \tag{2.13}$$

Example 2.4: Error Estimation: Let us consider the true values of temperature and heat transfer rate in a heat conduction problem as $T = 100°\text{C}$ and $q = 10\,000$ W. Estimate the error and percent relative error if the approximate computational solutions are $T = 999°\text{C}$ and $q = 9999$ W.

Solution: Error for temperature is

$$E_\text{t} = 100 - 99 = 1°\text{C}$$

$$\varepsilon_\text{t} = \frac{1}{100} \times 100\% = 1\%$$

Error for heat transfer rate is

$$E_\mathrm{t} = 10\,000 - 9999 = 1\ \mathrm{W}$$

$$\varepsilon_\mathrm{t} = \frac{1}{10\,000} \times 100\% = 0.01\%$$

It can be concluded that the computational solution is adequate for predicting the heat transfer rate. However, the computational solution for temperature could be improved.

2.4 APPROXIMATE ERROR

Computational errors cannot be estimated exactly for problems in which there are no analytical solutions. In such cases we have to rely on approximate errors. In cases where true values are not known, we estimate the error using the best available estimate of the true value, i.e., the approximate value itself, and express the ***approximate percent relative error*** as

$$\varepsilon_\mathrm{a} = \frac{E_\mathrm{a}}{\phi_\mathrm{av}} \times 100\% \tag{2.14}$$

where

E_a = approximate error
ϕ_av = approximate value

In these cases the major challenge is to determine the error in the absence of the true value. In most computational methods for fluid flows and heat transfer we employ an iterative scheme to compute answers. In this approach the present approximate value (ϕ^{k+1}) is estimated based on the previous approximate value (ϕ^k). This process is repeated or iterated to compute better and better approximation. In such cases we estimate ***approximate error*** and ***approximate percent relative error*** as

$$E_\mathrm{a} = \phi^{k+1} - \phi^k \tag{2.15}$$

and

$$\varepsilon_\mathrm{a} = \frac{\phi^{k+1} - \phi^k}{\phi^k} \tag{2.16}$$

2.5 CONVERGENCE CRITERIA

Signs of the approximate error can be positive or negative. However, we are interested in the absolute value of the error, i.e., $|\varepsilon_a|$. In an iterative scheme, convergence is assumed to have been reached when the absolute value of the error is less than a pre-specified tolerance limit, ε_s. In such a case computation is repeated or iterated until

$$|\varepsilon_a| \leq \varepsilon_s \tag{2.17a}$$

or

$$\left|\frac{\phi^{k+1} - \phi^k}{\phi^k}\right| \leq \varepsilon_s \tag{2.17b}$$

The pre-specified tolerance limit is selected based on the number of significant digits desired in the numerical solution. It should be noted that the smallest numerical error that can be achieved is limited by the fact that the pre-assigned tolerance limit has to be greater than the error bound given by Equation 2.7 or Equation 2.8, which relates the error to the number of significant figures or digits in the approximation. If this tolerance limit is met, then we can be assured that the computational result is correct to at least n significant figures or digits.

In the case of a vector array, $\{\phi\} = \{\phi_1, \phi_2 \cdots \phi_n\}$ or ϕ_i, Equation 2.17 is applied successively to all elements of the vector as

$$|\varepsilon_{ia}| \leq \varepsilon_s \tag{2.18a}$$

or

$$\left|\frac{\phi_i^{k+1} - \phi_i^k}{\phi_i^k}\right| \leq \varepsilon_s, i = 1, 2, 3 \ldots n \tag{2.18b}$$

In problems where the number of unknowns, n, or vector size is large, an alternate convergence criterion based on ***Euclidean norm*** is used as

$$\left\|\varepsilon_{ia}^{k+1}\right\|_e \leq \varepsilon_s \tag{2.19}$$

where the ***Euclidean norm*** is defined as

$$\left\|\varepsilon_{ia}^{k+1}\right\|_e = \sqrt{\sum_{i=1}^{n}\left(\varepsilon_{ia}^{k+1}\right)^2} = \sqrt{\left(\left(\varepsilon_{1a}^{k+1}\right)^2 + \left(\varepsilon_{2a}^{k+1}\right)^2 \cdots + \left(\varepsilon_{na}^{k+1}\right)\right)^2} \tag{2.20}$$

In problems where the unknowns are arranged in a two-dimensional vector array or matrix array ϕ_{ij}, the convergence criterion can be assigned in the form of a ***Frobenius norm*** as

$$\left\|\varepsilon_{ij}^{k+1}\right\|_{\mathrm{e}} \leq \varepsilon_{\mathrm{s}} \tag{2.21}$$

where the ***Frobenius norm*** is defined as

$$\left\|\varepsilon_{ij}^{k+1}\right\|_{\mathrm{e}} = \sqrt{\sum_{i=1}^{n}\sum_{j=1}^{n}\left(\varepsilon_{ij}^{k+1}\right)^2} \tag{2.22}$$

Another alternative convergence criterion is sometimes given in terms of uniform vector norm as follows. For a vector array

$$\left\|\varepsilon_{i}^{k+1}\right\|_{\infty} \leq \varepsilon_{\mathrm{s}} \tag{2.23a}$$

where

$$\left\|\varepsilon_{i}^{k+1}\right\|_{\infty} = \max_{1\leq i\leq n} |\varepsilon_i| \tag{2.23b}$$

and for a matrix array

$$\left\|\varepsilon_{ij}^{k+1}\right\|_{\infty} \leq \varepsilon_{\mathrm{s}} \tag{2.23c}$$

where

$$\left\|\varepsilon_{ij}^{k+1}\right\|_{\infty} = \max_{1\leq i\leq n} \sum_{j=1}^{n} |\varepsilon_{ij}| \tag{2.24}$$

Example 2.5: Total Error: Compute the numerical estimate of the derivative and the associated total numerical error for the function $y=x^4$ using an approximation scheme for the first derivative and progressively smaller step size. Estimate the true percent relative error and approximate percent relative error. Identify the optimum step size and error.

Solution: Let us estimate the derivative at a point $x_i = 10$ and $x_{i+1} = x_i + h$, and with progressively decreasing the step size by a factor of 10.

The exact derivative of the function is given as

$$\text{Der } E = y'(x_i) = 4x_i^3 = 4000 \tag{E.2.5.1}$$

We can approximate the first derivative of the function using the first-order approximation given by Equation 2.4b

$$\text{Der } A = y'(x_i) \cong \frac{y(x_{i+1}) - y(x_i)}{h} \tag{E.2.5.2}$$

Using the first estimate of the derivative with a step size $h = 0.1$, we get

$$\text{Der } A = \frac{(10 + 0.1)^4 - 10^4}{0.1} = 4060.401\,000\,000$$

We repeat this procedure with decreasing step sizes. Results for the estimate of the derivative and associated true and approximate errors are given in the table below.

Step size	Estimated derivative	True percent relative error	Approximate percent relative error
1.0E–01	4060.401 000 000	1.510 025 000	–
1.0E–02	4006.004 001 000	0.150 100 025	1.357 886 787
1.0E–03	4000.600 039 999	0.015 001 000	0.135 078 762
1.0E–04	4000.060 000 391	0.001 500 010	0.013 500 788
1.0E–05	4000.005 999 853	0.000 149 996	0.001 350 011
1.0E–06	4000.000 597 006	0.000 014 925	0.000 135 071
1.0E–07	4000.000 035 687	0.000 000 892	0.000 014 033
1.0E–08	4000.000 337 001	0.000 008 425	0.000 007 533
1.0E–09	4000.000 330 961	0.000 008 274	0.000 000 151
1.0E–10	4000.000 330 961	0.000 008 274	0.000 000 000
1.0E–11	3999.645 059 594	0.008 873 510	0.008 882 572
1.0E–12	4000.355 602 329	0.008 890 058	0.017 761 989
1.0E–13	3979.039 320 257	0.524 016 994	0.535 714 286
1.0E–14	4263.256 414 561	6.581 410 364	6.666 666 667
1.0E–15	7105.427 357 601	77.635 683 940	40.000 000 000

We can see that the optimum size for minimum total error is around $h = 1.0\,E - 09$ for true percent relative error. With further decrease in step size, the total error starts increasing. The approximate percent relative error follows a similar pattern.

PROBLEMS

2.1 What would be the assigned tolerance limit, (ε_s), if convergence with four significant digits is required in an iterative scheme.

2.2 How many significant digits are there in the number 0.6843×10^{-3}?

2.3 Round the number $0.786\,54 \times 10^3$ to four significant digits.

2.4 Convert the binary number 11011011 to an equivalent decimal number in an 8-bit computer.

2.5 Represent the following binary numbers in decimal form: (a) 1011101 and (b) 110101.

2.6 Round the following numbers correctly to six significant digits: (a) 35.674 57, (b) 0.000 231 45 and (c) 50 143.567

2.7 In a computer the value of $\pi = 3.141\,592\,654$ is approximated as 3.141 59 by chopping. What is the number of significant digits retained in the computer?

2.8 Write a computer program to estimate the value of the function $\phi(x) = \cos x$ at $x = \pi/3$ based on the values of the function $\phi(x)$ and its derivatives at $x = \pi/4$ by considering one to six terms in the Taylor series. Present your value of the function, true percent relative error, and approximate percent relative error at each step in a table.

2.9 The series expansion for $\sin x$ is given as

$$\sin x = x - \frac{x^3}{3!} + \frac{x^5}{5!} - \frac{x^7}{7!} + \cdots$$

Starting with the simplest version $\sin x \cong x$, add terms one at a time in order to estimate $\sin(\pi/3)$. Add terms until the absolute value of the approximate percent relative error falls below the assigned tolerance value of $\varepsilon_s = 0.5\%$.

2.10 Write a computer program to compute the numerical estimate of the derivative and total numerical error of the function $y = x^6$ using an approximation scheme for the first derivative and progressively smaller step size. Present your results for estimated derivate, true percent relative error and approximate percent relative error in a table. Decrease step sizes by a factor of 10, i.e., $\Delta x = 1/10^i$ and identify the optimum step size and error.

CHAPTER

3

NUMERICAL SOLUTIONS OF SYSTEMS OF EQUATIONS

In the subsequent chapters, we shall solve the system of simultaneous equations that are obtained by approximating the governing differential equations of heat and mass transfer problems, using discretization methods such as finite difference–control volume and finite element methods. The mathematical analysis of a linear physical system often results in a model consisting of simultaneous linear algebraic equations. The solution methods of nonlinear systems or differential equations use the technique of linearization of the models, thus requiring the repetitive solution of simultaneous linear algebraic equations.

These problems may range in complexity from a set of a few algebraic equations to a set of 1000 or more. For small and medium size problems, the time required for the solution of such a system of equations does not exceed that required for the formation of the system of equations using discretization methods. However, for very large problems, the solution time for such systems dominates the total computing time. Also, the overall effectiveness of a discretization method depends to a large extent on the numerical procedure used for the solution of system of equations. As will be noted in subsequent chapters, the accuracy of the discretization method, in general, can be improved if a finer discretization mesh distribution is used. This means that the cost and effectiveness of a discretization method depends to a considerable extent on the solution algorithm for solving the resulting system of equations. Therefore, use of inappropriate techniques for the solution of the system equations may lead to excessive computational time and even a numerically unstable solution algorithm.

The major objective of this chapter is to present various numerical methods for solving linear systems of simultaneous equations. Discretizations of differential equations give rise to various types of linear systems. Based on the physical problems under consideration, these systems pose special properties and structure, such as symmetric, nonsymmetric, banded and sparse, diagonally dominant, tridiagonal, block banded, block tridiagonal, and positive definite. The solution methods for solving these systems are described briefly in the following sections. For a detailed and further study of this topic, readers are suggested to refer to Ortega and Poole (1981), Datta (1995), Chapara and Canale (2002), Golub and Van Loan (1989), James et al. (1985), Hildebrand (1974), Forsythe et al. (1977), Saad (1996), and Rao (2002).

3.1 MATHEMATICAL BACKGROUND

3.1.1 Representation of the System of Equations

Consider the system of simultaneous linear algebraic equations represented as

$$\begin{aligned} &a_{11}x_1 + a_{12}x_2 + \cdots + a_{1n}x_n = c_1 \quad \text{(a)} \\ &a_{21}x_1 + a_{22}x_2 + \cdots + a_{2n}x_n = c_2 \quad \text{(b)} \\ &\vdots \\ &a_{n1}x_1 + a_{n2}x_2 + \cdots + a_{nn}x_n = c_n \quad \text{(n)} \end{aligned} \tag{3.1}$$

where the a_{ij} terms are the known coefficients of the equation, the x_j terms are the unknown variables, and the c_i terms are the known constants. Equation 3.1 represents a system of linear equations with n unknowns.

In developing a systematic approach for the solution of linear algebraic equations, we will make use of matrix-vector notations. For this reason, and for the benefit of the reader, a brief review of selected matrix and vector operations is given in Appendix A. In general, these systems of equations are often expressed in matrix notion as

$$[A]\{x\} = \{c\} \tag{3.2a}$$

or simply as

$$Ax = c \tag{3.2b}$$

where $[A]$ is the coefficient matrix written as

$$[A] = \begin{bmatrix} a_{11} & a_{12} & \cdots & \cdots & \cdots & a_{1n} \\ a_{21} & a_{22} & \cdots & \cdots & \cdots & a_{2n} \\ \vdots & & & & & \\ \vdots & & & & & \\ \vdots & & & & & \\ a_{n1} & a_{n2} & \cdots & \cdots & \cdots & a_{nn} \end{bmatrix} \tag{3.3}$$

The elements of the coefficient matrix are the known coefficients of the system of equations and are designated by a_{ij}. The first subscript i designates the row number in which the element lies. The second subscript j designates the corresponding column number. $\{c\}$ is a $(n \times 1)$ column vector of constants or known values

$$\{c\}^{\mathrm{T}} = [c_1 \quad c_2 \cdots c_n] \tag{3.4}$$

and $\{x\}$ is the $(n \times 1)$ column vector of unknowns:

$$\{x\}^{\mathrm{T}} = [x_1 \quad x_2 \cdots x_n] \tag{3.5}$$

We will consider cases when the matrix A is square; that is the number of unknowns equals the number of equations. Also, the nonhomogeneous set of linear equations (3.1) is assumed to be nonsingular; that is it has linearly independent rows (and columns) or the determinant of the coefficient matrix is nonzero, for which a unique nontrivial solution exists.

3.1.2 Cramer's Rule and the Elimination of Unknowns

The solution of a set of a few linear algebraic equations can be obtained by ***Cramer's rule*** or by the ***method of elimination of unknowns***. Cramer's rule is a solution technique that is used to calculate the solution of the nonhomogeneous linear algebraic equation (3.1) by expressing each unknown as the ratio of determinants of the coefficient matrix A and substituted matrix A_j as

$$x_j = \frac{|A_j|}{|A|}, \quad j = 1, 2, \ldots, n \tag{3.6}$$

The substituted matrix A_j is obtained by replacing jth columns of the matrix A with the known vector c as

$$[A_j] = \begin{bmatrix} a_{11} & a_{12} & \cdots & c_1 & \cdots & a_{1n} \\ a_{21} & a_{22} & \cdots & c_2 & \cdots & a_{2n} \\ \vdots & & & & & \\ \vdots & & & & & \\ \vdots & & & & & \\ a_{n1} & a_{n2} & \cdots & c_n & \cdots & a_{nn} \end{bmatrix} \tag{3.7}$$

The solution of a set of a few linear algebraic equations can be obtained easily by the application of Cramer's rule. However, for systems involving five or more equations, Cramer's rule requires a rapidly escalating number of arithmetic operations to evaluate the determinants, too large even for today's high-speed computers, and becomes impractical. The ***method of elimination of unknowns*** is an algebraic approach that involves multiplying the equations by constants in order that one of the unknowns will be eliminated when two equations are combined. This will result in a single equation that can be solved for the remaining unknown. The value of this unknown can then be substituted into one of the original equations to compute the other variables.

A solution procedure based on the elimination of unknowns appears to be simple to evaluate for systems with three or four equations. However, for larger systems, the elimination of unknown methods becomes too complex and extremely tedious to implement by hand calculations.

In the subsequent sections we will discuss more efficient alternative solution methods. There are two general types of numerical methods available for solving a nonhomogeneous set of linear algebraic equations. Methods of the first type are called ***direct methods*** and the methods of the second type are called ***iterative methods***.

3.2 DIRECT METHODS

The direct methods perform a finite number of steps to obtain the solution. Among these we will consider Gaussian elimination, the Gauss–Jordan elimination or factorization, Lu decomposition, Cholesky decomposition, and the diagonally dominant system.

3.2.1 Gaussian Elimination

The Gaussian elimination method uses systematically the procedure utilized in the method of elimination of unknowns. This method employs the scheme or algorithm in two steps, which are referred to as the ***elimination*** and ***back substitutions***. In the ***elimination step***, a set of n equations in n unknown is reduced to an equivalent upper triangular set, which is then easily solved in the ***back substitution step***. The procedure will be illustrated by solving the set of simultaneous equations given by Equation 3.1.

Elimination Step

1. This step is designed to reduce the set of equations to an upper triangular set. The initial step will be to eliminate the first unknown, x, from the second through the nth equation. To do this, divide Equation 3.1a by the coefficient of x_1, i.e., a_{11}, referred to as the ***pivot element***, to get

$$x_1 + \frac{a_{12}}{a_{11}}x_2 + \frac{a_{13}}{a_{11}}x_3 + \cdots + \frac{a_{1n}}{a_{11}}x_n = \frac{c_1}{a_{11}} \quad (3.8)$$

Next, multiply Equation 3.8 by the coefficient of x in Equation 3.1b, i.e., a_{21}, to get

$$a_{21}x_1 + \frac{a_{21}}{a_{11}}a_{12}x_2 + \frac{a_{21}}{a_{11}}a_{13}x_3 + \cdots + \frac{a_{21}}{a_{11}}a_{1n}x_n = \frac{a_{21}}{a_{11}}c_1 \tag{3.9}$$

Now this equation can be subtracted from Equation 3.1b to get

$$\left(a_{22} - \frac{a_{21}}{a_{11}}a_{12}\right)x_2 + \left(a_{23} - \frac{a_{21}}{a_{11}}a_{13}\right)x_3 + \cdots + \left(a_{2n} - \frac{a_{21}}{a_{11}}a_{1n}\right)x_n$$
$$= \left(c_2 - \frac{a_{21}}{a_{11}}c_1\right) \tag{3.10}$$

or

$$a'_{22}x_2 + a'_{23}x_3 + \cdots + a'_{2n}x_n = c'_2 \tag{3.11}$$

In a similar manner, x_1 is eliminated from all equations (3.1c–3.1n) of the set except Equation 3.1a, so that the set assumes the modified form

$$\begin{aligned}
a_{11}x_1 + a_{12}x_2 + a_{13}x_3 + \cdots + a_{1n}x_n &= c_n \quad \text{(a)}\\
a'_{22}x_2 + a'_{23}x_3 + \cdots + a'_{2n}x_n &= c'_2 \quad \text{(b)}\\
a'_{32}x_2 + a'_{33}x_3 + \cdots + a'_{3n}x_n &= c'_3 \quad \text{(c)}\\
&\vdots\\
&\vdots\\
a'_{n2}x_2 + a'_{n3}x_3 + \cdots + a'_{nn}x_n &= c'_n \quad \text{(n)}
\end{aligned} \tag{3.12}$$

In the foregoing steps, Equation 3.1a is used to eliminate the unknown in the subsequent equations and is called the ***pivot equation***. In the pivot equation, a_{11}, the coefficient of the unknown, x_1, which is to be eliminated from the subsequent equations, is called the ***pivot element***. The prime superscripts indicate that the coefficients are modified in the elimination process.

2. Repeat the above step in order to eliminate the second unknown from all equations following the pivot equation that is Equation 3.12b and using

a'_{22} as the pivot element. This elimination results in

$$\begin{aligned}
a_{11}x_1 + a_{12}x_2 + a_{13}x_3 + \cdots + a_{1n}x_n &= c_n \quad \text{(a)}\\
a'_{22}x_2 + a'_{23}x_3 + \cdots + a'_{2n}x_n &= c'_2 \quad \text{(b)}\\
a''_{33}x_3 + \cdots + a''_{3n}x_n &= c''_3 \quad \text{(c)}\\
&\vdots\\
&\vdots\\
a''_{n3}x_3 + \cdots + a''_{nn}x_n &= c''_n \quad \text{(n)}
\end{aligned} \tag{3.13}$$

3. This procedure is continued until the original set of equations has been transformed to an upper triangular system in the $(n-1)$th elimination step

$$\begin{aligned}
a_{11}x_1 + a_{12}x_2 + a_{13}x_3 + \cdots + a_{1n}x_n &= c_n \quad \text{(a)}\\
a'_{22}x_2 + a'_{23}x_3 + \cdots + a'_{2n}x_n &= c'_2 \quad \text{(b)}\\
a''_{33}x_3 + \cdots + a''_{3n}x_n &= c''_3 \quad \text{(c)}\\
&\vdots\\
a^{n-2}_{n-1,\,n-1}x_{n-1} + a^{n-2}_{n-1,\,n}x_n &= c^{n-2}_{n-1} \quad \text{(n − 1)}\\
a^{n-1}_{nn}x_n &= c^{n-1}_n \quad \text{(n)}
\end{aligned} \tag{3.14}$$

Back Substitution. Once the upper triangular system of equations is obtained, the unknown can be solved one at a time through back substitution. Equation 3.14n is solved directly for x_n as

$$x_n = \frac{c^{n-1}_n}{a^{n-1}_{nn}} \quad \text{(n)}$$

This value is then substituted into Equation 3.14n − 1 to solve for x_{n-1}

$$x_{n-1} = \frac{c^{n-2}_{n-1} - a^{n-2}_{n-1,n}x_n}{a^{n-2}_{n-1,n-1}} \quad \text{(n − 1)}$$

This back substitution procedure is continued to solve the remaining unknown as

$$\begin{aligned}
x_{n-2} &= \frac{c^{n-3}_{n-2} - a^{n-3}_{n-2,n-1}x_{n-1} - a^{n-3}_{n-2,n}x_n}{a^{n-3}_{n-2,n-2}} \quad \text{(n − 2)}\\
&\vdots\\
&\vdots\\
x_2 &= \frac{c'_2 - a'_{23}x_3 \cdots a'_{2,\,n-1}x_{n-1} - a'_{2,n}x_n}{a'_{22}} \quad \text{(b)}\\
x_1 &= \frac{c_1 - a_{12}x_2 - a_{13}x_3 \cdots a_{1,n-1}x_{n-1} - a_{1n}x_n}{a_{11}} \quad \text{(a)}
\end{aligned} \tag{3.15}$$

Generalized Formula. The procedure outlined in the elimination and back substitution steps can be generalized to express the modified elements and the unknown as

Elimination

$$a_{ij}^{k} = a_{ij}^{k-1} - \frac{a_{kj}^{k-1}}{a_{kk}^{k-1}} a_{ik}^{k-1} \quad \text{for } k+1 \leq j \leq m,\ k+1 \leq i \leq n \text{ and } k = 1, 2 \ldots n-1, \tag{3.16}$$

where

i = index identifying the row number of the matrix
j = index identifying the column number of the augmented matrix
k = index identifying pivot row or the elimination step
n = number of rows in the matrix
m = number of columns in the augmented matrix $= n + 1$

It can be noted that the lower limit for j is $k + 1$ instead of k. There is no need to calculate the anticipated initial zero values that appear during the elimination steps.

Back Substitution

$$x_n = \frac{a_{nm}}{a_{nn}} \tag{3.17a}$$

and

$$x_i = \frac{a_{im}^{i} - \sum_{j=i+1}^{n} a_{ij}^{i-1} x_j}{a_{ii}^{i-1}}, \quad \text{for } i = n-1, n-2, \ldots, 1 \tag{3.17b}$$

Some problems arise in the Gauss elimination procedure when a pivot element is zero or close to zero. This leads to division by zero or to large round-off errors. In order to avoid these problems, a procedure known as ***pivoting*** is used in the elimination step. Each elimination step should be made by choosing a row with the largest pivot element as the pivot row. This is achieved by searching for the row with the largest element, and switching the rows with the largest element. This procedure is known as ***partial pivoting***. If columns as well as rows are searched and switched for the largest element, then the procedure is called ***complete pivoting***. For solving a large system of equations, partial pivoting is always incorporated in the computer program. Complete pivoting is also desirable for some problems. However, it adds more complexity to the computer program.

As it has been mentioned before, there may be a considerable loss of accuracy in the results owing to the round-off error that accumulates in a large number of arithmetic operations. This may be quite significant when solving a large system of equations. So, it is important to check the accuracy by substituting the answers back into the original equations and compute the residuals as

$$
\begin{aligned}
r_1 &= a_{11}x_1 + a_{12}x_2 \cdots a_{1n}x_n - c_1 && \text{(a)}\\
r_2 &= a_{21}x_1 + a_{22}x_2 + \cdots a_{2n}x_n - c_2 && \text{(b)}\\
&\vdots\\
r_n &= a_{n1}x_1 + a_{n2}x_2 + \cdots + a_{nn}x_n - c_n && \text{(c)}
\end{aligned}
\tag{3.18}
$$

Example 3.1: Gauss Elimination Method: Use the Gauss elimination method to solve the following system of equations

$$
\begin{bmatrix} 4.855 & -4 & 1 & 0 \\ -4 & 5.855 & -4 & 1 \\ 1 & -4 & 5.855 & -4 \\ 0 & 1 & -4 & 4.855 \end{bmatrix}
\begin{pmatrix} x_1 \\ x_2 \\ x_3 \\ x_4 \end{pmatrix} =
\begin{pmatrix} -1.59 \\ 1 \\ 1 \\ -1.64 \end{pmatrix}
\tag{E.3.1.1}
$$

Solution: In this problem, we use the basic Gauss elimination algorithm given by Equation 3.16 for elimination steps and Equation 3.17 for the back substitution step. Following this algorithm, we get

Elimination step, $\boldsymbol{k} = 1$*, pivot row* $= 1$*, pivot element:* $a_{11} = 4.855$

$$
\begin{bmatrix} 4.855\,00 & -4 & 1 & 0 \\ 0 & 2.559\,44 & -3.176\,10 & 1 \\ 0 & -3.176\,10 & 5.649\,02 & -4 \\ 0 & 1 & -4 & 4.855\,00 \end{bmatrix}
\begin{pmatrix} x_1 \\ x_2 \\ x_3 \\ x_4 \end{pmatrix} =
\begin{bmatrix} -1.590\,00 \\ -0.309\,980 \\ 1.327\,19 \\ -1.640\,00 \end{bmatrix}
\tag{E.3.1.2}
$$

Elimination step: $\boldsymbol{k} = 2$*, pivot row* $= 2$*, pivot element:* $a'_{22} = 2.55944$

$$
\begin{bmatrix} 4.855\,00 & -4 & 1 & 0 \\ 0 & 2.559\,44 & -3.176\,10 & 1 \\ 0 & 0 & 1.707\,71 & -2.759\,07 \\ 0 & 0 & -2.759\,07 & 4.464\,29 \end{bmatrix}
\begin{pmatrix} x_1 \\ x_2 \\ x_3 \\ x_4 \end{pmatrix} =
\begin{pmatrix} -1.590\,00 \\ -0.309\,980 \\ 0.942\,827 \\ -1.518\,88 \end{pmatrix}
\tag{E.3.1.3}
$$

Elimination step: $\boldsymbol{k} = 3$*, pivot row* $= 3$*, pivot element:* $a''_{33} = 1.70771$

$$
\begin{bmatrix} 4.855\,00 & -4 & 1 & 0 \\ 0 & 2.559\,44 & -3.176\,10 & 1 \\ 0 & 0 & 1.707\,71 & -2.759\,07 \\ 0 & 0 & & 0.006\,600 \end{bmatrix}
\begin{pmatrix} x_1 \\ x_2 \\ x_3 \\ x_4 \end{pmatrix} =
\begin{pmatrix} -1.590\,00 \\ -0.309\,980 \\ 0.942\,827 \\ 0.004\,390 \end{pmatrix}
\tag{E.3.1.4}
$$

Back Substitution Step

$$x = \begin{pmatrix} 0.686\,706 \\ 1.637\,68 \\ 1.626\,74 \\ 0.665\,151 \end{pmatrix} \tag{E.3.1.5}$$

In order to check the accuracy of the answer, we now evaluate the residuals by substituting the answer back into the original system of equation

$$r = \begin{pmatrix} 0.000\,022\,37 \\ 0.000\,016\,60 \\ 0.000\,055\,30 \\ -0.000\,028\,105 \end{pmatrix} \tag{E.3.1.6}$$

3.2.2 Gauss–Jordan Elimination Method

This method is a variation of the Gaussian elimination method. The major difference is that when an unknown is eliminated it is eliminated from all the other equations preceding the pivot equation as well as those following it. Thus the elimination steps reduce the coefficient matrix to an identity matrix rather than a triangular matrix. The solution is then obtained directly without requiring a back substitution step. The solution vector becomes exactly equal to the modified right-hand side vector. The following steps illustrate the procedure for solving the system given by Equation 3.1.

Original System

$$\begin{aligned} a_{11}\,x_1 + a_{12}\,x_2 \cdots a_{1n}\,x_n &= c_1 \\ a_{21}\,x_1 + a_{22}\,x_2 \cdots a_{2n}\,x_n &= c_2 \\ &\vdots \\ a_{n1}\,x_1 + a_{n2}\,x_2 \cdots a_{nn}\,x_n &= c_n \end{aligned} \tag{3.1}$$

Elimination step = 1, pivot row = 1, pivot element = a_{11}

$$\begin{aligned} x_1 + a'_{12}x_2 + a'_{13}x_3 + \cdots + a'_{1n} &= c'_1 \\ a'_{22}x_2 + a'_{23}x_3 + \cdots + a'_{2n} &= c'_2 \\ &\vdots \\ &\vdots \\ &\vdots \\ a'_{n2}x_2 + a'_{n3}x_3 + \cdots + a'_{nn} &= c'_n \end{aligned} \tag{3.19}$$

Elimination step = 2, pivot row = 2, pivot element = a'_{22}

$$
\begin{aligned}
x_1 + 0x_2 + a''_{13}x_3 + \cdots + a''_{1n} &= c''_1 \\
0x_2 + a''_{23}x_3 + \cdots + a''_{2n} &= c''_2 \\
&\vdots \\
&\vdots \\
&\vdots \\
0x_2 + a''_{n3}x_3 + \cdots + a''_{nn} &= c''_n
\end{aligned}
\tag{3.20}
$$

Elimination step = n, pivot row = n, pivot element = a_{nn}^{n-1}

$$
\begin{aligned}
x_1 \qquad\qquad\qquad &= c_1^n \\
x_2 \qquad\quad &= c_2^n \\
&\vdots \\
&\vdots \\
x_n &= c_n^n
\end{aligned}
\tag{3.21}
$$

The solution vector $\{x\}$ is given directly by the modified $\{c\}$ vector without requiring any back substitution step.

Generalized Formula. The procedure outlined in the Gauss–Jordan method can be generalized to express modified elements for all elimination steps as

$$a_{kj}^k = \frac{a_{kj}^{k-1}}{a_{kk}} \tag{3.22a}$$

$$a_{ij}^k = a_{ij}^{k-1} - a_{ik}^{k-1} a_{kj}^k \tag{3.22b}$$

for $k = 1, 2, \ldots, n$, $1 \leq i \leq n$ except $i = k$, and $k + 1 \leq j \leq n + 1$.

Example 3.2: Gauss–Jordan Method: Use the Gauss–Jordan method to solve the following system of equations

$$
\begin{bmatrix} 3 & -6 & 7 \\ 9 & 0 & -5 \\ 5 & -8 & 6 \end{bmatrix}
\begin{Bmatrix} x_1 \\ x_2 \\ x_3 \end{Bmatrix}
=
\begin{Bmatrix} 3 \\ 3 \\ -4 \end{Bmatrix}
\tag{E.3.2.1}
$$

Solution: We combine the coefficient matrix and the right-hand side vector to write the augmented matrix, A, as

$$\begin{bmatrix} 3 & -6 & 7 & 3 \\ 9 & 0 & -5 & 3 \\ 5 & -8 & 6 & -4 \end{bmatrix} \quad \text{(E.3.2.2)}$$

We then use the Gauss–Jordan algorithm given by Equation 3.22 that reduces the original coefficient matrix to an identity matrix. Following this algorithm, we have

Elimination step, k = 1, pivot row = 1, pivot element $a_{11} = 3$

Elements of row 1 are calculated using Equation 3.22a, which is $a'_{ij} = a_{ij}/a_{11}$ for this elimination step. This gives the modified matrix as

$$\begin{bmatrix} 1 & a'_{12} & a'_{13} & a'_{14} \\ a_{21} & a_{22} & a_{23} & a_{24} \\ a_{31} & a_{32} & a_{33} & a_{34} \end{bmatrix} = \begin{bmatrix} 1 & -2 & 2.33 & 1 \\ 9 & 0 & -5 & 3 \\ 5 & -8 & 6 & -4 \end{bmatrix} \quad \text{(E.3.2.3)}$$

Equation 3.22b is then used to calculate elements of rows 2 and 3 as follows

For i = 2, j = 2, 3, 4

$$\begin{bmatrix} 1 & a'_{12} & a'_{13} & a'_{14} \\ 0 & a'_{22} & a'_{23} & a'_{24} \\ a_{31} & a_{32} & a_{33} & a_{34} \end{bmatrix} = \begin{bmatrix} 1 & -2 & 2.33 & 1 \\ 0 & 18 & -26 & -6 \\ 5 & -8 & 6 & -4 \end{bmatrix} \quad \text{(E.3.2.4)}$$

For i = 3, j = 2, 3, 4

$$\begin{bmatrix} 1 & a'_{12} & a'_{13} & a'_{14} \\ 0 & a'_{22} & a'_{23} & a'_{24} \\ 0 & a'_{32} & a'_{33} & a'_{34} \end{bmatrix} = \begin{bmatrix} 1 & -2 & 2.33 & 1 \\ 0 & 18 & -26 & -6 \\ 0 & 2 & -5.65 & -9 \end{bmatrix} \quad \text{(E.3.2.5)}$$

Elimination step, k = 2, pivot row = 2, pivot element $a'_{22} = 18$

Elements of row 2 are calculated using Equation 3.22a, which reduces to $a''_{2j} = a'_{2j}/a'_{22}$. This gives the modified matrix as

$$\begin{bmatrix} 1 & a'_{12} & a'_{13} & a'_{14} \\ 0 & 1 & a''_{23} & a''_{24} \\ 0 & a'_{32} & a'_{33} & a'_{34} \end{bmatrix} = \begin{bmatrix} 1 & -2 & 2.33 & 1 \\ 0 & 1 & 1.44 & -0.33 \\ 0 & 2 & -5.65 & -9 \end{bmatrix} \quad \text{(E.3.2.6)}$$

Equation 3.22b is then used to calculate elements of rows 3 and 1 as

***For i* = 3, *j* = 3, 4**

$$\begin{bmatrix} 1 & 0 & a'_{13} & a'_{14} \\ 0 & 1 & a''_{23} & a''_{24} \\ 0 & 0 & a''_{33} & a''_{34} \end{bmatrix} = \begin{bmatrix} 1 & -2 & 2.33 & 1 \\ 0 & 1 & -1.44 & -0.33 \\ 0 & 0 & -2.77 & -8.33 \end{bmatrix} \quad \text{(E.3.2.7)}$$

***For i* = 1, *j* = 3, 4**

$$\begin{bmatrix} 1 & 0 & a''_{13} & a''_{14} \\ 0 & 1 & a''_{23} & a''_{24} \\ 0 & 0 & a''_{33} & a''_{34} \end{bmatrix} = \begin{bmatrix} 1 & 0 & -0.55 & 0.33 \\ 0 & 1 & -1.44 & -0.33 \\ 0 & 0 & -2.77 & -8.33 \end{bmatrix} \quad \text{(E.3.2.8)}$$

***Elimination step, k* = 3, *pivot row* = 3, *pivot element* $a''_{33} = -2.77$**

Elements of row 3 are calculated using $a'''_{3j} = a''_{3j}/a''_{33}$. This modifies the augmented matrix to

$$\begin{bmatrix} 1 & 0 & a''_{13} & a''_{14} \\ 0 & 1 & a''_{23} & a''_{24} \\ 0 & 0 & 1 & a'''_{34} \end{bmatrix} = \begin{bmatrix} 1 & 0 & -0.55 & 0.33 \\ 0 & 1 & -1.44 & -0.33 \\ 0 & 0 & 1 & 3.01 \end{bmatrix} \quad \text{(E.3.2.9)}$$

Equation 3.22b is then used to calculate elements of row 1 and 2 as

***For i* = 1, *j* = 4**

$$\begin{bmatrix} 1 & 0 & 0 & a'''_{14} \\ 0 & 1 & a''_{23} & a''_{24} \\ 0 & 0 & 1 & a'''_{34} \end{bmatrix} = \begin{bmatrix} 1 & 0 & 0 & 1.99 \\ 0 & 1 & -1.44 & -0.33 \\ 0 & 0 & 1 & 3.01 \end{bmatrix} \quad \text{(E.3.2.10)}$$

***For i* = 2, *j* = 4**

$$\begin{bmatrix} 1 & 0 & 0 & a'''_{14} \\ 0 & 1 & 0 & a'''_{24} \\ 0 & 0 & 1 & a'''_{34} \end{bmatrix} = \begin{bmatrix} 1 & 0 & 0 & 1.99 \\ 0 & 1 & 0 & 4.00 \\ 0 & 0 & 1 & 3.01 \end{bmatrix} \quad \text{(E.3.2.11)}$$

The solution vector is then given by the modified elements of the right side vector as

$$\begin{Bmatrix} x_1 \\ x_2 \\ x_3 \end{Bmatrix} = \begin{Bmatrix} a'''_{14} \\ a'''_{24} \\ a'''_{34} \end{Bmatrix} = \begin{Bmatrix} 1.99 \\ 4.00 \\ 3.01 \end{Bmatrix} \quad \text{(E.3.2.12)}$$

□

All problems and improvements discussed for the Gauss elimination method are applicable to the Gauss–Jordan method. So, the method of pivoting is applicable to the Gauss–Jordan method to avoid zero division and to reduce round-off error. It can be mentioned here that the Gauss–Jordan method usually takes more operations than Gauss-elimination. Therefore, the Gauss-elimination method is usually the method of choice for obtaining the exact solutions of a system of linear equations. The Gauss–Jordan method is preferred for obtaining a matrix inverse (A^{-1}) that provides a convenient way of evaluating a set of simultaneous equations with multiple right-hand side force vectors, $\{c\}$.

The solution of the set of simultaneous equations given in Equation 2.1 can be obtained by using the matrix inverse A^{-1} as

$$\{x\} = [A]^{-1}\{c\} \tag{3.23}$$

Applications of such procedures occur when it is necessary to solve the same system of equations as in Equation 2.1 many times with different right-hand side vector $\{c\}$. The step for obtaining the matrix inverse is performed only once. This step involves augmenting the coefficient matrix $[A]$ with an identify matrix, $[I]$. Then the Gauss–Jordan algorithm is employed to reduce the coefficient matrix to an identify matrix. This procedure will provide the inverse matrix as the modified augmented unit matrix.

3.2.3 Decomposition or Factorization Methods

Another class of elimination methods that is sometimes more efficient than Gauss elimination or the Gauss–Jordan method is based on decomposition or factorization of the coefficient matrix. The main advantage of these methods is that a smaller number of arithmetic operations are needed and this results in a smaller round-off error and less computing time.

The Basic LU Decomposition. In the basic LU decomposition method, the system $Ax = c$ is solved in three-steps. In the ***first step***, the coefficient matrix, A, is decomposed into an upper triangular matrix, U, and a lower triangular matrix, L, such that

$$A = LU \tag{3.24}$$

where the upper triangular matrix U is given as

$$U = \begin{bmatrix} 1 & u_{12} & u_{13} & \cdots & \cdots & u_{1n} \\ 0 & 1 & u_{23} & \cdots & \cdots & u_{2n} \\ \vdots & & & & & \\ \vdots & & & & & \\ \vdots & & & & & \\ 0 & 0 & 0 & \cdots & \cdots & 1 \end{bmatrix} \tag{3.25}$$

and the lower triangular matrix is expressed as

$$L = \begin{bmatrix} l_{11} & 0 & 0 & \cdots & \cdots & 0 \\ l_{21} & l_{22} & 0 & \cdots & \cdots & 0 \\ \vdots & & & & & \\ \vdots & & & & & \\ \vdots & & & & & \\ l_{n1} & l_{n2} & & & & l_{n2} \end{bmatrix} \tag{3.26}$$

Once the decomposition or factorization of A is obtained, the solution of the system becomes equivalent to solving two systems $Ly = c$ and $Ux = y$.

In the ***second step***, the lower triangular matrix, L, and known right-hand side vector, c, is used to solve

$$Ly = c \tag{3.27}$$

using back substitution given by the generalized formula

$$y_1 = \frac{c_1}{l_{11}} \tag{3.28a}$$

and

$$y_i = \frac{c_i - \sum_{j=1}^{i-1} l_{ij} y_j}{l_{ii}}, \quad \text{for } i = 2, 3, \ldots, n \tag{3.28b}$$

In the ***third step***, the upper triangular matrix U is used along with vector y as the known right-hand side vector to solve the system

$$Ux = y \tag{3.29}$$

for the unknown vector x using the generalized formula for back substitution as

$$x_n = y_n \tag{3.30a}$$

$$x_i = y_i - \sum_{j=i+1}^{n} u_{ij} x_j, \quad i = n-1, n-2, \ldots, 1 \tag{3.30b}$$

Decomposition Procedure There are two different LU decomposition methods: ***Dolittle's*** method and ***Crout's*** method. In ***Dolittle's*** method, the matrix L has all the diagonal elements equal to 1, and the decomposed matrices are obtained through the use of the Gauss-elimination method. The upper triangular matrix U is obtained in the forward elimination step as

$$U = \begin{bmatrix} a_{11} & a_{12} & a_{13} & \cdots & \cdots & a_{1n} \\ 0 & a'_{22} & a'_{23} & \cdots & \cdots & a'_{2n} \\ \vdots & & & & & \\ \vdots & & & & & \\ \vdots & & & & & \\ 0 & 0 & 0 & \cdots & \cdots & a_{nn}^{n-1} \end{bmatrix} \tag{3.31}$$

The lower triangular matrix, L, is formed by retaining the multipliers, $m_{kj} = a_{kj}^{k-1}/a_{kk}^{k-1}$, for $j = k+1 \cdots n$, which are used during the elimination step and are used as elements l_{ij}

$$L = \begin{bmatrix} 1 & 0 & 0 & \cdots & \cdots & 0 \\ l_{21} = m_{21} & 1 & 0 & \cdots & \cdots & 0 \\ l_{31} = m_{31} & l_{32} = m_{32} & 1 & & & \\ \vdots & & & & & \\ \vdots & & & & & \\ l_{n1} = m_{n1} & l_{n2} = m_{n2} & l_{n3} = m_{n3} & \cdots & \cdots & 1 \end{bmatrix} \tag{3.32}$$

In ***Crout's method***, the U matrix has 1s in the diagonal elements. This is one of the most efficient decomposition procedures without requiring the time-consuming elimination steps. This involves a direct matrix multiplication of $[L]$ $[U]$ and equating the resulting terms with the corresponding elements of the coefficient matrix A. This procedure results in the following generalized formula for the elements in the L and U matrices

First column of L

$$l_{i1} = a_{i1}, \quad \text{for } i = 1, 2, 3, \ldots, n \tag{3.33a}$$

First row of U

$$u_{1j} = \frac{a_{1j}}{l_{11}}, \quad \text{for } j = 2, 3, 4, \ldots, n \tag{3.33b}$$

Rest of the columns in L

$$l_{ij} = a_{ij} - \sum_{k=1}^{j-1} l_{ik} u_{kj}, \quad \text{for } j = 2, 3, n-1 \text{ and } i = j, j+1, \ldots, n \tag{3.33c}$$

Rest of the rows in U

$$u_{ij} = \frac{a_{ij} - \sum_{k=1}^{i-1} l_{ik} u_{kj}}{l_{ii}}, \quad \text{for } i = 2, 3, 4, \ldots, n-1 \text{ and } j = i+1, i+2, \ldots, n$$

(3.33d)

Last diagonal element in U

$$l_{nn} = a_{nn} - \sum_{k=1}^{n-1} l_{nk} u_{kn} \tag{3.33e}$$

Since there is no need to store the 0s and 1s of the L and U matrices, considerable reduction in computer storage is obtained by storing both U and L matrices in the same storage location for the original coefficient matrix A. The original a_{ij} elements of the coefficient matrix are stored first. Then, calling both the U_{ij} and L_{ij} elements as a_{ij}, the old a_{ij} element values are replaced with u_{ij} and l_{ij} element values as the new a_{ij} element values.

Example 3.3: *LU* Decomposition: Use Crout's LU decomposition to solve

$$\begin{bmatrix} -4 & 1 & 1 & 0 \\ 1 & -4 & 0 & 1 \\ 1 & 0 & -4 & 1 \\ 0 & 1 & 1 & -4 \end{bmatrix} \begin{Bmatrix} x_1 \\ x_2 \\ x_3 \\ x_4 \end{Bmatrix} = \begin{Bmatrix} 0 \\ -90 \\ 0 \\ -90 \end{Bmatrix} \tag{E.3.3.1}$$

Solution: In the first step of the solution, we use the generalized formula given by Equation 3.33 to determine elements of matrices L and U.

Equation 3.33a gives elements of the first column of L

$$\text{For } j = 1 \text{ (first column) and}$$
$$i = 1, \quad l_{11} = a_{11} = -4$$
$$i = 2, \quad l_{21} = a_{21} = 1$$
$$i = 3, \quad l_{31} = a_{31} = 1$$

Equation 3.33b gives elements of the first row of U

$$\text{For } i = 1 \text{ (first row) and}$$
$$j = 2, \quad u_{12} = \frac{a_{12}}{l_{11}} = \frac{1}{-4} = -0.25$$
$$j = 3, \quad u_{13} = \frac{a_{13}}{l_{11}} = \frac{1}{-4} = -0.25$$
$$j = 4, \quad u_{14} = \frac{a_{14}}{l_{11}} = \frac{0}{-4} = 0.0$$

Equation 3.33c gives elements of the second column of L

For $j = 2$(column – 2) and

$$i = 2, \quad l_{22} = a_{22} - l_{21}u_{12} = -4 - (1)(-0.25) = -3.75$$

$$i = 3, \quad l_{32} = a_{32} - l_{31}u_{12} = 0 - (1)(-0.25) = 0.25$$

$$i = 4, \quad l_{42} = a_{42} - l_{41}u_{14} = 1 - (0)(0) = 1$$

Equation 3.33d gives elements of the second row of U

For $i = 2$(row – 2) and

$$j = 3, \quad u_{23} = \frac{a_{23} - l_{21}u_{13}}{l_{22}} = \frac{0 - (1)(-0.25)}{-3.75} = -0.066\,67$$

$$j = 4, \quad u_{24} = \frac{a_{24} - l_{21}u_{14}}{l_{22}} = \frac{1 - (1)(0)}{-3.75} = -0.266\,67$$

Equation 3.33c gives elements of the third column of L

For $j = 3$ and

$$i = 3, \quad l_{33} = a_{33} - (l_{31}u_{13} + l_{32}u_{23})$$
$$= -4 - [(1)(-0.25) + (0.25)(-0.066\,67)] = -3.733\,33$$

$$i = 4, \quad l_{43} = a_{43} - (l_{41}u_{13} + l_{42}u_{23})$$
$$= 1 - [(0)(-0.25) + (1)(-0.066\,67)] = 1.066\,67$$

Equation 3.33d gives elements of the third row of U

For $i = 3$(row – 3) and

$$j = 4, \quad u_{34} = \frac{a_{34} - [l_{31}u_{14} + l_{32}u_{24}]}{l_{33}}$$
$$= \frac{1 - [(1)(0) + (0.25)(-0.266\,67)]}{-3.733\,33} = -0.285\,71$$

Equation 3.33e gives the last diagonal element of L

For $j = 4$ and $i = 4$

$$l_{44} = a_{44} - (l_{41}u_{14} + l_{42}u_{24} + l_{43}u_{34})$$
$$= -4 - [(0)(0) + (1)(-0.266\,67) + (1.066\,67)(-0.285\,71)] = -3.428\,57$$

Assembling the elements, we have the decomposition of A into L and U as

$$L = \begin{bmatrix} -4 & 0 & 0 & 0 \\ 1 & -3.75 & 0 & 0 \\ 1 & 0.25 & -3.733\,33 & 0 \\ 0 & 1 & 1.0666 & -3.428\,57 \end{bmatrix} \qquad \text{(E.3.3.2)}$$

and

$$U = \begin{bmatrix} 1 & -0.25 & -0.25 & 0 \\ 0 & 1 & -0.066\,67 & -0.0266\,67 \\ 0 & 0 & 1 & -0.28571 \\ 0 & 0 & 0 & 1 \end{bmatrix} \qquad \text{(E.3.3.3)}$$

In the ***second step***, the system of equation $Ly = c$ is solved for the vector y using back substitution (Equation 3.28) as

$$y_1 = \frac{c_1}{l_{11}} = \frac{0}{-4} = 0$$

$$y_2 = \frac{c_2 - l_{21}y_1}{l_{22}} = \frac{-90 - (1)(0)}{-3.75} = 24$$

$$y_3 = \frac{c_3 - l_{31}y_1 - l_{32}y_2}{l_{33}} = \frac{0 - [(1)(0) - (0.25)(24)]}{-3.733\,33} = 1.607\,14$$

$$y_4 = \frac{c_4 - [l_{41}y_1 + l_{42}y_2 + l_{43}y_3]}{l_{44}}$$

$$= -90 - \frac{[(0)(0) + (1)(24) + (1.0666)(1.607\,14)]}{-3.428\,57} = 33.750\,01$$

Thus, the vector y is

$$y = \begin{Bmatrix} 0 \\ 24 \\ 1.607\,14 \\ 33.750\,01 \end{Bmatrix} \qquad \text{(E.3.3.4)}$$

In the ***third step***, the system $Ux = y$ is solved for the solution vector x using back substitution (Equation 3.30) as

$$x_4 = y_4 = 33.750\,01$$

$$x_3 = y_3 - u_{34}x_4 = 1.607\,14 - (-0.285\,71)(33.750\,01) = 11.249\,86$$

$$x_2 = y_2 - [u_{23}x_3 + u_{24}x_4]$$

$$= 24 - [(-0.066\,67)(11.249\,86) + (-0.266\,67)(33.750\,01)] = 33.750\,14$$

$$x_1 = y_1 - [u_{12}x_2 + u_{13}x_3 + u_{14}x_4]$$

$$= 0 - [(-0.25)(33.501\,4) + (-0.25)(11.249\,86) + (0)(33.750\,01)]$$

$$= 11.25$$

The final solution vector x is

$$x = \begin{Bmatrix} 11.25 \\ 33.750\,14 \\ 11.249\,86 \\ 33.750\,01 \end{Bmatrix} \tag{E.3.3.5}$$

□

Cholesky Factorization Method. A symmetric square matrix often arises in many engineering problems. For such a matrix, the decomposition can be more efficiently performed using a recursive procedure known as the ***Cholesky factorization method***. The method is based on factorizing the symmetric matrix A into a lower triangular matrix L and its transpose matrix L^{T} such that

$$A = LL^{\mathrm{T}} \tag{3.34}$$

where the lower triangular matrix L and its transpose matrix L^{T} are expressed as

$$L = \begin{bmatrix} l_{11} & 0 & 0 & \cdots & 0 \\ l_{21} & l_{22} & 0 & \cdots & 0 \\ \vdots & & & & \\ \vdots & & & & \\ l_{n1} & l_{n2} & 0 & \cdots & l_{nn} \end{bmatrix} \quad \text{and} \quad L^{\mathrm{T}} = \begin{bmatrix} l_{11} & l_{21} & \cdots & \cdots & l_{n1} \\ & l_{22} & \cdots & \cdots & l_{n2} \\ & & & & \vdots \\ & & & & \vdots \\ & & & & l_{nn} \end{bmatrix}$$

The elements of the Cholesky factorization are obtained directly from Equation 3.34 and elements of factorized matrices are given by the generalized formulas

$$l_{11} = \sqrt{a_{11}} \tag{3.35a}$$

$$l_{i1} = \frac{a_{i1}}{l_{11}}, \quad \text{for } i = 2, 3, \ldots, n \tag{3.35b}$$

$$l_{ii} = \sqrt{a_{ii} - \sum_{k=1}^{i-1} l_{ik}^2} \tag{3.35c}$$

$$l_{ij} = \frac{1}{l_{jj}}\left(a_{ij} - \sum_{k=1}^{j-1} l_{ik} l_{jk}\right), \quad \text{for } j = 2, 3, \ldots, i-1, \text{ and } j<i \tag{3.35d}$$

Once the factorization is obtained, the solution to the linear system $Ax=c$ is obtained by first solving the lower triangular system $Ly=c$ and followed by solving the upper transpose system $L^T x=y$.

The Cholesky decomposition method offers computational advantages because only half the storage is needed, and in most cases requires half the computational time as compared to the LU decomposition method.

Example 3.4: Cholesky Factorization Method: Use the Cholesky factorization method to decompose the symmetric matrix

$$A=\begin{bmatrix} 1 & -4 & 1 & 0 \\ -4 & 60 & -4 & 0 \\ 1 & -4 & 6 & -4 \\ 0 & 0 & -4 & 25 \end{bmatrix}$$

Solution:

$$l_{11}=\sqrt{a_{11}}=\sqrt{1}=1$$

The second row $(i=2)$

$$j=1,\quad l_{21}=\frac{a_{21}}{l_{11}}=\frac{-4}{1}=-4$$

$$j=2,\quad l_{22}=\sqrt{a_{22}-\sum_{k=1}^{2-1} l_{2k}^2}=\sqrt{60-(-4)^2}=\sqrt{44}$$

Third row $(i=3)$

$$j=1,\quad l_{31}=\frac{a_{31}}{l_{11}}=\frac{1}{1}=1$$

$$j=2,\quad l_{32}=\frac{\left(a_{32}-\sum_{k=1}^{2-1} l_{3k}l_{2k}\right)}{l_{22}}=\frac{[a_{32}-(l_{31}l_{21})]}{l_{22}}=\frac{[-4-(1)(-4)]}{\sqrt{44}}=0$$

$$j=3,\quad l_{33}=\sqrt{a_{33}-\sum_{k=1}^{3-1} l_{3k}^2}=\sqrt{6-(l_{31}^2+l_{32}^2)}=\sqrt{6-(1^2+0^2)}=\sqrt{5}$$

Fourth row ($i=4$)

$$j=1,\quad l_{41}=\frac{a_{41}}{l_{11}}=\frac{0}{1}=0$$

$$j=2,\quad l_{42}=\frac{\left(a_{42}-\sum_{k=1}^{2-1} l_{4k}l_{2k}\right)}{l_{22}}=\frac{0-[(0)(-4)]}{\sqrt{44}}=0$$

$$j=3,\quad l_{43}=\frac{\left(a_{43}-\sum_{k=1}^{3-1} l_{4k}l_{3k}\right)}{l_{33}}$$

$$=\frac{[a_{43}-(l_{41}l_{31}+l_{42}l_{32})]}{l_{33}}=\frac{[-4-((0)(1)+(0)(0))]}{\sqrt{5}}=-\frac{4}{\sqrt{5}}$$

$$j=4,\quad l_{44}=\sqrt{a_{44}^2-\sum_{k=1}^{4-1} l_{4k}^2}=\sqrt{a_{44}-[l_{41}^2+l_{42}^2+l_{43}^2]}$$

$$=\sqrt{25-\left[(0)^2+(0)^2+\left(-\frac{4}{\sqrt{5}}\right)^2\right]}=\sqrt{\frac{109}{5}}$$

The matrix L and its transpose matrix L^{T} can now be written as

$$L=\begin{bmatrix}1 & 0 & 0 & 0\\ -4 & \sqrt{44} & 0 & 0\\ 1 & 0 & \sqrt{5} & 0\\ 0 & 0 & -\frac{4}{\sqrt{5}} & \sqrt{\frac{109}{5}}\end{bmatrix} \quad \text{and} \quad L^{\mathrm{T}}=\begin{bmatrix}1 & -4 & 1 & 0\\ 0 & \sqrt{44} & 0 & 0\\ 0 & 0 & \sqrt{5} & -\frac{4}{\sqrt{5}}\\ 0 & 0 & 0 & \sqrt{\frac{109}{5}}\end{bmatrix}$$

□

***QR* Factorization by Householder Method.** In the QR factorization method, the matrix A is factorized into the product of an orthogonal matrix Q and an upper triangular matrix R such that

$$A=QR \tag{3.36}$$

There are different choices for QR factorization techniques such as the one based on the ***Householder method*** or based on the ***Given's method***. We will discuss here the QR factorization based on the Householder method in which the matrix A is transformed into a form for which the decomposition can be more rapidly obtained by the Householder method. One such desired form is the ***Hessenberg matrix***, which has the form

$$A_H = \begin{bmatrix} a_{11}^* & a_{12}^* & a_{13}^* & \cdots & \cdots & a_{1n}^* \\ a_{21}^* & a_{22}^* & a_{23}^* & \cdots & \cdots & a_{2n}^* \\ 0 & a_{32}^* & a_{33}^* & & & \\ 0 & 0 & a_{43}^* & \ddots & & \\ \vdots & \vdots & \vdots & \ddots & \ddots & \\ 0 & 0 & 0 & \cdots & a_{n-1,1}^* & a_{nn}^* \end{bmatrix} \tag{3.37}$$

Note that it has one nonzero diagonal below the main diagonal and the elements above the main diagonal are nonzero, in general. The reduction of the original matrix A to the Hessenburg form can be done by Householder transformations in $n-1$ steps. The procedure is outlined as follows:

Step 1: If we select a nonzero vector z_1 to be such that the Householder matrix can be defined as $H_1 = I - \left(2z_1z_1^T/z_1^Tz_1\right)$, it can be used to create 0s in a vector as

$$H_1 \begin{Bmatrix} a_{11} \\ a_{21} \\ a_{31} \\ \vdots \\ \vdots \\ a_{n1} \end{Bmatrix} = \begin{Bmatrix} a_{11}^1 \\ 0 \\ 0 \\ \vdots \\ \vdots \\ 0 \end{Bmatrix} \tag{3.38}$$

We transform the matrix A to the form

$$A_H^1 = H_1A = \begin{bmatrix} a_{11}^1 & a_{12}^1 & a_{13}^1 & \cdots & a_{1n}^1 \\ 0 & a_{22}^1 & a_{23}^1 & \cdots & a_{2n}^1 \\ 0 & a_{32}^1 & a_{33}^1 & \cdots & a_{3n}^1 \\ \vdots & \vdots & & & \\ \vdots & \vdots & & & \\ 0 & a_{n2}^1 & a_{n3}^1 & \cdots & a_{nn}^1 \end{bmatrix} \tag{3.39}$$

Note that the matrix $A_H^1 = H_1A$ has 0s below the diagonal element in the first column.

Step 2: We select a nonzero vector z_2 to be such that the Householder matrix can be defined as $H_2 = I - \left(2z_2z_2^T/z_2^Tz_2\right)$, and it can be used to transform the matrix to the form

$$A_H^2 = H_2A_H^1 = \begin{bmatrix} a_{11}^1 & a_{12}^1 & a_{13}^1 & \cdots & a_{1n}^1 \\ 0 & a_{22}^2 & a_{23}^2 & \cdots & a_{2n}^2 \\ 0 & 0 & a_{33}^2 & \cdots & a_{3n}^2 \\ \vdots & \vdots & & & \\ \vdots & \vdots & & & \\ 0 & 0 & a_{n3}^2 & \cdots & a_{nn}^2 \end{bmatrix} \tag{3.40}$$

Note that the matrix $A_H^2 = H_2A_H^1 = H_2H_1A$ has 0s in the first two columns below the main diagonal.

This procedure is repeated until in the $(n-1)$th step the Householder transformation leads to

$$A_H^{n-1} = H_{n-1}A_H^{n-2} = \begin{bmatrix} a_{11}^1 & a_{12}^1 & a_{13}^1 & \cdots & & a_{1n}^1 \\ 0 & a_{22}^2 & a_{23}^2 & \cdots & & a_{2n}^2 \\ 0 & 0 & a_{33}^3 & \cdots & & a_{3n}^3 \\ \vdots & \vdots & & \ddots & & \\ \vdots & \vdots & & & a_{n-1,n-1}^{n-1} & a_{n-1,n}^{n-1} \\ 0 & 0 & 0 & \cdots & 0 & a_{nn}^{n-1} \end{bmatrix} \tag{3.41}$$

which represents an upper triangular matrix R as

$$R = H_{n-1} \cdots H_2H_1A \tag{3.42}$$

Now, if we define

$$Q^T = H_{n-1} \cdots H_2H_1 \tag{3.43}$$

then Equation 3.42 can be written as $Q^TA = R$. Since each Householder matrix H is orthogonal, the product Q^T is also orthogonal. So, $Q^{-1} = Q^T$, and Equation 3.42 is written as

$$A = QR \tag{3.44}$$

Once the QR factorization is obtained, the system of linear equations $Ax = c$ is solved in the following two steps.

In the ***first step***, the vector y is obtained from

$$y = Q^{\mathrm{T}} c \tag{3.45a}$$

In the ***second step***, the upper triangular matrix R is used along with y as the right-hand side vector to solve for unknown vector x as

$$Rx = y \tag{3.45b}$$

3.2.4 Banded Systems

In many heat transfer and fluid flow problems, the discretization equations form a system in which not all of the unknowns are present in each equation, and hence consist of many zero elements in the coefficient matrix. A banded matrix is one such matrix that has all elements equal to zero with the exception of a band centered around the main diagonal as shown below.

$$\begin{bmatrix} a_{11} & a_{12} & a_{13} & 0 & 0 & 0 & \cdots & 0 \\ a_{21} & a_{22} & a_{23} & a_{24} & 0 & 0 & \cdots & 0 \\ a_{31} & a_{32} & a_{33} & a_{34} & a_{35} & 0 & \cdots & 0 \\ 0 & a_{42} & a_{43} & a_{44} & a_{45} & a_{46} & \cdots & 0 \\ 0 & 0 & a_{53} & a_{54} & & \leftarrow hbw & \rightarrow & 0 \\ \vdots & & & \leftarrow & \leftarrow & bw & \rightarrow & \rightarrow \\ \vdots & & & & & & & \\ 0 & 0 & 0 & 0 & 0 & a_{n6} & \cdots & a_{nn} \end{bmatrix} \tag{3.46}$$

The dimensions of such a banded system are quantified by two parameters: the bandwidth $\boldsymbol{bw}$ and the half bandwidth $\boldsymbol{hbw}$, as shown. These parameters are related to each other by $\boldsymbol{bw} = 2\ \boldsymbol{hbw} + 1$. In general, a system $Ax = c$ is defined as banded when elements of the coefficient matrix $a_{ij} = 0$ for $|i - j| > \boldsymbol{hbw}$.

Although all direct methods, such as the elimination and the factorization methods, can be used to solve such banded systems, they become quite inefficient for cases where pivoting is not required. A ***diagonally dominant matrix*** is one such case where no partial pivoting is required during triangularization procedures. For such matrices, the absolute value of the diagonal element in each of the equations must be larger than the sum of the absolute values of other elements in the equation. That is, $|a_{jj}| > \sum |a_{ij}|$, where the summation is taken from $i = 1$ to n, excluding the element for $i = j$. Considerable economics are achieved while solving such systems using alternate, more efficient algorithms that do not involve zero elements and require less expenditure of space and time in storing and manipulating zero elements.

Tridiagonal Systems and Tridiagonal-Matrix Algorithm (TDMA). One of the most commonly encountered banded systems is the tridiagonal system, which has a bandwidth of $\boldsymbol{bw} = 3$. Such a matrix has the form

$$A = \begin{bmatrix} a_1 & b_1 & 0 & 0 & 0 & \cdots & 0 \\ c_2 & a_2 & b_2 & 0 & 0 & \cdots & 0 \\ 0 & c_3 & a_3 & b_3 & 0 & \cdots & 0 \\ 0 & 0 & c_4 & a_4 & b_4 & \cdots & 0 \\ \vdots & & & & & & \\ & & & & c_{n-1} & a_{n-1} & b_{n-1} \\ & & & & & c_n & a_n \end{bmatrix} \tag{3.47}$$

Note that the notation for matrix elements are changed from a single two-dimensional matrix array, a_{ij}, to three one-dimensional vectors a_i, b_i, and c_i, that store only the elements in the three diagonals. Because of this simple form of the system, the elimination or factorization process leads into a convenient algorithm known as ***tridiagonal matrix algorithm (TDMA)*** or ***Thomas algorithm***.

Thomas Algorithm – A Gauss Elimination Method for Tridiagonal System. As with the conventional Gauss elimination method, the scheme utilized in this algorithm involves two steps: ***elimination*** and ***back substitution***. The procedure will be illustrated by considering the set of equations written as

$$\begin{bmatrix} a_1 & b_1 & 0 & 0 & 0 & \cdots & 0 \\ c_2 & a_2 & b_2 & 0 & 0 & \cdots & 0 \\ 0 & c_3 & a_3 & b_3 & 0 & \cdots & 0 \\ 0 & 0 & c_4 & a_4 & b_4 & \cdots & 0 \\ \vdots & & & & & & \\ 0 & 0 & 0 & \cdots & c_{n-1} & a_{n-1} & b_{n-1} \\ 0 & 0 & 0 & \cdots & 0 & c_n & a_n \end{bmatrix} \begin{Bmatrix} x_1 \\ x_2 \\ x_3 \\ x_4 \\ \vdots \\ x_{n-1} \\ x_n \end{Bmatrix} = \begin{Bmatrix} d_1 \\ d_2 \\ d_3 \\ d_4 \\ \vdots \\ d_{n-1} \\ d_n \end{Bmatrix} \tag{3.48}$$

In the ***elimination step***, as Gaussian elimination is applied to this system, only one of the c's is eliminated from the column containing the pivot element in each step, since the remaining elements below the diagonal element are zero.

The original zero elements remain unchanged and the matrix reduces to the form

$$\begin{bmatrix} a_1 & b_1 & 0 & 0 & 0 & \cdots & 0 \\ 0 & a_2' & b_2' & 0 & 0 & \cdots & 0 \\ 0 & & a_3' & b_3' & 0 & \cdots & 0 \\ 0 & 0 & & a_4 & b_4 & \cdots & 0 \\ \vdots & & & & & & \\ 0 & 0 & 0 & \cdots & 0 & a_{n-1}' & b_{n-1}' \\ 0 & 0 & 0 & \cdots & 0 & 0 & a_n' \end{bmatrix} \begin{Bmatrix} x_1 \\ x_2 \\ x_3 \\ x_4 \\ \vdots \\ x_{n-1} \\ x_n \end{Bmatrix} = \begin{Bmatrix} d_1 \\ d_2' \\ d_3' \\ d_4' \\ \vdots \\ d_{n-1}' \\ d_n' \end{Bmatrix} \tag{3.49}$$

From this reduced system of equations, the unknowns x_i may easily be obtained by back substitution.

The ***generalized formulas*** that express the modified elements in the system during each elimination step and the solution during the substitution steps are given by

Elimination

$$c_i' = 0, \quad a_i' = a_i - b_{i-1}\frac{c_i}{a_{i-1}'} \tag{3.50a}$$

$$b_i' = b_i, \quad d_i' = d_i - d_{i-1}'\frac{c_i}{a_{i-1}'} \quad \text{for } i = 2, 3, \ldots, n \tag{3.50b}$$

Back substitution

$$x_n = \frac{d_n'}{a_n'} \tag{3.51a}$$

$$x_i = \frac{d_i' - b_{i+1}'}{a_i'} \quad \text{for } i = n-1, n-2, \ldots, 3, 2, 1 \tag{3.51b}$$

Thomas Algorithm – An LU Decomposition Method for Tridiagonal System. As with the conventional LU decomposition method, this algorithm consists of three steps: ***decomposition***, ***forward***, and ***backward substitution***. In the first step, the LU decomposition of the tridiagonal matrix A gives upper bidiagonal and lower bidiagonal matrices such that

$$A = LU \tag{3.52a}$$

or

$$\begin{bmatrix} a_1 & b_1 & 0 & 0 & 0 & \cdots & 0 \\ c_2 & a_2 & b_2 & 0 & 0 & \cdots & 0 \\ 0 & c_3 & a_3 & b_3 & 0 & \cdots & 0 \\ 0 & 0 & c_4 & a_4 & b_4 & \cdots & 0 \\ \vdots & & & & & & \\ & & & & c_{n-1} & a_{n-1} & b_{n-1} \\ & & & & & c_n & a_n \end{bmatrix}$$

$$= \begin{bmatrix} 1 & & & & & & \\ l_2 & 1 & & & & & \\ 0 & l_3 & 1 & & & & \\ 0 & 0 & l_4 & 1 & & & \\ \vdots & & & & & & \\ & & & & l_{n-1} & 1 & \\ & & & & & l_n & 1 \end{bmatrix} \begin{bmatrix} u_1 & b_1 & & & & & \\ & u_2 & b_2 & & & & \\ & & u_3 & b_3 & & & \\ & & & u_4 & b_4 & & \\ & & & & & & \\ & & & & & u_{n-1} & b_{n-1} \\ & & & & & & u_n \end{bmatrix} \tag{3.52b}$$

The elements of decomposed matrices are obtained directly by equating the corresponding elements of matrices on both sides of Equation 3.47 and expressed by the formulas

$$u_1 = a_1 \tag{3.53a}$$

and

$$l_i = \frac{c_i}{u_{i-1}} \tag{3.53b}$$

$$u_i = a_i - l_i b_{i-1} \quad \text{for } i = 2, 3, \ldots, n \tag{3.53c}$$

Once the decomposition is obtained, the solution to the tridiagonal linear system $Ax = c$ is obtained by solving the lower bidiagonal system $Ly = c$ and then by solving the upper bidiagonal system $ux = y$.

TDMA Algorithm–A General Elimination Method for Tridiagonal System. This is one of the most popular algorithms (Patankar 1980) for solving a tridiagonal system. This algorithm employs a scheme in two steps: ***forward substitution*** and ***backward substitution***. In the forward substitution step x_i is expressed in terms of x_{i+1} until x_n is obtained as a numerical value.

In the subsequent back substitution step $x_{n-1}, x_{n-2}, \ldots, x_3, x_2, x_1$ are obtained. The procedure will be illustrated by solving the set of equations (3.48) written as

$$a_i x_i = -b_i x_{i+1} - c_i x_{i-1} + d_i \tag{3.54}$$

where $c_1 = 0$, $b_n = 0$, and $i = 1, 2, \ldots, n$.

The ***forward substitution*** step is designed to reduce the set of equations to the following form

$$x_i = P_i x_{i+1} + Q_i \tag{3.55}$$

where P_i and Q_i are obtained by substituting $x_{i-1} = P_{i-1} x_i + Q_{i-1}$ into Equation 3.54 as

$$a_i x_i = -b_i x_{i+1} - c_i(P_{i-1} x_i + Q_{i-1}) + d_i \tag{3.56}$$

Rearranging Equation (3.56), we get

$$x_i = -\frac{b_i}{a_i + c_i P_{i-1}} x_{i+1} + \frac{d_i - c_i Q_{i-1}}{a_i + c_i P_{i-1}} \tag{3.57}$$

Comparison of Equation 3.55 and Equation 3.57 leads to the recurrence relations

$$P_i = -\frac{b_i}{a_i + c_i P_{i-1}} \tag{3.58a}$$

$$Q_i = \frac{d_i - c_i Q_{i-1}}{a_i + c_i P_{i-1}} \tag{3.58b}$$

To start the recurrence process, P_1 and Q_1 are expressed by substituting $c_1 = 0$ into Equation 3.58, as

$$P_1 = -\frac{b_1}{a_1} \tag{3.59a}$$

$$Q_1 = \frac{d_1}{a_1} \tag{3.59b}$$

The recurrence process leads to $P_n = 0$ due to the fact that $b_n = 0$, and hence from Equation 3.55

$$x_n = Q_n \tag{3.60}$$

In the subsequent back substitution step x_{n-1} is obtained from Equation 3.55 as

$$x_{n-1} = P_{n-1}x_n + Q_{n-1} \tag{3.61a}$$

This back substitution procedure is continued to solve the remaining unknowns in the following manner

$$\begin{aligned} x_{n-2} &= P_{n-2}x_{n-1} + Q_{n-2} && \text{(b)} \\ &\vdots \\ x_2 &= P_2x_3 + Q_2 && \text{(n − 1)} \\ x_1 &= P_1x_2 + Q_1 && \text{(n)} \end{aligned} \tag{3.61b}$$

A pseudo-code for the TDMA algorithm is presented in Figure 3.1.

$P_1 = -b_1/a_1$

$Q_1 = d_1/a_1$

do for $i = 2, 3$ to n

$P_i = -b_i/(a_i + c_iP_{i-1})$

$Q_i = (d_i - c_iQ_{i-1})/(a_i + c_iP_{i-1})$

enddo

$x_n = Q_n$

do for $i = n-1, n-2, \ldots, 3, 2, 1$

$x_i = P_ix_{i+1} + Q_i$

enddo

Figure 3.1 Pseudo code for TDMA

Unlike general direct methods as applied to a dense coefficient matrix, the number of operations needed for solving a tridiagonal system using TDMA or the Thomas method is of $0(n)$ rather than $0(n^2)$ or $0(n^3)$. Therefore, much smaller times and round-off errors arise in the solutions of such systems while using these alternate algorithms.

Unfortunately, the Thomas algorithm based on the factorization procedure breaks down if any u_i is zero. In fact, the stability of the process in general cannot be guaranteed even if all u_i are nonzero. However, in many heat transfer and fluid flow problems the discretization procedure leads to a system that is tridiagonal as well as symmetric positive definite. In such cases the Thomas algorithm based on *LU* factorization is quite stable. In the general case, the TDMA algorithm based on the elimination method is preferred to ensure stability.

Example 3.5: Tridiagonal Matrix Algorithm (TDMA): Use the Tridiagonal matrix algorithm to solve the following system of equations

$$\begin{bmatrix} -2 & 1 & 0 & 0 & 0 & 0 & 0 \\ 1 & -2 & 1 & 0 & 0 & 0 & 0 \\ 0 & 1 & -2 & 1 & 0 & 0 & 0 \\ 0 & 0 & 1 & -2 & 1 & 0 & 0 \\ 0 & 0 & 0 & 1 & -2 & 1 & 0 \\ 0 & 0 & 0 & 0 & 1 & -2 & 1 \\ 0 & 0 & 0 & 0 & 0 & 1 & -2 \end{bmatrix} \begin{Bmatrix} x_1 \\ x_2 \\ x_3 \\ x_4 \\ x_5 \\ x_6 \\ x_7 \end{Bmatrix} = \begin{Bmatrix} -240 \\ -40 \\ -40 \\ -40 \\ -40 \\ -40 \\ -60 \end{Bmatrix} \tag{E.3.5.1}$$

Solution: We use Equation 3.59 for $i = 1$

$$P_1 = -\frac{b_1}{a_1} = -\frac{1}{-2} = 0.5, \quad Q_1 = \frac{d_1}{a_1} = \frac{-240}{-2} = 120$$

Use Equation 3.58 for $i > 1$

$$i = 2, \quad P_2 = \frac{-b_2}{a_2 + c_2 P_1} = \frac{-1}{-2 + 1 \times 0.5} = 0.666\,67,$$

$$Q_2 = \frac{d_2 - c_2 Q_1}{a_2 + c_2 P_1} = \frac{-40 - 1 \times 120}{-2 + 1 \times 0.5} = 106.666\,67$$

$$i = 3, \quad P_3 = \frac{-b_3}{a_3 + c_3 P_2} = \frac{-1}{-2 + 1 \times 0.666\,67} = 0.75,$$

$$Q_3 = \frac{d_3 - c_3 Q_2}{a_3 + c_3 P_2} = \frac{-40 - 1 \times 106.666\,67}{-2 + 1 \times 0.666\,67} = 110$$

$$i = 4, \quad P_4 = \frac{-b_4}{a_4 + c_4 P_3} = \frac{-1}{-2 + 1 \times 0.75} = 0.80,$$

$$Q_4 = \frac{d_4 - c_4 Q_3}{a_4 + c_4 P_3} = \frac{-40 - 1 \times 110}{-2 + 1 \times 0.75} = 120$$

$$i = 5, \quad P_5 = \frac{-b_5}{a_5 + c_5 P_4} = \frac{-1}{-2 + 1 \times 0.80} = 0.833\,33,$$

$$Q_5 = \frac{d_5 - c_5 Q_4}{a_5 + c_5 P_4} = \frac{-40 - 1 \times 120}{-2 + 1 \times 0.80} = 133.333\,33$$

$$i = 6, \quad P_6 = \frac{-b_6}{a_6 + c_6 P_5} = \frac{-1}{-2 + 1 \times 0.833\,33} = 0.857\,14,$$

$$Q_6 = \frac{d_6 - c_6 Q_5}{a_6 + c_6 P_5} = \frac{-40 - 1 \times 133.333\,3}{-2 + 1 \times 0.833\,33} = 148.571\,00$$

$$i = 7, \quad P_7 = 0,$$

$$Q_7 = \frac{d_7 - c_7 Q_6}{a_7 + c_7 P_6} = \frac{-40 - 1 \times 148.571\,00}{-2 + 1 \times 0.857\,14} = 148.571\,00$$

Use Equation 3.60 and Equation 3.61 in the back substitution step for the solution vector x as follows. For

$$\begin{aligned}
&i = 7, \quad x_7 = Q_7 = 182.5 \\
&i = 6, \quad x_6 = P_6 x_7 + Q_6 = 0.857\,14(182.5) + 148.571 = 305 \\
&i = 5, \quad x_5 = P_5 x_6 + Q_5 = 0.833\,33(305) + 133.333\,33 = 387.5 \\
&i = 4, \quad x_4 = P_4 x_5 + Q_4 = 0.8(387.5) + 120 = 430 \\
&i = 3, \quad x_3 = P_3 x_4 + Q_3 = 0.75(430) + 110 = 432.5 \\
&i = 2, \quad x_2 = P_2 x_3 + Q_2 = 0.666\,67(432.5) + 106.666\,67 = 395 \\
&i = 1, \quad x_1 = P_1 x_2 + Q_1 = 0.5(395) + 120 = 317.5
\end{aligned}$$

The solution vector is

$$\{x\} = \begin{Bmatrix} 317.5 \\ 395 \\ 432.5 \\ 430 \\ 387.5 \\ 305 \\ 182.5 \end{Bmatrix} \tag{E.3.5.2}$$

Block Tridiagonal System. Many physical problems in fluid flow and heat transfer give rise to a special system such as the block tridiagonal system when numerical discretization schemes are applied to the governing equations. For example, application of a finite difference–control volume discretization scheme to a two-dimensional elliptic field equation in the form a *Poisson's* equation with *Dirichlet* boundary conditions leads to a system of linear equations of form

$$\begin{bmatrix}
4 & -1 & 0 & 0 & \cdots & -1 & 0 & 0 & 0 & 0 \\
-1 & 4 & -1 & 0 & \cdots & 0 & -1 & 0 & 0 & 0 \\
0 & -1 & 4 & -1 & \cdots & & 0 & -1 & 0 & 0 \\
0 & 0 & -1 & 4 & \cdots & 0 & & 0 & -1 & 0 \\
 & & & & \vdots & & & & & \\
0 & 0 & 0 & 0 & \cdots & 4 & -1 & 0 & 0 & 0 \\
-1 & 0 & & & \cdots & -1 & 4 & -1 & 0 & 0 \\
0 & -1 & 0 & & \cdots & 0 & -1 & 4 & -1 & 0 \\
0 & 0 & -1 & 0 & \cdots & & 0 & -1 & 4 & -1 \\
0 & 0 & 0 & -1 & \cdots & & & 0 & -1 & 4
\end{bmatrix}
\begin{Bmatrix} x_1 \\ x_2 \\ x_3 \\ x_4 \\ \vdots \\ \\ \\ \\ \\ x_n \end{Bmatrix}
=
\begin{Bmatrix} d_1 \\ d_2 \\ d_3 \\ d_4 \\ \vdots \\ \\ \\ \\ \\ d_n \end{Bmatrix} \tag{3.62}$$

Equation 3.62 can be written in a block matrix form as

$$\begin{bmatrix} A_N & -I_N & & & \\ -I_N & A_N & & & \\ & -I_N & & & \\ & & & & -I_N \\ & & & -I_N & A_N \end{bmatrix} \begin{Bmatrix} x_{1N} \\ x_{2N} \\ \vdots \\ \\ x_{NN} \end{Bmatrix} = \begin{Bmatrix} d_{1N} \\ d_{2N} \\ \\ \\ x_{NN} \end{Bmatrix} \tag{3.63}$$

where A_N is a $N \times N$ matrix given as

$$A_N = \begin{bmatrix} 4 & -1 & & & \\ -1 & 4 & & & \\ & & & & \\ & & & 4 & -1 \\ & & & -1 & 4 \end{bmatrix} \tag{3.64}$$

and I_N is the $N \times N$ identity matrix.

Equation 3.63 can also be written in a block tridiagonal matrix form similar to Equation 3.48 as

$$\begin{bmatrix} A_1 & B_1 & & & \\ C_2 & A_2 & B_2 & & \\ & C_3 & & & \\ & & & & B_{N-1} \\ & & & C_N & A_N \end{bmatrix} \begin{Bmatrix} x_{1N} \\ x_{2N} \\ \vdots \\ x_{N-1,N} \\ x_{NN} \end{Bmatrix} = \begin{Bmatrix} d_{1N} \\ d_{2N} \\ \\ x_{N-1,N} \\ x_{NN} \end{Bmatrix} \tag{3.65}$$

So, Equation 3.65 represents a block tridiagonal system, which is an $n \times n$ square matrix, and in which each element is a $N \times N$ matrix, each solution vector element is a vector, and each constant vector element is a vector. The algorithm for the solution of a block tridiagonal system follows the same procedure as in the case of scalar tridiagonal algorithm: factorization and substitution. The basic technique based on LU factorization discussed in Section 3.2.3 can be used for solving a block tridiagonal system. This is called ***block LU factorization***.

Block LU Factorization In the ***first step***, the block tridiagonal matrix A_N is factorized into upper and lower triangular matrices such that

$$A_N = LU \tag{3.66a}$$

or

$$\begin{bmatrix} A_1 & B_1 & & & \\ C_2 & A_2 & B_2 & & \\ & C_3 & & & \\ & & & & B_{N-1} \\ & & & C_N & A_N \end{bmatrix} = \begin{bmatrix} I & & & & \\ L_2 & I & & & \\ & L_3 & & & \\ & & & & \\ & & & L_N & I \end{bmatrix} \begin{bmatrix} U_1 & B_1 & & & \\ & U_2 & B_2 & & \\ & & & & \\ & & & & B_{N-1} \\ & & & & U_N \end{bmatrix} \tag{3.66b}$$

The elements of the block L and U matrices are obtained directly by equating the corresponding elements on both sides of Equation 3.66b, and expressed by the formulas

$$U_1 = A_1 \tag{3.67a}$$

and

$$L_i = \frac{C_i}{U_{i-1}} \tag{3.67b}$$

$$U_i = A_i - L_i B_{i-1} \quad \text{for } i = 2 \text{ to } n \tag{3.67c}$$

In the ***second step***, the lower triangular matrix L and right-hand side vector d are used to solve the system $Ly = d$ by the block forward elimination algorithm as

$$y_1 = d_1 \tag{3.68a}$$

and

$$y_i = d_i - L_{i-1} y_{i-1}, \quad \text{for } i = 2 \text{ to } n \tag{3.68b}$$

In the ***third step***, the upper triangular matrix U and vector y are used to solve the system $Ux = y$ using the block back substitution algorithm

$$U_n x_n = y_n \tag{3.69a}$$

and

$$U_i x_i = y_i - B_i x_{i+1}, \quad \text{for } i = n-1, \ldots, 1 \tag{3.69b}$$

3.2.5 Error Equation and Iterative Refinement

Direct methods may lead to considerable error in the result due to round-off error. This is particularly significant in a large system of equations. In many cases, using an iterative refinement procedure that involves use of error equations can reduce this error. The procedure is outlined as follows.

Consider a system of the form

$$Ax = c \tag{3.2b}$$

If x^* is the computed solution vector of the system, then substitution of this result into Equation 3.2b gives

$$Ax^* = c^* \tag{3.70}$$

Now, suppose that $\hat{x}$ is the correction vector to the approximate solution vector x^*, then the exact solution vector x is expressed as

$$x = x^* + \hat{x} \tag{3.71}$$

Substituting Equation 3.71 into Equation 3.2b the following system results in

$$A(x^* + \hat{x}) = c \tag{3.72}$$

Now subtracting Equation 3.70 from Equation 3.72 results in

$$A\hat{x} = c - c^* \tag{3.73a}$$

or

$$A\hat{x} = e \tag{3.73b}$$

where e is the *residual vector* and can be rewritten

$$e = c - Ax^* \tag{3.74}$$

Equation 3.73 is a system of *error equations*, which can also be solved using the direct solver, and the correction vector, $\hat{x}$, can be applied to refine the solution as specified by Equation 3.71.

This refinement procedure is continued until a desired accuracy is achieved. The desired accuracy can be set by a criterion such as

$$\|\hat{x}\|_e \leq \varepsilon_s \tag{3.75a}$$

or

$$\frac{\|x^*\|_e}{\|x\|_e} \leq \varepsilon_s \tag{3.75b}$$

where ε_s is the assigned tolerance limit. It is relatively simple to implement this iterative refinement algorithm into computer programs for any direct methods. However, it is especially effective for those direct methods for which triangularization of the matrix exist, such as in *LU* decomposition or matrix inversion by the Gauss–Jordan elimination method.

Figure 3.2 shows a pseudo code to implement this iterative refinement procedure.

---

initialize $x_i = x^*$

do for

  compute residual vector e from $e = c - Ax_i$

  Solve the correction vector $\hat{x}$ from $A\hat{x} = e$

  Refine solution vector x from $x_i = x_i + \hat{x}$

  If $(\|\hat{x}\|_e < \varepsilon_s)$ then

    Solution converged

    Stop

  Else

End do

Figure 3.2 Pseudo code for iterative refinement

3.3 ITERATIVE METHODS

The direct methods discussed in the preceding sections are usually effective for a small number of equations, typically of the order of several hundreds. This number can often be expanded if the system is well conditioned, the matrix is sparse, and iterative refinement is employed. However, because of the round-off error, computer time, and storage, direct methods become inadequate for larger systems. The discretization of governing equations in many fluid flows and heat transfer problems results in a system as large as several thousands or more. For these kinds of problems, ***iterative methods*** can be used more effectively.

The solution of a large matrix system requires a large memory as well as a large portion of solution time along with the assembly of the global matrix. Direct methods have been successfully applied to the solution of such systems of equations, running on high-speed computers including super computers. However, memory limitations on super computers, as well as high cost/performance ratios, severely limit the routine application of direct solvers for large-scale three-dimensional flow problems. On the other hand, because of their lower storage requirements and operation costs, iterative methods are well adapted to engineering workstations, which are characterized by large memory, fast scale performance, and low cost/performance ratio.

Iterative methods are approximate methods, which start with an initial guess solution and iterate to converged solutions with some pre-specified tolerance limits. There are many iterative methods that can be used to solve large systems. These include: (1) Gauss–Seidel method, (2) Jacobi method, (3) successive over-relaxation (SOR) method, (4) conjugate gradient (CG) method, and (5) generalized minimal residual (GMRES) method.

The ***Gauss–Seidel*** method is the most commonly used iterative method. It is a simple modification of the ***Jacobi*** method and is a special case of the SOR method. The Gauss–Seidel method converges for diagonally dominant matrices and for symmetric positive definite matrices. The conjugate gradient method is highly efficient for symmetric problems. The choice of iterative methods for the system is, however, complicated by the nonsymmetry and nonpositive definitiveness of the matrix. Classical iterative methods such as Gauss–Seidel, SOR, and CG methods become inapplicable. Under such cases, several other methods look more attractive. These include: (1) generalized minimal residual (GMRES), (2) conjugate gradient squared (CGS), (3) bi-conjugate gradient stabilized (Bi-CGSTAB), and (4) transpose free quasi-minimal residual (TFQMR).

3.3.1 Jacobi Method

Let us consider the system of equations given by Equation 3.1, i.e.

$$Ax = c \tag{3.2b}$$

This system of equations can be rewritten explicitly for the unknown x_i as

$$\begin{aligned}
x_i &= \frac{c_1 - a_{12}x_2 - a_{13}x_3 \cdots a_{1n}x_n}{a_{11}} \\
x_2 &= \frac{c_2 - a_{21}x_1 - a_{23}x_3 \cdots a_{2n}x_n}{a_{22}} \\
&\vdots \\
&\vdots \\
x_n &= \frac{c_n - a_{n1}x_1 - a_{n2}x_2 \cdots a_{n,n-1}x_{n-1}}{a_{nn}}
\end{aligned} \tag{3.76a}$$

Equation 3.76a can be written in a compact form as

$$x_i = \frac{c_i - \sum_{j=1, j\neq i}^{n} a_{ij}x_j}{a_{ii}}, \quad \text{for } i = 1, 2, \ldots, n \tag{3.76b}$$

The sequence of steps constituting the Jacobi method is as given below.

In the ***first step***, assign an initial guess value for each unknown appearing in the system. If no reasonable guess of these values can be made, then any arbitrarily chosen values such as zeros can be made, i.e., $x_1^0 = 0$, $x_2^0 = 0, \ldots, x_n^0 = 0$.

In the ***second step***, these initial guess values are substituted into Equation 3.76 to compute a new value of x_i^1 as

$$\begin{aligned} x_1^1 &= \frac{c_1 - a_{12}x_2^0 - a_{13}x_3^0 \cdots a_{1n}x_n^0}{a_{11}} \\ x_2^1 &= \frac{c_2 - a_{21}x_1^0 - a_{23}x_3^0 \cdots a_{2n}x_n^0}{a_{22}} \\ &\vdots \\ x_n^1 &= \frac{c_n - a_{n1}x_1^0 - a_{n2}x_2^0 \cdots a_{n,n-1}x_{n-1}^0}{a_{nn}} \end{aligned} \tag{3.77a}$$

and in a compact form as

$$x_i^1 = \frac{c_i - \sum_{j=1, j\neq i}^{n} a_{ij}x_j^0}{a_{ii}}, \quad \text{for } i = 1, 2, \ldots, n \tag{3.77b}$$

When all values are estimated by the above equations, the first iteration ($k=1$) is said to have been completed.

In ***step three***, the iteration procedure outlined in step two is continued by substituting the values of each unknown determined in a particular iteration, k, into the right-hand side of Equation 3.76 as

$$x_i^{k+1} = \frac{c_i - \sum_{j=1, j\neq i}^{n} a_{ij}x_j^k}{a_{ii}}, \quad \text{for } i = 1, 2, \ldots, n \tag{3.78}$$

The iteration process is continued until the unknown values obtained in the present iteration $\boldsymbol{k+1}$ differs from its respective values obtained in the preceding iteration k by a specified tolerance limit, ε_s. This convergence criterion is expressed as

$$\left| \frac{x_i^{k+1} - x_i^k}{x_i^{k+1}} \right| \le \varepsilon_s \tag{3.79}$$

The convergence can also be met using an alternative criterion based on the Euclidean norm such as

$$\frac{\left\| x_i^{k+1} - x_i^k \right\|_e}{\left\| x_i^{k+1} \right\|_e} \le \varepsilon_s \tag{3.80}$$

This alternate convergence criterion is particularly suitable for large systems to save computational time.

3.3.2 Gauss–Seidel Method

It can be observed in the ***Jacobi method*** that all unknown values are computed using values estimated in the previous iteration step, rather than using most recent values of the other unknowns. Considerable improvement in the storage requirements and the rate of convergence are achieved in many problems by employing an alternative iterative scheme known as the ***Gauss–Seidel method***, in which as unknowns are computed in an iteration step; they are subsequently employed in the computation of the rest of the unknown in the same iteration step. This iterative scheme can be expressed as

$$x_i^{k+1} = \frac{c_i - \sum_{j=1}^{i-1} a_{ij}\, x_j^{k+1} - \sum_{j=i+1}^{n} a_{ij}\, x_j^k}{a_{ii}}, \quad \text{for } i = 1, 2, \ldots, n \tag{3.81}$$

There are many problems where the Jacobi method converges faster. However, in most cases, the Gauss–Seidel method is the method of choice.

3.3.3 Convergence Criterion for Iterative Methods

One major disadvantage of the iteration method is that it does not always converge. There are problems in which the solution may oscillate and eventually diverge. A ***sufficient*** condition for the convergence of the Gauss–Seidel and Jacobi methods is that the solution is guaranteed to converge if the coefficient matrix A of the system of Equation 1.1 is diagonally dominant, that is

$$\left| a_{ij} \right| > \sum_{j=1, j \ne i}^{n} \left| a_{ij} \right| \tag{3.82}$$

It is important to note that such a condition is a sufficient condition and not a ***necessary*** condition. That is, at times, the system may violate the criterion and still attain converged solutions. It is quite desirable to have a discretization scheme that leads to a diagonally dominant system.

3.3.4 The Successive Over-relaxation (SOR) Method

There are many problems in which the Gauss–Seidel method converges at a very slow rate, especially when the system is very large. However, the rate of convergence can be enhanced by using a parameter, Λ, and using a solution scheme given by

$$x_i^{k+1} = \Lambda x_i^{k+1} + (1 - \Lambda)x_i^k \tag{3.83}$$

where Λ is a constant and termed as the ***relaxation parameter***, and x_i^{k+1} on the right-hand side of the equation is the value of x_i obtained in the $(k+1)$th iterative step by using the Gauss–Seidel iteration scheme given by Equation 3.81. Substituting Equation 3.81 into Equation 3.83, we obtain the relaxation scheme as

$$x_i^{k+1} = \Lambda \left[\frac{c_i - \sum_{j=1}^{i-1} a_{ij}\, x_j^{k+1} - \sum_{j=i+1}^{n} a_{ij}\, x_j^k}{a_{ii}} \right] + (1 - \Lambda)x_i^k \tag{3.84}$$

If $1 < \Lambda < 2$, then this modified Gauss–Seidel iterative scheme is known as the ***successive over-relaxation (SOR) method***. It can be noted from Equation 3.84 that the SOR iteration method reduces to Gauss–Seidel method when $\Lambda = 1$. If $0 < \Lambda < 1$, the iteration scheme is known as ***successive under-relaxation (SUR) method***. A successive under-relaxation method is generally employed to make a nonconvergent Gauss–Seidel iteration method to converge or to facilitate convergence by dampening out any small-scale oscillation. An optimum value of the relaxation parameter Λ is established through numerical experimentation.

Example 3.6: Gauss–Seidel Method: Use the Gauss–Seidel algorithm to solve the following system of equations

$$\begin{aligned} 3x_1 + x_2 - 2x_3 &= 9 \\ -x_1 + 4x_2 - 3x_3 &= -8 \\ x_1 - x_2 + 4x_3 &= 1 \end{aligned} \tag{E.3.6.1}$$

Continue iteration until all unknowns are accurate to three decimal places, i.e., converged to specified tolerance of $\varepsilon_s = 0.0009$.

Solution: In the ***first step***, we use Equation 3.76a to express each equation for its unknown on the diagonal in an explicit form as

$$x_1^{k+1} = \frac{9 - x_2^k + 2x_3^k}{3} \tag{E.3.6.2a}$$

$$x_2^{k+1} = \frac{-8 + x_1^{k+1} + 3x_3^k}{4} \tag{E.3.6.2b}$$

$$x_3^{k+1} = \frac{1 - x_1^{k+1} + x_2^{k+1}}{4} \tag{E.3.6.2c}$$

In the ***second step***, we assume the initial guess values as

$$x_1^0 = 0 \tag{E.3.6.3a}$$

$$x_2^0 = 0 \tag{E.3.6.3b}$$

$$x_3^0 = 0 \tag{E.3.6.3c}$$

In the ***third step***, we start the iteration process. The first iteration is completed by substituting the initial guess values into Equation E.3.6.2 as

***Iteration, k* = 1**

$$x_1^1 = \frac{9 - x_2^0 + 2x_3^0}{3} = \frac{9 - (0) + 2(0)}{3} = 3.000\,000\,00 \tag{E.3.6.4a}$$

$$x_2^1 = \frac{-8 + x_1^1 + 3x_3^0}{4} = \frac{-8 + (3.000\,000\,00 + 3(0)}{4} = -1.250\,000\,00 \tag{E.3.6.4b}$$

$$x_3^1 = \frac{1 - x_1^1 + x_2^1}{4} = \frac{1 - (3.000\,000\,00) + (-1.250\,000\,00)}{4} = -0.812\,5 \tag{E.3.6.4c}$$

Percent relative error

$$|\varepsilon_{a1}| = \left|\frac{x_1^1 - x_1^0}{x_1^1}\right| = 1.000\,000\,00 \tag{E.3.6.5a}$$

$$|\varepsilon_{a2}| = \left|\frac{x_2^1 - x_2^0}{x_2^1}\right| = 1.000\,000\,00 \tag{E.3.6.5b}$$

$$|\varepsilon_{a3}| = \left|\frac{x_3^1 - x_3^0}{x_3^1}\right| = 1.000\,000\,00 \tag{E.3.6.5c}$$

Iteration, k = 2

$$x_1^2 = \frac{9 - x_2^1 + 2x_3^1}{3} = \frac{9 - (-1.250\,000\,00) + 2(-0.812\,5)}{3} = 2.875 \quad \text{(E.3.6.6a)}$$

$$x_2^2 = \frac{-8 + x_1^2 + 3x_3^1}{4} = \frac{-8 + (2.875) + 3(-0.812\,5)}{4} = -1.890\,625 \quad \text{(E.3.6.6b)}$$

$$x_3^2 = \frac{1 - x_1^2 + x_2^2}{4} = \frac{1 - (2.875) + (-1.890\,625)}{4} = -0.941\,406\,25 \quad \text{(E.3.6.6c)}$$

Percent relative error

$$|\varepsilon_{a1}| = \left|\frac{x_1^2 - x_1^1}{x_1^2}\right| = 0.043\,478\,261 \quad \text{(E.3.6.7a)}$$

$$|\varepsilon_{a2}| = \left|\frac{x_2^2 - x_2^1}{x_2^2}\right| = 0.338\,842\,975 \quad \text{(E.3.6.7b)}$$

$$|\varepsilon_{a3}| = \left|\frac{x_3^2 - x_3^1}{x_3^2}\right| = 0.136\,929\,461 \quad \text{(E.3.6.7c)}$$

For the subsequent iterations, a similar procedure is repeated until a solution vector converges to the specified tolerance of $\varepsilon_s = 0.000\,09$. The results are summarized in the table below.

Iteration	x_1	x_2	x_3	ε_{a1}	ε_{a2}	ε_{a3}
1	3.000 000 0	−1.250 000 0	−0.812 5	1.000 000 0	1.000 000 0	1.000 000 0
2	2.875	−1.890 625	−0.941 406 25	0.043 478 21	0.338 842 9	0.136 929 4
3	3.002 604 1	−1.955 403 64	−0.989 501 95	0.042 497 83	0.033 128 0	0.048 605 9
4	2.992 133 24	−1.994 093 15	−0.996 556 6	0.346 485 10	0.019 402 0	0.007 079 0
5	3.000 326 65	−1.997 335 78	−0.999 415 61	0.002 730 83	0.003 242 6	0.002 860 6
6	2.999 501 52	−1.999 686 32	−0.999 796 19	0.000 275 08	0.001 175 4	0.000 380 66
7	3.000 031 31	−1.999 839 32	−0.999 967 65	0.000 176 59	0.000 076 5	0.000 171 46
8	2.999 968 00	−1.999 983 74	−0.999 987 93	0.000 021 10	0.000 072 2	0.000 020 27

3.3.5 Conjugate Gradient Method

One difficulty associated with the SOR method is that it is not always guaranteed to converge and often the convergence depends upon the relaxation parameter Λ that is quite difficult to choose without numerical experimentation. The ***conjugate gradient (CG)*** method, on the other hand, is guaranteed to converge, particularly for a large and sparse symmetric positive definite system of equations. In this section, we present the basic conjugate gradient method of Hestenes and Stiefel (1952) that has its origin in optimization theory. The conjugate gradient method can be considered as an acceleration technique that does not require estimation of acceleration or relaxation parameter as in the case of other iteration techniques such as the SOR method.

As stated, the conjugate gradient method can be viewed as an optimization problem. If $\boldsymbol{A}$ is a symmetric and positive definite, then it can be shown that the problem of solving a linear system $Ax=c$ is equivalent to minimizing a quadratic function

$$\Pi(x) = \tfrac{1}{2} x^{\mathrm{T}} A x - x^{\mathrm{T}} c \tag{3.85}$$

The quadratic function, $\Pi(x)$, attains its minimum value of $-\frac{1}{2} c^{\mathrm{T}} A^{-1} c$ at $x = A^{-1} C$. Our objective is not only to decrease the function Π^{k+1} efficiently, but also to estimate x^{k+1} efficiently and attain a rapid convergence process for converging the quadratic function to a minimum.

In the basic conjugate gradient method, the minimization problem is solved iteratively by the recursive scheme

$$x^{k+1} = x^k + \alpha_p^k p^k \tag{3.86}$$

where the scalar quantity α_p^k is chosen such as to minimize the function $\Pi(x)$, and it is expressed as

$$\alpha_p^k = \frac{r^{(k)^{\mathrm{T}}} r^k}{p^{(k)^{\mathrm{T}}} A p^{(k)}} \tag{3.87}$$

The residual r^k is estimated as

$$r^{k+1} = r^k - \alpha_p^k A p^k \tag{3.88}$$

where the vector p^k is called the direction vector and is linearly independent. The quadratic function is minimized in the space spanned by these vectors, which are generated in the preceding iterative step as

$$p^{k+1} = r^{k+1} + \beta_p^k p^k \tag{3.89}$$

where

$$\beta^k = \frac{r^{(k+1)^{\mathrm{T}}} r^k}{r^{(k)^{\mathrm{T}}} r^k} \tag{3.90}$$

The iterative process starts with an initial guess value of x^1 and by setting $p^1 = r^1 = c - Ax^1$ the iterative process is continued until convergence.

The basic conjugate gradient algorithm can be summarized as follows:

Choose the initial guess x^1 (often a null vector)

Choose convergence tolerance ε_s

Calculate the residual $r^1 = c - Ax^1$

Set $p^1 = r^1$

for $k = 1, 2, \ldots,$ calculate

$$\alpha_p^k = \frac{r^{(k)^T} r^k}{p^{(k)^T} A p^k}$$

$$x^{k+1} = x^k = \alpha_p^k p^k$$

$$r^{k+1} = r^k - \alpha_p^k A p^k$$

$$\beta_p^k = \frac{r^{(k+1)^T} r^{k+1}}{r^{(k)^T} r^k}$$

$$p^{(k+1)} = r^{k+1} + \beta_p^k p^k$$

Continue iteration until $\|r^k\|_e \leq \varepsilon_s$

A convergence criterion on $\left|\frac{x^{k+1} - x^k}{x^{k+1}}\right|$ could also be used. A pseudo code for basic conjugate gradient algorithm is illustrated in Figure 3.3.

The rate of convergence of the conjugate gradient method depends on the condition number of the matrix A. The larger the condition number, the slower the convergence, and in cases where the matrix A is ill-conditioned, the convergence can be extremely slow. The convergence speed of the conjugate gradient method can be increased using ***preconditioning***. In fact, the conjugate gradient method is rarely used without any preconditioning.

Example 3.7: Conjugate Gradient Method: Use the conjugate gradient method to solve the following system of equations. Continue iteration until converged to a tolerance limit of $\varepsilon_s = 0.0009$ using the Euclidean norm

$$\begin{bmatrix} 12 & 1 & 7 \\ 1 & 12 & 3 \\ 7 & 3 & 12 \end{bmatrix} \begin{Bmatrix} x_1 \\ x_2 \\ x_{31} \end{Bmatrix} = \begin{Bmatrix} 27 \\ -24 \\ 3 \end{Bmatrix}$$

We use the basic conjugate gradient algorithm as follows.

Step 1

Choose x^1, ε_s and itmax

Choose $k = 0$ and err $= 0$

Calculate the residual $r^1 = c - Ax^1$

Set $p^1 = r^1$

Step 2

do while (($k <$ itmax) and err $= 0$)

err $= 1$

dofor $k = 1, 2, \ldots$

$$\alpha_p^k = \frac{r^{(k)^{\mathrm{T}}} r^k}{p^{(k)^{\mathrm{T}}} A p^k}$$

$$x^{k+1} = x^k + \alpha_p^k p^k$$

$$r^{k+1} = r^k - \alpha_p^k A p^k$$

$$\beta_p^k = \frac{r^{(k+1)^{\mathrm{T}}} r^{k+1}}{r^{(k)^{\mathrm{T}}} r^k}$$

$$p^{(k+1)} = r^{k+1} + \beta_p^k p^k$$

If ((err $= 1$) and $r^{k+1} \neq 0$)

$\varepsilon_a = \|r^k\|$

If ($\varepsilon_a > \varepsilon_s$) then err $= 0$

end if

enddo

Figure 3.3 A pseudo code for CG algorithm

Solution:
Step 1

Choose initial guess as

$$x^1 = \begin{Bmatrix} 0 \\ 0 \\ 0 \end{Bmatrix}$$

$$r^1 = C - Ax^1 = \begin{Bmatrix} 27 \\ -24 \\ 3 \end{Bmatrix} - \begin{bmatrix} 12 & 1 & 7 \\ 1 & 12 & 3 \\ 7 & 3 & 12 \end{bmatrix} \begin{Bmatrix} 0 \\ 0 \\ 0 \end{Bmatrix} = \begin{Bmatrix} 27 \\ -24 \\ 3 \end{Bmatrix}$$

$$p^1 = \begin{Bmatrix} 27 \\ -24 \\ 3 \end{Bmatrix}, \quad p^{(1)^{\mathrm{T}}} = \{27 \quad -24 \quad 3\}$$

Step 2

$$\alpha_p^1 = \frac{r^{(k)^{\mathrm{T}}} r^k}{p^{(k)^{\mathrm{T}}} A p^k} = \frac{\left(\|r^1\|\right)^2}{p^{(1)^{\mathrm{T}}} A p^1} = 0.087$$

$$x^2 = x^1 + \alpha_p^1 p^1 = \begin{Bmatrix} 0 \\ 0 \\ 0 \end{Bmatrix} + 0.087 \begin{Bmatrix} 27 \\ -24 \\ 3 \end{Bmatrix}, \quad x^2 = \begin{pmatrix} 2.349 \\ -2.088 \\ 0.261 \end{pmatrix}$$

$$r^2 = r^1 - \alpha_p^1 A p^1 = \begin{Bmatrix} 27 \\ -24 \\ 3 \end{Bmatrix} - 0.087 \begin{bmatrix} 12 & 1 & 7 \\ 1 & 12 & 3 \\ 7 & 3 & 12 \end{bmatrix} \begin{Bmatrix} 27 \\ -24 \\ 3 \end{Bmatrix}, \quad r^2 = \begin{pmatrix} -0.927 \\ -2.076 \\ -10.311 \end{pmatrix}$$

$$\beta_p^1 = \frac{r^{(2)^{\mathrm{T}}} r^2}{r^{(1)^{\mathrm{T}}} r^1} = \frac{(-0.927)^2 + (-2.076)^2 + (-10.311)^2}{(27)^2 + (-24)^2 + (3)^2}, \quad \beta_p^1 = 0.084\,845$$

$$p^2 = r^2 + \beta_p^1 p^1 = \begin{pmatrix} -0.927 \\ -2.076 \\ -10.311 \end{pmatrix} + 0.084\,845 \begin{Bmatrix} 27 \\ -24 \\ 3 \end{Bmatrix}, \quad p^2 = \begin{pmatrix} 1.368 \\ -4.116 \\ -10.056 \end{pmatrix}$$

Check for convergence using the ***Euclidean norm***, i.e., $\left\|r_i^{k+1}\right\|_e \le \varepsilon_s$:

$$\left(\|r^2\|\right)^2 = \left[(1.368)^2 + (-4.116)^2 + (-10.056)^2\right] = 111.485\,826$$

or

$$\|r^2\| = 10.558$$

Since $(\|r^2\|) > \varepsilon_s$, we continue the iteration steps and the results are summarized in the table below.

Iteration No.	r_1^k	r_2^k	r_3^k	x_1^k	x_2^k	x_3^k	$(\|r^2\|)$
1	1.368	−4.116	−10.056	2.349	−2.088	0.261	10.558
2	3.4299	3.7884	−1.0527	2.4516	−2.3967	0.4932	5.217 69
3	0.036 341	−0.076 39	0.017 14	2.997 335	−1.993 00	−1.001 62	0.086 31
4	0.006 9231	−0.003 26	−0.003 22	3.000 661	−1.999 73	−1.000 18	0.008 30
5	−0.001 065	0.000 290	0.002 092	3.000 076	−2.000 09	−1.000 45	0.002 23

Solution converged in five iteration steps.

3.3.6 Preconditioned Conjugate Gradient Method

The basic conjugate gradient (CG) method works poorly for an ill-conditioned system. Convergence speed of the conjugate gradient method can be increased using preconditioning. Preconditioning means that we change variables to obtain a new system $\tilde{A}x = c$ whose coefficient matrix $\tilde{A}$ has a much improved eigenvalue distribution or condition number, condition $(\tilde{A})$, than that of the original system $Ax = c$. The new system is obtained by substituting $x = L^T\tilde{x}$ into the original system as

$$AL^T\tilde{x} = c \tag{3.91}$$

Premultiplying Equation 3.91 by L yields the equivalent system

$$LAL^T\tilde{x} = Lc$$

or

$$\tilde{A}\tilde{x} = \tilde{c} \tag{3.92}$$

where

$$\tilde{A} = LAL^T \tag{3.93}$$

and

$$\tilde{c} = Lc \tag{3.94}$$

The conjugate gradient method is now applied to the new system and the original variable x is then recovered from the variable $\tilde{x}$ using $x = L^T\tilde{x}$.

The nonsingular matrix $A_p = (LL^T)^{-1}$ is called the ***preconditioning matrix or preconditioner***. The main objective with this transformation is to obtain a matrix $\tilde{A}$ that has a much improved condition number. The preconditioned conjugate gradient algorithm is expressed as follows.

Choose the initial guess x^1

Choose convergence tolerance ε_s

Calculate the residual $r^1 = c - Ax^1$

Calculate $z^1 = A_p^{-1} r^1$

Set $p^1 = z^1$

Calculate for $k = 2, 3, \ldots$

$$\alpha_p^k = \frac{z^{(k)^T} r^k}{p^{(k)^T} A p^k}$$

$$x^{k+1} = x^k + \alpha_p^k p^k$$

$$r^{k+1} = r^k - \alpha_p^k A p^k$$

$$z^{k+1} = A_p^{-1} r^{k+1}$$

$$\beta_p^k = \frac{z^{(k+1)^T} r^{k+1}}{z^{(k)^T} r^k}$$

$$p^{(k+1)} = z^{k+1} + \beta_p^k p^k$$

Continue iteration until $\|r^k\|_e \leq \varepsilon_s$

It can be noted that in this algorithm an intermediate vector z^k has been introduced. If no preconditioning is used, i.e., if $A_p = I$, then this intermediate vector z^k is equal to r^k and the precondition conjugate gradient algorithm reduces to the basic conjugate gradient algorithm. Significant savings in computer storage and time are gained by using the precondition conjugate gradient method.

One of the important steps that has to be considered before using the precondition conjugate gradient method is to determine a preconditioner, A_p. Many different preconditioners have been proposed and used in solving fluid flow and heat transfer problems. Among them, the most popular ones are the incomplete Cholesky factorization, the polynomial preconditioner, the diagonal scaling preconditioner, and the block factorization preconditioner. The choice of a good preconditioner can have a dramatic effect upon the rate of convergence. For a detailed description of preconditioners see Golub and Van Loan (1989), Ortega and Poole (1981), and Saad (1996).

PROBLEMS

3.1 Consider the following matrix

$$[A] = \begin{bmatrix} 4 & -1 & 2 \\ 4 & -8 & 2 \\ -2 & 1 & 6 \end{bmatrix}$$

Determine (a) the determinant of the matrix $|A|$, (b) the inverse of the matrix A^{-1}, and (c) the condition number ***Cond* (*A*)** using the row-sum norm.

3.2 Develop a code ***Gauss*** to implement the Gauss elimination algorithm with partial pivoting to solve the following system of equations.

$$\begin{bmatrix} 4 & -1 & 0 & 0 & 0 & 0 \\ 0 & 4 & -1 & 0 & 0 & 0 \\ 0 & -1 & 4 & -1 & 0 & 0 \\ 0 & 0 & -1 & 4 & -1 & 0 \\ 0 & 0 & 0 & -1 & 4 & 0 \\ 0 & 0 & 0 & 0 & -1 & -4 \end{bmatrix} \begin{Bmatrix} x_1 \\ x_2 \\ x_3 \\ x_4 \\ x_5 \\ x_6 \end{Bmatrix} = \begin{Bmatrix} -90 \\ 90 \\ 0 \\ 0 \\ 90 \\ 90 \end{Bmatrix}$$

Print the augmented matrix and the pivot row and pivot element in every elimination step. Use the results to estimate the residual terms.

3.3 Develop a code ***Gauss–Jordan*** implementing the Gauss–Jordan algorithm to determine the inverse of the coefficient matrix and solve the following system of equations.

$$\begin{bmatrix} 2 & -1 & 0 & 0 & 0 & 0 \\ 0 & 2 & -1 & 0 & 0 & 0 \\ 0 & -1 & 2 & -1 & 0 & 0 \\ 0 & 0 & -1 & 2 & -1 & 0 \\ 0 & 0 & 0 & -1 & 2 & 0 \\ 0 & 0 & 0 & 0 & -1 & -2 \end{bmatrix} \begin{Bmatrix} x_1 \\ x_2 \\ x_3 \\ x_4 \\ x_5 \\ x_6 \end{Bmatrix} = \begin{Bmatrix} -30 \\ 30 \\ 0 \\ 0 \\ 30 \\ 30 \end{Bmatrix}$$

3.4 Determine the Euclidean norm $\|A\|_e$ and the row-sum norm $\|A\|_\infty$ for the coefficient matrix A of the system given in Problem 3.2. Determine the condition number ***Cond*** **(*A*)** for the system in Problem 3.2 using row-sum norm.

3.5 Develop a computer code ***ludecom*** based on *LU* decomposition algorithms to solve the following system of equations and present results in each elimination step.

$$\begin{bmatrix} -4 & 1 & 0 & 4 & 0 & 0 & 0 & 0 & 0 \\ 1 & -4 & 1 & 0 & 1 & 0 & 0 & 0 & 0 \\ 0 & 1 & -4 & 0 & 0 & 1 & 0 & 0 & 0 \\ 1 & 0 & 0 & -4 & 1 & 0 & 1 & 0 & 0 \\ 0 & 1 & 0 & 1 & -4 & 1 & 0 & 1 & 0 \\ 0 & 0 & 1 & 0 & 1 & -4 & 0 & 0 & 1 \\ 0 & 0 & 0 & 1 & 0 & 0 & -4 & 0 & 0 \\ 0 & 0 & 0 & 0 & 1 & 0 & 1 & -4 & 1 \\ 0 & 0 & 0 & 0 & 0 & 1 & 0 & 1 & -4 \end{bmatrix} \begin{Bmatrix} T_1 \\ T_2 \\ T_3 \\ T_4 \\ T_5 \\ T_6 \\ T_7 \\ T_8 \\ T_9 \end{Bmatrix} = \begin{Bmatrix} -420 \\ -400 \\ -420 \\ -20 \\ 0 \\ -20 \\ -40 \\ -20 \\ -40 \end{Bmatrix}$$

Use partial pivoting if necessary.

3.6 Solve the following system of equations using the TDMA algorithm.

$$\begin{bmatrix} 1.0835 & -0.04175 & 0 & 0 \\ -0.04175 & 1.0835 & -0.04175 & 0 \\ 0 & -0.04175 & 1.0835 & -0.04175 \\ 0 & 0 & -0.04175 & 1.0835 \end{bmatrix} \begin{Bmatrix} x_1 \\ x_2 \\ x_3 \\ x_4 \end{Bmatrix} = \begin{Bmatrix} 4.175 \\ 0 \\ 0 \\ 2.0875 \end{Bmatrix}$$

3.7 Develop a computer code ***tdma*** implementing the TDMA algorithm and solve the system of equations in Problem 3.1.

3.8 Use the Cholesky factorization method to solve the system of equations given in Problem 3.4.

3.9 Use the Gauss–Seidel method to solve the following system of equations.

$$\begin{aligned} 12x_1 + x_2 + 7x_3 &= 27 \\ x_1 + 12x_2 + 3x_3 &= -24 \\ 7x_1 + 3x_2 + 12x_3 &= 3 \end{aligned}$$

Show three iterations in hand calculations.

3.10 Consider the following system of equations

$$\begin{aligned} 3x_1 - 0.1x_2 - 0.2x_3 &= 7.85 \\ 0.1x_1 + 7x_2 - 0.3x_3 &= -19.3 \\ 0.3x_1 - 0.2x_2 + 10x_3 &= 71.4 \end{aligned}$$

(a) Determine whether the coefficient matrix is diagonally dominant.

(b) Use the Gauss–Seidel method to obtain the solution of the system. Show three iterations and percent relative error in each step of hand calculations.

(c) Use the SOR method with a relaxation factor $\lambda = 1.1$ to obtain the solution of the system. Show three iterations and percent relative error in each step of hand calculations.

3.11 Develop a computer code ***sor*** by using the SOR iterative algorithm and solve the system of equations given in Problem 3.3. Continue iteration until convergence is reached for all variables with a tolerance limit of $\varepsilon_s = 0.01\%$. Present your results in a summary table for unknown values and associated approximate percent relative error in each iteration.

3.12 Use the conjugate gradient algorithm to solve the following system of equations.

$$\begin{bmatrix} 3 & -3 & 0 & 0 & 0 & 0 \\ -3 & 10 & -1 & -4 & -2 & 0 \\ 0 & -1 & 44 & -2 & 19 & 0 \\ 0 & -4 & -2 & 10 & -4 & 0 \\ 0 & -2 & 19 & -4 & 90 & -17 \\ 0 & 0 & 0 & 0 & -17 & 43 \end{bmatrix} \begin{Bmatrix} x_1 \\ x_2 \\ x_3 \\ x_4 \\ x_5 \\ x_6 \end{Bmatrix} = \begin{Bmatrix} 2 \\ 77 \\ 75 \\ 79 \\ 77 \\ 2 \end{Bmatrix}$$

(a) Show hand calculations in two iterations.

(b) Develop a computer code **cg** *implementing the conjugate gradient algorithm and solve the system of equations.* Continue iterations until all unknowns are accurate to three decimal places, i.e., converged to a specified tolerance of $\varepsilon_s = 0.0009$. Present the results in a summary table showing values and percent relative error in each iteration step.

CHAPTER

4

NUMERICAL INTEGRATION

Quite often we will be confronted with problems of integrating functions for which no convenient integrals are available or which cannot be integrated in terms of elementary function. Some examples of such integrals that we will see in fluid flow and heat transfer problems are as follows.

For flow in any arbitrary flow section

Average velocity   Mass flow rate

$$\bar{V} = \frac{1}{A}\int_A \vec{V}\,\vec{d}A \qquad m = \int_A \rho\vec{V}\cdot\vec{d}A$$

Average fluid temperature at a section of fluid flow in a circular pipe

$$\bar{T} = \frac{\int_A \rho c_p u T\,dA}{\int_A \rho u\,dA} = \frac{\int_0^R \rho c_p u T\,2\pi r\,dr}{\int_0^R \rho u 2\pi r\,dr}$$

Average heat transfer coefficient for convective heat transfer from a surface

$$\bar{h} = \frac{1}{A}\int_A h\,dA$$

Numerical integrations are also encountered in the evaluation element characteristic matrices and source vectors in finite element formulations. Some examples of such integrals are:

$$\text{(a)}\quad I=\int_{x_i}^{x_j}\Gamma(x)\frac{d[N]^{\mathrm{T}}}{dx}\frac{d\phi}{dx}dx,\quad I=\int_{x_i}^{x_j}S(x)[N]^{\mathrm{T}}\,dx,\quad I=\int_{x_i}^{x_j}q(x)([N]^{\mathrm{T}}[N]\,dx$$

where

$$[N]^{\mathrm{T}}=\begin{Bmatrix}N_i(x)\\N_j(x)\end{Bmatrix}=\begin{Bmatrix}\dfrac{x_j-x}{x_j-x_i}\\[2ex]\dfrac{x-x_i}{x_j-x_i}\end{Bmatrix},\quad \phi^*=N_i\phi_i+N_j\phi_j$$

$$\text{(b)}\quad I=\int_{-1}^{1}q(x)[N]^{\mathrm{T}}[N]\,dx$$

where

$$[N]^{\mathrm{T}}=\begin{Bmatrix}N_i\\N_j\\N_k\end{Bmatrix}=\begin{Bmatrix}\dfrac{x}{2}(x-1)\\[1ex]-(x^2-1)\\[1ex]\dfrac{x}{2}(x+1)\end{Bmatrix}$$

Note that the following integral represents a matrix whose each element involves an integral given as

$$I=\int_{-1}^{1}q(x)\begin{bmatrix}N_i^2 & N_iN_j & N_iN_k\\N_jN_i & N_j^2 & N_jN_k\\N_kN_i & N_kN_j & N_k^2\end{bmatrix}dx$$

$$\text{(c)}\quad I=\int_A\{D\}^{\mathrm{T}}[\Gamma]\{D\}\,dA$$

where

$$[\Gamma]=\text{transport property matrix}=\begin{bmatrix}\Gamma_x & 0\\0 & \Gamma_y\end{bmatrix}$$

$$\{D\}=\text{column vector with derivatives}=\begin{Bmatrix}\dfrac{\partial[N]}{\partial x}\\[2ex]\dfrac{\partial[N]}{\partial y}\end{Bmatrix}$$

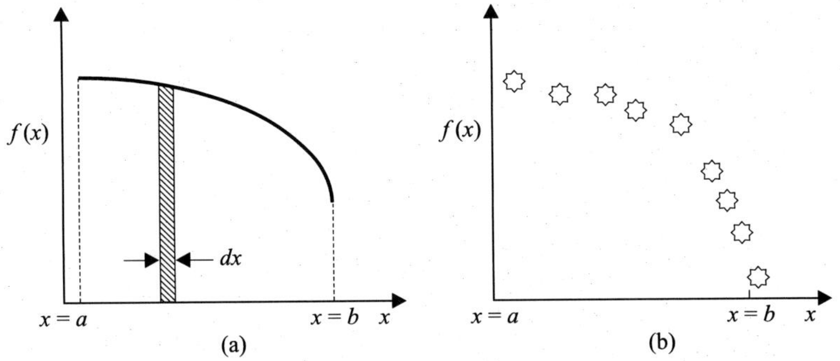

Figure 4.1 Function variation

Numerical integration, also referred to as ***quadrature formulas***, involves approximation of the integrand by a function that can be evaluated exactly. Numerical integration essentially finds a close approximation to the area under a curve represented by the function $f(x)$

$$I = \int_a^b f(x)\,dx \tag{4.1}$$

The function $f(x)$ could be a continuous function, as in Figure 4.1a, or a distribution of discrete values such as velocity and temperature at discrete points, as in Figure 4.1b. The integration methods can be classified into two basic groups, the Newton–Cotes formulas, and the Gauss quadrature formula.

4.1 NEWTON–COTES INTEGRATION FORMULAS

This is the most common numerical integration scheme. The basic strategy is to replace a complicated function or distribution of data with some approximating function such as

$$I = \int_a^b f(x)\,dx = \int_a^b f_n(x)\,dx \tag{4.2}$$

where $f_n(x)$ is a polynomial of the form

$$f_n(x) = a_0 + a_1 x + a_2 x^2 + \cdots + a_n x^n \tag{4.3}$$

Here $\boldsymbol{n}$ represents the order of the polynomial or order of the approximation. For example:

$\boldsymbol{n} = 1$ represents first-order approximation or a straight-line approximation given as $f_n(x) = a_0 + a_1 x$ and as depicted in Figure 4.2.

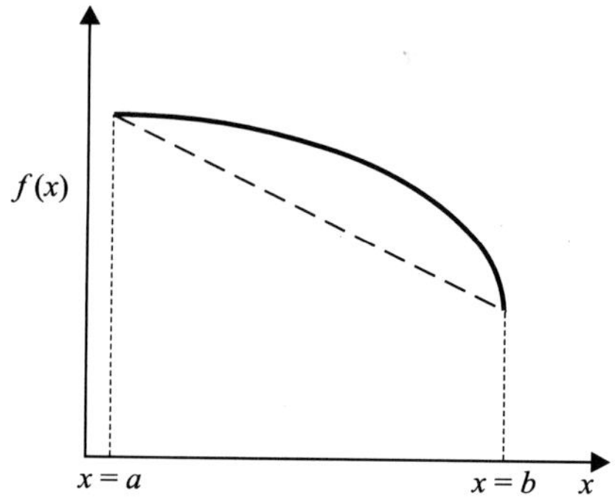

Figure 4.2 First-order linear approximation

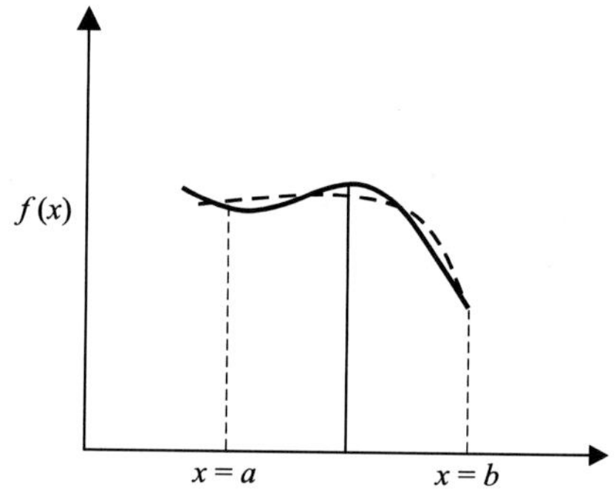

Figure 4.3 Second-order quadratic approximation

$n = 2$ represents second order approximation that uses a single parabola requiring three consecutive points, as given by the function $f_n(x) = a_0 + a_1x + a_2x^2$ and as depicted in Figure 4.3.

4.1.1 The Trapezoidal Rule

This is the first of the ***Newton–Cotes*** integration formulas and is the first-order or straight-line approximation shown in Figure 4.2. The integration is approximated by

$$I = \int_a^b f(x)\,dx \cong \int_a^b f_1(x)\,dx \tag{4.4}$$

where the function $f_1(x) = a_0 + a_1x$ is expressed in terms of the function values at the two limits as

$$f_1(x) = f(x_i) + \frac{f(x_{i+1}) - f(x_i)}{h}(x - x_i) \tag{4.5}$$

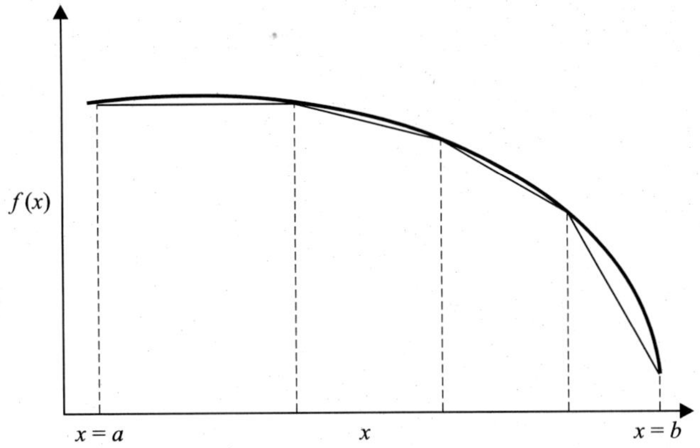

Figure 4.4 Multiple use of the trapezoidal rule with multiple strips

Substituting Equation 4.5 into Equation 4.4, we obtain the first-order estimate of the integral as

$$I \cong \frac{h}{2}[f(x_i) + f(x_{i+1})] \tag{4.6}$$

Equation 4.6 also represents the trapezoidal area under the straight line, which approximates the area under the curve. So, Equation 4.6 is referred to as the trapezoidal rule.

In order to obtain greater accuracy the area under the curve is divided into multiple areas, as shown in Figure 4.4, and the trapezoidal rule is applied many times to estimate the total area under the curve.

Multiple application of the trapezoidal rule leads to a general ***trapezoidal quadrature formula***

$$I \cong \frac{h}{2}\left[f(x_0) + 2\sum_{i=2}^{n-1} f(x_i) + f(x_n)\right] \tag{4.7}$$

where

$$n = \text{number of strips}$$
$$h = \frac{b-a}{n} = \text{step size}$$

It can be shown by using the Taylor series expansion that the truncation error associated with the single strip trapezoidal rule is given as

$$E_{t1} = -\frac{h^3}{12} f''(\xi_1) \tag{4.8}$$

where ξ_1 is a point in the interval between the two limiting points ***a*** and ***b***.

Evaluating the truncation errors for all the strips in a similar manner, the total truncation error for the general trapezoidal formula can be written as

$$E_t = -\frac{h^3}{12}[f''(\xi_1) + f''(\xi_2) + \cdots + f''(\xi_n)] \tag{4.9}$$

Equation (4.9) is further simplified by introducing an average value for the second derivative of the function as

$$E_t = -\tfrac{1}{12}(nf''(\xi)_{\text{av}}h^3 \tag{4.10}$$

Substituting $h = (b - a)/n$, Equation 4.10 reduces to

$$E_t = -\tfrac{1}{12}(b - a)f''(\xi)_{\text{av}}h^2 \tag{4.11}$$

Since the point ξ is difficult to estimate and $f''(\xi)_{\text{av}}$ may not be known, Equation 4.11 is written in an alternate form as

$$E_t \cong ch^2 \tag{4.12a}$$

or

$$E_t \cong O(h^2) \tag{4.12b}$$

Equation 4.12 is particularly useful for comparative evaluation of integral estimate using different step sizes, h, or number of strips, n.

Example 4.1: Trapezoidal Rule: Numerical results for the axial velocity distribution in a circular pipe are given at discrete points at a cross-section in the table below. The pipe is 1 ft in diameter. Compute the volume flow rate and the average velocity in the pipe using the trapezoidal rule.

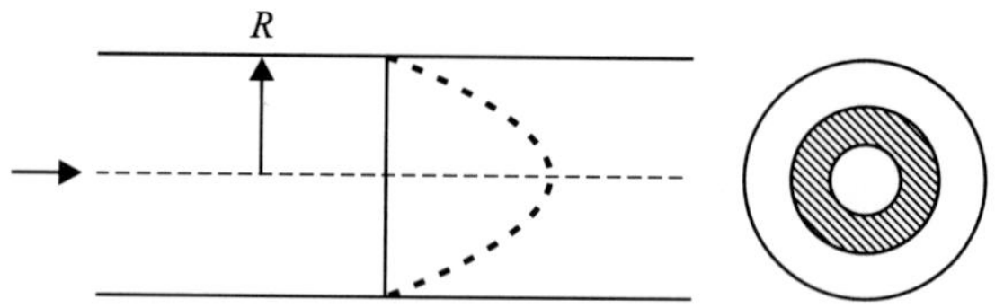

i	r (ft)	u_z (ft/s)
1	0	10.000
2	1/12	9.722
3	1/6	8.889
4	1/4	7.500
5	1/3	5.556
6	5/12	3.056
7	1/2	0

Solution: The volume flow rate of an incompressible fluid is given by the integral

$$\forall = \int_A \vec{V} \cdot \vec{dA} \tag{E.4.1.1}$$

If the axial velocity distribution at any cross-section in a circular pipe is known, then Equation E.4.1.1 reduces to

$$\forall = \int_0^R u_z 2\pi r \, dr \tag{E.4.1.2}$$

The trapezoidal rule can be used to estimate the value of the integral

$$\forall \cong \frac{h}{2}\left[f(x_0) + 2\sum_{i=2}^{n-1} f(x_i) + f(x_n)\right]$$

Using $f(x) = u_z 2\pi r$, $h = \Delta r = \frac{1}{12}$ and $n = 7$, we have

$$\forall \cong \frac{\Delta r}{2}\left[(u_{z0} 2\pi r_0) + 2\sum_{i=2}^{6} (u_{z_i} 2\pi r_i) + (u_{zR} 2\pi R)\right]$$

or

$$\forall \cong \frac{2\pi\left(\frac{1}{12}\right)}{2}\left[10(0) + 2\left\{9.722\left(\tfrac{1}{12}\right) + 8.889\left(\tfrac{1}{6}\right) + 7.500\left(\tfrac{1}{4}\right)\right.\right.$$

$$\left.\left. + 5.556\left(\tfrac{1}{3}\right) + 3.056\left(\tfrac{5}{6}\right)\right\} + 0\left(\tfrac{1}{2}\right)\right]$$

or

$$\forall \cong 3.818 \frac{\text{ft}^3}{\text{sec}}$$

The average velocity in the pipe is estimated as

$$\overline{V} = \frac{\int_0^R u_z 2\pi r \, dr}{\pi R^2} = \frac{3.818}{\pi(1/2)^2}$$

or

$$\overline{V} = 1.1259$$

4.1.2 Simpson's Integration Formula

This is the second ***Newton–Cotes*** integration formula and second-order or quadratic approximation using a parabola with three consecutive points. Approximation of the integral is obtained by connecting successive groups of three points by parabolas and summing the areas under the parabolas to approximate the area under the curve. Consider an area under two strips with three data points $(x_{i-1}, f(x_{i-1}))$, $(x_i, f(x_i))$, and $(x_{i+1}, f(x_{i+1}))$, as shown in Figure 4.5.

The integration is approximated by

$$I = \int_{x_{i-1}}^{x_{i+1}} f(x)\,dx \cong \int_{i-1}^{i+1} f_2(x)\,dx \tag{4.13}$$

where the second-order quadratic function $f_2(x) = a_0 + a_1 x + a_2 x^2$ is assumed to pass through the three points. The coefficients a_0, a_1, and a_2 are determined by substituting the co-ordinate of the three points as

$$a_0 = f(x_i), \quad a_1 = \frac{f(x_{i+1}) - f(x_{i-1})}{2h}, \quad \text{and} \quad a_2 = \frac{f(x_{i-1} - 2f(x_i) + f(x_{i+1})}{2h^2} \tag{4.14}$$

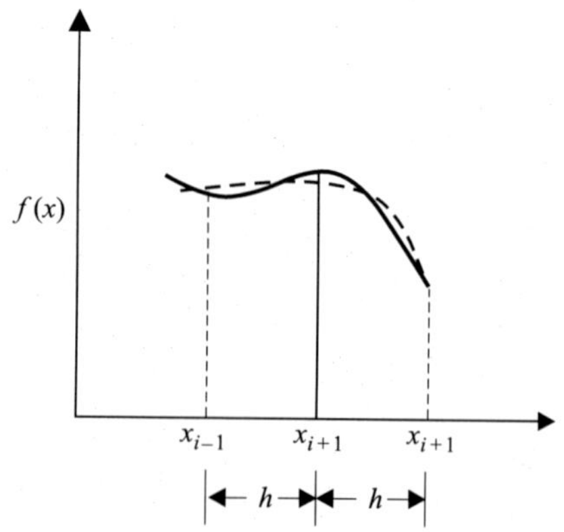

Figure 4.5 Simpson's 1/3rd rule using a three-point quadratic function

The integral given by Equation 4.13 or the area under the function can now be evaluated by substituting the function as

$$I \cong \int_{x_{i-1}}^{x_{i+1}} f_2(x) = \int_{x_{i-1}}^{x_{i+1}} (a_0 + a_1 x + a_2 x^2)\, dx$$

$$\cong a_0 x|_{-h}^{h} + \tfrac{1}{2} a_1 x^2 \big|_{-h}^{h} + \tfrac{1}{3} a_2 x^3 \big|_{-h}^{h}$$

$$\cong 2a_0 h + \tfrac{2}{3} a_2 h^3$$

Now, by substituting the expressions of a_0, a_1 and a_2, we obtain the formula for the approximation of the integral or the area under the curve that passes through the three points as

$$I \cong \frac{h}{3}(f(x_{i-1}) + 4f(x_i) + f(x_{i+1})) \tag{4.15}$$

This formula is referred to as the ***Simpson's 1/3rd rule*** and it is based on the estimation of the area of two segments or three points. For the estimation of the integral, Equation 4.1, in the range (a, b), we need to divide the regions into multiple strips. For $\boldsymbol{n}$ even number of strips of width $h = (b - a)/n$, Equation 4.15 is applied many times in a group of two strips, and the general formula by Simpson's 1/3rd rule is given as

$$I \cong \frac{h}{3}\left[f(x_0) + 4 \sum_{i=1,3,5} f(x_i) + 2 \sum_{i=2,4,6} f(x_i) + f(x_n) \right] \tag{4.16}$$

Note that the *Simpson's 1/3rd rule* formula is applicable when the area is divided into $\boldsymbol{n}$ even number of strips of width h and $\boldsymbol{n} + 1$ odd number of data points. In order to overcome the situation that involves odd number of strips and even number of data points, the *Simpson's 3/8th rule* is developed. This is a second-order approximation obtained by considering a third-order polynomial, and fitting it with four consecutive data points.

Consider an area under three strips with four data points $(x_{i-1}, f(x_{i-1}))$, $(x_i, f(x_i))$, $(x_{i+1}, f(x_{i+1}))$ and $(x_{i+2}, f(x_{i+2}))$ as shown in the Figure 4.6. The integration is approximated by

$$I = \int_{x_{i-1}}^{x_{i+2}} f(x)\, dx \cong \int_{x_{i-1}}^{x_{i+2}} f_3(x)\, dx \tag{4.17}$$

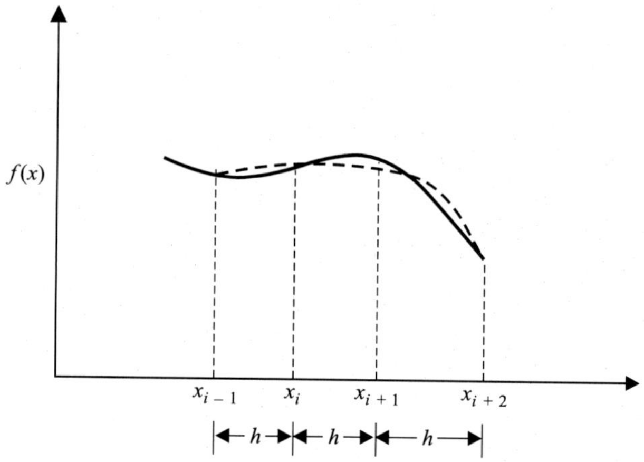

Figure 4.6 Simpson's 3/8th rule using a four-point quadratic function

where the second-order quadratic function $f_3(x) = a_0 + a_1 x + a_2 x^2 + a_3 x^3$ is assumed to pass through the four points. The coefficients a_0, a_1, a_2, and a_3 are determined by substituting the coordinate of the three points into the function $f_3(x) = a_0 + a_1 x + a_2 x^2 + a_3 x^3$ and solving the system of equations for

$$a_0 = f(x_i), \qquad a_1 = \frac{-f(x_{i+2}) + 6f(x_{i+1}) - 3f(x_i) - 2f(x_{i-1})}{6h}$$

$$a_2 = \frac{f(x_{i-1}) - 2f(x_i) + f(x_{i+1})}{2h^2}, \quad \text{and} \quad a_3 = \frac{f(x_{i+2}) - f(x_{i+1}) - 3f(x_i) - f(x_{i-1})}{6h^3}$$

The integral given by Equation 4.17 or the area under the function can now be evaluated by substituting the function as

$$I \cong \int_{x_{i-1}}^{x_{i+2}} f_3(x) = \int_{-h}^{2h} (a_0 + a_1 x + a_2 x^2 + a_3 x^3)\, dx$$

$$\cong a_0 x\Big|_{-h}^{2h} + \tfrac{1}{2} a_1 x^2\Big|_{-h}^{2h} + \tfrac{1}{3} a_2 x^3\Big|_{-h}^{2h} + \tfrac{1}{4} a_3 x^4\Big|_{-h}^{2h}$$

$$\cong 3a_0 h + \tfrac{1}{2} a_1 h^3 + 3a_2 h^3 + \tfrac{15}{3} a_3 h^4$$

Now, by substituting the expressions of a_0, a_1, a_2, and a_3, we obtain the formula for the approximation of the integral or the area under the curve that passes through the four points as

$$I \cong \frac{3h}{8}(f(x_{i-1}) + 3f(x_i) + 3f(x_{i+1}) + f(x_{i+2})) \qquad (4.18)$$

It can be noted that that the Simpson's 3/8th formula, Equation 4.18, is used along with three strips or four points. For the estimation of the intergral given by Equation 4.1 in the range (a, b), we need to divide the regions into $\boldsymbol{n}$ multiple strips with width $h=(b-a)/n$, where Equation 4.18 is applied many times in a group of three strips. The general formula by *Simpson's 3/8th rule* is derived as

$$I \cong \frac{3h}{8}\left[f(x_0) + 3\sum_{i=1,4,7}\{f(x_i) + f(x_{i+1})\} + 2\sum_{i=3,6,9} f(x_i) + f(x_n)\right] \tag{4.19}$$

It can be shown by using the Taylor series expansion that the truncation error associated with the multi-strip Simpson's 1/3rd rule is

$$E_t = -\frac{1}{180}(b-a)f^{iv}(\xi)_{\text{av}}h^4 \tag{4.20}$$

Since the point ξ is difficult to estimate and $f^{iv}(\xi)_{\text{av}}$ may not be known, Equation 4.20 is written in an alternate form as

$$E_t \cong ch^4 \tag{4.21a}$$

or

$$E_t \cong O(h^4) \tag{4.21b}$$

This indicates that truncation error in multi-strip Simpson's 1/3rd rule is proportional to h^4, the error decreases with a decrease in step size at the fourth power.

Example 4.2: Simpson's 1/3rd Rule: Repeat Problem 4.1 using Simpson's 1/3rd rule.

Solution: We start with Equation E.4.1.1 for the volume flow rate in terms of the axial velocity distribution as

$$\forall = \int_0^R u_z 2\pi r \, dr \tag{E.4.2.1}$$

Applying Simpson's 1/3rd rule to the data given in the table, we obtain the following estimate of the flow rate

$$I \cong \frac{h}{3}\left[f(x_0) + 4\sum_{i=1,3,5} f(x_i) + 2\sum_{i=2,4,6} f(x_i) + f(x_n)\right]$$

Using $f(x) = u_z 2\pi r$, $h = \Delta r = \frac{1}{12}$ and $n = 7$, we have

$$\forall \cong \frac{\Delta r}{3}\left[(u_{z0}2\pi r_0) + 4\sum_{i=1,3,5}^{6}(u_{zi}2\pi r_i) + 2\sum_{i=2,4}^{6}(u_{zi}2\pi r_i) + (u_{z6}2\pi R)\right]$$

$$\forall \cong \frac{\Delta r}{3}\left[(u_{z0}2\pi r_0) + 4\{(u_{z1}2\pi r_1) + (u_{z3}2\pi r_3) + (u_{z5}2\pi r_5)\}\right.$$

$$\left. + 2\{(u_{z2}2\pi r_2) + (u_{z4}2\pi r_4)\} + (u_{z6}2\pi r_6)\right]$$

or

$$\forall \cong \frac{\pi}{18}\left[10(0) + 4\{9.722(\tfrac{1}{12}) + 7.500(\tfrac{1}{4}) + 3.056(\tfrac{5}{12})\}\right.$$

$$\left. + 2\{8.889(\tfrac{1}{6}) + 5.556(\tfrac{1}{3}) +\} + 0(\tfrac{1}{2})\right]$$

or

$$\forall \cong 3.927 \frac{\text{ft}^3}{\text{sec}}$$

The average velocity in the pipe is estimated as

$$\bar{V} = \frac{\int_0^R u_z 2\pi r\, dr}{\pi R^2} = \frac{3.927}{\pi(1)^2}$$

or

$$\bar{V} = 1.1506\,\text{ft/s}$$

4.1.3 Summary of Newton–Cotes Integration Formulas

A summary of some of the common Newton–Cotes integration formulas are given in Table 4.1.

Simpson's 1/3rd and 3/8th rules as well as the five-point and six-point formulas have the same order of error. Even segments/odd points formulas such as the 1/3rd rule and Boole's rule are usually the methods of preference. Higher-order formulas (greater than four points) are rarely used. Simpson's rule is sufficient for most applications. Accuracy can be improved by multiple applications schemes (***such as the Romberg integration algorithm***) rather than using higher-order formulas. If the function is known, and higher accuracy is needed, methods such as the ***Romberg integration algorithm*** or ***Gauss quadrature*** are used.

Table 4.1 Newton–Cotes Integration Formulas

Strips (n)	Points ($n+1$)	Name	Integration Formula	Truncation Error
1	2	Trapezoidal rule	$I \cong \frac{h}{2} f(x_i) + f(x_{i+1})$	$-\frac{1}{12}(b-a)f''(\xi)_{av}h^2$
2	3	Simpson's 1/3rd rule	$I \cong \frac{h}{3}(f(x_{i-1}) + 4f(x_i) + f(x_{i+1}))$	$-\frac{1}{180}(b-a)f^{iv}(\xi)_{av}h^4$
3	4	Simpson's 3/8th rule	$I \cong \frac{3h}{8}(f(x_{i-1}) + 3f(x_i) + 3f(x_{i+1} + f(x_{i+2}))$	$-\frac{3}{240}(b-a)f^{iv}(\xi)_{av}h^4$
4	5	Boole's rule	$I \cong \frac{2h}{45}(7f(x_{i-2}) + 32f(x_{i-1}) + 12f(x_i) + 32f(x_{i+1}) + 7f(x_{i+2}))$	$-\frac{2}{945}(b-a)f^{vi}(\xi)_{av}h^6$
5	6		$I \cong \frac{5h}{288}(19f(x_{i-2}) + 75f(x_{i-1}) + 50f(x_i) + 50f(x_{i+1}) + 75f(x_{i+2}) + 19f(x_{i+3}))$	$-\frac{55}{2096}(b-a)f^{vi}(\xi)_{av}h^6$

4.2 ROMBERG INTEGRATION

The Romberg integration formula is the computational algorithm based on the Richardson's extrapolation, which combines two first-order numerical integration estimates and obtains a higher-order more accurate numerical estimate. The first-order estimates are obtained using the trapezoidal rule or Simpson's rule, or any other Newton–Cotes integration formula given in Table 4.1. Let us consider two approximate solutions I_{h1} and I_{h2} based on step sizes h_1 and h_2, respectively, and using the trapezoidal rule. The truncation error associated with these two first-order estimates can be written based on Equation 4.12 as

$$E_T \cong c_1 h^2 \quad or \quad o(h^2) \tag{4.22a}$$

or

$$E_T = c_1 h^2 + o(h^4) \tag{4.22b}$$

Based on this error estimate, we can express the improved integration estimate as

$$I = I_{1,1} = I_{h1} + c_1 h_1^2 \tag{4.23a}$$

and

$$I = I_{2,1} = I_{h2} + c_1 h_2^2 \tag{4.23b}$$

Solving Equations 4.23a and 4.23b for the constant, c_1, we have

$$c_1 = \frac{I_{h2} - I_{h1}}{(h_1/h_2)^2 - 1} \tag{4.24}$$

Substituting Equation 4.24 into Equation 4.23b, we obtain a second-order improved estimate as

$$I_{1,2} = I_{h2} + \frac{I_{h2} - I_{h1}}{(h_1/h_2)^2 - 1} \tag{4.25}$$

If we consider increasing the number of strips such that $(h_1/h_2) = 2$, then Equation 4.25 reduces to

$$I_{1,2} = I_{h2} + \frac{I_{h2} - I_{h1}}{4 - 1} \tag{4.26a}$$

or

$$I_{1,2} = \frac{4^1 I_{h2} - I_{h1}}{4^1 - 1} \tag{4.26b}$$

It can be noted that this second-order estimate is obtained by eliminating the error term $c_1 h^2$ or $o(h^2)$. With the elimination of this error term, the truncation error is expressed as

$$E_T = c_2 h^4 + o(h^6) \tag{4.27}$$

In order to eliminate the error term $c_2 h^4$ or $o(h^4)$ and obtain a third-order estimate, we need to have another second-order estimate, $I_{2,2}$. By considering two first-order estimates I_{h2} and I_{h3}, based on step sizes h_2 and h_3, respectively, and applying a similar procedure, we can obtain another second-order estimate as

$$I_{2,2} = I_{h3} + \frac{I_{h3} - I_{h2}}{(h_2/h_3)^2 - 1} \tag{4.28}$$

By considering $(h_2/h_3) = 2$, we can express Equation 4.28 as

$$I_{2,2} = \frac{4^1 I_{h3} - I_{h2}}{4^1 - 1} \tag{4.29}$$

Now, from the final two improved integrations given by Equation 4.26 and Equation 4.29, we can compute a new improved integral value by eliminating the error term $c_2 h^4$. This process of removing error terms is continued until the desired accuracy is obtained. The integration estimate at any accuracy level can be expressed by a general formula given as

$$I_{i,j} \cong \frac{4^{j-1} I_{i+1,j-1} - I_{i,j-1}}{4^{j-1} - 1} \tag{4.30}$$

where $\boldsymbol{j}$ is the index for the present order of integration and $\boldsymbol{j} - 1$ represents the previous lower order of integration. The indices $\boldsymbol{i}$ and $\boldsymbol{i} + 1$ represent the less and more accurate integration estimates. So, the integration $I_{i,j}$ represents a new improved integration estimation at the jth order of integration approximation, whereas $I_{i+1,j-1}$ and $I_{i,j-1}$ represent the more accurate and less accurate integration estimates at the $(j-1)$th order of integration approximations, respectively. A pseudo code for the Romberg integration formula is given in Table 4.2.

Table 4.2 Pseudo Code for Romberg Integration

Input a, b:	integration limits
Input: ε_s:	pre-specified tolerance limit
$n = 1$:	number of strips
call trapz (n, a, b, int):	use trapezoidal rule for first-order estimate
$I_{1,1} = \text{int}$	
$\varepsilon_a = 1.05\varepsilon_s$	
$k = 0$	
dowhile $(\varepsilon_a > \varepsilon_s)$	
$k = k + 1$	
$n = 2^k$	
call trapz (n, a, b, int)	
$I_{k+1,1} = \text{int}$	
dofor $j = 2$ to $k + 1$	
$i = 2 + k - j$	
$I_{i,j} \cong \dfrac{4^{j-1} I_{i+1,j-1} - I_{i,j-1}}{4^{j-1} - 1}$	
enddo	
$\varepsilon_a = \left\lvert \dfrac{I_{1,k+1} - I_{1,k}}{I_{1,k+1}} \right\rvert$	
enddo	

4.3 GAUSS QUADRATURE

One of the important characteristics of the Newton–Cotes integration formulas is that the integral estimate is based on evenly-spaced function values or the values are known at fixed interior points and fixed weights. The number of points on each strip are selected to define a nth degree polynomial that can be integrated exactly. However, the positions of the interior points and weights are not optimized so as to minimize the truncation error. Accuracy is improved by considering an increased number of strips or using a higher-order polynomial. Also, since the formula is based on using the function values at the limit points, it will lead to inaccurate results for selecting a strip by joining any two points, such as shown in Figure 4.7.

It can be seen that such an approximation will involve both a positive error and a negative error. We could, however, select these points and straight line in such a way that these errors will cancel each other and give an improved estimate. ***Gauss quadrature*** is one such class of techniques that implement such a strategy. One particular gauss quadrature that has become very popular in engineering applications is the ***Gauss–Legendre formula***, which is derived using the method of undetermined coefficients. The development of this formula is done in two steps. First, the integral is transformed from Equation 4.1 into a general form as

$$I = \int_{-1}^{1} g(x')\,dx' \tag{4.31}$$

It can be noted that the integral limits have been changed from (a, b) to $(-1, 1)$. This is done in order to make the formulation simple and general. However,

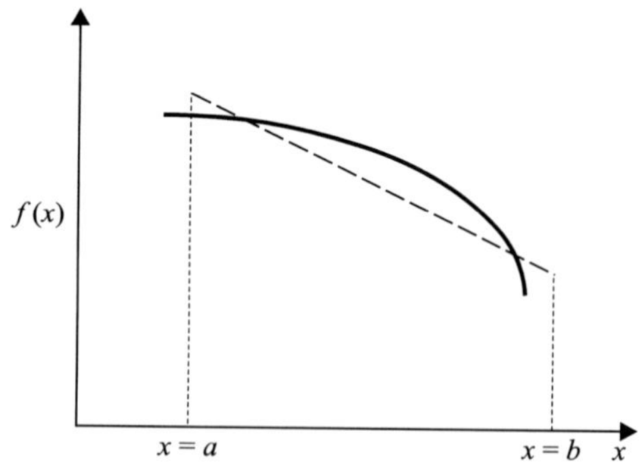

Figure 4.7 Integral estimate based on the area under a straight line passing through two points other than the integration limits

we need to make a change of variable in order to translate other limits of integration into this general form. This is achieved by substituting the following two equations

$$x = \frac{b+a}{2} + \frac{b-a}{2}x' \tag{4.32a}$$

and

$$dx = \frac{b-a}{2}dx' \tag{4.32b}$$

into the integral given by Equation 4.1.

In the second step, the integral is approximated by a quadrature formula that is expressed in a general form as

$$I = \int_{-1}^{1} g(x')\,dx' \cong \sum_{i=0}^{n-1} w_i g(x') \tag{4.33a}$$

where ***w*** are the unknown co-efficients called ***weighting factors***; x' are the unknown points, called ***function arguments***, and ***n*** is the number of points. The major objective is to determine unknown weighting factors, ***w***, and function arguments, x'.

$$I \cong w_0 g(x'_0) + w_1 g(x'_1) + w_2 g(x'_2) + \cdots + w_{n-1} g(x'_{n-1}) \tag{4.33b}$$

Let us now consider the development of a two-point ***Gauss–Legendre formula*** and present a summary of other commonly used higher-order **Gauss–Legendre formulas** in the following section.

4.3.1 Two-Point Gauss–Legendre Formula

In the two-point formula, the objective is to approximate the integral as

$$I \cong w_0 g(x'_0) + w_1 g(x'_1) \tag{4.34}$$

where the ***w*** are the unknown coefficients that need to be determined. The x' are the selected points that also have to be determined. Thus, we have four unknowns and we require four conditions to determine them exactly. These four conditions are established by selecting four expressions for the function that will yield exact results. Four simple equations that represent these cases are that it fits the integral of a constant function $g(x') = 1$, a straight line

$g(x') = x'$, a parabola $g(x') = x'^2$, and a cubic function $g(x') = x'^3$. In order to satisfy these conditions we need to solve equations

$$I \cong w_0 g(x'_0) + w_1 g(x'_1) = \int_{-1}^{1} 1\, dx' = 2 \tag{4.35a}$$

$$I \cong w_0 g(x'_0) + w_1 g(x'_1) = \int_{-1}^{1} x\, dx' = 0 \tag{4.35b}$$

$$I \cong w_0 g(x'_0) + w_1 g(x'_1) = \int_{-1}^{1} x'^2\, dx' = \frac{2}{3} \tag{4.35c}$$

$$I \cong w_0 g(x'_0) + w_1 g(x'_1) = \int_{-1}^{1} x'^3\, dx' = 0 \tag{4.35d}$$

Solving these equations simultaneously, we get

$$w_0 = 1 \tag{4.36a}$$

$$w_1 = 1 \tag{4.36b}$$

$$x'_0 = -\frac{1}{\sqrt{3}} = -0.577\,350\,269 \tag{4.36c}$$

$$x'_1 = \frac{1}{\sqrt{3}} = 0.577\,350\,269 \tag{4.36d}$$

4.3.2 Higher-Point Gauss–Legendre Formulas

Higher-point formulas are developed based on the equation

$$I \cong w_0 g(x'_0) + w_1 g(x'_1) + w_2 g(x'_2) + \cdots + w_{n-1} g(x'_{n-1}) \tag{4.37}$$

The weighting factors and the function arguments are derived using a similar procedure, and the values are summarized in Table 4.3.

One limitation of the Gauss–Legendre formulas is that it requires function evaluations at nonuniformly spaced points within the integration limits. For problems where the function is known and requires repetitive evaluations

Table 4.3 Function Arguments and Weight Factors in Gauss–Legendre Quadrature Formula

(n)	(w_j)	(x'_j)
2	$w_0=1.000\,000\,000$	$x_0=-0.577\,350\,269$
	$w_1=1.000\,000\,000$	$x_1=0.577\,350\,269$
3	$w_0=0.555\,555\,556$	$x_0=-0.774\,596\,669$
	$w_1=0.888\,888\,889$	$x_1=0.0$
	$w_2=0.555\,555\,556$	$x_2=0.774\,596\,669$
4	$w_0=0.347\,854\,845$	$x_0=-0.861\,136\,312$
	$w_1=0.652\,145\,155$	$x_1=-0.339\,981\,044$
	$w_2=0.652\,145\,155$	$x_2=0.339\,981\,044$
	$w_3=0.347\,854\,845$	$x_3=-0.861\,136\,312$
5	$w_0=0.236\,926\,885$	$x_0=-0.906\,179\,846$
	$w_1=0.478\,628\,670$	$x_1=-0.538\,469\,310$
	$w_2=0.568\,888\,889$	$x_2=0.0$
	$w_3=0.478\,628\,670$	$x_3=0.538\,469\,310$
	$w_4=0.236\,926\,885$	$x_4=0.906\,179\,846$
6	$w_0=0.171\,324\,492$	$x_0=-0.932\,469\,514$
	$w_1=0.360\,761\,573$	$x_1=-0.661\,209\,386$
	$w_2=0.467\,913\,935$	$x_2=-0.238\,619\,186$
	$w_3=0.467\,913\,935$	$x_3=0.238\,619\,186$
	$w_4=0.360\,761\,573$	$x_4=0.661\,209\,386$
	$w_5=0.171\,324\,492$	$x_5=0.932\,469\,514$
8	$w_0=0.101\,228\,536$	$x_0=-0.960\,289\,856$
	$w_1=0.222\,381\,034$	$x_1=-0.796\,666\,477$
	$w_2=0.313\,706\,646$	$x_2=-0.525\,532\,409$
	$w_3=0.362\,683\,783$	$x_3=-0.183\,434\,642$
	$w_4=0.362\,683\,783$	$x_4=0.183\,434\,642$
	$w_5=0.313\,706\,646$	$x_5=0.525\,532\,409$
	$w_6=0.222\,381\,034$	$x_6=0.796\,666\,477$
	$w_7=0.101\,228\,536$	$x_7=0.960\,289\,856$
10	$w_0=0.066\,671\,344$	$x_0=-0.973\,906\,529$
	$w_1=0.149\,451\,349$	$x_1=-0.865\,063\,367$
	$w_2=0.219\,086\,363$	$x_2=-0.679\,409\,568$
	$w_3=0.269\,266\,719$	$x_3=-0.433\,395\,394$
	$w_4=0.295\,524\,225$	$x_4=-0.148\,874\,339$
	$w_5=0.295\,524\,225$	$x_5=0.148\,874\,339$
	$w_6=0.269\,266\,719$	$x_6=0.433\,395\,394$
	$w_7=0.219\,086\,363$	$x_7=0.679\,409\,568$
	$w_8=0.149\,451\,349$	$x_8=0.865\,063\,367$
	$w_9=0.066\,671\,344$	$x_9=0.973\,906\,529$

of the integrals, such formulas are usually the method of choice for efficient evaluation of the integrals. However, these formulas are not suitable for problems that involve discrete values of the variables at selected points such as velocity or temperature values at nodal points given by the finite difference–control volume or finite element methods. In such cases Newton–Cotes formulas are preferred.

Example 4.4: Gauss–Legendre Quadrature: Use three-point, four-point and six-point Gauss–Legendre formulas to evaluate the integral

$$I = \int_{-1}^{1} [N]^{\mathrm{T}}[N]\,dx \tag{E.4.4.1}$$

where

$$[N]^{\mathrm{T}} = \begin{Bmatrix} N_i \\ N_j \\ N_k \end{Bmatrix} = \begin{Bmatrix} \frac{x}{2}(x-1) \\ -(x^2-1) \\ \frac{x}{2}(x+1) \end{Bmatrix} \tag{E.4.4.2}$$

Solution: Note that the integral represents a matrix whose each element involves an integral given as

$$I = \int_{-1}^{1} \begin{bmatrix} N_i \\ N_j \\ N_k \end{bmatrix} [N_i \quad N_j \quad N_k] \tag{E.4.4.3}$$

or

$$I = \int_{-1}^{1} \begin{bmatrix} N_i^2 & N_iN_j & N_iN_k \\ N_jN_i & N_j^2 & N_jN_k \\ N_kN_i & N_kN_j & N_k^2 \end{bmatrix} dx \tag{E.4.4.4}$$

Substituting the expressions for shape functions, we get

$$I = \int_{-1}^{1} \begin{bmatrix} \frac{x^2}{4}(x-1)^2 & -\frac{x}{2}(x-1)(x^2-1) & \frac{x^2}{4}(x^2-1) \\ -\frac{x}{2}(x-1)(x^2-1) & (x^2-1)^2 & -\frac{x}{2}(x+1)(x^2-1) \\ \frac{x^2}{4}(x^2-1) & -\frac{x}{2}(x+1)(x^2-1) & \frac{x^2}{4}(x+1)^2 \end{bmatrix} dx \tag{E.4.4.5}$$

Note that the matrix is symmetric and we just need to estimate the elements in the upper triangular part. Let us consider evaluation of the first integral element as

$$I_{11} = \int_{-1}^{1} N_i^2\, dx = \int_{-1}^{1} \left[\frac{x}{2}(x-1)\right]^2 dx \tag{E.4.4.6}$$

or

$$I_{11} = \int_{-1}^{1} \left(\frac{x^4}{4} - \frac{2x^3}{4} + \frac{x^2}{4}\right) dx$$

or

$$I_{11} = \int_{-1}^{1} f(x)\, dx$$

where

$$f(x) = \frac{x^4}{4} - \frac{2x^3}{4} + \frac{x^2}{4} \tag{E.4.4.7}$$

Three-point Gauss–Legendre formula

$$I_{11} \cong w_0 f(x_0) + w_1 f(x_1) + w_2 f(x_2) \tag{E.4.4.8}$$

where

$$w_0 = 0.555\,555\,556, \quad w_1 = 0.888\,888\,889, \quad w_2 = 0.555\,555\,556$$
$$x_0 = -0.774\,596\,669, \quad x_1 = 0.0, \quad x_0 = 0.774\,596\,669$$

The function is evaluated at the function points as

$$\begin{aligned} f(x_0) &= \frac{(-0.774\,596\,669)^4}{4} - \frac{(-0.774\,596\,669)^3}{2} + \frac{(0.774\,596\,669)^2}{4} \\ &= 0.089\,999\,999 + 0.232\,379 + 0.149\,999\,999 \\ &= 0.472\,378\,998 \\ f(x_1) &= 0.0 \\ f(x_2) &= 0.089\,999\,999 - 0.232\,379 + 0.149\,999\,999 \\ &= 0.007\,620\,998 \end{aligned}$$

Substituting the function values, we have the first integral element estimated from equation (E.4.4.8) as

$$
\begin{aligned}
I_{11} &\cong w_0 f(x_0) + w_1 f(x_1) + w_2 f(x_2) \\
&= 0.555\,555\,556 \times 0.472\,378\,998 + 0.888\,888\,889 \times 0 \\
&\quad + 0.555\,555\,556 \times 0.007\,620\,998 \\
&= 0.262\,432\,776 + 0 + 0.004\,233\,888 \\
&= 0.266\,666\,664
\end{aligned}
$$

The rest of the integral elements are computed in a similar manner and the integral based on the three-point formula is given by

$$
I \cong \begin{bmatrix} 0.266\,666\,666\,8 & -1.200\,000\,01 & -0.066\,666\,666\,7 \\ -1.200\,000\,01 & 6.400\,000\,002 & 0.133\,333\,333\,5 \\ -0.066\,666\,666\,7 & 0.133\,333\,333\,5 & 0.266\,666\,666\,8 \end{bmatrix}
$$

Four-point Gauss–Legendre formula

$$
I_{11} \cong w_0 f(x_0) + w_1 f(x_1) + w_2 f(x_2) + w_3 f(x_3) \tag{E.4.4.9}
$$

where

$$
\begin{aligned}
w_0 &= 0.347\,854\,845, \quad w_1 = 0.652\,145\,155, \\
w_2 &= 0.652\,145\,155, \quad w_3 = 0.347\,854\,845 \\
x_0 &= -0.861\,136\,312, \quad x_1 = -0.339\,981\,044, \\
x_2 &= 0.339\,981\,044, \quad x_3 = 0.861\,1363\,12
\end{aligned}
$$

$$
\begin{aligned}
f(x_0) &= \frac{(-0.861\,136\,312)^4}{4} - \frac{(-0.861\,136\,312)^3}{2} + \frac{(-0.861\,136\,312)^2}{4} \\
&= 0.137\,476\,231 + 0.319\,290\,29 + 0.185\,388\,937 \\
&= 0.642\,155\,458
\end{aligned}
$$

$$
\begin{aligned}
f(x_1) &= \frac{(-0.339\,981\,044)^4}{4} - \frac{(-339\,981\,044)^3}{2} + \frac{(-0.339\,981\,044)}{4} \\
&= 0.003\,340\,095 + 0.019\,648\,713 + 0.028\,896\,777 \\
&= 0.051\,885\,585
\end{aligned}
$$

$$
\begin{aligned}
f(x_2) &= \frac{(0.339\,981\,044)^4}{4} - \frac{(339\,981\,044)^3}{2} + \frac{(0.339\,981\,044)}{4} \\
&= 0.003\,340\,095 - 0.019\,648\,713 + 0.028\,896\,777 \\
&= 0.012\,588\,159
\end{aligned}
$$

$$f(x_3) = \frac{(0.861\,136\,312)^4}{4} - \frac{(0.861\,136\,312)^3}{2} + \frac{(0.861\,136\,312)^2}{4}$$
$$= 0.137\,476\,231 - 0.319\,290\,29 + 0.185\,388\,937$$
$$= 0.003\,574\,878$$

Substituting the function values, we have the first integral element estimated from equation (E.4.4.9) as

$$I_{11} \cong w_0 f(x_0) + w_1 f(x_1) + w_2 f(x_2) + w_3 f(x_3)$$
$$= 0.223\,376\,887 + 0.038\,369\,32 + 0.007\,996\,768 + 0.002\,233\,133\,8$$
$$= 0.267\,541\,925$$

The rest of the integral elements are computed in a similar manner and the integral based on four-point formula is given by

$$I \cong \begin{bmatrix} 0.266\,666\,667\,0 & -1.200\,000\,01 & -0.066\,666\,666\,6 \\ -1.200\,000\,01 & 6.400\,000\,004 & 0.133\,333\,333\,5 \\ -0.066\,666\,666\,6 & 0.133\,333\,333\,3 & 0.266\,666\,667\,0 \end{bmatrix}$$

Six-point Gauss–Legendre formula

$$I_{11} \cong w_0 f(x_0) + w_1 f(x_1) + w_2 f(x_2) + w_3 f(x_3) + w_4 f(x_4) + w_5 f(x_5) \tag{E.4.4.10}$$

where

$$w_0 = 0.171\,324\,492, \quad w_1 = 0.360\,761\,573, \quad w_2 = 0.467\,913\,935$$
$$w_3 = 0.467\,913\,935, \quad w_4 = 0.360\,761\,573, \quad w_5 = 0.171\,324\,492$$
$$x_0 = -0.932\,469\,514, \quad x_1 = -0.661\,209\,386, \quad x_2 = -0.238\,619\,186$$
$$x_3 = 0.238\,619\,186, \quad x_4 = 0.661\,209\,386, \quad x_5 = 0.932\,469\,514$$
$$f(x_0) = \frac{(-0.932\,469\,514)^4}{4} - \frac{(-0.932\,469\,514)^3}{2} + \frac{(-0.932\,469\,514)^2}{4}$$
$$= 0.189\,007\,299 + 0.405\,390\,838 + 0.217\,374\,848$$
$$= 0.8117\,729\,85$$
$$f(x_1) = \frac{(-0.661\,209\,386)^4}{4} - \frac{(-0.661\,209\,386)^3}{2} + \frac{(-0.661\,209\,386)^2}{4}$$
$$= 0.477\,854\,9 + 0.144\,539\,661 + 0.109\,299\,463$$
$$= 0.301\,624\,614$$

$$f(x_2) = \frac{(-0.238\,619\,186)^4}{4} - \frac{(-0.238\,619\,186)^3}{2} + \frac{(-0.238\,619\,186)^2}{4}$$

$$= 0.000\,810\,515 + 0.006\,793\,382 + 0.014\,234\,778$$

$$= 0.002\,183\,867\,5$$

$$f(x_3) = \frac{(0.238\,619\,186)^4}{4} - \frac{(0.238\,619\,186)^3}{2} + \frac{(0.238\,619\,186)^2}{4}$$

$$= 0.000\,810\,515 - 0.006\,793\,382 + 0.014\,234\,778$$

$$= 0.008\,257\,911$$

$$f(x_4) = \frac{(0.661\,209\,386)^4}{4} - \frac{(0.661\,209\,386)^3}{2} + \frac{(0.661\,209\,386)^2}{4}$$

$$= 0.477\,854\,9 - 0.144\,539\,661 + 0.109\,299\,463$$

$$= 0.012\,545\,292$$

$$f(x_5) = \frac{(0.932\,469\,514)^4}{4} - \frac{(0.932\,469\,514)^3}{2} + \frac{(0.932\,469\,514)^2}{4}$$

$$= 0.189\,007\,299 - 0.405\,390\,838 + 0.217\,374\,848$$

$$= 0.000\,913\,09$$

Substituting the function values, we have the first integral element estimated from equation (E.4.4.10) as

$$I_{11} \cong w_0 f(x_0) + w_1 f(x_1) + w_2 f(x_2) + w_3 f(x_3) + w_4 f(x_4) + w_5 f(x_5)$$

$$= 0.139\,076\,590 + 0.108\,814\,57 + 0.010\,218\,62 + 0.003\,861\,184$$

$$+ 0.004\,525\,859 + 0.000\,169\,835$$

$$= 0.266\,666\,662$$

The rest of the integral elements are computed in a similar manner and the integral based on the six-point formula is given by

$$I \cong \begin{bmatrix} 0.266\,666\,666\,4 & -1.199\,999\,997 & -0.066\,666\,666\,6 \\ -1.199\,999\,997 & 6.399\,999\,992 & 0.133\,333\,331\,1 \\ -0.066\,666\,666\,6 & 0.133\,333\,331\,1 & 0.266\,666\,666\,4 \end{bmatrix}$$

4.4 MULTI-DIMENSIONAL NUMERICAL INTEGRATION

Let us consider multi-dimensional integrals of the following forms:

$$I = \int_c^d \int_a^b f(x, y)\, dy\, dx \quad \text{for two dimensions} \tag{4.38a}$$

and

$$I = \int_e^f \int_c^d \int_a^b f(x, y, z)\, dx\, dy\, dz \quad \text{(for three dimensions)} \tag{4.38b}$$

Multi-dimensional integrals can be evaluated numerically using the methods discussed for one-dimensional case, i.e. by ***Newton–Cotes formulas*** or by ***Gauss–Legendre formula***. The method is first applied in the first dimension while keeping each value of the second variable held constant. This will give an array of integrated values. Then the method is applied again for the second dimension using the array of integrated values. The procedure is described as follows:

Integration using Newton–Cotes formulas

$$I = \int_c^d \int_a^b f(x, y)\, dy\, dx = \int_c^d \left[\int_a^b f(x, y)\, dy \right] dx \tag{4.39}$$

The inner integral for y-dimension is evaluated using one-dimensional integration formula for different x-values. This gives the array of integrated values as

$$I(x) = \int_a^b f(x, y)\, dy \tag{4.40}$$

One-dimensional integration is then applied for the x-direction using the array of integrated values as

$$I = \int_a^b I(x)\, dx \tag{4.41}$$

Let us consider the double integral in the general form as

$$I = \int_{-1}^{1} \int_{-1}^{1} g(x', y')\, dy'\, dx' \tag{4.42}$$

Again, this two-dimensional integral is solved using the one-dimensional formula for each coordinate system. For example, we can use an n-term

Gauss–Legendre quadrature in the y-direction to obtain

$$I = \int_{-1}^{1} \left(\sum_{j=1}^{n} w_j g(x', y'_j) \right) dx' \tag{4.43}$$

This is followed by the use of another n-term Gauss–Legendre quadrature in the x-direction that gives

$$I \cong \sum_{i=1}^{n} \sum_{j=1}^{n} w_i w_j g(x'_i, y'_j) \tag{4.44}$$

The weight factors w_i and w_j, and the corresponding Gauss points x_i and x_j, are the same as those given in Table 4.3, and $\boldsymbol{n}$ is the number of Gauss points.

In a similar manner, we can evaluate a triple integral by repeatedly applying the one-dimensional Gauss–Legendre quadrature as

$$I \cong \int_{-1}^{1} \int_{-1}^{1} \int_{-1}^{1} g(x', y', z')\, dz'\, dy'\, dx' \cong \sum_{i=1}^{n} \sum_{j=1}^{n} \sum_{k=1}^{n} w_i w_j w_k g(x'_i, y'_j, z'_k) \tag{4.45}$$

where w_i, w_j, and w_k are weight factors, and x_i, x_j, and x_k are corresponding Gauss points as given in Table 4.3.

PROBLEMS

4.1 Local Nusselt numbers for developing flow in a circular tube are computed using a numerical scheme, and results are shown in the table below.

Location (x)	Local Nusselt Number (Nu_D)
1.0	13.2
2.0	11.1
3.0	8.4
4.0	6.7
5.0	7.3
6.0	6.8
7.0	5.9
8.0	5.0
9.0	4.3
10.0	3.9

Estimate the average Nusselt number using the trapezoidal rule.

4.2 Use the trapezoidal rule to evaluate the integral

$$I = \int_{1}^{3} \left(\frac{x+2}{x}\right)^2 dx$$

with eight segments.

(a) Develop a computer code ***Trapz*** using the quadrature formula for the trapezoidal rule.

(b) Use the code and progressively increase the number of segments as $n = 2$, 4, 8 and 16 to calculate the integral. Estimate approximate percent relative error for each integration estimate.

(c) Use first-order estimates obtained in part (b) to estimate the higher-order estimates using the Romberg integration algorithm.

4.3 Power intensity of a high energy laser beam is given by

$$I_x = I_0 e^{-x^2/R_0^2}$$

where

$$I_0 = 2.30 \times 10^8 \text{ W/m}^2$$

$$R_0 = 0.001 \text{ m}$$

Determine the average power of the laser beam using Simpson's 1/3rd rule.

4.4 Consider the integral

$$I = \int_{0}^{8} \left(12 + 2x - 6x^2 + 5x^3\right) dx$$

(a) Evaluate the integral with ***Simpson's 1/3rd rule*** and using eight segments.

(b) Develop a computer code ***simp*** using the algorithm for Simpson's rule.

(c) Use the code and progressively increase the number of segments as $n = 2$, 4, 8, and 16 to calculate the integral. Estimate approximate percent relative error for each integration estimate.

(d) Use results obtained in part (c) and obtain improved estimates of the integral using the ***Romberg integration formula***. Show approximate percent relative error in each step.

4.5 Develop a computer code to implement the Romberg integration formula to integrate the integral

$$I = \int_0^{\pi/2} \cos x \, dx$$

with $\varepsilon_s = 0.000\,001$ and a trapezoidal rule for the first-order estimates.

4.6 Use three-point ***Gauss–Legendre*** quadrature formula to evaluate the integral given in Problem 4.2.

4.7 Evaluate the following integral using the four-point Gauss–Legendre quadrature formula

$$I = \int_{-1}^{1} N_1 N_3$$

where

$$N_1 = \frac{x}{2}(x-1) \quad \text{and} \quad N_2 = \frac{x}{2}(x+1)$$

4.8 Write a computer code to implement three-, four-, six-, and eight- point Gauss–Legendre quadrature formula and evaluate the integral

$$I = \int_{-1}^{1} 0.1x[N]^{\mathrm{T}}[N]\, dx$$

where

$$[N]^{\mathrm{T}} = \begin{bmatrix} N_1 \\ N_2 \\ N_3 \end{bmatrix} = \begin{bmatrix} \frac{x}{2}(x-1) \\ -(x^2-1) \\ \frac{x}{2}(x+1) \end{bmatrix}$$

Present your result including percent relative error in a summary table.

4.9 Develop a computer code based on the Romberg integration algorithm using the subroutine ***trapz*** to evaluate

$$\int_0^2 \frac{e^x}{1+x^4} \, dx$$

to an accuracy of 0.1%. Present your results in each step including the percent relative errors.

4.10 Develop a computer code using two through six-point Gauss–Legendre formulas to solve

$$\int_{-2}^{2} \frac{3}{1+3x^3}\, dx$$

Present your result in a table.

(a) Evaluate the following integral using four- and six-point Gauss–Legendre quadrature

$$I = \int_{-1}^{1} q(x)([N]^{\mathrm{T}}[N]\, dx$$

where

$$[N]^{\mathrm{T}} = \begin{Bmatrix} N_i \\ N_j \\ N_k \end{Bmatrix} = \begin{Bmatrix} \frac{x}{2}(x-1) \\ -(x^2-1) \\ \frac{x}{2}(x+1) \end{Bmatrix}$$

$$q(x) = e^{-0.2x}$$

4.11 Use Simpson's algorithm and computer program to evaluate the average temperature distribution in a square slab. The temperature distribution at selected nodal points is given below:

300	430	545	625	710
220	280	410	515	605
125	192	254	335	440
100	157	184	241	307

Note that you have to use quadrature formula for two dimensional integrals.

4.12 Develop a computer code to evaluate the double integral

$$\int_{-4}^{4}\int_{0}^{8} (x^3 - 4y^2 + 3xy^3)\, dx\, dy$$

using subroutine ***simp***.

Part II
FINITE DIFFERENCE–CONTROL VOLUME METHOD

CHAPTER

5

BASICS

The finite difference method has been in use for a long time. The method is popular due to its simplicity in the discretization procedure and the relative ease of implementation into a computer code. In this chapter we will present the basic steps in the formulation and application of the finite difference method, using a general steady-state equation containing diffusion and source terms only. Also, we will restrict our formulation procedure to one-dimensional problems. In order to comprehend the basic concepts and steps clearly, a number of simple classical problems will be considered. Detailed formulation and application to multi-dimensional steady state problems will be given in Chapter 6. Treatment of the unsteady state or storage term and convection terms will be considered in the subsequent chapters.

5.1 INTRODUCTION AND BASIC STEPS IN FINITE DIFFERENCE METHOD

The mathematical formulation or model of a physical process is developed based on using fundamental laws of science and engineering, subject to some assumptions related to the process, and with the application of relevant mathematical methods. Such formulation results in mathematical statements, often differential equations, relating quantities of interest in the understanding and/or design of physical processes. The complexity in the mathematical model and in the geometry of the problem very often precludes the use of analytical solution methods. In such cases, we rely on numerical solution methods such as the finite difference and finite element methods. In a numerical simulation, we use an approximate method to evaluate the mathematical model and estimate the characteristics of the process using computers.

In the ***finite difference method***, the governing equations are approximated by a point-wise discretization scheme where derivatives are replaced by difference equations that involve the values of the solution at the nodal points. In this method, the discretization process involves first dividing the solution region into a network of grid or mesh of intersecting lines, which are drawn parallel to the coordinate axes. The discrete intersecting points of these gridlines are called the grid or nodal points. The number or the distances between the gridlines affect the number of the resulting grid points or grid size, and hence affect the accuracy of the numerical solution. The accuracy improves with the increase of grid points or decrease of the grid size. The time coordinate is also discretized in a similar manner and it is assumed that solution progresses in a sequence of time steps. Once the grid is generated, the governing equations and boundary conditions are then transformed into discretization equations at each nodal point to obtain a set of algebraic equations that involve the unknown values at the grid points. Finally, this set of algebraic equations is solved for the unknown nodal values.

The basic steps in obtaining a numerical solution using finite difference are categorized as follows.

1. State the mathematical statement of the problem in terms of governing equations, boundary conditions and initial conditions.
2. Disretize the solution domain into a network of discrete nodal points. The unknown values are sought only at those discrete points rather than obtaining a continuous solution in the domain.
3. Obtain discretization equations for all nodal points by approximating the governing differential equations and boundary conditions. This discretization procedure leads to a set of algebraic equations involving the unknown values at the nodal points.
4. Use an appropriate solution algorithm to solve the set of algebraic equations involving the unknown values at the nodal points.
5. Post-processing of the data to evaluate secondary quantities.

We will outline the basic principles of the finite difference method in the following sections.

5.2 DISCRETIZATION OF THE DOMAIN

In this discretization method, the two-dimensional calculation domain is divided into equal or nonequal small regions with increments of Δx and Δy

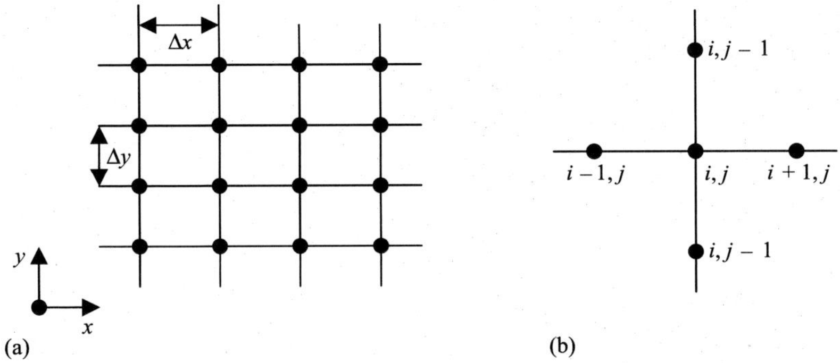

Figure 5.1 Nodal network for finite difference method

in the x and y directions, respectively, as shown in Figure 5.1. The values of the dependent variables $\phi(x, y)$ are calculated at a finite number of discrete points in the solution domain as shown in the figure. These discrete points are referred to as the grid points or nodal points.

Each nodal point is designated by a numbering scheme i and j, where the i and j locations indicate the x and y increments defined by $x = i\Delta x$ and $y = j\Delta y$, respectively. So, the continuous function behavior of the dependent variables $\phi(x, y)$ contained in the analytical solution of the governing equations is replaced by the discrete values $\phi(x_i, y_j)$ at these grid points. The discrete value of the function ϕ at a location (x_i, y_j) is designated by $\phi_{i,j}$. The discrete values at selected grid points are established by a system of algebraic equations, which we will refer to as discretization equations or finite difference equations. When the number grid points are small, the resulting discretization or numerical solution is approximate, which may deviate considerably from the analytical solution of the governing equations. The accuracy of the numerical solution increases as we increase the number of grid points.

This designation procedure can be extended to three-dimensional and unsteady-state problems with the discrete values represented as $\phi^l_{i,j,k} = \phi(x_i, y_j, z_k, t_l)$. The index k represents z space locations with increments of Δz, and l represents the time level with time increments of Δt. We will describe the systematic approach for obtaining these discretization equations in the following sections.

5.3 DISCRETIZATION OF THE MATHEMATICAL MODEL

The discretization equation or finite difference equations can be derived by using several approaches, the most popular ones being the ***Taylor series method*** and the ***control volume method***. The governing equations for fluid flows, heat and mass transfer involve first- and second-order partial derivatives of the independent variables such as velocities, pressure, concentration and temperature. In the Taylor series method, each derivative is approximated with the help of a difference formula that is derived from the truncated Taylor series expansion as discussed in the following section, whereas in a control volume method the governing equation is discretized by integrating it over an appropriately chosen control volume. In the following section, we present a brief summary of these discretization methods.

5.3.1 The Taylor Series Method

As we have already mentioned, in the Taylor series method each space and time derivatives are approximated by appropriate finite difference approximation formulas, which are derived by using the Taylor series expansions. So, this method involves two steps. First, appropriate numerical ***finite difference formulas*** are selected and then the discretization equation is obtained by substituting these formulas in the governing equation. Let us briefly discuss availability of such finite difference formulas followed by the derivation of discretization equations in the following section.

5.3.1.1 Numerical Differentiation–Finite Difference Formulas. Numerical differentiation of a governing differential equation is carried out by approximating the derivatives with the finite difference formulas, which are derived utilizing the truncated Taylor series expansions. A function $\phi(x, y)$ at the point $(i+1, j)$ can be expressed in terms of the function value and its derivatives at a neighboring point (i, j) using forward Taylor series expansion

$$\phi_{i+1,j} = \phi_{i,j} + \phi'_{i,j}\Delta x + \frac{1}{2!}\phi''_{i,j}(\Delta x)^2 + \frac{1}{3!}\phi'''_{i,j}(\Delta x)^3 + \cdots + \frac{1}{n!}\phi^{(n)}_{i,j}(\Delta x)^n + R_n \quad (5.1)$$

where $\Delta x = x_{i+1} - x_i$ is the step size in the x direction, and R_n is the remainder term that is included to account for all the remaining terms from $n+1$ to infinity. The remainder term is basically a representation of the truncation error, and it is expressed as

$$R_n = \frac{1}{(n+1)!}\phi_i^{(n+1)}(\Delta x)^{n+1} \quad (5.2)$$

Similarly, a function $\phi(x, y)$ at the point $(i-1, j)$ can be expressed in terms of the function value and its derivatives at a neighboring point (i, j) using the backward Taylor series expansion

$$\phi_{i-1,j} = \phi_{i,j} - \phi'_{i,j}\,\Delta x + \frac{1}{2!}\phi''_{i,j}(\Delta x)^2 - \frac{1}{3!}\phi'''_{i,j}(\Delta x)^3 + \cdots + \frac{1}{n!}\phi^{(n)}_{i,j}(\Delta x)^n + R_n \quad (5.3)$$

Equation 5.2 and Equation 5.3 can be used to derive many different finite difference approximation formulas. Let us discuss here the derivation of some basic approximation formulas for the first derivatives.

Forward Difference Approximation This approximation formula is derived by neglecting all the terms of $o(\Delta x)^2$ and higher in the forward Taylor series expansion given by Equation 5.1, and solving for the first derivative term as

$$\phi'_{i,j} = \frac{\phi_{i+1,j} - \phi_{i,j}}{\Delta x} + \frac{R_1}{\Delta x} \tag{5.4}$$

where the second term represents the truncation error associated with the approximation of the first derivative, and is expressed as

$$\frac{R_1}{\Delta x} = \frac{\phi''}{2!}\Delta x \tag{5.5a}$$

or

$$\frac{R_1}{\Delta x} = o(\Delta x) \tag{5.5b}$$

Using Equation 5.5, we can rewrite the approximation, Equation 5.4, as

$$\phi'_{i,j} = \frac{\phi_{i+1,j} - \phi_{i,j}}{\Delta x} + o(\Delta x) \tag{5.6a}$$

or

$$\phi'_{i,j} \cong \frac{\phi_{i+1,j} - \phi_{i,j}}{\Delta x} \tag{5.6b}$$

This approximation is called the ***forward difference approximation*** because it estimates the derivative at the point (i,j) using the values at the point (i,j) and a forward point $(i+1,j)$ (Figure 5.1a), and it has an error of $o(\Delta x)$. Physically this approximation means that the derivative at x_i, which represents the slope or the tangent at x_i, is assumed to be equal to the slope of the straight line joining points x_i and x_{i+1} as shown in Figure 5.2a.

Backward Difference Approximation Similarly, the backward difference approximation formula is derived by neglecting all the terms of $o(\Delta x)^2$ and higher in the backward Taylor series expansion given by Equation 5.3, and solving for the first derivative term as

$$\phi'_{i,j} = \frac{\phi_{i,j} - \phi_{i-1,j}}{\Delta x} + o(\Delta x) \tag{5.7a}$$

or

$$\phi'_{i,j} \cong \frac{\phi_{i,j} - \phi_{i-1,j}}{\Delta x} \tag{5.7b}$$

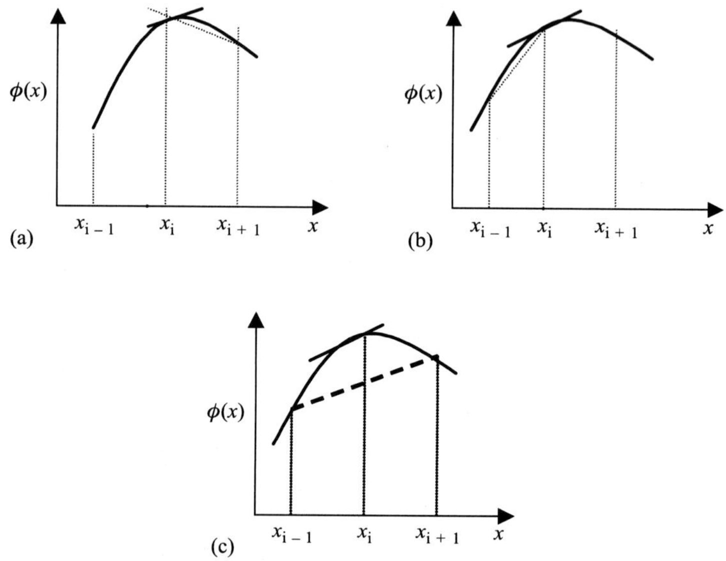

Figure 5.2 Graphical representation of the difference approximations: (a) forward difference, (b) backward difference, and (c) central difference

This approximation is called the ***backward difference approximation*** because it estimates the derivative at the point (i, j) using the values at the point (i, j) and a backward point $(i\!-\!1, j)$ (Figure 5.1b), and it has an error of $o(\Delta x)$. Physically this approximation means that the derivative at x_i is assumed to be equal to the slope of the straight line joining points x_{i-1} and x_i as shown in Figure 5.2b.

Central Difference Approximation In order to derive this approximation formula, Equation 5.3 is subtracted from Equation 5.1 to obtain

$$\phi_{i+1,j} - \phi_{i-1,j} = 2\phi'_{i,j}\Delta x + \frac{2\phi'''_{i,j}}{3!}(\Delta x)^3 + \cdots + R_n \tag{5.8}$$

Now, neglecting the terms of $o(\Delta x)^3$ and higher, we have

$$\phi_{i+1,j} - \phi_{i-1,j} = 2\phi'_{i,j}\Delta x + R_2 \tag{5.9}$$

Solving for the first derivative we get the approximation as

$$\phi'_i = \frac{\phi_{i+1} - \phi_{i-1}}{2\Delta x} + \frac{R_2}{2\Delta x} \tag{5.10a}$$

or

$$\phi'_i = \frac{\phi_{i+1} - \phi_{i-1}}{2\Delta x} + 0\left(\Delta x^2\right) \tag{5.10b}$$

or

$$\phi_i' \cong \frac{\phi_{i+1} - \phi_{i-1}}{2\Delta x} \tag{5.10c}$$

This approximation is called the ***central difference approximation*** as it expresses the derivative at the point (i, j) in terms of a forward point $(i+1, j)$ and a backward point $(i-1, j)$ (Figure 5.1c). Basically, the derivative at x_i is assumed to be equal to the slope of the straight line joining points x_{i-1} and x_{i+1}. It can be noted that this central difference approximation of the first derivative has an error of the order of $o(\Delta x)^2$ as compared to the forward and backward differences that have an error of $o(\Delta x)$. The level of accuracy depends on the number of terms retained in the Taylor series expansion during the derivation of these formulas. By including higher-order terms of the Taylor series we can develop more accurate approximations of the first derivative. In a similar manner, we can also develop approximations for second- and higher-order derivatives. Also, the above equations can be derived for derivatives in other space and time coordinates in a similar manner. A brief summary of these formulas is given in Table 5.1.

The list of derivatives given in Table 5.1 uses values at only two, three or five grid points. We generally try to use two- or three-point formulas to represent partial derivatives in the governing equations of fluid flow and heat transfer. Finite difference approximations for derivatives involving more than three grid points are used when higher accuracy is required by reducing the truncation error.

Discretization Equations. We will describe the procedure for deriving the discretization equation using a two-dimensional general governing equation of the form

$$\frac{\partial}{\partial x}\left(\Gamma_x \frac{\partial \phi}{\partial x}\right) + \frac{\partial}{\partial y}\left(\Gamma_y \frac{\partial \phi}{\partial y}\right) + S = 0 \tag{5.11a}$$

and

$$\frac{\partial^2 \phi}{\partial x^2} + \frac{\partial^2 \phi}{\partial y^2} + \frac{5}{\Gamma} = 0 \tag{5.11b}$$

for isotropic materials with

$$\Gamma_x = \Gamma_y = \Gamma$$

This equation contains diffusion terms and a source term. Such a mathematical equation represents a mathematical statement of many different physical processes including heat conduction, mass transfer, and fully developed flows in channels. In order to derive the discretization equation for any interior nodes (i, j) in the solution domain, we replace the derivatives by a

Table 5.1 Two, Three and Five Point Finite Difference Formulas for First and Second Derivatives

Derivatives	Number of Points	Difference Formula	Order of Error
		First derivative	
Forward difference			
$\left.\frac{d\phi}{dx}\right)_i$	2	$\phi_i' = \frac{\phi_{i+1} - \phi_i}{\Delta x}$	$O(\Delta x)$
$\left.\frac{d\phi}{dx}\right)_i$	3	$\phi_i' = \frac{-\phi_{i+2} + 4\phi_{i+1} - 3\phi_i}{2\Delta x}$	$O(\Delta x^2)$
Backward difference			
$\left.\frac{d\phi}{dx}\right)_i$	2	$\phi_i' = \frac{\phi_i - \phi_{i-1}}{\Delta x}$	$O(\Delta x)$
$\left.\frac{d\phi}{dx}\right)_i$	3	$\phi_i' = \frac{3\phi_i - 4\phi_{i-1} + \phi_{i-2}}{2\Delta x}$	$O(\Delta x^2)$
Central difference			
$\left.\frac{\partial\phi}{\partial x}\right)_i$	2	$\phi_i' = \frac{\phi_{i+1} - \phi_{i-1}}{2\Delta x}$	$O(\Delta x^2)$
$\left.\frac{\partial\phi}{\partial x}\right)_i$	4	$\phi_i' = \frac{-\phi_{i+2} + 8\phi_{i+1} - 8\phi_{i-1} - \phi_{i-2}}{12\Delta x}$	$O(\Delta x^4)$
		Second derivative	
Forward difference			
$\frac{d^2\phi}{dx^2}$	3	$\phi_i'' = \frac{\phi_{i+2} - 2\phi_{i+1} + \phi_i}{\Delta x^2}$	$O(\Delta x)$
$\frac{d^2\phi}{dx^2}$	5	$\phi_i'' = \frac{-\phi_{i+3} + 4\phi_{i+2} - 5\phi_{i+1} + 2\phi_i}{\Delta x^2}$	$O(\Delta x^2)$
Backward difference			
$\frac{d^2\phi}{dx^2}$	3	$\phi_i'' = \frac{\phi_i - 2\phi_{i-1} + \phi_{i-2}}{\Delta x^2}$	$O(\Delta x)$
$\frac{d^2\phi}{dx^2}$	5	$\phi_i'' = \frac{2\phi_i - 5\phi_{i-1} + 4\phi_{i-2} - \phi_{i-3}}{\Delta x^2}$	$O(\Delta x^2)$
Central difference			
$\frac{d^2\phi}{dx^2}$	3	$\phi_i'' = \frac{\phi_{i+1} - 2\phi_i + \phi_{i-1}}{\Delta x^2}$	$O(\Delta x^2)$
$\frac{d^2\phi}{dx^2}$	5	$\phi_i'' = \frac{-\phi_{i+2} + 16\phi_{i+1} - 30\phi_i + 16\phi_{i-1} - \phi_{i-2}}{12\Delta x^2}$	$O(\Delta x^4)$

suitable finite difference formula given in Table 5.1. For example, the second derivative at the node (i, j) can first be approximated by using a two-point central difference formula for the first derivatives

$$\frac{\partial}{\partial x}\left(\Gamma_x \frac{\partial \phi}{\partial x}\right)\bigg|_{i,j} \approx \frac{\Gamma_x \frac{\partial \phi}{\partial x}\Big)_{i+\frac{1}{2},j} - \Gamma_x \frac{\partial \phi}{\partial x}\Big)_{i-\frac{1}{2},j}}{\Delta x} \tag{5.12a}$$

$$\frac{\partial}{\partial y}\left(\Gamma_y \frac{\partial \phi}{\partial y}\right)\bigg|_{i,j} \approx \frac{\Gamma_y \frac{\partial \phi}{\partial y}\Big)_{i,j+\frac{1}{2}} - \Gamma_y \frac{\partial \phi}{\partial y}\Big)_{i,j-\frac{1}{2}}}{\Delta y} \tag{5.12b}$$

Next, the first derivatives located at

$$i+\tfrac{1}{2}, j, \quad i-\tfrac{1}{2}, j, \quad i, j=\tfrac{1}{2} \text{ and } i, j-\tfrac{1}{2}$$

are replaced by two-point forward difference formulas

$$\Gamma_x \frac{\partial \phi}{dx}\Big)_{i+\frac{1}{2},j} \cong \Gamma_x \frac{\phi_{i+1,i} - \phi_{i,j}}{\Delta x} \tag{5.13a}$$

$$\Gamma_x \frac{\partial \phi}{dx}\Big)_{i-\frac{1}{2},j} \cong \Gamma_x \frac{\phi_{i,i} - \phi_{i-1,j}}{\Delta x} \tag{5.13b}$$

$$\Gamma_y \frac{\partial \phi}{\partial y}\Big)_{i,j+\frac{1}{2}} \cong \Gamma_y \frac{\phi_{i,j+1} - \phi_{i,j}}{\Delta y} \tag{5.13c}$$

$$\Gamma_y \frac{\partial \phi}{\partial y}\Big)_{i,j-\frac{1}{2}} \cong \Gamma_y \frac{\phi_{i,j} - \phi_{i,j-1}}{\Delta y} \tag{5.13d}$$

where Γ_x and Γ_y are assumed to be constant.

Substituting Equations 5.12 and 5.13 into Equation 5.11a, we have

$$\frac{\Gamma_x\left[\frac{\phi_{i+1,j} - \phi_{i,j}}{\Delta x}\right] - \Gamma_x\left[\frac{\phi_{i,j} - \phi_{i-1,j}}{\Delta x}\right]}{\Delta x} + \frac{\Gamma_y\left[\frac{\phi_{i,j+1} - \phi_{i,j}}{\Delta y}\right] - \Gamma_y\left[\frac{\phi_{i,j} - \phi_{i,j-1}}{\Delta y}\right]}{\Delta y} + S = 0$$

$$\frac{\Gamma_x(\phi_{i+1,j} - 2\phi_{i,j} + \phi_{i-1,j})}{\Delta x^2} + \frac{\Gamma_y(\phi_{i,j+1} - 2\phi_{i,j} + \phi_{i,j-1})}{\Delta y^2} + S = 0 \tag{5.14a}$$

or

$$\frac{\Gamma_x}{\Delta x^2}\left(\phi_{i+1,j}+\phi_{i-1,j}\right)+\frac{\Gamma_y}{(\Delta y)^2}\left(\phi_{i,j+1}+\phi_{i,j-1}\right)-\left(\frac{2\Gamma_x}{\Delta x^2}+\frac{2\Gamma_y}{(\Delta y)^2}\right)\phi_{i,j}+S=0 \tag{5.14b}$$

For isotropic materials with $\Gamma_x = \Gamma_y = \Gamma$, we get

$$\left(\phi_{i+1,j}+\phi_{i-1,j}\right)+\left(\frac{\Delta x}{\Delta y}\right)^2\left(\phi_{i,j+1}+\phi_{i,j-1}\right)-2\left[1+\left(\frac{\Delta x}{\Delta y}\right)^2\right]\phi_{i,j}+\frac{S}{\Gamma}(\Delta x)^2 \tag{5.15}$$

Note that we can also derive the discretization Equation 5.15 directly by replacing the second derivative terms in Equation 5.11b by the central difference formula given in Table 5.1 as

$$\left.\frac{\partial^2\phi}{\partial x^2}\right)_{i,j}\approx\frac{\phi_{i+1,j}-2\phi_{i,j}+\phi_{i-1,j}}{(\Delta x)^2} \tag{5.16a}$$

$$\left.\frac{\partial^2\phi}{\partial y^2}\right)_{i,j}=\frac{\phi_{i,j+1}-2\phi_{i,j}+\phi_{i,j-1}}{(\Delta y)^2} \tag{5.16b}$$

Substituting Equation 5.16 into Equation 5.11b, we get

$$\frac{\phi_{i+1,j}-2\phi_{i,j}+\phi_{i-1,j}}{(\Delta x)^2}+\frac{\phi_{i,j+1}-2\phi_{i,j}+\phi_{i,j-1}}{(\Delta y)^2}+\frac{S}{\Gamma}=0$$

$$\phi_{i+1,j}+\phi_{i-1,j}+\left(\frac{\Delta x}{\Delta y}\right)^2\left[\phi_{i,j+1}+\phi_{i,j-1}\right]-2\phi_{i,j}\left[1+\left(\frac{\Delta x}{\Delta y}\right)^2\right]+\frac{S}{\Gamma}(\Delta x)^2=0 \tag{5.17}$$

If we choose $\Delta x = \Delta y$, then Equation 5.17 reduces to

$$4\phi_{i,j}=\phi_{i+1,j}+\phi_{i-1,j}+\phi_{i,j+1}+\phi_{i,j-1}+\frac{S}{\Gamma}(\Delta x)^2 \tag{5.18}$$

Equations 5.17 and 5.18 represents finite difference approximations of the Poisson equation for any interior nodal points. This equation can be reduced to some limiting case problems as follows.

Two-dimensional steady-state problems without any source term

$$4\phi_{i,j} = \phi_{i+1,j} + \phi_{i-1,j} + \phi_{i,j+1} + \phi_{i,j-1} \tag{5.19}$$

One-dimensional steady state with the source term

$$2\phi_i = \phi_{l+1} + \phi_{i-1} + \frac{S}{\Gamma}(\Delta x)^2 \tag{5.20}$$

5.3.2 Control Volume Method

Another approach to derive the finite difference equation is by the control volume method, in which the solution domain is discretizized into a number of non-overlapping control volumes. Integration is then carried out over a control volume surrounding the grid point (i, j). The integral formulation will satisfy the overall balance of an extensive property such as energy in a heat-transfer problem. For each grid point, we consider a control volume bounded by the dashed lines located at

$$\left(i-\tfrac{1}{2}, j\right), \quad \left(i+\tfrac{1}{2}, j\right), \quad \left(i, j+\tfrac{1}{2}\right) \quad \text{and} \quad \left(i, j-\tfrac{1}{2}\right) \tag{5.21}$$

as shown in Figure 5.3.

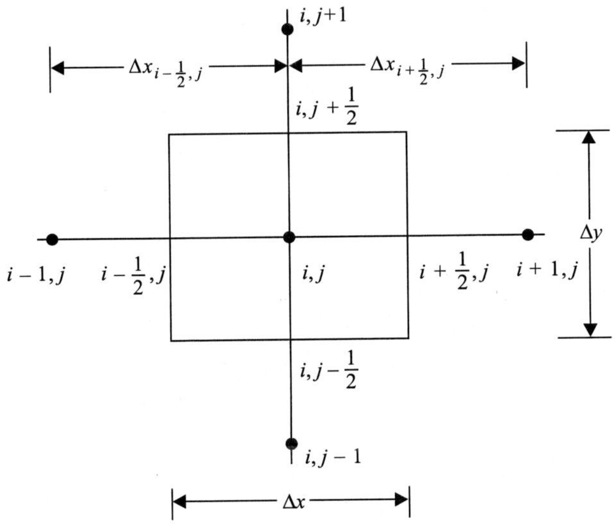

Figure 5.3 A typical control volume for an interior node

Integrating Equation 5.11 over the control volume, we get

$$\int_{i,\,j-\frac{1}{2}}^{i,\,j+\frac{1}{2}} \int_{i-\frac{1}{2},\,j}^{i+\frac{1}{2},\,j} \frac{\partial}{\partial x}\left(\Gamma_x \frac{\partial \phi}{\partial x}\right) dx\, dy + \int_{i,\,j-\frac{1}{2}}^{i,\,j+\frac{1}{2}} \int_{i-\frac{1}{2},\,j}^{i+\frac{1}{2},\,j} \frac{\partial}{\partial y}\left(\Gamma_y \frac{\partial \phi}{\partial y}\right) dx\, dy$$

$$+ \int_{i,\,j-\frac{1}{2}}^{i,\,j+\frac{1}{2}} \int_{i-\frac{1}{2},\,j}^{i+\frac{1}{2},\,j} S\, dx\, dy = 0 \tag{5.22}$$

$$\left(\Gamma_x \frac{\partial \phi}{\partial x}\right)_{i+\frac{1}{2},\,j} \Delta y - \left(\Gamma_x \frac{\partial \phi}{\partial x}\right)_{i-\frac{1}{2},\,j} \Delta y + \left(\Gamma_y \frac{\partial \phi}{\partial y}\right)_{i,\,j+\frac{1}{2}} \Delta x - \left(\Gamma_y \frac{\partial \phi}{\partial y}\right)_{i,\,j-\frac{1}{2}}$$

$$\times\, \Delta x + \int_{i,\,j-\frac{1}{2}}^{i,\,j+\frac{1}{2}} \int_{i-\frac{1}{2},\,j}^{i+\frac{1}{2}} S\, dx\, dy = 0 \tag{5.23}$$

In the next step the derivative terms are approximated with a profile assumption using an interpolation formula. Two very common profile assumptions are a step-wise constant profile and a piece-wise linear profile. Figure 5.4 shows such profiles for a variable ϕ in the x direction.

A step-wise constant profile, shown in Figure 5.4(a), can be used for approximating material transport or thermo-physical properties such as Γ and also for a dependent variable ϕ. For example, the temperature, T, heat generation, and thermal conductivity, k, at a node i can be assumed to be a constant value over the entire control volume that surrounds the node i. A piece-wise linear profile, shown in Figure 5.4(b), is suitable for approximating a flux quantity that is continuous across the control volume surface and involves a first derivative. For example, the heat transfer rate or mass transfer rate functions involve first derivatives that have to be continuous across the control volume surfaces.

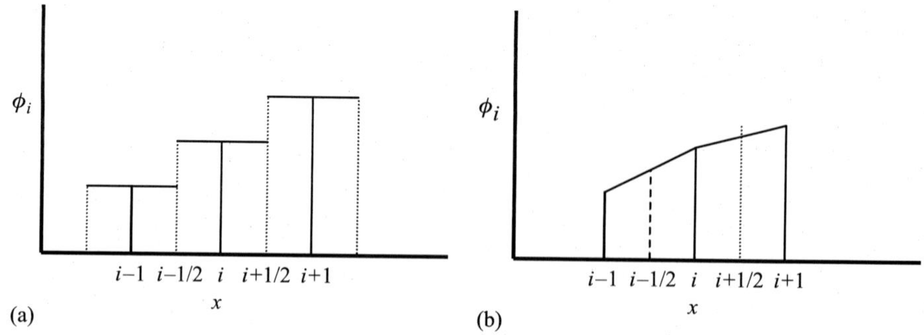

Figure 5.4 Profile assumptions in a control volume: (a) step-wise constant profile; (b) piece-wise linear profile

Considering a step-wise constant profile for the source term we can express the source term integral as

$$\int_{i,\, j-\frac{1}{2}}^{i,\, j+\frac{1}{2}} \int_{i-\frac{1}{2},\, j}^{i+\frac{1}{2},\, j} S\, dx\, dy = \overline{S}\, \Delta x\, \Delta y \tag{5.24}$$

where $\overline{S}$ is the mean or average heat generation rate in the control volume.

Also, considering a piece-wise constant profile for the transport properties Γ and a piece-wise linear profile to approximate the first derivative terms, we can express the flux quantities across the control volume surfaces in Equation 5.23 as

$$\left.\Gamma_x \frac{\partial \phi}{\partial x}\right)_{i+\frac{1}{2},\, j} \approx \frac{\Gamma_{xi+\frac{1}{2}}\left(\phi_{i+1,\, j} - \phi_{i,\, j}\right)}{\Delta x_{i+\frac{1}{2}}} \tag{5.25a}$$

$$\left.\Gamma_x \frac{\partial \phi}{\partial x}\right)_{i-\frac{1}{2},\, j} \approx \frac{\Gamma_{xi-\frac{1}{2}}\left(\phi_{i,\, j} - \phi_{i-1,\, j}\right)}{\Delta x_{i-\frac{1}{2}}} \tag{5.25b}$$

$$\left.\Gamma_y \frac{\partial \phi}{\partial y}\right)_{i,\, j+\frac{1}{2}} = \frac{\Gamma_{yj+\frac{1}{2}}\left(\phi_{i,\, j+1} - \phi_{i,\, j}\right)}{\Delta y_{j+\frac{1}{2}}} \tag{5.25c}$$

$$\left.\Gamma_y \frac{\partial \phi}{\partial y}\right)_{i,\, j-\frac{1}{2}} = \frac{\Gamma_{yj-\frac{1}{2}}\left(\phi_{i,\, j} - \phi_{i,\, j-1}\right)}{\Delta y_{j-\frac{1}{2}}} \tag{5.25d}$$

Substituting these approximations (Equations 5.24 and 5.25) into Equation 5.23, we get

$$\frac{\Gamma_{xi+\frac{1}{2}}\left(\phi_{i+1,\, j} - \phi_{i,\, j}\right)}{\Delta x_{i+\frac{1}{2}}} \Delta y - \frac{\Gamma_{xi-\frac{1}{2}}\left(\phi_{i,\, j} - \phi_{i-1,\, j}\right)}{\Delta x_{i-\frac{1}{2}}} \Delta y + \frac{\Gamma_{yj+\frac{1}{2}}\left(\phi_{i,\, j+1} - \phi_{i,\, j}\right)}{\Delta y_{j+\frac{1}{2}}} \Delta x$$
$$- \frac{\Gamma_{yj-\frac{1}{2}}\left(\phi_{i,\, j} - \phi_{i,\, j-1}\right)}{\Delta y_{j-\frac{1}{2}}} \Delta x + \overline{S}\, \Delta x\, \Delta y = 0 \tag{5.26}$$

Rearranging, we get the finite difference equation as

$$\frac{\Gamma_{xi+\frac{1}{2}}}{\Delta x_{i+\frac{1}{2}}} \Delta y \phi_{i+1} + \frac{\Gamma_{xi-\frac{1}{2}}}{\Delta x_{i-\frac{1}{2}}} \Delta y \phi_{i-1} + \frac{\Gamma_{yj+\frac{1}{2}}}{\Delta y_{j+\frac{1}{2}}} \Delta x \phi_{j+1} + \frac{\Gamma_{yj-\frac{1}{2}}}{\Delta y_{j-\frac{1}{2}}} \Delta x \phi_{j-1}$$
$$- \left(\frac{\Gamma_{xi+\frac{1}{2}}}{\Delta x_{i+\frac{1}{2}}} \Delta y + \frac{\Gamma_{xi-\frac{1}{2}}}{\Delta x_{i-\frac{1}{2}}} \Delta y + \frac{\Gamma_{yj+\frac{1}{2}}}{\Delta y_{j+\frac{1}{2}}} \Delta x + \frac{\Gamma_{yj-\frac{1}{2}}}{\Delta y_{j-\frac{1}{2}}} \Delta x \right) \phi_{i,\, j} + \overline{S}\, \Delta x\, \Delta y = 0$$
$$\tag{5.27a}$$

Equation 5.27a can be written in a compact form as

$$a_{i,j}\phi_{i,j} = a_{i+1,j}\phi_{i+1,j} + a_{i-1,j}\phi_{i-1,j} + a_{i,j+1}\phi_{i,j+1} + a_{i,j-1}\phi_{i,j-1} + d \quad (5.27b)$$

where

$$a_{i,j} = a_{i+1,j} + a_{i-1,j} + a_{i,j+1} + a_{i,j-1} \quad (5.28a)$$

$$a_{i+1,j} = \frac{\Gamma_{xi+\frac{1}{2}}}{\Delta x_{i+\frac{1}{2}}}\Delta y \quad (5.28b)$$

$$a_{i-1,j} = \frac{\Gamma_{xi-\frac{1}{2}}}{\Delta x_{i-\frac{1}{2}}}\Delta y \quad (5.28c)$$

$$a_{i,j+1} = \frac{\Gamma_{yj+\frac{1}{2}}}{\Delta y_{j+\frac{1}{2}}}\Delta x \quad (5.28d)$$

$$a_{i,j-1} = \frac{\Gamma_{yj-\frac{1}{2}}}{\Delta y_{j-\frac{1}{2}}}\Delta x \quad (5.28e)$$

$$d = \overline{S}\,\Delta x\,\Delta y \quad (5.28f)$$

For $\Gamma_x = \Gamma_y = \Gamma$ and for uniform grid size distributions in x and y directions, i.e. $\Delta x_{i+\frac{1}{2}} = \Delta x_{i-\frac{1}{2}} = \Delta x$ and $\Delta y_{j+\frac{1}{2}} = \Delta y_{j-\frac{1}{2}} = \Delta y$, Equation 5.27b reduces to

$$a_{i,j}\phi_{i,j} = a_{i+1,j}\phi_{i+1,j} + a_{i-1,j}\phi_{i-1,j} + a_{i,j+1}\phi_{i,j+1} + a_{i,j-1}\phi_{i,j-1} + d \quad (5.29)$$

where the coefficients are given as

$$a_{i,j} = a_{i+1,j} + a_{i-1,j} + a_{i,j+1} + a_{i,j-1} = \frac{2\Delta y}{\Delta x}\left(1 + \frac{\Delta x^2}{\Delta y^2}\right) \quad (5.30a)$$

$$a_{i+1,j} = \frac{1}{\Delta x}\Delta y \quad (5.30b)$$

$$a_{i-1,j} = \frac{1}{\Delta x}\Delta y \quad (5.30c)$$

$$a_{i,j+1} = \frac{1}{\Delta y}\Delta x \quad (5.30d)$$

$$a_{i,j-1} = \frac{1}{\Delta y}\Delta x \quad (5.30e)$$

$$d = \frac{\overline{S}\,\Delta x\,\Delta y}{\Gamma} \quad (5.30f)$$

For $\Delta x = \Delta y$, Equation 5.30 reduces to

$$a_{i,j} = 4$$
$$a_{i+1,j} = a_{i-1,j} = a_{i,j+1} = a_{i,j-1} = 1$$
$$d = \frac{\overline{s}\Delta x^2}{\Gamma} \tag{5.31}$$

and discretization Equation 5.29 for interior nodes becomes

$$4\phi_{i,j} = \phi_{i+1,j} + \phi_{i-1,j} + \phi_{i,j+1} + \phi_{i,j-1} + \frac{\overline{S}\Delta x^2}{\Gamma} \tag{5.32}$$

It can be mentioned here that the assumption of the piecewise linear profile is exactly the same as that made in the derivation of two-point approximate formulas for derivatives using Taylor's series. However, the primary advantage in control volume formulation is that we can choose many other profile assumptions within the control volume. Also, we are free to make different profile assumptions for different derivative terms in the governing equations and for different quantities such as material properties, Γ, and source term, *S*.

5.4 ONE-DIMENSIONAL STEADY STATE DIFFUSION

For a one-dimensional steady state diffusion with uniform source, the appropriate governing equation is

$$\frac{d}{dx}\left(\Gamma_x \frac{d\phi}{dx}\right) + \overline{S} = 0 \tag{5.33}$$

The discretization equation for interior nodes can be written from Equation 5.29 as

$$a_i\phi_i = a_{i+1}\phi_{i+1} + a_{i-1}\phi_{i-1} + d \tag{5.34}$$

where

$$a_i = a_{i+1} + a_{i-1} \tag{5.35a}$$

$$a_{i+1} = \frac{\Gamma_{xi+\frac{1}{2}}}{\Delta x_{i+\frac{1}{2}}} \tag{5.35b}$$

$$a_{i-1} = \frac{\Gamma_{xi-\frac{1}{2}}}{\Delta x_{i-\frac{1}{2}}} \tag{5.35c}$$

$$d = \overline{S}\Delta x \tag{5.35d}$$

Example 5.1: One-dimensional Steady State Conduction: Consider a one-dimensional steady state conduction without heat generation in a plane slab, shown in the figure below. The boundary surfaces at $x=0$ and $x=L$ are maintained at constant temperatures T_0 and T_L respectively. Determine the temperature distribution in the slab and heat transfer rates at the surfaces. Use $k = 120\,\mathrm{W/cm \cdot C}$, $L = 10\,\mathrm{cm}$, $T_0 = 100°\mathrm{C}$, $T_L = 200°\mathrm{C}$.

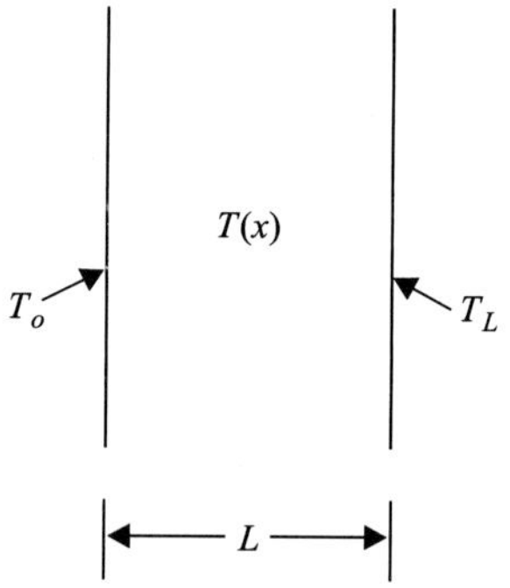

The mathematical statement of the problem is

Governing equation

$$\frac{d}{dx}\left(k\frac{dT}{dx}\right) = 0 \tag{E.5.1.1}$$

Boundary condition

1. $x = 0, \qquad T = T_0$ (E.5.1.2)
2. $x = L, \qquad T = T_L$ (E.5.1.3)

Let us obtain the numerical solution with the given data.

Solution: In the ***first step***, let us discretize the domain using nine uniform divisions as shown in the figure.

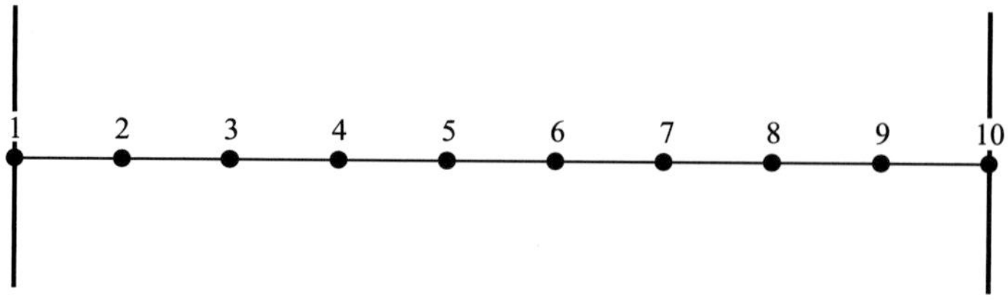

With this grid size distribution, we have $\Delta x_{i+\frac{1}{2}} = \Delta x_{i-\frac{1}{2}} = 1.111\,\mathrm{cm}$ and number of grid points $N = 10$ with grids $i = 1$ and $i = 11$ located at $x = 0$ and $x = 10$. In the ***second step***, we select an appropriate discretization equation.

As the boundary temperature is known, we set

$$T_1 = T_0 = 100°\text{C} \quad \text{and} \quad T_{10} = T_L = 200°\text{C}$$

The rest of the nodes are interior nodes and we can apply the discretization Equation 5.34

$$a_i T_i = a_{i+1} T_{i+1} + a_{i-1} T_{i-1} + d \tag{E.5.1.4}$$

where

$$a_i = \frac{2k}{\Delta x} = 216$$

$$a_{i+1} = a_{i-1} = \frac{k}{\Delta x} = 108$$

$$d = \dot{q}\,\Delta x = 0$$

Substituting these coefficients into Equation E.5.1.4, we have

$$216T_i = 108T_{i+1} + 108T_{i-1} \tag{E.5.1.5a}$$

Applying this equation succeeding to all interior nodes 2, 3, 4 and 5, we get

Node 2 $216\text{T}_2 = 108T_3 + T_1$

$$2T_2 - T_3 = 100 \tag{E.5.1.5b}$$

Node 3 $216T_3 = 108T_4 + 108T_2$

$$-108T_2 + 216T_3 - 108T_4 = 0 \tag{E.5.1.5c}$$

Node 4 $216T_4 = 108T_5 + 108T_3$

$$-108T_3 + 216T_4 - 108T_5 = 0 \tag{E.5.1.5d}$$

Node 5 $216T_5 = 108T_6 + 108T_4$

$$-108T_4 + 216T_s - 108T_6 = 0 \tag{E.5.1.5e}$$

Node 6 $216T_6 = 108T_7 + 108T_5$

$$-108T_5 + 216T_6 - 108T_7 = 0 \tag{E.5.1.5f}$$

Node 7 $216T_7 = 108T_8 + 108T_6$

$$-108T_6 + 216T_7 - 108T_8 = 0 \tag{E.5.1.5g}$$

Node 8 $216T_8 = 108T_9 + 108T_7$

$$-108T_7 + 216T_8 - 108T_9 = 0 \tag{E.5.1.5h}$$

Node 9 $216T_9 = 108T_{10} + 108T_8$

$$-108T_8 + 216T_9 = 108 \times 200 \tag{E.5.1.5i}$$

The system of algebraic Equation E.5.1.5a to Equation E.5.1.5g can be written in matrix form as

$$\begin{bmatrix} 2 & -1 & 0 & 0 & 0 & 0 & 0 & 0 \\ -1 & 2 & -1 & 0 & 0 & 0 & 0 & 0 \\ 0 & -1 & 2 & -1 & 0 & 0 & 0 & 0 \\ 0 & 0 & -1 & 2 & -1 & 0 & 0 & 0 \\ 0 & 0 & 0 & -1 & 2 & -1 & 0 & 0 \\ 0 & 0 & 0 & 0 & -1 & 2 & -1 & 0 \\ 0 & 0 & 0 & 0 & 0 & -1 & 2 & -1 \\ 0 & 0 & 0 & 0 & 0 & 0 & -1 & 2 \end{bmatrix} \begin{Bmatrix} T_2 \\ T_3 \\ T_4 \\ T_4 \\ T_6 \\ T_7 \\ T_8 \\ T_9 \end{Bmatrix} = \begin{Bmatrix} 100 \\ 0 \\ 0 \\ 0 \\ 0 \\ 0 \\ 0 \\ 200 \end{Bmatrix}$$

In the ***final step***, we choose an appropriate solver for this system of equations. Using the Gaussian elimination program, we get the numerical solution as

$$\begin{aligned} T_2 &= 111.11°\text{C} & T_3 &= 122.22°\text{C} \\ T_4 &= 133.33°\text{C} & T_5 &= 144.44°\text{C} \\ T_6 &= 155.55°\text{C} & T_7 &= 166.66°\text{C} \\ T_8 &= 177.77°\text{C} & T_9 &= 188.88°\text{C} \end{aligned}$$

The heat transfer rate at $x = 0$ is estimated as

$$q''_{x=0} = k\frac{T_1 - T_2}{\Delta x} = 120\frac{100 - 111.11}{1.111}$$
$$q''_{x=0} = -1199.89\ \text{W/cm}^2$$

The heat transfer rate at $x = L$ is estimated as

$$q''_{x=L} = k\frac{T_9 - T_{10}}{\Delta x} = 120 \times \frac{188.88 - 200}{1.111}$$
$$q''_{x=L} = -1200.96\ \text{W/cm}^2$$

The negative sign indicates that the heat transfer rate is in the negative x-direction.

Example 5.2: Fully Developed Flow in a Channel: Consider the fully developed laminar steady flow of Newtonian viscous fluid through a two-dimensional channel formed by two infinite parallel plates as shown below. The lower plate is stationary and the upper plate is moving in the x direction with a constant speed, U_w, in the positive x-direction.

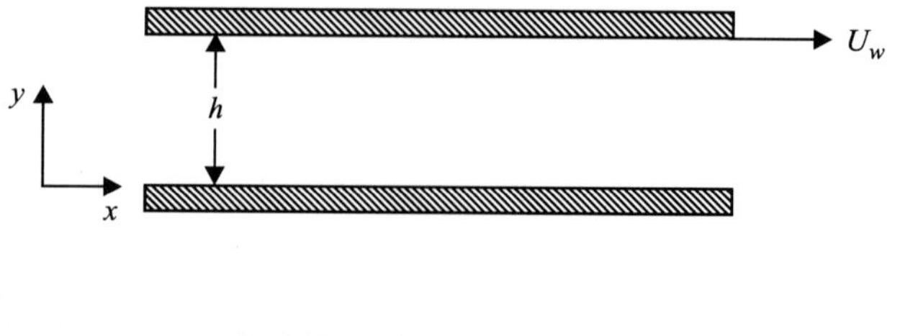

The mathematical statement of the problem can be derived from mass and momentum equations of fluid flow and it is given as follows.

Governing equation

$$\frac{d^2u}{dy^2} = \frac{1}{\mu}\frac{dp}{dx} \tag{E.5.2.1}$$

Boundary conditions

1. $y = 0, \quad u = 0$ (E.5.2.2)
2. $y = h, \quad u = U_w$ (E.5.2.3)

Determine (a) the velocity distribution, (b) volume flow rate, and (c) the shear stress at the plate and use the following data for the calculation: $h = 6.00$ mm, $\mu = 0.02$ kg/m · s, $\partial P/\partial x = \overline{S} = -500$ Pa/m, and $U_w = 10$ mm/s.

Solution: In the ***first step***, let us discretize the domain using six uniform divisions as shown in the figure.

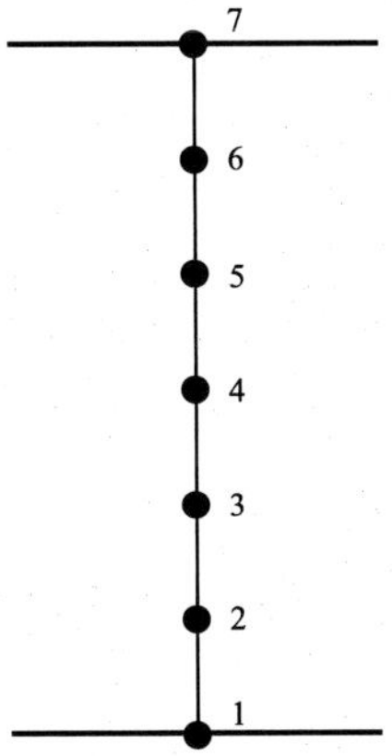

With this grid size distribution, we have $\Delta y_{i+\frac{1}{2}} = \Delta y_{i-\frac{1}{2}} = \Delta y = 0.001$ m and number of grid points $N = 7$ with grids $i = 1$ and $i = 7$ located at $y = 0$ and $y = 6$ mm, respectively. In the ***second step***, we select an appropriate discretization equation. As the boundary velocities are known, we set

$$u_1 = 0 = 0 \quad \text{and} \quad u_7 = U_w = 10 \text{ mm/s}$$

The rest of the nodes are interior nodes and we select the discretization Equation 5.34 for a one-dimensional problem as

$$2u_{i,j} = u_{i+1} + u_{i-1} + \frac{\overline{S}\Delta y^2}{\mu} \tag{E.5.2.4}$$

where

$$S = -\frac{\overline{S}\Delta y^2}{\mu} = -\frac{-500\,\text{N/m}^2/\text{m}\cdot(0.001)^2\,\text{m}^2}{0.02\,(\text{kg}\cdot\text{m/s}^2)} = 0.025\,\text{m/s} \qquad \text{(E.5.2.5)}$$

Substituting these coefficients into the equation, we have

$$2u_i = u_{i+1} + u_{i-1} + S \qquad \text{(E.5.2.6)}$$

Applying this equation succeeding to all interior nodes 2, 3, 4, 5, 6 and 7, we get

Node 2 $2u_2 = u_3 + u_1 + S$

$2u_2 - u_3 = S + u_1$

Node 3 $2u_3 = u_4 + u_2 + S$

$-u_2 + 2u_3 - u_4 = S$

Node 4 $2u_4 = u_5 + u_3 + S$

$-u_3 + 2u_4 - u_5 = S$

Node 5 $2u_5 = u_6 + u_4 + S$

$-u_4 + 2u_5 - u_6 = S$

Node 6 $2u_6 = u_7 + u_5$

$-u_5 + 2u_6 = u_7 + S$

Assembly of all nodal equations form the system of equations as

$$\begin{bmatrix} 2 & -1 & 0 & 0 & 0 \\ -1 & 2 & -1 & 0 & 0 \\ 0 & -1 & 2 & -1 & 0 \\ 0 & 0 & -1 & 2 & -1 \\ 0 & 0 & 0 & -1 & 2 \end{bmatrix} \begin{Bmatrix} u_2 \\ u_3 \\ u_4 \\ u_5 \\ u_6 \end{Bmatrix} = \begin{Bmatrix} S + u_1 \\ S \\ S \\ S \\ S + u_7 \end{Bmatrix} \qquad \text{(E.5.2.7)}$$

Substituting $u_1 = 0$, $u_7 = U_w = 0.01$ m/s and $S = 0.025$ m/s, we get

$$\begin{bmatrix} 2 & -1 & 0 & 0 & 0 \\ -1 & 2 & -1 & 0 & 0 \\ 0 & -1 & 2 & -1 & 0 \\ 0 & 0 & -1 & 2 & -1 \\ 0 & 0 & 0 & -1 & 2 \end{bmatrix} \begin{Bmatrix} u_2 \\ u_3 \\ u_4 \\ u_5 \\ u_6 \end{Bmatrix} = \begin{Bmatrix} 0.025 \\ 0.025 \\ 0.025 \\ 0.025 \\ 0.035 \end{Bmatrix}$$

In the ***final step***, we choose an appropriate solver for this system of equations. The coefficient is tri-diagonal and so we can use the tri-diagonal matrix algorithm to get the numerical solution

$$\begin{Bmatrix} u_2 \\ u_3 \\ u_4 \\ u_5 \\ u_6 \end{Bmatrix} = \begin{Bmatrix} 0.064\,166 \\ 0.103\,333 \\ 0.117\,500 \\ 0.106\,666 \\ 0.070\,833 \end{Bmatrix}$$

5.5 VARIABLE SOURCE TERM

So far in our derivation of discretization equations we have treated the source term S as a constant. We now consider the source term to be a function of the dependent variable, i.e., $S(\phi)$. In order to include this dependence in our formulation, let us assume that for the grid shown in Figure 5.5 ϕ_i can be assumed to be the representative dependent variable of the control volume surrounding the grid point i.

If we like to solve the set of discretization equations by the techniques of linear algebraic equations, then it is essential to express $S(\phi)$ by a linear relation. The procedure for linearization will be discussed in a later section. Here we assume a linear variation of average heat source $\overline{S}$ with ϕ_i as

$$\overline{S} = S_0 + S_1\phi_i \tag{5.36}$$

where S_0 and S_1 are constants in the linear expression.

Substituting Equation 5.36 into Equation 5.23, we get the discretization equation and the coefficients as

$$a_{i,j}\phi_{i,j} = a_{i+1,j}\phi_{i+1,j} + a_{i-1,j}\phi_{i-1,j} + a_{i,j+1}\phi_{i,j+1} + a_{i,j-1}\phi_{i,j-1} + d \tag{5.37}$$

where

$$a_{i+1,j} = \frac{\Gamma_{i+\frac{1}{2}}}{\Delta x_{i+\frac{1}{2}}}\Delta y \tag{5.38a}$$

$$a_{i-1,j} = \frac{\Gamma_{i-\frac{1}{2}}}{\Delta x_{i-\frac{1}{2}}}\Delta y \tag{5.38b}$$

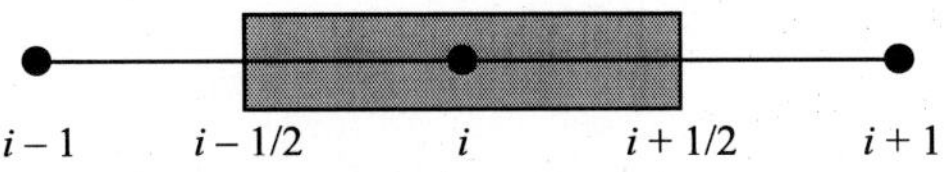

Figure 5.5 Control volume for an internal node

$$a_{i,\,j+1} = \frac{\Gamma_{j+\frac{1}{2}}}{\Delta y_{j+\frac{1}{2}}} \Delta x \tag{5.38c}$$

$$a_{i,\,j-1} = \frac{\Gamma_{j-\frac{1}{2}}}{\Delta y_{j-\frac{1}{2}}} \Delta x \tag{5.38d}$$

$$a_{i,\,j} = a_{i+1,\,j} + a_{i-1,\,j} + a_{i,\,j+1} + a_{i,\,j-1} - S_1\,\Delta x\,\Delta y \tag{5.38e}$$

$$d = S_0\,\Delta x\,\Delta y \tag{5.38f}$$

When the source form does not depend on the temperature, S_1 becomes zero and $S_0 = \overline{S}$, and Equation 5.38 reduces to Equation 5.28.

5.6 BOUNDARY CONDITIONS

When the boundary condition is of the ***first kind***, i.e., of ***constant surface value***, we can simply assign the known value to the boundary nodes and solve the system of equations as was done in the previous example. However, when the boundary conditions are of the ***second*** or ***third kind***, we need to use appropriate discretizations equations for boundary nodes, which are obtained by applying overall balance of conservation quantities such as mass, heat, and momentum, or applying control volume formulation over a half control volume surrounding the boundary nodes. Let us show the derivation of such boundary equations for a one-dimensional problem for which the boundary control volume is a half control volume surrounding boundary node i as shown in Figure 5.6.

Let us first consider the boundary nodes with constant surface heat flux in a one-dimensional problem. The governing equation is

$$\frac{d}{dx}\left(\Gamma\frac{d\phi}{dx}\right) + S = 0 \tag{5.39}$$

Boundary condition

$$\text{at } x = 0, \quad \left(-\Gamma\frac{\partial\phi}{\partial x} = f''_{ls}\right) \tag{5.40}$$

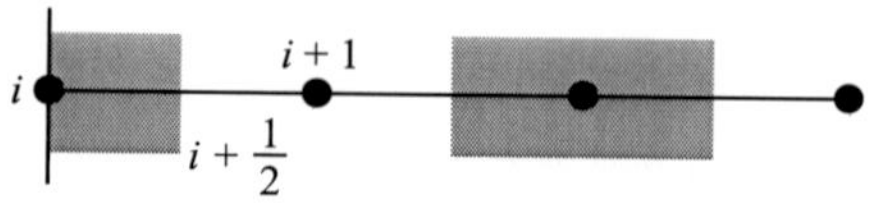

Figure 5.6 Half control volume for boundary nodes

where f_{ls}'' is constant surface flux at the left boundary surface. Integrating Equation 5.39 over the half control volume

$$\int_{i}^{i+\frac{1}{2}} \frac{d}{dx}\left(\Gamma \frac{d\phi}{dx}\right) dx + \int_{i}^{i+\frac{1}{2}} S\, dx = 0 \tag{5.41a}$$

or

$$\left[\left(\Gamma \frac{d\phi}{dx}\right)_{i+\frac{1}{2}} - \left(\Gamma \frac{d\phi}{dx}\right)_{i}\right] + \overline{S}\frac{\Delta x_{i+\frac{1}{2}}}{2} \tag{5.41b}$$

Let us assume the piecewise linear profile to approximate the temperature gradient at the face of the control volume as

$$\left(\frac{d\phi}{dx}\right)_{i+\frac{1}{2}} = \frac{(\phi_{i+1} - \phi_i)}{\Delta x_{i+\frac{1}{2}}} \tag{5.42}$$

The source term $\overline{S}$ is linearized with a stepwise profile for the dependent variable as

$$\overline{S} = S_0 + S_1\phi_i \tag{5.43}$$

The gradient at the boundary node is described by the constant flux boundary condition as

$$\left(-\Gamma \frac{d\phi}{dx}\right)_{i} = f_{ls}'' \tag{5.44}$$

Substituting Equations 5.42 to 5.44 into Equation 5.41

$$\frac{\Gamma_{i+\frac{1}{2}}(\phi_{i+1} - \phi_i)}{\Delta x_{i+\frac{1}{2}}} + f_{ls}'' + S_0\frac{\Delta x_{i+\frac{1}{2}}}{2} + S_1\phi_i\frac{\Delta x_{i+\frac{1}{2}}}{2} = 0 \tag{5.45}$$

Rearranging, we have the discretization equation for the boundary node as

$$\left(\frac{\Gamma_{i+\frac{1}{2}}}{\Delta x_{i+\frac{1}{2}}} - S_1\frac{\Delta x_{i+\frac{1}{2}}}{2}\right)\phi_i = \frac{\Gamma_{i+\frac{1}{2}}\phi_{i+1}}{\Delta x_{i+\frac{1}{2}}} + S_0\frac{\Delta x_{i+\frac{1}{2}}}{2} + f_{ls}'' \tag{5.46a}$$

or

$$a_i\phi_i = a_{i+1}\phi_{i+1} + d \tag{5.46b}$$

where

$$a_i = \left(\frac{\Gamma_{i+\frac{1}{2}}}{\Delta x_{i+\frac{1}{2}}} - S_1\frac{\Delta x_{i+\frac{1}{2}}}{2}\right) \tag{5.47a}$$

$$a_{i+1} = \frac{\Gamma_{i+\frac{1}{2}}}{\Delta x_{i+\frac{1}{2}}} \tag{5.47b}$$

$$d = S_0 \frac{\Delta x_{i+\frac{1}{2}}}{2} + f''_{ls} \tag{5.47c}$$

We can see that boundary flux term f''_{ls} has been lumped together with the source term d.

Special Cases

1. For ***zero surface flux or at a symmetric boundary*** $f''_{ls} = 0$, so the discretization equation for boundary nodes become

$$a_i \phi_i = a_{i+1} \phi_{i+1} + d \tag{5.48}$$

where

$$a_i = \frac{\Gamma_{i+\frac{1}{2}}}{\Delta x_{i+\frac{1}{2}}} - S_1 \frac{\Delta x_{i+\frac{1}{2}}}{2} \tag{5.49a}$$

$$a_{i+1} = \frac{\Gamma_{i+\frac{1}{2}}}{\Delta x_{i+\frac{1}{2}}} \tag{5.49b}$$

$$d = S_0 \frac{\Delta x_{i+\frac{1}{2}}}{2} \tag{5.49c}$$

2. Another possible boundary condition is the boundary condition of the third kind or the mixed boundary conditions given as

$$-\Gamma \frac{d\phi}{dx} = h(\phi - \phi_\infty) \tag{5.50a}$$

or

$$f''_{ls} = h(\phi_\infty - \phi) \tag{5.51}$$

In heat transfer problems, such boundary conditions usually arise from convective boundary conditions such as

$$k \frac{dT}{dx} = h_c(\mathrm{T}_\infty - T_i) \tag{5.52}$$

or

$$q''_{ls} = h_c(T_\infty - T_i) \tag{5.53}$$

Now, if we substitute Equation 5.51 into Equation 5.48, we get the discretization equation for nonzero heat flux as

$$a_i\phi_i = a_{i+1}\phi_{i+1} + d \tag{5.54}$$

where

$$a_i = \frac{\Gamma_{i+\frac{1}{2}}}{\Delta x_{i+\frac{1}{2}}} - S_1\frac{\Delta x_{i+\frac{1}{2}}}{2} + h \tag{5.55a}$$

$$a_{i+1} = \frac{\Gamma_{i+\frac{1}{2}}}{\Delta x_{i+\frac{1}{2}}} \tag{5.55b}$$

$$d = s_0\frac{\Delta x_{i+\frac{1}{2}}}{2} + h\phi_\infty \tag{5.55c}$$

After we solve the grid-point values, the flux at the boundary can be checked by performing an overall integral balance as

$$f''_{ls} = \frac{\Gamma_{i+\frac{1}{2}}(\phi_i - \phi_{i+1})}{\Delta x_{i+\frac{1}{2}}} - (S_0 + S_1\phi_i)\frac{\Delta x_{i+\frac{1}{2}}}{2} \tag{5.56}$$

Similarly, we can derive the discretization equations for other boundary nodes and boundary conditions. A summary of these discretization equations is given in Table 5.2.

We can also use discretization, Equations 5.42 and 5.43, for problems when boundary heat flux is a nonlinear function of ϕ_i. In such cases, we express the nonlinear function in the form of Equation 5.43 through linearization techniques.

Example 5.3. One-dimension Steady State Fin: Consider a straight fin of uniform cross-sectional area A, length $L = 2.0\,\text{cm}$ and a thickness $t = 1.4\,\text{mm}$. The fin thermal conductivity is $k = 60\,\text{W/m}\cdot\text{C}$ and it is exposed to a convection environment at $T_\infty = 20°\text{C}$ and $h = 500\,\text{W/m}^2\cdot\text{C}$. The base of the fin is at a constant temperature $T_0 = 150°\text{C}$ and the tip of the fin is assumed to be convective.

Table 5.2 Summary of Finite Difference Formulas for Interior and Boundary Nodes in One-dimensional Steady State Problems

Nodes	Finite Difference Formula
Interior nodes	
(a) Variable transport property and non-uniform grid 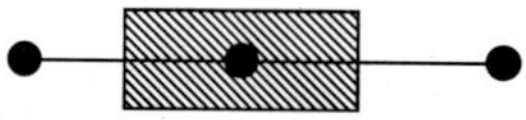	$a_i\phi_i = a_{i+1}\phi_{i+1} + a_{i-1}\phi_{i-1} + d$ $a_{i+1} = \Gamma_{i+\frac{1}{2}}/\Delta x_{i+\frac{1}{2}}, \quad a_{i-1} = \Gamma_{i-\frac{1}{2}}/\Delta x_{i-\frac{1}{2}}$ $a_i = a_{i+1} + a_{i-1}, \quad d = \bar{S}\Delta x$
(b) Constant transport property and uniform grid in x and y 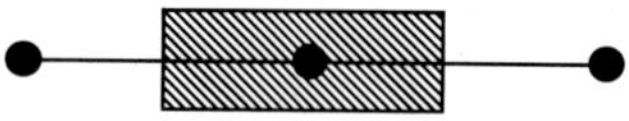	$a_i\phi_i = a_{i+1}\phi_{i+1} + a_{i-1}\phi_{i-1} + d$ $a_{i+1} = a_{i-1} = \Gamma/\Delta x$ $a_i = a_{i+1} + a_{i-1}$ $d = \bar{S}\Delta x$
(c) Variable transport property, variable source ($S = S_0 + S_1\, T_i$), and non-uniform grid 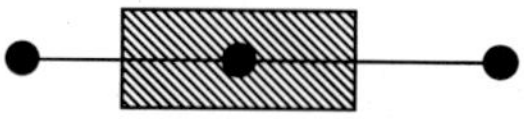	$a_i\phi_i = a_{i+1}\phi_{i+1} + a_{i-1}\phi_{i-1} + d$ $a_{i+1} = \Gamma_{i+\frac{1}{2}}/\Delta x_{i+\frac{1}{2}}, \quad a_w = \Gamma_{i-\frac{1}{2}}/\Delta x_{i-\frac{1}{2}}$ $a_i = a_{i+1} + a_{i-1} - S_1\Delta x$ $d = S_0\Delta x$
Boundary nodes	
(d) Generalized form 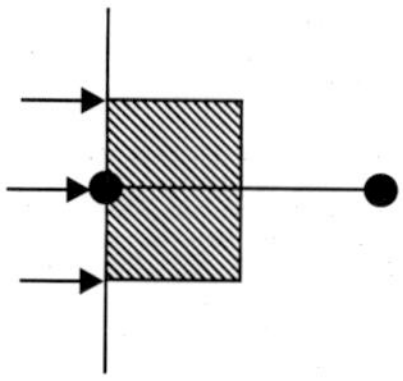	$a_i\phi_i = a_{i+1}\phi_{i+1} + d$ $a_{i+1} = \Gamma_{i+\frac{1}{2}}/\Delta x_{i+\frac{1}{2}}, \quad a_i = \left(\Gamma_{i+\frac{1}{2}}/\Delta x_{i+\frac{1}{2}}\right) - S_1\left(\Delta x_{i+\frac{1}{2}}/2\right)$ $d = S_0\left(\Delta x_{i+\frac{1}{2}}/2\right) + f''_{ls}$
(e) Zero flux or symmetric boundary: $f''_{ls} = 0$ 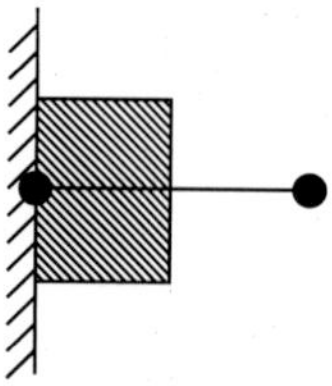	$a_i\phi_i = a_{i+1}\phi_{i+1} + d$ $a_{i+1} = \Gamma_{i+\frac{1}{2}}/\Delta x_{i+\frac{1}{2}}, \quad a_i = \left(\Gamma_{i+\frac{1}{2}}/\Delta x_{i+\frac{1}{2}}\right) - S_1\left(\Delta x_{i+\frac{1}{2}}/2\right)$ $d = S_0\left(\Delta x_{i+\frac{1}{2}}/2\right)$
(f) Variable surface flux: $f''_{ls} = f_0 + f_1\phi_i$ 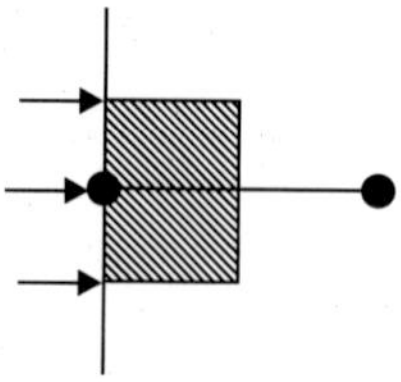	$a_i = a_{i+1}\phi_{i+1} + d$ $a_{i+1} = \Gamma_{i+\frac{1}{2}}/\Delta x_{i+\frac{1}{2}}, \quad a_i = \left(\Gamma_{i+\frac{1}{2}}/\Delta x_{i+\frac{1}{2}}\right) - S_1\left(\Delta x_{i+\frac{1}{2}}/2\right) - f_1,$ $d = S_0\left(\Delta x_{i+\frac{1}{2}}/2\right) + f_0$

Table 5.2 Continued

Nodes	Finite Difference Formula
(g) Mixed boundary condition: a special case of (c) 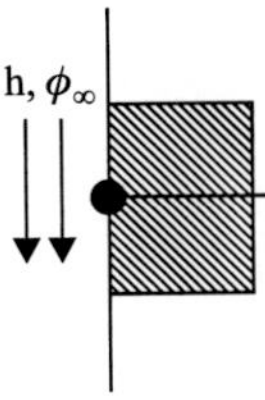	$a_i\phi_i = a_{i+1}\phi_{i+1} + d$ $a_{i+1} = \Gamma_{i+\frac{1}{2}}/\Delta x_{i+\frac{1}{2}}$ $a_i = \left(\Gamma_{i+\frac{1}{2}}/\Delta x_{i+\frac{1}{2}}\right) - S_1\left(\Delta x_{i+\frac{1}{2}}/2\right) + h$ $d = S_0\left(\Delta x_{i+\frac{1}{2}}/2\right) + hT_\infty$

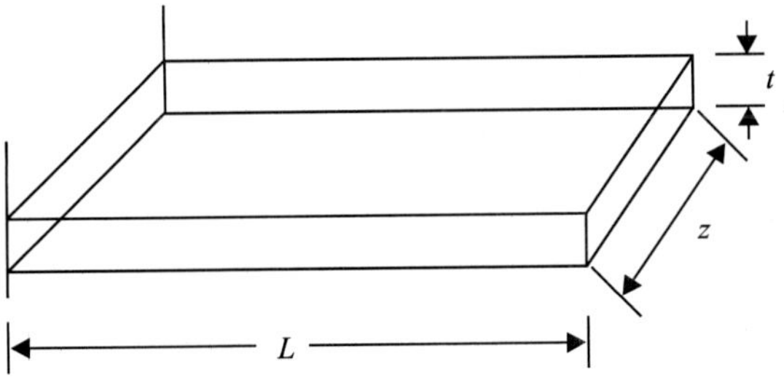

The mathematical statement of the problem is:

Governing equation

$$\frac{d}{dx}\left(k\frac{dT}{dx}\right) + \frac{hP}{A}(T_\infty - T) = 0 \qquad \text{(E.5.3.1)}$$

Boundary conditions

1. $x = 0, \qquad T = T_0 \qquad$ (E.5.3.2a)
2. $x = L, \qquad -k\dfrac{dT}{dx} = h(T - T_\infty) \qquad$ (E.5.3.2b)

Determine the temperature distribution in the fin and fin heat loss using one-dimensional heat conduction analysis.

Solution: In this problem the fin is rectangular with uniform cross-section. We can estimate the cross-sectional area and the perimeter as

$$P = 2(z + t) = 2\left(1.0 \times 10^{-2} + 1.4 \times 10^{-3}\right) = 0.0228 \text{ m}$$
$$A = zt = \left(1.0 \times 10^{-2}\right)\left(1.4 \times 10^{-3}\right) = 1.4 \times 10^{-5} \text{ m}^2$$

The source term $(hP/A)(T_\infty - T)$ can be linearized as

$$S_0 = \frac{hP}{A}T_\infty = \frac{(500)(2.28 \times 10^{-2})}{(1.4 \times 10^{-5})} \times (20) = 16\,285\,714.29$$

$$S_1 = -\frac{hP}{A} = -\frac{(500)(2.28 \times 10^{-2})}{(1.4 \times 10^{-5})} = -814\,285.7145$$

For constant conductivity

$$k_{i+\frac{1}{2}} = k_{i-\frac{1}{2}} = k = 60\,\mathrm{W/m \cdot C}$$

Now the solution of the problem by the finite difference method is given as follows.

In the ***first step***, we discretize the domain using a uniform grid size and number of division $N_{\text{div}} = 8$ as shown below. This leads to

Number of grid points, $N = 9$

Grid size, $\Delta x_{i+\frac{1}{2}} = \Delta x_{i-\frac{1}{2}} = \Delta x = \dfrac{L}{N_{\text{div}}} = \dfrac{0.002}{8} = 2.5 \times 10^{-3}\,\mathrm{m}$

In the ***second step***, we select the following discretization equation for the node.

For ***node 1***, we use the known constant temperature, i.e.

$$T_1 = T_0 = 200 \tag{E.5.3.3}$$

For the ***interior nodes***, we use Equation (f) from Table 5.2.

$$a_i T_i = a_{i+1} T_{i+1} + a_{i-1} T_{i-1} + d \tag{E.5.3.4}$$

where

$$a_{i+1} = \frac{k_{i+\frac{1}{2}}}{\Delta x_{i+\frac{1}{2}}}, \quad a_{i-1} = \frac{k_{i-\frac{1}{2}}}{\Delta x_{i-\frac{1}{2}}}$$

$$a_i = a_{i+1} + a_{i-1} - s_1 \Delta x$$

$$d = s_0 \Delta x$$

Substituting the input variables we get the coefficients as

$$a_{i+1} = \frac{k_{i+\frac{1}{2}}}{\Delta x_{i+\frac{1}{2}}} = \frac{60}{2.5 \times 10^{-3}} = 24\,000$$

$$a_{i-1} = \frac{k_{i-\frac{1}{2}}}{\Delta x_{i-\frac{1}{2}}} = \frac{60}{2.5 \times 10^{-3}} = 24\,000$$

$$a_i = a_{i+1} + a_{i-1} - S_1\Delta x = 24\,000 + 24\,000$$
$$+ \left(814\,285.7145 \times 2.5 \times 10^{-3}\right) = 50\,035.714\,29$$
$$b = S_0\Delta x = \left(16\,285\,714.29 \times 2.5 \times 10^{-3}\right) = 40\,714.285\,73$$

Applying this discretization equation in turn to all interior nodes, we have

Node 2 $\quad 50\,035.714\,29T_2 = 24\,000T_1 + 24\,000T_3 + 40\,714.285\,73$

Node 3 $\quad 50\,035.714\,29T_3 = 24\,000T_2 + 24\,000T_4 + 40\,714.285\,73$

Node 4 $\quad 50\,035.714\,29T_4 = 24\,000T_3 + 24\,000T_5 + 40\,714.285\,73$

Node 5 $\quad 50\,035.714\,29T_5 = 24\,000T_4 + 24\,000T_6 + 40\,714.285\,73$

Node 6 $\quad 50\,035.714\,29T_6 = 24\,000T_5 + 24\,000T_7 + 40\,714.285\,73$

Node 7 $\quad 50\,035.714\,29T_7 = 24\,000T_6 + 24\,000T_8 + 40\,714.285\,73$

Node 8 $\quad 50\,035.714\,29T_8 = 24\,000T_7 + 24\,000T_9 + 40\,714.285\,73$

Node 9 is a convective boundary node on the right side boundary, and we use Equation 5.54 as the discretization equation, which is given as

$$a_iT_i = a_{i-1}T_{i-1} + d \qquad \text{(E.5.3.5)}$$

where the coefficients are given as

$$hT_\infty = 500 \times 20, \qquad h = 500$$

$$a_{i-1} = \frac{k_{i-1}}{\Delta x_{i-1}} = \frac{k}{\Delta x} = 2400$$
$$a_i = \frac{k_{i-1}}{\Delta x_{i-1}} - S_1\frac{\Delta x_{i-\frac{1}{2}}}{2} + h = 24\,000 + (814\,285.7145)\left(1.25 \times 10^{-3}\right)$$
$$+ 500 = 25\,517.857\,14$$
$$b = S_0\frac{\Delta x_{i-1}}{2} + hT_\infty = (16\,285\,714.29)\left(1.25 \times 10^{-3}\right) + 10\,000 = 30\,357.142\,86$$

Substituting the coefficients we get the discretization equation for the boundary node as

Node 9 $\quad 25\,517.857\,14T_9 = 24\,000T_8 + 30\,357.142\,86$

Now by assembling all nodal equations we obtain the system of algebraic equation as

$$\begin{bmatrix} 1 & 0 & 0 & 0 & 0 & 0 & 0 & 0 & 0 \\ -24\,000 & 5\,003\,571\,429 & -24\,000 & 0 & 0 & 0 & 0 & 0 & 0 \\ 0 & -24\,000 & 5\,003\,571\,429 & -24\,000 & 0 & 0 & 0 & 0 & 0 \\ 0 & 0 & -24\,000 & 5\,003\,571429 & -24\,000 & 0 & 0 & 0 & 0 \\ 0 & 0 & 0 & -24\,000 & 5\,003\,571\,429 & -24\,000 & 0 & 0 & 0 \\ 0 & 0 & 0 & 0 & -24\,000 & 5\,003\,571\,429 & -24\,000 & 0 & 0 \\ 0 & 0 & 0 & 0 & 0 & -24\,000 & 5\,003\,571\,429 & -24\,000 & 0 \\ 0 & 0 & 0 & 0 & 0 & 0 & -24\,000 & 5\,003\,571\,429 & -24\,000 \\ 0 & 0 & 0 & 0 & 0 & 0 & 0 & -24\,000 & 2\,551\,785\,714 \end{bmatrix} \times \begin{Bmatrix} T_1 \\ T_2 \\ T_3 \\ T_4 \\ T_5 \\ T_6 \\ T_7 \\ T_8 \\ T_9 \end{Bmatrix} = \begin{Bmatrix} 150 \\ 40\,714.285\,73 \\ 40\,714.285\,73 \\ 40\,714.285\,73 \\ 40\,714.285\,73 \\ 40\,714.285\,73 \\ 40\,714.285\,73 \\ 40\,714.285\,73 \\ 30\,357.142\,86 \end{Bmatrix} \tag{E.5.3.6}$$

In the ***third step***, we use the ***triadiagonal matrix algorithm*** given in Figure 3.1 to solve the system of equations. The solution given by this algorithm is given in the table below.

Node	a_i	b_i	c_i	d_i	P_i	Q_i	T_i
1	1	0	0	150	0	150	150
2	50 035.714 29	24 000	24 000	40 714.285 73	0.479 66	72.762 31	117.886 70
3	50 035.714 29	24 000	24 000	40 714.285 73	0.622 99	46.387 07	94.075 79
4	50 035.714 29	24 000	24 000	40 714.285 73	0.684 07	32.892 64	76.548 14
5	50 035.714 29	24 000	24 000	40 714.285 73	0.713 90	24.693 22	63.817 29
6	50 035.714 29	24 000	24 000	40 714.285 73	0.729 44	19.249 58	54.803 30
7	50 035.714 29	24 000	24 000	40 714.285 73	0.737 80	15.454 00	48.747 11
8	50 035.714 29	24 000	24 000	40 714.285 73	0.742 38	12.732 08	45.116 71
9	25 517.857 14	0	24 000	30 357.142 86	0	43.622 72	43.622 72

In the ***fourth step***, we estimate the fin heat loss and check the overall balance of heat.

Fin heat loss
Heat loss per unit area is estimated as

$$q_f = q_i\big|_{x=0} = \frac{k_{i+\frac{1}{2}}(T_i - T_{i+1})}{\Delta x_{i+\frac{1}{2}}} - (S_o + S_1 T_i)\frac{\Delta x_{i+\frac{1}{2}}}{2}$$

$$= \frac{60(150 - 117.89)}{2.5 \times 10^{-3}}$$

$$- (16\,285\,714.29 - 814\,285.714\,56150) \times 1.25 \times 10^{-3}$$

$$= 902\,961.4286\,\text{w/m}^2$$

Total heat loss

$$Q_f = q_f \times A = 902\,061.4286 \times 1.4 \times 10^{-5} = 12.641\,46\,\text{W}$$

Check for Overall Integral Heat Balance Heat transfer rate at the base = heat generation or heat transferred by convection

$$Q_f = A\int_0^L (S_0 + S_1 T_i)\,dx = A\sum (S_0 + S_1 T_i)\Delta x$$

Substituting the temperature values on the right-hand side of the equation, we get

$$
\begin{aligned}
\sum (S_o + S_1 T_i)\Delta x &= [16\,285\,714.29 - 8.7145(150)] \times 1.25 \times 10^{-3} \\
&+ [16\,285\,714.29 - 8.7145(117.89)] \times 2.5 \times 10^{-3} \\
&+ [16\,285\,714.29 - 8.7145(94.08)] \times 2.5 \times 10^{-3} \\
&+ [16\,285\,714.29 - 8.7145(76.55)] \times 2.5 \times 10^{-3} \\
&+ [16\,285\,714.29 - 8.7145(63.82)] \times 2.5 \times 10^{-3} \\
&+ [16\,285\,714.29 - 8.7145(54.80)] \times 2.5 \times 10^{-3} \\
&+ [16\,285\,714.29 - 8.7145(48.74)] \times 2.5 \times 10^{-3} \\
&+ [16\,285\,714.29 - 8.7145(45.12)] \times 2.5 \times 10^{-3} \\
&+ [16\,285\,714.29 - 8.7145(43.62)] \times 1.25 \times 10^{-3} \\
\sum (S_o + S_1 T_i)\Delta x &= -891\,256.0717\ \mathrm{W/m^2}
\end{aligned}
$$

The total heat convection heat loss is

$$A \sum (S_o + S_1 T_i)\Delta x = -891\,256.0717 \times 1.4 \times 10^{-3} = -12.477\,585$$

Comparison with the Exact Solution The analytical solutions for temperature distribution and fin heat loss are given as

Temperature distribution: $$\frac{T - T_\infty}{T_0 - T_\infty} = \frac{\cosh m(L - x) + (h/mk)\sinh m(L - x)}{\cosh mL + (h/mk)\sinh mL}$$

Fin heat loss: $$\sqrt{hPkA}(T_0 - T_\infty)\frac{\sinh mL + (h/mk)\cosh mL}{\cosh mL + (h/mk)\sinh mL}$$

where $m^2 = hP/kA$.

The percent error for temperature can be estimated as

$$\varepsilon_{\mathrm{T}} = \left| \frac{(T_{\mathrm{num}} - T_{\mathrm{exact}})}{(\mathrm{T}_0 - T_\infty)} \right| \times 100\%$$

The percent error for heat flux can be estimated as

$$\varepsilon_q = \left| \frac{q_{f,\mathrm{num}} - q_{f,\mathrm{exact}}}{q_{f,\mathrm{exact}}} \right| \times 100\%$$

A comparison of the computational solution with the exact solution is presented in the table and figure below.

Node	T_i (Exact)	T_i (Numerical)	Error (ε_T)
1	150	150	0
2	112.78	117.89	0.000 846 2
3	93.91	94.08	0.001 307 7
4	76.35	76.55	0.001 538 5
5	63.61	63.82	0.001 615 4
6	54.60	54.80	0.001 538 5
7	48.54	48.74	0.001 538 5
8	44.91	45.12	0.001 615 4
9	43.42	43.62	0.001 538 5

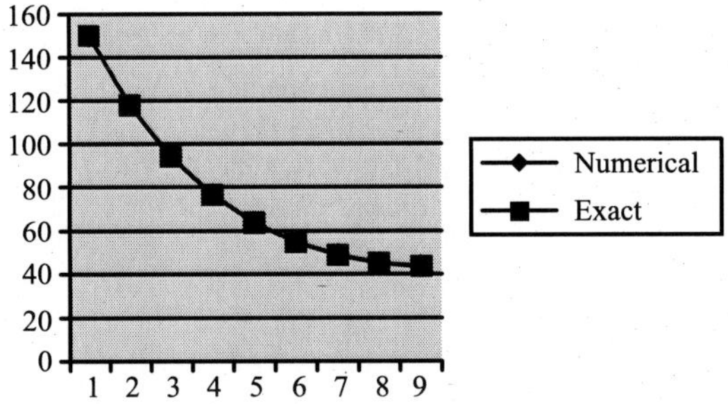

5.7 GRID SIZE DISTRIBUTION

In the preceding section we have used uniform grid size distributions, which is convenient for computation. However, non-uniform grid size distribution may become more effective in achieving desired accuracy for some problems that involve regions of steep variation of dependent variables. For such problems, we can use finer grid size distribution in regions when there is a very sharp variation of the dependent variables, and coarser grid size distribution in regions where there is very gradual variation. Let us consider the temperature variation in the straight fin as shown in Figure 5.7. One possible grid size distribution that was used in the previous example problem is the uniform grid size distribution as shown in Figure 5.8a.

However, a more effective grid size would be the nonuniform grid size as shown in Figure 5.8b. For this problem we could make this decision because of our prior knowledge of the problem and availability of the analytical solution.

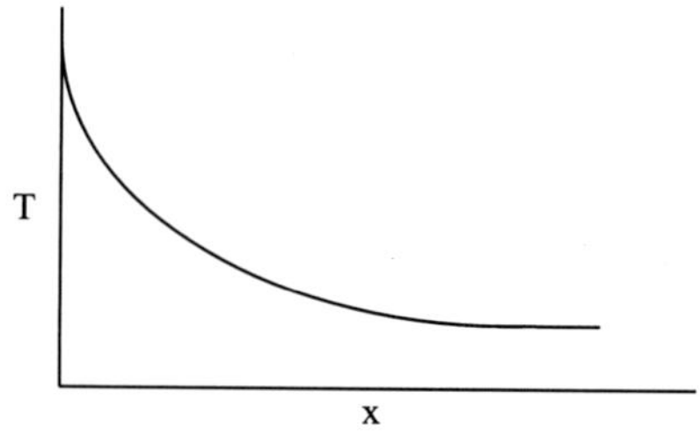

Figure 5.7 Typical temperature distribution in a fin

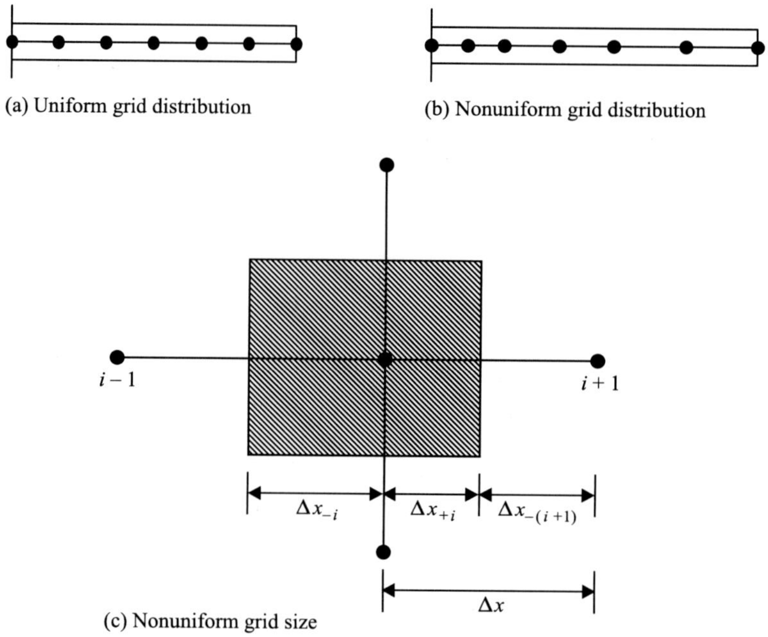

Figure 5.8 Grid size distributions: (a) uniform grid, (b) nonuniform grid and (c) nonunifrom grid size

In most problems we have to conduct some preliminary numerical experimentation to gain some physical knowledge before making such a decision. A number of parametric studies are to be made before deciding the number of grids and the degree of nonuniformity of the grid. A sufficient number of grid divisions and nonuniformity are to be used to capture the sharp variation near the boundaries and/or at locations near the source and sink. Some initial calculation with various degree of nonuniformity will show the extent of regions of sharp velocity, temperature, and concentration gradients. A sufficient number of grid divisions are then used to capture the sharp variations in these regions.

5.8 NONUNIFORM TRANSPORT PROPERTY

Nonuniform transport property can result due to the presence of nonhomogeneous materials such as composite materials, and even in homogeneous materials with strong variation of transport property with dependent variable. In order to consider this nonuniform transport property, we have to estimate the transport properties $\Gamma_{i+\frac{1}{2}}$ and $\Gamma_{i-\frac{1}{2}}$ at the interfaces $i+\frac{1}{2}$ and $i-\frac{1}{2}$ of two adjacent control volumes as shown in the Figure 5.9.

As was discussed in Section 5.3.2, a step-wise profile, i.e., a constant value, is usually assumed for the transport and thermo-physical properties in a control volume surrounding a node, *i*. One common procedure for obtaining interface transport property, $\Gamma_{i+\frac{1}{2}}$, is to assume a linear variation of Γ between nodes i and $i+1$, and this leads to the expression

$$\Gamma_{i+\frac{1}{2}} = \frac{\Delta x_{+i}}{\Delta x_{i+\frac{1}{2}}}\Gamma_i + \left(1 - \frac{\Delta x_{-(i+1)}}{\Delta x_{i+\frac{1}{2}}}\right)\Gamma_{i+1} \tag{5.57}$$

For the case of a uniform grid size distribution, interface transport property would be given by the ***arithmetic mean*** of Γ_i and Γ_{i+1} as

$$\Gamma_{i+\frac{1}{2}} = \left(\frac{\Gamma_i + \Gamma_{i+1}}{2}\right) \tag{5.58}$$

This approach is quite suitable for a homogeneous material with strong temperature dependent transport properties. However, these equations may lead to erroneous results for composite materials that involve abrupt changes in transport properties between two adjacent materials. In such cases an alternate equation is derived based on evaluating the surface flux using a one-dimensional steady state analysis as

$$f_s'' = \Gamma_i \frac{\phi_i - \phi_{i+\frac{1}{2}}}{\Delta x_{+i}} \tag{5.59a}$$

$$f_s'' = \Gamma_{i+1} \frac{\phi_{i+\frac{1}{2}} - \phi_{i+1}}{\Delta x_{-(i+1)}} \tag{5.59b}$$

$$f_s'' = \Gamma_{i+\frac{1}{2}} \frac{\phi_i - \phi_{i+1}}{\Delta x_{i+\frac{1}{2}}} \tag{5.59c}$$

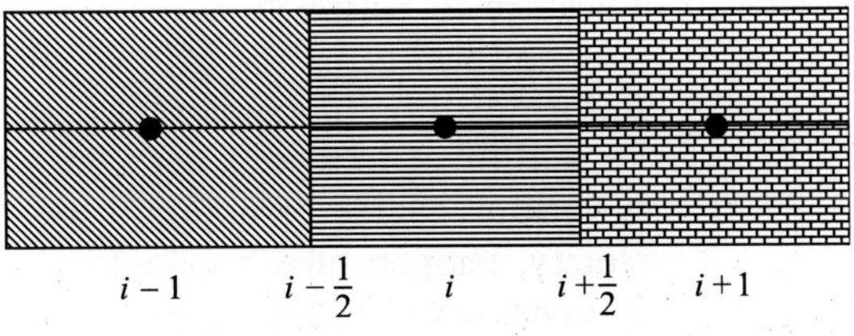

Figure 5.9 Grid distribution and interfaces in a composite material

Eliminating $\phi_{i+\frac{1}{2}}$ from Equations 5.59a and 5.59b, and equating with Equation 5.59c, we can derive the interface transport property as

$$\Gamma_{i+\frac{1}{2}} = \frac{1}{\dfrac{\Delta x_{+i}/\Delta x_{i+\frac{1}{2}}}{\Gamma_i} + \dfrac{\Delta x_{-(i+1)}/\Delta x_{i+\frac{1}{2}}}{\Gamma_{i+1}}} \tag{5.60}$$

Again, for a uniform grid size distribution, Equation 5.60 reduces to a simplified form, which estimates the interface transport property as a ***harmonic mean*** of Γ_i and Γ_{i+1} as

$$\Gamma_{i+\frac{1}{2}} = \frac{2\Gamma_i\Gamma_{i+1}}{\Gamma_i + \Gamma_{i+1}} \tag{5.61}$$

Equations 5.65 and 5.57 are desirable for maintaining the continuity of surface flux at the interface of two dissimilar materials such as in composite materials without requiring an excessively fine grid size distribution. These equations are also desirable to ensure a zero surface flux at the interface of two materials, one of which has zero transport property ($\Gamma = 0$) such as in an impermeable material or in an insulating material.

Example 5.4: One-dimensional Steady State Conduction in Composite Wall: Consider steady state heat conduction in a composite wall as shown below.

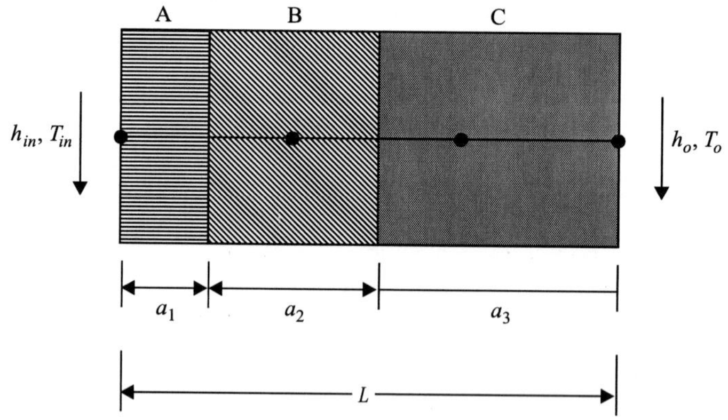

The mathematical statement of the problem is given by:

Governing equation

$$\frac{d}{dx}\left(k(x)\frac{dT_n}{dx}\right) = 0 \tag{E.5.4.1}$$

where T_n is the temperature distribution in the nth layer.

Boundary conditions

$$1. \quad x = 0, \quad k\frac{dT}{dx} = h_\text{i}(T - T_\text{i}) \tag{E.5.4.2}$$

$$2. \quad x = L, \quad -k\frac{dT}{dx} = h_0(T - T_0) \tag{E.5.4.3}$$

Calculate the nodal temperature values within the composite wall and evaluate the heat flow per unit width through each material based on the following data: $a_1 = 5\,\text{cm}$, $a_2 = 10\,\text{cm}$, $a_3 = 15\,\text{cm}$, $k_\text{A} = 2.0\,\text{W/m°C}$, $k_\text{B} = 30\,\text{W/m°C}$, $k_\text{C} = 0.1\,\text{W/m°C}$, $h_\text{i} = 20\,\text{W/cm}^{2}\text{°C}$, $h_0 = 20\,\text{W/cm}^{2}\text{°C}$, $T_{\infty,i} = 100°\text{C}$ and $T_{\infty,0} = 20°\text{C}$.

Solution: Let us consider a uniform grid system shown in the figure below. The step size is given as $\Delta x = 1.0\,\text{cm}$.

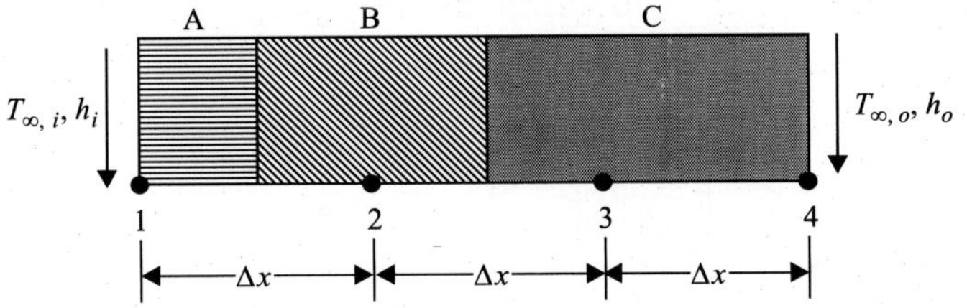

For the ***interior nodes***, we use Equation 5.34

$$a_i T_i = a_{i+1} T_{i+1} + a_{i-1} T_{i-1} + b \tag{E.5.4.4}$$

where

$$a_{i+1} = \frac{k_{i+\frac{1}{2}}}{\Delta x_{i+\frac{1}{2}}}, \qquad a_{i-1} = \frac{k_{i-\frac{1}{2}}}{\Delta x_{i-\frac{1}{2}}}$$

$$a_i = a_{i+1} + a_{i-1}$$

$$d = \overline{S}_\text{o} \Delta x$$

Nodes 2 and 3 have control volume surfaces that are the interface of two dissimilar materials. The thermal conductivity at the interface of a uniform grid is given as a ***harmonic mean*** Equation 5.61 as

$$k_{i+\frac{1}{2}} = \frac{2k_i k_{i+1}}{k_i + k_{i+1}} \tag{E.5.4.5}$$

Substituting material conductivities, the conductivities at the interfaces are computed as

$$k_{2-\frac{1}{2}} = \frac{2k_A k_B}{k_A + k_{B+1}} = \frac{2 \times 2.0 \times 30.0}{2.0 + 30.0} = 3.75$$

and

$$k_{2+\frac{1}{2}} = k_{3-\frac{1}{2}} = \frac{2k_B k_C}{k_B + k_C} = \frac{2 \times 30.0 \times 0.1}{30.0 + 0.1} = 0.1993$$

$$k_{3+\frac{1}{2}} = k_{4-\frac{1}{2}} = k_C = 0.1$$

For ***node 2***, the discretization equation becomes

$$a_2 T_2 = a_3 T_3 + a_1 T_1 + d$$

where

$$a_3 = \frac{k_{2+\frac{1}{2}}}{\Delta x} = \frac{0.1993}{0.1} = 1.933, \qquad a_1 = \frac{k_{2-\frac{1}{2}}}{\Delta x} = \frac{3.75}{0.1} = 37.5$$

$$a_2 = a_3 + a_1 = 1.933 + 37.5 = 39.433, \qquad d = 0$$

$$39.433 T_2 = 1.933 T_3 + 37.5 T_1 \tag{E.5.4.6a}$$

For ***node 3***, the discretization equation becomes

$$a_3 T_3 = a_2 T_2 + a_4 T_4 + d$$

where

$$a_4 = \frac{k_{3+\frac{1}{2}}}{\Delta x} = \frac{0.1}{0.1} = 1.0, \qquad a_2 = \frac{k_{3-\frac{1}{2}}}{\Delta x} = \frac{0.1993}{0.1} = 1.933$$

$$a_3 = a_4 + a_2 = 1.0 + 1.933 = 2.933, \qquad d = 0$$

With the substitution of the coefficients, the equation becomes

$$2.933 T_3 = 1.933 T_2 + 1.0 T_4 \tag{E.5.4.6b}$$

For boundary ***node 1***, we use the convective boundary equation from Table 5.2 with $S_0 = 0$ and $S_1 = 0$ as

$$a_1 T_1 = a_2 T_2 + d$$

where

$$k_{1+\frac{1}{2}} = k_{2-\frac{1}{2}} = \frac{2k_A k_B}{k_A + k_{B+1}} = 3.75$$

$$a_1 = \frac{k_{1+\frac{1}{2}}}{\Delta x_{1+\frac{1}{2}}} + h_i = \frac{3.75}{0.1} + 20 = 57.5$$

$$a_2 = \frac{k_{1+\frac{1}{2}}}{\Delta x_{1+\frac{1}{2}}} = \frac{3.75}{0.1} = 37.5$$

$$d = h_i T_{\infty,1} = 20.0 \times 100 = 2000$$

and

$$k_{1+\frac{1}{2}} = k_{2-\frac{1}{2}} = \frac{2k_A k_B}{k_A + k_{B+1}} = 3.75$$

With the substitution of the coefficients, the equation becomes

$$57.5T_1 = 37.5T_2 + 2000 \qquad \text{(E.5.4.6c)}$$

Similarly, for the boundary ***node 4*** we use the convective boundary equation

$$a_4 T_4 = a_3 T_3 + d$$

where

$$a_3 = \frac{k_{4-\frac{1}{2}}}{\Delta x_{4-\frac{1}{2}}} = \frac{0.1}{0.1} = 1.0$$

$$a_4 = \frac{k_{4-\frac{1}{2}}}{\Delta x_{4-\frac{1}{2}}} + h_{\infty.0} = \frac{0.1}{0.1} + 50 = 51$$

$$d = h_0 T_{\infty,0} = 50.0 \times 20.0 = 1000$$

With the substitution of the coefficients, we have the equation

$$51T_4 = T_3 + 1000 \qquad \text{(E.5.4.6d)}$$

Assembling all nodal equations, we have the system of equations

$$\begin{bmatrix} 57.5 & -37.5 & 0 & 0 \\ -37.5 & 39.433 & -1.933 & 0 \\ 0 & -1.933 & 2.933 & -1.0 \\ 0 & 0 & -1.0 & 51 \end{bmatrix} \begin{Bmatrix} T_1 \\ T_2 \\ T_3 \\ T_4 \end{Bmatrix} = \begin{Bmatrix} 2000 \\ 0 \\ 0 \\ 1000 \end{Bmatrix} \qquad \text{(E.5.4.7)}$$

Solving the system of equations, we get

$$\begin{Bmatrix} T_1 \\ T_2 \\ T_3 \\ T_4 \end{Bmatrix} = \begin{Bmatrix} 97.5217°\text{C} \\ 96.1999°\text{C} \\ 70.5577°\text{C} \\ 20.9913°\text{C} \end{Bmatrix} \tag{E.5.4.8}$$

Let us now estimate heat transfer rates at two surfaces as follows.

The heat transfer rate at the left surface is estimated as

$$q_L = h_{\text{i}}\left(T_{\infty,\text{i}} - T_1\right) = 20.0(100 - 97.5217) = 49.566 \text{ W/m}^2$$

The heat transfer rate at the right surface is estimated as

$$q_R'' = h_0\left(T_4 - T_{\infty,0}\right) = 50.0(20.9913 - 20.0) = 49.565 \text{ W/m}^2$$

As expected, the overall energy balance under steady state is satisfied even with the coarse grid considered.

5.9 NONLINEARITY

So far we have considered the governing equations as linear, and so discretization of these equations leads to a system of linear algebraic equations, which is solved by a linear solver. Governing differential equations could become nonlinear due to the presence of the convective terms and/or due to a variable material or fluid properties or a variable source term. Nonlinearity due to the inclusion of convective terms will be addressed in subsequent chapters on convection problems.

Nonlinearity due to the dependence of the transport and thermo-physical properties or dependence of the source term or film transfer coefficient on the dependent variable such as $\Gamma(\phi)$, $S(\phi)$ or $h(T)$ are usually tackled by an iterative process. The iteration process starts with a guess for all nodal values of dependent variables. Using these guess values, the coefficients a and d of the discretization equations that includes the properties and source term are estimated. This is followed by solving the resulting linear system of algebraic equations by the usual linear equation solver. The process is repeated with the latest iterated nodal values of dependent variables until convergence is reached.

5.10 LINEARIZATION OF A VARIABLE SOURCE TERM

In Section 5.5 we have mentioned that, in order to solve the set discretization equations by the solution techniques of linear algebraic equations, it is essential to express a variable source $S(\phi)$ by a linear relation. We have assumed a linear variation of an average source $\overline{S}$ with ϕ_i as $\overline{S} = S_0 + S_1\phi_i$ in deriving discretization equations 5.37 and 5.38 for the two-dimensional steady state problem with source term. In many situations the selections of S_0 and S_1 are quite straightforward. However, it is the situations where source functions are such that we need to decide whether we would simply evaluate the source function based on the previous iterated value of the dependent variable or try to linearize the function through rearrangement or using expansion. In this section let us describe one of the ways of linearizing a source term as given by Patankar (1991). In this approach the source function $\overline{S}(\phi)$ is expressed using a truncated Taylor series in the following manner

$$\overline{S}(\phi) = \overline{S}(\phi^m) + \left(\frac{\partial \overline{S}}{\partial \phi}\right)^m (\phi_i - \phi_i^m) \tag{5.62}$$

where the superscript m represents the previous iteration number, and ϕ_i^m represents the value of the dependent variable at a previous iteration. Equation 5.62 now is expressed in linear form as

$$\overline{S}(\phi) = S_0 + S_1\phi_i \tag{5.63}$$

where

$$S_0 = \overline{S}(\phi^m) - \left(\frac{\partial \overline{S}}{\partial \phi}\right)^m \phi_i^m \tag{5.64a}$$

$$S_1 = \left(\frac{\partial \overline{S}}{\partial \phi}\right)^m \tag{5.64b}$$

The selection of S_0 and S_1 through Equations 5.64 is only done when it gives a negative value of S_1. If this is not achievable, then we simply select $S_0 = \overline{S}(\phi)$ and $S_1 = 0$.

PROBLEMS

5.1 Consider one-dimensional steady state conduction without heat generation in a plane wall shown in the figure below. The boundary surfaces at $x=0$ and $x=L$ are exposed to indoor and outdoor air environment, respectively. Determine the temperature distribution in the slab and heat transfer rates at the surfaces using six equally-spaced nodes. Use $k=20\ \text{W/cm}^\circ\text{C}$, $L=10\ \text{cm}$, $T_i=25^\circ\text{C}$, $T_0=40^\circ\text{C}$, $h_i=50\ \text{W/m}^{2\circ}\text{C}$ and $h_0=100\ \text{W/m}^{2\circ}\text{C}$. Solve the system of equations using ***TDMA*** solver.

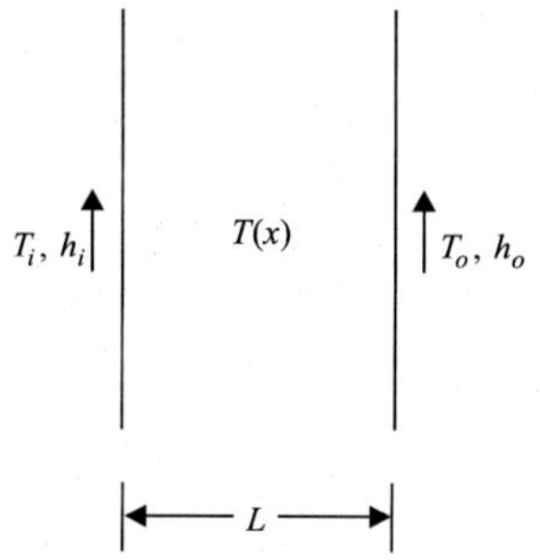

5.2 Consider one-dimensional steady state heat conduction without heat generation in a plane wall with surface at $x=0$ subjected to incident radiation q_0'' as shown in the figure below. The boundary surface at $x=L$ is exposed to a convective fluid environment with $T_0=40°C$, $h_0=100\,W/m^{2}°C$. Thermal conductivity of the material is $k=20$ W/cm°C. Use a finite difference–control volume scheme to derive the system of equations for temperatures using six nodal points. Solve the system of equations for temperature distribution in the wall and heat transfer rates at the surfaces using Gauss-elimination solver.

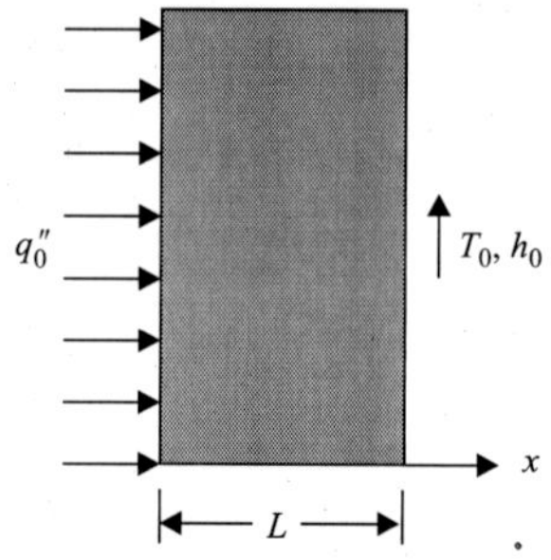

5.3 An exterior wall of a building is a composite wall made of four layers as shown in the figure below.

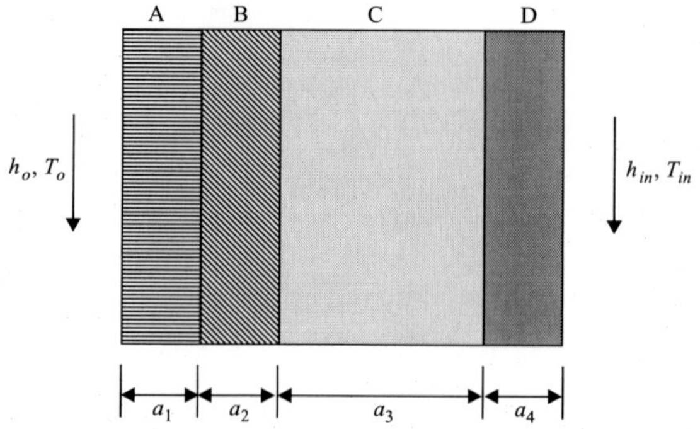

Consider the layers as homogeneous with the following data: $a_1 = 2\,\text{cm}$, $a_2 = 1\,\text{cm}$, $a_3 = 5\,\text{cm}$, $a_4 = 1\,\text{cm}$, $k_A = 0.72\,\text{W/m K}$, $k_B = 0.12\,\text{W/m K}$ (W/cm°C), $k_C = 0.026\,\text{W/m K}$ (W/cm°C), and $k_D = 0.22\,\text{W/m K}$. Calculate the nodal temperature values within the composite wall and evaluate the heat flow per unit width based on assuming winter design indoor and outdoor conditions as: $T_{in} = 25°\text{C}$, $h_{in} = 10\,\text{w/cm}^{2}°\text{C}$ and $T_{out} = -20°\text{C}$, $h_0 = 50\,\text{W/cm}^{2}°\text{C}$. Also, compare these numerical solutions to the analytical solution given by:

Temperature distribution

$$T_n = T_{in} - \frac{(T_{in} - T_0)\left[h_{in}\left(\dfrac{x - x_{n-1}}{k_{n-1}} + \sum_{m=1}^{n-1}\dfrac{a_m}{k_m}\right) + 1\right]}{h_{in}\sum_{m=1}^{N}\dfrac{a_m}{k_m} + \dfrac{h_{in}}{h_0} + 1}$$

Heat flux

$$q'' = \frac{T_{in} - T_0}{\frac{1}{h_{in}} + \sum_{m=1}^{N}\dfrac{a_m}{k_m} + \dfrac{1}{h_0}}$$

5.4 The exposed surface ($x = 0$) of a plane wall of thermal conductivity k is subjected to thermal radiation that causes volumetric heat generation in the media to vary as

$$\dot{q}(x) = \dot{q}_0\left(1 - \frac{x}{L}\right)$$

where $\dot{q}_0$(W/m^3) is a constant. The boundary at $x = L$ is convectively cooled, while the exposed surface is maintained at a constant temperature T_0.

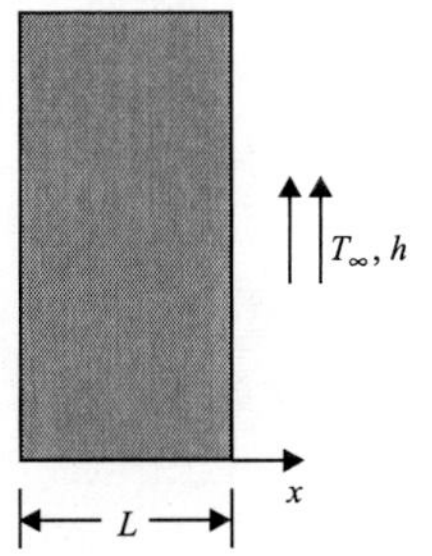

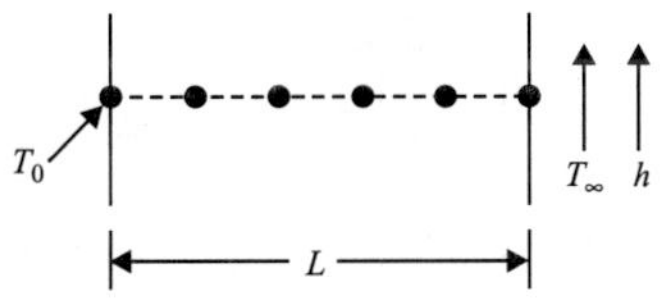

The mathematical statement of the problem is given as:

Governing equation

$$\frac{\partial}{\partial x}\left(k\frac{\partial T}{\partial x}\right)+\dot{q}=0$$

Boundary conditions

1. $x=0, \quad T(x,0)=T_0$

2. $x=L, \quad -k\frac{\partial T}{\partial x}=h(T-T_\infty)$

Consider the grid size distribution shown in the figure with six equally-spaced grid points and determine the temperature distribution in the slab. Use the following data for computation

$$\dot{q}=1.0\times 10^5\,\text{W/m}^3, \quad L=12\,\text{cm}, \quad T_0=30°\text{C},$$

$$h=50\,\text{W/m}^{2°}\text{C}, \quad T_\infty=25°\text{C}$$

Solve the system of equations using the Gauss-elimination solver.

5.5 A semi-transparent wall is irradiated at the surface at $x=0$ by an incident radiation such that the absorbed radiation in the media results in an internal heat generation given as

$$\dot{q}=q_0''ae^{-ax}$$

where a is the coefficient absorption of the material and q_0'' is the incident radiation flux. The wall at $x=0$ is also exposed to a convective environment of h_0 and T_0, and the surface at $x=L$ is maintained at constant temperature T_L. Derive discretization equations for interior nodes and for the convective boundary node for the nonuniform grid system shown in the figure below. Derive the system of equations for the solution of nodal temperatures and heat transfer rates at the inner and outer surfaces.

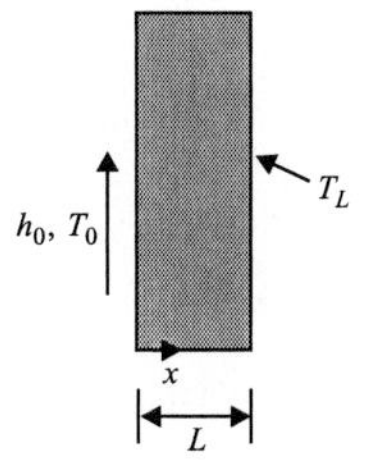

Solve the system of equations by (a) the Gauss-elimination solver, and (b) the Gauss–Seidel solver.

5.6 Consider one-dimensional steady state analysis for the cylindrical fin with a diameter of 10 mm and material thermal conductivity of 30 W/m K. The surrounding fluid temperature and convection heat transfer coefficient are 30°C and 40 W/m^2 K. Obtain the system of equations for the grid system shown in the figure below. Solve the equations for the temperature distribution and fin heat loss. Make an energy balance to check the accuracy.

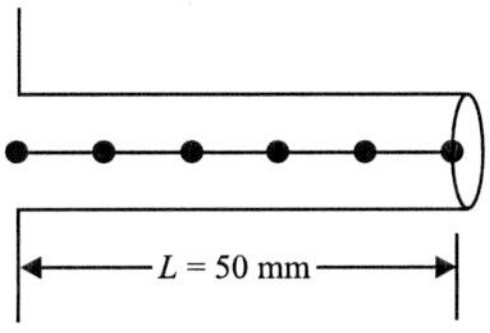

5.7 An aluminum pin fin having a diameter $D = 1$ cm and length $L = 6$ cm is exposed to the surrounding fluid with temperature $T_\infty = 10$°C, and convection heat transfer coefficient $h = 8.0$ W/m^2°C. The fin base temperature is $T_0 = 100$°C. Determine the temperature distribution and fin heat loss using the finite difference method and using the grid system shown in the figure below. Assume thermal conductivity of the material is $k = 200$ W/m°C. Show overall-energy balance.

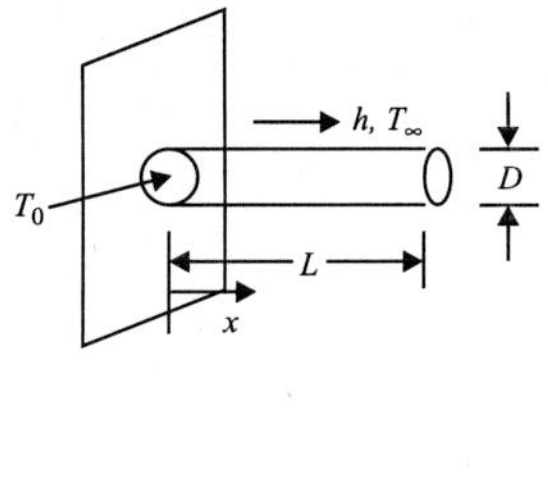

Solve the system of equations using the ***tdma*** solver.

5.8 Consider the fully developed laminar steady flow of Newtonian viscous fluid through a two-dimensional channel formed by two infinite parallel plates as shown in the figure below. The lower plate is moving in the negative x-direction with a constant speed U_{w1} and the upper plate is moving in the positive x-direction with a constant speed U_{w2} in the positive x-direction.

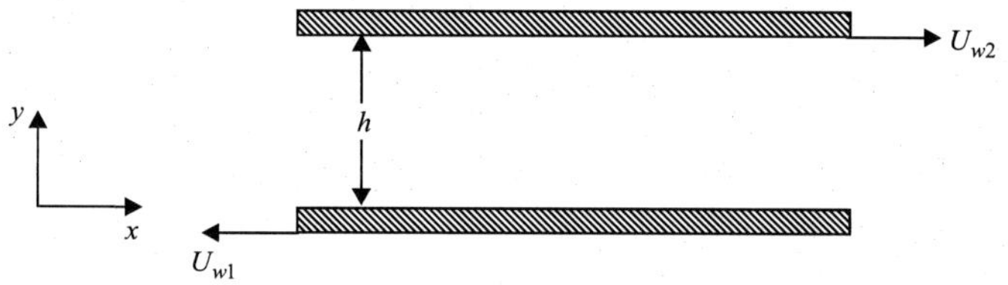

Determine (a) the velocity distribution, (b) volume flow rate, and (c) the shear stress at the plate and use the following data for the calculation: $h = 3.00$ mm, $\mu = 0.02$ kg/m s, $\partial P/\partial x = \overline{S} = -1000$ Pa/m, $U_{w1} = 10$ mm/s and $U_{w2} = 20$ mm/s.
Determine the shear stresses at the bottom and top surfaces.

5.9 Consider a one-dimensional steady state diffusion of hydrogen in an anode electrode made of a porous diffusion layer with integrated catalyst. A gas flow channel maintains a convective mass transfer with constant hydrogen concentration of C_∞ at the left surface of the electrode. The right surface of the electrode is assumed to be impermeable to hydrogen. Assume a constant rate of hydrogen consumption, $\dot{m}_{H_2}$, due to the catalytic reaction throughout the porous layer, and determine the hydrogen concentration profile in the electrode.

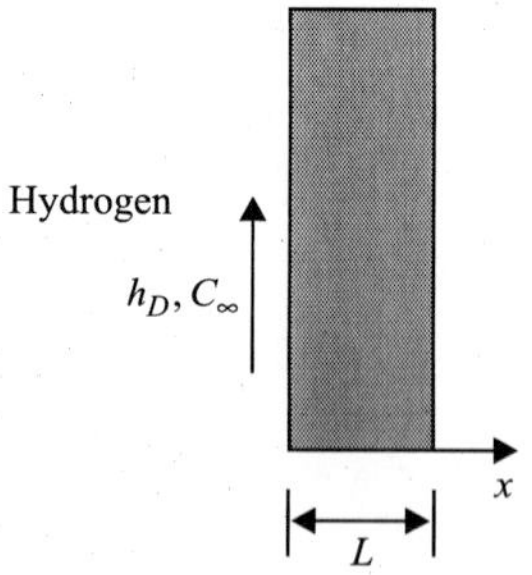

The mathematical statement of the problem is given as:

Governing equation

$$\frac{d}{dx}\left(D_{H_2}\frac{dC}{dx}\right) - \dot{m}_{H_2} = 0$$

Boundary conditions

1. $x = 0, \qquad D_{H_2}\dfrac{dC}{dx} = h_D(C - C_\infty)$

2. $x = L, \qquad \dfrac{dC}{dx} = 0$ due to impermeable right surface

Use a nonuniform grid size distribution shown in the figure to derive the system of equations for the nodal hydrogen concentration.

5.10 Consider a one-dimensional steady state diffusion of hydrogen in an anode electrode made of porous diffusion layer with a thin catalyst layer placed at the right side as shown in the figure below. The catalyst layer provides a heterogeneous chemical reaction for the consumption of hydrogen at a rate m''_{H_2}, defined as the consumption per unit area of the catalyst. Assume the reaction rate as given by a ***first-order reaction***, i.e., $m''_{H_2} = -k_1''C(0)$, where k_1'' is the reaction rate constant. A gas flow channel maintains a constant hydrogen concentration of C_0 at the left surface of the electrode.

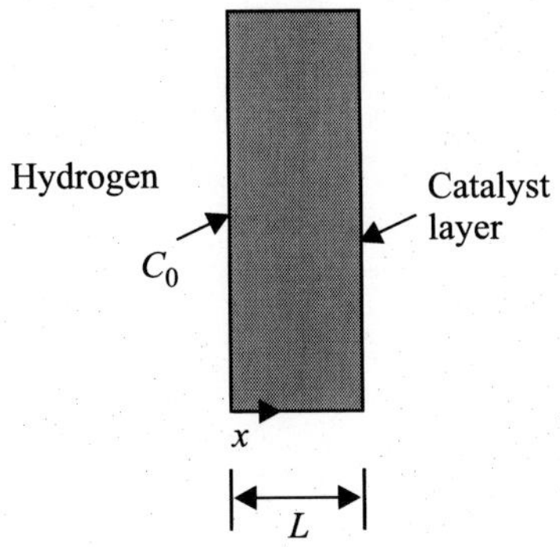

The mathematical statement of the problem is given as:

Governing equation

$$\frac{d}{dx}\left(D\frac{dC}{dx}\right) = 0$$

Boundary conditions

1. $x = 0,\qquad C(0) = C_0$
2. $x = L,\qquad -D\frac{dC}{dx}\bigg|_{x=L} = -k_1''C(L)$

Use a uniform grid size distribution shown in the figure to derive the system of equations for the nodal hydrogen concentration.

5.11 Develop a computer code for the solution one-dimensional diffuse equation in a plane slab of thickness L. The code generates the system of equations for unknown nodal values using discretization equations for interior nodes and that for the boundary nodes at the left and right surfaces. Solve the system of equations using the Gauss-elimination solver.

a. Use the code for the solution of the problem given in Problem 5.4.
b. Repeat the problem by progressively increasing the number of nodes as $N_x=8$, $N_x=10$, and $N_x=12$. Present an approximate percent relative error for the estimation of heat transfer rate at the left surface at $x=0$ with increasing number of nodes.

CHAPTER

6

MULTI-DIMENSIONAL PROBLEMS

The objective of this chapter is to consider the application of the finite difference method to multidimensional problems. Basics for the finite difference method were presented in Chapter 5 dealing with one-dimensional steady state problems. Even though the introduction of the discretization procedure was given using a two-dimensional steady state equation, Chapter 5 was primarily involved with applications to one-dimensional problems. In this chapter, we present detailed formulation for obtaining discretization equations for two-dimensional boundary nodes on plane surfaces and corners. The objective is to build enough knowledge to formulate discretization equations for nodes involving more complex physical phenomena.

6.1 TWO-DIMENSIONAL STEADY STATE PROBLEMS

For two-dimensional steady state problems, the appropriate governing equation is

$$\frac{\partial}{\partial x}\left(\Gamma \frac{\partial \phi}{\partial x}\right) + \frac{\partial}{\partial y}\left(\Gamma \frac{\partial \phi}{\partial y}\right) + S = 0 \quad (6.1)$$

The discretization equation for interior nodes was derived in Chapter 5 as

$$a_{i,j}\phi_{i,j} = a_{i+1,j}\phi_{i+1,j} + a_{i-1,j}\phi_{i-1,j} + a_{i,j+1}\phi_{i,j+1} + a_{i,j-1}\phi_{i,j-1} + d \quad (6.2a)$$

where

$$a_{i+1,j} = \frac{\Gamma_{i+\frac{1}{2},j}}{\Delta x_{i+\frac{1}{2}}} \Delta y \quad a_{i-1,j} = \frac{\Gamma_{i-\frac{1}{2},j}}{\Delta x_{i-\frac{1}{2},j}} \Delta y \quad a_{i,j+1} = \frac{\Gamma_{i,j+\frac{1}{2}}}{\Delta y_{i,j+\frac{1}{2}}} \Delta x$$

$$a_{i,j-1} = \frac{\Gamma_{i,j-\frac{1}{2}}}{\Delta y_{i,j-\frac{1}{2}}} \Delta x \quad d = S_0 \, \Delta x \, \Delta y$$

$$a_{i,j} = a_{i+1,j} + a_{i-1,j} + a_{i,j+1} + a_{i,j-1} - S_1 \, \Delta x \, \Delta y \tag{6.2b}$$

Equation 6.2 can be further simplified for many different cases. A summary of finite difference formulas for such cases is given in Table 6.1.

Example 6.1: Two-dimensional Steady State Conduction: Consider a 2 m by 2 m square slab with the top surface maintained at 400°C and all other surfaces maintained at 20°C. Apply the finite difference method to calculate the two-dimensional temperature distribution using a grid with $\Delta x = \Delta y = 0.5$ m as shown in the figure. Estimate the heat transfer rates at all surfaces, and show the overall energy balance.

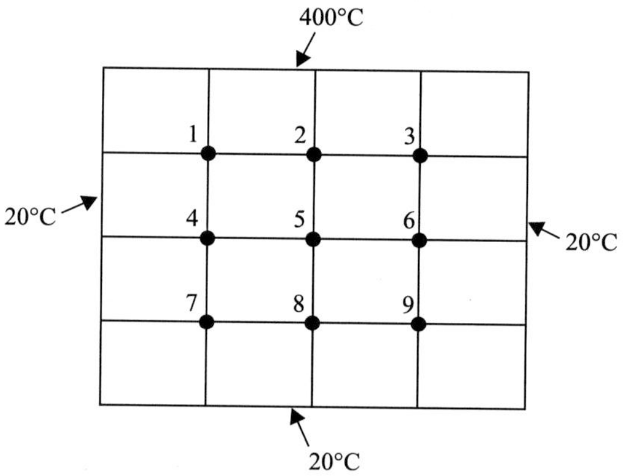

Solution: In the prescribed grid, the boundary nodes can be assigned with the prescribed surface temperatures. There are nine interior nodes for which the appropriate finite difference equation is given by

$$T_{i+1,j} + T_{i-1,j} + T_{i,j+1} + T_{i,j-1} - 4T_{i,j} = 0$$

Table 6.1 A Summary of Finite Difference Formulas for Interior Nodes in Two-dimensional Steady State Problems

Physical Situation	Finite-Difference Equation
(a) Non-uniform grid size 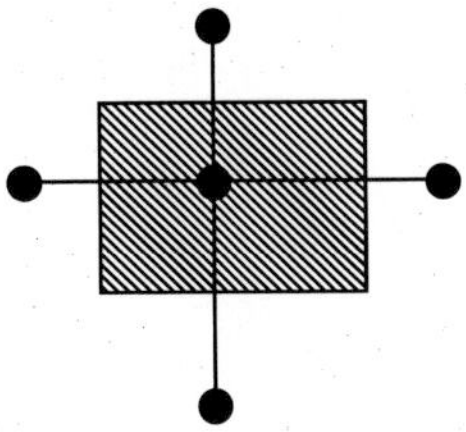	$a_{i,j}\phi_{i,j} = a_{i+1,j}\phi_{i+1,j} + a_{i-1,j}\phi_{i-1,j} + a_{i,j+1}\phi_{i,j+1} + a_{i,j-1}\phi_{i,j-1} + d$ $a_{i+1,j} = \frac{\Gamma_{i+\frac{1}{2}}}{\Delta x_{i+\frac{1}{2},j}}\Delta y, \quad a_{i-1,j} = \frac{\Gamma_{i-\frac{1}{2},j}}{\Delta x_{i-\frac{1}{2},j}}\Delta y$ $a_{i,j+1} = \frac{\Gamma_{i,j+\frac{1}{2}}}{\Delta y_{i,j+\frac{1}{2}}}\Delta x, \quad a_{i,j-1} = \frac{\Gamma_{i,j-\frac{1}{2}}}{\Delta y_{i,j-\frac{1}{2}}}\Delta x \quad d = S_0\,\Delta x\,\Delta y$ $a_{i,j} = a_{i+1,j} + a_{i-1,j}, a_{i,j+1} + a_{i,j-1} - S_1\,\Delta x\,\Delta y$
(b) For constant transport property, constant source, and uniform grid size distribution in x and y 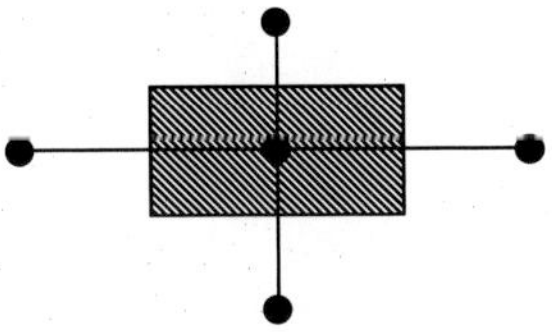	$a_{i,j}\phi_{i,j} = a_{i+1,j}\phi_{i+1,j} + a_{i-1,j}\phi_{i-1,j} + a_{i,j+1}\phi_{i,j+1} + a_{i,j-1}\phi_{i,j-1} + d$ $a_{i+1,j} = a_{i-1,j} = \frac{\Delta y}{\Delta x}$ $a_{i,j+1} = a_{i,j-1} = \frac{\Delta x}{\Delta y}$ $a_{i,j} = a_{i+1,j} + a_{i-1,j} + a_{i,j-1} + a_{i,j-1} = 2\frac{\Delta y}{\Delta x}\left(1 + \frac{\Delta x^2}{\Delta y^2}\right)$ $d = \bar{S}\,\Delta x\,\Delta y/\Gamma$
(c) Constant transport property and $\Delta x = \Delta y$ 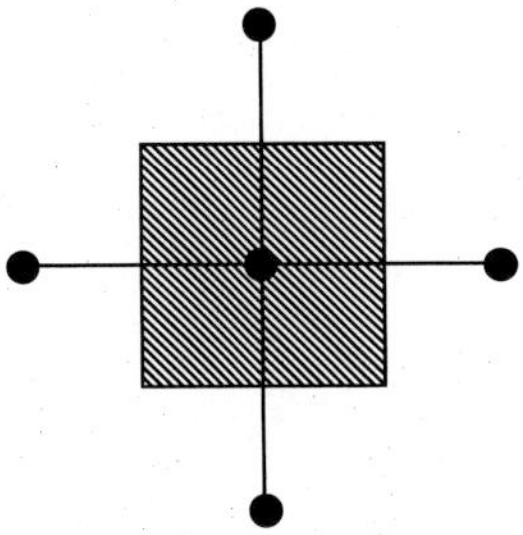	$a_{i,j}\phi_{i,j} = a_{i+1,j}\phi_{i+1,j} + a_{i-1,j}\phi_{i-1,j} + a_{i,j+1}\phi_{i,j+1} + a_{i,j-1}\phi_{i,j-1} + d$ $a_{i+1,j} = a_{i-1,j} = a_{i,j+1} = a_{i,j-1} = 1, a_{i,j} = 4$ $d = \bar{S}\Delta x^2/\Gamma$
(d) Constant transport property, $\Delta x = \Delta y$, and no source 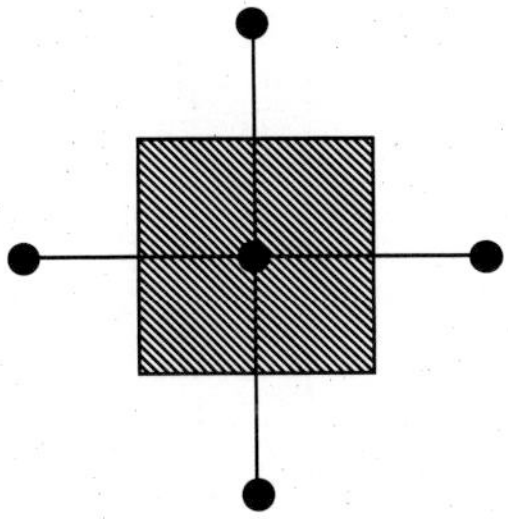	$4\phi_{i,j} = \phi_{i+1,j} + \phi_{i-1,j} + \phi_{i,j+1} + \phi_{i,j-1}$

Using this equation, the finite difference equations for the nine interior nodes are obtained as

$$\begin{aligned}
&\textit{Node 1} \quad T_2 + T_4 + 20 + 400 - 4T_1 = 0\\
&\textit{Node 2} \quad T_1 + T_5 + T_3 + 400 - 4T_2 = 0\\
&\textit{Node 3} \quad T_2 + T_6 + 20 + 400 - 4T_3 = 0\\
&\textit{Node 4} \quad 20 + T_1 + T_5 + T_7 - 4T_4 = 0\\
&\textit{Node 5} \quad T_2 + T_4 + T_8 + T_6 - 4T_5 = 0\\
&\textit{Node 6} \quad T_3 + T_5 + T_9 + 20 - 4T_6 = 0\\
&\textit{Node 7} \quad T_4 + 20 + 20 + T_8 - 4T_7 = 0\\
&\textit{Node 8} \quad T_5 + T_7 + 20 + T_9 - 4T_8 = 0\\
&\textit{Node 9} \quad T_6 + T_8 + 20 + 20 - 4T_9 = 0
\end{aligned}$$

Rearranging these equations, we get the system of equations as

$$\begin{aligned}
&-4T_1 + T_2 + 0 + T_4 + 0 + 0 + 0 + 0 + 0 = -420\\
&T_1 - 4T_2 + T_3 + 0 + T_5 + 0 + 0 + 0 + 0 = -400\\
&0 + T_2 - 4T_3 + 0 + 0 + T_6 + 0 + 0 + 0 = -420\\
&T_1 + 0 + 0 - 4T_4 + T_5 + 0 + T_7 + 0 + 0 = -20\\
&0 + T_2 + 0 + T_4 - 4T_5 + T_6 + 0 + T_8 + 0 = 0\\
&0 + 0 + T_3 + 0 + T_5 - 4T_6 + 0 + 0 + T_9 = -20\\
&0 + 0 + 0 + T_4 + 0 + 0 - 4T_7 + T_8 + 0 = -40\\
&0 + 0 + 0 + 0 + T_5 + 0 + T_7 - 4T_8 + T_9 = -20\\
&0 + 0 + 0 + 0 + 0 + T_6 + 0 + T_8 - 4T_9 = -40
\end{aligned}$$

The system of equations can be expressed in matrix notation as

$$\begin{bmatrix}
-4 & 1 & 0 & 1 & 0 & 0 & 0 & 0 & 0\\
1 & -4 & 1 & 0 & 1 & 0 & 0 & 0 & 0\\
0 & 1 & -4 & 0 & 0 & 1 & 0 & 0 & 0\\
1 & 0 & 0 & -4 & 1 & 0 & 1 & 0 & 0\\
0 & 1 & 0 & 1 & -4 & 1 & 0 & 1 & 0\\
0 & 0 & 1 & 0 & 1 & -4 & 0 & 0 & 1\\
0 & 0 & 0 & 1 & 0 & 0 & -4 & 0 & 0\\
0 & 0 & 0 & 0 & 1 & 0 & 1 & -4 & 1\\
0 & 0 & 0 & 0 & 0 & 1 & 0 & 1 & -4
\end{bmatrix}
\begin{Bmatrix} T_1\\ T_2\\ T_3\\ T_4\\ T_5\\ T_6\\ T_7\\ T_8\\ T_9 \end{Bmatrix}
=
\begin{Bmatrix} -420\\ -400\\ -420\\ -20\\ 0\\ -20\\ -40\\ -20\\ -40 \end{Bmatrix}$$

Using the Gauss elimination routine, the solution to this system of equations is obtained as

$$\begin{Bmatrix} T_1 \\ T_2 \\ T_3 \\ T_4 \\ T_5 \\ T_6 \\ T_7 \\ T_8 \\ T_9 \end{Bmatrix} = \begin{Bmatrix} 182.86^\circ\text{C} \\ 220.16^\circ\text{C} \\ 182.86^\circ\text{C} \\ 91.25^\circ\text{C} \\ 115.00^\circ\text{C} \\ 91.25^\circ\text{C} \\ 47.14^\circ\text{C} \\ 57.32^\circ\text{C} \\ 47.14^\circ\text{C} \end{Bmatrix}$$

The heat transfer rate from the top surface is computed from the expression

$$\begin{aligned} q_{\text{T}} = \sum kA_y \frac{\Delta T}{\Delta y} &= k\Delta x \cdot 1 \frac{(400 - T_1)}{\Delta y} + k\Delta x \cdot 1 \frac{(400 - T_2)}{\Delta y} + k\Delta x \cdot 1 \frac{(400 - T_3)}{\Delta y} \\ &= 40.0 \times 0.5 \times \frac{(400 \quad 182.86)}{0.5} + 40.0 \times 0.5 \times \frac{(400 - 220.16)}{0.5} \\ &\quad + 40.0 \times 0.5 \times \frac{(400 - 182.86)}{0.5} \\ &= 24\,564.8\ \text{W} \end{aligned}$$

The heat transfer rate from the left surface is computed from the expression

$$\begin{aligned} q_{\text{L}} = \sum kA_x \frac{\Delta T}{\Delta x} &= k\Delta y \cdot 1 \frac{(20 - T_1)}{\Delta x} + k\Delta y \cdot 1 \frac{(20 - T_4)}{\Delta x} + k\Delta y \cdot 1 \frac{(20 - T_7)}{\Delta x} \\ &= 40.0 \times 0.5 \times \frac{(20 - 182.86)}{0.5} + 40.0 \times 0.5 \times \frac{(20 - 91.25)}{0.5} \\ &\quad + 40.0 \times 0.5 \times \frac{(20 - 47.14)}{0.5} \\ &= -10\,450\ \text{W} \end{aligned}$$

The heat transfer rate from the right surface is computed from the expression

$$\begin{aligned} q_{\text{R}} = \sum kA_x \frac{\Delta T}{\Delta x} &= k\Delta y \cdot 1 \frac{(20 - T_3)}{\Delta x} + k\Delta y \cdot 1 \frac{(20 - T_6)}{\Delta x} + k\Delta y \cdot 1 \frac{(20 - T_9)}{\Delta x} \\ &= 40.0 \times 0.5 \times \frac{(20 - 182.86)}{0.5} + 40.0 \times 0.5 \times \frac{(20 - 91.25)}{0.5} \\ &\quad + 40.0 \times 0.5 \times \frac{(20 - 47.14)}{0.5} \\ &= -10\,450\ \text{W} \end{aligned}$$

The heat transfer rate from the bottom surface is computed from the expression

$$\begin{aligned} q_{\mathrm{B}} &= \sum kA_y\frac{\Delta T}{\Delta y} = k\Delta x\cdot 1\frac{(20-T_7)}{\Delta y} + k\Delta x\cdot 1\frac{(20-T_8)}{\Delta y} + k\Delta x\cdot 1\frac{(20-T_9)}{\Delta y} \\ &= 40.0\times 0.5\times\frac{(20-47.14)}{0.5} + 40.0\times 0.5\times\frac{(20-57.32)}{0.5} \\ &\quad + 40.0\times 0.5\times\frac{(20-47.14)}{0.5} \\ &= -3\,664\ \mathrm{W} \end{aligned}$$

It can be noted that the heat transfer rates at the left and right surfaces are identical. This is due to the symmetric nature of the boundary conditions and geometry of the problem. The negative heat transfer rates indicate that heat is lost from these surfaces. The overall energy balance for the two-dimensional steady state problem states that rate of energy inflow be balanced by the rate of energy outflow, i.e

$$E_{\mathrm{in}} = E_{\mathrm{out}}$$

In this problem energy inflow is given by the heat transfer rate at the top surface

$$E_{\mathrm{in}} = q_{\mathrm{T}} = 24\,564.8\ \mathrm{W}$$

The energy outflow is given by the summation of all heat transfer rates at the left, right, and bottom surfaces

$$\begin{aligned} E_{\mathrm{out}} &= q_{\mathrm{L}} + q_{\mathrm{R}} + q_{\mathrm{B}} = -10\,450 - 10\,450 - 3,664 \\ E_{\mathrm{out}} &= -24\,564.0\ \mathrm{W} \end{aligned}$$

The percent relative error in the overall energy balance is given as

$$\varepsilon_{\mathrm{ae}} = \frac{E_{\mathrm{in}} - E_{\mathrm{out}}}{E_{\mathrm{in}}} = \frac{24\,564.8 - 24\,564.0}{24\,564.8}\times 100$$

$$\varepsilon_{\mathrm{ae}} = 0.003\%$$

It can be noted that there is an excellent agreement in the overall energy balance as expected, and this indicates the accuracy in the finite difference formulation and the solution of the system of equations. However, it should be remembered here that the even though the computed temperatures at the nodal points satisfy the overall energy balance, they do not necessarily give us the correct temperature distribution. This approximate temperature distribution can be improved further by performing a *grid refinement* study by reducing the grid size or using increased number of nodal points. Such grid refinement study will lead to a converged and more accurate solution. Another way of checking the accuracy of the approximate temperature distribution is by comparing this solution with that given by an exact solution for a simpler or limiting case problem for which an exact solution exists.

6.2 BOUNDARY CONDITIONS

When the boundary condition is of the ***first kind***, i.e. ***of constant surface value***, we can simply assign the known value of temperature to the boundary nodes and solve the system of equations as was done in the previous example. However, for boundary conditions of the ***second*** or ***third kind***, we need to use discretization equations derived for those boundary nodes. Figure 6.1 shows typical two dimensional boundary nodes, such as ***boundary surface node***, ***exterior corner node*** and ***interior corner node***.

Figure 6.2 show control volumes for boundary nodes on all four sides of a two-dimensional co-ordinate system. It can be noted that only one-half of

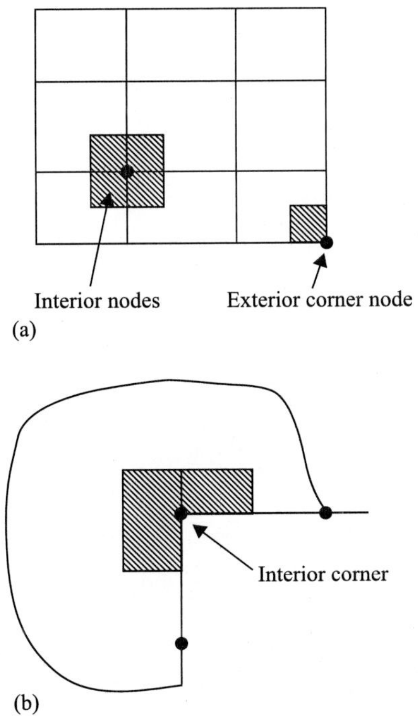

Figure 6.1 Control volumes for boundary nodes in two-dimensional problems

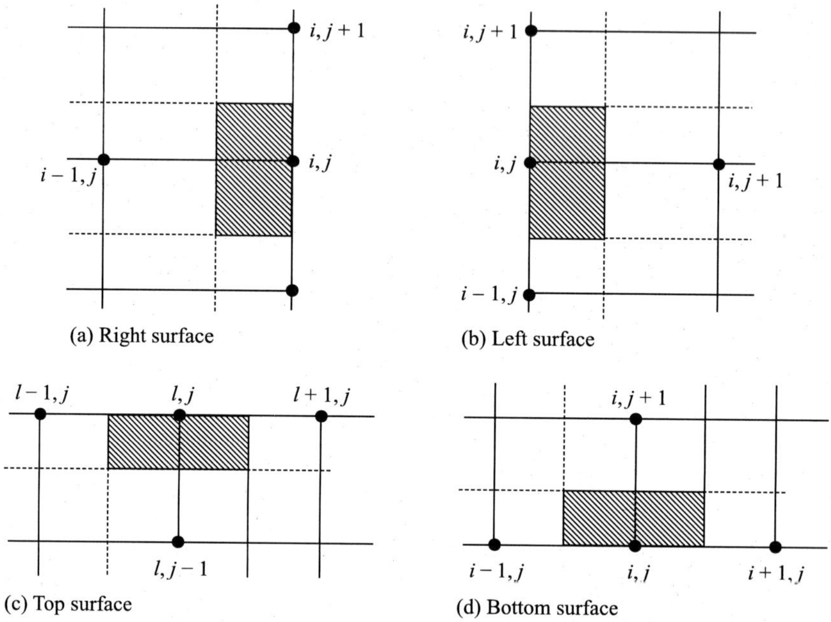

Figure 6.2 Control volumes for boundary nodes on two-dimensional plane surfaces

a control volume surrounds the boundary node ***i***, as marked by the shading in the figures.

Let us show the derivation of one such boundary node based on the control volume formulation. Consider the grid point located on the ***right-hand side boundary*** as shown in Figure 6.2a. Integrating the governing Equation (Equation 6.1) over this half control volume, we get

$$\int_{i,j-\frac{1}{2}}^{i,j+\frac{1}{2}}\int_{i-\frac{1}{2},j}^{i,j}\frac{\partial}{\partial x}\left(\Gamma\frac{\partial\phi}{\partial x}\right)dx\,dy+\int_{i,j-\frac{1}{2}}^{i,j+\frac{1}{2}}\int_{i-\frac{1}{2},j}^{i,j}\frac{\partial}{\partial y}\left(\Gamma\frac{\partial\phi}{\partial y}\right)dxdy+\int_{i,j-\frac{1}{2}}^{i,j+\frac{1}{2}}\int_{i-\frac{1}{2},j}^{i,j}S\,dx\,dy$$

or

$$\left[\left(\Gamma\frac{\partial\phi}{\partial x}\right)_{i,j}-\left(\Gamma\frac{\partial\phi}{\partial x}\right)_{i-\frac{1}{2},j}\right]\Delta y+\left[\left(\Gamma\frac{\partial\phi}{\partial y}\right)_{i,j+\frac{1}{2}}-\left(\Gamma\frac{\partial\phi}{\partial y}\right)_{i,j-\frac{1}{2}}\right]\frac{\Delta x}{2}+\Delta y\left(\int_{i-\frac{1}{2},j}^{i,j}S\Delta x\right) \tag{6.3}$$

The first term in the equation is described by the constant surface flux condition as

$$\left(\Gamma\frac{\partial\phi}{\partial x}\right)_{i,j} = f''_{rs} \tag{6.4}$$

We assume piecewise-linear profiles to approximate gradients at the faces of the control volume as follows:

$$\left(\Gamma\frac{\partial\phi}{\partial y}\right)_{i,j+\frac{1}{2}} = \Gamma_{i,j+\frac{1}{2}}\frac{\phi_{i,j+1}-\phi_{i,j}}{\Delta y} \tag{6.5a}$$

$$\left(\Gamma\frac{\partial\phi}{\partial y}\right)_{i,j-\frac{1}{2}} = \Gamma_{i,j-\frac{1}{2}}\frac{\phi_{i,j}-\phi_{i,j-1}}{\Delta y} \tag{6.5b}$$

$$\left(\Gamma\frac{\partial\phi}{\partial x}\right)_{i-\frac{1}{2},j} = \Gamma_{i-\frac{1}{2},j}\frac{\phi_{i,j}-\phi_{i-1,j}}{\Delta x} \tag{6.5c}$$

A step-wise constant profile for the dependent variable is assumed in the linearized form of the source term

$$S = S_0 + S_1\phi_i \tag{6.6}$$

Substituting Equation 6.4 to Equation 6.6 into Equation 6.3, we get

$$\left(\Gamma_{i-\frac{1}{2},j}\frac{\Delta y}{\Delta x} + \frac{\Gamma_{i,j+\frac{1}{2}}}{2}\frac{\Delta x}{\Delta y} + \frac{\Gamma_{i,j-\frac{1}{2}}}{2}\frac{\Delta x}{\Delta y} - \frac{S_1}{2}\Delta x\,\Delta y\right)\phi_{i,j} = \Gamma_{i-\frac{1}{2},j}\frac{\Delta y}{\Delta x}\phi_{i-1,j}$$

$$+\frac{\Gamma_{i,j+\frac{1}{2}}}{2}\frac{\Delta x}{\Delta y}\phi_{i,j+1} + \frac{\Gamma_{i,i-\frac{1}{2}}}{2}\frac{\Delta x}{\Delta y}\phi_{i,j-1} + \frac{S_0}{2}\Delta x\,\Delta y + f''_{rs}\Delta y \tag{6.7}$$

Simplifying, we get

$$a_{i,j}\phi_{i,j} = a_{i-1,j}\phi_{i-1,j} + a_{i,j+1}\phi_{i,j+1} + a_{i,j-1}\phi_{i,j-1} + d \tag{6.8}$$

where

$$a_{i-1,j} = \Gamma_{i-\frac{1}{2},j}\frac{\Delta y}{\Delta x} \tag{6.9a}$$

$$a_{i,j+1} = \Gamma_{i,j+\frac{1}{2}}\frac{\Delta x}{2\Delta y} \tag{6.9b}$$

$$a_{i,j-1} = \Gamma_{i,j-\frac{1}{2}}\frac{\Delta x}{2\Delta y} \tag{6.9c}$$

$$a_{i,j} = a_{i-1,j} + a_{i,j+1} + a_{i,j-1} - \frac{S_1}{2}\Delta x\,\Delta y \tag{6.9d}$$

$$d = \frac{S_0}{2}\Delta x \Delta y + f''_{rs} \tag{6.9e}$$

Zero Surface Flux For the case of a ***zero surface flux*** or ***along a line of symmetry***, the appropriate boundary condition is defined as

$$\Gamma\frac{\partial \phi}{\partial x} = 0 \quad \text{or} \quad f''_{rs} = 0 \tag{6.10}$$

We can obtain the appropriate finite difference for this case by setting $f''_{rs} = 0$ in Equation 6.10.

Similarly, we can derive the discretization equations for other boundary nodes as shown in Figure 6.2. A summary of these equations is given in Table 6.2.

Mixed Boundary Condition. Another possible boundary condition is the boundary condition of the ***third kind*** or ***the mixed boundary condition*** defined as

$$-\Gamma\frac{\partial \phi}{\partial x} = h(\phi - \phi_\infty) \tag{6.11a}$$

or

$$f''_{rs} = h(\phi_\infty - \phi) \tag{6.11b}$$

Substituting Equation 6.11b into Equation 6.7, we get

$$\left(\Gamma_{i-\frac{1}{2},j}\frac{\Delta y}{\Delta x} + \frac{\Gamma_{i,j+\frac{1}{2}}}{2}\frac{\Delta x}{\Delta y} + \frac{\Gamma_{i,j-\frac{1}{2}}}{2}\frac{\Delta x}{\Delta y} - \frac{S_1}{2}\Delta x\,\Delta y\right)\phi_{i,j} = \Gamma_{i-\frac{1}{2},j}\frac{\Delta y}{\Delta x}\phi_{i-1,j}$$

$$+\frac{\Gamma_{i,j+\frac{1}{2}}}{2}\frac{\Delta x}{\Delta y}\phi_{i,j+1} + \frac{\Gamma_{i,j-\frac{1}{2}}}{2}\frac{\Delta x}{\Delta y}\phi_{i,j-1} + \frac{S_0}{2}\Delta x\,\Delta y + h(\phi_\infty - \phi_{i,j})\,\Delta y$$

Rearranging

$$\left(\Gamma_{i-\frac{1}{2},j}\frac{\Delta y}{\Delta x} + \frac{\Gamma_{i,j+\frac{1}{2}}}{2}\frac{\Delta x}{\Delta y} + \frac{\Gamma_{i,j-\frac{1}{2}}}{2}\frac{\Delta x}{\Delta y} - \frac{S_1}{2}\Delta x\,\Delta y + h\,\Delta y\right)\phi_{i,j}$$

$$= \Gamma_{i-\frac{1}{2},j}\frac{\Delta y}{\Delta x}\phi_{i-1,j} + \frac{\Gamma_{i,j+\frac{1}{2}}}{2}\frac{\Delta x}{\Delta y}\phi_{i,j+1} + \frac{\Gamma_{i,j-\frac{1}{2}}}{2}\frac{\Delta x}{\Delta y}\phi_{i,j-1} + \frac{S_0}{2}\Delta x\,\Delta y + h\phi_\infty\,\Delta y$$

Simplifying, we get

$$a_{i,j}\phi_{i,j} = a_{i-1,j}\phi_{i-1,j} + a_{i,j+1}\phi_{i,j+1} + a_{i,j-1}\phi_{i,j-1} + d \tag{6.12}$$

Table 6.2 A Summary of Finite Difference Equations for Boundary Nodes with Constant Surface Flux

Physical Situations	Finite Difference Formula
(a) Grid point located at the right-hand side boundary 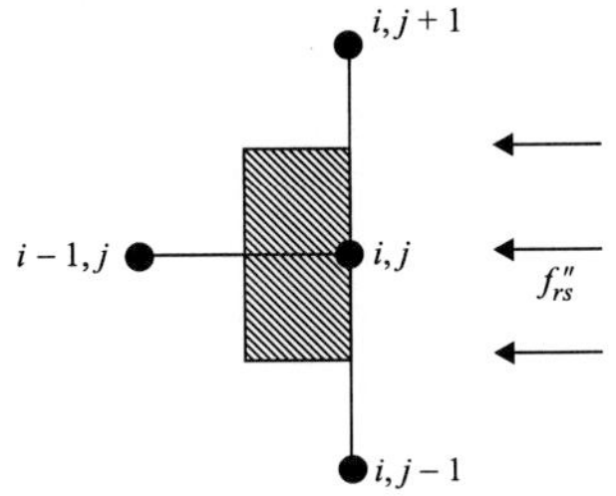	$a_{i,j}\phi_{i,j} = a_{i-1,j}\phi_{i-1,j} + a_{i,j+1}\phi_{i,j+1} + a_{i,j-1}\phi_{i,j-1} + d$ $a_{i-1,j} = \Gamma_{i-\frac{1}{2},j}\frac{\Delta y}{\Delta x}, \quad a_{i,j+1} = \frac{\Gamma_{i,j+\frac{1}{2}}}{2}\frac{\Delta x}{\Delta y}, a_{i,j-1} = \frac{\Gamma_{i,j-\frac{1}{2}}}{2}\frac{\Delta x}{\Delta y}$ $a_{i,j} = a_{i-1,j} + a_{i,j+1} + a_{i,j-1} - \frac{S_1}{2}\Delta x\,\Delta y$ $d = \frac{S_0}{2}\Delta x\,\Delta y + f''_{rs}\,\Delta y$
(b) Grid point located at the left-hand side boundary 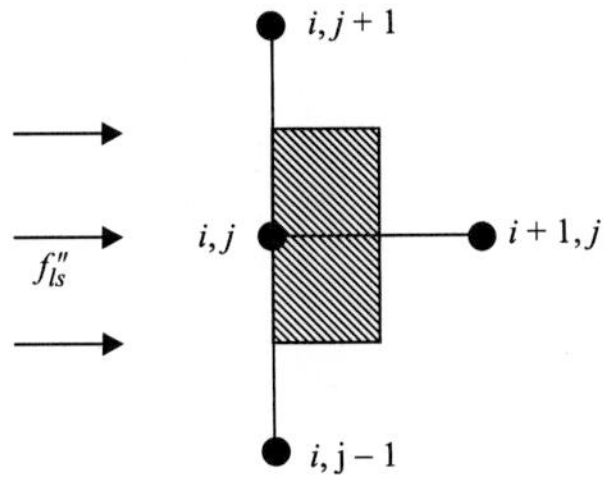	$a_{i,j}\phi_{i,j} = a_{i+1,j}\phi_{i+1,j} + a_{i,j+1}\phi_{i,j+1} + a_{i,j-1}\phi_{i,j-1} + d$ $a_{i+1,j} = \Gamma_{i+\frac{1}{2},j}\frac{\Delta y}{\Delta x}, \quad a_{i,j+1} = \frac{\Gamma_{i,j+\frac{1}{2}}}{2}\frac{\Delta x}{\Delta y}, \quad a_{i,j-1} = \frac{\Gamma_{i,j-\frac{1}{2}}}{2}\frac{\Delta x}{\Delta y}$ $a_{i,j} = a_{i+1,j} + a_{i,j+1} + a_{i,j-1} - \frac{S_1}{2}\Delta x\,\Delta y$ $d = \frac{S_0}{2}\Delta x\,\Delta y + f''_{ls}\,\Delta y$
(c) Grid point located at the top boundary 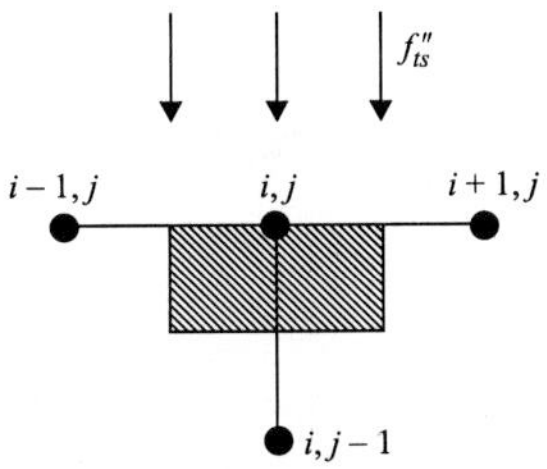	$a_{i,j}\phi_{i,j} = a_{i+1,j}\phi_{i+1,j} + a_{i-1,j}\phi_{i-1,j} + a_{i,j-1}\phi_{i,j-1} + d$ $a_{i+1,j} = \frac{\Gamma_{i+\frac{1}{2},j}}{2}\frac{\Delta y}{\Delta x}, \quad a_{i-1,j} = \frac{\Gamma_{i-\frac{1}{2},j}}{2}\frac{\Delta y}{\Delta x}, \quad a_{i,j-1} = \Gamma_{i,j-\frac{1}{2}}\frac{\Delta x}{\Delta y}$ $a_{i,j} = a_{i+1,j} + a_{i-1,j} + a_{i,j-1} - \frac{S_1}{2}\Delta x\,\Delta y$ $d = \frac{S_0}{2}\Delta x\,\Delta y + f''_{ts}\,\Delta y$
(d) Grid point located at bottom boundary 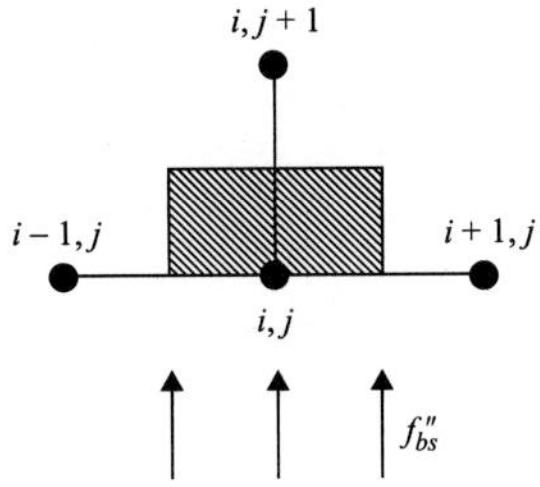	$a_{i,j}\phi_{i,j} = a_{i+1,j}\phi_{i+1,j} + a_{i-1,j}\phi_{i-1,j} + a_{i,j-1}\phi_{i,j-1} + d$ $a_{i+1,j} = \frac{\Gamma_{i+\frac{1}{2},j}}{2}\frac{\Delta y}{\Delta x}, \quad a_{i-1,j} = \frac{\Gamma_{i-\frac{1}{2},j}}{2}\frac{\Delta y}{\Delta x}, \quad a_{i,j+1} = \Gamma_{i,j+\frac{1}{2}}\frac{\Delta x}{\Delta y}$ $a_{i,j} = a_{i+1,j} + a_{i-1,j+1} + a_{i,j+1} - \frac{S_1}{2}\Delta x\,\Delta y$ $d = \frac{S_0}{2}\Delta x\,\Delta y + f''{}_{bs}\,\Delta y$

where

$$a_{i-1,j} = \Gamma_{i-\frac{1}{2},j}\frac{\Delta y}{\Delta x} \tag{6.13a}$$

$$a_{i,j+1} = \Gamma_{i,j+\frac{1}{2}}\frac{\Delta x}{2\Delta y} \tag{6.13b}$$

$$a_{i,j-1} = \Gamma_{i,j-\frac{1}{2}}\frac{\Delta x}{2\Delta y} \tag{6.13c}$$

$$a_{i,j} = a_{i-1,j} + a_{i,j+1} + a_{i,j-1} - \frac{S_1}{2}\Delta x\,\Delta y + h\,\Delta y \tag{6.13d}$$

$$d = \frac{S_0}{2}\Delta x\,\Delta y + h\phi_\infty\,\Delta y \tag{6.13e}$$

For uniform grid size distributions, $\Delta x = \Delta y$, and constant transport property, Γ, Equation 6.12 and Equation 6.13 are simplified to

$$\left(\frac{h\Delta x}{\Gamma} + 2 - \frac{S_1}{2\Gamma}\Delta x^2\right)\phi_{i,j} = \frac{1}{2}(2\phi_{i-1,j} + \phi_{i,j+1} + \phi_{i,j-1}) + \frac{S_0}{2\Gamma}\Delta x^2 + \frac{h\Delta x}{\Gamma}\phi_\infty \tag{6.14}$$

For the case with no source term, we can set $S_1 = 0$ and $S_0 = 0$, and Equation 6.14 transforms into

$$\left(\frac{h\Delta x}{\Gamma} + 2\right)\phi_{i,j} = \frac{1}{2}(2\phi_{i-1,j} + \phi_{i,j+1} + \phi_{i,j-1}) + \frac{h\Delta x}{\Gamma}\phi \tag{6.15}$$

Similarly, we can derive the discretization equations for other boundary nodes as shown in Figure 6.2. A summary of these equations is given in Table 6.3.

6.2.1 Corner Boundary Nodes

The procedure outlined for boundary nodes on a plane boundary surface can be used to derive the discretization equations for corner boundary nodes. Let us show the derivation of one such equation for the northeast exterior corner node shown below.

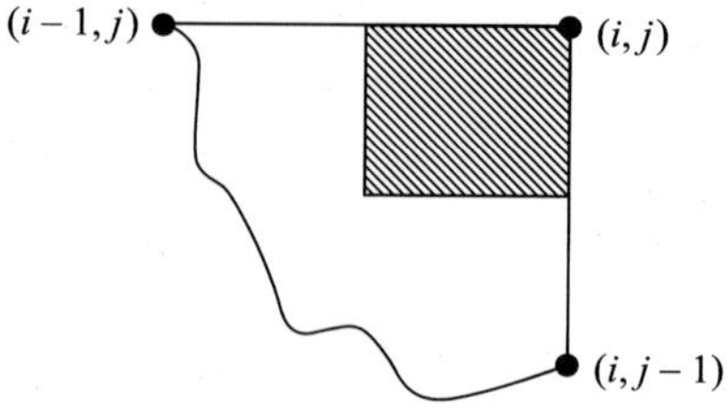

Figure 6.3 Control volume for an exterior corner node

Table 6.3 A Summary of Finite Difference Equations for Boundary Nodes with Mixed Condition

Physical Situations	Finite Difference Formula
(a) Grid point located at the right-hand side boundary 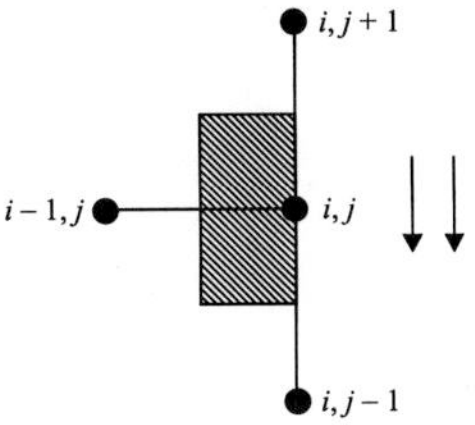	$\left(\frac{h\Delta x}{\Gamma}+2\right)\phi_{i,j}=\frac{1}{2}(2\phi_{i-1,j}+\phi_{i,j+1}+\phi_{i,j-1})+\frac{h\Delta x}{\Gamma}\phi_{\infty}$
(b) Grid point located at the left-hand side boundary 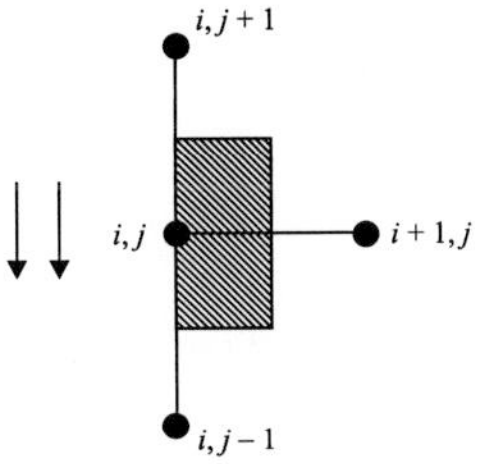	$\left(\frac{h\Delta x}{\Gamma}+2\right)\phi_{i,j}=\frac{1}{2}(2\phi_{i+1,j}+\phi_{i,j+1}+\phi_{i,j-1})+\frac{h\Delta x}{\Gamma}\phi_{\infty}$
(c) Grid point located at the top boundary 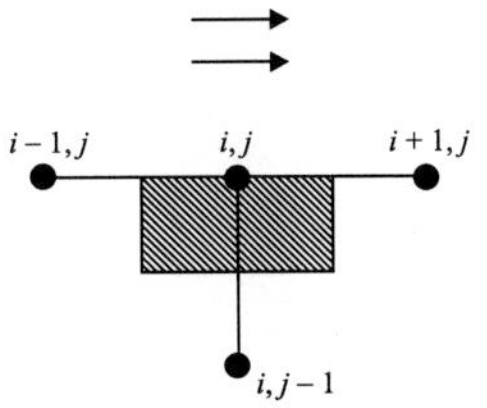	$\left(\frac{h\Delta x}{\Gamma}+2\right)\phi_{i,j}=\frac{1}{2}(2\phi_{i,j-1}+\phi_{i+1,j}+\phi_{i-1,j})+\frac{h\Delta x}{\Gamma}\phi_{\infty}$
(d) Grid point located at bottom boundary 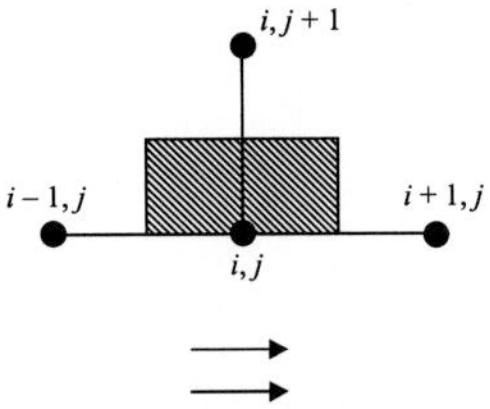	$\left(\frac{h\Delta x}{\Gamma}+2\right)\phi_{i,j}=\frac{1}{2}(2\phi_{i,j+1}+\phi_{i+1,j}+\phi_{i-1,j})+\frac{h\Delta x}{\Gamma}\phi_{\infty}$
(e) Convective boundary node as above with uniform heat generation	$\left(\frac{h\Delta x}{\Gamma}+2\right)\phi_{i,j}=\frac{1}{2}(2\phi_{i,j+1}+\phi_{i+1,j}+\phi_{i-1,j})+\frac{h\Delta x}{\Gamma}\phi_{\infty}+\frac{\bar{S}(\Delta x)^2}{\Gamma}$

Integrating the governing equation (Equation 6.1) over the quarter control volume surrounding the node (i, j), we get

$$\int_{i,j-\frac{1}{2}}^{i,j}\int_{i-\frac{1}{2},j}^{i,j}\frac{\partial}{\partial x}\left(\Gamma\frac{\partial\phi}{\partial x}\right)dx\,dy+\int_{i,j-\frac{1}{2}}^{i,j}\int_{i-\frac{1}{2},j}^{i,j}\frac{\partial}{\partial y}\left(\Gamma\frac{\partial\phi}{\partial y}\right)dx\,dy+\int_{i,j-\frac{1}{2}}^{i,j}\int_{i-\frac{1}{2},j}^{i,j}S\;dx\,dy \tag{6.16}$$

or

$$\left[\left(\Gamma\frac{\partial\phi}{\partial x}\right)_{i,j}-\left(\Gamma\frac{\partial\phi}{\partial x}\right)_{i-\frac{1}{2},j}\right]\frac{\Delta y}{2}+\left[\left(\Gamma\frac{\partial\phi}{\partial y}\right)_{i,j}-\left(\Gamma\frac{\partial\phi}{\partial y}\right)_{i,j-\frac{1}{2}}\right]\frac{\Delta x}{2}+\frac{\Delta y}{2}\left(\int_{i-\frac{1}{2},j}^{i,j}S\,\Delta x\right) \tag{6.17}$$

The first and the third terms in the equation are described by the constant surface flux condition as

$$\left(\Gamma\frac{\partial\phi}{\partial x}\right)_{i,j}=f''_{rs} \tag{6.18a}$$

and

$$\left(\Gamma\frac{\partial\phi}{\partial y}\right)_{i,j}=f''_{ts} \tag{6.18b}$$

We assume piecewise-linear profiles to approximàte gradients at the faces of the control volume as

$$\left(\Gamma\frac{\partial\phi}{\partial y}\right)_{i,j-\frac{1}{2}}=\Gamma_{i,j-\frac{1}{2}}\frac{\phi_{i,j}-\phi_{i,j-1}}{\Delta y} \tag{6.19a}$$

$$\left(\Gamma\frac{\partial\phi}{\partial x}\right)_{i-\frac{1}{2},j}=\Gamma_{i-\frac{1}{2},j}\frac{\phi_{i,j}-\phi_{i-1,j}}{\Delta x} \tag{6.19b}$$

A step-wise constant profile for the dependent variable is assumed in the linearized form of the source term

$$S=S_0+S_1\phi_i \tag{6.20}$$

Substituting Equations 6.18 to 6.20 into Equation 6.17, we get

$$\left(\Gamma_{i-\frac{1}{2},j}\frac{\Delta y}{2\Delta x}+\Gamma_{i,j-\frac{1}{2}}\frac{\Delta x}{2\Delta y}-S_1\frac{\Delta x\Delta y}{4}\right)\phi_{i,j}=\Gamma_{i-\frac{1}{2},j}\frac{\Delta y}{2\Delta x}\phi_{i-1,j}$$
$$+\frac{\Gamma_{i,i-\frac{1}{2}}}{2}\frac{\Delta x}{\Delta y}\phi_{i,j-1}+\frac{S_0}{4}\Delta x\,\Delta y+f''_{rs}\frac{\Delta y}{2}+f''_{ts}\frac{\Delta x}{2} \tag{6.21}$$

Simplifying, we get

$$a_{i,j}\phi_{i,j}=a_{i-1,j}\phi_{i-1,j}+a_{i,j-1}\phi_{i,j-1}+d \tag{6.22}$$

where

$$a_{i-1,j}=\Gamma_{i-\frac{1}{2},j}\frac{\Delta y}{2\Delta x} \tag{6.23a}$$

$$a_{i,j-1}=\Gamma_{i,j-\frac{1}{2}}\frac{\Delta x}{2\Delta y} \tag{6.23b}$$

$$a_{i,j}=a_{i-1,j}+a_{i,j-1}-\frac{S_1}{4}\Delta x\,\Delta y \tag{6.23c}$$

$$d=\frac{S_0}{4}\Delta x\Delta y+f''_{rs}\frac{\Delta y}{2}+f''_{ts}\frac{\Delta x}{2} \tag{6.23d}$$

Zero Surface Flux. For the case of a ***zero surface flux*** or ***along a line of symmetry***, the appropriate boundary condition is defined as

$$\Gamma\frac{\partial\phi}{\partial x}=0 \quad \text{or} \quad f''_{rs}=0 \tag{6.24a}$$

$$\Gamma\frac{\partial\phi}{\partial y}=0 \quad or \quad f''_{ts}=0 \tag{6.24b}$$

We can obtain the appropriate finite difference equation for this case by setting $f''_{rs}=0, f''_{ts}=0$ in Equation 6.23.

Mixed Boundary Condition. Another possible boundary condition is the boundary condition of the ***third kind*** or ***the mixed boundary condition*** defined as

$$-\Gamma\frac{\partial\phi}{\partial x}=h(\phi-\phi_\infty) \tag{6.25a}$$

or

$$f''_{rs}=h(\phi_\infty-\phi) \tag{6.25b}$$

and

$$-\Gamma\frac{\partial\phi}{\partial y} = h(\phi - \phi_\infty) \qquad (6.26a)$$

or

$$f''_{ts} = h(\phi_\infty - \phi) \qquad (6.26b)$$

Substituting Equations 6.25 and 6.26 into Equation 6.21, we get

$$\left(\Gamma_{i-\frac{1}{2},j}\frac{\Delta y}{2\Delta x} + \frac{\Gamma_{i,i-\frac{1}{2}}}{2}\frac{\Delta x}{\Delta y} - \frac{S_1}{4}\Delta x\Delta y\right)\phi_{i,j} = \Gamma_{i-\frac{1}{2},j}\frac{\Delta y}{2\Delta x}\phi_{i-1,j} + \frac{\Gamma_{i,i-\frac{1}{2}}}{2}\frac{\Delta x}{\Delta y}\phi_{i,j-1}$$
$$+ \frac{S_0}{4}\Delta x\,\Delta y + h_{rs}(\phi_\infty - \phi_{i,j})\Delta y$$
$$+ h_{ts}(\phi_\infty - \phi_{i,j})\Delta x \qquad (6.27)$$

Rearranging

$$\left(\Gamma_{i-\frac{1}{2},j}\frac{\Delta y}{2\Delta x} + \frac{\Gamma_{i,i-\frac{1}{2}}}{2}\frac{\Delta x}{\Delta y} - \frac{S_1}{4}\Delta x\,\Delta y + h_{rs}\frac{\Delta y}{2} + h_{ts}\frac{\Delta x}{2}\right)\phi_{i,j}$$
$$= \Gamma_{i-\frac{1}{2},j}\frac{\Delta y}{2\Delta x}\phi_{i-1,j} + \frac{\Gamma_{i,i-\frac{1}{2}}}{2}\frac{\Delta x}{\Delta y}\phi_{i,j-1} + \frac{S_0}{4}\Delta x\Delta y + h''_{rs}\phi_\infty\frac{\Delta y}{2} + h''_{ts}\phi_\infty\frac{\Delta x}{2}$$
$$(6.28)$$

Simplifying, we get

$$a_{i,j}\phi_{i,j} = a_{i-1,j}\phi_{i-1,j} + a_{i,j-1}\phi_{i,j-1} + d \qquad (6.29)$$

where

$$a_{i-1,j} = \Gamma_{i-\frac{1}{2},j}\frac{\Delta y}{2\Delta x} \qquad (6.30a)$$

$$a_{i,j-1} = \Gamma_{i,j-\frac{1}{2}}\frac{\Delta x}{2\Delta y} \qquad (6.30b)$$

$$a_{i,j} = a_{i-1,j} + a_{i,j-1} - \frac{S_1}{4}\Delta x\,\Delta y + h_{rs}\frac{\Delta y}{2} + h_{ts}\frac{\Delta x}{2} \qquad (6.30c)$$

$$d = \frac{S_0}{4}\Delta x\Delta y + h_{rs}\frac{\Delta y}{2}\phi_\infty + h_{ts}\frac{\Delta x}{2}\phi_\infty \qquad (6.30d)$$

For uniform grid size distributions, $\Delta x = \Delta y$, and using constant transport property Γ and constant coefficients $h''_{rs} = h''_{ts} = h$, Equation 6.30 is simplified to

$$a_{i-1,j} = \frac{\Gamma}{2} \tag{6.31a}$$

$$a_{i,j-1} = \frac{\Gamma}{2} \tag{6.31b}$$

$$a_{i,j} = a_{i-1,j} + a_{i,j-1} - \frac{S_1}{4}\Delta x^2 + h\Delta x \tag{6.31c}$$

$$d = \frac{S_0}{4}\Delta x^2 + h\phi_\infty \Delta x \tag{6.31d}$$

For the case with no source term, we can set $S_1 = 0$ and $S_0 = 0$ to get the simplified form

$$2\left(\frac{h\Delta x}{\Gamma} + 1\right)\phi_{i,j} = \phi_{i-1,j} + \phi_{i,j-1} + \frac{2h\Delta x}{\Gamma}\phi_\infty \tag{6.32}$$

Similarly, we can derive the discretization equations for other corner boundary nodes. A summary of these equations is given in Table 6.4.

Example 6.2: Two-dimensional Steady State Heat Conduction with Mixed Boundary Conditions: Consider a 1.5 m by 1.5 m square slab with the top and the left surfaces maintained at a high temperature, $T_H = 600°C$, and a low temperature, $T_C = 20°C$, respectively. The right and the bottom surfaces are subjected to convection conditions with $h_c = 30\ W/m^2°C$ and $T_\infty = 25°C$. The material conductivity is $k = 30\ W/m°C$. Apply the finite difference method to calculate the two-dimensional temperature distribution in the slab based on the grid distribution shown in the figure. Estimate the heat transfer rates at all surfaces, and show the overall energy balance.

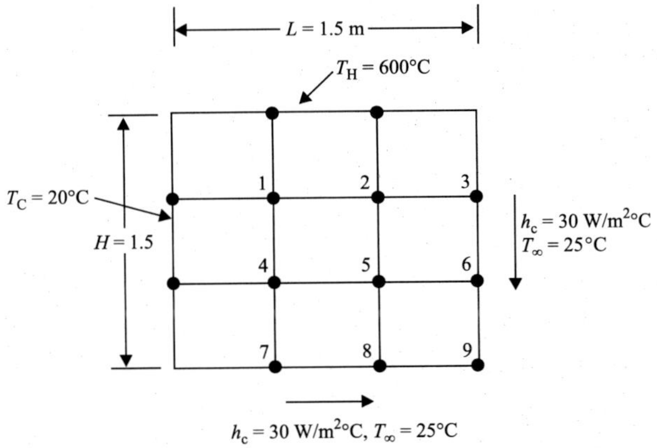

Table 6.4 A Summary of Finite Difference Equations for Corner Boundary Nodes with Mixed Condition

Physical Situations	Finite Difference Formula

Exterior corner nodes

(a) Northeast corner

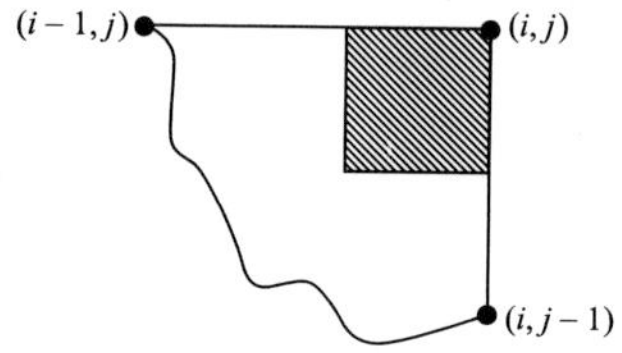

$$2\left(\frac{h\Delta x}{\Gamma}+1\right)\phi_{i,j}=\left(\phi_{i-1,j}+\phi_{i,j-1}\right)+2\frac{h\Delta x}{k}\phi_{\infty}$$

(b) Southeast corner

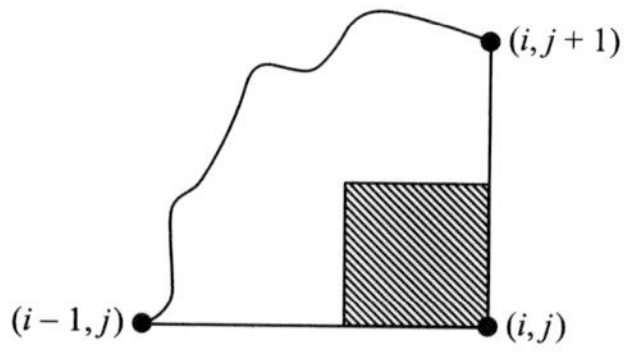

$$2\left(\frac{h\Delta x}{\Gamma}+1\right)\phi_{i,j}=\left(\phi_{i,j+1}+\phi_{i-1,j}\right)+2\frac{h\Delta x}{k}\phi_{\infty}$$

(c) Northwest corner

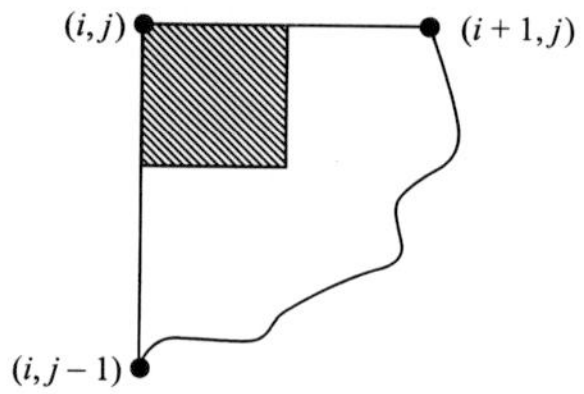

$$2\left(\frac{h\Delta x}{k}+1\right)\phi_{i,j}=\left(\phi_{i+1,j}+\phi_{i,j-1}\right)+2\frac{h\Delta x}{k}\phi_{\infty}$$

(d) Southwest corner

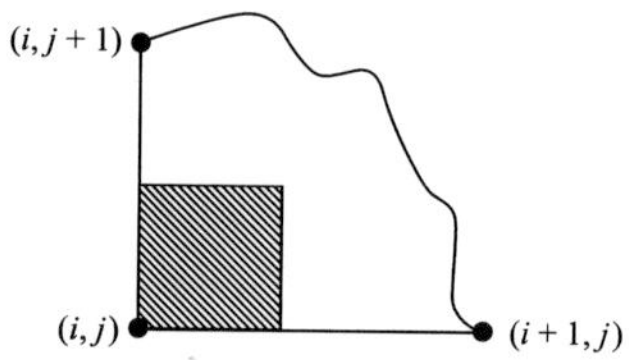

$$2\left(\frac{h\Delta x}{\Gamma}+1\right)\phi_{i,j}=\left(\phi_{i,j+1}+\phi_{i+1,j}\right)+2\frac{h\Delta x}{k}\phi_{\infty}$$

Interior corner node

(e) Northwest corner

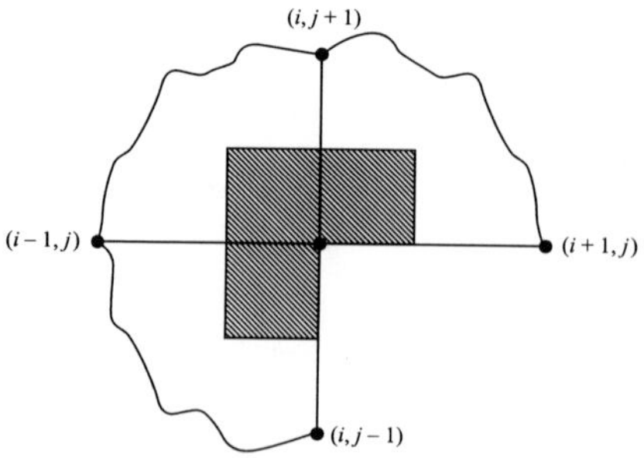

$$2\left(3+\frac{h\Delta x}{\Gamma}+1\right)\phi_{i,j}$$
$$=\phi_{i+1,j}+2\phi_{i-1,j}+2\phi_{i,j+1}+\phi_{i,j-1}+2\frac{h\Delta x}{k}\phi_{\infty}$$

The mathematical statement of the problem is as follows

Governing equation

$$\frac{\partial^2 T}{\partial x^2} + \frac{\partial^2 T}{\partial x^2} = 0$$

Boundary conditions

1. $x = 0, \qquad T(0, y) = T_C$
2. $x = L, \qquad -k\left.\frac{\partial T}{\partial x}\right|_{x=L} = h_c(T|_{x=L} - T_\infty)$
3. $y = 0, \qquad k\left.\frac{\partial T}{\partial y}\right|_{y=0} = h(T|_{y=0} - T_\infty)$
4. $y = H, \qquad T(x, H) = T_H$

Solution: For the prescribed grid size distribution with uniform grid sizes and number of division $N_{div} = 3$, we can estimate the following parameters as

$$\Delta x = \frac{L}{N_{x\text{div}}} = \frac{1.5\,\text{m}}{3} = 0.5\,\text{m}, \quad \Delta y = \frac{H}{N_{y\,\text{div}}} = \frac{1.5\,\text{m}}{3} = 0.5\,\text{m}$$

$$\frac{h_c \Delta x}{k} = \frac{30.0 \times 0.5}{30.0} = 0.5$$

In the next step, we select appropriate discretization equations for each node. There are four interior nodes 1, 2, 4 and 5 for which the appropriate finite difference equation is given by

$$a_{i,j} = a_{i+1,j}T_{i+1,j} + a_{i-1,j}T_{i-1,j} + a_{i,j+1}T_{i,j+1} + a_{i,j-1}T_{i,j-1} + d$$

$$a_{i+1,j} = a_{i-1,j} = a_{i,j+1} = a_{i,j-1} = 1, \quad a_{i,j} = 4$$

$$d = \bar{S}\Delta x^2/k$$

With no heat source term, $\bar{S} = 0$, and $d = 0$, the equation is expressed in a simplified form as

$$4T_{i,j} = T_{i+1,j} + T_{i-1,j} + T_{i,j+1} + T_{i,j-1}$$

Using this equation, the finite difference equations for the interior nodes 1, 2, 4 and 5 are obtained as

Node 1 $\quad 4T_1 = T_2 + T_C + T_H + T_4$

Node 2 $\quad 4T_2 = T_3 + T_1 + T_H + T_5$

Node 4 $\quad 4T_4 = T_5 + T_C + T_1 + T_7$

Node 5 $\quad 4T_5 = T_6 + T_4 + T_2 + T_8$

Nodes 3 and 6 are convective boundary nodes on right side surface, and we use the appropriate discretization equation which is given as

$$a_{i,j}T_{i,j} = a_{i-1,j}T_{i-1,j} + a_{i,j+1}T_{i,j+1} + a_{i,j-1}T_{i,j-1} + d$$

$$a_{i-1,j} = k_{i-\frac{1}{2},j}\frac{\Delta y}{\Delta x}, \quad a_{i,j+1} = \frac{k_{i,j+\frac{1}{2}}}{2}\frac{\Delta x}{\Delta y}, \quad a_{i,j-1} = \frac{k_{i,j-\frac{1}{2}}}{2}\frac{\Delta x}{\Delta y}$$

$$a_{i,j} = a_{i-1,j} + a_{i,j+1} + a_{i,j-1} - \frac{S_1}{2}\Delta x\,\Delta y + h\,\Delta y$$

$$d = \frac{S_0}{2}\Delta x\,\Delta y + hT_\infty\,\Delta y$$

Again, for uniform grid $\Delta x = \Delta y$, constant conductivity, $k_{i-\frac{1}{2},j} = k_{i+\frac{1}{2},j} = k_{i,j+\frac{1}{2}} = k_{i,j-\frac{1}{2}} = k$, and no heat source, $S_0 = S_1 = 0$, the equation reduces to

$$\left(\frac{h\Delta x}{k} + 2\right)T_{i,j} = \frac{1}{2}\left(2T_{i-1,j} + T_{i,j+1} + T_{i,j-1}\right) + \frac{h\Delta x}{k}T_\infty$$

Applying this equation to nodes 3 and 5, we get

$$\textit{Node 3}\ (0.5 + 2)T_3 = \frac{1}{2}(2T_2 + T_{\mathrm{H}} + T_6) + 0.5T_\infty$$

$$5.0T_3 = 2T_2 + T_{\mathrm{H}} + T_6 + T_\infty$$

$$\textit{Node 6}\ (0.5 + 2)T_6 = \frac{1}{2}(2T_5 + T_3 + \mathrm{T}_9) + 0.5T_\infty$$

$$5.0T_6 = 2T_5 + T_3 + T_9 + T_\infty$$

Nodes 7 and 8 are convective boundary nodes on the bottom surface, and we select the discretization as

$$a_{i,j}T_{i,j} = a_{i+1,j}T_{i+1,j} + a_{i-1,j}T_{i-1,j} + a_{i,j-1}T_{i,j-1} + d$$

$$a_{i+1,j} = \frac{\Gamma_{i+\frac{1}{2},j}}{2}\frac{\Delta y}{\Delta x}, \quad a_{i-1,j} = \frac{\Gamma_{i-\frac{1}{2},j}}{2}\frac{\Delta y}{\Delta x}, a_{i,j+1} = \Gamma_{i,j+\frac{1}{2}}\frac{\Delta x}{\Delta y}$$

$$a_{i,j} = a_{i+1,j} + a_{i-1,j+1} + a_{i,j-1} - \frac{S_1}{2}\Delta x\,\Delta y + h\,\Delta x$$

$$d = \frac{S_0}{2}\Delta x\,\Delta y + hT_\infty\,\Delta x$$

This equation is simplified in a similar manner for a uniform grid, constant conductivity, and no heat source of the form

$$\left(h\frac{\Delta x}{k} + 2\right)T_{i,j} = \frac{1}{2}\left(2T_{i,j+1} + T_{i+1,j} + T_{i-1,j}\right) + \frac{h\Delta x}{k}T_\infty$$

Applying this equation to nodes 7 and 8, we get

$$\textit{Node 7} \quad (0.5+2)T_7 = \frac{1}{2}(2T_4 + T_8 + T_c) + 0.5T_\infty$$
$$5.0T_7 = 2T_4 + T_8 + T_c + T_\infty$$

$$\textit{Node 8} \quad (0.5+2)T_8 = \frac{1}{2}(2T_s + T_9 + T_7) + 0.5T_\infty$$
$$5.0T_8 = 2T_s + T_9 + T_7 + T_\infty$$

Node 9 is an exterior corner node, and the discretization equation is selected for a southeast corner node as

$$2\left(\frac{h\Delta x}{k} + 1\right)T_{i,j} = \left(T_{i,j+1} + T_{i-1,i}\right) + 2\frac{h\Delta x}{k}T_\infty$$

Applying this equation to corner node 9, we get

$$\textit{Node 9} \quad 2(0.5+1)T_9 = T_6 + T_8 + T_\infty$$
$$3T_9 = T_6 + T_8 + T_\infty$$

Rearranging and assembling all nodal equations, we get the following system of equations

$$-4T_1 + T_2 + 0 + T_4 + 0 + 0 + 0 + 0 + 0 = -(T_C + T_H)$$
$$T_1 - 4T_2 + T_3 + 0 + T_5 + 0 + 0 + 0 + 0 = -T_H$$
$$0 + 2T_2 - 5T_3 + 0 + 0 + T_6 + 0 + 0 + 0 = -(T_H + T_\infty)$$
$$T_1 + 0 + 0 - 4T_4 + T_5 + 0 + T_7 + 0 + 0 = -T_C$$
$$0 + T_2 + 0 + T_4 - 4T_5 + T_6 + 0 + T_8 + 0 = 0$$
$$0 + 0 + T_3 + 0 + 2T_5 - 5T_6 + 0 + 0 + T_9 = -T_\infty$$
$$0 + 0 + 0 + 2T_4 + 0 + 0 - 5T_7 + T_8 + 0 = -(T_C + T_\infty)$$
$$0 + 0 + 0 + 0 + 2T_5 + 0 + T_7 - 5T_8 + T_9 = -T_\infty$$
$$0 + 0 + 0 + 0 + 0 + T_6 + 0 + T_8 - 3T_9 = -T_\infty$$

Now, substituting $T_H = 600°C$, $T_C = 20°C$ and $T_\infty = 25°C$, and writing this in matrix form, we get the system of equations as

$$\begin{bmatrix} -4 & 1 & 0 & 1 & 0 & 0 & 0 & 0 & 0 \\ 1 & -4 & 1 & 0 & 1 & 0 & 0 & 0 & 0 \\ 0 & 2 & -5 & 0 & 0 & 1 & 0 & 0 & 0 \\ 1 & 0 & 0 & -4 & 1 & 0 & 1 & 0 & 0 \\ 0 & 1 & 0 & 1 & -4 & 1 & 0 & 1 & 0 \\ 0 & 0 & 1 & 0 & 2 & -5 & 0 & 0 & 1 \\ 0 & 0 & 0 & 2 & 0 & 0 & -5 & 1 & 0 \\ 0 & 0 & 0 & 0 & 2 & 0 & 1 & -5 & 1 \\ 0 & 0 & 0 & 0 & 0 & 1 & 0 & 1 & -3 \end{bmatrix} \begin{Bmatrix} T_1 \\ T_2 \\ T_3 \\ T_4 \\ T_5 \\ T_6 \\ T_7 \\ T_8 \\ T_9 \end{Bmatrix} = \begin{Bmatrix} -620 \\ -600 \\ -625 \\ -20 \\ 0 \\ -25 \\ -45 \\ -25 \\ -25 \end{Bmatrix}$$

Using the Gauss elimination routine, the solution to this system of equations is obtained as

$$\begin{Bmatrix} T_1 \\ T_2 \\ T_3 \\ T_4 \\ T_5 \\ T_6 \\ T_7 \\ T_8 \\ T_9 \end{Bmatrix} = \begin{Bmatrix} 276.1977°C \\ 340.1860°C \\ 293.1860°C \\ 144.6047°C \\ 191.3605°C \\ 160.5581°C \\ 90.8605°C \\ 120.0930°C \\ 101.8837°C \end{Bmatrix}$$

The heat transfer rate from the top surface is computed based on the *Fourier* law as

$$q_T = \sum kA_y \frac{\Delta T}{\Delta y} = k\Delta x \cdot 1 \frac{(600 - T_1)}{\Delta y} + k\Delta x \cdot 1 \frac{(600 - T_2)}{\Delta y} + k\frac{\Delta x}{2} \cdot 1 \frac{(600 - T_3)}{\Delta y}$$

$$= 30.0 \times 0.5 \times \frac{(600 - 276.1977)}{0.5} + 30.0 \times 0.5 \times \frac{(600 - 340.1860)}{0.5} + 30.0 \times \frac{0.5}{2} \times \frac{(600 - 293.1860)}{0.5}$$

$$= 9714.069 + 7794.42 + 4602.2$$

$$= 22\,110.699\ \text{W}$$

The heat transfer rate from the left surface is computed based on the *Fourier* law as

$$
\begin{aligned}
q_L &= \sum kA_x \frac{\Delta T}{\Delta x} = k\Delta y \cdot 1 \frac{(T_1 - 20)}{\Delta x} + k\Delta y \cdot 1 \frac{(T_4 - 20)}{\Delta x} + k\Delta y \cdot 1 \frac{(T_7 - 20)}{\Delta x} \\
&= 30.0 \times 0.5 \times \frac{(276.1977 - 20)}{0.5} + 30.0 \times 0.5 \times \frac{(144.6047 - 20)}{0.5} + 30.0 \\
&\quad \times \frac{0.5}{2} \times \frac{(90.8605 - 20)}{0.5} \\
&= 7685.931 + 3738.141 + 1062.9075 \\
&= 12\,486.9795\,\text{W}
\end{aligned}
$$

The heat transfer rate from the right surface is computed based on *Newton's* law of cooling as

$$
\begin{aligned}
q_R &= \Sigma h\,\Delta y\big(T_{i,j} - T_\infty\big) = h\Delta y.1(T_3 - T_\infty) + h\Delta y.1(T_6 - T_\infty) + h\frac{\Delta y}{2}.1(T_9 - T_\infty) \\
&= 30.0 \times 0.5 \times (293.1860 - 25) + 30.0 \times 0.5 \times (160.5581 - 25) + 30.0 \times \frac{0.5}{2} \\
&\quad \times (101.8837 - 25) \\
&= 4022.79 + 2033.3715 + 576.62775 \\
&= 6632.789\,25\,\text{W}
\end{aligned}
$$

The heat transfer rate from the bottom surface is computed based on *Newton's* law of cooling as

$$
\begin{aligned}
q_R &= \Sigma h\,\Delta x\big(T_{i,j} - T_\infty\big) = h\Delta x.1(T_7 - T_\infty) + h\Delta x.1(T_8 - T_\infty) + h\frac{\Delta x}{2}.1(T_9 - T_\infty) \\
&= 30.0 \times 0.5 \times (90.8605 - 25) + 30.0 \times 0.5 \times (120.0930 - 25) + 30.0 \times \frac{0.5}{2} \\
&\quad \times (101.8837 - 25) \\
&= 987.9075 + 1426.395 + 576.62775 \\
&= 2990.930\,25\,\text{W}
\end{aligned}
$$

The overall energy balance for the two-dimensional steady state problems states that rate of energy inflow is balanced by the rate of energy outflow, i.e.

$$E_{\text{in}} = E_{\text{out}}$$

In this problem energy inflow is given by the heat transfer rate at the top surface

$$E_{\text{in}} = q_T = 22\,110.699\,\text{W}$$

The energy outflow is the sum of heat transfer rates at the left, right and bottom surfaces, i.e.

$$E_{out} = q_L + q_R + q_B = 12\,486.9795 + 6632.789\,25 + 2990.930\,25\,\text{W}$$
$$E_{out} = 22\,110.699\,\text{W}$$

Example 6.3: Two-dimensional Steady State Conduction in a Slab Irradiated by a High Energy Laser Beam at The Surface: Use the finite difference method to solve a two-dimensional steady state conduction in a rectangular aluminum ($k = 200\,\text{W/m°C}$) slab subjected to a constant surface heat flux irradiated by a high-energy laser beam at the top surface. For simplicity, assume the heat flux distribution to be a constant average value, $I_0 = 2\times10^8\,\text{W/m}^2$, acting over a section of the surface equal to the beam diameter, $d = 4\,\text{mm}$ as shown in the figure. The remaining portion of the top surface is subjected to convection with $h_c = 100\,\text{W/m}^{2\circ}\text{C}$. All other surfaces are assumed to be maintained at constant temperature of $T_\infty = 25°\text{C}$.

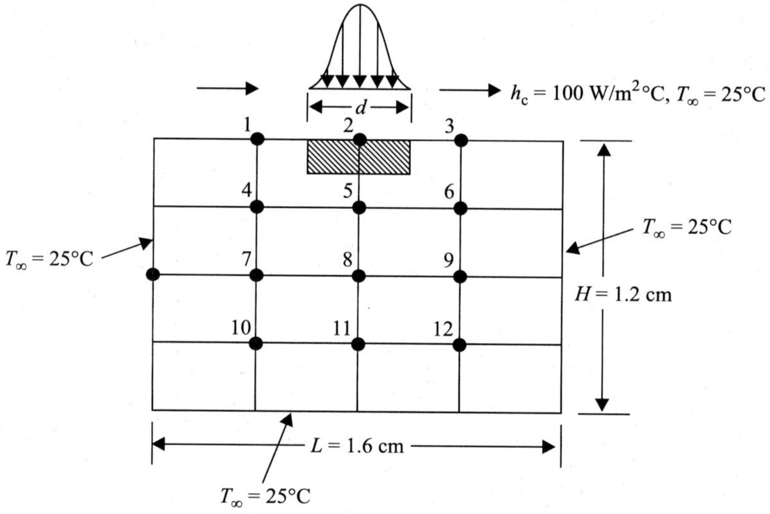

The mathematical statement of the problem is given as:

Governing heat equation

$$\frac{\partial^2 T}{\partial x^2} + \frac{\partial^2 T}{\partial y^2} = 0$$

Boundary conditions

1. at the left surface, $T(-\frac{L}{2}, y) = T_\infty$
2. at the right surface, $T(\frac{L}{2}, y) = T_\infty$

3. at the bottom surface, $T(x, H) = T_\infty$
4. at the top surface, the boundary condition is specified in two regions

Region 1

$$-\frac{L}{2} < x < -\frac{d}{2} \quad \text{and} \quad \frac{d}{2} < x < \frac{L}{2} \qquad k\left.\frac{\partial T}{\partial y}\right|_{y=0} = h(T|_{y=0} - T_\infty)$$

Region 2

$$-\frac{d}{2} < x < \frac{d}{2} \qquad -k\left.\frac{\partial T}{\partial y}\right|_{y=0} = I_0$$

Use the finite difference method to calculate the two-dimensional temperature distribution in the slab based on grid distribution shown in the figure. Estimate the heat transfer rates at all surfaces, and check the overall energy balance.

Solution: For the prescribed 4×4 grid size distribution with $N_{x\text{div}} = N_{y\text{div}} = 4$, we can estimate the parameters as

$$\Delta x = \frac{L}{N_{x\text{div}}} = \frac{1.6\,\text{cm}}{4} = 0.4\,\text{cm}, \quad \Delta y = \frac{L}{N_{y\text{div}}} = \frac{1.2\,\text{cm}}{4} = 0.3\,\text{cm}$$

$$\frac{\Delta x}{\Delta y} = \frac{0.4}{0.3} = 1.333, \quad \frac{\Delta y}{\Delta x} = \frac{0.3}{0.4} = 0.75, \quad \frac{h_c \Delta x}{k} = \frac{100 \times 0.004}{200} = 0.002$$

In the next step, we select appropriate discretization equations for each node. There are six interior nodes $4, 5, 6, 7, 8, 9, 10, 11$ and 12 for which the appropriate finite difference equation is given by the Equation in Table 6.1b

$$a_{i,j}\phi_{i,j} = a_{i+1,j}\phi_{i+1,j} + a_{i-1,j}\phi_{i-1,j} + a_{i,j+1}\phi_{i,j+1} + a_{i,j-1}\phi_{i,j-1} + d$$

$$a_{i+1,j} = a_{i-1,j} = \frac{\Delta y}{\Delta x} = 0.75$$

$$a_{i,j+1} = a_{i,j-1} = \frac{\Delta x}{\Delta y} = 1.333$$

$$a_{i,j} = a_{i+1,j} + a_{i-1,j} + a_{i,j=1} + a_{i,j-1} = 2\frac{\Delta y}{\Delta x}\left(1 + \frac{\Delta x^2}{\Delta y^2}\right)$$

$$= 2 \times 0.75 \times (1 + (1.333)^2) = 4.165$$

$$d = \bar{S}\,\Delta x\,\Delta y / k = 0$$

With no heat source term, $\bar{S} = 0$, and $d = 0$, and the equation is expressed in a simplified form as

$$4.165T_{i,j} = 0.75T_{i+1,j} + 0.75T_{i-1,j} + 1.333T_{i,j+1} + 1.333T_{i,j-1}$$

Using this equation, the finite difference equation for the interior nodes 4 through 12 are written as

$$\textit{Node 4} \quad 4.165T_4 = 0.75T_5 + 0.75T_0 + 1.333T_1 + 1.333T_7$$

$$\textit{Node 5} \quad 4.165T_5 = 0.75T_6 + 0.75T_4 + 1.333T_2 + 1.333T_8$$

$$\textit{Node 6} \quad 4.165T_6 = 0.75T_0 + 0.75T_5 + 1.333T_3 + 1.333T_9$$

$$\textit{Node 7} \quad 4.165T_7 = 0.75T_8 + 0.75T_0 + 1.333T_4 + 1.333T_{10}$$

$$\textit{Node 8} \quad 4.165T_8 = 0.75T_9 + 0.75T_7 + 1.333T_5 + 1.333T_{11}$$

$$\textit{Node 9} \quad 4.165T_9 = 0.75T_0 + 0.75T_8 + 1.333T_6 + 1.333T_{12}$$

$$\textit{Node 10} \quad 4.165T_{10} = 0.75T_{11} + 0.75T_0 + 1.333T_7 + 1.333T_0$$

$$\textit{Node 11} \quad 4.165T_{11} = 0.75T_{12} + 0.75T_{10} + 1.333T_8 + 1.333T_0$$

$$\textit{Node 12} \quad 4.165T_{12} = 0.75T_0 + 0.75T_{11} + 1.333T_9 + 1.333T_0$$

Nodes 1 and 3 are convective boundary nodes on the top surface, and we use Equation 6.12 and Equation 6.13 as the appropriate discretization equation, which is given as

$$a_{i,j}T_{i,j} = a_{i-1,j}T_{i-1,j} + a_{i+1,j}T_{i+1,j} + a_{i,j-1}T_{i,j-1} + d$$

$$a_{i,j-1} = k_{i,j-\frac{1}{2}}\frac{\Delta x}{\Delta y}, \quad a_{i+1,j} = \frac{k_{i+\frac{1}{2},j}}{2}\frac{\Delta y}{\Delta x}, \quad a_{i-1,j} = \frac{k_{i-\frac{1}{2},j}}{2}\frac{\Delta y}{\Delta x}$$

$$a_{i,j} = a_{i-1,j} + a_{i+1,j} + a_{i,j-1} - \frac{S_1}{2}\Delta x\,\Delta y + h\,\Delta x$$

$$d = \frac{S_0}{2}\Delta x\,\Delta y + hT_\infty\,\Delta x$$

Again, for constant conductivity, $k_{i-\frac{1}{2},j} = k_{i+\frac{1}{2},j} = k_{i,j+\frac{1}{2}} = k_{i,j-\frac{1}{2}} = k$, and no heat source, $S_0 = S_1 = 0$, the equation reduces to

$$\left(\frac{h_c\Delta x}{k} + 2.833\right)T_{i,j} = \frac{1}{2}(2.666T_{i,j-1} + 0.75T_{i+1,j} + 0.75T_{i-1,j}) + \frac{h_c\Delta x}{k}T_\infty$$

or

$$(0.002 + 2.833)T_{i,j} = \frac{1}{2}(2.666T_{i,j-1} + 0.75T_{i+1,j} + 0.75T_{i-1,j}) + 0.002T_\infty$$

or

$$5.67T_{i,j} = (2.666T_{i,j-1} + 0.75T_{i+1,j} + 0.75T_{i-1,j}) + 0.004T_\infty$$

Applying this equation to nodes 1 and 3, we get

$$\textit{Node 1} \quad 5.67T_1 = (2.666T_4 + 0.75T_2 + 0.75T_0) + 0.004T_\infty$$

or

$$5.67T_1 = 2.666T_4 + 0.75T_2 + 0.75T_0 + 0.754T_\infty$$

$$\textit{Node 3} \quad 5.67T_3 = (2.666T_6 + 0.75T_0 + 0.75T_2) + 0.004T_\infty$$

or

$$5.67T_3 = 2.666T_6 + 0.75T_2 + 0.75T_0 + 0.004T_\infty$$

Node 2 is a boundary node on the top surface with constant surface heat flux, and we select the discretization equation as

$$a_{i,j}T_{i,j} = a_{i+1,j}T_{i+1,j} + a_{i-1,j}T_{i-1,j} + a_{i,j-1}T_{i,j-1} + d$$

$$a_{i+1,j} = \frac{k_{i+\frac{1}{2},j}}{2}\frac{\Delta y}{\Delta x}, \quad a_{i-1,j} = \frac{k_{i-\frac{1}{2},j}}{2}\frac{\Delta y}{\Delta x}, \quad a_{i,j-1} = k_{i,j-\frac{1}{2}}\frac{\Delta x}{\Delta y}$$

$$a_{i,j} = a_{i+1,j} + a_{i-1,j+1} + a_{i,j-1} - \frac{S_1}{2}\Delta x\,\Delta y$$

$$d = \frac{S_0}{2}\Delta x\,\Delta y + f''_{ts}\,\Delta x$$

This equation is simplified in a similar manner for constant conductivity and no heat source to the form

$$\left(\frac{\Delta x}{\Delta y} + \frac{\Delta y}{\Delta x}\right)T_{i,j} = \frac{1}{2}\left(2\frac{\Delta x}{\Delta y}T_{i,j-1} + \frac{\Delta y}{\Delta x}T_{i+1,j} + \frac{\Delta y}{\Delta x}T_{i-1,j}\right) + \frac{I_0\Delta x}{k}$$

Substituting the numerical values, we get

$$(1.333 + 0.75)T_{i,j} = \frac{1}{2}(2.666T_{i,j-1} + 0.75T_{i+1,j} + 0.75T_{i-1,j}) + \frac{I_0 \times 0.004}{200}$$

or

$$4.166T_{i,j} = 2.666T_{i,j-1} + 0.75T_{i+1,j} + 0.75T_{i-1,j} + 4 \times 10^{-5}I_0$$

Applying this equation to node 2, we get

$$\textit{Node 2} \quad 4.166T_2 = 2.666T_5 + 0.75T_3 + 0.75T_1 + 4.0 \times 10^{-5}I_0$$

Rearranging and assembling all nodal equations, we get the following system of equations.

$$5.67T_1 - 0.75T_2 + 0T_3 - 2.666T_4 + 0T_5 + 0T_6 + 0T_7 + 0T_8 + 0T_9 + 0T_{10} + 0T_{11} + 0T_{12} = 0.75T_0 + 0.754T_\infty$$

$$-0.75T_1 + 4.166T_2 - 0.75T_3 + 0T_4 - 2.666T_5 + 0T_6 + 0T_7 + 0T_8 + 0T_9 + 0T_{10} + 0T_{11} + 0T_{12} = 4.0 \times 10^{-5} I_0$$

$$0T_1 - 0.75T_2 + 5.67T_3 + 0T_4 + 0T_5 - 2.666T_6 + 0T_7 + 0T_8 + 0T_9 + 0T_{10} + 0T_{11} + 0T_{12} = 0.75T_0 + 0.004T_\infty$$

$$-1.333T_1 + 0T_2 + 0T_3 + 4.165T_4 - 0.75T_5 + 0T_6 - 1.333T_7 + 0T_8 + 0T_9 + 0T_{10} + 0T_{11} + 0T_{12} = 0.75T_0$$

$$0T_1 - 1.333T_2 + 0T_3 - 0.75T_4 + 4.165T_5 - 0.75T_6 + 0T_7 - 1.333T_8 + 0T_9 + 0T_{10} + 0T_{11} + 0T_{12} = 0$$

$$0T_1 + 0T_2 - 1.333T_3 + 0T_4 - 0.75T_5 + 4.165T_6 + 0T_7 + 0T_8 - 1.333T_9 + 0T_{10} + 0T_{11} + 0T_{12} = 0.75T_0$$

$$0T_1 + 0T_2 + 0T_3 - 1.333T_4 + 0T_5 + 0T_6 + 4.165T_7 - 0.75T_8 + 0T_9 - 1.333T_{10} + 0T_{11} + 0T_{12} = 0.75T_0$$

$$0T_1 + 0T_2 + 0T_3 + 0T_4 - 1.333T_5 + 0T_6 - 0.75T_7 + 4.165T_8 - 0.75T_9 + 0T_{10} - 1.333T_{11} + 0T_{12} = 0$$

$$0T_1 + 0T_2 + 0T_3 + 0T_4 + 0T_5 - 1.333T_6 + 0T_7 - 0.75T_8 + 4.165T_9 + 0T_{10} + 0T_{11} - 1.333T_{12} = 0.75T_0$$

$$0T_1 + 0T_2 + 0T_3 + 0T_4 + 0T_5 + 0T_6 - 1.333T_7 + 0T_8 + 0T_9 + 4.165T_{10} - 0.75T_{11} + 0T_{12} = 0.75T_0 + 1.333T_0$$

$$0T_1 + 0T_2 + 0T_3 + 0T_4 + 0T_5 + 0T_6 + 0T_7 - 1.333T_8 + 0T_9 - 0.75T_{10} + 4.165T_{11} - 0.75T_{12} = 1.333T_0$$

$$0T_1 + 0T_2 + 0T_3 + 0T_4 + 0T_5 + 0T_6 + 0T_7 + 0T_8 - 1.333T_9 + 0T_{10} - 0.75T_{11} + 4.165T_{12} = 0.75T_0 + 1.333T_0$$

Now, substituting $I_0 = 2 \times 10^8\,\mathrm{W/m^2}$, $T_0 = 25°\mathrm{C}$ and $T_\infty = 25°\mathrm{C}$, and writing this in matrix form, we get the system of equations as

$$\begin{bmatrix} 5.67 & -0.75 & 0 & 2.666 & 0 & 0 & 0 & 0 & 0 & 0 & 0 & 0 \\ -0.75 & 4.166 & 0.75 & 0 & -2.666 & 0 & 0 & 0 & 0 & 0 & 0 & 0 \\ 0 & -0.75 & 5.67 & 0 & 0 & -2.666 & 0 & 0 & 0 & 0 & 0 & 0 \\ -1.333 & 0 & 0 & 4.165 & -0.75 & 0 & -1.333 & 0 & 0 & 0 & 0 & 0 \\ 0 & -1.333 & 0 & -0.75 & 4.165 & -0.75 & 0 & -1.333 & 0 & 0 & 0 & 0 \\ 0 & 0 & -1.333 & 0 & -0.75 & 4.165 & 0 & 0 & -1.333 & 0 & 0 & 0 \\ 0 & 0 & 0 & -1.333 & 0 & 0 & 4.165 & -0.75 & 0 & -1.333 & 0 & 0 \\ 0 & 0 & 0 & 0 & -1.333 & 0 & -0.75 & 4.165 & -0.75 & 0 & -1.333 & 0 \\ 0 & 0 & 0 & 0 & 0 & -1.333 & 0 & -0.75 & 4.165 & 0 & 0 & -1.333 \\ 0 & 0 & 0 & 0 & 0 & 0 & -1.333 & 0 & 0 & 4.165 & -0.75 & 0 \\ 0 & 0 & 0 & 0 & 0 & 0 & 0 & -1.333 & 0 & -0.75 & 4.165 & -0.75 \\ 0 & 0 & 0 & 0 & 0 & 0 & 0 & 0 & -1.333 & 0 & -0.75 & 4.165 \end{bmatrix} \begin{Bmatrix} T_1 \\ T_2 \\ T_3 \\ T_4 \\ T_5 \\ T_6 \\ T_7 \\ T_8 \\ T_9 \\ T_{10} \\ T_{11} \\ T_{12} \end{Bmatrix} = \begin{Bmatrix} 18.85 \\ 8000 \\ 18.85 \\ 18.75 \\ 0 \\ 18.75 \\ 18.75 \\ 0 \\ 18.75 \\ 52.075 \\ 33.325 \\ 52.075 \end{Bmatrix}$$

Using the Gauss elimination routine, the solution to this system of equations is obtained as

$$\begin{Bmatrix} T_1 \\ T_2 \\ T_3 \\ T_4 \\ T_5 \\ T_6 \\ T_7 \\ T_8 \\ T_9 \\ T_{10} \\ T_{11} \\ T_{12} \end{Bmatrix} = \begin{Bmatrix} 100.533^\circ\text{C} \\ 401.784^\circ\text{C} \\ 100.533^\circ\text{C} \\ 93.712^\circ\text{C} \\ 196.186^\circ\text{C} \\ 93.712^\circ\text{C} \\ 67.825^\circ\text{C} \\ 105.754^\circ\text{C} \\ 67.825^\circ\text{C} \\ 44.640^\circ\text{C} \\ 57.924^\circ\text{C} \\ 44.640^\circ\text{C} \end{Bmatrix}$$

Note that the solution shows symmetric temperature distribution in the workpiece. This is true because the geometry as well as the boundary conditions are symmetric.

Example 6.4: Cooling of Electronic Chips: Consider the problem of cooling an electronic chip that is mounted on a substrate. The electric power generation in the chip is $\dot{q} = 1.6 \times 10^8\ \text{W/m}^3$ and the size of the chip is 1 cm × 0.5 cm. The top surface is assumed to be convectively cooled with $h_C = 50\ \text{W/m}^{2\circ}\text{C}$ and $T_\infty = 20^\circ\text{C}$. The left, right, and bottom surfaces are assumed to be at constant temperature $T_0 = 25^\circ\text{C}$. Assume the conductivity of the chip and substrate materials as $k = 15\ \text{W/m·K}$.

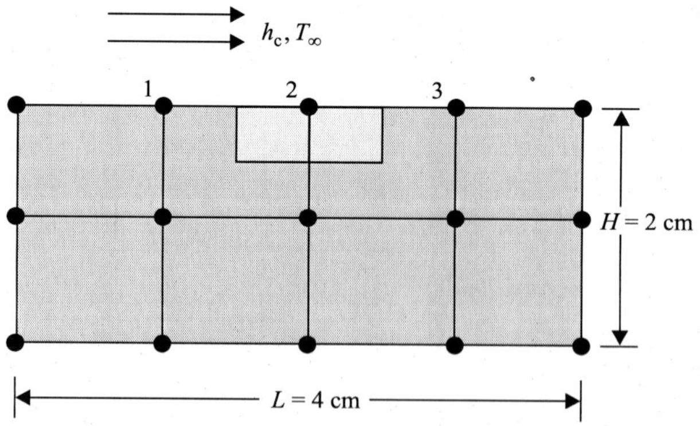

(a) Derive the finite difference equation for node 2, (b) select the finite difference equations for the rest of the nodes 1, 3, 4, 5, and 6, (c) assemble the equations to form the system of equations, and (d) obtain the solution for the temperature distribution.

Solution: For the selected uniform grid system, we have

$$\Delta x = \frac{L}{N_{\text{div}x}} = \frac{4}{4} = 1\,\text{cm} = 0.01\,\text{m}, \qquad \Delta y = \frac{H}{N_{\text{div}y}} = \frac{2}{2} = 1\,\text{cm} = 0.01\,\text{m}$$

$$\frac{h\Delta x}{k} = \frac{50 \times 0.01}{15} = 0.0333$$

The following discretization equations are selected from Table 6.4.

Node 2 Convective boundary nodes with uniform heat generation

$$\left(\frac{h\Delta x}{k} + 2\right)T_{i,j} = \frac{1}{2}(2T_{i,j-1} + T_{i+1,j} + T_{i-1,j}) + \frac{h\Delta x}{\Gamma}T_\infty + \frac{q(\dot{\Delta} x)^2}{k}$$

or

$$2\left(\frac{h\Delta x}{k} + 2\right)T_2 = (2T_5 + T_3 + T_1) + \frac{2h\Delta x}{\Gamma}T_\infty + \frac{q(\dot{\Delta} x)^2}{k}$$

$$2(0.0333 + 2)T_2 = (2T_5 + T_3 + T_1) + 2 \times 0.0333 \times 20 + \frac{1.6 \times 10^8(0.01)^2}{15}$$

$$T_1 - 4.0666T_2 + T_3 + 2T_5 = -1.332 - \frac{2 \times 8 \times 10^7(0.01)^2}{15} \tag{E.6.4.1}$$

Nodes 1 and 3 Convective boundary nodes

$$\left(\frac{h\Delta x}{k} + 2\right)T_{i,j} = \frac{1}{2}(2T_{i,j-1} + T_{i+1,j} + T_{i-1,j}) + \frac{h\Delta x}{k}T_\infty$$

Node 1 $2(0.0333 + 2)T_1 = (2T_4 + T_2 + 25) + 2 \times 0.0333 \times 20$

or

$$-4.0666T_1 + T_2 + 2T_4 = -26.332 \tag{E.6.4.2}$$

Node 3 $2(0.0333 + 2)T_3 = (2T_6 + 25 + T_2) + 2 \times 0.0333 \times 20$

or

$$T_2 - 4.0666T_3 + 2T_6 = -26.332 \tag{E.6.4.3}$$

Nodes 4, 5, and 6 Interior nodes

$$4T_{i,j} = T_{i+1,j} + T_{i-1,j} + T_{i,j+1} + T_{i,j-1}$$

Node 4 $4T_4 = T_5 + 25 + T_1 + 25$

or

$$T_1 - 4T_4 + T_5 = -50 \tag{E.6.4.4}$$

Node 5 $4T_5 = T_6 + T_4 + T_2 + 25$

or

$$T_2 + T_4 - 4T_5 + T_6 = -25 \tag{E.6.4.5}$$

Node 6 $4T_6 = 25 + T_5 + T_3 + 25$

or

$$T_3 + T_5 - 4T_6 = -50 \tag{E.6.4.6}$$

Assembly of all nodal equations (E.6.4.1) to (E.6.4.6) leads to the system of equations as

$$\begin{bmatrix} -4.0666 & 1 & 0 & 2 & 0 & 0 \\ 0 & -4.0666 & 1 & 0 & 2 & 0 \\ 0 & 1 & -4.0666 & 0 & 0 & 2 \\ 1 & 0 & 0 & -4.0 & 1 & 0 \\ 0 & 1 & 0 & 1 & -4.0 & 1 \\ 0 & 0 & 1 & 0 & 1 & -4 \end{bmatrix} \begin{Bmatrix} T_1 \\ T_2 \\ T_3 \\ T_4 \\ T_5 \\ T_6 \end{Bmatrix} = \begin{Bmatrix} -26.332 \\ -1067.992 \\ -26.332 \\ -50.0 \\ -25.0 \\ -50.0 \end{Bmatrix}$$

Solution of the system of equations gives the temperature distribution as

$$\begin{Bmatrix} T_1 \\ T_2 \\ T_3 \\ T_4 \\ T_5 \\ T_6 \end{Bmatrix} = \begin{Bmatrix} 135.6302 \\ 363.6288 \\ 135.6302 \\ 80.7964 \\ 137.5554 \\ 80.7964 \end{Bmatrix}$$

Example 6.5: Fully Developed Flow in a Rectangular Channel with Top Surface Moving: Consider a fully-developed flow in a rectangular channel with top surface moving with an axial velocity U_0 and the pressure drop in the axial direction assumed to be $dP/dz = S$. Determine the axial-component velocity distribution.

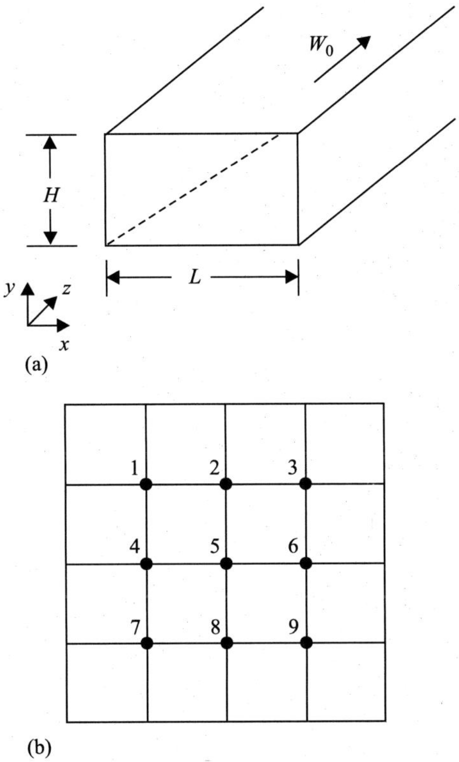

A mathematical statement of the problem can be derived from the continuity and momentum equations for two-dimensional steady incompressible flow equations given as:

Governing equation

$$\frac{\partial^2 u}{\partial x^2} + \frac{\partial^2 u}{\partial y^2} = \frac{1}{\mu}\frac{dp}{dx} \tag{E.6.5.1}$$

Boundary conditions

1. $y = 0, \quad w = 0$
2. $y = H, \quad w = W_0$
3. $x = 0, \quad w = 0$
4. $x = L, \quad w = 0$

The finite difference equation reduces to

$$w_{i+1,j} + w_{i-1,j} + w_{i,j+1} + w_{i,j-1} - 4w_{i,j} + S\Delta x^2 = 0 \tag{E.6.5.2}$$

where

$$S = -\frac{1}{\mu}\frac{dP}{dz} \tag{E.6.5.3}$$

Applying this equation successively to all nodes, we get

$$\begin{aligned}
&\textit{Node 1} \quad w_2 + w_L + w_0 + w_4 - 4w_1 + S\Delta x^2 = 0\\
&\textit{Node 2} \quad w_3 + w_1 + w_0 + w_5 - 4w_2 + S\Delta x^2 = 0\\
&\textit{Node 3} \quad w_R + w_2 + w_0 + w_6 - 4w_3 + S\Delta x^2 = 0\\
&\textit{Node 4} \quad w_5 + w_L + w_1 + w_7 - 4w_4 + S\Delta x^2 = 0\\
&\textit{Node 5} \quad w_6 + w_4 + w_2 + w_8 - 4w_5 + S\Delta x^2 = 0\\
&\textit{Node 6} \quad w_R + w_5 + w_3 + w_9 - 4w_6 + S\Delta x^2 = 0\\
&\textit{Node 7} \quad w_8 + w_L + w_4 + w_B - 4w_7 + S\Delta x^2 = 0\\
&\textit{Node 8} \quad w_9 + w_7 + w_5 + w_B - 4w_8 + S\Delta x^2 = 0\\
&\textit{Node 9} \quad w_R + w_8 + w_6 + w_B - 4w_9 + S\Delta x^2 = 0
\end{aligned}$$

Rearranging and writing in matrix form, we get the system of equations as

$$\begin{bmatrix}
-4 & 1 & 0 & 1 & 0 & 0 & 0 & 0 & 0\\
1 & -4 & 1 & 0 & 1 & 0 & 0 & 0 & 0\\
0 & 1 & -4 & 0 & 0 & 1 & 0 & 0 & 0\\
1 & 0 & 0 & -4 & 1 & 0 & 1 & 0 & 0\\
0 & 1 & 0 & 1 & -4 & 1 & 0 & 1 & 0\\
0 & 0 & 1 & 0 & 1 & -4 & 0 & 0 & 1\\
0 & 0 & 0 & 1 & 0 & 0 & -4 & 1 & 0\\
0 & 0 & 0 & 0 & 1 & 0 & 1 & -4 & 1\\
0 & 0 & 0 & 0 & 0 & 1 & 0 & 1 & -4
\end{bmatrix}
\begin{Bmatrix} w_1\\ w_2\\ w_3\\ w_4\\ w_5\\ w_6\\ w_7\\ w_8\\ w_9 \end{Bmatrix}
=
\begin{Bmatrix} -(w_0 + S\Delta x^2)\\ -(w_0 + S\Delta x^2)\\ -(w_0 + S\Delta x^2)\\ -S\Delta x^2\\ -S\Delta x^2\\ -S\Delta x^2\\ -S\Delta x^2\\ -S\Delta x^2\\ -S\Delta x^2 \end{Bmatrix} \tag{E.6.5.4}$$

6.3 IRREGULAR GEOMETRIES

So far, we have considered geometries involving only regular boundaries, which match well with the grid system. However, there are many physical problems that involve complex boundaries including slant and curved surfaces. As was discussed in Chapter 5, a nonuniform grid size distribution is essential for problems that have regions of steep variation in dependent variables, and thus finer grids are used to capture the necessary resolution. Such nonuniform grids are also necessary near an irregular surface. For example, a uniform grid size distribution is used at the interior regions, away from the curved surface,

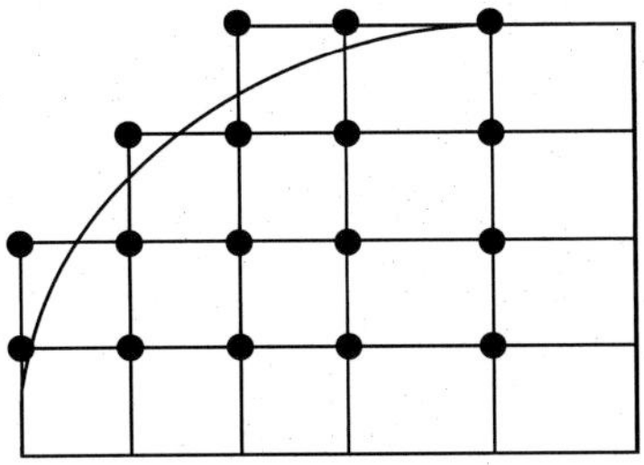

Figure 6.4 Grid distribution in a region with an irregular boundary surface

as shown in Figure 6.3. However, the spacing became nonuniform for spaces between the boundary nodes and the neighboring interior nodes due to the irregular shape of the boundary surface. Another option could be using a nonuniform grid with finer grids near the irregular surface and coarser ones away from the boundary.

The procedure for developing the finite difference discretization equations for such a nonuniform grid near an irregular surface is, however, the same as those discussed for regular geometries. Let us show the development of one such finite difference equation for an interior node near a curved surface as well as nodes on such boundaries as shown in Figure 6.4. As can be noted, a uniform grid size distribution with grid sizes Δx and Δy are used in the x and y directions for most of the regions. The grid sizes near the curved boundary are defined arbitrarily as $\Delta x' = \alpha \Delta x$ between the interior node (i, j) and boundary node b_1, $\Delta y' = \beta \Delta y$ between the interior node (i, j) and the boundary node b_2, and $\Delta y'' = \gamma \Delta y$ between boundary nodes b_2 and b_3 as shown in Figure 6.5.

The control volume method can be applied to develop the finite difference equation for interior nodes near, as well as on, the irregular surfaces by considering overlapping control volumes that surround a node and extend halfway to its neighboring nodes.

Interior Nodes Near a Curved Surface In order to derive the discretization equation for the interior node (i, j) close to the boundary, we consider a control volume bounded by the dashed lines located at $-\Delta x'/2$, $\Delta x/2$, $-\Delta y'/2$ and $\Delta y/2$ as shown in Figure 6.4. Integrate the two-dimensional steady state governing Equation 6.1 over the control volume surrounding the grid point (i, j) as

$$\int_{-\Delta x'/2}^{\Delta x/2} \int_{-\Delta y/2}^{\Delta y'/2} \frac{\partial}{\partial x}\left(\Gamma_x \frac{\partial \phi}{\partial x}\right) dx\, dy + \int_{-\Delta x'/2}^{\Delta x/2} \int_{-\Delta y/2}^{\Delta y'/2} \frac{\partial}{\partial y}\left(\Gamma_y \frac{\partial \phi}{\partial y}\right) dx\, dy$$
$$+ \int_{-\Delta x'/2}^{\Delta x/2} \int_{-\Delta y/2}^{\Delta y'/2} S\, dx\, dy = 0 \tag{6.33}$$

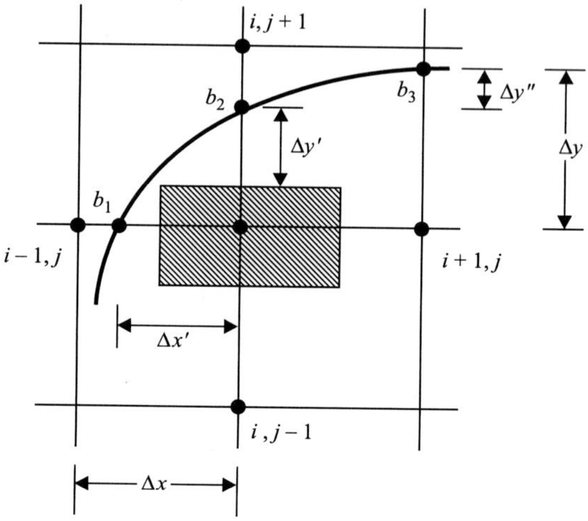

Figure 6.5 Nodes near and on a curved boundary

Now, the source term is approximated with a step-wise profile and integrated over x and y. The diffusion terms in x and y are integrated over x and y, respectively, to reduce the equation to

$$\left[\int_{-\Delta x'/2}^{\Delta x/2} \frac{\partial}{\partial x}\left(\Gamma_x \frac{\partial \phi}{\partial x}\right) dx \left(y\Big|_{-\Delta y/2}^{\Delta y'/2}\right)\right] + \left[\int_{-\Delta y/2}^{\Delta y'/2} \frac{\partial}{\partial y}\left(\Gamma_y \frac{\partial \phi}{\partial y}\right) dy \left(x\Big|_{-\Delta x'/2}^{\Delta x/2}\right)\right]$$
$$+ \bar{S}\left(y\Big|_{-\Delta y/2}^{\Delta y'/2}\right)\left(x\Big|_{-\Delta x'/2}^{\Delta x/2}\right) = 0$$

Applying the limit values, we get

$$\left[\int_{-\Delta x'/2}^{\Delta x/2} \frac{\partial}{\partial x}\left(\Gamma_x \frac{\partial \phi}{\partial x}\right) dx \left(\frac{\Delta y'}{2} + \frac{\Delta y}{2}\right)\right] + \left[\int_{-\Delta y/2}^{\Delta y'/2} \frac{\partial}{\partial y}\left(\Gamma_y \frac{\partial \phi}{\partial y}\right) dy \left(\frac{\Delta x}{2} + \frac{\Delta x'}{2}\right)\right]$$
$$+ \bar{S}\left(\frac{\Delta y'}{2} + \frac{\Delta y}{2}\right)\left(\frac{\Delta x}{2} + \frac{\Delta x'}{2}\right) = 0$$

Substituting $\Delta x' = \alpha \Delta x$ and $\Delta y' = \beta \Delta y$, we get

$$\left[\int_{-\Delta x'/2}^{\Delta x/2} \frac{\partial}{\partial x}\left(\Gamma_x \frac{\partial \phi}{\partial x}\right) dx \left(\frac{\Delta y}{2}(\beta + 1)\right)\right] + \left[\int_{-\Delta y/2}^{\Delta y'/2} \frac{\partial}{\partial y}\left(\Gamma_y \frac{\partial \phi}{\partial y}\right) dy \left(\frac{\Delta x}{2}(1 + \alpha)\right)\right]$$
$$+ \bar{S}\left(\frac{\Delta y}{2}(1 + \beta)\right)\left(\frac{\Delta x}{2}(1 + \alpha)\right) = 0$$

Now, integrating over x, we get

$$\left[\Gamma_x \frac{\partial\phi}{\partial x}\bigg|_{\Delta x/2} - \Gamma_x \frac{\partial\phi}{\partial x}\bigg|_{-\Delta x'/2}\right]\frac{\Delta y}{2}(1+\beta) + \left[\Gamma_y \frac{\partial\phi}{\partial y}\bigg|_{\Delta y'/2} - \Gamma_y \frac{\partial\phi}{\partial y}\bigg|_{-\Delta y/2}\right]\frac{\Delta x}{2}(1+\alpha)$$

$$+ \bar{S}(1+\alpha)(1+\beta)\frac{\Delta x\, \Delta y}{4} = 0 \tag{6.34}$$

Also, considering a step-wise constant profile for the transport properties Γ and a piece-wise linear profile to approximate the first derivative terms, we can express the flux quantities across the control volume surfaces in Equation 6.23 as

$$\Gamma_x \frac{\partial\phi}{\partial x}\bigg|_{-\Delta x'/2} = \Gamma_x \frac{\phi_{i,j} - \phi_{b1}}{\Delta x'} = \Gamma_x \frac{\phi_{i,j} - \phi_{b1}}{\alpha \Delta x} \tag{6.35a}$$

$$\Gamma_x \frac{\partial\phi}{\partial x}\bigg|_{\Delta x/2} = \Gamma_x \frac{\phi_{i+1,j} - \phi_{i,j}}{\Delta x} \tag{6.35b}$$

$$\Gamma_y \frac{\partial\phi}{\partial y}\bigg|_{-\Delta y/2} = \Gamma_y \frac{\phi_{i,j} - \phi_{i,j-1}}{\Delta y} \tag{6.35c}$$

$$\Gamma_y \frac{\partial\phi}{\partial y}\bigg|_{-\Delta y'/2} = \Gamma_y \frac{\phi_{b2} - \phi_{i,j}}{\Delta y'} = \Gamma_y \frac{\phi_{b2} - \phi_{i,j}}{\beta \Delta y} \tag{6.35d}$$

Substituting Equation 6.35 into Equation 6.34, we get

$$\left[\Gamma_x \frac{\phi_{i+1,j} - \phi_{i,j}}{\Delta x} - \Gamma_x \frac{\phi_{i,j} - \phi_{b1}}{\alpha \Delta x}\right]\frac{\Delta y}{2}(1+\beta)$$

$$+ \left[\Gamma_y \frac{\phi_{b2} - \phi_{i,j}}{\beta \Delta y} - \Gamma_y \frac{\phi_{i,j} - \phi_{i,j-1}}{\Delta y}\right]\frac{\Delta x}{2}(1+\alpha) + \bar{S}(1+\alpha)(1+\beta)\frac{\Delta x \Delta y}{4} = 0$$

Rearranging

$$\left[\Gamma_x \frac{\Delta y}{2\Delta x}(1+\beta)\left(\frac{1}{\alpha}+1\right) + \Gamma_y \frac{\Delta x}{2\Delta y}(1+\alpha)\left(1+\frac{1}{\beta}\right)\right]\phi_{i,j} - \Gamma_x \frac{\Delta y}{2\Delta x}\frac{1+\beta}{\alpha}\phi_{b1}$$

$$- \Gamma_x \frac{\Delta y}{2\Delta x}(1+\beta)\phi_{i+1,j} - \Gamma_y \frac{\Delta x}{2\Delta y}(1+\alpha)\phi_{i,j-1} - \Gamma_y \frac{\Delta x}{2\Delta y}\frac{1+\alpha}{\beta}\phi_{b2}$$

$$- \bar{S}(1+\alpha)(1+\beta)\frac{\Delta x \Delta y}{4} = 0 \tag{6.36}$$

We can write this in a compact form as

$$a_{i,j}\phi_{i,j} = a_{i+1,j}\phi_{i+1,j} + a_{i,j-1}\phi_{i,j-1} + d \tag{6.37}$$

where

$$\begin{aligned}
a_{i+1,j} &= \Gamma_x \frac{\Delta y}{2\Delta x}(1+\beta) \\
a_{i,j-1} &= \Gamma_y \frac{\Delta x}{2\Delta y}(1+\alpha) \\
a_{i,j} &= a_{i+1,j}\left(1+\frac{1}{\alpha}\right) + a_{i,j-1}\left(1+\frac{1}{\beta}\right) \\
d &= \Gamma_x \frac{\Delta y}{2\Delta x}\frac{1+\beta}{\alpha}\phi_{b1} + \Gamma_y \frac{\Delta x}{2\Delta y}\frac{1+\alpha}{\beta}\phi_{b2} + \bar{S}(1+\alpha)(1+\beta)\frac{\Delta x \Delta y}{4}
\end{aligned}$$

For constant transport property $\Gamma_x = \Gamma_y$ and for uniform grid size in x and y directions, i.e., $\Delta x = \Delta y$, Equation (6.37) reduces to

$$a_{i,j}\phi_{i,j} = a_{i+1,j}\phi_{i+1,j} + a_{i,j-1}\phi_{i,j-1} + d \tag{6.38}$$

where

$$\begin{aligned}
a_{i+1,j} &= \tfrac{1}{2}(1+\beta) \\
a_{i,j-1} &= \tfrac{1}{2}(1+\alpha) \\
a_{i,j} &= a_{i+1,j}\left(1+\frac{1}{\alpha}\right) + a_{i,j-1}\left(1+\frac{1}{\beta}\right) \\
d &= \frac{1+\beta}{2\alpha}T_{b1} + \frac{1+\alpha}{2\beta}T_{b2} + \bar{S}(1+\alpha)(1+\beta)\frac{\Delta x^2}{4\Gamma}
\end{aligned}$$

Boundary Nodes on a Curved Surface. A similar procedure can be applied to derive finite difference discretization equations for other boundary nodes with ***mixed boundary*** or ***Newman's*** condition. By considering a half control volume surrounding the boundary node, b_2, we can derive the finite difference equation as

$$\begin{aligned}
&\left(\Gamma\frac{\beta\Delta y}{\sqrt{\alpha^2\Delta x^2+\beta^2\Delta y^2}} + \Gamma\frac{\beta\Delta y}{\sqrt{\gamma^2\Delta y^2+\Delta x^2}} + \Gamma\frac{\Delta x}{\Delta y}\frac{\alpha+1}{\beta} + h_c\sqrt{\gamma^2\Delta y^2+\Delta x^2}\right. \\
&\left. + h_c\sqrt{\alpha^2\Delta x^2+\beta^2\Delta y^2}\right)\phi_{b2} \\
&= \Gamma\frac{\beta\Delta y}{\sqrt{\alpha^2\Delta x^2+\beta^2\Delta y^2}}\phi_{b1} + \Gamma\frac{\beta\Delta y}{\sqrt{\gamma^2\Delta y^2+\Delta x^2}}\phi_{b3} + \Gamma\frac{\Delta x}{\Delta y}\frac{\alpha+1}{\beta}T_{i,j} \\
&+ h_c\left(\sqrt{\gamma^2\Delta y^2+\Delta x^2} + \sqrt{\alpha^2\Delta x^2+\beta^2\Delta y^2}\right)\phi_\infty + \frac{1}{2}\bar{S}(\alpha+1)\beta\Delta x\Delta y
\end{aligned} \tag{6.39}$$

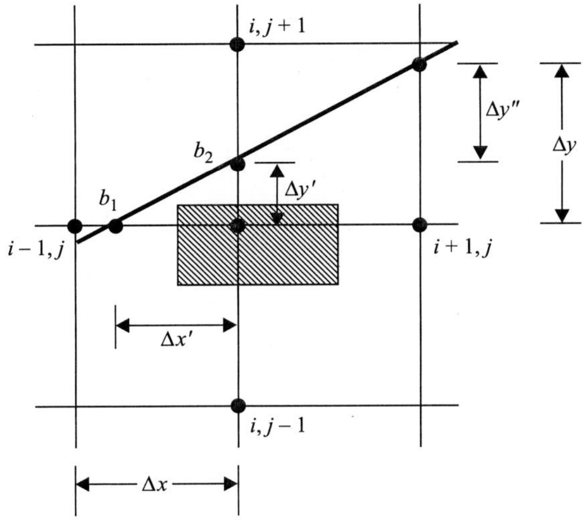

Figure 6.6 Interior nodes near a slanted boundary surface

Boundary Nodes with No Flux Conditions. The finite difference equations for nodes on a boundary surface with no flux conditions, such as an insulated surface or an impermeable surface, can be obtained by setting $h_c = 0$ in the Equation 6.39.

Slanted Surface. In the case of a slanted surface (Figure 6.6), we can deduce the finite difference equations for interior nodes and boundary nodes from Equations 6.38 and 6.39, respectively, by setting $\alpha = \frac{1}{2}, \beta = \frac{1}{2}$ and $\gamma = \frac{3}{2}$.

6.4 THREE-DIMENSIONAL STEADY STATE PROBLEMS

So far, we have considered only two-dimensional problems, assuming that the dimension in the z direction is significantly larger than those in the x and y directions, and thus net flux quantity in the z direction is negligible. Let us now consider three-dimensional problems in which dependent variables are a function of three space variables. A discretization procedure outlined for the two-dimensional problem can be extended and applied in a similar manner to the three-dimensional problem. The three-dimensional calculation domain is divided into equal or nonequal small regions with increments of Δx, Δy, and Δz in x, y, and z directions, respectively, as shown in Figure 6.7. The values of the dependent variables $\phi(x, y, z)$ are calculated at a finite number of discrete points in the solution domain as shown in the figure.

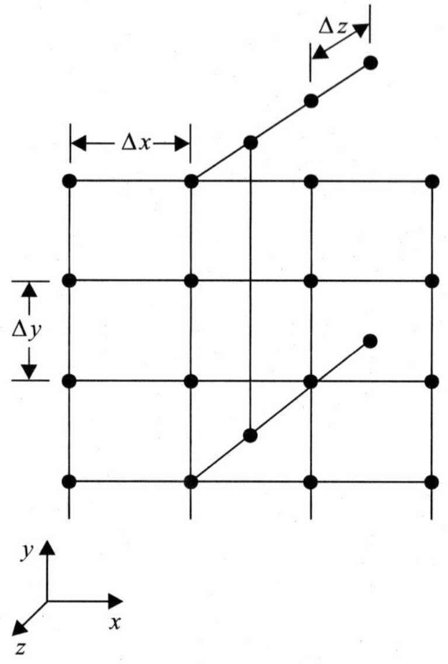

Figure 6.7 Nodal network for a three-dimensional region

Each nodal point is designated by an additional index k, which indicates z increments defined $z = k\Delta z$. The discrete value of the function ϕ at a location (x_i, y_j, z_k) is designated as $\phi_{i,j,k}$.

For three-dimensional steady state problems, the appropriate governing equation is

$$\frac{\partial}{\partial x}\left(\Gamma\frac{\partial \phi}{\partial x}\right) + \frac{\partial}{\partial y}\left(\Gamma\frac{\partial \phi}{\partial y}\right) + \frac{\partial}{\partial z}\left(\Gamma\frac{\partial \phi}{\partial z}\right) + S = 0 \tag{6.40}$$

Let us again consider control volume method to obtain the discretization equation for the interior nodes. For each grid point we consider a control volume bounded by the dashed lines localized at

$$\left(i-\tfrac{1}{2}, j, k\right),\ \left(i+\tfrac{1}{2}, j, k\right),\ \left(i, j+\tfrac{1}{2}, k\right),\ \left(i, j-\tfrac{1}{2}, k\right),\ \left(i, j, k+\tfrac{1}{2}\right) \text{ and } \left(i, j, k-\tfrac{1}{2}\right)$$

as shown in Figure 6.8. Integrate the governing equation over a three-dimensional control volume surrounding the grid point (i, j, k) as shown.

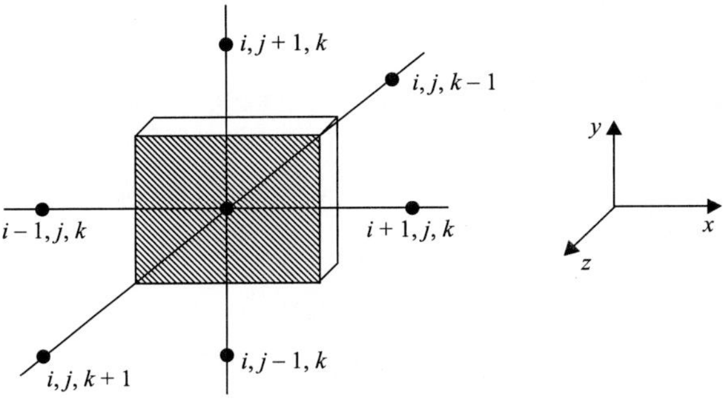

Figure 6.8 Details of grid points in a three-dimensional grid system

Integrating Equation 6.40 over the control volume, we get

$$\int_{i,j-\frac{1}{2},k}^{i,j+\frac{1}{2},k}\int_{i-\frac{1}{2},j,k}^{i+\frac{1}{2},j,k}\int_{i,j,k-\frac{1}{2}}^{i,j,k+\frac{1}{2}}\frac{\partial}{\partial x}\left(\Gamma_x\frac{\partial\phi}{\partial x}\right)dx\,dy\,dz+\int_{i,j-\frac{1}{2}}^{i,j+\frac{1}{2}}\int_{i-\frac{1}{2},j}^{i+\frac{1}{2},j}\int_{i,j,k-\frac{1}{2}}^{i,j,k+\frac{1}{2}}\frac{\partial}{\partial y}\left(\Gamma_y\frac{\partial\phi}{\partial y}\right)dx\,dy\,dz$$
$$+\int_{i,j-\frac{1}{2},k}^{i,j+\frac{1}{2},k}\int_{i-\frac{1}{2},j,k}^{i+\frac{1}{2},j,k}\int_{i,j,k-\frac{1}{2}}^{i,j,k+\frac{1}{2}}\frac{\partial}{\partial z}\left(\Gamma_z\frac{\partial\phi}{\partial z}\right)dx\,dy\,dz+\int_{i,j-\frac{1}{2},k}^{i,j+\frac{1}{2},k}\int_{i-\frac{1}{2},j,k}^{i+\frac{1}{2},j,k}\int_{i,j,k-\frac{1}{2}}^{i,j,k+\frac{1}{2}}S\,dx\,dy\,dz=0 \tag{6.41}$$

$$\left(\Gamma_x\frac{\partial\phi}{\partial x}\right)_{i+\frac{1}{2},j,k}\Delta y\Delta z-\left(\Gamma_x\frac{\partial\phi}{\partial x}\right)_{i-\frac{1}{2},j,k}\Delta y\,\Delta z+\left(\Gamma_y\frac{\partial\phi}{\partial y}\right)_{i,j+\frac{1}{2},k}\Delta x\,\Delta z$$
$$-\left(\Gamma_y\frac{\partial\phi}{\partial y}\right)_{i,j-\frac{1}{2},k}\Delta x\,\Delta z+\left(\Gamma_z\frac{\partial\phi}{\partial z}\right)_{i,j,k+\frac{1}{2}}\Delta x\,\Delta y-\left(\Gamma_z\frac{\partial\phi}{\partial z}\right)_{i,j,k-\frac{1}{2}}\Delta x\,\Delta y$$
$$+\int_{i,j-\frac{1}{2}}^{i,j+\frac{1}{2}}\int_{i-\frac{1}{2},j}^{i+\frac{1}{2}}S\,dx\,dy=0 \tag{6.42}$$

Considering a step-wise constant profile for the source term we can express the source term integral as

$$\int_{i,j-\frac{1}{2},k}^{i,j+\frac{1}{2},k}\int_{i-\frac{1}{2},j,k}^{i+\frac{1}{2},j,k}\int_{i,j,k-\frac{1}{2}}^{i,j,k+\frac{1}{2}}S\,dx\,dy=\bar{S}\,\Delta x\,\Delta y\,\Delta z \tag{6.43}$$

where $\bar{S}$ is the mean or average heat generation rate in the control volume, and it can be linearized as discussed before.

Also, considering a step-wise constant profile for the transport properties Γ and a piece-wise linear profile to approximate the first derivative terms, we can express the flux quantities across the control volume surfaces in Equation 6.42 as

$$\left.\Gamma_x \frac{\partial \phi}{\partial x}\right)_{i+\frac{1}{2},j,k} \approx \frac{\Gamma_{x i+\frac{1}{2}}\left(\phi_{i+1,j,k} - \phi_{i,j,k}\right)}{\Delta x_{i+\frac{1}{2}}} \tag{6.44a}$$

$$\left.\Gamma_x \frac{\partial \phi}{\partial x}\right)_{i-\frac{1}{2},j,k} \approx \frac{\Gamma_{x i-\frac{1}{2}}\left(\phi_{i,j,k} - \phi_{i-1,j,k}\right)}{\Delta x_{i-\frac{1}{2}}} \tag{6.44b}$$

$$\left.\Gamma_y \frac{\partial \phi}{\partial y}\right)_{i,j+\frac{1}{2},k} = \frac{\Gamma_{y j+\frac{1}{2}}\left(\phi_{i,j+1,k} - \phi_{i,j,k}\right)}{\Delta y_{j+\frac{1}{2}}} \tag{6.44c}$$

$$\left.\Gamma_y \frac{\partial \phi}{\partial y}\right)_{i,j-\frac{1}{2},k} = \frac{\Gamma_{y j-\frac{1}{2}}\left(\phi_{i,j,k} - \phi_{i,j-1,k}\right)}{\Delta y_{j-\frac{1}{2}}} \tag{6.44d}$$

$$\left.\Gamma_z \frac{\partial \phi}{\partial z}\right)_{i,j,k+\frac{1}{2}} = \frac{\Gamma_{z k+\frac{1}{2}}\left(\phi_{i,j,k+1} - \phi_{i,j,k}\right)}{\Delta z_{k+\frac{1}{2}}} \tag{6.44e}$$

$$\left.\Gamma_z \frac{\partial \phi}{\partial z}\right)_{i,j,k-\frac{1}{2}} = \frac{\Gamma_{z k-\frac{1}{2}}\left(\phi_{i,j,k} - \phi_{i,j,k-1}\right)}{\Delta z_{k-\frac{1}{2}}} \tag{6.44f}$$

Substituting Equations 6.43 and 6.44 into Equation 6.42 and simplifying, we get

$$\begin{aligned} a_{i,j,k}\phi_{i,j,k} = {} & a_{i+1,j,k}\phi_{i+1,j,k} + a_{i-1,j,k}\phi_{i-1,j,k} + a_{i,j+1,k}\phi_{i,j+1,k} + a_{i,j-1,k}\phi_{i,j-1,k} \\ & + a_{i,j,k+1}\phi_{i,j,k+1} + a_{i,j,k-1}\phi_{i,j,k-1} + d \end{aligned} \tag{6.45}$$

where

$$a_{i+1,j,k} = \frac{\Gamma_{i+\frac{1}{2},j,k}}{\Delta x_{i+\frac{1}{2},j,k}} \Delta y \, \Delta z \quad a_{i-1,j,k} = \frac{\Gamma_{i-\frac{1}{2},j,k}}{\Delta x_{i-\frac{1}{2},j,k}} \Delta y \, \Delta z \quad a_{i,j+1,k} = \frac{\Gamma_{i,j+\frac{1}{2},k}}{\Delta y_{i,j+\frac{1}{2},k}} \Delta x \, \Delta z$$

$$a_{i,j-1,k} = \frac{\Gamma_{i,j-\frac{1}{2},k}}{\Delta y_{i,j-\frac{1}{2},k}} \Delta x \, \Delta z \quad a_{i,j,k+1} = \frac{\Gamma_{i,j,k+\frac{1}{2}}}{\Delta z_{i,j,k+\frac{1}{2}}} \Delta x \, \Delta y \quad a_{i,j,k-1} = \frac{\Gamma_{i,j,k-\frac{1}{2}}}{\Delta y_{i,j,k-\frac{1}{2}}} \Delta x \Delta y$$

$$a_{i,j,k} = a_{i+1,j,k} + a_{i-1,j,k} + a_{i,j+1,k} + a_{i,j-1,k} + a_{i,j,k+1} + a_{i,j,k-1} - S_1 \, \Delta x \, \Delta y \, \Delta z$$

$$d = S_0 \, \Delta x \, \Delta y \, \Delta z$$

Equation 6.45 can be simplified further for many different cases as was shown for the two-dimensional case.

Boundary Nodes. The finite difference equations for boundary nodes on x and y planes are the same as those derived for the two-dimensional case and presented in Table 6.2 to Table 6.4. Equations for the boundary nodes on z planes, i.e., the front and the back planes, can be deduced from those derived for x or y planes.

6.5 SOLUTION TECHNIQUES AND COMPUTER IMPLEMENTATION

It can be noted from Examples 6.1 to 6.5 worked out in the previous section that the system of equations resulting from a two-dimensional diffusion equation is no longer tridiagonal, as it was in the case of a one-dimensional diffusion problem. The reason for this is the fact that, while the nodal discretization equation can be rearranged so that three of the terms are at or adjacent to the main diagonal, the remaining two terms are displaced. Hence, unlike the unidirectional diffusion problem, the resulting system of equations for the two-dimensional heat diffusion cannot be solved by TDMA. The resulting system of equations is in fact very sparse, containing a large number of zeros. We can certainly use a direct solver such as the Gauss elimination or *LU* decomposition method discussed in Chapter 3. However, with more refined grid size distributions such direct methods are subject to increased computer memory limitation and become computationally expansive. As we have discussed in Chapter 3, that iterative point-by-point methods such as Gauss–Seidel are the method of choice for an increased number of nodes and for such a sparse system. Additional discussion about the solution techniques for multidimensional problems is given in Chapter 7 for multidimensional unsteady state problems. In the following section we present the solution algorithm or pseudo code for solving a two-dimensional diffusion equation based on the Gauss–Seidel algorithm.

6.5.1 Solution Algorithm Based on the Gauss–Seidel Method

This is the simplest to implement of all iterative methods as discussed in Chapter 3. It generally involves the following steps.

1. Before assembling the discretizations for all nodes, it is desirable to reorder each equation such that the resulting system of equations is diagonally dominant, a sufficient condition for the convergence of the Gauss–Seidel method.
2. Write each equation in the system in explicit form for the nodal value associated with the diagonal element as

$$\phi_i^{k+1} = \frac{c_i - \sum_{j=1}^{i-1} a_{ij}\,\phi_j^{k+1} - \sum_{j=i+1}^{n} a_{ij}\,\phi_j^{k}}{a_{ii}}, \quad \text{for } i = 1, 2, \ldots, n \tag{6.46}$$

   where k is the index that represents the iteration number.
3. An initial set of values for ϕ_i is assumed as ϕ_i^0.
4. New improved iterative values ϕ_i^1 for $k=1$ are calculated by substituting initial guess value ϕ_i^0 at $k=0$ and any recent calculated value ϕ_i^1 at $k=1$ into the right-hand side of Equation 6.46. This concludes the first iteration step.
5. Step 3 is repeated for all new iterated values of ϕ_i^{k+1} using the most recent values of ϕ.
6. The iteration process is continued until a prescribed convergence criterion such as

$$\left|\frac{\phi_i^{k+1} - \phi_i^k}{\phi_i^{k+1}}\right| \le \varepsilon_s \tag{6.47a}$$

   or

$$\frac{\left\|\phi_i^{k+1} - \phi_i^k\right\|_e}{\left\|\phi_i^{k+1}\right\|_e} \le \varepsilon_s \tag{6.47b}$$

   is satisfied.

A pseudo code for solving a two-dimensional diffusion equation by the Gauss–Seidel method is given in Table 6.5.

Table 6.5 Pseudo Code for Steady State Solution of Two-dimensional Diffusion Equation using the Finite Difference Scheme and the Gauss-Seidel Method

Input dimensions, properties, select number of divisions, maximum number of iteration parameters

```
  Input dimension L and H
  Input transport property Γ
  Input number of divisions NDIVX and NDIVY
  Number of node in x-direction NX = NDIVX + 1
  Number of node in y-direction NY = NDIVY + 1
  Input maximum number of iteration, itmax
```

Set initial guess values

```
  dofor I = 1, NX
    dofor J = 1, NY
      φ(I, J) = φ_I
    enddo
  enddo

  iflag = 1
  iter = 0
  dowhile (it < itmax and iflag = 1)
    iter = iter+1
```

Set new values to old values

```
    dofor I = 1, NX
    dofor J=1, NY
        φ0(I, J) = φ(I, J)
    enddo
```

Calculate new iterated nodal values

```
    dofor I = 1, NX
      do for j = 1, NY
        Solve for φ(I, J)
      enddo
    enddo
```

Estimate approximate percent relative error and check for convergence

```
    dofor I = 1, NX
      dofor J=1, NY
```

$$\varepsilon_a = ABS\left(\frac{\phi(I,J) - \phi_0(I,J)}{\phi(I,J)}\right)$$

```
        if (ε_a > ε_s) then
          iflag = 1
        else
          iflag = 0
        endif
      enddo
    enddo
  enddo
```

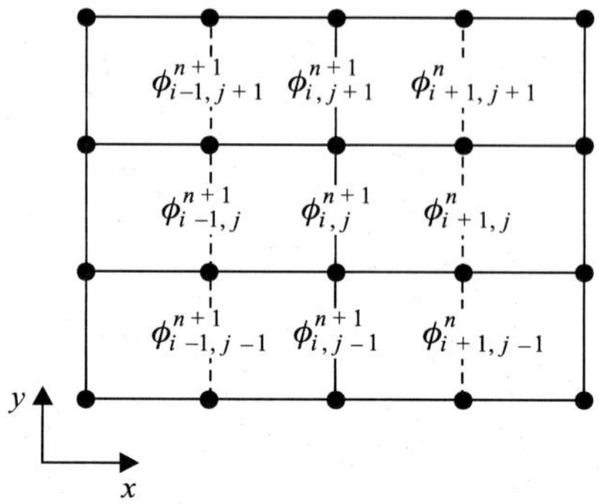

Figure 6.9 Computational grid in the line-by-line method with sweeping in the x direction

6.5.2 Solution by Combination of TDMA and Gauss–Seidel Method (Line-By-Line Method)

This procedure is also referred to as the ***line-by-line method*** as suggested by Patankar (1980). The ***line-by-line method*** is primarily a combination of a TDMA algorithm for a one-dimensional problem and a Gauss–Seidel iteration method. The basic approach involves either sweeping in the x direction while determining all nodal values along a vertical line using TDMA or sweeping in the y direction while determining all nodal values along a horizontal line using TDMA. The values of all nodal points ($\phi_{i,j}^{n+1}$, $j = 1,\ldots, N$) along a vertical line (solid line in Figure 6.9) are determined in the present iteration (n+1) as a one-dimensional problem by using the TDMA algorithm while assuming known values for the nodal points along two adjacent vertical lines, shown as dotted lines at x_{i-1} and x_{i+1} in Figure 6.9. The nodal values in the x_{i-1} line are assumed to be known in the present iteration (n+1), while the nodal values in the x_{i+1} line are known based on the previous iteration (n).

This is repeated for all vertical lines in the x direction, and one iteration is said to be complete when all vertical lines are scanned. In the next iteration, one can choose to sweep in the y direction and estimate all nodal values along all horizontal lines.

For example, in order to sweep all vertical lines in the x direction, the discretization equation for the two-dimensional diffusion equation (5.27) can be rearranged in the following manner

$$a_{i,,j}\phi_{i,j} = a_{i,j+1}\phi_{i,j+1} + a_{i,j-1}\phi_{i,j-1} + a_{i+1,j}\phi_{i+1,j} + a_{i-1,j}\phi_{i-1,j} + d \tag{6.48}$$

In order to apply the TDMA algorithm at any y line, we can first rewrite the discretization equation in the form of a tridiagonal matrix format as

$$A_j\phi_{i,j} = B_j\phi_{i,j+1} + C_j\phi_{i,j-1} + D_j \tag{6.49}$$

where

$$A_j = a_{i,j}, \quad B_j = a_{i,j+1}, \quad C_i = a_{i,j-1}$$
$$D_i = a_{i+1,j}\phi_{i+1,j} + a_{i-1,j}\phi_{i-1,j} + b + b$$

Equation 6.49 can now be solved using the TDMA algorithm. Let us demonstrate the development of a pseudo code for solving a two-dimensional diffusion problem by considering two different cases involving different types of boundary conditions in a rectangular slab.

Case I: With Boundary Condition of the First Kind

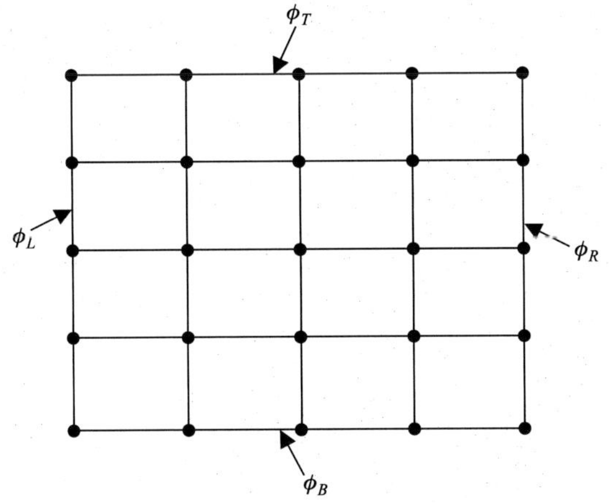

Table 6.6 Pseudo-Code for a Line-by-line Method with Sweeping in the x Direction and with Boundary Condition of the First Kind at the Top and Bottom Surfaces

```
Dofor I = 2, (NX−1)
   A(1) = 1.0
   B(1) = 0.0
   C(1) = 0.0
   D(1) = φ_T
   A(NY) = 1.0
   B(NY) = 0.0
   C(NY) = 0.0
   D(NY) = φ_B
   Dofor J = 2, (NY−1)
      A(J) = a_{i,j}
      B(J) = a_{i,j+1}
      C(J) = a_{i,j−1}
      D(J) = a_{i+1,j}φ_{i+1,j} + a_{i−1,j}φ_{i−1,j} + d
   Enddo
   Call tdma (NX, A, B, C, D, φ)
Enddo
```

Case II: With Boundary Condition of the First Kind at the Top and No Flux at the Bottom

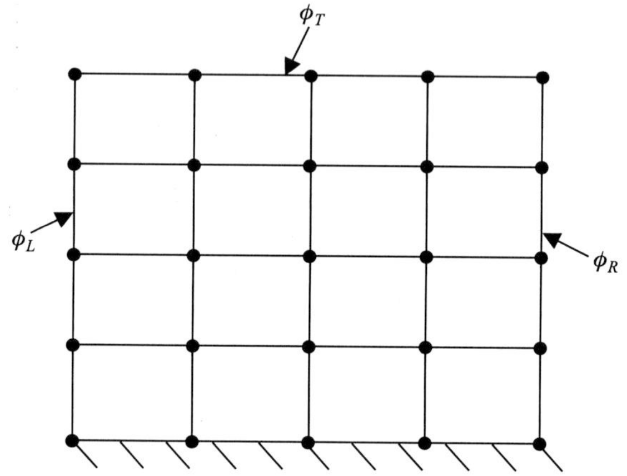

Table 6.7 Pseudo-Code for a Line-by-line Method Sweeping in the x Direction and with Boundary Condition of the First Kind at the Top Surface and No Flux at the Bottom Surface

```
Dofor I = 2, (NX − 1)
  A(1) = 1.0
  B(1) = 0.0
  C(1) = 0.0
  D(1) = φ_T
  A(NY) = 1.0
  B(NY ) = 1.0
  C(NY) = 0.0
  D(NY ) = 0.0
  Dofor J = 2, (NY − 1)
    A(J) = a_{i,j}
    B(J) = a_{i,j+1}
    C(J) = a_{i,j−1}
    D(J) = a_{i+1,j}φ_{i+1,j} + a_{i−1,j}φ_{i−1,j} + d
  Enddo
  Call tdma (NY in NX, A, B, C, D, φ)
Enddo
```

We can see from the pseudo codes that the boundary condition information is brought in to calculate interior nodes values. This results in an improved convergence rate in a line-by-line method compared to the Gauss–Seidel method. Alternating the sweeping directions essentially improves the convergence rate. However, selection of the preferred sweeping direction may be dictated, in many problems, by the geometry and magnitude of the boundary conditions, as well as magnitudes of the coefficient values. Alternating the sweeping directions essentially improves the convergence rate.

PROBLEMS

6.1 Consider a 1.5 m by 1.5 m square slab with the top and left surfaces maintained at high temperature, $T_H = 600°C$. The left, right, and bottom surfaces are subjected to a convection condition with $h_c = 30\ W/m^{2}°C$ and $T_{\infty} = 25°C$. The material conductivity is $k = 30\ W/m°C$. Apply the finite difference method to calculate the two-dimensional temperature distribution in the slab based on the grid distribution shown in the figure. Estimate the heat transfer rates at all surfaces, and check for the overall energy balance.

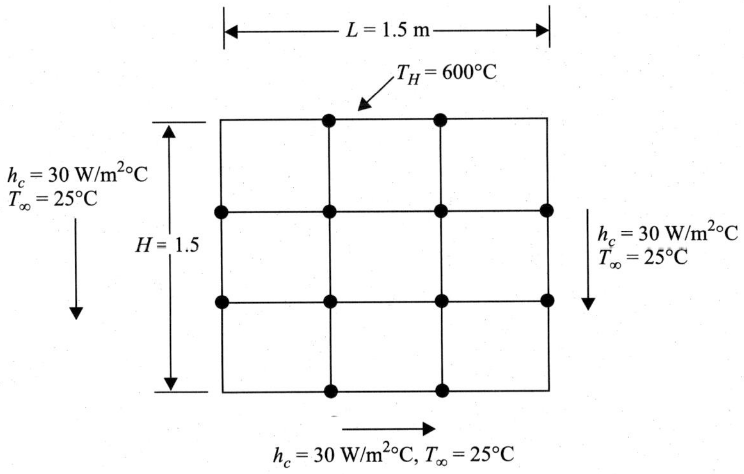

The mathematical statement of the problem is as follows.

Governing equation

$$\frac{\partial^2 T}{\partial x^2} + \frac{\partial^2 T}{\partial x^2} = 0$$

Boundary conditions

1. $x = 0, \quad k\left.\frac{\partial T}{\partial x}\right|_{x=0} = h_c(T|_{x=0} - T_{\infty})$

2. $x = L, \quad -k\left.\frac{\partial T}{\partial x}\right|_{x=L} = h_c(T|_{x=L} - T_{\infty})$

3. $y = 0, \quad k\left.\frac{\partial T}{\partial y}\right|_{y=0} = h(T|_{y=0} - T_{\infty})$

4. $y = H, \quad T(x, H) = T_H$

Solve the system of equations by using (a) the Gauss-elimination method and (b) the Gauss–Seidel method.

6.2 Develop a computer code for solving a two-dimensional diffusion equation in a rectangular slab using the pseudo code given in Table 6.5 based on using the Gauss–Seidel algorithm. Use the computer code to solve Problem 6.1 with refining the grid as 3×3, 4×4, 6×6, and 8×8. Check the convergence of the solution by comparing the temperature distribution at the mid-section with the progressively refined grids.

6.3 Consider a 0.75 m by 1.0 m rectangular porous slab with the top and bottom surfaces maintained at a constant moisture concentration of $C_0 = 0.002\,\mathrm{kg/m^3}$. The left surface is exposed to moving air with a mass transfer coefficient of $h_D = 0.02\,\mathrm{m/s}$ and a moisture concentration of $C_\infty = 0.02\,\mathrm{kg/m^3}$. The right surface is attached to an impermeable aluminum foil. The material is assumed to be homogeneous with diffusion coefficient $D_0 = 4.0 \times 10^{-5}\,\mathrm{m^2/s}$.

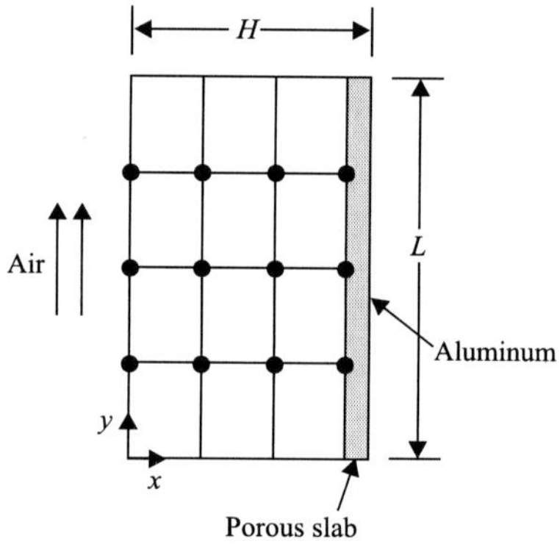

The mathematical statement of the problem is as follows.

Governing equation

$$\frac{\partial^2 C}{\partial x^2} + \frac{\partial^2 C}{\partial y^2} = 0$$

Boundary conditions

1. $x = 0, \quad -D_0 \left.\frac{\partial C}{\partial x}\right|_{x=0} = h_D(C_\infty - C|_{x=0})$
2. $x = H, \quad \left.\frac{\partial C}{\partial x}\right|_{x=H} = 0$
3. $y = 0, \quad C(x, 0) = C_0$
4. $y = L, \quad C(x, L) = C_0$

Apply a finite difference method to calculate the two-dimensional moisture concentration distribution in the slab based on the grid distribution shown in the figure. Estimate the moisture transfer rates at all surfaces, and show the overall moisture balance.

6.4 Consider temperature rise in a metal cutting tool. Heat is generated in primary and secondary shear zones of a metallic workpiece and metal chips during metal cutting processes. This heat is transmitted by conduction into the cutting tools. In order to study the thermal effects on the cutting tool, it is necessary to evaluate the temperature distribution in the tools.

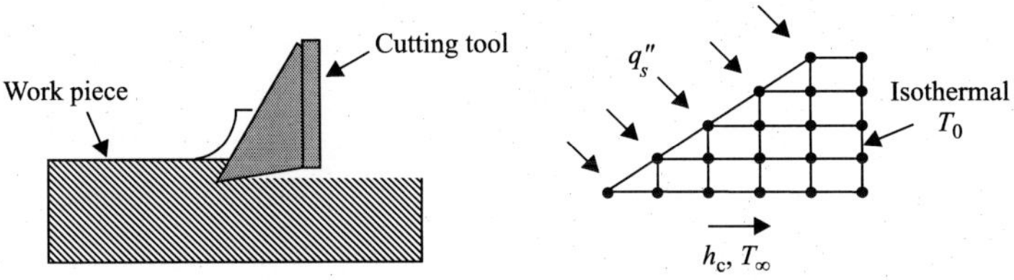

A simplified heating condition is considered in this problem by considering a constant surface heat flux at the top rake surface and the bottom surface exposed to forced convective cooling. Further interior regions, i.e., the right surface, are assumed to be isothermal or at a constant temperature T_0. Consider the grid distribution, write finite difference equations for each node, and obtain the system of equations for the unknown nodal values.

6.5 Consider the problem of cooling an electronic chip that is mounted on a substrate. The electric power generation in the chip is $q = 40$ W and the size of the chip is 10 mm × 5 mm. The top and bottom surfaces are assumed to be convectively cooled with $h_c = 50$ W/m^2°C and $T_\infty = 20$°C. The two side surfaces are assumed to be at a constant temperature $T_0 = 25$°C. The conductivities of the substrate and chip materials are $K_s = 4$ W/m °C and $K_c = 60$ W/m °C, respectively.

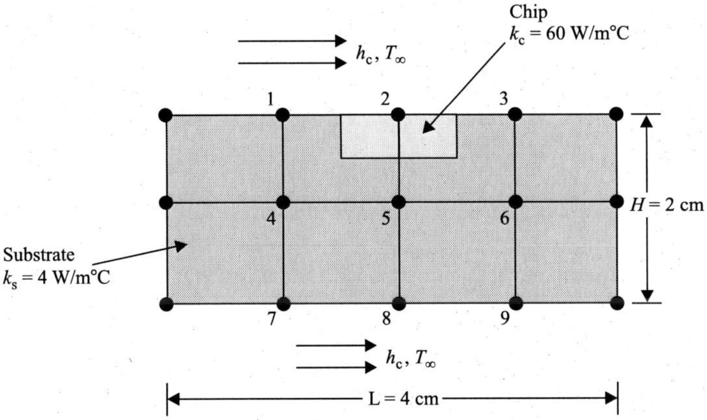

Use the finite difference method/control volume to calculate the two-dimensional temperature distribution in the chip and the substrate using the grid distribution shown in the figure. Check the overall energy balance.

6.6 Use the finite difference method to solve the two-dimensional steady state conduction in a rectangular carbon steel ($k = 200\,\text{W/m}\,°\text{C}$) slab subjected to a constant surface heat flux irradiated by a high-energy laser beam at the top surface. For simplicity, assume the heat flux distribution to be a constant average value, $q''_0 = 2 \times 10^8\,\text{W/m}^2$, acting over a section of the surface. The top surface is also subjected to forced convection with $h_c = 500\,\text{W/m}^2\,°\text{C}$. The left surface is assumed to be adiabatic as a line of symmetry. All other surfaces are assumed to be subjected to free convection with $T_\infty = 25°\text{C}$ and $h = 40\,\text{W/m}^2\,°\text{C}$. Derive the finite difference equations for all nodes and represent the system in matrix form.

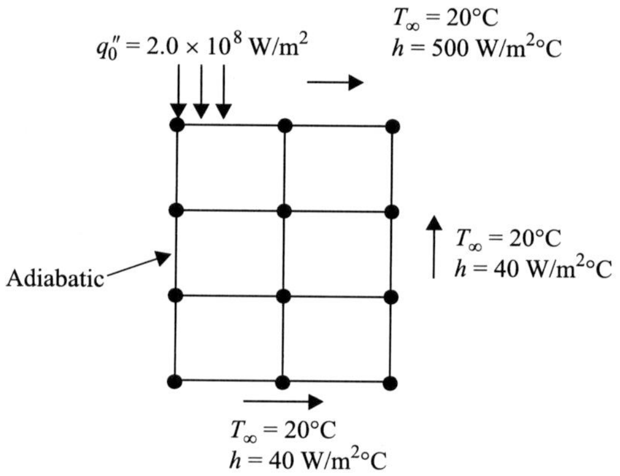

(a) Derive the finite difference equations for all nodes and represent the system in matrix form.

(b) Solve the system of equations using the Gauss elimination solver and check for the overall energy balance.

6.7 Solve the velocity distribution for the example Problem 6.5 using the computer code based on the Gauss–Seidel method in Problem 6.1. Estimate the average velocity and volume flow rate using two-dimensional quadrature formulas presented in Chapter 4. Use the following data points for the computation. $H = L = 8.00\,\text{mm}$, $\mu = 0.02\,\text{kg/m/s}$, $\partial P/\partial x = \bar{S} = -750\,\text{Pa/m}$, and $U_0 = 10\,\text{mm/s}$.

Repeat the computation by further refining the grid to 3×3, 4×4, 6×6, and 8×8. Check the convergence by observing velocity distribution at the mid section along the y direction and by observing percent relative error in average velocity or volume flow rate. Use the assigned tolerance limit, $\varepsilon_s = 0.001$.

6.8 Develop a computer code for solving a two-dimensional diffusion equation in a rectangular slab using the pseudo code given in Table 6.7, based on using the line-by-line algorithm. Use the computer code to solve the problem presented in Example 6.1 by progressively refining the grid as 4×4, 6×6, and 8×8. Check the convergence of the solution by comparing temperature distribution at the mid section with progressively refined grids. Use the assigned tolerance limit, $\varepsilon_s = 0.001$.

CHAPTER

7

UNSTEADY STATE DIFFUSION EQUATION

In this chapter we will introduce the basic treatment of an unsteady state term in the finite difference method using the one-dimensional steady state problem. Different temporal approximations that are needed to discretize time derivative are introduced. The procedure will be extended to two- and three-dimensional problems.

In previous chapters on steady state problems, we have demonstrated the use of both Taylor series approach and control volume approach in deriving the finite difference equations. In our treatment of the unsteady term, we also employ both these approaches to demonstrate the discretization procedure.

7.1 TIME DISCRETIZATION PROCEDURE

Let us demonstrate the time discretization procedure by considering the governing equation as the one-dimensional diffusion equation with one diffusion term in the x direction and a storage term or unsteady state term

$$C\frac{\partial \phi}{\partial t} = \frac{\partial}{\partial x}\left(\Gamma_x \frac{\partial \phi}{\partial x}\right) \tag{7.1}$$

Also, we will demonstrate the procedure using the control volume approach and considering an interior node shown in Figure 7.1.

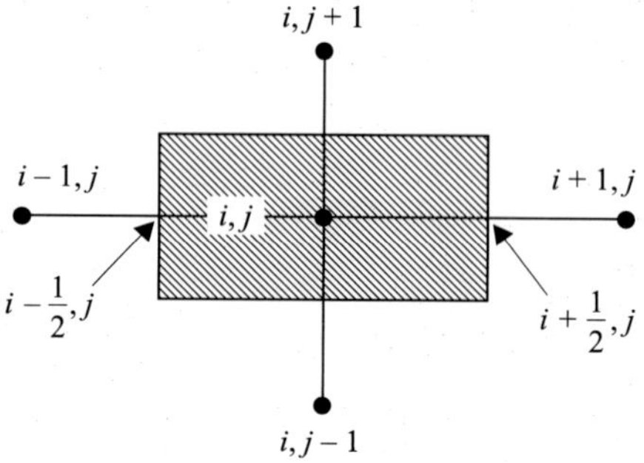

Figure 7.1 Control volume for an interior node

As time is a one-way coordinate we obtain the solution by marching in time, i.e., for a given ϕ value at time step t, we find ϕ at $t+\Delta t$. Integrating Equation 7.1 over the control volume (Figure 7.1) from $i-\frac{1}{2}$ to $i+\frac{1}{2}$ and over the time interval t to $t+\Delta t$, we get

$$\int_{t}^{t+\Delta t}\int_{i-\frac{1}{2}}^{i+\frac{1}{2}} C\frac{\partial \phi}{\partial t}\,dx\,dt = \int_{t}^{t+\Delta t}\int_{i-\frac{1}{2}}^{i+\frac{1}{2}} \frac{\partial}{\partial x}\left(\Gamma_x \frac{\partial T}{\partial x}\right)dx\,dt \tag{7.2}$$

The left-hand side of the integral is evaluated by assuming that the value of ϕ at the node i, i.e., ϕ_i, prevails over the entire control volume surrounding the node

$$C\int_{i-\frac{1}{2}}^{i+\frac{1}{2}}\int_{t}^{t+\Delta t} \frac{\partial \phi}{\partial t}\,dt\,dx = C\Delta x_i \int_{t}^{t+\Delta t} \frac{\partial \phi_i}{\partial t}\,dt = C\Delta x_i \phi_i\big|_t^{t+\Delta t} = C(\phi_i^{l+1} - \phi_i^l)\Delta x_i \tag{7.3}$$

where

$$\phi_i^{l+1} = \text{new value at time } t+\Delta t$$

and

$$\phi_i^{l} = \text{old value at time } t$$

The right-hand side integral of Equation 7.2 is evaluated based on the discretization procedure outlined in Chapter 5 for the diffusion term as

$$\int_{t}^{t+\Delta t}\int_{i-\frac{1}{2}}^{i+\frac{1}{2}} \frac{\partial}{\partial x}\left(\Gamma_x \frac{\partial \phi}{\partial x}\right)dx\,dt = \int_{t}^{t+\Delta t}\left[\frac{\Gamma_{i+\frac{1}{2}}(\phi_{i+1}-\phi_i)}{(\Delta x)_{i+\frac{1}{2}}} - \frac{\Gamma_{i-\frac{1}{2}}(\phi_i-\phi_{i-1})}{(\Delta x)_{i-\frac{1}{2}}}\right]dt \tag{7.4}$$

Now, we need a profile assumption for the variation of ϕ over the time step t to $t+\Delta t$. Let us assume a general profile such as

$$\phi_i = \left[f\phi_i^{l+1} + (1-f)\phi_i^l\right] \tag{7.5}$$

where f is called the ***time approximation weighting factor***.

Using this general temporal profile assumption, we can write each one of the integral terms in Equation 7.3 as

$$\int_t^{t+\Delta t} \phi_i \, dt = \left[f\phi_i^{l+1} + (1-f)\phi_i^l\right]\Delta t \tag{7.6a}$$

$$\int_t^{t+\Delta t} \phi_{i+1} \, dt = \left[f\phi_{i+1}^{l+1} + (1-f)\phi_{i+1}^l\right]\Delta t \tag{7.6b}$$

$$\int_t^{t+\Delta t} \phi_{i-1} \, dt = \left[f\phi_{i-1}^{l+1} + (1-f)\phi_{i-1}^l\right]\Delta t \tag{7.6c}$$

Substituting Equation 7.6 into Equation 7.4, we get

$$\begin{aligned}\int_t^{t+\Delta t}\int_{i-\frac{1}{2}}^{i+\frac{1}{2}} \frac{\partial}{\partial x}\left(\Gamma_x \frac{\partial \phi}{\partial x}\right) dx\, dt = {} & f\left[\frac{\Gamma_{i+\frac{1}{2}}\left(\phi_{i+1}^{l+1} - \phi_i^{l+1}\right)}{(\Delta x)_{i+\frac{1}{2}}} - \frac{\Gamma_{i-\frac{1}{2}}\left(\phi_i^{l+1} - \phi_{i-1}^{l+1}\right)}{(\Delta x)_{i-\frac{1}{2}}}\right]\Delta t \\ & + (l-f)\left[\frac{\Gamma_{i+\frac{1}{2}}\left(\phi_{i+1}^{l} - \phi_i^{l}\right)}{(\Delta x)_{i+\frac{1}{2}}} - \frac{\Gamma_{i-\frac{1}{2}}\left(\phi_i^{l} - \phi_{i-1}^{l}\right)}{(\Delta x)_{i-\frac{1}{2}}}\right]\Delta t\end{aligned} \tag{7.7}$$

Substituting Equations 7.3 and 7.7 into Equation 7.2, we have the discretization equation for an interior node as

$$\begin{aligned}C\left(\phi_i^{l+1} - \phi_i^l\right)\Delta x_i = {} & f\left[\frac{\Gamma_{i+\frac{1}{2}}\left(\phi_{i+1}^{l+1} - \phi_i^{l+1}\right)}{(\Delta x)_{i+\frac{1}{2}}} - \frac{\Gamma_{i-\frac{1}{2}}\left(\phi_i^{l+1} - \phi_{i-1}^{l+1}\right)}{(\Delta x)_{i-\frac{1}{2}}}\right]\Delta t \\ & + (1-f)\left[\frac{\Gamma_{i+\frac{1}{2}}\left(\phi_{i+1}^{l} - \phi_i^{l}\right)}{(\Delta x)_{i+\frac{1}{2}}} - \frac{\Gamma_{i-\frac{1}{2}}\left(\phi_i^{l} - \phi_{i-1}^{l}\right)}{(\Delta x)_{i-\frac{1}{2}}}\right]\Delta t\end{aligned} \tag{7.8}$$

Rearranging and regrouping terms, we have

$$\begin{aligned}a_i\phi_i^{l+1} = {} & a_{i+1}\left[f\phi_{i+1}^{l+1} + (1-f)\phi_{i+1}^l\right] + a_{i-1}\left[f\phi_{i-1}^{l+1} + (1-f)\phi_{i-1}^l\right] \\ & + \left[a_i^l - (1-f)a_{i+1} - (1-f)a_{i-1}\right]\phi_i^l\end{aligned} \tag{7.9a}$$

where

$$a_i = fa_{i+1} + fa_{i-1} + a_i^l \tag{7.9b}$$

$$a_{i+1} = \frac{\Gamma_{i+\frac{1}{2}}}{(\Delta x)_{i+\frac{1}{2}}} \tag{7.9c}$$

$$a_{i-1} = \frac{\Gamma_{i-\frac{1}{2}}}{(\Delta x)_{i-\frac{1}{2}}} \tag{7.9d}$$

$$a_i^l = \frac{C\Delta x}{\Delta t} \tag{7.9e}$$

The selection of the time approximation weighting factor leads to different time approximation schemes, which are named as the ***explicit scheme*** with $f=0$, the ***Crank–Nicolson scheme*** with $f=0.5$, and ***fully implicit scheme*** with $f=1$. A brief description along with the physical meaning of these schemes is given below.

7.2 EXPLICIT SCHEME

7.2.1 Discretization Equation by Control Volume Approach

An explicit form of the discretization equation is obtained substituting $f=0$ into Equation (7.9) as

$$a_i\phi_i^{l+1} = a_{i+1}\phi_{i+1}^l + a_{i-1}\phi_{i-1}^l + \left[a_i^l - a_{i+1} - a_{i-1}\right]\phi_i^l \tag{7.10a}$$

where

$$a_i = a_i^l \tag{7.10b}$$

$$a_{i+1} = \frac{\Gamma_{i+\frac{1}{2}}}{(\Delta x)_{i+\frac{1}{2}}} \tag{7.10c}$$

$$a_{i-1} = \frac{\Gamma_{i-\frac{1}{2}}}{(\Delta x)_{i-\frac{1}{2}}} \tag{7.10d}$$

$$a_i^l = \frac{C\Delta x_i}{\Delta t} \tag{7.10e}$$

Physically, in the ***explicit scheme*** it is assumed that the old value of ϕ_i^l prevails throughout the entire time step except at time $t+\Delta t$, as depicted in Figure 7.2, which show temporal variation of ϕ with time.

Substituting the coefficient values in Equation 7.10, we get

$$\frac{C\Delta x_i}{\Delta t}\phi_i^{l+1} = \frac{\Gamma_{i+\frac{1}{2}}}{\Delta x_{i+\frac{1}{2}}}\phi_{i+1}^l + \frac{\Gamma_{i-\frac{1}{2}}}{\Delta x_{i-\frac{1}{2}}}\phi_{i-1}^l + \left(\frac{C\Delta x_i}{\Delta t} - \frac{\Gamma_{i+\frac{1}{2}}}{\Delta x_{i+\frac{1}{2}}} - \frac{\Gamma_{i-\frac{1}{2}}}{\Delta x_{i-\frac{1}{2}}}\right)\phi_i^l \tag{7.11}$$

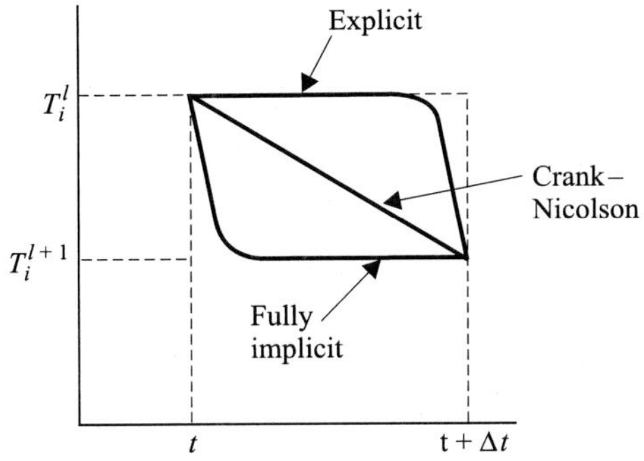

Figure 7.2 Time approximation schemes

For a special case of constant transport property Γ and uniform grid size distribution, i.e., $\Delta x_{i+\frac{1}{2}} = \Delta x_{i-\frac{1}{2}} = \Delta x_i = \Delta x$, we get

$$\phi_i^{l+1} = \frac{\Gamma \Delta t}{C \Delta x^2}\left(\phi_{i+1}^l + \phi_{i-1}^l\right) + \left(1 - \frac{2\Gamma \Delta t}{C \Delta x^2}\right)\phi_i^l \tag{7.12}$$

Similarly, for a two-dimensional problem, we get

$$\phi_{i,j}^{l+1} = \frac{\Gamma \Delta t}{C(\Delta x)^2}\left(\phi_{i+1,j}^l + \phi_{i-1,j}^l + \phi_{i,j+1}^l + \phi_{i,j-1}^l\right) + \left[1 - \frac{4\Gamma \Delta t}{C(\Delta x)^2}\right]\Gamma_{i,j}^l \tag{7.13}$$

The finite difference equations presented here are called the ***explicit form,*** because variable ϕ_i^{l+1} at time step $(l+1)$ can be determined directly from the knowledge of the variables ϕ_{i+1}^l, ϕ_i^l and ϕ_{i-1}^l at the previous time step l.

7.2.2 Finite Difference Equation by Taylor Series Expansion

As was demonstrated before, we can derive the discretization equation directly by applying finite difference approximations to the derivatives in the unsteady state diffusion equation (7.1). Let us demonstrate this by considering the one-dimensional, time dependent diffusion equation of the form

$$C\frac{\partial \phi}{\partial t} = \Gamma \frac{\partial^2 \phi}{\partial x^2} \tag{7.14}$$

Using the second-order central-difference formula for the second derivative diffusion term and the forward difference formula for the first-order time derivative, we get

$$C\frac{\phi_i^{l+1} - \phi_i^l}{\Delta t} = \Gamma \frac{\phi_{i+1}^l - 2\phi_i^l + \phi_{i-1}^l}{(\Delta x)^2} \tag{7.15}$$

with error of the $0(\Delta t) + 0(\Delta x)^2$.

Rearranging, we obtain

$$\phi_i^{l+1} = \frac{\Gamma \Delta t}{C(\Delta x)^2}\left[\phi_{i+1}^l + \phi_{i-1}^l\right] + \left[1 - \frac{2\Gamma \Delta t}{C(\Delta x)^2}\right]\phi_i^l \tag{7.16}$$

Similarly, for the two-dimensional unsteady diffusion equation of the form

$$C\frac{\partial \phi}{\partial t} = \Gamma\left(\frac{\partial^2 \phi}{\partial x^2} + \frac{\partial^2 \phi}{\partial y^2}\right) \tag{7.17}$$

we obtain

$$C\frac{\phi_{i,j}^{l+1} - \phi_{i,j}^l}{\Delta t} = \Gamma\frac{\phi_{i+1,j}^l - 2\phi_{i,j}^l + \phi_{i-1,j}^l}{(\Delta x)^2} + \Gamma\frac{\phi_{i,j+1}^l - 2\phi_{i,j}^l + \phi_{i,j-1}^l}{\Delta y^2} \tag{7.18}$$

For uniform grid size distribution, i.e., $\Delta x = \Delta y$

$$\phi_{i,j}^{l+1} = \frac{\Gamma \Delta t}{C(\Delta x)^2}\left[\phi_{i+1,j}^l + \phi_{i-1,j}^l + \phi_{i,j+1}^l + \phi_{i,j-1}^l\right] + \left[1 - \frac{4\Gamma \Delta t}{C(\Delta x)^2}\right]\phi_{i,j}^l \tag{7.19}$$

The explicit finite difference scheme discussed here is also referred to as the ***FTCS (forward in time central in space) scheme.*** This explicit form of finite-difference representation provides a relatively straightforward procedure for computer implementation. However, the explicit scheme suffers serious limitation from a ***stability*** point of view.

7.2.3 Stability Consideration

An important issue that needs to be examined before using finite difference equations based on an explicit scheme is that of stability. Although it is possible for any fluctuations resulting from round-off errors in numerical solutions to dampen out and reach convergence, this is not always guaranteed to occur. This error may grow at any stage of the computation and cause termination of the program. It may be indicated that the round-off error is the one that is relevant to the stability of the numerical solution. There are several methods for stability analysis that examine the behavior of the round-off errors. The most common one is the ***von Newmann*** method.

von Newmann Stability Criterion. The von Newmann method is based on the prediction of the growth or decay of the errors identified earlier as the difference between the exact solution and the numerical solution. A limitation of this method is that it yields necessary and sufficient conditions for the stability of mathematical models that can be described as linear, initial value problems with constant coefficients. Otherwise, this method provides only necessary conditions for stability. Even though the von Newman method pertains to interior nodes of the domain, it can provide useful approximate information on the effect of the boundary conditions on the stability solution if it is applied to the boundaries of the solution domain.

The method involves solving an error equation using the Fourier series and the method of separation of variables. The method yields the ***von Newman*** stability criterion, which needs to be satisfied for the stability of the solution. For a two-dimensional problem this criterion is derived as

$$\frac{\Gamma\Delta t}{C\Delta x^2} + \frac{\Gamma\Delta t}{C\Delta y^2} < \frac{1}{2} \tag{7.20}$$

For uniform grid size distribution, i.e., $\Delta x = \Delta y$, the criterion becomes

$$\frac{\Gamma\Delta t}{C\Delta x^2} < \frac{1}{4} \tag{7.21}$$

and for one-dimensional problems

$$\frac{\Gamma\Delta t}{C\Delta x^2} < \frac{1}{2} \tag{7.22}$$

The parameter $\gamma_0 = \Gamma\Delta t/C\Delta x^2$ is defined as the stability parameter. The stability criterion tells us that for given values of Γ, C, and selected grid size, Δx, there is a maximum permissible time step Δt, which cannot be exceeded for stability of the solution. In other words, if the time step Δt exceeds the limit imposed by the criterion, then the numerical computations become unstable resulting from the amplification of errors. Figure 7.3 illustrates typical numerical results with different stability parameters.

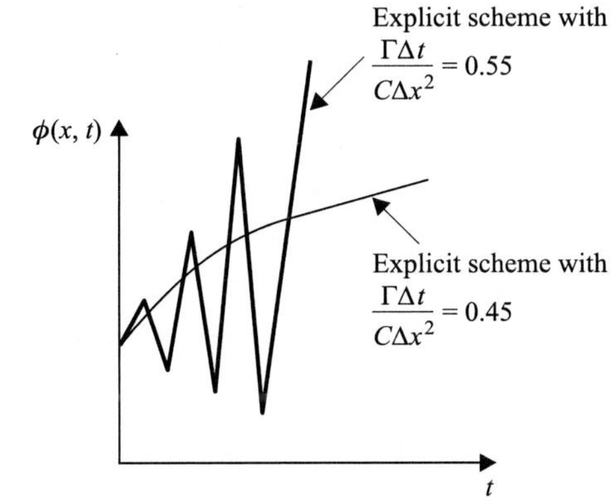

Figure 7.3 Effect of stability parameter in finite difference solution using the explicit scheme

Table 7.1 Pseudo Code for an Explicit Finite Difference Scheme in a Plane Slab with Constant Surface Values

```
Input dimensions, properties, select number of divisions, maximum time, and stability parameters
   Input dimension L_x
   Input thermo physical properties Γ and C
   Select number of division in x direction NDIVX
   Select stability parameter γ0
   Number of node NX = NDIVX+1
   Calculate step size DX = L_x/NDIVX

   Calculate time step from stability parameter
   Dt = γ0 C(DX)^2/Γ

Initialize time
   time = 0.0

Set initial condition
   dofor I = 1, NX
   φ(I) = φ_I
enddo

Set boundary conditions
   φ(1) = φ_L
   φ(NX) = φ_R
dowhile (time < tmax)
   time = time +Dt

Calculate nodal values at interior nodes
   dofor I = 2, NX−1
      φ(I) = γ0 * (φ(I + 1) + φ(I − 1)) + (1 − 2 * γ0) * φ(I)
   enddo

Estimate flux quantity at right surface
   f_x'' = Γ(φ(NX) − φ(NX − 1)/DX)
enddo
```

Again it may be noted here that the above criterion provides the necessary and sufficient stability condition for linear problems with constant coefficients. For more complex problems with nonlinearities and variable coefficients it usually provides only the necessary (but not always sufficient) stability condition. For more detailed discussion on stability analysis, refer to Jaluria and Tortance (1986), Smith (1978), Fletcher (1991), and Poulikakos (1994).

A pseudo code for solving one-dimensional transient problems based on an explicit finite difference scheme is given in Table 7.1.

Example 7.1: Explicit Scheme with Constant Surface Temperatures: Consider a one-dimensional unsteady state heat conduction in a plane slab of thickness $L = 4\,\text{mm}$ as shown in the figure. Initially the slab is at a temperature $T_I = 0°\text{C}$. The right side of the plate is suddenly brought to a temperature $T_L = 200°\text{C}$. The left side is maintained at $T_0 = 0°\text{C}$. The material's thermo-physical properties are $k = 50\,\text{W/m°C}$ and $\alpha = 2 \times 10^{-5}\,\text{m}^2/\text{s}$.

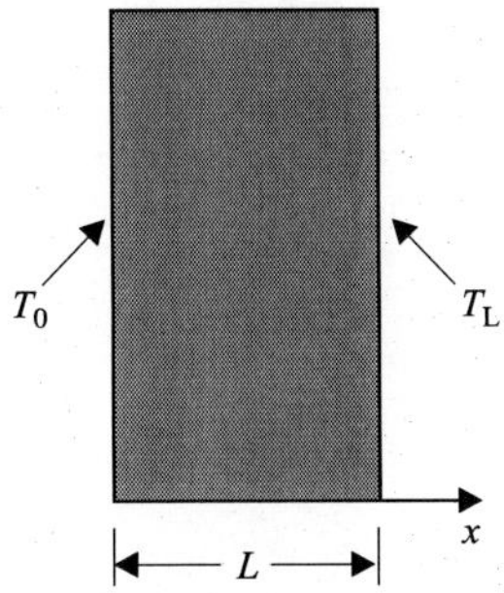

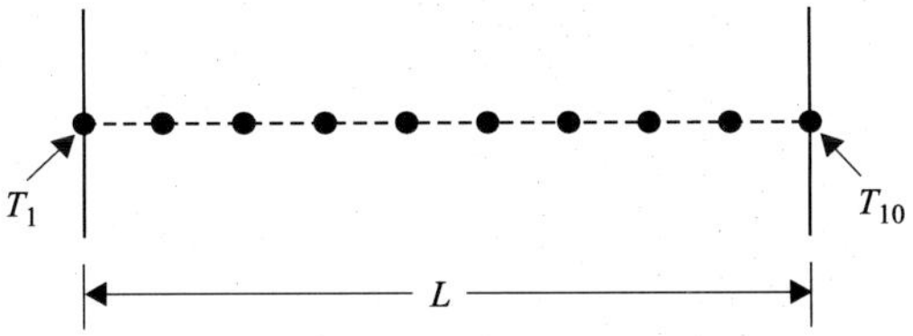

Governing equation

$$\frac{\partial T}{\partial t} = \alpha \frac{\partial^2 T}{\partial x^2}$$

Boundary conditions

1. $x = 0, \quad T(0, t) = \mathrm{T_L}$
2. $x = L, (L, t) = T_R$

Initial condition

$$T = 0, \quad T(x, 0) = T_I$$

Solution: For a number of nodes $n = 11$, the spatial step size is

$$\Delta x = \frac{L}{n-1} = \frac{4 \times 10^{-3}}{10} = 4 \times 10^{-4}\,\mathrm{m}$$

Noting that for the heat transfer problems $\Gamma = k$ and $C = \rho c$, we can write

$$\frac{\Gamma \Delta t}{C \Delta x^2} = \frac{\alpha \Delta t}{\Delta x^2} = f_0$$

For satisfying the stability criterion, let us assume $\left(\Gamma \Delta t / C \Delta x^2\right) = \left(\alpha \Delta t / \Delta x^2\right) = 0.4$ and this gives a time step estimate of

$$\Delta t = \frac{0.4 \times \Delta x^2}{2 \times 10^{-5}} = 0.003\,95\,\mathrm{s}$$

We select the explicit discretization equation (7.15) for interior nodes as

$$T_i^{l+1} = f_0\left(T_{i+1}^l + T_{i-1}^l\right) + (1 - 2f_0)T_i^l$$

Applying this equation to all interior nodes, we get

$$\begin{aligned}
&\textit{Node } 2 \quad T_2^{l+1} = f_0\left(T_3^l + T_1^l\right) + (1 - 2f_0)T_2^l \\
&\textit{Node } 3 \quad T_3^{l+1} = f_0\left(T_4^l + T_2^l\right) + (1 - 2f_0)T_3^l \\
&\textit{Node } 4 \quad T_4^{l+1} = f_0\left(T_5^l + T_3^l\right) + (1 - 2f_0)T_4^l \\
&\textit{Node } 5 \quad T_5^{l+1} = f_0\left(T_6^l + T_4^l\right) + (1 - 2f_0)T_5^l \\
&\textit{Node } 6 \quad T_6^{l+1} = f_0\left(T_7^l + T_5^l\right) + (1 - 2f_0)T_6^l \\
&\textit{Node } 7 \quad T_7^{l+1} = f_0\left(T_8^l + T_6^l\right) + (1 - 2f_0)T_7^l \\
&\textit{Node } 8 \quad T_8^{l+1} = f_0\left(T_9^l + T_7^l\right) + (1 - 2f_0)T_8^l \\
&\textit{Node } 9 \quad T_9^{l+1} = f_0\left(T_{10}^l + T_8^l\right) + (1 - 2f_0)T_9^l
\end{aligned}$$

Since the nodal equations are explicit, the system of equations can be solved at the present time with all temperature values on the right-hand side of the equations known at the previous time. The heat transfer rate at the right surface is then estimated as

$$q_L'' = k\frac{T(N) - T(N-1)}{\Delta X}$$

A computer code based on the pseudo code given in Table 7.1 is used to solve the problem. Results for transient temperature distribution at selected time intervals are summarized in the table and figure below.

	Time					
Node	0.01 s	0.03 s	0.05 s	0.07 s	0.09 s	0.1
T_1	0	0	0	0	0	0
T_2	0	0	0.4876	3.8687	8.1739	11.8623
T_3	0	0	2.4662	10.4488	18.8527	25.6856
T_4	0	0	8.099	21.4315	33.2186	42.3091
T_5	0	2.048	20.2003	38.1834	52.0149	62.2003
T_6	0	12.4928	41.1502	61.362	75.4229	85.4112
T_7	0	12.8	71.7258	90.6736	103.0272	111.5885
T_8	0	55.04	110.5805	124.8823	133.8867	140.0408
T_9	80	123.776	154.6249	162.0634	166.6964	169.8489
T_{10}	200	200	200	200	200	200

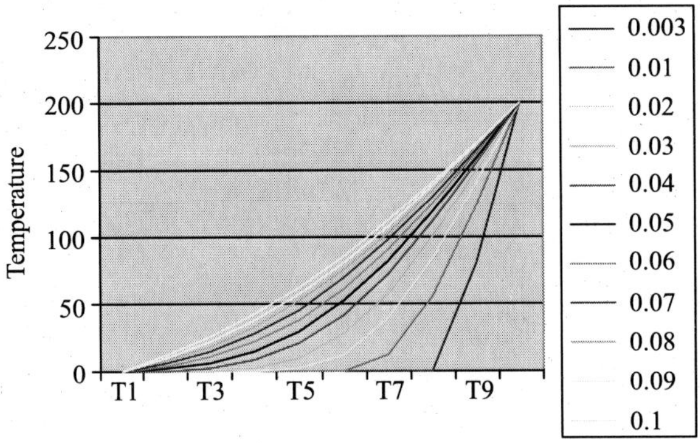

The heat transfer rates at the right surface are also estimated at selected time intervals and presented in the table and figure below:

Time	q''_L
0	13 500 000
0.01	8 575 200
0.02	6 765 777
0.03	5 763 110
0.04	5 104 696
0.05	4 630 217
0.06	4 267 871
0.07	3 980 437
0.08	3 746 659
0.09	3 553 408
0.1	3 391 994

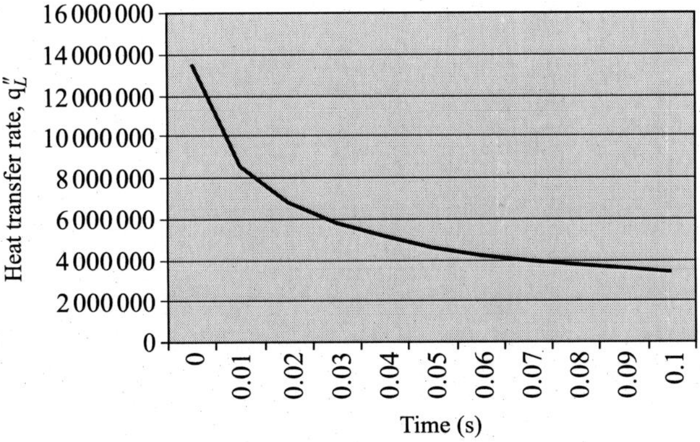

7.2.4 Other Explicit Schemes

The explicit scheme discussed so far is called a ***FTCS*** (forward in time central in space) scheme, and it uses a first-order-accurate forward-difference approximation for the time derivative. There are many other explicit schemes that are proposed to improve the accuracy and stability using higher-order finite difference formulas. A few of them are discussed here.

Richardson Scheme. In this scheme a second-order-accurate central-difference approximation for the time derivative is used in order to improve the accuracy. The discretized equation is

$$C\frac{\phi_{i,j}^{l+1} - \phi_{i,j}^{i-1}}{2\Delta t} = \Gamma\frac{\phi_{i+1,j}^{l} - 2\phi_{i,j}^{l} + \phi_{i-1,j}^{l}}{(\Delta x)^2} + \Gamma\frac{\phi_{i,j+1}^{l} - 2\phi_{i,j}^{l} + \phi_{i,j-1}^{l}}{(\Delta y)^2} \tag{7.23a}$$

Rearranging

$$\phi_{i,j}^{l+1} = \phi_{i,j}^{l-1} + \frac{2\Gamma\Delta t}{C(\Delta x)^2}\left[\phi_{i+1,j}^{l} + \phi_{i-1,j}^{l}\right] + \frac{2\Gamma\Delta t}{C(\Delta y)^2}\left[\phi_{i,j+1}^{l} + \phi_{i,j-1}^{l}\right]$$
$$- \left[\frac{2\Gamma\Delta t}{C(\Delta x)^2} + \frac{2\Gamma\Delta t}{C(\Delta y)^2}\right]\phi_{i,j}^{l}$$

For uniform grid size distributions, i.e., $\Delta x = \Delta y$, we have

$$\phi_{i,j}^{l+1} = \phi_{i,j}^{l-1} + \frac{2\Gamma\Delta t}{C(\Delta x)^2}\left[\phi_{i+1,j}^{l} + \phi_{i-1,j}^{l} + \phi_{i,j+1}^{l} + \phi_{i,j-1}^{l} - 2\phi_{i,j}^{l}\right] \tag{7.23b}$$

The equation is second-order accurate of the order of $O(\Delta x^2) + O(\Delta t^2)$. However, a stability analysis indicates that it is unconditionally unstable. Therefore, it is usually not used for heat and fluid flow problems.

DuFort–Frankel Scheme. In this scheme $\phi_{l,j}^{l}$ in above Richardson's equation is replaced by $\frac{1}{2}\left(\phi_{l,j}^{l-1} + \phi_{i,j}^{l+1}\right)$ to avoid the stability problem. The resulting scheme is

$$C\frac{\phi_{i,j}^{l+1} - \phi_{i,j}^{i-1}}{2\Delta t} = \Gamma_x\frac{\phi_{i+1,j}^{l} - \left(\phi_{i,j}^{l-1} + \phi_{i,j}^{l+1}\right) + \phi_{i-1,j}^{l}}{(\Delta x)^2} + \Gamma_y\frac{\phi_{i,j+1}^{l} - \left(\phi_{i,j}^{l-1} + \phi_{i,j}^{l+1}\right) + \phi_{i,j-1}^{l}}{(\Delta y)^2}$$

or

$$\left(1 + \frac{2\Gamma_x\Delta t}{C(\Delta x)^2} + \frac{2\Gamma_y\Delta t}{C(\Delta y)^2}\right)\phi_i^{l+1}$$
$$= \frac{2\Gamma_x\Delta t}{C(\Delta x)^2}\left(\phi_{i+1,j}^{l} + \phi_{i-1,j}^{l}\right) + \frac{2\Gamma_y\Delta t}{C(\Delta y)^2}\left(\phi_{i,j+1}^{l} + \phi_{i-1,j-1}^{l}\right)$$
$$+ \left(1 - \frac{2\Gamma_x\Delta t}{C(\Delta x)^2} - \frac{2\Gamma_y\Delta t}{C(\Delta y)^2}\right)\phi_{i,j}^{l-1}$$

Simplifying

$$\phi_{i,j}^{l+1} = \frac{1 - [2\Gamma_x \Delta t/C(\Delta x)^2] - [2\Gamma_y \Delta t/C(\Delta y)^2]}{1 + [2\Gamma_x \Delta t/C(\Delta x)^2] + [2\Gamma_y \Delta t/C(\Delta y)^2]} \phi_{i,j}^{l-1}$$
$$+ \frac{2[2\Gamma_x \Delta t/C(\Delta x)^2]}{1 + [2\Gamma_x \Delta t/C\Delta x^2] + [2\Gamma_y \Delta t/C\Delta y^2]} \left(\phi_{i+1,j}^{l} + \phi_{i-1,j}^{l}\right)$$
$$+ \frac{2[2\Gamma_y \Delta t/C(\Delta y)^2]}{1 + [2\Gamma_x \Delta t/C\Delta x^2] + [2\Gamma_y \Delta t/C\Delta y^2]} \left(\phi_{i,j+1}^{l} + \phi_{i,j-1}^{l}\right) \quad (7.24)$$

For uniform grid size distribution and for an isotropic media, Equation 7.24 reduces to

$$\phi_{i,j}^{l+1} = \frac{1 - [4\Gamma \Delta t/C(\Delta x)^2]}{1 + [4\Gamma \Delta t/C(\Delta x)^2]} \phi_{i,j}^{l-1} + \frac{[2\Gamma \Delta t/C(\Delta x)^2]}{1 + [4\Gamma \Delta t/C(\Delta x^2)]} \left(\phi_{i+1,j}^{l} + \phi_{i-1,j}^{l}\right)$$
$$+ \frac{[2\Gamma \Delta t/C\Delta x^2]}{1 + [4\Gamma \Delta t/C(\Delta x)^2]} \left(\phi_{i,j+1}^{l} + \phi_{i,j-1}^{l}\right)$$

For ***one-dimensional problems***, Equation 7.25 reduces to

$$\phi_{i}^{l+1} = \frac{1 - [2\Gamma \Delta t/C\Delta x^2]}{1 + [2\Gamma \Delta t/C(\Delta x)^2]} \phi_{i}^{l-1} + \frac{[2\Gamma \Delta t/C\Delta x^2]}{1 + [2\Gamma \Delta t/C\Delta x^2]} \left(\phi_{i+1,j}^{l} + \phi_{i-1,j}^{l}\right) \quad (7.26)$$

It can be shown by application of stability analysis that the DuFort–Frankel scheme is unconditionally stable. It can also be noted that the DuFort–Frankel scheme utilizes three levels in time: $l-1$, l and $l+1$, and hence brings in additional complexity in computations.

7.2.5 Boundary Conditions

The discretization equations given in the previous section are applicable for determining the temperature of interior nodes of a solid as a function of space and time. These equations can be solved along with the boundary conditions of constant surface temperature or first kind. For other types of boundary conditions, we need to use appropriate discretization equations for boundary nodes, which are derived based on assuming a half control volume surrounding the boundary nodes.

Let us briefly discuss the derivation of one such equation for the grid point located at the ***right-hand side boundary*** shown below.

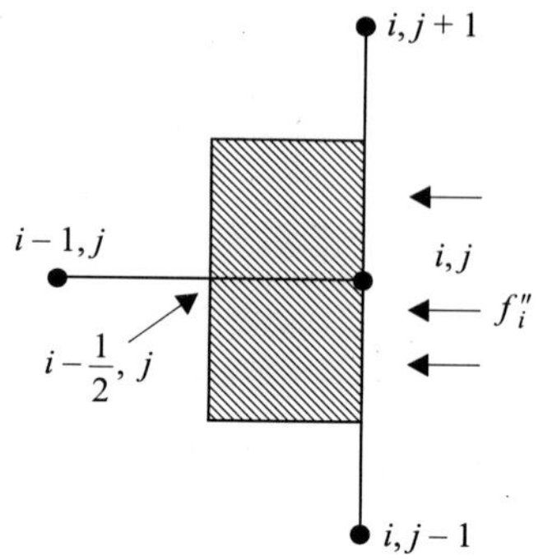

Figure 7.4 Control volume for a boundary node

Let us demonstrate derivation of the boundary node equation by considering the two-dimensional unsteady state diffusion equation given as

$$\frac{\partial}{\partial x}\left(\Gamma\frac{\partial\phi}{\partial x}\right)+\frac{\partial}{\partial x}\left(\Gamma\frac{\partial\phi}{\partial y}\right)=C\frac{\partial\phi}{\partial t} \tag{7.27}$$

To proceed, we integrate the equation over the half control volume and over the time interval t to $t+\Delta t$:

$$\int_{i-\frac{1}{2}}^{i}\int_{j-\frac{1}{2}}^{j+\frac{1}{2}}\int_{t}^{t+\Delta t}\frac{\partial}{\partial x}\left(\Gamma_x\frac{\partial\phi}{\partial x}\right)dx\,dy\,dt+\int_{i-\frac{1}{2}}^{i}\int_{j-\frac{1}{2}}^{j+\frac{1}{2}}\int_{t}^{t+\Delta t}\frac{\partial}{\partial y}\left(\Gamma_y\frac{\partial\phi}{\partial y}\right)dx\,dy\,dt$$
$$=C\int_{i-\frac{1}{2}}^{i}\int_{j-\frac{1}{2}}^{j+\frac{1}{2}}\int_{t}^{t+\Delta t}\frac{\partial\phi}{\partial t}dx\,dy\,dt \tag{7.28}$$

$$\left.\Gamma_x\frac{\partial\phi}{\partial x}\right)_i\Delta y\Delta t-\left.\Gamma_y\frac{\partial\Gamma}{\partial x}\right)_{i-\frac{1}{2}}\Delta y\Delta t+\left.\Gamma_y\frac{\partial\phi}{\partial y}\right)_{j+\frac{1}{2}}\frac{\Delta x\Delta t}{2}-\left.\Gamma_y\frac{\partial\phi}{\partial y}\right)_{j-\frac{1}{2}}\frac{\Delta x\Delta t}{2}$$
$$=C\frac{\Delta x\Delta y}{2}\int_{t}^{t+\Delta t}\frac{\partial\phi}{\partial t}dt \tag{7.29}$$

The first term in the equation is described by the heat flux boundary conditions as

$$\left.-\Gamma_x\frac{\partial\phi}{\partial x}\right)_i=-f_i'' \tag{7.30a}$$

or

$$\Gamma_x \frac{\partial\phi}{\partial x}\Big)_i = +f_i'' \tag{7.30b}$$

Next, we assume a piecewise-linear profile or central difference formula to approximate temperature gradients at points $j+\frac{1}{2}$, $j-\frac{1}{2}$, and $i-\frac{1}{2}$, and forward difference formula for first-order time derivative

$$\Gamma_x \frac{\partial\phi}{\partial x}\Big)_{i-\frac{1}{2}} = \Gamma_{x\,i-\frac{1}{2}} \frac{\phi_{i,j}-\phi_{i-1,j}}{\Delta x}, \quad \Gamma_y \frac{\partial\phi}{\partial y}\Big)_{j+\frac{1}{2}} = \Gamma_{yj+\frac{1}{2}} \frac{\phi_{i,j+1}-\phi_{i,j}}{\Delta y} \tag{7.31a}$$

$$\Gamma_y \frac{\partial\phi}{\partial y}\Big)_{j-\frac{1}{2}} = \Gamma_{yj-\frac{1}{2}} \frac{(\phi_{i,j}-\phi_{i,j-1})}{\Delta y}, \quad \frac{\partial\phi}{\partial t} = \frac{\phi_{i,j}^{l+1}-\phi_{i,j}^{l}}{\Delta t} \tag{7.31b}$$

Substituting these approximations, we get

$$-f_i''\Delta y \Delta t - \frac{\Gamma_{i-\frac{1}{2}}\left(\phi_{i,j}^{l}-\phi_{i-1,j}^{l}\right)}{\Delta x}\Delta y\,\Delta t + \frac{\Gamma_{j+\frac{1}{2}}\left(\phi_{i,j+1}^{l}-\phi_{i,j}^{l}\right)}{\Delta y}\frac{\Delta x}{2}\Delta t$$

$$- \frac{\Gamma_{j-\frac{1}{2}}\left(\phi_{i,j}^{l}-\phi_{i,j-1}^{l}\right)}{\Delta y}\frac{\Delta x}{2}\Delta t = C\frac{\Delta x}{2}\Delta y\left(\phi_{i,j}^{l+1}-\phi_{i,j}^{l}\right) \tag{7.32}$$

$$\frac{C\Delta x\,\Delta y}{2\Delta t}\phi_{i,j}^{l+1} = \frac{\Gamma_{i-\frac{1}{2}}\Delta y}{\Delta x}\phi_{i-1}^{l} + \frac{\Gamma_{j+\frac{1}{2}}\Delta x}{2\Delta y}\phi_{i,j+1}^{l} + \frac{\Gamma_{j-\frac{1}{2}}\Delta x}{2\Delta y}\phi_{i,j-1}^{l}$$

$$+\left(\frac{C\Delta x\,\Delta y}{2\Delta t} - \frac{\Gamma_{i-\frac{1}{2}}\Delta y}{\Delta x} - \frac{\Gamma_{i,j+\frac{1}{2}}\Delta x}{2\Delta y} - \frac{\Gamma_{j-\frac{1}{2}}\Delta x}{2\Delta y}\right)\phi_{i,j}^{l} + f_i''\Delta y \tag{7.33}$$

Simplifying

$$a_i\phi_{i,j}^{l+1} = a_{i-1}\phi_{i-1}^{l} + a_{j+1}\phi_{i,j+1}^{l} + a_{j-1}\phi_{i,j-1}^{l} + \left(a_i^{l} - a_{i-1} - a_{j+1} - a_{j-1}\right)\phi_{i,j}^{l} + d \tag{7.33}$$

where

$$a_i = \frac{C\,\Delta x\,\Delta y}{2\Delta t}, \quad a_{i-1} = \frac{\Gamma_{i-\frac{1}{2}}\Delta y}{\Delta x}, \quad a_{j+1} = \frac{\Gamma_{j+\frac{1}{2}}\Delta x}{2\Delta y}, \quad a_{j-1} = \frac{\Gamma_{j-\frac{1}{2}}\Delta x}{2\Delta y}$$

$$a_i^{l} = \frac{C\,\Delta x\,\Delta y}{2\Delta t}, \quad d = -f_i''\Delta y \tag{7.34}$$

For constant transport property, Γ, and for uniform grid size distributions, i.e., $\Delta x = \Delta y$, we can write the above equation in an alternative form as

$$\phi_{i,j}^{l+1} = \frac{\Gamma \Delta t}{C \Delta x^2}\left[2\phi_{i-1,j}^{l} + \phi_{i,j+1}^{l} + \phi_{i,j-1}^{l}\right] + \left[1 - 4\frac{\Gamma \Delta t}{C \Delta x^2}\right]\phi_i^l + \frac{2\Delta t}{C \Delta x} f_i'' \quad (7.35)$$

For the ***one-dimensional case***, the equation reduces to

$$\phi_i^{l+1} = \frac{\Gamma \Delta t}{C \Delta x^2}\left[2\phi_{i-1}^{l}\right] + \left[1 - \frac{2\Gamma \Delta t}{C \Delta x^2}\right]\phi_i^l + \frac{2\Delta t}{C \Delta x} f_i'' \quad (7.36)$$

Let us now consider two special cases

1. ***Adiabatic boundary condition or a symmetric boundary***
   For an adiabatic surface, the surface flux is zero, i.e., $f_i'' = 0$

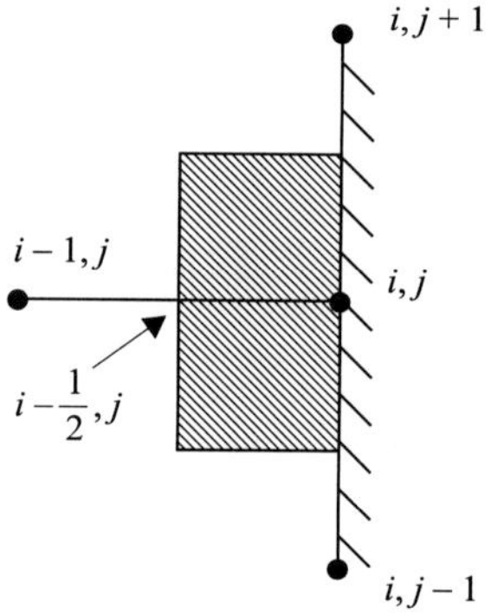

Figure 7.5 Adiabatic boundary node

Substituting Equations 7.35 and 7.36, we get

$$\phi_{i,j}^{l+1} = \frac{\Gamma \Delta t}{C \Delta x^2}\left[2\phi_{i-1,j}^{l} + \phi_{i,j+1}^{l} + \phi_{i,j-1}^{l}\right] + \left[1 - \frac{4\Gamma \Delta t}{C \Delta x^2}\right]\phi_{i,j}^l \quad (7.37)$$

for two-dimensional problems

and

$$\phi_i^{l+1} = \frac{\Gamma \Delta t}{C \Delta x^2}\left[2\phi_{i-1}^{l}\right] + \left[1 - \frac{2\Gamma \Delta t}{C \Delta x^2}\right]\phi_i^l \quad (7.38)$$

for one-dimensional problems

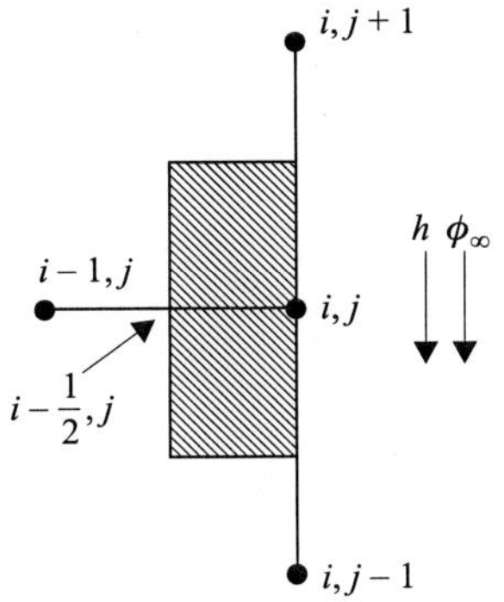

Figure 7.6 Convective boundary node

2. ***Convective boundary***
   For a convective boundary condition

$$-\Gamma \frac{\partial \phi}{\partial x} = h(\phi - \phi_\infty) \tag{7.39a}$$

or

$$f_i'' = h\left(\phi_{i,j} - \phi_\infty\right) \tag{7.39b}$$

Substituting Equation 7.39 into Equation 7.35, we get

$$\phi_{i,j}^{l+1} = \frac{\Gamma \Delta t}{C \Delta x^2}\left[2\phi_{i-1,j}^{l} + \phi_{i,j+1}^{l} + \phi_{i,j-1}^{l} + \frac{2h\Delta x}{\Gamma}\phi_\infty\right] + \left[1 - \frac{4\Gamma \Delta t}{C \Delta x^2} - \frac{2h\Delta t}{C \Delta x}\right]\phi_{i,j}^{l} \tag{7.40}$$

A summary of some explicit discretization equations for boundary nodes is given in Table 7.2.

The explicit method discussed here is simple computationally but, due to its stability requirement, it may require a very small time increment Δt, and hence requires a large number of time steps for problems that need to be analyzed over a longer period of time. In such situations, other finite difference schemes in which the time increment is not restricted by the stability criterion are used. Two such widely used schemes are the ***fully implicit scheme*** and the ***Crank–Nicholson scheme***. These are discussed in the subsequent section.

A pseudo code for a transient solution in a plane slab with a convective boundary condition on the right side and using the explicit finite difference scheme is given in Table 7.3.

Table 7.2 A Summary of Explicit Discretization Equations with Uniform Grid Size Distributions

Physical Situation	Discretization Equation	Stability Requirement
(a) ***Interior node*** 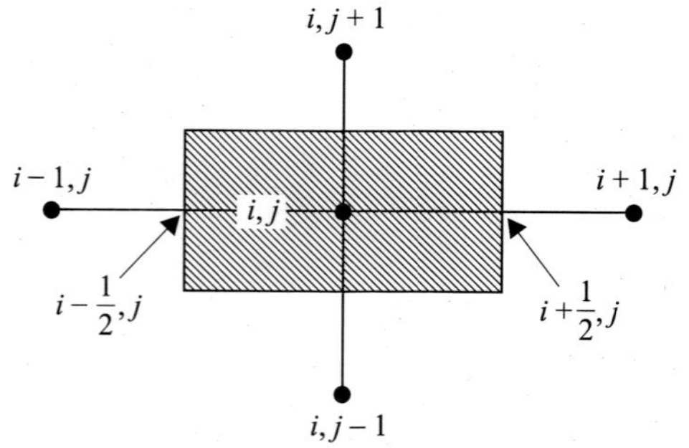	$T_{i,j}^{l+1} = \frac{\Gamma\Delta t}{C\Delta x^2}\left[\phi_{i+1,j}^l + \phi_{i-1,j}^l + \phi_{i,j+1}^l + \phi_{i,j-1}^l\right] + \left[1 - \frac{4\Gamma\Delta t}{C\Delta x^2}\right]\phi_{i,j}^l$	$\frac{\Gamma\Delta t}{\Delta x^2} \le \frac{1}{4}$
(b) ***Convective boundary*** 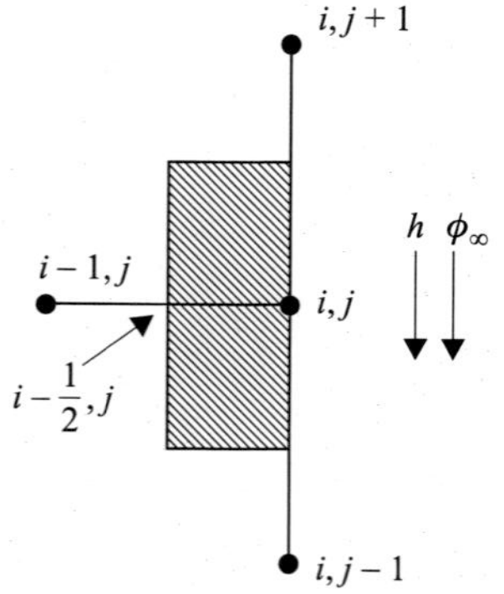	$\phi_{i,j}^{l+1} = \frac{\Gamma\Delta t}{C\Delta x^2}\left[2\phi_{i-1,j}^l + \phi_{i,j+1}^l + \phi_{i,j-1}^l + 2h\Delta x/\Gamma\phi_\infty\right] + \left[1 - \frac{4\Gamma\Delta t}{C\Delta x^2} - \frac{2h\Delta t}{C\Delta x}\right]\phi_{i,j}^l$	$\frac{\Gamma\Delta t}{C\Delta x^2}\left(2 + \frac{h\Delta x}{\Gamma}\right) \le \frac{1}{2}$

(c) ***Exterior corner with convection boundary***

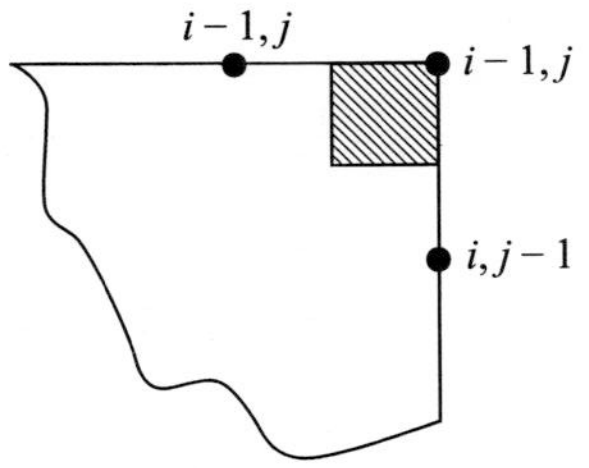

$$\phi_{i,j}^{l+1} = \frac{2\Gamma\Delta t}{C\Delta x^2}\left[\phi_{i-1,j}^{l} + \phi_{i,j-1}^{l} + \frac{2h\Delta x}{\Gamma}T_\infty\right] + \left[1 - \frac{4\Gamma\Delta t}{C\Delta x^2} - \frac{4\Gamma\Delta t}{C\Delta x^2}\frac{h\Delta x}{\Gamma}\right]\phi_{i,j}^{l}$$

$$\frac{\Gamma\Delta t}{C\Delta x^2}\left(1 + \frac{h\Delta x}{\Gamma}\right) \leq \frac{1}{4}$$

(d) ***Interior corner with convection boundary***

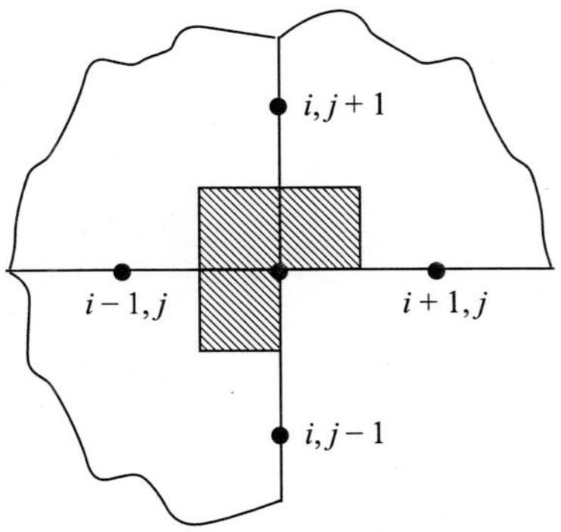

$$\phi_{i,j}^{l+1} = \frac{\frac{2}{3}\Gamma\Delta t}{C\Delta x^2}\left[2\phi_{i,j+1}^{l} + \phi_{i+1,j}^{l} + 2\phi_{i-1,j}^{l} + \phi_{i,j-1}^{l} + 2\frac{h\Delta x}{K}T_\infty\right] + \left[1 - 4\frac{\Gamma\Delta t}{C\Delta x^2} - \frac{4}{3}\frac{\Gamma\Delta t}{C\Delta x^2}\frac{h\Delta x}{\Gamma}\right]\phi_{i,j}^{l}$$

$$\frac{\Gamma\Delta t}{C\Delta x^2}\left(3 + \frac{h\Delta x}{\Gamma}\right) \leq \frac{3}{4}$$

(Continued)

Table 7.2 Continued

Physical Situation	Discretization Equation	Stability Requirement
(e) ***Adiabatic boundary*** 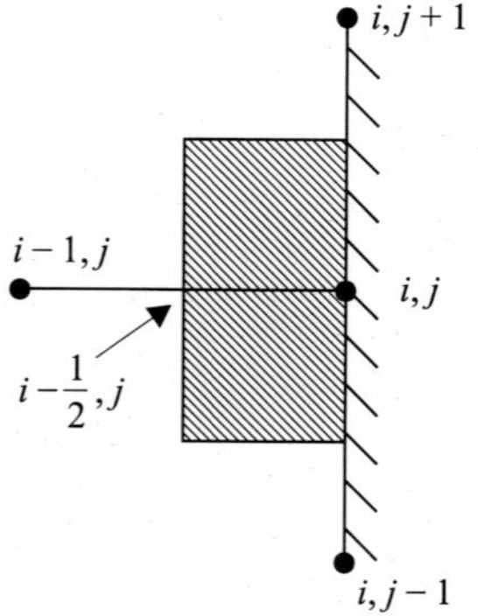	$\phi_{i,j}^{l+1} = \frac{\Gamma\Delta t}{C\Delta x^2}\left[2\phi_{i-1,j}^{l} + \phi_{i,j+1}^{l} + \phi_{i,j-1}^{l}\right] + \left[1 - 4\frac{\Gamma\Delta t}{C\Delta x^2}\right]\phi_{i,j}^{l}$	$\frac{\Gamma\Delta t}{C\Delta x^2} \le \frac{1}{4}$
(f) ***Constant surface heat flux*** 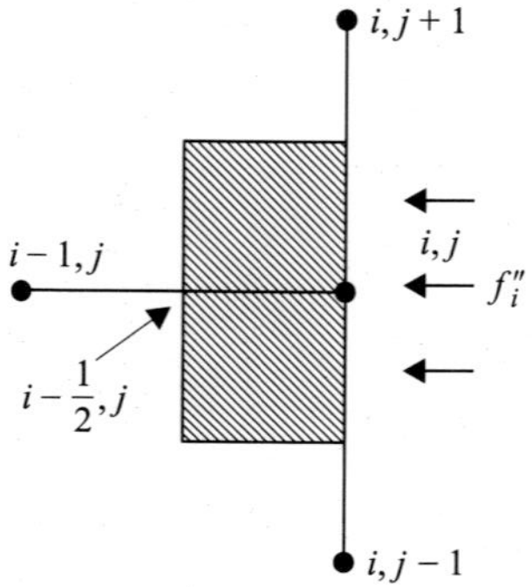	$\phi_{i,j}^{l+1} = \frac{\Gamma\Delta t}{C\Delta x^2}\left[2\phi_{i-1,j}^{l} + \phi_{i,j+1}^{l} + \phi_{i,j-1}^{l}\right]_i'' + \left[1 - 4\frac{\Gamma\Delta t}{C\Delta x^2}\right]\phi_i^{l} - \frac{2\Delta t}{C\Delta x}f_i''$	

Table 7.3 Pseudo Code for a Transient Solution in a Plane Slab with a Convective Boundary on One Side and Using the Explicit Finite Difference Scheme

Input dimensions, properties, select number of divisions, maximum time, and stability parameters
Input dimension L_x
Input thermo physical properties Γ and C
Input convective boundary parameters $h$and ϕ_∞
Select number of division in x-direction $NDIVX$
Select stability parameter γ_0
Number of node $NX = NDIVX{+}1$
Calculate step size $DX = L_x/NDIVX$

Calculate time step using stability parameter
$Dt = \gamma_0 C(DX)^2/\Gamma$

Initialize time
time = 0.0

Set initial condition
dofor i = 1, NX
$\phi(I) = \phi_I$
enddo

Set boundary conditions
$\phi(1) = \phi_L$
dowhile (time < tmax)
time = time + Dt

Calculate nodal values at interior nodes
dofor I = 2, NX−1
$\phi(I) = \gamma_0 * (\phi(I+1) + \phi(I-1)) + (1 - 2 * \gamma_0) * \phi(I)$
enddo

Calculate nodal values at the right boundary nodes
$\phi(NX) = \gamma_0 * (2\phi(NX-1) + (2hDX/\Gamma)\phi_\infty) + (1 - 2 * \gamma_0 - 2 * \gamma_0 * (hDX)/\Gamma) * \phi(NX)$

Estimate flux quantity at right surface
$f_x'' = h * (\phi(NX) - \phi_\infty)$
enddo

Example 7.2: Explicit Scheme with Convective Boundary: Consider a one-dimensional unsteady state conduction in a plane slab of thickness $L = 4\,\text{mm}$. Initially the slab is at a temperature $T_I = 20°\text{C}$. The right side of the plate is suddenly brought to a convective environment with temperature $T_\infty = 100°\text{C}$ and $h = 80\,\text{W/m}^2{}°\text{C}$. The other side is maintained at 20°C. The material's thermo-physical properties are $k = 50\,\text{W/m°C}$ and $\alpha = 2 \times 10^{-5}\,\text{m}^2/\text{s}$.

Governing equation

$$\frac{\partial T}{\partial t} = \alpha \frac{\partial^2 T}{\partial x^2}$$

Boundary conditions

$$1.\quad x = 0, \quad T(x, 0) = \mathrm{T_L}$$

$$2.\quad x = L, \quad -k\frac{\partial T}{\partial x} = h(T - T_\infty)$$

Initial condition

$$T = 0, \quad T(x, 0) = T_\mathrm{I}$$

Consider the grid size distribution shown with ten grid points and examine the effect of the time step or the stability parameter. Determine the transient temperature distribution in the slab.

Solution: For number of nodes $n = 11$, spatial step size is

$$\Delta x = \frac{L}{n-1} = \frac{4 \times 10^{-3}}{10} = 4 \times 10^{-4}\,\mathrm{m}$$

Noting that for heat transfer problems $\Gamma = k$ and $C = \rho c$, we can write

$$\frac{\Gamma \Delta t}{C \Delta x^2} = \frac{\alpha \Delta t}{\Delta x^2} = f_0$$

and we select the discretization equation for the nodes as follows.

Interior nodes: *1, 2, 3, 4, 5, 6, 7, 8, 9*

$$T_i^{l+1} = f_0\left(T_{i+1}^l + T_{i-1}^l\right) + (1 - 2f_0)T_i^l$$

Node 1 $\quad T_1^{l+1} = f_0\left(T_2^l + T_0^l\right) + (1 - 2f_0)T_1^l$

Node 2 $\quad T_2^{l+1} = f_0\left(T_3^l + T_1^l\right) + (1 - 2f_0)T_2^l$

Node 3 $\quad T_3^{l+1} = f_0\left(T_4^l + T_2^l\right) + (1 - 2f_0)T_3^l$

Node 4 $\quad T_4^{l+1} = f_0\left(T_5^l + T_3^l\right) + (1 - 2f_0)T_4^l$

Node 5 $\quad T_5^{l+1} = f_0\left(T_6^l + T_4^l\right) + (1 - 2f_0)T_5^l$

Node 6 $\quad T_6^{l+1} = f_0\left(T_7^l + T_5^l\right) + (1 - 2f_0)T_6^l$

Node 7 $\quad T_7^{l+1} = f_0\left(T_8^l + T_6^l\right) + (1 - 2f_0)T_7^l$

Node 8 $\quad T_8^{l+1} = f_0\left(T_9^l + T_7^l\right) + (1 - 2f_0)T_8^l$

Node 9 $\quad T_9^{l+1} = f_0\left(T_{10}^l + T_8^l\right) + (1 - 2f_0)T_9^l$

Right-hand convective node: *10*

$$T_i^{l+1} = f_0\left(2T_{i-1}^l + \frac{2h\Delta x}{k}T_\infty\right) + \left(1 - 2f_0 - 2f_0\frac{h\Delta x}{k}\right)T_i^l$$

$$\text{Node 10} \quad T_{10}^{l+1} = f_0\left(2T_9^l + \frac{2h\Delta x}{k}T_\infty\right) + \left(1 - 2f_0 - 2f_0\frac{h\Delta x}{k}\right)T_{10}^l$$

The stability condition for interior nodes is given by Equation 7.22 as

$$f_0 = \frac{\alpha\Delta t}{(\Delta x)^2} < 0.5$$

The time step is estimated based on the value of f_0 used

$$\Delta t < \frac{(\Delta x)^2}{2\alpha} = 0.004\text{ s}$$

A pseudo-code for this solution methodology is shown in Table 7.3.

In order to show the effect of the time step, the heat transfer rate at the right surface can be evaluated with increase in time for different Fourier numbers and results could be plotted for visual observation of the convergence and transient temperature distribution in the slab.

7.3 IMPLICIT SCHEME

7.3.1 Discretization Equation by Control Volume Approach

The implicit form of the finite difference equation for the governing differential equation (7.1) can be derived directly from generalized formulation given before using the control volume approach. By considering the ***time approximation weighting factor*** $f = 1$, the discretization equation (7.9) becomes

$$a_i\phi_i^{l+1} = a_{i+1}\phi_{i+1}^{l+1} + a_{i-1}\phi_{i-1}^{l+1} + a_i^l\phi_i^l \tag{7.41}$$

where

$$a_i = a_{i+1} + a_{i-1} + a_i^l \tag{7.42a}$$

$$a_{i+1} = \frac{\Gamma_{i+\frac{1}{2}}}{(\Delta x)_{i+\frac{1}{2}}} \tag{7.42b}$$

$$a_{i-1} = \frac{\Gamma_{i-1\frac{1}{2}}}{(\Delta x)_{i-\frac{1}{2}}} \tag{7.42c}$$

$$a_i^l = \frac{C\Delta x_i}{\Delta t} \tag{7.42d}$$

Physically, the fully ***implicit scheme*** assumes that, at time t, ϕ changes from ϕ_i^l to ϕ_i^{l+1} and then remains constant at this value over the entire time step as depicted in Figure 7.2.

Substituting the coefficient values in Equation 7.41, we get

$$\left(\frac{\Gamma_{i+\frac{1}{2}}}{\Delta x_{i+\frac{1}{2}}}+\frac{\Gamma_{i-\frac{1}{2}}}{\Delta x_{i-\frac{1}{2}}}+\frac{C\Delta x}{\Delta t}\right)\phi_i^{l+1}=\frac{\Gamma_{i+\frac{1}{2}}}{\Delta x_{i+\frac{1}{2}}}\phi_{i+1}^{l+1}+\frac{\Gamma_{i-\frac{1}{2}}}{\Delta x_{i-\frac{1}{2}}}\phi_{i-1}^{l+1}+\frac{C\Delta x}{\Delta t}\phi_i^l \tag{7.43}$$

For the special case of constant transport property Γ and uniform grid size distribution, i.e., $\Delta x_{i+\frac{1}{2}}=\Delta x_{i-\frac{1}{2}}=\Delta x$, we get

$$\left(\frac{2\Gamma\Delta t}{C\Delta x^2}+1\right)\phi_i^{l+1}=\frac{\Gamma\Delta t}{C\Delta x^2}\left(\phi_{i+1}^{l+1}+\phi_{i-1}^{l+1}\right)+\phi_i^l \tag{7.44a}$$

Substituting $\gamma_0=\Gamma\Delta t/C\Delta x^2$

$$(2\gamma_o+1)\phi_i^{l+1}=\gamma_o\left(\phi_{i+1}^{l+1}+\phi_{i-1}^{l+1}\right)+\phi_i^l \tag{7.44b}$$

This is called an implicit form of the finite difference representation, because to determine the unknown dependent variable ϕ at the $(l+1)$th time step, a system of simultaneous algebraic equations are to be solved. Based on ***von Newmann stability analysis***, it can be shown that the fully implicit scheme is ***unconditionally stable***. The main advantage of the implicit scheme is that it is stable for all time steps Δt.

The outlined procedure can easily be extended to two-dimensional and three-dimensional cases as follows.

Two-dimensional Fully Implicit Discretization Equation. For the *two-dimensional diffusion equation with a source term*

$$C\frac{\partial\phi}{\partial t}=\frac{\partial}{\partial x}\left(\Gamma\frac{\partial\phi}{\partial x}\right)+\frac{\partial}{\partial y}\left(\Gamma\frac{\partial\phi}{\partial y}\right)+S \tag{7.45}$$

the fully implicit form of the finite difference equations is

$$a_{i,j}\phi_{i,j}^{l+1}=a_{i+1}\phi_{i+1,j}^{l+1}+a_{i-1}\phi_{i-1,j}^{l+1}+a_{j+1}\phi_{i,j+1}^{l+1}+a_{j-1}\phi_{i,j-1}^{l+1}+d \tag{7.46}$$

where

$$a_{i+1}=\frac{\Gamma_{i+\frac{1}{2}}\Delta y}{\Delta x_{i+\frac{1}{2}}},\quad a_{i-1}=\frac{\Gamma_{i-\frac{1}{2}}\Delta y}{\Delta x_{i-\frac{1}{2}}},\quad a_{j+1}=\frac{\Gamma_{j+\frac{1}{2}}\Delta x}{\Delta y_{j+\frac{1}{2}}},\quad a_{j-1}=\frac{\Gamma_{j-\frac{1}{2}}\Delta x}{\Delta y_{j-\frac{1}{2}}}$$

$$a_i=a_{i+1}+a_{i-1}+a_{j+1}+a_{j-1}+a_i^l-S_1\Delta x\,\Delta y$$

$$a_i^l=\frac{C\Delta x\Delta y}{\Delta t},\quad d=S_0\Delta x\Delta y+a_i^l\phi_i^l \tag{7.47}$$

For the special case of a constant transport property Γ, uniform grid size distribution and uniform source, i.e.,

$$\Gamma_{i+\frac{1}{2}} = \Gamma_{i-\frac{1}{2}} = \Gamma_{j+\frac{1}{2}} = \Gamma_{j-\frac{1}{2}} = \Gamma, \quad \Delta x_{i+\frac{1}{2}} = \Delta x_{i-\frac{1}{2}} = \Delta y_{j+\frac{1}{2}} = \Delta y_{j-\frac{1}{2}} = \Delta x$$

$$S_0 = \overline{S}, \qquad S_1 = 0$$

the coefficients of Equation 7.47 reduce to

$$a_{i+1} = a_{i-1} = a_{j+1} = a_{j-1} = \Gamma$$

$$a_i^l = \frac{C\Delta x^2}{\Delta t}, \quad b = a_i^l \phi_i^l + S\Delta x^2$$

$$a_i = 4\Gamma + \frac{C\Delta x^2}{\Delta t} \tag{7.48}$$

Substituting the coefficients, we get the simplified form

$$\left(4\Gamma + \frac{C\Delta x^2}{\Delta t}\right)\phi_{i,j}^{l+1} - \Gamma\left[\phi_{i+1,j}^{l+1} + \phi_{i-1,j}^{l+1} + \phi_{i,j+1}^{l+1} + \phi_{i,j-1}^{l+1}\right] + \frac{C\Delta x^2}{\Delta t}\phi_i^l + \overline{S}\Delta x^2 \tag{7.49a}$$

or

$$\left[4 + \frac{C\Delta x^2}{\Gamma\Delta t}\right]\phi_{i,j}^{l+1} = \left[\phi_{i+1,j}^{l+1} + \phi_{i-1,j}^{l+1} + \phi_{i,j+1}^{l+1} + \phi_{i,j-1}^{l+1}\right] + \frac{C\Delta x^2}{\Gamma\Delta t}\phi_i^l + \frac{\overline{S}\Delta x^2}{\Gamma} \tag{7.49b}$$

Substituting

$$\gamma_0 = \frac{\Gamma\Delta t}{C\Delta x^2} \tag{7.50}$$

We have

$$[1 + 4\gamma_0]\phi_{i,j}^{l+1} - \gamma_0\left[\phi_{i+1,j}^{l+1} + \phi_{i-1,j}^{l+1} + \phi_{i,j+1}^{l+1} + \phi_{i,j-1}^{l+1}\right] - \phi_{i,j}^l = \frac{\overline{S}\Delta x^2\gamma_0}{\Gamma} \tag{7.51}$$

Three-dimensional Fully Implicit Finite Difference Equation. Consider the three-dimensional governing equation as

$$C\frac{\partial\phi}{\partial t} = \frac{\partial}{\partial x}\left(\Gamma\frac{\partial\phi}{\partial x}\right) + \frac{\partial}{\partial y}\left(\Gamma\frac{\partial\phi}{\partial y}\right) + \frac{\partial}{\partial z}\left(\Gamma\frac{\partial\phi}{\partial z}\right) + S \tag{7.52}$$

with a three-dimensional grid shown in Figure 7.7 as

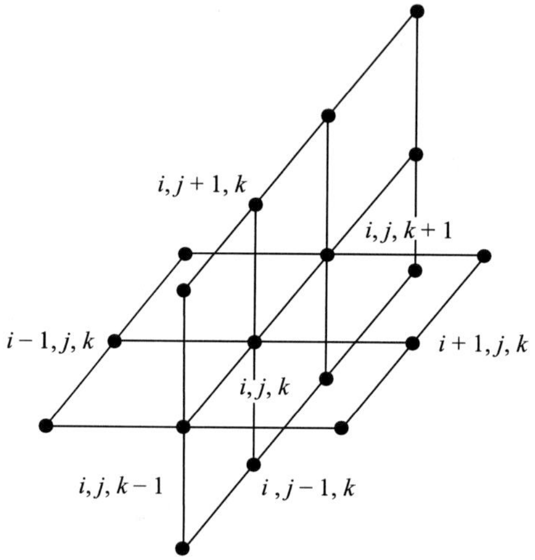

Figure 7.7 Three-dimensionl grid

Integration over the control volume gives

$$
\begin{aligned}
C\int_{i-\frac{1}{2}}^{i+\frac{1}{2}}\int_{j-\frac{1}{2}}^{j+\frac{1}{2}}\int_{k-\frac{1}{2}}^{k+\frac{1}{2}}\int_{t}^{t+\Delta t}\frac{\partial\phi}{\partial t}\,dx\,dy\,dz\,dt &= \int_{t}^{\Delta t}\int_{i-\frac{1}{2}}^{i+\frac{1}{2}}\int_{j-\frac{1}{2}}^{j+\frac{1}{2}}\int_{k-\frac{1}{2}}^{k+\frac{1}{2}}\frac{\partial\phi}{\partial x}\,dx\,dy\,dz\,dt \\
&= \int_{t}^{t+\Delta t}\int_{i-\frac{1}{2}}^{i+\frac{1}{2}}\int_{j-\frac{1}{2}}^{j+\frac{1}{2}}\int_{k-\frac{1}{2}}^{k+\frac{1}{2}}\frac{\partial}{\partial y}\left(\Gamma\frac{\partial\phi}{\partial y}\right)dx\,dy\,dz\,dt \\
&\quad + \int_{t}^{t+\Delta t}\int_{i-\frac{1}{2}}^{i+\frac{1}{2}}\int_{j-\frac{1}{2}}^{j+\frac{1}{2}}\int_{k-\frac{1}{2}}^{k+\frac{1}{2}}\frac{\partial}{\partial z}\left(\Gamma\frac{\partial\phi}{\partial z}\right)dx\,dy\,dz\,dt \\
&\quad + \int_{t}^{t+\Delta t}\int_{i-\frac{1}{2}}^{i+\frac{1}{2}}\int_{j-\frac{1}{2}}^{j+\frac{1}{2}}\int_{k-\frac{1}{2}}^{k+\frac{1}{2}} S\,dx\,dy\,dz\,dt
\end{aligned}
\tag{7.53}
$$

The fully implicit form of the finite difference equation

$$
\begin{aligned}
a_i\phi_{i,j,k} &= a_{i+1}\phi_{i+1,j,k} + a_{i-1}\phi_{i-1,j,k} + a_{j+1}\phi_{i,j+1,k} + a_{j-1}\phi_{i,j-1,k} \\
&\quad + a_{k+1}\phi_{i,j,k+1} + a_{k-1}\phi_{i,j,k-1} + d
\end{aligned}
\tag{7.54a}
$$

where

$$a_{i+1} = \frac{\Gamma_{i+\frac{1}{2}}\Delta y\,\Delta z}{\Delta x_{i+\frac{1}{2}}}, \quad a_{i-1} = \frac{\Gamma_{i-\frac{1}{2}}\Delta y\,\Delta z}{\Delta x_{i-\frac{1}{2}}} \tag{7.54b}$$

$$a_{j+1} = \frac{\Gamma_{j+\frac{1}{2}}\Delta z \Delta x}{\Delta z_{j+\frac{1}{2}}}, \quad a_{j-1} = \frac{\Gamma_{j-\frac{1}{2}}\Delta z \Delta x}{\Delta z_{j-\frac{1}{2}}} \tag{7.54c}$$

$$a_{k+1} = \frac{\Gamma_{k+\frac{1}{2}}\Delta x \Delta y}{\Delta z_{k+\frac{1}{2}}}, \quad a_{k-1} = \frac{\Gamma_{k-\frac{1}{2}}\Delta x \Delta y}{\Delta z_{k-\frac{1}{2}}} \tag{7.54d}$$

$$a_i^l = \frac{C\Delta x \Delta y \Delta z}{\Delta t}, \quad b = S_0 \Delta x \Delta y \Delta z + a_i^l \phi_{i,j,k}^l \tag{7.54e}$$

$$a_i = a_{i+1} + a_{i-1} + a_{j+1} + a_{j-1} + a_{k+1} + a_{k-1} + a_i^l - S_1 \Delta x\,\Delta y\,\Delta z \tag{7.54e}$$

7.3.2 Finite Difference Equation by Taylor Series Expansion

Let us consider the two-dimensional, time dependent heat equation given as

$$C\frac{\partial \phi}{\partial t} = \Gamma\left[\frac{\partial^2 \phi}{\partial x^2} + \frac{\partial^2 \phi}{\partial y^2}\right] \tag{7.55}$$

Using the central-difference formula for the second derivative at the $(l+1)$th time step as

$$\left.\frac{\partial^2 \phi}{\partial x^2}\right)_{i,j} = \frac{\phi_{i+l,j}^{l+l} - 2\phi_{i,j}^{l+1} + \phi_{i-1,j}^{l+1}}{\Delta x^2} \tag{7.56a}$$

$$\left.\frac{\partial^2 \phi}{\partial y^2}\right)_{l,i} = \frac{\phi_{i,j+1}^{l+1} - 2\phi_{i,j}^{l+1} + \phi_{l,j-1}^{l+1}}{\Delta y^2} \tag{7.56b}$$

and backward-difference formula at the $(l+1)$th time step for the first-order time derivative from Table 5.1 as

$$\left.\frac{\partial \phi}{\partial t}\right|_{i,j} = \frac{\phi_{i,j}^{l+1} - \phi_{i,j}^{l}}{\Delta t} \tag{7.56c}$$

we get

$$\begin{aligned} \phi_{i,j}^{l+1} &= \frac{\Gamma \Delta t / C\Delta x^2}{1 + 2[\alpha(\Delta t/\Delta x^2)] + 2[\alpha \Delta t/\Delta y^2]}\left[\phi_{i+1,j}^{l+1} + \phi_{i-l,j}^{l+1}\right] \\ &+ \frac{\alpha \Delta t/\Delta y^2}{1 + 2[\alpha \Delta t/\Delta x^2] + 2[\alpha \Delta t/\Delta y^2]}\left(\phi_{l,j+1}^{l+1} + \phi_{i,j-1}^{l+1}\right) \\ &+ \frac{1}{1 + 2[\alpha \Delta t/\Delta x^2] + 2[\alpha \Delta t/\Delta y^2]}\phi_{i,j}^{l} \end{aligned} \tag{7.57}$$

The above equation can be simplified to the following special cases.

For uniform grid size distribution, i.e., $\Delta x = \Delta y$

$$\phi_{i,j}^{l+1} = \frac{\Gamma \Delta t/C\Delta x^2}{1 + 4\Gamma \Delta t/C\Delta x^2}\left[\phi_{i+1,j}^{l+1} + \phi_{i-1,j}^{l+1}\right] + \frac{\Gamma \Delta t/C\Delta x^2}{1 + 4\Gamma \Delta t/C\Delta x^2}\left[\phi_{l,j+1}^{l+1} + \phi_{i,j-1}^{l+1}\right]$$
$$+ \frac{1}{1 + 4\Gamma \Delta t/C\Delta x^2}\phi_{l,i}^{l}$$

or

$$[1 + 4\gamma_0]\phi_{i,j}^{l+1} - \gamma_o\left[\phi_{i+1,j}^{l+1} + \phi_{i-l,j}^{l+1} + \phi_{i,j+1}^{l+1} + \phi_{i,j-1}^{l+1}\right] = \phi_{i,j}^{l} \tag{7.58}$$

For one-dimensional unsteady state case

$$\phi_i^{l+1} = \frac{\Gamma \Delta t/C\Delta x^2}{1 + 2\Gamma \Delta t/C\Delta x^2}\left[\phi_{i+1}^{l+1} + \phi_{i-1}^{l+1}\right] + \frac{1}{1 + 2\Gamma \Delta t/C\Delta x^2}\phi_i^{l}$$

or

$$[1 + 2\gamma_o]\phi_i^{l+1} - \gamma_o\left[\phi_{i+1}^{l+1} + \phi_{i-l}^{l+1}\right] = \phi_i^{l} \tag{7.59}$$

7.3.3 A General Formulation of Fully Implicit Scheme for One-dimensional Problems

Let us consider a general one-dimensional problem with the grid system shown in the figure and boundary conditions of the first kind at the left- and right-hand side of the solution domain.

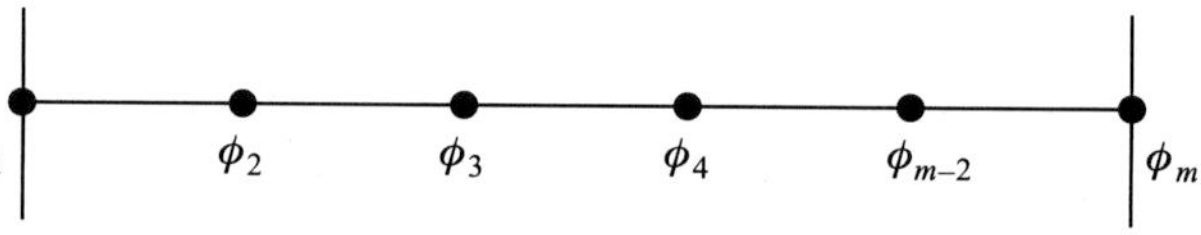

Figure 7.8 One dimensional grid

Assuming that the boundary conditions are of constant temperature (i.e., boundary temperatures ϕ_1^{l+1} and ϕ_m^{l+1} are known), the following system of equations can be constructed using Equation 7.59.

$$\begin{bmatrix} 1+2\gamma_0 & -\gamma_0 & & & \\ -\gamma_0 & 1+2\gamma_0 & -\gamma_0 & & \\ & -\gamma_0 & 1+2\gamma_0 & -\gamma_0 & \\ \vdots & \vdots & \vdots & \vdots & \vdots \\ & & & -\gamma_0 & 1+2\gamma_0 \end{bmatrix} \begin{bmatrix} \phi_2^{l+1} \\ \phi_3^{l+1} \\ \vdots \\ \phi_{m-2}^{l+1} \\ \phi_{m-1}^{l+1} \end{bmatrix} = \begin{Bmatrix} \phi_2^{l} + \gamma_o\phi_1^{l+1} \\ \phi_3^{l} \\ \vdots \\ \phi_{m-2}^{l} \\ \phi_{m-1}^{l} + \gamma_o\phi_m^{l+1} \end{Bmatrix} \tag{7.60}$$

It can be noted the coefficient matrix is tridiagonal and all its elements are known. So, the unknown solution vector

$$\begin{Bmatrix} \phi_2^{l+1} \\ \phi_3^{l+1} \\ \vdots \\ \phi_{m-1}^{l+1} \end{Bmatrix}$$

can be solved by using a direct or iterative solution method.

Since the matrix is tridiagonal, the TDMA algorithm is quite suitable as a direct solver. Another convenient way to solve the system with increase in time steps is to evaluate the inverse of the matrix A^{-1} and solve the system as

$$\begin{bmatrix} \phi_2^{l+1} \\ \phi_3^{l+1} \\ \vdots \\ \phi_{m-2}^{l+1} \\ \phi_{m-1}^{l+1} \end{bmatrix} = \begin{bmatrix} 1+2\gamma_0 & -\gamma_0 & & & \\ -\gamma_0 & 1+2\gamma_0 & -\gamma_0 & & \\ & -\gamma_0 & 1+2\gamma_0 & -\gamma_0 & \\ \vdots & \vdots & \vdots & \vdots & \vdots \\ & & & -\gamma_0 & 1+2\gamma_0 \end{bmatrix}^{-1} \begin{Bmatrix} \phi_2^l + \gamma_o \phi_1^{l+1} \\ \phi_3^l \\ \vdots \\ \phi_{m-2}^l \\ \phi_{m-1}^l + \gamma_o \phi_m^{l+1} \end{Bmatrix} \tag{7.61}$$

The above equation can be used to determine the solution vector with marching in time.

Example 7.3: One-dimensional Unsteady State with Fully Implicit Scheme: Let us consider a one-dimensional unsteady state heat conduction in a plane slab of thickness $L = 10\,\text{cm}$. Initially the slab is a uniform temperature of $T_\text{I} = 0°\text{C}$. At time $t > 0$, the left and right sides of the slab are simultaneously brought to a temperature of 100°C and 50°C, respectively, as shown in the figure.

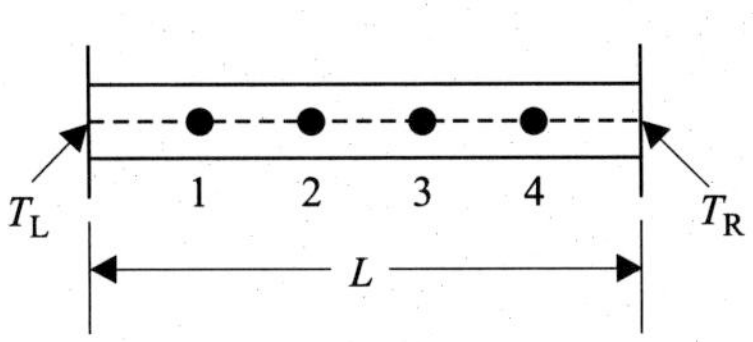

Governing equation

$$\frac{\partial T}{\partial t} = \alpha \frac{\partial^2 T}{\partial x^2}$$

Boundary conditions

1. $x = 0, \quad T = T_L = 100°C$
2. $x = L, \quad T = T_R = 50°C$

Initial condition

$t = 0, \quad T(x) = T_I = 0$

The rod is aluminum with thermal diffusivity $\alpha = 0.835\,cm^2/s$. Use a fully implicit scheme to obtain the system of equations for the grid shown in the figure and a time step $\Delta t = 0.1$ s. Obtain temperature distribution in the rod at $t = 0.1$ s and $t = 0.2$ s.

Solution: For the selected grid system, the spatial step

$$\Delta x = \frac{L}{Ndiv_x} = \frac{10}{5} = 2\,cm$$

and the dimensionless time $\gamma_0 = f_0 =$ Fourier number is

$$f_0 = \frac{\alpha \Delta t}{\Delta x^2} = \frac{0.835(0.1)}{(2)^2} = 0.041\,75$$

The appropriate discretization equation for all interior nodes is the fully implicit Equation 7.59, which for the heat diffusion equation becomes

$$[1 + 2f_o]T_i^{l+1} - f_0\left[T_{i+1}^{l+1} + T_{i-1}^{l+1}\right] = T_i^l$$

Applying this equation successively to all nodes, we get

Node 1
$$[1 + 2f_0]T_1^{l+1} - f_0\left[T_2^{l+1} + T_L^{l+1}\right] = T_1^l$$
$$[1 + 2(0.041\,75]T_1^1 - 0.041\,75\left[T_2^1 + T_L^{l+1}\right] = T_1^0$$
$$[1 + 2(0.041\,75]T_1^1 - 0.041\,75\left[T_2^1 + 100\right] = 0$$
$$1.0835T_1^1 - 0.041\,75T_2^1 = 4.175$$

Node 2
$$[1 + 2f_0]T_2^{l+1} - f_0\left[T_3^{l+1} + T_1^{l+1}\right] = T_2^l$$
$$[1 + 2(0.041\,75]T_2^1 - 0.041\,75\left[T_3^{l+1} + T_1^{l+1}\right] = T_2^0$$
$$-\,0.041\,75T_1^1 - 1.0835T_2^1 - 0.041\,75T_3^1 = 0$$

Node 3
$$[1 + 2f_0]T_3^{l+1} - f_0\left[T_4^{l+1} + T_2^{l+1}\right] = T_3^l$$
$$[1 + 2(0.041\,75]T_3^1 - 0.041\,75\left[T_4^{l+1} + T_2^{l+1}\right] = T_3^0$$
$$-\,0.041\,75T_2^1 - 1.0835T_3^1 - 0.041\,75T_4^1 = 0$$

Node 4
$$[1 + 2f_0]T_4^{l+1} - f_0\left[T_R^{l+1} + T_3^{l+1}\right] = T_4^l$$
$$[1 + 2(0.041\,75]T_4^1 - 0.041\,75\left[T_R^{l+1} + T_3^{l+1}\right] = T_4^0$$
$$-\,0.041\,75T_3^1 - 1.0835T_4^1 = 2.0875$$

Assembling all nodal equations, we get the system of equations in matrix form as

$$\begin{bmatrix} 1.0835 & -0.04175 & 0 & 0 \\ -0.04175 & 1.0835 & -0.04175 & 0 \\ 0 & -0.04175 & 1.0835 & -0.04175 \\ 0 & 0 & -0.04175 & 1.0835 \end{bmatrix} \begin{Bmatrix} T_1^1 \\ T_2^1 \\ T_3^1 \\ T_4^1 \end{Bmatrix} = \begin{Bmatrix} T_1^0 + 4.175 \\ T_2^0 \\ T_3^0 \\ T_4^0 + 2.0875 \end{Bmatrix}$$

Setting $T_1^0 = T_2^0 = T_3^0 = T_4^0 = 0$ as initial condition and solving the system of equations, we have the temperature distribution at time $t = 0.1\text{s}$ as

$$\begin{Bmatrix} T_1^1 \\ T_2^1 \\ T_3^1 \\ T_4^1 \end{Bmatrix} = \begin{Bmatrix} 3.85910^\circ\text{C} \\ 0.15179^\circ\text{C} \\ 0.08021^\circ\text{C} \\ 1.92972^\circ\text{C} \end{Bmatrix}$$

Similarly at time $t = 2\Delta t$ the system of equations for nodal temperatures is obtained as

$$\begin{bmatrix} 1.0835 & -0.04175 & 0 & 0 \\ -0.04175 & 1.0835 & -0.04175 & 0 \\ 0 & -0.04175 & 1.0835 & -0.04175 \\ 0 & 0 & -0.04175 & 1.0835 \end{bmatrix} \begin{Bmatrix} T_1^2 \\ T_2^2 \\ T_3^2 \\ T_4^2 \end{Bmatrix} = \begin{Bmatrix} T_1^1 + 4.175 \\ T_2^1 \\ T_3^1 \\ T_4^1 + 2.0875 \end{Bmatrix}$$

or

$$\begin{bmatrix} 1.0835 & -0.04175 & 0 & 0 \\ -0.04175 & 1.0835 & -0.04175 & 0 \\ 0 & -0.04175 & 1.0835 & -0.04175 \\ 0 & 0 & -0.04175 & 1.0835 \end{bmatrix} \begin{Bmatrix} T_1^2 \\ T_2^2 \\ T_3^2 \\ T_4^2 \end{Bmatrix} = \begin{Bmatrix} 3.85910 + 4.175 \\ 0.15179 \\ 0.08021 \\ 1.92972 + 2.0875 \end{Bmatrix}$$

or

$$\begin{bmatrix} 1.0835 & -0.04175 & 0 & 0 \\ -0.04175 & 1.0835 & -0.04175 & 0 \\ 0 & -0.04175 & 1.0835 & -0.04175 \\ 0 & 0 & -0.04175 & 1.0835 \end{bmatrix} \begin{Bmatrix} T_1^2 \\ T_2^2 \\ T_3^2 \\ T_4^2 \end{Bmatrix} = \begin{Bmatrix} 8.0341 \\ 0.15179 \\ 0.08021 \\ 4.01722 \end{Bmatrix}$$

The solution of the system gives temperature distribution at time $t = 0.2\,\text{s}$ as

$$\begin{Bmatrix} T_1^2 \\ T_2^2 \\ T_3^2 \\ T_4^2 \end{Bmatrix} = \begin{Bmatrix} 7.43173^\circ\text{C} \\ 0.43547^\circ\text{C} \\ 0.23402^\circ\text{C} \\ 3.71665^\circ\text{C} \end{Bmatrix}$$

7.3.4 A General Formulation of Fully Implicit Scheme for Two-dimensional Problems

Let us consider a general two-dimensional problem with the grid system and boundary conditions as shown in the figure.

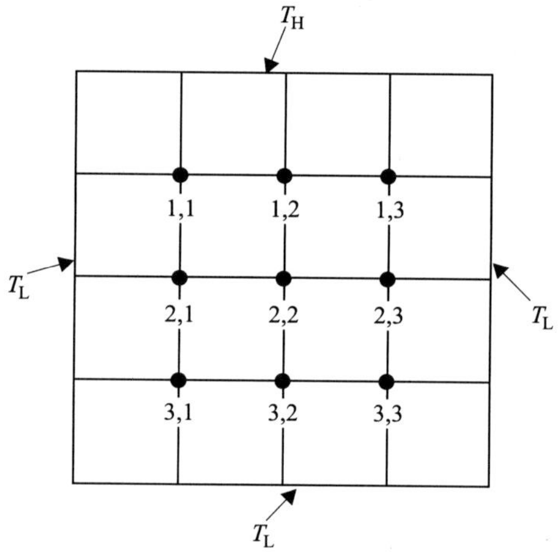

Figure 7.9 Two-dimensional grid

The appropriate discretization equation for the interior node is given by Equation 7.58 as

$$[1 + 4\gamma_0]\phi_{i,j}^{l+1} - \gamma_o\left[\phi_{i+1,j}^{l+1} + \phi_{i-l,j}^{l+1} + \phi_{i,j+1}^{l+1} + \phi_{i,j-1}^{l+1}\right] = \phi_{i,j}^{l} \tag{7.62}$$

Let us apply this equation to all interior nodes as follows:

Node (1, 1) $[1 + 4\gamma_0]T_{1,1}^{l+1} - \gamma_o\left[T_{1,2}^{l+1} + T_{\text{L}}^{l+1} + T_{\text{H}}^{l+1} + T_{2,1}^{l+1}\right] = T_{1,1}^{l}$

Node (1, 2) $[1 + 4\gamma_0]T_{1,2}^{l+1} - \gamma_o\left[T_{1,3}^{l+1} + T_{1,1}^{l+1} + T_{\text{H}}^{l+1} + T_{2,2}^{l+1}\right] = T_{1,2}^{l}$

Node (1, 3) $[1 + 4\gamma_0]T_{1,3}^{l+1} - \gamma_o\left[T_{\text{L}}^{l+1} + T_{1,2}^{l+1} + T_{\text{H}}^{l+1} + T_{2,3}^{l+1}\right] = T_{1,3}^{l}$

$$\text{Node } (2,1) \quad [1+4\gamma_0]T_{2,1}^{l+1} - \gamma_o\left[T_{2,2}^{l+1} + T_{L}^{l+1} + T_{1,1}^{l+1} + T_{3,1}^{l+1}\right] = T_{2,1}^{l}$$

$$\text{Node } (2,2) \quad [1+4\gamma_0]T_{2,2}^{l+1} - \gamma_o\left[T_{2,3}^{l+1} + T_{2,1}^{l+1} + T_{1,2}^{l+1} + T_{3,2}^{l+1}\right] = T_{2,2}^{l}$$

$$\text{Node } (2,3) \quad [1+4\gamma_0]T_{2,3}^{l+1} - \gamma_o\left[T_{L}^{l+1} + T_{2,2}^{l+1} + T_{1,3}^{l+1} + T_{3,3}^{l+1}\right] = T_{2,3}^{l}$$

$$\text{Node } (3,1) \quad [1+4\gamma_0]T_{3,1}^{l+1} - \gamma_o\left[T_{3,2}^{l+1} + T_{L}^{l+1} + T_{2,1}^{l+1} + T_{L}^{l+1}\right] = T_{3,1}^{l}$$

$$\text{Node } (3,2) \quad [1+4\gamma_0]T_{3,2}^{l+1} - \gamma_o\left[T_{3,3}^{l+1} + T_{3,1}^{l+1} + T_{2,2}^{l+1} + T_{L}^{l+1}\right] = T_{3,2}^{l}$$

$$\text{Node } (3,3) \quad [1+4\gamma_0]T_{3,3}^{l+1} - \gamma_o\left[T_{L}^{l+1} + T_{3,2}^{l+1} + T_{2,3}^{l+1} + T_{L}^{l+1}\right] = T_{3,3}^{l}$$

Writing the system of equations in matrix form, we have

$$\begin{bmatrix} 1+4\gamma_0 & -\gamma_0 & 0 & -\gamma_0 & 0 & 0 & 0 & 0 & 0 \\ -\gamma_0 & 1+4\gamma_0 & -\gamma_0 & 0 & -\gamma_0 & 0 & 0 & 0 & 0 \\ 0 & -\gamma_0 & 1+4\gamma_0 & 0 & 0 & -\gamma_0 & 0 & 0 & 0 \\ -\gamma_0 & 0 & 0 & 1+4\gamma_0 & -\gamma_0 & 0 & -\gamma_0 & 0 & 0 \\ 0 & -\gamma_0 & 0 & -\gamma_0 & 1+4\gamma_0 & -\gamma_0 & 0 & -\gamma_0 & 0 \\ 0 & 0 & -\gamma_0 & 0 & -\gamma_0 & 1+4\gamma_0 & 0 & 0 & -\gamma_0 \\ 0 & 0 & 0 & -\gamma_0 & 0 & 0 & 1+4\gamma_0 & -\gamma_0 & 0 \\ 0 & 0 & 0 & 0 & -\gamma_0 & 0 & -\gamma_0 & 1+4\gamma_0 & -\gamma_0 \\ 0 & 0 & 0 & 0 & 0 & -\gamma_0 & 0 & -\gamma_0 & 1+4\gamma_0 \end{bmatrix} \begin{Bmatrix} T_{1,1} \\ T_{1,2} \\ T_{1,3} \\ T_{2,2} \\ T_{2,3} \\ T_{3,3} \\ T_{3,1} \\ T_{3,2} \\ T_{3,3} \end{Bmatrix}$$

$$= \begin{Bmatrix} T_{1,1} + \gamma_0(T_L + T_H) \\ T_{1,2} + \gamma_0 T_L \\ T_{1,3} + \gamma_0(T_L + T_H) \\ T_{2,1} + \gamma_0 T_L \\ T_{2,2} \\ T_{2,3} + \gamma_0 T_L \\ T_{3,1} + 2\gamma_0 T_L \\ T_{3,2} + \gamma_0 T_L \\ T_{3,3} + \gamma_0 T \end{Bmatrix} \qquad (7.63)$$

7.3.5 Solution Methods for a Two-dimensional Implicit Scheme

It can be noted that Equation 7.63 resulting from a two-dimensional diffusion equation is no longer tridiagonal, and hence, unlike the unidirectional transient diffusion problem, the discretization equations for the two-dimensional transient heat conduction cannot be solved by TDMA. The reason for this is the fact that while Equation 7.58 can be rearranged so that three of the terms are at or adjacent to the main diagonal, the remaining two terms are displaced.

We can certainly use a direct solver such as the Gauss elimination or *LU* decomposition method or iterative point-by-point methods such as Gauss–Seidel or SOR methods. However, there are two other alternate methods, ***line-by-line method*** and ***block tridiagonal method*** that is effectively used for some class of problems.

Application of Gauss–Seidel Method (Point-by-Point Method). This is the simplest of all iterative methods to implement, as discussed in Chapter 2 and Chapter 6. It stores the most recent calculated values of nodal temperatures. An iterative process at each time step is executed before proceeding to the next time step. For example, for the two-dimensional diffusion problems, the temperature at each grid point in a body is obtained from Equation 7.58 by writing it in an explicit form as

$$\phi_{i,j}^{n+1} = \frac{\gamma_0\left(\phi_{i+1,j}^{n+1} + \phi_{i-1,j}^{n+1}\right) + \gamma_0\left(\phi_{i,j+1}^{n+1} + \phi_{i,j-1}^{n+1}\right) + \phi_{i,j}^{n}}{1 + 4\gamma_0} \tag{7.64}$$

where, $\phi_{i+l,j}^{n+1}$, $\phi_{i-l,j}^{n+1}$, $\phi_{i,j+1}^{n+1}$, $\phi_{i,j-1}^{n+1}$ are temperatures at four grid points that are currently stored in the memory. The iterative scheme is continued until all nodal values converge with assigned tolerance at each time step before increasing. This procedure can also be extended to the SOR algorithm.

Line-By-Line Method. The ***line-by-line method*** is primarily a combination of a TDMA algorithm for a one-dimensional problem and a Gauss–Seidel iteration method. Instead of estimating one nodal value at a time, as in a Gauss–Seidel method, all nodal values in a vertical line or in a horizontal line are computed directly by using a TDMA algorithm and sweeping from left to right or from bottom to top. For example, in sweeping all vertical lines in an x direction, the values of all nodal points ($\phi_{i,j}^{n+1}$, $j = 1, \ldots, N$) along a vertical line or an x line (x_i) are determined in a present iteration ($n+1$) by using a TDMA algorithm while assuming the nodal values along two adjacent vertical or x lines (x_{i-1} and x_{i+1}) are known, as shown in Figure 7.10. The nodal values in the x_{i-1} line are assumed to be known in the present iteration ($n+1$), while the nodal values in the x_{i+1} line are obtained in the previous iteration (n).

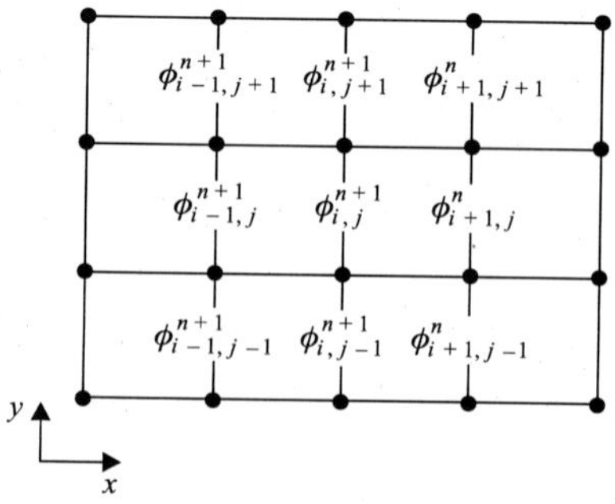

Figure 7.10 Line-by-line method with sweep in the x direction

This is repeated for all vertical lines in an x direction, and one iteration is said to be completed when all vertical lines are scanned. In the next iteration, we can choose to sweep in a y direction and estimate all nodal values along all horizontal lines. Alternating the sweeping directions essentially improves the convergence rate. Once convergence is reached at the present given time, solution procedure is repeated for the next subsequent time steps.

Block Tridiagonal Matrix Algorithm. The second approach, known as the ***block tridiagonal matrix algorithm*** as discussed in Chapter 2, is also applicable to the system of Equation 7.58 by writing it in a $N^2 \times N^2$ coefficient matrix in a block tridiagonal form as

$$\gamma_0 \begin{bmatrix} T_N & -I_N & & & & & & & \\ -I_N & T_N & & & & & & & \\ & & T_N & & & & & & \\ & & & T_N & & & & & \\ & & & & T_N & & & & \\ & & & & & T_N & & & \\ & & & & & & T_N & & \\ & & & & & & & T_N & -I_N \\ & & & & & & & -I_N & T_N \end{bmatrix} \left\{ \quad \right\} = \left\{ \quad \right\} \tag{7.65}$$

where

$$T_N = \begin{bmatrix} \gamma_0^{-1}+4 & -1 & & & & & & & \\ -1 & \gamma_0^{-1}+4 & -1 & & & & & & \\ & -1 & \gamma_0^{-1}+4 & -1 & & & & & \\ & & -1 & \gamma_0^{-1}+4 & -1 & & & & \\ & & & -1 & \gamma_0^{-1}+4 & -1 & & & \\ & & & & -1 & \gamma_0^{-1}+4 & -1 & & \\ & & & & & -1 & \gamma_0^{-1}+4 & -1 & \\ & & & & & & -1 & \gamma_0^{-1}+4 & -1 \\ & & & & & & & -1 & \gamma_0^{-1}+4 \end{bmatrix}$$

and I_N is a $N \times N$ identity matrix.

7.3.6 Boundary Conditions for Implicit Scheme

Implicit discretization equations for boundary nodes can be derived using the procedure outlined in the case of an explicit scheme.

Let us briefly discuss the derivation of one such equation for the grid point located at the ***right-hand side boundary*** shown below.

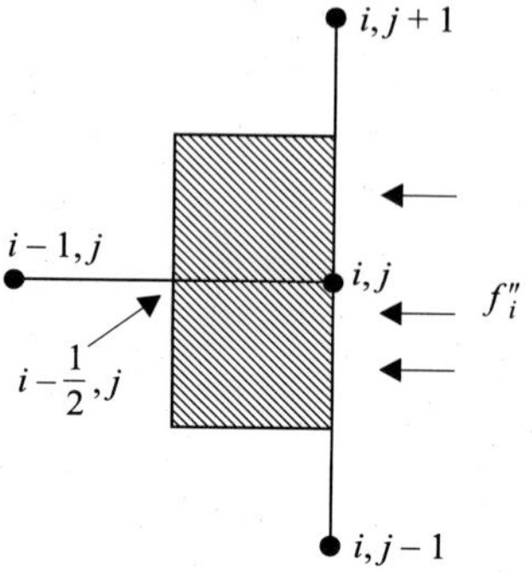

Figure 7.11 Control volume for a boundary node

Let us demonstrate derivation of the boundary node equation by considering the two-dimensional unsteady state diffusion equation given as

$$\frac{\partial}{\partial x}\left(\Gamma_x \frac{\partial \phi}{\partial x}\right) + \frac{\partial}{\partial x}\left(\Gamma_y \frac{\partial \phi}{\partial y}\right) = C\frac{\partial \phi}{\partial t} \tag{7.66}$$

To proceed, we integrate the equation over the half control volume and over the time interval t to $t + \Delta t$:

$$\int\limits_{i-\frac{1}{2}}^{i}\int\limits_{j-\frac{1}{2}}^{j+\frac{1}{2}}\int\limits_{t}^{t+\Delta t} \frac{\partial}{\partial x}\left(\Gamma_x \frac{\partial \phi}{\partial x}\right) dx\, dy\, dt + \int\limits_{i-\frac{1}{2}}^{i}\int\limits_{j-\frac{1}{2}}^{j+\frac{1}{2}}\int\limits_{t}^{t+\Delta t} \frac{\partial}{\partial y}\left(\Gamma_y \frac{\partial \phi}{\partial y}\right) dx\, dy\, dt$$

$$= C\int\limits_{i-\frac{1}{2}}^{i}\int\limits_{j-\frac{1}{2}}^{j+\frac{1}{2}}\int\limits_{t}^{t+\Delta t} \frac{\partial \phi}{\partial t} dx\, dy\, dt \tag{7.67}$$

$$\left.\Gamma_x \frac{\partial \phi}{\partial x}\right)_i \Delta y\, \Delta t - \left.\Gamma_y \frac{\partial \Gamma}{\partial x}\right)_{i-\frac{1}{2}} \Delta y\, \Delta t + \left.\Gamma_y \frac{\partial \phi}{\partial y}\right)_{j+\frac{1}{2}} \frac{\Delta x\, \Delta t}{2} - \left.\Gamma_y \frac{\partial \phi}{\partial y}\right)_{j-\frac{1}{2}} \frac{\Delta x\, \Delta t}{2}$$

$$= C\frac{\Delta x\, \Delta y}{2}\int\limits_{t}^{t+\Delta t} \frac{\partial \phi}{\partial t} dt \tag{7.68}$$

The first term in the equation is described by the heat flux boundary conditions as

$$\left.-\Gamma_x \frac{\partial \phi}{\partial x}\right)_i = -f_i{}'' \tag{7.69a}$$

or

$$\left.\Gamma_x \frac{\partial \phi}{\partial x}\right)_i = f_i{}'' \tag{7.69b}$$

Next, we assume a piecewise-linear profile or a central difference formula to approximate temperature gradients at points $j+\frac{1}{2}$, $j-\frac{1}{2}$, and $i-\frac{1}{2}$, and a forward difference formula for first-order time derivative

$$\Gamma_x \frac{\partial \phi}{\partial x}\bigg)_{i-\frac{1}{2}} = \Gamma_{x i-\frac{1}{2}} \frac{\phi_{i,j}^{l+1} - \phi_{i-1,j}^{l+1}}{\Delta x}, \quad \Gamma_y \frac{\partial \phi}{\partial y}\bigg)_{j+\frac{1}{2}} = \Gamma_{y j+\frac{1}{2}} \frac{\phi_{i,j+1}^{l+1} - \phi_{i,j}^{l+1}}{\Delta y} \tag{7.70a}$$

$$\Gamma_y \frac{\partial \phi}{\partial y}\bigg)_{j-\frac{1}{2}} = \Gamma_{y j-\frac{1}{2}} \frac{\left(\phi_{i,j}^{l+1} - \phi_{i,j-1}^{l+1}\right)}{\Delta y}, \quad \frac{\partial \phi}{\partial t} = \frac{\phi_{i,j}^{l+1} - \phi_{i,j}^{l}}{\Delta t} \tag{7.70b}$$

Substituting these approximations, we get

$$+f_i{}'' \Delta y \, \Delta t - \frac{\Gamma_{i-\frac{1}{2}}\left(\phi_{i,j}^{l+1} - \phi_{i-1,j}^{l+1}\right)}{\Delta x} \Delta y \, \Delta t + \frac{\Gamma_{j+\frac{1}{2}}\left(\phi_{i,j+1}^{l+1} - \phi_{i,j}^{l+1}\right)}{\Delta y} \frac{\Delta x}{2} \Delta t$$

$$- \frac{\Gamma_{j-\frac{1}{2}}\left(\phi_{i,j}^{l+1} - \phi_{i,j-1}^{l+1}\right)}{\Delta y} \frac{\Delta x}{2} \Delta t = C \frac{\Delta x}{2} \Delta y \left(\phi_{i,j}^{l+1} - \phi_{i,j}^{l}\right) \tag{7.71}$$

$$\frac{\Gamma_{i-\frac{1}{2}} \Delta y}{\Delta x} \phi_{i-1,j}^{l+1} + \frac{\Gamma_{j+\frac{1}{2}} \Delta x}{2 \Delta y} \phi_{i,j+1}^{l+1} + \frac{\Gamma_{j-\frac{1}{2}} \Delta x}{2 \Delta y} \phi_{i,j-1}^{l+1}$$

$$+ \left(-\frac{C \Delta x \Delta y}{2 \Delta t} - \frac{\Gamma_{i-\frac{1}{2}} \Delta y}{\Delta x} - \frac{\Gamma_{i,j+\frac{1}{2}} \Delta x}{2 \Delta y} - \frac{\Gamma_{j-\frac{1}{2}} \Delta x}{2 \Delta y}\right) \phi_{i,j}^{l+1} = -\frac{C \Delta x \Delta y}{2 \Delta t} \phi_{i,j}^{l} + f_i{}'' \Delta y$$

Simplifying

$$a_{i-1} \phi_{i-1,j}^{l+1} + a_{j+1} \phi_{i,j+1}^{l+1} + a_{j-1} \phi_{i,j-1}^{l+1} - \left(a_i^l + a_{i-1} + a_{j+1} + a_{j-1}\right) \phi_{i,j}^{l+1} = a_i^l \phi_{i,j}^{l} + d \tag{7.72}$$

where

$$a_i = a_i^l + a_{i-1} + a_{j+1} + a_{j-1}, \quad a_{i-1} = \frac{\Gamma_{i-\frac{1}{2}} \Delta y}{\Delta x}, \quad a_{j+1} = \frac{\Gamma_{j+\frac{1}{2}} \Delta x}{2 \Delta y}, \quad a_{j-1} = \frac{\Gamma_{j-\frac{1}{2}} \Delta x}{2 \Delta y}$$

$$a_i^l = -\frac{C \Delta x \Delta y}{2 \Delta t}, \quad d = f_i{}'' \Delta y \tag{7.73}$$

For constant transport property, Γ, and for uniform grid size distributions, i.e., $\Delta x = \Delta y$, we can write the above equation in an alternate form as

$$\phi_{i,j}^{l+1} = \frac{\Gamma \Delta t}{C \Delta x^2}\left[2\phi_{i-1,j}^{l+1} + \phi_{i,j+1}^{l+1} + \phi_{i,j-1}^{l+1}\right] + \left[1 - 4\frac{\Gamma \Delta t}{C \Delta x^2}\right]\phi_i^l + \frac{2 \Delta t}{C \Delta x} f_i{}'' \tag{7.74}$$

For a ***one-dimensional case***, the equation reduces to

$$\phi_i^{l+1} = \frac{\Gamma \Delta t}{C \Delta x^2}\left[2\phi_{i-1}^{l+1}\right] + \left[1 - \frac{2\Gamma \Delta t}{C \Delta x^2}\right]\phi_i^l + \frac{2\Delta t}{C \Delta x} f_i'' \tag{7.75}$$

Let us now consider special cases.

1. ***Adiabatic boundary condition or a symmetric boundary***
   For an adiabatic surface, the surface flux is zero, i.e., $f_i'' = 0$

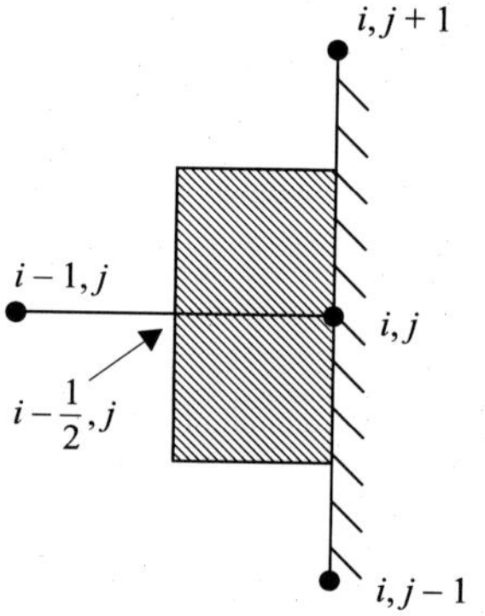

Figure 7.12 Convective boundary node

The Equations 7.74 and 7.75 reduce to

$$\phi_{i,j}^{l+1} = \frac{\Gamma \Delta t}{C \Delta x^2}\left[2\phi_{i-1,j}^{l+1} + \phi_{i,j+1}^{l+1} + \phi_{i,j-1}^{l+1}\right] + \left[1 - \frac{4\Gamma \Delta t}{C \Delta x^2}\right]\phi_{i,j}^l \tag{7.76}$$

for two-dimensional problems
and

$$\phi_i^{l+1} = \frac{\Gamma \Delta t}{C \Delta x^2}\left[2\phi_{i-1}^{l+1}\right] + \left[1 - \frac{2\Gamma \Delta t}{C \Delta x^2}\right]\phi_i^l \tag{7.77}$$

for one-dimensional problems

2. ***Convective boundary***
   For convective boundary condition

$$-\Gamma \frac{\partial \Gamma}{\partial x} = h(\phi - \phi_\infty) \tag{7.78a}$$

or

$$f_i'' = h(\phi_{i,j} - \phi_\infty) \tag{7.78b}$$

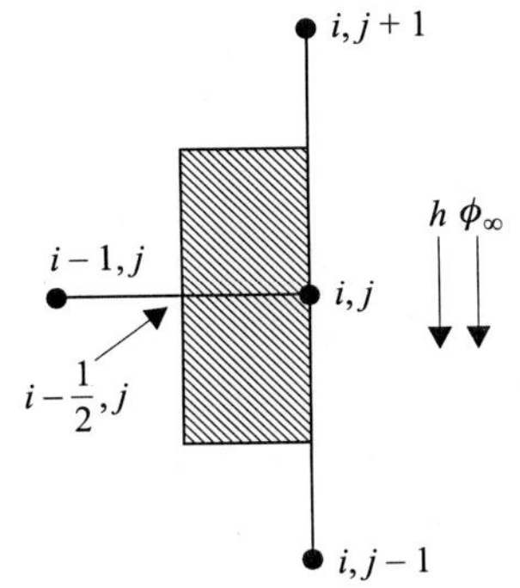

Figure 7.13 Convective boundary node

Substituting Equation 7.78b into Equation 7.74, we get

$$\phi_{i,j}^{l+1} = \frac{\Gamma \Delta t}{C \Delta x^2}\left[2\phi_{i-1,j}^{l+1} + \phi_{i,j+1}^{l+1} + \phi_{i,j-1}^{l+1} + \frac{2h\Delta x}{\Gamma}\phi_\infty\right] + \left[1 - \frac{4\Gamma\Delta t}{C\Delta x^2} - \frac{2h\Delta t}{C\Delta x}\right]\phi_{i,j}^{l} \tag{7.79}$$

A summary of these implicit discretization formulas is given in Table 7.4.

Example 7.4: Two-dimensional Unsteady State Problem with Fully Implicit Scheme: A fin shown below is initially at a uniform temperature $T_\mathrm{I} = 300°\mathrm{C}$ and then suddenly exposed to the convection environment at a temperature of $T_\infty = 20°\mathrm{C}$ and $h = 40\,\mathrm{W/m^{2}°C}$. The base of the fin is maintained at a temperature of $T_0 = 300°\mathrm{C}$. Consider two-dimensional unsteady state conduction in the fin and determine temperature as a function of time using an 8×2 grid. The material thermo-physical properties are $\rho = 7800\,\mathrm{kg/m^3}$, $c = 800\,\mathrm{J/kg°C}$, and $k = 10\,\mathrm{W/m°C}$. The length and thickness of the fin are $L = 8\,\mathrm{cm}$ and $H = 2\,\mathrm{cm}$, respectively.

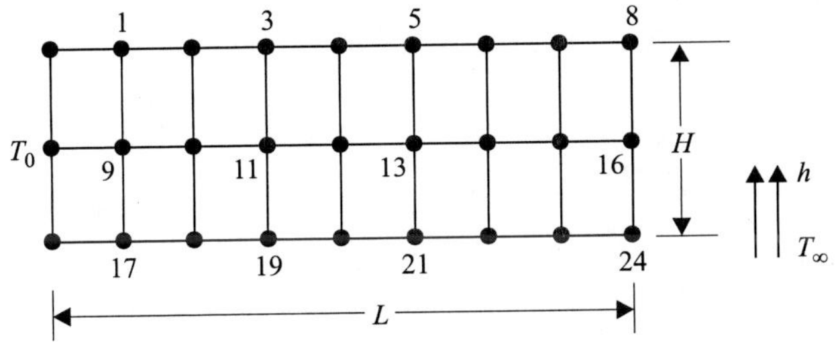

Table 7.4 A Summary of Implicit Discretization Equations with $\Delta x = \Delta y$

Physical Situation	Discretization Equation
(a) ***Interior node*** 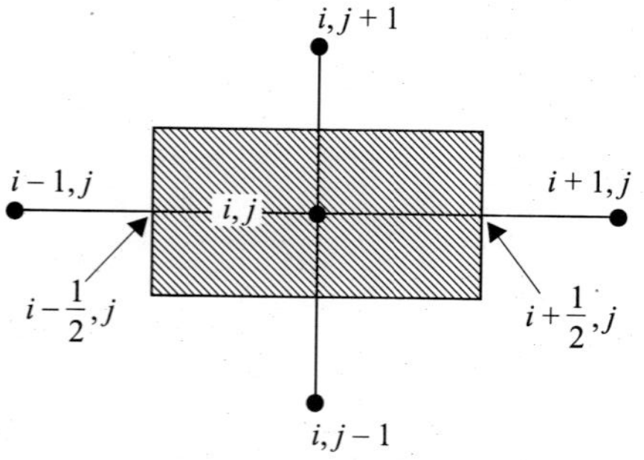	$[1+4\gamma_o]\phi^{l+1}_{i,j} - \gamma_o\left[\phi^{l+1}_{i+1,j} + \phi^{l+1}_{i-1,j} + \phi^{l+1}_{i,j+1} + \phi^{l+1}_{i,j-1}\right] = \phi^{l}_{i,j}$
(b) ***Convective boundary node*** 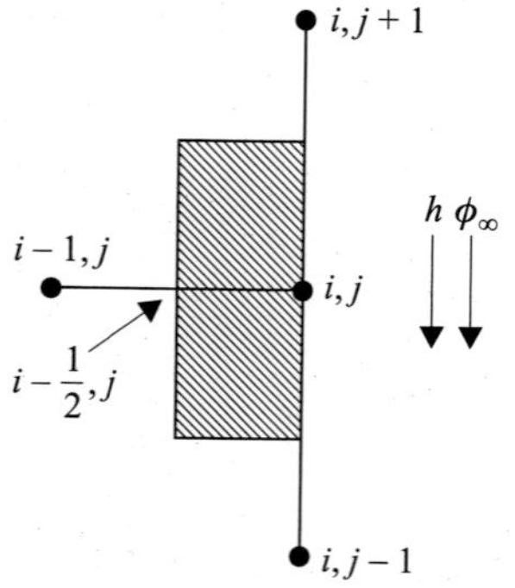	$\left[1 + 2\gamma_0\left(2 + \frac{h\Delta x}{\Gamma}\right)\right]\phi^{l+1}_{i,j} - \gamma_0\left[2\phi^{l+1}_{i-l,j} + \phi^{l+1}_{l,j-1} + \phi^{l+1}_{i,j+1} + 2\frac{h\Delta x}{\Gamma}\phi_\infty\right] = \phi^{l}_{i,j}$

(c) ***Exterior corner with convection boundary***

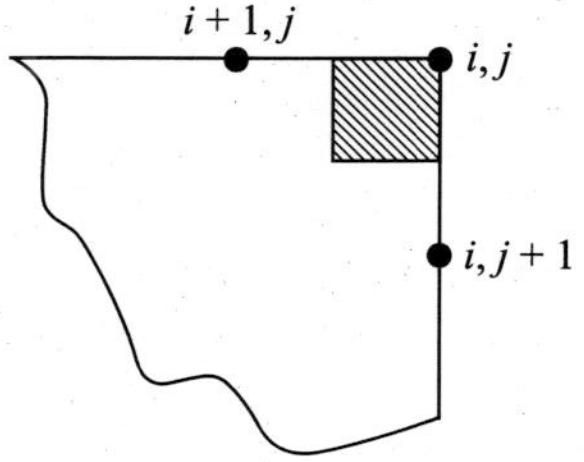

$$\left[1 + 4\gamma_0\left(1 + \frac{h\Delta x}{\Gamma}\right)\right]\phi_{i,j}^{l+1} - 2\gamma_0\left[\phi_{i-1,j}^{l+1} + \phi_{i,j-1}^{l+1} + 2\frac{h\Delta x}{\Gamma}\phi_\infty\right] = \phi_{i,j}^{l}$$

(d) ***Interior corner with convection boundary***

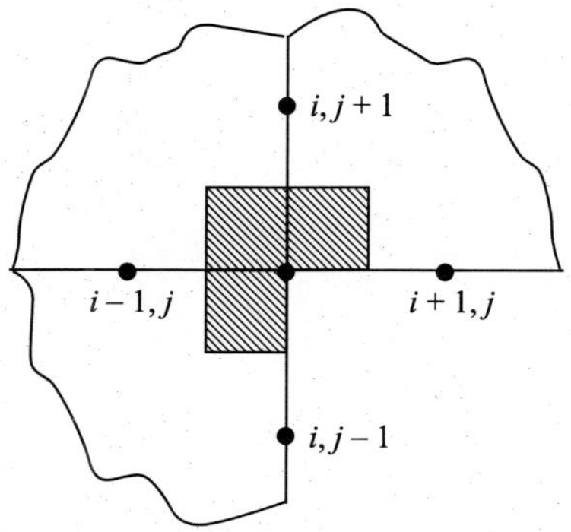

$$\left[1+4\gamma_o\left(1 + \frac{h\Delta x}{3\Gamma}\right)\right]\phi_{i,j}^{l+1} - 2\left(\frac{\gamma_0}{3}\right)\left[2\phi_{i-1,j}^{l+1} + \phi_{i,j-1}^{l+1} + 2\phi_{i,j+1}^{l+1} + \phi_{i+1,j}^{l+1} + 2\frac{h\Delta x}{\Gamma}\phi_\infty\right] = \phi_{i,j}^{l}$$

(Continued)

Table 7.4 Continued

Physical Situation	Discretization Equation
(e) ***Adiabatic boundary*** 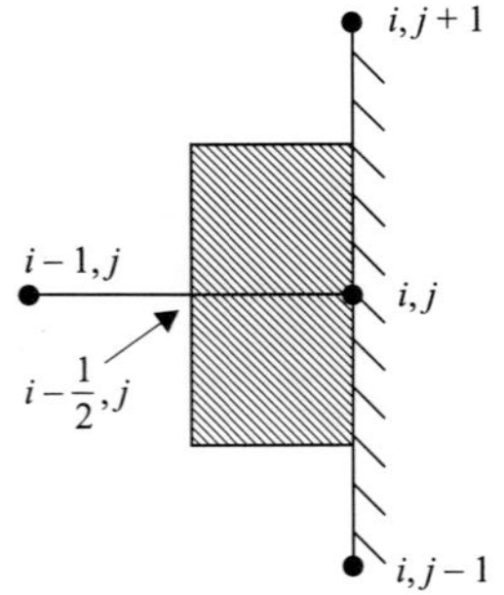	$[1+4\gamma_o]\phi_{i,j}^{l+1} - \frac{2\gamma_o}{3}\left[2\phi_{l-1,j}^{l+1} + \phi_{i,j+1}^{l+1} + \phi_{i,j-1}^{l+1}\right] = \phi_{i,j}^{l}$
(f) ***Constant surface heat flux*** 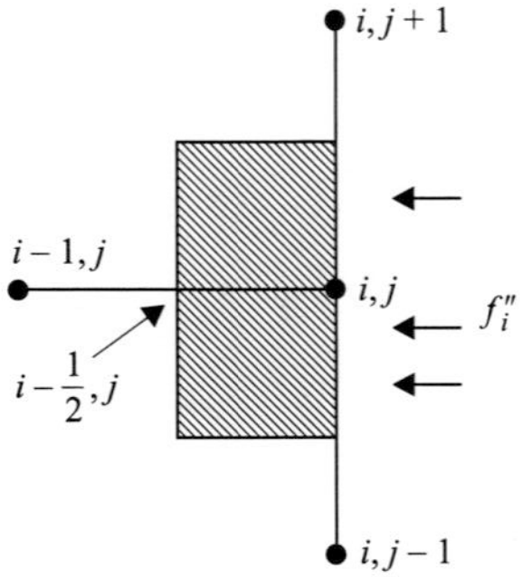	$\phi_{i,j}^{l+1} - \frac{\Gamma\Delta t}{C\Delta x^2}\left[2\phi_{i-1,j}^{l+1} + \phi_{i,j+1}^{l+1} + \phi_{i,j-1}^{l+1}\right] = \left[1 - 4\frac{\Gamma\Delta t}{C\Delta x^2}\right]\phi_i^l + \frac{2\Delta t}{C\Delta x}f_i''$

Note: $\gamma_0 = \Gamma\Delta t/C\Delta x^2$.

Solution:

Governing equation

$$\frac{\partial^2 T}{\partial x^2} + \frac{\partial^2 T}{\partial y^2} = \frac{1}{\alpha}\frac{\partial T}{\partial t}$$

Initial condition: $t = 0,\ T(x, y, 0) = T_0 = 300°\text{C}$

Boundary Conditions

1. $x = 0, \quad T(0, y, t) = T_0 = 300°\text{C}$
2. $x = L$ and on boundary surface, $-k(\partial T/\partial n) = h(T - T_\infty)$

The grid sizes in the x and y directions are

$$\Delta x = \frac{L}{NDIV_x} = \frac{8}{8} = 1\text{ cm} = 0.01\text{ m}, \quad \Delta y = \frac{L}{NDIV_y} = \frac{2}{2} = 1\text{ cm} = 0.01\text{ m}.$$

Discretization equations for the nodes are selected as follows:

Interior nodes: *9, 10, 11, 12, 13, 14, 15*

$$(1 + f_0)T_{i,j}^{l+1} - f_0\left(T_{i+1,j}^{l+1} + T_{i-1,j}^{l+1} + T_{i,j+1}^{l+1} + T_{i,j-1}^{l+1}\right) = T_{i,j}^{l}$$

Node 9

$$(1 + f_0)T_9^{l+1} - f_0\left(T_{10,}^{l+1} + T_0 + T_1^{l+1} + T_{17}^{l+1}\right) = T_9^{l}$$

⋮

⋮

Node 15

$$(1 + f_0)T_{15}^{l+1} - f_0\left(T_{16,}^{l+1} + T_{14} + T_7^{l+1} + T_{23}^{l+1}\right) = T_{15}^{l}$$

Convective nodes on top surface: *1, 2, 3, 4, 5, 6, 7*

$$[1 + 2f_0(2 + B_i]T_{i,j}^{l+1} - f_0\left[2T_{i,j-1}^{l+1} + T_{i-1,j}^{l+1} + T_{i+1,j}^{l+1} + 2B_iT_\infty\right] = T_{i,j}^{l}$$

Node 1

$$[1 + 2f_0(2 + B_i]T_1^{l+1} - f_0[2T_9^{l+1} + T_0 + T_2^{l+1} + 2B_iT_\infty] = T_1^{l}$$

⋮

⋮

Node 7

$$[1 + 2f_0(2 + B_i]T_7^{l+1} - f_0[2T_{15}^{l+1} + T_6 + T_8^{l+1} + 2B_iT_\infty] = T_7^{l}$$

Convective nodes on bottom surface: *17, 18, 19, 20, 21, 22, 23*

$$[1+2f_0(2+B_i]T_{i,j}^{l+1} - f_0\left[2T_{i,j+1}^{l+1} + T_{i-1,j}^{l+1} + T_{i+1,j}^{l+1} + 2B_iT_\infty\right] = T_{i,j}^{l}$$

Node 17

$$[1+2f_0(2+B_i]T_{17}^{l+1} - f_0\left[2T_{9}^{l+1} + T_0 + T_{18}^{l+1} + 2B_iT_\infty\right] = T_{17}^{l}$$

⋮

⋮

Node 23

$$[1+2f_0(2+B_i]T_{23}^{l+1} - f_0\left[2T_{15}^{l+1} + T_{22}^{l+1} + T_{24}^{l+1} + 2B_iT_\infty\right] = T_{23}^{l}$$

Convective node right surface: *16*

$$[1+2f_0(2+B_i]T_{i,j}^{l+1} - f_0\left[2T_{i-1,j}^{l+1} + T_{i,j-1}^{l+1} + T_{i,j+1}^{l+1} + 2B_iT_\infty\right] = T_{i,j}^{l}$$

Node 16

$$[1+2f_0(2+B_i]T_{16}^{l+1} - f_0\left[2T_{15}^{l+1} + T_{24}^{l+1} + T_{8}^{l+1} + 2B_iT_\infty\right] = T_{16}^{l}$$

North-east corner with convective node: *8*

$$[1+4f_0(1+B_i]T_{i,j}^{l+1} - 2f_0\left[2T_{i-1,j}^{l+1} + T_{i,j-1}^{l+1} + 2B_iT_\infty\right] = T_{i,j}^{l}$$

Node 8

$$[1+4f_0(1+B_i]T_{8}^{l+1} - 2f_0\left[2T_{7}^{l+1} + T_{16}^{l+1} + 2B_iT_\infty\right] = T_{8}^{l}$$

South-east corner with convective node: *24*

$$[1+4f_0(1+B_i]T_{i,j}^{l+1} - 2f_0\left[2T_{i-1,j}^{l+1} + T_{i,j+1}^{l+1} + 2B_iT_\infty\right] = T_{i,j}^{l}$$

Node 24

$$[1+4f_0(1+B_i]T_{24}^{l+1} - 2f_0\left[2T_{23}^{l+1} + T_{16}^{l+1} + 2B_iT_\infty\right] = T_{24}^{l}$$

Node	Temperature
Top surface	
1	261.73332029
2	233.95998474
3	213.05863823
4	197.13343750
5	184.78988480
6	175.09333805
7	167.27218777
8	160.95470105

(Continued)

Node	Temperature
	Mid-section
9	265.63788848
10	238.09027107
11	216.90011772
12	200.66013908
13	188.09017058
14	178.17752625
15	170.21842068
16	163.77223699
	Bottom surface
17	261.70592749
18	233.91145214
19	212.99992949
20	197.05738540
21	184.72851089
22	175.01658512
23	167.21506168
24	160.90214644

Table 7.5 Pseudo Code for Transient Solution in a Rectangular Slab Using Implicit Finite Difference Scheme and Gauss–Seidel Method

Input dimensions, properties, select number of divisions, maximum time, and stability parameters

```
  Input dimension L and H
  Input thermophysical properties Γ and C
  Input parameters γ0
  Select number of divisions NDIVX and NDIVY
  Number of node NX = NDIVX + 1
  Number of node NY = NDIVY + 1
  Time step Dt
```

Initialize time

```
  time = 0.0
```

Set initial condition

```
  dofor I = 1, NX
    dofor J = 1, NY
      φ(I, J) = φI
    enddo
  enddo
dowhile (time < tmax)
  time = time + Dt
    iflag = 1
    iter = 0
  dowhile (it < itmax and iflag = 1)
    iter = iter + 1
```

Set new values to old values

```
      dofor I = 2, NX−1
        dofor J=2, NY−1
          φo(I, J) = φ(I, J)
      enddo
```

(*Continued*)

Table 7.5 Continued

```
Calculate nodal values at interior nodes
    dofor I = 2, NX−1
      dofor J=2, NY−1
        φ(I, J) = (γ0(φ(I + 1, J) + φ(I − 1, j) + φ(I, J + 1) + φ(I, J − 1)) + φo(I, J))/(1 + 4γ0)
    enddo
Similarly calculate nodal values at other boundary nodes or assign boundary values
Estimate approximate percent relative error
    dofor I = 1, NX
      dofor J = 1, NY
        εa = ABS(φ(I, J) − φo(I, J)/φ(I, J))
        If (εa > εs) then
          iflag = 1
        else
          iflag = 0
        endif
      enddo
    enddo
  enddo
enddo
```

A pseudo-code implementing these discretizations and solving the system of equation using the Gauss–Seidel method is given in Table 7.5. A solution to the problem is obtained using a computer code based on the pseudo code and results for temperature distribution at time $t = 10\Delta t$ are given below.

As expected the top and bottom surface temperature distributions are almost identical. The discrepancy is due to the round off error.

7.4 CRANK–NICOLSON SCHEME

Crank and Nicolson suggested a modified implicit method. In this scheme, the space derivatives are averaged between the $(l+1)$th time step and lth time step. A forward-difference formula is used to approximate the time derivative. The resulting discretization equation for two-dimensional unsteady case with constant thermophysical properties and without heat generation is

$$C\frac{\phi_{i,j}^{l+1}-\phi_{l,j}^{l}}{\Delta t}=\Gamma\left[\frac{1}{2}\frac{\phi_{i+l,j}^{l+1}-2\phi_{i,j}^{l+1}+\phi_{i-1,j}^{l+1}}{\Delta x^2}+\frac{1}{2}\frac{\phi_{i+1,j}^{l}-2\phi_{i,j}^{l}+\phi_{i-1,j}^{l}}{\Delta x^2}\right]$$

$$+\Gamma\left[\frac{1}{2}\frac{\phi_{i,j+1}^{l+1}-2\phi_{i,j}^{l+1}+\phi_{i,j-1}^{l+1}}{\Delta y^2}+\frac{1}{2}\frac{\phi_{i,j+1}^{l}-2\phi_{i,j}^{l}+\phi_{i,j-1}^{l}}{\Delta y^2}\right]$$

Solving for the $\phi_{i,j}^{l+1}$, we obtain

$$\phi_{i,j}^{l+1} = \frac{1-[\Gamma\Delta t/C\Delta x^2]-[\Gamma\Delta t/C\Delta y^2]}{1+[\Gamma\Delta t/C\Delta x^2+\Gamma\Delta t/C\Delta y^2]}\phi_{i,j}^{l}$$
$$+\frac{[\alpha\Delta t/2\Delta x^2]}{1+[\alpha\Delta t/\Delta x^2]+[\alpha\Delta t/\Delta y^2]}\left[\phi_{i-1,j}^{l+1}+\phi_{i+1,j}^{l+1}+\phi_{i-1,j}^{l}+\phi_{i+1,j}^{l}\right]$$
$$+\frac{[\Gamma\Delta t/2C\Delta x^2]}{1+[\Gamma\Delta t/C\Delta x^2]+[\Gamma\Delta t/C\Delta y^2]}\left[\phi_{i,j+1}^{l+1}+\phi_{i,j-1}^{l+1}+\phi_{i,j+1}^{l}+\phi_{i,j-1}^{l}\right] \quad (7.80)$$

It can be mentioned here that the discretization equation 7.80 can also be obtained by setting the weighting factor $f=0.5$ in general form of discretization equation (7.9) obtained by using the control volume method.

The truncation error is of the order of $O(\Delta x^2+\Delta t^2)$. The advantage of this method is that, for given values of space and time steps Δx and Δt, the resulting solution involve less truncation error due to Δt than the explicit and implicit schemes. A von Neumann stability analysis shows that the *Crank–Nicolson scheme* is *unconditionally stable*. However, the Crank–Nicolson scheme involves additional computation. If there are N internal grid points over the region, then this method involves the solution of N simultaneous algebraic equations for each time step.

7.4.1 Solution Methods for Crank–Nicolson Method

Regarding the solution of the system of algebraic equations, the methodologies utilized in the fully implicit scheme can be used in this case as well. For example, in the case of a one-dimension heat conduction, the above equation can be rearranged as

$$-\frac{1}{2}\frac{\Gamma\Delta t}{C\Delta x^2}\phi_{i-1}^{l+1}+\left[1+\frac{\Gamma\Delta t}{C\Delta x^2}\right]\phi_i^{l+1}-\frac{1}{2}\frac{\Gamma\Delta t}{C\Delta x^2}\phi_{i+1}^{l+1}$$
$$=\frac{1}{2}\frac{\Gamma\Delta t}{C\Delta x^2}\phi_{i-1}^{l}+\left[1-\frac{\Gamma\Delta t}{C\Delta x^2}\right]\phi_i^{l}+\frac{1}{2}\frac{\Gamma\Delta t}{C\Delta x^2}\phi_{i+1}^{l} \quad (7.81)$$

For two-dimensional problems, the resulting system of discretization equations is not tridiagonal.

7.5 SPLITTING METHODS

The difficulties encountered in solving two-dimensional diffusion equations by a conventional implicit scheme led to the development of splitting algorithms such as the ***alternating-direction implicit method (ADI method)*** and the ***alternating direction explicit method (ADE)***. In the ADI method, marching of the solution in a time step takes place in two half steps with the implicit scheme being used alternatively between the x and y directions. This leads to the formation of a tridiagonal system at each time step and the system can be solved by a direct solver such as TDMA.

7.5.1 ADI Method

In the alternating direction implicit (ADI) method given by Peaceman and Rachford (1955), the two-step procedure for the solution of Equation 7.27 is given as follows.

Step 1. In this step the equation is solved at intermediate time step $\Delta t/2$ with an implicit scheme used only for the derivative in the x direction and an explicit scheme for the derivative in y direction

$$C\frac{\phi_{i,j}^{l+\frac{1}{2}}-\phi_{i,j}^{l}}{\Delta t/2}=\Gamma\frac{\phi_{i+l,j}^{l+\frac{1}{2}}-2\phi_{i,j}^{l+\frac{1}{2}}+\phi_{i-1,j}^{l+\frac{1}{2}}}{\Delta x^2}+\Gamma\frac{\phi_{i,j+1}^{l}-2\phi_{i,j}^{l}+\phi_{l,j-1}^{l}}{\Delta y^2} \tag{7.82}$$

Rearranging

$$\left(2+\frac{2C\Delta x^2}{\Gamma\Delta t}\right)\phi_{i,j}^{l+\frac{1}{2}}-\phi_{i+1,j}^{i+\frac{1}{2}}-\phi_{i-1,j}^{i+\frac{1}{2}}=\left(\frac{2C\Delta x^2}{\Gamma\Delta t}-2\frac{\Delta x^2}{\Delta y^2}\right)\phi_{i,j}^{l}+\frac{\Delta x^2}{\Delta y^2}\left(\phi_{i,j+1}^{l}+\phi_{i,j-1}^{l}\right) \tag{7.83}$$

Equation 7.83 represents a tridiagonal system for a fixed j location and as approximate solution is obtained at this intermediate time step for nodes i = $1, 2, \ldots, N$.

Step 2. The solution is now advanced to time step $l+1$ with $\Delta t/2$ with an implicit scheme used only for the derivative in the y direction and an explicit scheme for the derivative in the x direction as follows:

$$C\frac{\phi_{i,j}^{l+1}-\phi_{i,j}^{l+\frac{1}{2}}}{\Delta t/2}=\Gamma\frac{\phi_{i+l,j}^{l+\frac{1}{2}}-2\phi_{i,j}^{l+\frac{1}{2}}+\phi_{i-1,j}^{l+\frac{1}{2}}}{\Delta x^2}+\Gamma\frac{\phi_{i,j+1}^{l+1}-2\phi_{i,j}^{l+1}+\phi_{l,j-1}^{l+1}}{\Delta y^2} \tag{7.84}$$

Rearranging

$$\left(2+\frac{2C\Delta x^2}{\Gamma\Delta t}\right)\phi_{i,j}^{l+1}-\phi_{i,j+1}^{i+1}-\phi_{i,j-1}^{i+1}=\left(\frac{2C\Delta x^2}{\Gamma\Delta t}-2\frac{\Delta x^2}{\Delta y^2}\right)\phi_{i,j}^{l+\frac{1}{2}}+\frac{\Delta x^2}{\Delta y^2}\left(\phi_{i+1,j}^{l+\frac{1}{2}}+\phi_{i-1,j}^{l+\frac{1}{2}}\right) \tag{7.85}$$

Equation 7.85 represents a tridiagonal system for a fixed i location and a solution is obtained at the time step for all nodes $j = 1, 2, \ldots, N$.

The solution is marched forward by alternating between the rows and columns with the implicit scheme being used only in one direction and the explicit scheme in the other direction. The ADI method is second-order accurate with an error of the order $O\left[(\Delta t)^2, (\Delta x)^2, (\Delta y)^2\right]$. A stability analysis leads to the conclusion that the ADI scheme is unconditionally stable. The method can also be employed along with the Crank–Nicolson scheme with two intermediate time steps and for three-dimensional problems.

7.5.2 ADE Method

In the alternating direction explicit (ADE) method, the two-step procedure for the solution of Equation 7.27 is used alternately for x and y directions, and the resulting system is solved in an explicit manner. For example, the two-step marching procedure for the x direction is used as follows.

Step 1. In this step the equation is solved at an intermediate time step $\Delta t/2$ with the solution obtained explicitly by marching from left to right (i.e., $i = 1, 2, \ldots, N$). The scheme is shown as

$$C\frac{\phi_{i,j}^{l+\frac{1}{2}} - \phi_{i,j}^{l}}{\Delta t/2} = \Gamma\frac{\phi_{i+l,j}^{l} - \phi_{i,j}^{l} - \phi_{i,j}^{l+\frac{1}{2}} + \phi_{i-1,j}^{l+\frac{1}{2}}}{\Delta x^2} + \Gamma\frac{\phi_{i,j+1}^{l} - 2\phi_{i,j}^{l} + \phi_{l,j-1}^{l}}{\Delta y^2} \tag{7.86}$$

Rearranging

$$\left(1 + \frac{2C\Delta x^2}{\Gamma\Delta t}\right)\phi_{i,j}^{l+\frac{1}{2}} = \phi_{i-1,j}^{i+\frac{1}{2}} + \phi_{i+1,j}^{i} + \left(\frac{2C\Delta x^2}{\Gamma\Delta t} - 2\frac{\Delta x^2}{\Delta y^2} - 1\right)\phi_{i,j}^{l} + \frac{\Delta x^2}{\Delta y^2}\left(\phi_{i,j+1}^{l} + \phi_{i,j-1}^{l}\right) \tag{7.87}$$

Note that in marching from left to right ($i = 1, 2, \ldots, N$), all items including $\phi_{i-1,j}^{i+\frac{1}{2}}$ on the right-hand side are known.

Step 2. In this step the equation is solved at the time step Δt with solution obtained explicitly by marching from right to left, i.e., $i = N$, $(N-1) \ldots 2, 1$

$$C\frac{\phi_{i,j}^{l+1} - \phi_{i,j}^{l+\frac{1}{2}}}{\Delta t/2} = \Gamma\frac{\phi_{i+l,j}^{l+1} - \phi_{i,j}^{l+1} - \phi_{i,j}^{l+\frac{1}{2}} + \phi_{i-1,j}^{l+\frac{1}{2}}}{\Delta x^2} + \Gamma\frac{\phi_{i,j+1}^{l+\frac{1}{2}} - 2\phi_{i,j}^{l+\frac{1}{2}} + \phi_{l,j-1}^{l+\frac{1}{2}}}{\Delta y^2} \tag{7.88}$$

Rearranging

$$\left(1 + \frac{2C\Delta x^2}{\Gamma\Delta t}\right)\phi_{i,j}^{l+1} = \phi_{i+1,j}^{l+1} + \phi_{i-1,j}^{i+\frac{1}{2}} + \left(\frac{2C\Delta x^2}{\Gamma\Delta t} - 2\frac{\Delta x^2}{\Delta y^2} - 1\right)\phi_{i,j}^{l+\frac{1}{2}} + \frac{\Delta x^2}{\Delta y^2}\left(\phi_{i,j+1}^{l+\frac{1}{2}} + \phi_{i,j-1}^{l+\frac{1}{2}}\right) \tag{7.89}$$

Note that in marching from left to right ($i = 1, 2, \ldots, N$), all items including $\phi_{i+1,j}^{i+1}$ on the right-hand side are known. The two-step procedure is repeated for the y direction by alternating the marching direction from top to bottom and then from bottom to top.

For a one-dimensional problem, Equations 7.86 and 7.88 reduce to

$$\left(1+\frac{2C\Delta x^2}{\Gamma\Delta t}\right)\phi_i^{l+\frac{1}{2}} = \phi_{i-1}^{i+\frac{1}{2}} + \phi_{i+1}^{i} + \left(\frac{2C\Delta x^2}{\Gamma\Delta t} - 1\right)\phi_i^{l} \tag{7.90}$$

and

$$\left(1+\frac{2C\Delta x^2}{\Gamma\Delta t}\right)\phi_i^{l+1} = \phi_{i+1}^{l+1} + \phi_{i-1}^{i+\frac{1}{2}} + \left(\frac{2C\Delta x^2}{\Gamma\Delta t} - 1\right)\phi_i^{l+\frac{1}{2}} \tag{7.91}$$

PROBLEMS

7.1 Consider a one-dimensional unsteady state conduction in a plane slab [$k = 50\,\text{W/m°C}$, $\alpha = 2\times10^{-5}\,\text{m}^2/\text{s}$] of thickness L = 4 mm. Initially, the slab is at a 20°C. The right side of the plate is suddenly brought to a temperature of 200°C. The left side is maintained at 20°C.

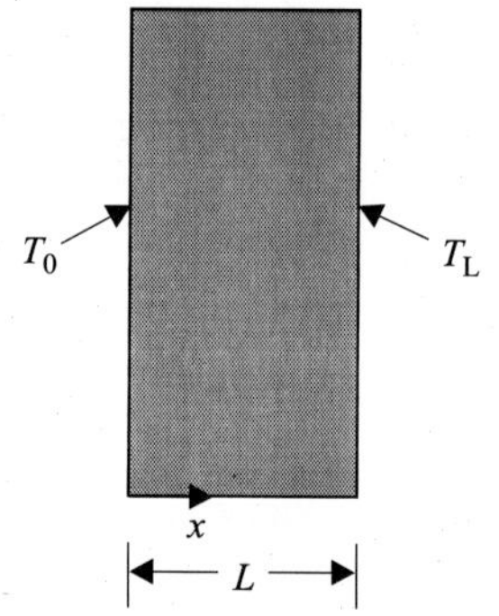

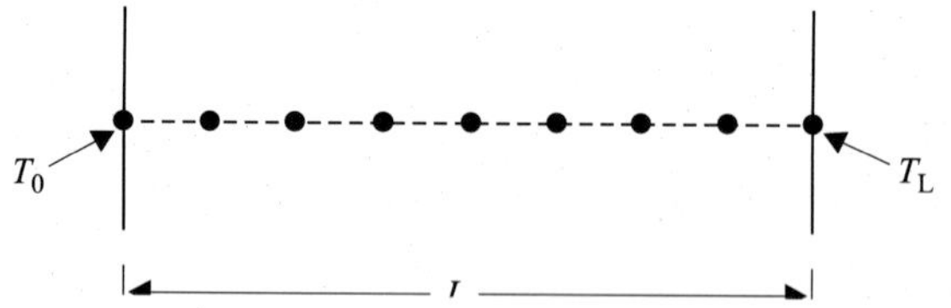

Solve the problem using an explicit finite difference scheme to study the transient temperature distribution and rate of heat transfer from the slab. Use the number of nodes as 10 and the stability parameters as 0.4 and 0.55. Carry out the computation for four consecutive time steps and plot heat flux at the right surface as a function of time with each stability parameter.

7.2 Consider a one-dimensional unsteady state conduction in a plane slab [$k = 50\,\mathrm{W/m^\circ C}$, $\alpha = 2 \times 10^{-5}\,\mathrm{m^2/s}$] of thickness $L = 4\,\mathrm{mm}$. Initially the slab is at a temperature of 20°C. The right side of the plate is suddenly brought to a convective environment with a temperature of $T_\infty = 200^\circ\mathrm{C}$ and $h = 80\,\mathrm{W/m^{2\circ}C}$. The left side is maintained at 20°C. Consider the grid size distribution shown with 10 grid points and determine the transient temperature distribution in the slab.

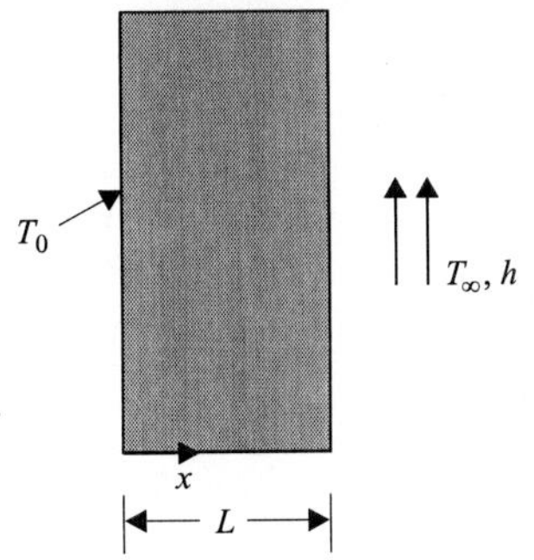

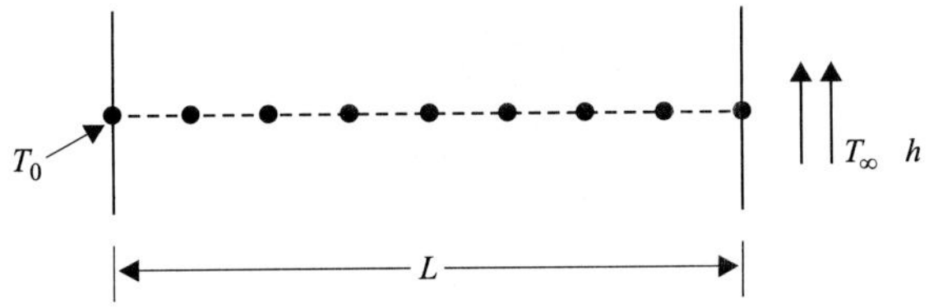

Develop a computer program using an explicit finite difference scheme and the pseudo code presented in Table 7.3 to study the transient temperature distribution and rate of heat transfer from the slab.

(a) Examine the effect of stability parameter, $\gamma_0 = 0.4$, 0.45, 0.5, 0.55, and 0.6. Plot the heat flux at $x = L$ as a function of time for different values of stability parameter.

(b) Select the appropriate stability parameter and determine the transient temperature distribution with increase in time. Present temperature distributions at 10 different time steps up to 5 s.

7.3 Consider a one-dimensional unsteady state heat conduction in a plane slab of thickness $L = 10\,\mathrm{cm}$. Initially, the slab is at a uniform temperature of $T_\mathrm{I} = 20^\circ\mathrm{C}$. At time $t > 0$, the left side of the slab is subjected to a uniform constant surface heat flux $q''_s = 2.0 \times 10^4\,\mathrm{W/cm^2}$ and the right surface is maintained at $T_\mathrm{R} = 20^\circ\mathrm{C}$.

The slab is carbon steel with $\alpha = 0.835\,\mathrm{cm^2/s}$. Use the fully implicit finite difference scheme to obtain the system of equations with the number of nodes as six and a time step of $\Delta t = 0.2\,\mathrm{s}$. Show computations for two time steps.

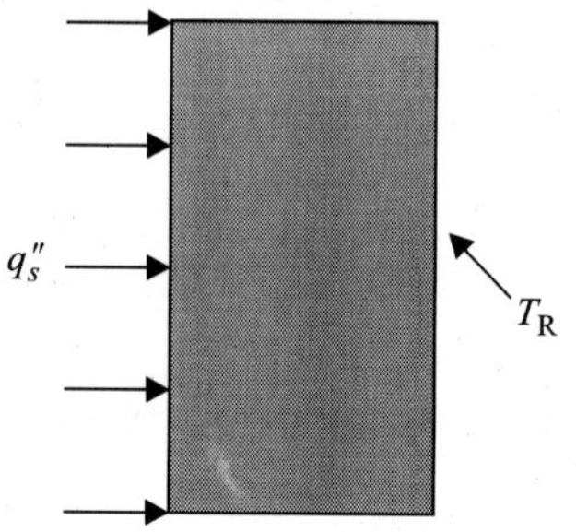

7.4 Develop a computer program using a fully implicit finite difference scheme and TDMA solver to study the transient temperature distribution and rate of heat transfer for Problem 7.3. Present results for Problem 7.3 in terms of temperature distribution with the number of nodes as 10 and a time step of $\Delta t = 0.2\,\text{s}$ up to 5 s.

7.5 Redo Problem 7.4 based on developing a computer code that uses a matrix inversion procedure with a Gauss–Jordan solver.

7.6 Let us consider a general two-dimensional unsteady state problem with the grid system and boundary conditions as shown in the figure.

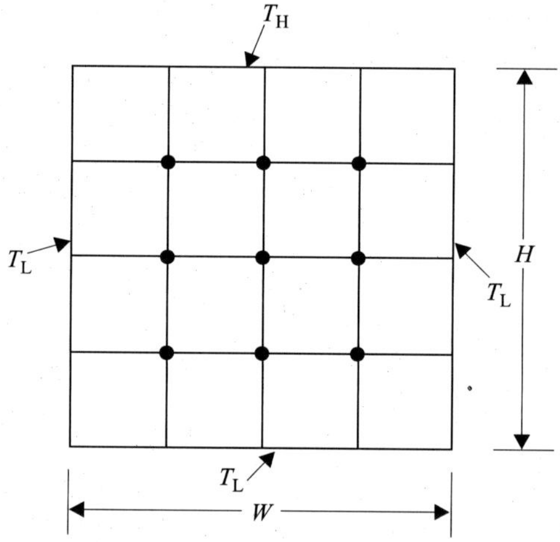

Derive the system of equations using a fully implicit scheme and the following data: $W = 8\,\text{cm}$, $H = 8\,\text{cm}$, $T_I = 25°\text{C}$, $T_H = 400°\text{C}$, $T_L = 20°\text{C}$, and $f_0 = 0.05$. Solve the system for two time steps with $\Delta t = 0.2\,\text{s}$ and using (a) Gauss elimination, (b) a matrix inversion with Gauss–Jordan, (c) LU decomposition, and (d) a Gauss–Seidel method.

7.7 A two-dimensional $8\,\text{cm} \times 8\,\text{cm}$ square slab, initially at a uniform temperature of 300°C, is suddenly exposed to a convection environment at a temperature of 20°C and $h = 40\,\text{W/m}^2\text{°C}$ all around. Consider a

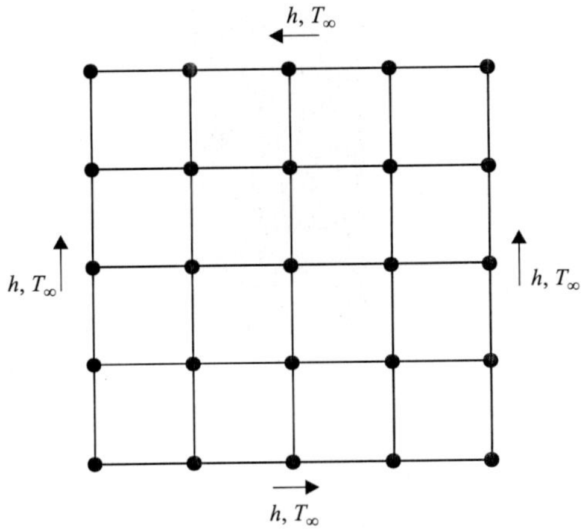

two-dimensional unsteady state conduction in the slab and determine temperature distribution and rate of heat loss as a function of time.

(a) Use the 4×4 grid and derive the system of equations for grid temperatures using a fully implicit scheme.

(b) Select an appropriate Δt and calculate the grid point temperature for five time increments. Material thermo-physical properties are $\rho = 7900\,\text{kg/m}^3$, $c = 477\,\text{J/kg K}$ and $k = 14.9\,\text{W/m K}$.

7.8 Repeat Problem 7.7 using the Crank–Nicolson scheme.

7.9 Develop a computer program using the fully implicit scheme and Gauss–Seidel method. Redo Problem 7.7 with refined grid size distributions as 4×4, 6×6 and 8×8 using this code.

(a) Present temperature distribution along with percent relative error for the vertical center line with refined grid size distributions at the fifth time step.

(b) Show the temperature distribution for the vertical center line with increase in time for the grid of 8×8.

7.10 Develop a computer program using the fully implicit scheme and line-by-line method. Repeat Problem 7.9 using this code.

7.11 The rectangular slab shown below is initially uniform in temperature at 300°C and then suddenly exposed to the convection environment at a temperature of 20°C and $h = 40\,\text{W/m}^2\,\text{K}$ all around except the left surface, which is maintained at $T_0 = 300°\text{C}$. Consider a two-dimensional unsteady state conduction in the slab and determine

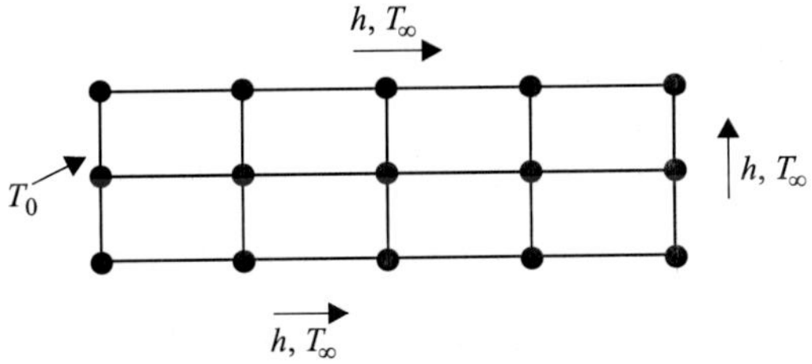

temperature distribution and total heat loss as a function of time. Material thermo-physical properties are $\rho = 2700\,\text{kg/m}^3$, $c = 900\,\text{J/kg K}$, and $k = 230\,\text{W/m K}$. The length and height of the slab are 4 cm and 2 cm respectively. Use the Gauss elimination solver.

7.12 Consider the diffusion of moisture from an airstream in a two-dimensional porous adsorbing felt material lined on the side of the air flow channel as shown. The other three sides of the felt are lined with thin aluminum foil, thus making it impermeable to mass diffusion. Initial moisture concentration in the felt is $C_\text{I} = 0.001\,\text{kgH}_2\text{O/m}^3$. The left surface of the felt is exposed to moving air in the channel with mass transfer coefficient $h_\text{D} = 0.02\,\text{m/s}$ and moisture concentration $C_\infty = 0.01\,\text{kg H}_2\text{O/m}^3$. The material is assumed to be homogeneous with a constant moisture adsorption rate of $\dot{m}_{ad} = 0.000\,04\,\text{kg H}_2\text{O/kg s}$.

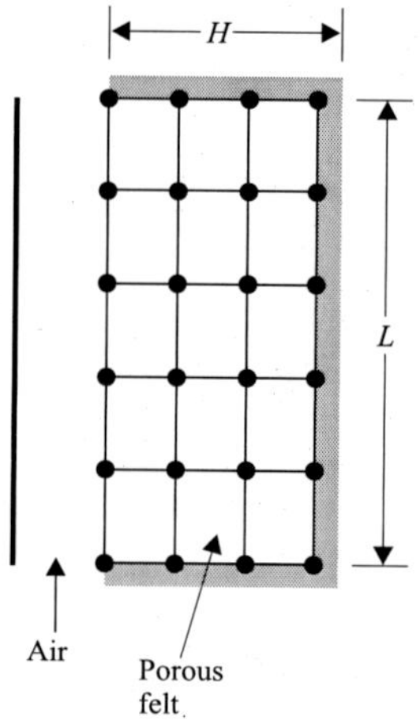

The mathematical statement of the problem is as follows.

Governing equation

$$\frac{\partial C}{\partial t} = \frac{\partial}{\partial x}\left(D\frac{\partial C}{\partial x}\right) + \frac{\partial}{\partial y}\left(D\frac{\partial C}{\partial y}\right) + \frac{(1-\varepsilon)}{\varepsilon}\rho\dot{m}_\text{ad}$$

Boundary conditions

1. $y = 0, \quad -D\left.\frac{\partial C}{\partial x}\right|_{y=0} = h_\text{D}(C_\infty - C|_{y=0})$

2. $y = H, \quad \left.\frac{\partial C}{\partial x}\right|_{x=H} = 0$

3. $x = 0, \quad \left.\frac{\partial C}{\partial x}\right|_{x=0} = 0$

4. $x = L, \quad \left.\frac{\partial C}{\partial x}\right|_{x=L} = 0$

where ε = porosity and ρ_b = material bulk density.

(a) Apply a finite difference–control volume method using an explicit scheme and derive the system of equations to calculate the two-dimensional transient moisture concentration distribution in the slab based on the grid distribution shown in the figure. Assume diffusion coefficient as $D = 4.0 \times 10^{-6}\,m^2/s$ and use the following data for computation: $H = 4\,mm$, $L = 0.1\,m$ $\rho_b = 450\,kg/m^3$, $\varepsilon = 0.5$.

(b) Solve the system of equations using the Gauss–Seidel method and present results along the four vertical lines as a function of time.

7.13 Use a finite difference–control volume method to solve a two-dimensional unsteady state conduction in a rectangular carbon steel ($k = 200\,W/m°C$) slab subjected to a constant surface heat flux irradiated by a continuous high-energy laser beam at the top surface. Assume the heat flux distribution to be a constant average value, $q''_0 = 2 \times 10^8\,W/m^2$, acting over a section of the surface. The top surface is also subjected to forced convection with $h_c = 500\,W/m^2°C$. The left surface is assumed to be adiabatic as a line of symmetry. All other surfaces are assumed to be subjected to free convection with $T_\infty = 25°C$ and $h = 40\,W/m°C$.

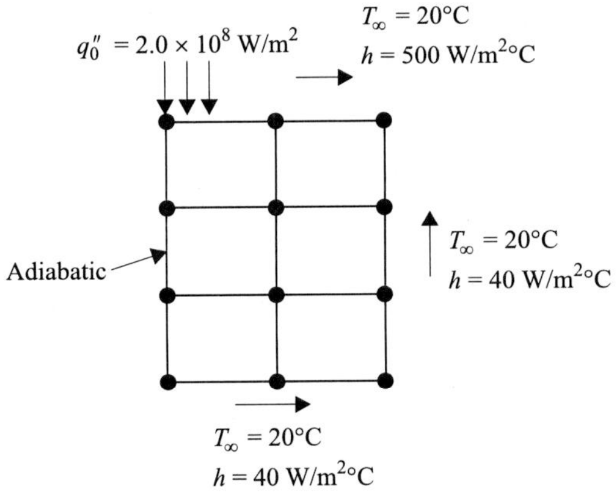

(a) Derive the finite difference equations for all nodes using a fully implicit scheme and represent the system in matrix form.

(b) Solve the system of equations using the Gauss elimination solver and present results for temperature distribution as a function of time.

7.14 A three-dimensional block ($0.4\,m \times 0.4\,m \times 0.4\,m$) is heated in a furnace to a uniform temperature of $T_I = 600°C$ and then suddenly submerged in cooling fluid with convection environment $h = 40\,W/m^2\,K$ and $T_\infty = 20°C$. Develop a computer program using a fully implicit scheme and line-by-line method to determine the temperature distribution in the slab with increase in time. Use TDMA with alternate sweeping on the x–y plane and marching scheme in the z direction. Use thermophysical properties of the material as $\rho = 7854\,\text{kg/m}^3$, $c = 434\,\text{J/kg K}$ and $k = 60\,\text{W/m K}$.

(a) Determine temperature distribution using a grid of $10 \times 10 \times 10$ and assigned tolerance limit of $\varepsilon_s = 0.001$, and plot the temperature distribution at the mid section along the z direction with increase in time.

(b) Repeat step (a) with an assigned tolerance limit of $\varepsilon_s = 0.00001$.

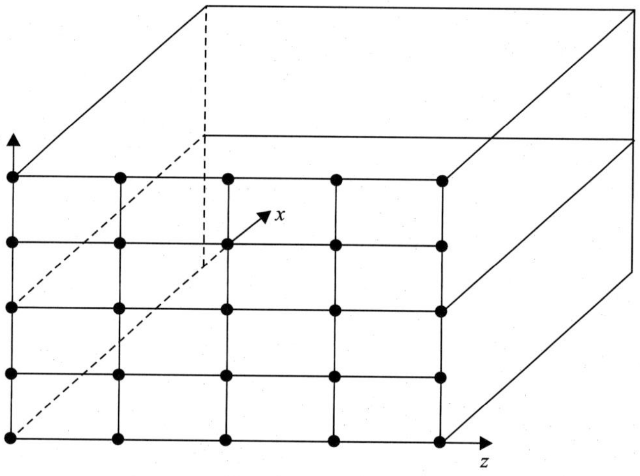

CHAPTER

8

CONVECTION PROBLEMS

Convection transport mechanism is primarily influenced by the associated flow field, and hence the solution of convection problems involves solution of conservation of mass and momentum equations for the flow field, followed by the solution of energy or mass concentration equations. The governing equations for convection problems differ from the diffusion equations discussed in previous chapters with the inclusion of additional terms such as the convective and pressure derivative terms. Major difficulties in numerical solution of convection problems arise from the discretization of the convective and pressure derivative terms in momentum equations. In this chapter, we will begin with the discussion of discretization procedures for convective and pressure derivative terms. This will be followed by the discussion of solution algorithms for solving coupled systems of partial differential equations for velocity field, pressure, and temperature and/or mass concentration fields.

8.1 SPATIAL DISCRETIZATION USING CONTROL VOLUME METHOD

Let us consider a one-dimensional steady state problem involving convection and diffusion terms as

$$\frac{d}{dx}(\rho u \phi) = \frac{d}{dx}\left(\Gamma_x \frac{d\phi}{dx}\right) \tag{8.1}$$

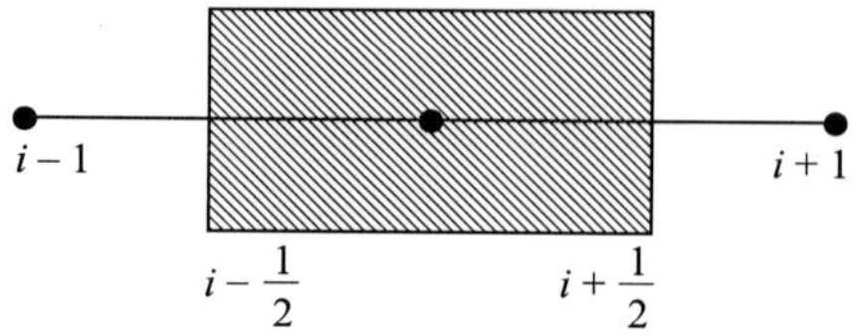

Figure 8.1 One-dimensional control volume

In the control volume method, we integrate the equation over a control volume shown in figure

$$\int_{i-\frac{1}{2}}^{i+\frac{1}{2}} \frac{d}{dx}(\rho u\phi)\,dx = \int_{i-\frac{1}{2}}^{i+\frac{1}{2}} \frac{d}{dx}\left(\Gamma_x \frac{d\phi}{dx}\right)dx \tag{8.2a}$$

or

$$(\rho u\phi)\big|_{i+\frac{1}{2}} - (\rho u\phi)\big|_{i-\frac{1}{2}} = \left(\Gamma_x \frac{d\phi}{dx}\right)_{i+\frac{1}{2}} - \left(\Gamma_x \frac{d\phi}{dx}\right)_{i-\frac{1}{2}} \tag{8.2b}$$

In the next step, quantities such as $\rho u, \phi$, Γ_x and $d\phi/dx$ are approximated with a profile assumption using interpolation formulas. We have mentioned before that the two common profile assumptions are (1) step-wise constant profile and (2) piece-wise linear profile, as depicted in Figure 5.4. Assuming a step-wise constant profile for the convection quantity, ρu, and transport property, Γ_x, and a piece-wise linear profile for the first derivative terms, we get

$$(\rho u)_{i+\frac{1}{2}}\phi_{i+\frac{1}{2}} - (\rho u)_{i-\frac{1}{2}}\phi_{i-\frac{1}{2}} = \Gamma_{xi+\frac{1}{2}} \frac{\phi_{i+1} - \phi_i}{\Delta x_{i+\frac{1}{2}}} - \Gamma_{xi-\frac{1}{2}} \frac{\phi_i - \phi_{i-1}}{\Delta x_{i-\frac{1}{2}}} \tag{8.3}$$

A profile assumption for the dependent variable, ϕ, in the convection term leads to so-called different discretization schemes for the convection term. Some of these schemes are: (1) ***central difference scheme***, (2) ***upwind scheme***, (3) ***exponential scheme***, (4) ***hybrid scheme***, and (5) ***power law scheme***. A brief discussion of these schemes is now given.

8.1.1 Central Difference Scheme

In this scheme, the dependent variable ϕ in the convection term is approximated by a piece-wise linear profile, such as the ***central difference formula***, as

$$\phi_{i+\frac{1}{2}} = \frac{1}{2}(\phi_{i+1} + \phi_i) \tag{8.4a}$$

and

$$\phi_{i-\frac{1}{2}} = \frac{1}{2}(\phi_i + \phi_{i-1}) \tag{8.4b}$$

Substituting Equations 8.4a and 8.4b into Equation 8.3, we get

$$\frac{1}{2}(\rho u)_{i+\frac{1}{2}}(\phi_{i+1} + \phi_i) - \frac{1}{2}(\rho u)_{i-\frac{1}{2}}(\phi_i + \phi_{i-1}) = \frac{\Gamma_{xi+\frac{1}{2}}(\phi_{i+1} - \phi_i)}{\Delta x_{i+\frac{1}{2}}} - \frac{\Gamma_{xi-\frac{1}{2}}(\phi_i - \phi_{i-1})}{\Delta x_{i-\frac{1}{2}}}$$

Rearranging and grouping terms, we get the discretization equation

$$a_i\phi_i = a_{i+1}\phi_{i+1} + a_{i-1}\phi_{i-1} \tag{8.5a}$$

where

$$a_{i+1} = \frac{\Gamma_{xi+\frac{1}{2}}}{\Delta x_{i+\frac{1}{2}}} - \frac{1}{2}(\rho u)_{i+\frac{1}{2}} = \frac{\Gamma_{xi+\frac{1}{2}}}{\Delta x_{i+\frac{1}{2}}}\left(1 - \frac{P_{i+\frac{1}{2}}}{2}\right) \tag{8.5b}$$

$$a_{i-1} = \frac{\Gamma_{xi-\frac{1}{2}}}{\Delta x_{i-\frac{1}{2}}} + \frac{1}{2}(\rho u)_{i-\frac{1}{2}} = \frac{\Gamma_{xi-\frac{1}{2}}}{\Delta x_{i-\frac{1}{2}}}\left(1 + \frac{P_{i-\frac{1}{2}}}{2}\right) \tag{8.5c}$$

$$a_i = a_{i+1} + a_{i-1} + \left\{(\rho u)_{i+\frac{1}{2}} - (\rho u)_{i-\frac{1}{2}}\right\} \tag{8.5d}$$

$$P = \frac{(\rho u)}{\Gamma/\Delta x} \tag{8.5e}$$

Application of this three-node ***central difference scheme*** to all internal nodes leads to a tridiagonal system of linear algebraic equations. An inspection of the coefficient terms $\boldsymbol{a}$ in this scheme reveals that the condition for the stable and converging solution of the linear system, i.e., $a_i \geq (|a_{i+1}| + |a_{i-1}|)$, may not always be satisfied because the net mass velocity term $\left\{(\rho u)_{i+\frac{1}{2}} - (\rho u)_{i-\frac{1}{2}}\right\}$ may become negative during computations. Under this situation, this scheme becomes computationally unstable. A central difference scheme works well for problems where diffusion dominates over convection, i.e., for low values of $P = \rho u/(\Gamma/\Delta x)$ or for low control volume Reynolds number, $\mathrm{Re}_{\Delta x} = ((\rho u \Delta x)/\mu)$.

8.1.2 Upwind Scheme

In the upwind scheme, the dependent variable at the interface of a control volume is assumed to be equal to the upstream or upwind side nodal values for positive values of the mass velocity at the interface and equal to the downstream-side nodal value for negative values of mass velocity at the interface. This is expressed as

$$\phi_{i+\frac{1}{2}} = \phi_i \quad \text{for } (\rho u)_{i+\frac{1}{2}} > 0 \tag{8.6a}$$

and

$$\phi_{i+\frac{1}{2}} = \phi_{i+1} \quad \text{for } (\rho u)_{i+\frac{1}{2}} < 0 \tag{8.6b}$$

Similarly

$$\phi_{i-\frac{1}{2}} = \phi_{i-1} \quad \text{for } (\rho u)_{i-\frac{1}{2}} > 0 \tag{8.6c}$$

and

$$\phi_{i-\frac{1}{2}} = \phi_i \quad \text{for } (\rho u)_{i-\frac{1}{2}} < 0 \tag{8.6d}$$

Using a compact notation given by Patankar (1980), we expressed the upwind scheme as

$$(\rho u)_{i+\frac{1}{2}}\phi_{i+\frac{1}{2}} = \phi_i\left\langle\left((\rho u)_{i+\frac{1}{2}}\right),\ 0\right\rangle - \phi_{i+1}\left\langle\left(-\ (\rho u)_{i+\frac{1}{2}}\right),\ 0\right\rangle \tag{8.7a}$$

and

$$(\rho u)_{i-\frac{1}{2}}\phi_{i-\frac{1}{2}} = \phi_{i-1}\left\langle\left((\rho u)_{i-\frac{1}{2}}\right),\ 0\right\rangle - \phi_i\left\langle\left(-\ (\rho u)_{i-\frac{1}{2}}\right),\ 0\right\rangle \tag{8.7b}$$

where the ***operator*** $\langle a,\ b\rangle$ represents the maximum of ***a*** and ***b***.

Substituting the upwind scheme given by Equation 8.7 into Equation 8.3, we get

$$\begin{aligned}\phi_i\left\langle\left((\rho u)_{i+\frac{1}{2}}\right),\ 0\right\rangle - \phi_{i+1}\left\langle\left((-\rho u)_{i+\frac{1}{2}}\right),\ 0\right\rangle - \phi_{i-1}\left\langle\left((\rho u)_{i-\frac{1}{2}}\right),\ 0\right\rangle - \phi_i\left\langle\left(-\ (\rho u)_{i-\frac{1}{2}}\right),\ 0\right\rangle \\ = \frac{\Gamma_{xi+\frac{1}{2}}(\phi_{i+1} - \phi_i)}{\Delta x_{i+\frac{1}{2}}} - \frac{\Gamma_{xi-\frac{1}{2}}(\phi_i - \phi_{i-1})}{\Delta x_{i-\frac{1}{2}}}\end{aligned} \tag{8.8}$$

Simplifying and rearranging, we get

$$a_i\phi_i = a_{i+1}\phi_{i+1} + a_{i-1}\phi_{i-1} \tag{8.9}$$

where

$$a_{i+1} = \frac{\Gamma_{xi+\frac{1}{2}}}{\Delta x_{i+\frac{1}{2}}} + \left\langle -(\rho u)_{i+\frac{1}{2}},\ 0 \right\rangle$$

$$a_{i-1} = \frac{\Gamma_{xi-\frac{1}{2}}}{\Delta x_{i-\frac{1}{2}}} + \left\langle (\rho u)_{i-\frac{1}{2}},\ 0 \right\rangle$$

$$a_i = a_{i+1} + a_{i-1} + (\rho u)_{i+\frac{1}{2}} - (\rho u)_{i-\frac{1}{2}}$$

It can be noted from these expressions for the coefficents ***a*** that they do not become negative during computation, thus ensuring stability.

8.1.3 Exponential Scheme

In the exponential scheme, the convection and diffusion terms are combined and expressed as a single flux term as

$$j = \rho u \phi - \Gamma \frac{d\phi}{dx} \tag{8.10}$$

Substituting Equation 8.10 into Equation 8.1, we get

$$\frac{dJ}{dx} = 0 \tag{8.11}$$

Integrating Equation 8.11 over the control volume, we get

$$J_{i+\frac{1}{2}} - J_{i-\frac{1}{2}} = 0 \tag{8.12}$$

Considering the analytical solution for the one-dimensional problem given by Equation 8.1 in a plane slab with boundary conditions of the first kind, the combined ***J*** flux is approximated as

$$J_{i+\frac{1}{2}} = (\rho u)_{i+\frac{1}{2}} \left(\phi_i + \frac{\phi_i - \phi_{i+1}}{\exp\left(P_{i+\frac{1}{2}}\right) - 1} \right) \tag{8.13a}$$

where

$$P_{i+\frac{1}{2}} = \frac{(\rho u)_{i+\frac{1}{2}}}{\Gamma_{i+\frac{1}{2}} / \Delta x_{i+\frac{1}{2}}} \tag{8.13b}$$

and

$$J_{i-\frac{1}{2}} = (\rho u)_{i-\frac{1}{2}} \left(\phi_{i-1} + \frac{\phi_{i-1} - \phi_i}{\exp\left(P_{i-\frac{1}{2}}\right) - 1} \right) \tag{8.14a}$$

where

$$P_{i-\frac{1}{2}} = \frac{(\rho u)_{i-\frac{1}{2}}}{\Gamma_{i-\frac{1}{2}}/\Delta x_{i-\frac{1}{2}}} \tag{8.14b}$$

Substituting Equations 8.13 and 8.14 into Equation 8.12, we get

$$(\rho u)_{i+\frac{1}{2}}\left(\phi_i + \frac{\phi_i - \phi_{i+1}}{\exp\left(P_{i+\frac{1}{2}}\right) - 1}\right) - (\rho u)_{i-\frac{1}{2}}\left(\phi_{i-1} + \frac{\phi_{i-1} - \phi_i}{\exp\left(P_{i-\frac{1}{2}}\right) - 1}\right) = 0 \tag{8.15a}$$

Rearranging

$$a_i\phi_i = a_{i+1}\phi_{i+1} + a_{i-1}\phi_{i-1} \tag{8.15b}$$

where

$$a_{i+1} = \frac{(\rho u)_{i+\frac{1}{2}}}{\exp\left(P_{i+\frac{1}{2}}\right) - 1} = \frac{\Gamma_{x_{i+\frac{1}{2}}}}{\Delta x_{i+\frac{1}{2}}} \frac{P_{i+\frac{1}{2}}}{\exp\left(P_{i+\frac{1}{2}}\right) - 1}$$

$$a_{i-1} = \frac{(\rho u)_{i-\frac{1}{2}} \exp\left(P_{i-\frac{1}{2}}\right)}{\exp\left(P_{i-\frac{1}{2}}\right) - 1} = \frac{\Gamma_{x_{i-\frac{1}{2}}}}{\Delta x_{i-\frac{1}{2}}} \frac{P_{i-\frac{1}{2}} \exp\left(P_{i-\frac{1}{2}}\right)}{\exp\left(P_{i-\frac{1}{2}}\right) - 1}$$

$$a_i = a_{i+1} + a_{i-1} + \left\{(\rho u)_{i+\frac{1}{2}} - (\rho u)_{i-\frac{1}{2}}\right\}$$

The main disadvantage of this scheme is that it involves exponential functions, which are computationally more expensive and time consuming, particularly for multi-dimensional problems involving a large system of equations.

8.1.4 Hybrid Scheme

The hybrid scheme is a combination of the central difference scheme and the upwind scheme. The exponential form of the analytical solution is approximated by three straight lines. It is identical with the central difference scheme in the mid range for $-2 \leq P_{i+\frac{1}{2}} < 2$, and outside this range it reduces to the upwind scheme with zero diffusion as

for $P_{i+\frac{1}{2}} < -2$,
$$a_{i+1} = -\frac{\Gamma_{x_{i+\frac{1}{2}}}}{\Delta x_{i+\frac{1}{2}}}\left(P_{i+\frac{1}{2}}\right) \tag{8.16a}$$

for $-2 \leq P_{i+\frac{1}{2}} < 2$,
$$a_{i+1} = \frac{\Gamma_{x_{i+\frac{1}{2}}}}{\Delta x_{i+\frac{1}{2}}}\left(1 - \frac{P_{i+\frac{1}{2}}}{2}\right) \tag{8.16b}$$

for $P_{i+\frac{1}{2}} > 2$,
$$a_{i+1} = 0 \tag{8.16c}$$

In compact notation, Equation 8.16 is written as

$$
\begin{aligned}
a_{i+1} &= \frac{\Gamma_{x_{i+\frac{1}{2}}}}{\Delta x_{i+\frac{1}{2}}}\left\langle -P_{i+\frac{1}{2}},\ 1-\frac{P_{i+\frac{1}{2}}}{2},\ 0\right\rangle \\
&= \frac{\Gamma_{x_{i+\frac{1}{2}}}}{\Delta x_{i+\frac{1}{2}}}\left\langle 1-\frac{P_{i+\frac{1}{2}}}{2},\ 0\right\rangle + \left\langle 0,\ -(\rho u)_{i+\frac{1}{2}}\right\rangle
\end{aligned}
\tag{8.17a}
$$

Similarly

$$
\begin{aligned}
a_{i-1} &= \frac{\Gamma_{x_{i-\frac{1}{2}}}}{\Delta x_{i-\frac{1}{2}}}\left\langle P_{i-\frac{1}{2}},\ 1+\frac{P_{i-\frac{1}{2}}}{2},\ 0\right\rangle \\
&= \frac{\Gamma_{x_{i+\frac{1}{2}}}}{\Delta x_{i+\frac{1}{2}}}\left\langle 1+\frac{P_{i-\frac{1}{2}}}{2},\ 0\right\rangle + \left\langle 0,\ (\rho u)_{i-\frac{1}{2}}\right\rangle
\end{aligned}
\tag{8.17b}
$$

Based on this approximation, the discretization scheme for the convection-diffusion equation given by Equation 8.1 can be derived as

$$a_i\phi_i = a_{i+1}\phi_{i+1} + a_{i-1}\phi_{i-1} \tag{8.18}$$

where

$$
\begin{aligned}
a_{i+1} &= \frac{\Gamma_{x_{i+\frac{1}{2}}}}{\Delta x_{i+\frac{1}{2}}}\left\langle 1-\frac{P_{i+\frac{1}{2}}}{2},\ 0\right\rangle + \left\langle 0,\ -(\rho u)_{i+\frac{1}{2}}\right\rangle \\
a_{i-1} &= \frac{\Gamma_{x_{i-\frac{1}{2}}}}{\Delta x_{i-\frac{1}{2}}}\left\langle 1+\frac{P_{i-\frac{1}{2}}}{2},\ 0\right\rangle + \left\langle 0,\ (\rho u)_{i-\frac{1}{2}}\right\rangle \\
a_i &= a_{i+1} + a_{i-1} + \left\{(\rho u)_{i+\frac{1}{2}} - (\rho u)_{i-\frac{1}{2}}\right\}
\end{aligned}
$$

8.1.5 Power Law Scheme

In the power law scheme, the exponential function is approximated by straight lines and power law profiles, which are computationally less expensive in the following way

for $P_{i+\frac{1}{2}} < -10$,
$$a_{i+1} = -\frac{\Gamma_{x_{i+\frac{1}{2}}}}{\Delta x_{i+\frac{1}{2}}}\left(P_{i+\frac{1}{2}}\right) \tag{8.19a}$$

for $-10 \le P_{i+\frac{1}{2}} < 0$,
$$a_{i+1} = \frac{\Gamma_{x_{i+\frac{1}{2}}}}{\Delta x_{i+\frac{1}{2}}}\left\{\left(1+0.1P_{i+\frac{1}{2}}\right)^5 - P_{i+\frac{1}{2}}\right\} \tag{8.19b}$$

for $0 \le P_{i+\frac{1}{2}} < 10,$ $$a_{i+1} = \frac{\Gamma_{x_{i+\frac{1}{2}}}}{\Delta x_{i+\frac{1}{2}}}\left(1 - 0.1P_{i+\frac{1}{2}}\right)^5 \tag{8.19c}$$

for $P_{i+\frac{1}{2}} > 10,$ $$a_{i+1} = 0 \tag{8.19d}$$

Equation 8.19 can also be written in the compact form as

$$a_{i+1} = \frac{\Gamma_{x_{i+\frac{1}{2}}}}{\Delta x_{i+\frac{1}{2}}}\left\langle 0,\ \left(1 - 0.1\left|P_{i+\frac{1}{2}}\right|\right)^5\right\rangle + \frac{\Gamma_{x_{i+\frac{1}{2}}}}{\Delta x_{i+\frac{1}{2}}}\left\langle 0,\ -P_{i+\frac{1}{2}}\right\rangle$$

or

$$a_{i+1} = \frac{\Gamma_{x_{i+\frac{1}{2}}}}{\Delta x_{i+\frac{1}{2}}}\left\langle 0,\ \left(1 - 0.1\left|P_{i+\frac{1}{2}}\right|\right)^5\right\rangle + \left\langle 0,\ -(\rho u)_{i+\frac{1}{2}}\right\rangle \tag{8.20a}$$

Similarly

$$a_{i-1} = \frac{\Gamma_{x_{i-\frac{1}{2}}}}{\Delta x_{i-\frac{1}{2}}}\left\langle 0,\ \left(1 - 0.1\left|P_{i-\frac{1}{2}}\right|\right)^5\right\rangle + \frac{\Gamma_{x_{i-\frac{1}{2}}}}{\Delta x_{i-\frac{1}{2}}}\left\langle 0,\ P_{i-\frac{1}{2}}\right\rangle$$

or

$$a_{i-1} = \frac{\Gamma_{x_{i-\frac{1}{2}}}}{\Delta x_{i-\frac{1}{2}}}\left\langle 0,\ \left(1 - 0.1\left|P_{i-\frac{1}{2}}\right|\right)^5\right\rangle + \left\langle 0,\ (\rho u)_{i-\frac{1}{2}}\right\rangle \tag{8.20b}$$

The discretized form of Equation 8.1 based on the power law scheme is given as

$$a_i\phi_i = a_{i+1}\phi_{i+1} + a_{i-1}\phi_{i-1} \tag{8.21}$$

where

$$a_{i+1} = \frac{\Gamma_{x_{i+\frac{1}{2}}}}{\Delta x_{i+\frac{1}{2}}}\left\langle 0,\ \left(1 - 0.1\left|P_{i+\frac{1}{2}}\right|\right)^5\right\rangle + \left\langle 0,\ -\ (\rho u)_{i+\frac{1}{2}}\right\rangle$$

$$a_{i-1} = \frac{\Gamma_{x_{i-\frac{1}{2}}}}{\Delta x_{i-\frac{1}{2}}}\left\langle 0,\ \left(1 - 0.1\left|P_{i-\frac{1}{2}}\right|\right)^5\right\rangle + \left\langle 0,\ (\rho u)_{i-\frac{1}{2}}\right\rangle$$

$$a_i = a_{i+1} + a_{i-1} + \left\{(\rho u)_{i+\frac{1}{2}} - (\rho u)_{i-\frac{1}{2}}\right\}$$

Table 8.1 Expression for the Function $A|P|$ in Different Discretization Schemes for Convection–Diffusion

Approximation Scheme	Function $A\|P\|$
Central difference	$1 - \frac{\|P\|}{2}$
Upwind	1
Hybrid	$\left\langle 0, 1 - \frac{\|P\|}{2} \right\rangle$
Exponential	$\left\langle \frac{\|P\|}{\exp(\|P\|)} - 1 \right\rangle$
Power	$\langle 0, (1 - 0.1\|P\|)^5 \rangle$

8.1.6 Generalized Convection–Diffusion Scheme

A generalized form of all approximation schemes for the convection–diffusion terms is given by Patankar (1980) as

$$a_i\phi_i = a_{i+1}\phi_{i+1} + a_{i-1}\phi_{i-1} \tag{8.22}$$

where

$$a_{i+1} = \frac{\Gamma_{x_{i+\frac{1}{2}}}}{\Delta x_{i+\frac{1}{2}}} A\left|P_{i+\frac{1}{2}}\right| + \left\langle 0, \; -(\rho u)_{i+\frac{1}{2}} \right\rangle$$

$$a_{i-1} = \frac{\Gamma_{x_{i-\frac{1}{2}}}}{\Delta x_{i-\frac{1}{2}}} A\left|P_{i-\frac{1}{2}}\right| + \left\langle 0, \; (\rho u)_{i-\frac{1}{2}} \right\rangle$$

$$a_i = a_{i+1} + a_{i-1} + \left\{ (\rho u)_{i+\frac{1}{2}} - (\rho u)_{i-\frac{1}{2}} \right\}$$

It can be seen that the only difference among difference approximation schemes is the expression for the function $A|P|$, which is listed in Table 8.1.

8.2 DISCRETIZATION OF A GENERAL TRANSPORT EQUATION

Let us now consider an unsteady state problem consisting of the continuity or conservation of a mass equation and a transport equation, which represents a general form for the momentum, energy, and other transport equations.

8.2.1 One-dimensional Unsteady State Problems

Let us now consider a one-dimensional unsteady state problem with mass conservation and transport equation including a linearized source term that is described by the following governing differential equation.

Mass conservation

$$\frac{\partial \rho}{\partial t} + \frac{\partial (\rho u)}{\partial x} = 0 \tag{8.23a}$$

Transport equations

$$\frac{\partial (\rho \phi)}{\partial t} + \frac{\partial}{\partial x}(\rho u \phi) = \frac{\partial}{\partial x}\left(\Gamma_x \frac{\partial \phi}{\partial x}\right) + (S_0 + S_1 \phi) \tag{8.23b}$$

With the representation of the convection and the diffusion terms in the form of a total flux given by Equation 8.10, then Equation (8.23b) reduces to

$$\frac{\partial (\rho \phi)}{\partial t} + \frac{\partial J_x}{\partial x} = (S_0 + S_1 \phi) \tag{8.24}$$

Integrating the transport equation (Equation 8.24) over the control volume (Figure 8.1) from $i - \frac{1}{2}$ to $i + \frac{1}{2}$ and over the time interval t to $t + \Delta t$, we get

$$\int_{i-\frac{1}{2}}^{i+\frac{1}{2}} \int_{t}^{t+\Delta t} \frac{\partial (\rho \phi)}{\partial t}\, dx\, dt + \int_{i-\frac{1}{2}}^{i+\frac{1}{2}} \int_{t}^{t+\Delta t} \frac{\partial J_x}{\partial x}\, dx\, dt = \int_{t}^{t+\Delta t} \int_{i-\frac{1}{2}}^{i+\frac{1}{2}} (S_0 + S_1 \phi)\, dx\, dt$$

Using the ***implicit time integration scheme***

$$(\rho_i \phi_i)\Big|_{t}^{t+\Delta t} \Delta x + J_x^{m+1}\Big|_{i-\frac{1}{2}}^{i+\frac{1}{2}} \Delta t = S_0\, \Delta x\, \Delta t + S_1 \phi_i^{m+1}\, \Delta x\, \Delta t$$

or

$$\frac{\rho_i^{m+1} \phi_i^{m+1} - \rho_i^m \phi_i^m}{\Delta t} \Delta x + J_{x_{i+\frac{1}{2}}}^{m+1} - J_{x_{i-\frac{1}{2}}}^{m+1} = S_0 \Delta x + S_1 \phi_i^{m+1} \Delta x \tag{8.25}$$

where the superscript $\boldsymbol{m} + 1$ represents the new values at time $t + \Delta t$ and $\boldsymbol{m}$ represents the old value at time t.

In a similar manner, we integrate the mass conservation equation over the control volume as

$$\int_{i-\frac{1}{2}}^{i+\frac{1}{2}} \int_{t}^{t+\Delta t} \frac{\partial(\rho)}{\partial t}\, dx\, dt + \int_{i-\frac{1}{2}}^{i+\frac{1}{2}} \int_{t}^{t+\Delta t} \frac{\partial(\rho u)}{\partial x}\, dx\, dt = 0$$

or

$$\frac{\rho_i^{m+1} - \rho_i^m}{\Delta t}\Delta x + (\rho u)_{i+\frac{1}{2}}^{m+1} - (\rho u)_{i-\frac{1}{2}}^{m+1} = 0 \tag{8.26}$$

Multiplying Equation 8.26 by ϕ_i^{m+1} and subtracting from Equation 8.25, we get

$$\left(\phi_i^{m+1} - \phi_i^m\right)\frac{\rho_i^m \Delta x}{\Delta t} + \left(J_{x_{i+\frac{1}{2}}}^{m+1} - (\rho u)_{i+\frac{1}{2}}^{m+1}\phi_i^{m+1}\right) - \left(J_{x_{i-\frac{1}{2}}}^{m+1} - (\rho u)_{i-\frac{1}{2}}^{m+1}\phi_i^{m+1}\right)$$
$$= S_0 \Delta x + S_1 \phi_i^{m+1} \Delta x \tag{8.27}$$

Let us now use the approximations (Patankar [1980])

$$\left(J_{x_{i+\frac{1}{2}}}^{m+1} - (\rho u)_{i+\frac{1}{2}}^{m+1}\phi_i^{m+1}\right) = a_{i+1}\left(\phi_i^{m+1} - \phi_{i+1}^{m+1}\right) \tag{8.28a}$$

and

$$\left(J_{x_{i-\frac{1}{2}}}^{m+1} - (\rho u)_{i-\frac{1}{2}}^{m+1}\phi_i^{m+1}\right) = a_{i-1}\left(\phi_{i-1}^{m+1} - \phi_i^{m+1}\right) \tag{8.28b}$$

where a_{i+1} and a_{i-1} are given by Equation 8.25 as

$$a_{i+1} = \frac{\Gamma_{x_{i+\frac{1}{2}}}}{\Delta x_{i+\frac{1}{2}}} A\left|P_{i+\frac{1}{2}}\right| + \left\langle 0,\ -(\rho u)_{i+\frac{1}{2}}\right\rangle \tag{8.25b}$$

$$a_{i-1} = \frac{\Gamma_{x_{i-\frac{1}{2}}}}{\Delta x_{i-\frac{1}{2}}} A\left|P_{i-\frac{1}{2}}\right| + \left\langle 0,\ (\rho u)_{i-\frac{1}{2}}\right\rangle \tag{8.25c}$$

Substituting Equation 8.28 into Equation 8.27, we get

$$a_i \phi_i^{m+1} = a_{i+1}\phi_{i+1}^{m+1} + a_{i-1}^{m+1}\phi_{i-1}^{m+1} + d \tag{8.29}$$

where

$$a_i = a_{i+1} + a_{i-1} + a_i^m - S_1 \Delta x$$

$$a_{i+1} = \frac{\Gamma_{x_{i+\frac{1}{2}}}}{\Delta x_{i+\frac{1}{2}}} A \left| P_{i+\frac{1}{2}} \right| + \left\langle 0, \; -(\rho u)_{i+\frac{1}{2}} \right\rangle$$

$$a_{i-1} = \frac{\Gamma_{x_{i-\frac{1}{2}}}}{\Delta x_{i-\frac{1}{2}}} A \left| P_{i-\frac{1}{2}} \right| + \left\langle 0, \; (\rho u)_{i-\frac{1}{2}} \right\rangle$$

$$a_i^m = \frac{\rho_i^m \Delta x}{\Delta t}$$

$$d = S_0 \Delta x + a_i^m \phi_i^m$$

Equation 8.29 represents an implicit form of the discretization equation for a general transport equation. This equation can now be applied to the x, y, and z components of the momentum equation, and to heat and mass transport equations.

8.2.2 Two-dimensional Unsteady State Problem

A two-dimensional unsteady state problem with mass conservation and a general form of transport equation is given as

Mass conservation

$$\frac{\partial \rho}{\partial t} + \frac{\partial (\rho u)}{\partial x} + \frac{\partial (\rho v)}{\partial y} = 0 \quad . \tag{8.30a}$$

Transport equations

$$\frac{\partial (\rho \phi)}{\partial t} + \frac{\partial}{\partial x}(\rho u \phi) + \frac{\partial}{\partial y}(\rho v \phi) = \frac{\partial}{\partial x}\left(\Gamma_x \frac{\partial \phi}{\partial x}\right) + \frac{\partial}{\partial y}\left(\Gamma_y \frac{\partial \phi}{\partial y}\right) + (S_0 + S_1 \phi) \tag{8.30b}$$

The implicit discretization of these equations is obtained by integrating over the two-dimensional control volume, shown in Figure 8.2, and expressed as

$$a_i \phi_i^{m+1} = a_{i+1} \phi_{i+1}^{m+1} + a_{i-1}^{m+1} \phi_{i-1}^{m+1} + a_{j+1} \phi_{j+1}^{m+1} + a_{j-1}^{m+1} \phi_{j-1}^{m+1} + d \tag{8.31}$$

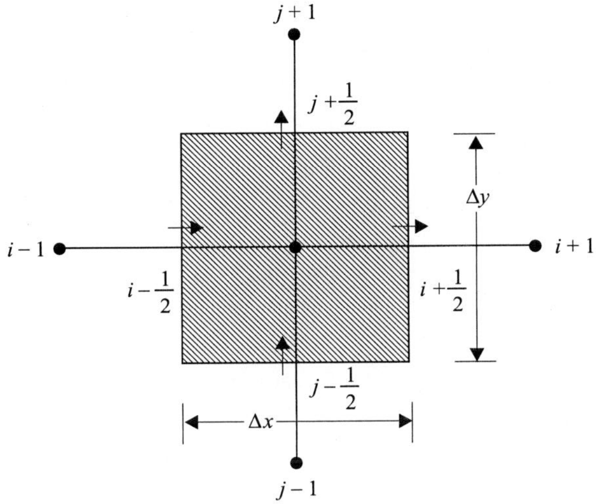

Figure 8.2 Two-dimensional control volume

where

$$a_i = a_{i+1} + a_{i-1} + a_{j+1} + a_{j-1} + a_i^m - S_1\,\Delta x\,\Delta y$$

$$a_{i+1} = \frac{\Gamma_{x_{i+\frac{1}{2}}}}{\Delta x_{i+\frac{1}{2}}} A\left|P_{i+\frac{1}{2}}\right| + \left\langle 0,\ -\ (\rho u)_{i+\frac{1}{2}}\right\rangle$$

$$a_{i-1} = \frac{\Gamma_{x_{i-\frac{1}{2}}}}{\Delta x_{i-\frac{1}{2}}} A\left|P_{i-\frac{1}{2}}\right| + \left\langle 0,\ (\rho u)_{i-\frac{1}{2}}\right\rangle$$

$$a_{j+1} = \frac{\Gamma_{y_{j+\frac{1}{2}}}}{\Delta y_{i+\frac{1}{2}}} A\left|P_{j+\frac{1}{2}}\right| + \left\langle 0,\ -\ (\rho v)_{j+\frac{1}{2}}\right\rangle$$

$$a_{j-1} = \frac{\Gamma_{y_{j-\frac{1}{2}}}}{\Delta y_{i-\frac{1}{2}}} A\left|P_{j-\frac{1}{2}}\right| + \left\langle 0,\ -\ (\rho v)_{j-\frac{1}{2}}\right\rangle$$

$$a_i^m = \frac{\rho_i^m \Delta x \Delta y}{\Delta t}$$

$$d = S_0 \Delta x \Delta y + a_i^m \phi_i^m$$

8.2.3 Three-dimensional Unsteady State Problem

A three-dimensional unsteady state problem with mass conservation and a general form of transport equation is given as

Mass conservation

$$\frac{\partial \rho}{\partial t} + \frac{\partial(\rho u)}{\partial x} + \frac{\partial(\rho v)}{\partial y} + \frac{\partial(\rho w)}{\partial z} = 0 \tag{8.32a}$$

Transport equations

$$\frac{\partial(\rho\phi)}{\partial t}+\frac{\partial}{\partial x}(\rho u\phi)+\frac{\partial}{\partial y}(\rho v\phi)+\frac{\partial}{\partial z}(\rho w\phi)$$
$$=\frac{\partial}{\partial x}\left(\Gamma_x\frac{\partial\phi}{\partial x}\right)+\frac{\partial}{\partial y}\left(\Gamma_y\frac{\partial\phi}{\partial y}\right)+\frac{\partial}{\partial z}\left(\Gamma_z\frac{\partial\phi}{\partial z}\right)+(S_0+S_1\phi) \quad (8.32b)$$

The implicit discretization of these equations leads to

$$a_i\phi_i^{m+1}=a_{i+1}\phi_{i+1}^{m+1}+a_{i-1}^{m+1}\phi_{i-1}^{m+1}+a_{j+1}\phi_{j+1}^{m+1}+a_{j-1}^{m+1}\phi_{j-1}^{m+1}$$
$$+a_{k+1}\phi_{k+1}^{m+1}+a_{k-1}^{m+1}\phi_{k-1}^{m+1}+c \quad (8.33a)$$

or

$$a_i\phi_i^{m+1}=\sum a_{nb}\phi_{nb}^{m+1}+d \quad (8.33b)$$

where

$$a_i=a_{i+1}+a_{i-1}+a_{j+1}+a_{j-1}+a_{k+1}+a_{k-1}+a_i^m-S_1\,\Delta x\,\Delta y\,\Delta z$$

$$a_{i+1}=\frac{\Gamma_{x_{i+\frac{1}{2}}}}{\Delta x_{i+\frac{1}{2}}}A\left|P_{i+\frac{1}{2}}\right|+\left\langle 0,\ -(\rho u)_{i+\frac{1}{2}}\right\rangle$$

$$a_{i-1}=\frac{\Gamma_{x_{i-\frac{1}{2}}}}{\Delta x_{i-\frac{1}{2}}}A\left|P_{i-\frac{1}{2}}\right|+\left\langle 0,\ (\rho u)_{i-\frac{1}{2}}\right\rangle$$

$$a_{j+1}=\frac{\Gamma_{y_{j+\frac{1}{2}}}}{\Delta y_{i+\frac{1}{2}}}A\left|P_{j+\frac{1}{2}}\right|+\left\langle 0,\ -(\rho v)_{j+\frac{1}{2}}\right\rangle$$

$$a_{j-1}=\frac{\Gamma_{y_{j-\frac{1}{2}}}}{\Delta y_{i-\frac{1}{2}}}A\left|P_{j-\frac{1}{2}}\right|+\left\langle 0,\ (\rho v)_{j-\frac{1}{2}}\right\rangle$$

$$a_{k+1}=\frac{\Gamma_{z_{k+\frac{1}{2}}}}{\Delta z_{k+\frac{1}{2}}}A\left|P_{k+\frac{1}{2}}\right|+\left\langle 0,\ -(\rho w)_{k+\frac{1}{2}}\right\rangle$$

$$a_{k-1}=\frac{\Gamma_{z_{k-\frac{1}{2}}}}{\Delta z_{k-\frac{1}{2}}}A\left|P_{k-\frac{1}{2}}\right|+\left\langle 0,\ (\rho w)_{k-\frac{1}{2}}\right\rangle$$

$$a_i^m=\frac{\rho_i^m\,\Delta x\,\Delta y\,\Delta z}{\Delta t}$$

$$d=S_0\Delta x\,\Delta y\,\Delta z+a_i^m\phi_i^m$$

8.3 SOLUTION OF FLOW FIELD

Solution of a flow field is obtained in terms of the velocity field $\vec{V} = \hat{i}u + \hat{j}v + \hat{k}w$ and pressure field $p(x, y, z)$, which are determined by solving continuity or conservation of mass and momentum. As we have discussed in Chapter 1, the conservation of mass and momentum for Newtonian viscous fluid flow is given by the Navier–Stokes equation

Continuity

$$\frac{\partial \rho}{\partial t} + \frac{\partial(\rho u)}{\partial x} + \frac{\partial(\rho v)}{\partial y} + \frac{\partial(\rho w)}{\partial z} = 0 \tag{8.34a}$$

x momentum

$$\rho\left(\frac{\partial u}{\partial t} + u\frac{\partial u}{\partial x} + v\frac{\partial u}{\partial y} + w\frac{\partial u}{\partial z}\right) = \rho g_x - \frac{\partial p}{\partial x} + \mu\left(\frac{\partial^2 u}{\partial x^2} + \frac{\partial^2 u}{\partial y^2} + \frac{\partial^2 u}{\partial z^2}\right) \tag{8.34b}$$

y momentum

$$\rho\left(\frac{\partial v}{\partial t} + u\frac{\partial v}{\partial x} + v\frac{\partial v}{\partial y} + w\frac{\partial v}{\partial z}\right) = \rho g_y - \frac{\partial p}{\partial y} + \mu\left(\frac{\partial^2 v}{\partial x^2} + \frac{\partial^2 v}{\partial y^2} + \frac{\partial^2 v}{\partial z^2}\right) \tag{8.34c}$$

z momentum

$$\rho\left(\frac{\partial w}{\partial t} + u\frac{\partial w}{\partial x} + v\frac{\partial w}{\partial y} + w\frac{\partial w}{\partial z}\right) = \rho g_z - \frac{\partial p}{\partial z} + \mu\left(\frac{\partial^2 w}{\partial x^2} + \frac{\partial^2 w}{\partial y^2} + \frac{\partial^2 w}{\partial z^2}\right) \tag{8.34d}$$

This set of coupled partial differential equations is nonlinear due to the presence of convective terms on the left-hand side of Equations 8.34b and 8.34c, and the equations are solved in an iterative manner. The major difficulty is that there is no equation that we can solve directly for the pressure. There are two major approaches for solving the flow field: (1) ***the direct solution involving the primitive variable***, (2) ***the stream function/vorticity-based method***.

In the direct solution method involving the primitive variables, the three momentum and continuity equations are solved directly for three velocity components u, v and w with a pre-assumed pressure field. The iteration process is continued with an updated pressure field. The correct pressure field is established when the estimated velocity components satisfy the continuity equation (8.34a). In the subsequent sections, we will present the development of a pressure equation by using this indirect information of pressure in the continuity equation. This procedure of solving the flow field directly in terms of four primitive variables, three velocity components and pressure, poses considerable complexity and computational difficulties.

8.3.1 Stream Function–Vorticity-based Method

The vorticity/stream function-based method is primarily used for two-dimensional problems. The set of equations for the two-dimensional incompressible fluid flow with constant viscosity is deduced from Equation 8.34 as

Continuity

$$\frac{\partial u}{\partial x} + \frac{\partial v}{\partial y} = 0 \tag{8.35a}$$

x momentum

$$\rho\left(\frac{\partial u}{\partial t} + u\frac{\partial u}{\partial x} + v\frac{\partial u}{\partial y}\right) = \rho g_x - \frac{\partial P}{\partial x} + \mu\left(\frac{\partial^2 u}{\partial x^2} + \frac{\partial^2 u}{\partial y^2}\right) \tag{8.35b}$$

y momentum

$$\rho\left(\frac{\partial v}{\partial t} + u\frac{\partial v}{\partial x} + v\frac{\partial v}{\partial y}\right) = \rho g_y - \frac{\partial P}{\partial y} + \mu\left(\frac{\partial^2 v}{\partial x^2} + \frac{\partial^2 v}{\partial y^2}\right) \tag{8.35c}$$

In order to transform the two momentum equations for u and v into a single equation we introduce ***vorticity*** as the ***curl*** of velocity vector or $\zeta = \nabla \times \vec{V}$, which reduces to the z-component vorticity as

$$\zeta = \zeta_z = \frac{\partial u}{\partial y} - \frac{\partial v}{\partial x} \tag{8.36}$$

Now by taking a ***curl*** of the momentum equations (Schlichting, 1955; Yuan, 1967) and eliminating pressure based on the z-component vorticity we have

$$\rho\left(\frac{\partial \zeta}{\partial t} + u\frac{\partial \zeta}{\partial x} + v\frac{\partial \zeta}{\partial y}\right) = \mu\left(\frac{\partial^2 \zeta}{\partial x^2} + \frac{\partial^2 \zeta}{\partial y^2}\right) \tag{8.37}$$

Equation 8.37 is referred to as the ***vorticity transport equation***, which is similar to the general transport Equation 8.28 considered in an earlier section. The terms on the left-hand side of the equation represent the local and convective variation of the vorticity, and the terms on the right-hand side represent the diffusion dissipation of vorticity. It can be noted that the pressure no longer appears in the equation as normal forces do not have any effect on the angular rotation of the fluid. The vorticity transport Equation 8.37, along with the continuity Equation 8.35a, forms a set of two equations for two unknowns u and v without the need for a third equation for the pressure. The numerical approximation schemes for convective and diffusion terms and the solution procedure described for the two-dimensional transport equation in Section 8.2.2 is also applicable for the solution of this set.

The number of equations can further be reduced by introducing the stream function, ψ, as

$$u = \frac{\partial \psi}{\partial y} \quad \text{and} \quad v = -\frac{\partial \psi}{\partial x} \tag{8.38}$$

This satisfies the continuity Equation 8.35a automatically, and transforms the vorticity given by Equation 8.36 and vorticity transport Equation 8.37 into

$$\frac{\partial^2 \psi}{\partial x^2} + \frac{\partial^2 \psi}{\partial y^2} = \zeta \tag{8.39}$$

and

$$\rho \left\{ \frac{\partial (\nabla^2 \psi)}{\partial t} + \frac{\partial \psi}{\partial y} \frac{\partial (\nabla^2 \psi)}{\partial x} - \frac{\partial \psi}{\partial x} \frac{\partial (\nabla^2 \psi)}{\partial y} \right\} = \mu \Delta^4 \psi \tag{8.40}$$

In this form, the vorticity transport equation contains only one unknown variable, ψ, but has the form of a nonlinear fourth-order differential equation. Equation 8.40, referred to as the ***biharmonic–stream function formulation***, can be solved directly by applying finite difference formulas for the derivatives. Such a formulation and solution is given in detail by Shih (1984).

A major advantage of this vorticity-based method is that there is no need to solve for the pressure. This is, however, also a disadvantage in some class of problems where it is essential to know the pressure distribution. Since the stream function can only be defined for two dimensions, the vorticity-based method is primarily restricted to two-dimensional problems. It will pose considerable difficulties if it were to be extended to three-dimensions where it involves solving a set of six differential equations: three for the three components of vorticity vector, $\varsigma = \hat{i}\varsigma_x + \hat{j}\varsigma_y + \hat{k}\varsigma_z$, and velocity potentials, ϕ. Numerical solution methods for the vorticity/stream function-based method are discussed by Jaluria and Torrance (1986) and Roaches (1982). As we have previously mentioned, for problems where estimation of pressure is essential and for three-dimensional problems, it is more appropriate to solve continuity and momentum equations directly for the primitive variables. In fact, in recent times this approach has become increasingly popular. In the following sections we will focus on the direct solution method involving the primitive variables, i.e., three velocity components u, v, w, and pressure, p.

8.3.2 Direct Solution with the Primitive Variable

We have previously mentioned that, in the direct solution method involving the primitive variable, the three momentum equations are solved directly for three velocity components u, v, and w, with an estimated value of the pressure field. The iteration process is continued with an updated pressure field. The correct pressure field is established when estimated velocity components satisfy continuity Equation 8.31a. There was considerable effort in developing solution algorithms based on this basic criterion. Among the most popular algorithms are the semi-implicit time algorithm such as ***SIMPLE*** (Patankar and Spalding, 1972; Patankar, 1980) and its variations, and the explicit time algorithm such as ***MAC*** (Harlow and Welch, 1965; Hirt and Cook, 1972; Majumdar and Deb, 2003). We will describe these two algorithms in the following sections.

In deriving the discretization equations for the momentum equations we note that the momentum Equations 8.34b to 8.34d are identical to the general form of the transport Equation 8.21b with the exception of the pressure derivative terms. The discretization schemes for convective and diffusion terms described for the general transport equation in Section 8.2.2 are also applicable for deriving the discretization equation for the momentum equation. However, before we proceed to the solution of the flow field, we need to present the discretization scheme for the pressure derivative terms in the momentum equation as well as the derivation of a separate equation for the pressure variable.

Discretization of Pressure Derivative. Let us consider the one-dimensional x-momentum equation by simply substituting $\phi = u$ and $\Gamma_x = \mu$ into Equation 8.26b, and including the pressure derivative term as

$$\frac{\partial(\rho u)}{\partial t} + \frac{\partial}{\partial x}(\rho uu) = -\frac{\partial p}{\partial x} + \frac{\partial}{\partial x}\left(\mu \frac{\partial u}{\partial x}\right) + (S_0 + S_1\phi) \tag{8.41}$$

Integration of the pressure gradient term over the control volume shown in Figure.8.1 gives

$$\int_{t}^{t+\Delta t}\int_{i-\frac{1}{2}}^{i+\frac{1}{2}} \frac{dP}{dx}dx = \left(P_{i+\frac{1}{2}}^{m+1} - P_{i-\frac{1}{2}}^{m+1}\right)\Delta t \tag{8.42}$$

Assuming a piece-wise linear profile for pressure at interfaces of the control volume, Equation 8.42 reduces to

$$\int_{t}^{t+\Delta t}\int_{i-\frac{1}{2}}^{i+\frac{1}{2}} \frac{dP}{dx}dx = \left(\frac{P_{i+1}^{m+1} + P_{i}^{m+1}}{2} - \frac{P_{i}^{m+1} + P_{i-1}^{m+1}}{2}\right)\Delta t = \frac{1}{2}\left(P_{i+1}^{m+1} - P_{i-1}^{m+1}\right)\Delta t \tag{8.43}$$

With the inclusion of this pressure derivative term in the momentum equation, velocity at the node $\boldsymbol{i}$ will be calculated based on Equation 8.29

$$a_i u_i^{m+1} = a_{i+1} u_{i+1}^{m+1} + a_{i-1}^{m+1} u_{i-1}^{m+1} + c - \frac{1}{2}(P_{i+1} - P_{i-1}) \tag{8.44a}$$

or

$$a_i u_i^{m+1} = \sum a_{nb} u_{nb}^{m+1} + d - \frac{1}{2}(P_{i+1} - P_{i-1}) \tag{8.44b}$$

We can see that the velocity component $\boldsymbol{u}$ at a node $\boldsymbol{i}$ is calculated based on difference of pressure values at two alternate nodes $\boldsymbol{i}-1$ and $\boldsymbol{i}+1$, and does not involve the pressure values at the node i. As a consequence, this velocity discretization equation may cause instability during the iteration process and may not converge or it may converge to unrealistic values. A detailed discussion of such a behavior is given by Patankar (1980). In order to avoid such unrealistic situations a recommended practice is to use a ***staggered grid***.

Staggered Grid. In a staggered grid the velocity components are calculated using a grid system that is staggered from the grid system that is being used for the estimation of pressure or any other scalar variables such as temperature or mass concentration. The nodal points of a staggered grid system, represented by broken lines and smaller bullets, lies at the faces of the control volumes for the regular grid as represented by solid lines and bigger bullets in Figure 8.3.

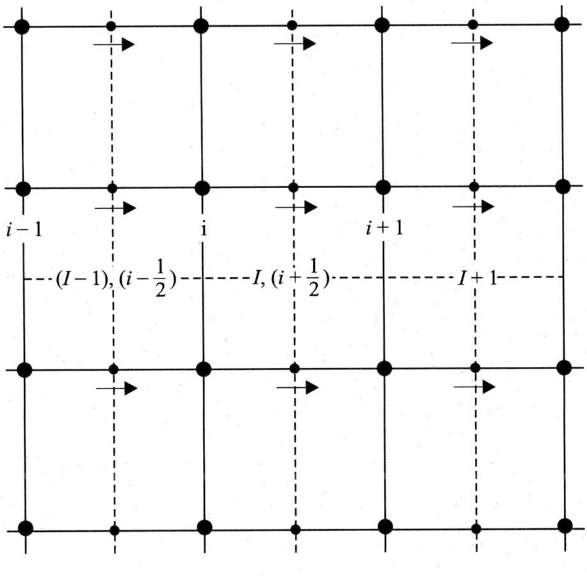

Figure 8.3 A staggered grid

Also, the nodal points in the staggered grid are designated by ***i***, ***i*** + 1 and ***i*** − 1 to differentiate from the regular grid (indicated by I, $I-1$, $I+1$) for the scalar quantities such as pressure. Using such a grid system and using Equation 8.42, the discretized equation for the x-component velocity is

$$a_i u_i^{m+1} = a_{i+1} u_{i+1}^{m+1} + a_{i-1}^{m+1} u_{i-1}^{m+1} + d - \left(P_{i+\frac{1}{2}}^{m+1} - P_{i-\frac{1}{2}}^{m+1}\right) \tag{8.45a}$$

or

$$a_i u_i^{m+1} = a_{i+1} u_{i+1}^{m+1} + a_{i-1}^{m+1} u_{i-1}^{m+1} + d - \left(P_I^{m+1} - P_{I-1}^{m+1}\right) \tag{8.45b}$$

Note that with this discretization equation, based on the staggered grid, the velocity at the faces of the regular grid is calculated based on the pressure difference between two adjacent nodal points of the regular grid. In a similar manner, this procedure is extended to derive y-component velocity and z-component velocity for two- and three-dimensional problems.

Discretization Equation for Velocity Component. The discretization equations for the three velocity components ***u***, ***v***, and ***w*** momentum can now be written in terms of guessed pressure values, following the procedures outlined in Section 8.2.1 for the general form of the transport equation and in Section 8.3.2.2 for the pressure derivative term, and using a staggered grid system for each of the velocity components. For the u-component velocity, we can consider a control volume in a staggered grid system as shown in Figure 8.4

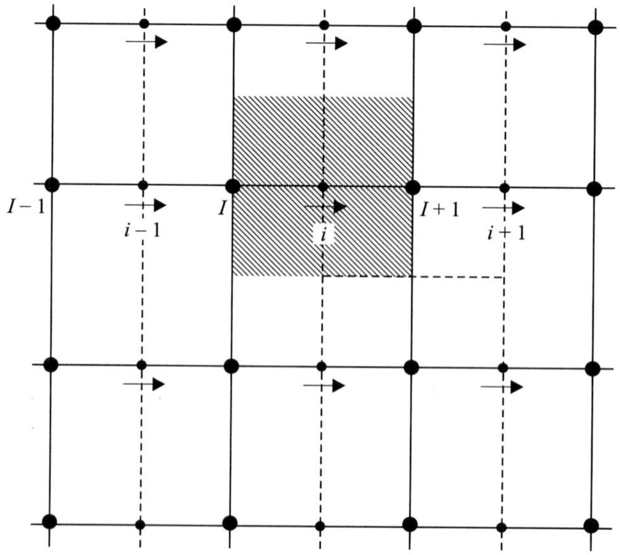

Figure 8.4 Control volume for u-component velocity

and discretize the x-momentum Equation 8.34b using Equation 8.32 and Equation 8.45 as

$$a_i u_i^{m+1} = a_{i+1} u_{i+1}^{m+1} + a_{i-1} u_{i-1}^{m+1} + a_{j+1} u_{j+1}^{m+1} + a_{j-1} u_{j-1}^{m+1} + a_{k+1} u_{k+1}^{m+1} + a_{k-1} u_{k-1}^{m+1} + d + (p_{I-1} - p_I)\Delta y\, \Delta z \tag{8.46a}$$

or

$$a_i u_i^{m+1} = \sum a_{nb} u_{nb}^{m+1} + d + (p_{I-1} - p_I)\, \Delta y\, \Delta z \tag{8.46b}$$

where

$$a_i = a_{i+1} + a_{i-1} + a_{j+1} + a_{j-1} + a_{k+1} + a_{k-1} + a_i^m \tag{8.47a}$$

$$a_{i+1} = \frac{\mu}{\Delta x_{i+\frac{1}{2}}} A\left|P_{i+\frac{1}{2}}\right| + \left\langle 0,\ -\ (\rho u)_{i+\frac{1}{2}}\right\rangle \tag{8.47b}$$

$$a_{i-1} = \frac{\mu}{\Delta x_{i-\frac{1}{2}}} A\left|P_{i-\frac{1}{2}}\right| + \left\langle 0,\ (\rho u)_{i-\frac{1}{2}}\right\rangle \tag{8.47c}$$

$$a_{j+1} = \frac{\mu}{\Delta y_{i+\frac{1}{2}}} A\left|P_{j+\frac{1}{2}}\right| + \left\langle 0,\ -\ (\rho v)_{j+\frac{1}{2}}\right\rangle \tag{8.47d}$$

$$a_{j-1} = \frac{\mu}{\Delta y_{i-\frac{1}{2}}} A\left|P_{j-\frac{1}{2}}\right| + \left\langle 0,\ (\rho v)_{j-\frac{1}{2}}\right\rangle \tag{8.47e}$$

$$a_{k+1} = \frac{\mu}{\Delta z_{k+\frac{1}{2}}} A\left|P_{k+\frac{1}{2}}\right| + \left\langle 0,\ -\ (\rho w)_{k+\frac{1}{2}}\right\rangle \tag{8.47f}$$

$$a_{k-1} = \frac{\mu}{\Delta z_{k-\frac{1}{2}}} A\left|P_{k-\frac{1}{2}}\right| + \left\langle 0,\ (\rho w)_{k-\frac{1}{2}}\right\rangle \tag{8.47g}$$

$$a_i^m = \frac{\rho\, \Delta x\, \Delta y\, \Delta z}{\Delta t} \tag{8.47h}$$

$$d = \rho g_x\, \Delta x\, \Delta y\, \Delta z + a_i^m u_i^m \tag{8.47i}$$

$$P = \frac{(\rho u)}{\mu/\Delta x} \tag{8.47j}$$

For the v-component velocity, we consider a control volume in a staggered grid system as shown in Figure 8.5 and discretize the y-momentum Equation 8.34c as

$$a_j v_j^{m+1} = a_{i+1} v_{i+1}^{m+1} + a_{i-1} v_{i-1}^{m+1} + a_{j+1} v_{j+1}^{m+1} + a_{j-1} v_{j-1}^{m+1} + a_{k+1} v_{k+1}^{m+1} + a_{k-1} v_{k-1}^{m+1} + d + (p_{J-1} - p_J)\, \Delta x\, \Delta z \tag{8.48a}$$

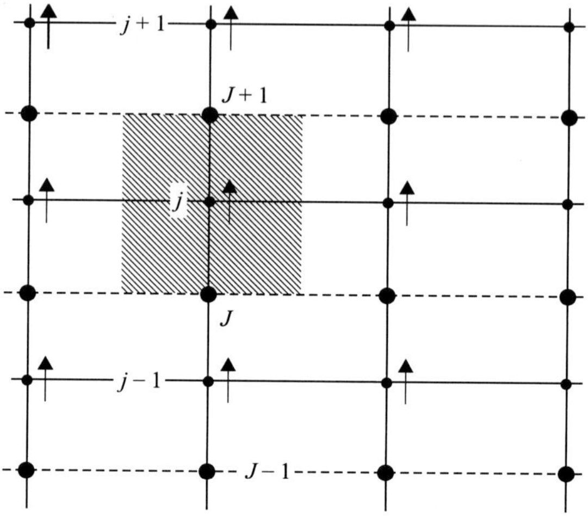

Figure 8.5 Control volume for v-component velocity

or

$$a_j v_j^{m+1} = \sum a_{nb} v_{nb}^{m+1} + d + (p_{J-1} - p_J)\,\Delta x\,\Delta z \tag{8.48b}$$

where

$$a_j = a_{i+1} + a_{i-1} + a_{j+1} + a_{j-1} + a_{k+1} + a_{k-1} + a_j^m \tag{8.49a}$$

$$a_{i+1} = \frac{\mu}{\Delta x_{i+\frac{1}{2}}} A\left|P_{i+\frac{1}{2}}\right| + \left\langle 0,\ -\ (\rho u)_{i+\frac{1}{2}}\right\rangle \tag{8.49b}$$

$$a_{i-1} = \frac{\mu}{\Delta x_{i-\frac{1}{2}}} A\left|P_{i-\frac{1}{2}}\right| + \left\langle 0,\ (\rho u)_{i-\frac{1}{2}}\right\rangle \tag{8.49c}$$

$$a_{j+1} = \frac{\mu}{\Delta y_{i+\frac{1}{2}}} A\left|P_{j+\frac{1}{2}}\right| + \left\langle 0,\ -\ (\rho v)_{j+\frac{1}{2}}\right\rangle \tag{8.49d}$$

$$a_{j-1} = \frac{\mu}{\Delta y_{i-\frac{1}{2}}} A\left|P_{j-\frac{1}{2}}\right| + \left\langle 0,\ (\rho v)_{j-\frac{1}{2}}\right\rangle \tag{8.49e}$$

$$a_{k+1} = \frac{\mu}{\Delta z_{k+\frac{1}{2}}} A\left|P_{k+\frac{1}{2}}\right| + \left\langle 0,\ -\ (\rho w)_{k+\frac{1}{2}}\right\rangle \tag{8.49f}$$

$$a_{k-1} = \frac{\mu}{\Delta z_{k-\frac{1}{2}}} A\left|P_{k-\frac{1}{2}}\right| + \left\langle 0,\ (\rho w)_{k-\frac{1}{2}}\right\rangle \tag{8.49g}$$

$$a_j^m = \frac{\rho\,\Delta x\,\Delta y\,\Delta z}{\Delta t} \tag{8.49h}$$

$$d = \rho g_y\,\Delta x\,\Delta y\,\Delta z + a_j^m v_j^m \tag{8.49i}$$

$$P = \frac{(\rho v)}{\mu/\Delta y} \tag{8.49j}$$

In a similar manner, the discretization equation for the w-component velocity is written as

$$a_k w_k^{m+1} = a_{i+1} w_{i+1}^{m+1} + a_{i-1} w_{i-1}^{m+1} + a_{j+1} w_{j+1}^{m+1} + a_{j-1} w_{j-1}^{m+1} + a_{k+1} w_{k+1}^{m+1} + a_{k-1} w_{k-1}^{m+1} + d + (p_{K-1} - p_K)\,\Delta x\,\Delta y \tag{8.50a}$$

or

$$a_k w_k^{m+1} = \sum a_{nb} w_{nb}^{m+1} + d + (p_{K-1} - p_K)\,\Delta x\,\Delta y \tag{8.50b}$$

where

$$a_k = a_{i+1} + a_{i-1} + a_{j+1} + a_{j-1} + a_{k+1} + a_{k-1} + a_k^m \tag{8.51a}$$

$$a_{i+1} = \frac{\mu}{\Delta x_{i+\frac{1}{2}}} A\left|P_{i+\frac{1}{2}}\right| + \left\langle 0,\ -\ (\rho u)_{i+\frac{1}{2}}\right\rangle \tag{8.51b}$$

$$a_{i-1} = \frac{\mu}{\Delta x_{i-\frac{1}{2}}} A\left|P_{i-\frac{1}{2}}\right| + \left\langle 0,\ (\rho u)_{i-\frac{1}{2}}\right\rangle \tag{8.51c}$$

$$a_{j+1} = \frac{\mu}{\Delta y_{i+\frac{1}{2}}} A\left|P_{j+\frac{1}{2}}\right| + \left\langle 0,\ -\ (\rho v)_{j+\frac{1}{2}}\right\rangle \tag{8.51d}$$

$$a_{j-1} = \frac{\mu}{\Delta y_{i-\frac{1}{2}}} A\left|P_{j-\frac{1}{2}}\right| + \left\langle 0,\ (\rho v)_{j-\frac{1}{2}}\right\rangle \tag{8.51e}$$

$$a_{k+1} = \frac{\mu}{\Delta z_{k+\frac{1}{2}}} A\left|P_{k+\frac{1}{2}}\right| + \left\langle 0,\ -\ (\rho w)_{k+\frac{1}{2}}\right\rangle \tag{8.51f}$$

$$a_{k-1} = \frac{\mu}{\Delta z_{k-\frac{1}{2}}} A\left|P_{k-\frac{1}{2}}\right| + \left\langle 0,\ (\rho w)_{k-\frac{1}{2}}\right\rangle \tag{8.51g}$$

$$a_k^m = \frac{\rho\,\Delta x\,\Delta y\,\Delta z}{\Delta t} \tag{8.51h}$$

$$d = \rho g_z\,\Delta x\,\Delta y\,\Delta z + a_k^m w_k^m \tag{8.51i}$$

$$P = \frac{(\rho w)}{\mu/\Delta z} \tag{8.51j}$$

Equations 8.46, 8.48, and 8.50 are the discretization equations for estimated values u, v, and w based on the guess values of the pressure field. In order to differentiate these estimated values from the corrected values during the iteration process, let us designate these estimated values of velocity components and guess values of pressure with a superscript * and write the discretization equations as

$$a_i u_i^* = a_{i+1} u_{i+1}^* + a_{i-1} u_{i-1}^* + a_{j+1} u_{j+1}^* + a_{j-1} u_{j-1}^* + a_{k+1} u_{k+1}^* + a_{k-1} u_{k-1}^* + d + \left(p_{I-1}^* - p_I^*\right)\Delta y\,\Delta z \tag{8.52a}$$

$$a_j v_j^* = a_{i+1} v_{i+1}^* + a_{i-1} v_{i-1}^* + a_{j+1} v_{j+1}^* + a_{j-1} v_{j-1}^* + a_{k+1} v_{k+1}^* + a_{k-1} v_{k-1}^* + d + \left(p_{J-1}^* - p_J^*\right) \Delta x \, \Delta z \tag{8.52b}$$

$$a_k w_k^* = a_{i+1} w_{i+1}^* + a_{i-1} w_{i-1}^* + a_{j+1} w_{j+1}^* + a_{j-1} w_{j-1}^* + a_{k+1} w_{k+1}^* + a_{k-1} w_{k-1}^* + d + \left(p_{K-1}^* - p_K^*\right) \Delta x \, \Delta y \tag{8.52c}$$

The Velocity Correction Equations. We have mentioned that the estimated starred velocity components given by Equation 8.52 have to be updated using corrected pressure values given as

$$p = p^* + p' \tag{8.53}$$

where p' is the pressure correction, which will be derived in the following section using the continuity equation.

The iterative procedure for updating the velocity components is similar to the iterative refinement procedure outlined in Section 3.2.5, Chapter 3, for a linear system of algebraic equations. Let us designate the updated velocity components as

$$u = u^* + u' \tag{8.54a}$$
$$v = v^* + v' \tag{8.54b}$$
$$w = w^* + w' \tag{8.54c}$$

where u', v', and w' are the velocity corrections. Substituting Equations 8.54 into Equations 8.46, 8.48, and 8.50, and subtracting from Equation 8.52, we obtain the correction equations from the velocity components as

$$a_i u_i' = a_{i+1} u_{i+1}' + a_{i-1} u_{i-1}' + a_{j+1} u_{j+1}' + a_{j-1} u_{j-1}' + a_{k+1} u_{k+1}' + a_{k-1} u_{k-1}' + \left(p_{I-1}' - p_I'\right) \Delta y \, \Delta z \tag{8.55a}$$

$$a_j v_j' = a_{i+1} v_{i+1}' + a_{i-1} v_{i-1}' + a_{j+1} v_{j+1}' + a_{j-1} v_{j-1}' + a_{k+1} v_{k+1}' + a_{k-1} v_{k-1}' + \left(p_{J-1}' - p_J'\right) \Delta x \, \Delta z \tag{8.55b}$$

$$a_k w_k' = a_{i+1} w_{i+1}' + a_{i-1} w_{i-1}' + a_{j+1} w_{j+1}' + a_{j-1} w_{j-1}' + a_{k+1} w_{k+1}' + a_{k-1} w_{k-1}' + \left(p_{K-1}' - p_K'\right) \Delta x \, \Delta y \tag{8.55c}$$

The system of equations given by Equations 8.55a, 8.55b, and 8.55c are similar to that given by Equation 8.52 with the exception of the body force term, and can be solved by the same linear solver discussed in Chapter 2. However, one of the major features of the ***SIMPLE*** algorithm, as discussed by Patankar (1980), is to simplify Equation 8.55 by dropping the velocity component of the surrounding nodes on the right-hand side of Equation 8.55, i.e., the term $\sum a_{nb}u'_{nb}$, and write the correction equations as

$$u'_i = \left(p'_{I-1} - p'_I\right)\frac{\Delta y\,\Delta z}{a_i} \tag{8.56a}$$

$$v'_j = \left(p'_{J-1} - p'_J\right)\frac{\Delta x\,\Delta z}{a_j} \tag{8.56b}$$

$$w'_k = \left(p'_{K-1} - p'_K\right)\frac{\Delta x\,\Delta y}{a_k} \tag{8.56c}$$

We can see that the use of Equation 8.56 instead of Equation 8.55 considerably reduces the computational difficulties, but sacrifices the accuracy in velocity corrections to some extent. However, accuracy of the estimated velocity components are achieved through iterations and assigned tolerance limits.

Substituting the velocity correction values given by either Equation 8.55 or Equation 8.56 into Equation 8.54, we get the expressions for corrected velocity components. If we use Equation 8.56 for velocity corrections, then the expressions for thc ***corrected velocity*** components are given as

$$u_i = u_i^* + \left(p'_{I-1} - p'_I\right)\frac{\Delta y\,\Delta z}{a_i} \tag{8.57a}$$

$$v_j = v_j^* + \left(p'_{J-1} - p'_J\right)\frac{\Delta x\,\Delta z}{a_j} \tag{8.57b}$$

$$w_k = w_k^* + \left(p'_{K-1} - p'_K\right)\frac{\Delta x\,\Delta y}{a_k} \tag{8.57c}$$

The Pressure Correction Equation. The pressure correction equation is derived by substituting the estimated velocity components into the continuity equation. Let us first integrate the continuity Equation 8.34a over the control volume in the regular grid system (Figure 8.6) for the pressure

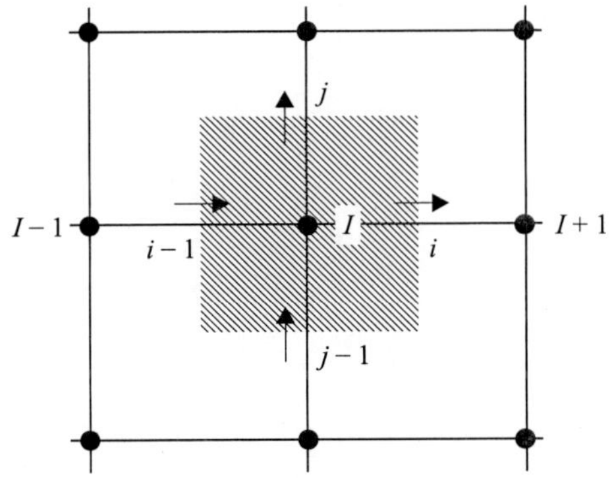

Figure 8.6 Control volume for pressure

$$\int_{t}^{t+\Delta t}\int_{i-1}^{i}\int_{j-1}^{j}\int_{k-1}^{k} \frac{\partial \rho}{\partial t}\, dt\, dx\, dy\, dz + \int_{t}^{t+\Delta t}\int_{i-1}^{i}\int_{j-1}^{j}\int_{k-1}^{k} \frac{\partial(\rho u)}{\partial x}\, dt\, dx\, dy\, dz$$
$$+ \int_{t}^{t+\Delta t}\int_{i-1}^{i}\int_{j-1}^{j}\int_{k-1}^{k} \frac{\partial(\rho v)}{\partial y}\, dt\, dx\, dy\, dz$$
$$+ \int_{t}^{t+\Delta t}\int_{i-1}^{i}\int_{j-1}^{j}\int_{k-1}^{k} \frac{\partial(\rho w)}{\partial z}\, dt\, dx\, dy\, dz = 0 \quad (8.58)$$

or

$$\rho_I|_t^{t+\Delta t}\, \Delta x\, \Delta y\, \Delta z + \left\{(\rho u)_i - (\rho u)_{i-1}\right\} \Delta y\, \Delta z\, \Delta t + \left\{(\rho v)_j - (\rho v)_{j-1}\right\} \Delta x\, \Delta z\, \Delta t$$
$$+ \left\{(\rho w)_k - (\rho w)_{k-1}\right\} \Delta x\, \Delta y\, \Delta t = 0 \quad (8.59)$$

Using the ***implicit time integration scheme***, i.e., values of density and velocity components at time $t + \Delta t$ are assumed to prevail over the entire time step Δt, and substituting the expressions for the velocity field given by Equation 8.57, we get

$$\frac{\left(\rho_I^{m+1} - \rho_I^{m}\right) \Delta x\, \Delta y\, \Delta z}{\Delta t} + \left[\left\{(\rho u^*)_i + \frac{\left(P'_I - P'_{I+1}\right) \Delta y\, \Delta z}{a_i}\right\}\right.$$
$$\left. - \left\{(\rho u^*)_{i-1} + \frac{\left(P'_{I-1} - P'_I\right) \Delta y\, \Delta z}{a_{i-1}}\right\}\right] \Delta y\, \Delta z$$
$$+ \left[\left\{(\rho v^*)_j + \frac{\left(P'_J - P'_{J+1}\right) \Delta x\, \Delta z}{a_j}\right\} - \left\{(\rho v^*)_{j-1} + \frac{\left(P'_{J-1} - P'_J\right) \Delta x\, \Delta z}{a_{j-1}}\right\}\right] \Delta x\, \Delta z$$
$$+ \left[\left\{(\rho w^*)_k + \frac{\left(P'_K - P'_{K+1}\right) \Delta x\, \Delta y}{a_k}\right\}\right.$$
$$\left. - \left\{(\rho w^*)_{k-1} + \frac{\left(P'_{K-1} - P'_K\right) \Delta x\, \Delta y}{a_{k-1}}\right\}\right] \Delta x\, \Delta y = 0 \quad (8.60)$$

In evaluating the integrals we assumed that the fluid is not strongly influenced by the pressure, and the equation is applicable to incompressible or weak compressible flows. Rearranging Equation 8.60, we get the ***pressure correction equation*** as

$$a_I p'_I = a_{I+1} p'_{I+1} + a_{I-1} p'_{I-1} + a_{J+1} p'_{J+1} + a_{J-1} p'_{J-1} + a_{K+1} p'_{K+1} + a_{K-1} p'_{K-1} + d \quad (8.61)$$

where

$$a_{I+1} = \rho_i \frac{(\Delta y\, \Delta z)^2}{a_i} \tag{8.62a}$$

$$a_{I-1} = \rho_{i-1} \frac{(\Delta y\, \Delta z)^2}{a_{i-1}} \tag{8.62b}$$

$$a_{J+1} = \rho_j \frac{(\Delta z\, \Delta x)^2}{a_j} \tag{8.62c}$$

$$a_{J-1} = \rho_{j-1} \frac{(\Delta z\, \Delta x)^2}{a_{j-1}} \tag{8.62d}$$

$$a_{K+1} = \rho_k \frac{(\Delta x\, \Delta y)^2}{a_k} \tag{8.62e}$$

$$a_{K-1} = \rho_{k-1} \frac{(\Delta x\, \Delta y)^2}{a_{k-1}} \tag{8.62f}$$

$$a_I = a_{I+1} + a_{I-1} + a_{J+1} + a_{J-1} + a_{K+1} + a_{K-1} \tag{8.62g}$$

$$d = \frac{\left(\rho_I^m - \rho_I^{m+1}\right)\Delta x\, \Delta y\, \Delta z}{\Delta t} + \left[(\rho u^*)_{i-1} - (\rho u^*)_i\right] \Delta y\, \Delta z$$
$$+ \left[(\rho v^*)_{j-1} - (\rho v^*)_j\right] \Delta z\, \Delta x + \left[(\rho w^*)_{k-1} - (\rho w^*)_k\right] \Delta x\, \Delta y \tag{8.62h}$$

It can be noted that the source term ***d*** in the pressure correction Equation 8.61 is a residual, which is obtained while the estimated starred velocity components given by Equation 8.52 are substituted into the continuity equation. So, as this mass source term, ***d***, approaches zero or a specified tolerance limit, a solution procedure leads to correct velocity and pressure fields.

Under-relaxation. The iterative solution of velocity and pressure fields may encounter instability, particularly due to the simplification made in the derivation of velocity correction Equation 8.56. This necessitates the need for underrelaxation parameters for the estimated starred velocity components in the momentum equations and pressure with respect to the previous iterated values as

$$u = u^* + \alpha_V u' \tag{8.63a}$$

$$v = v^* + \alpha_V v' \tag{8.63b}$$

$$w = w^* + \alpha_V w' \tag{8.63c}$$

and

$$p = p^* + \alpha_p p' \tag{8.64}$$

Where α_V and α_p are the underrelaxation parameters for velocity and pressure, respectively. The optimum values of these relaxation parameters are problem dependent and are usually established through numerical experimentations.

SIMPLE Algorithm. The sequence of calculation to be performed using the equations derived for velocity, pressures, and their correction values constitute an algorithm known as ***semi-implicit method for pressure-linked equation (SIMPLE)*** as proposed by Patankar and Spalding (Patankar and Spalding, 1972; Patankar, 1980). The algorithm is given as follows.

1. Guess the pressure field p^*.
2. Solve discretization Equations 8.52 for u^*, v^*, and w^*.
3. Solve pressure correction Equation 8.61 for p'.
4. Calculate corrected pressure from Equation 8.53 or Equation 8.64 with under-relaxation.
5. Calculate velocity components u, v, and w from their starred and correction components using Equation 8.57 or Equation 8.63 with under-relaxation.
6. Solve the discretization equation for other scalar quantities, such as turbulence quantities, temperature, and mass concentration if they influence the flow field though dependence, through fluid properties such as density, viscosity, and source terms. If a particular scalar quantity does not affect the velocity field, then that quantity is calculated after a converged velocity field is achieved.
7. Reset the corrected pressure p as the new guessed pressure p^*.
8. Repeat 2 to 7 until the solution converges within a specified tolerance limit.

Example 8.1. Consider a one-dimensional flow in a straight nozzle shown in the figure. The governing equations are given as

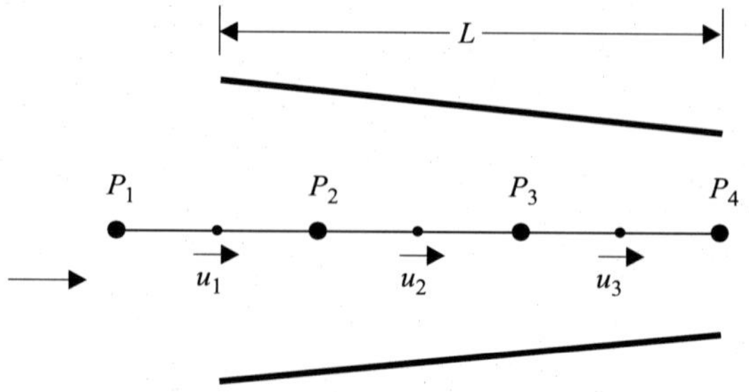

Continuity

$$\frac{\partial(uA)}{\partial x} = 0$$

x-momentum

$$\frac{\partial(\rho uu)}{\partial x} = -\frac{\partial p}{\partial x}$$

where A is the cross-sectional area of the nozzle. Consider the data

$$\rho = 1.00\,\text{kg/m}^3, \quad A_1 = 5\,\text{m}^2, \quad A_2 = 4\,\text{m}^2, \quad A_3 = 3\,\text{m}^2$$
$$p_1 = 300\,\text{kPa}, \quad p_4 = 100\,\text{kPa and } \alpha_v = 0.5$$

Show discretization equations for velocity and pressure and use the **SIMPLE** algorithm to compute velocity and pressure at the nodes for one iteration. Use initial guess for mass flow rate as $\dot{m} = \rho uA = 100\,\text{kg/s}$.

Solution. For the initial guess value of the mass flow rate, we can compute the initial guess values for the u-component velocities at the staggered grid points as

$$u_1 = \frac{100}{5} = 20\,\text{m/s}, \quad u_2 = \frac{100}{4} = 25\,\text{m/s} \quad \text{and} \quad u_3 = \frac{100}{3} = 33.333\,\text{m/s}$$

The **SIMPLE** algorithm can be used with following steps:

1. Guess the pressure field p^* at regular grid points as

$$p_1^* = 300\,\text{kPa}, \quad p_2^* = 250\,\text{kPa}, \quad p_3^* = 200\,\text{kPa}, \quad p_4^* = 100\,\text{kPa}$$

2. Discretization equations for u^* are obtained from Equation (8.52) for a one-dimensional problem as

$$a_i u_i^* = a_{i+1}u_{i+1}^* + a_{i-1}u_{i-1}^* + d + \left(p_I^* - p_{I+1}^*\right)A_i$$
$$a_i = a_{i+1} + a_{i-1}$$
$$a_{i+1} = \frac{\mu}{\Delta x_{i+\frac{1}{2}}} A\left|P_{i+\frac{1}{2}}\right| + \left\langle 0, -(\rho u)_{i+\frac{1}{2}}\right\rangle$$
$$a_{i-1} = \frac{\mu}{\Delta x_{i-\frac{1}{2}}} A\left|P_{i-\frac{1}{2}}\right| + \left\langle 0, (\rho u)_{i-\frac{1}{2}}\right\rangle$$
$$P = \frac{(\rho u)}{\mu/\Delta x}, \quad d = 0$$

For steady state and for zero gravity force as in this problem, and with $\mu = 0$, the discretization equation for the u-component velocity reduces to

$$a_i u_i^* = a_{i+1}u_{i+1}^* + a_{i-1}u_{i-1}^* + \left(p_{I-1}^* - p_I^*\right)A_i$$
$$a_i = a_{i+1} + a_{i-1}$$
$$a_{i+1} = \left\langle 0, -(\rho u)_{i+\frac{1}{2}}\right\rangle$$
$$a_{i-1} = \left\langle 0, (\rho u)_{i-\frac{1}{2}}\right\rangle$$

Applying this equation successively to nodes 1, 2, and 3 in a staggered system, we obtain the system of equations for the u component velocities as

Node 1

$$a_1 u_1^* = a_2 u_2^* + a_0 u_0^* + (p_1^* - p_2^*) A_1$$

$$a_2 = \left\langle 0, \; - \; (\rho u)_{i+\frac{1}{2}} \right\rangle = \left\langle 0, \; - \; (\rho u)_{1+\frac{1}{2}} \right\rangle = 0$$

$$a_0 = \left\langle 0, \; (\rho u)_{i-\frac{1}{2}} \right\rangle = \left\langle 0, \; (\rho u)_{\frac{1}{2}} \right\rangle = \frac{0 + u_1}{2} = \frac{20}{2} = 10$$

$$a_1 = a_2 + a_0 = 0 + 10 = 10$$

$$10u_1^* = 0u_2^* + 10 \times 0 + (300 - 250) \times 5$$

$$u_1^* = 25\,\mathrm{m/s}$$

Node 2

$$a_2 u_2^* = a_3 u_3^* + a_1 u_1^* + (p_2^* - p_3^*) A_2$$

$$a_i = a_{i+1} + a_{i-1}$$

$$a_3 = \left\langle 0, \; - \; (\rho u)_{i+\frac{1}{2}} \right\rangle = \left\langle 0, \; - \; (\rho u)_{2+\frac{1}{2}} \right\rangle = 0$$

$$a_1 = \left\langle 0, \; (\rho u)_{1+\frac{1}{2}} \right\rangle = \frac{u_1 + u_2}{2} = \frac{20 + 25}{2} = 22.5$$

$$a_2 = a_3 + a_1 = 22.5$$

$$22.5u_2^* = 0u_3^* + 22.5 \times 25 + (250 - 200) \times 4$$

$$u_2^* = 33.888\,\mathrm{m/s}$$

Node 3

$$a_3 u_3^* = a_2 u_2^* + a_4 u_4^* + (p_3^* - p_4^*) A_3$$

$$a_4 = \left\langle 0, \; - \; (\rho u)_{3+\frac{1}{2}} \right\rangle = 0$$

$$a_2 = \left\langle 0, \; (\rho u)_{3-\frac{1}{2}} \right\rangle = \frac{u_2 + u_3}{2} = \frac{25 + 33.333}{2} = 29.166$$

$$a_3 = 0 + 29.166 = 29.166$$

$$29.166u_3^* = 29.166 \times 33.888 + 0u_4^* + (200 - 100) \times 3$$

$$u_3^* = 44.174\,\mathrm{m/s}$$

3. Solve for pressure correction Equation 8.61 for p'

$$a_I p'_I = a_{I+1} p'_{I+1} + a_{I-1} p'_{I-1} + c$$

where

$$a_{I+1} = \rho_{I+\frac{1}{2}} \frac{\left(A_{I+\frac{1}{2}}\right)^2}{a_{I+\frac{1}{2}}}$$

$$a_{I-1} = \rho_{I-\frac{1}{2}} \frac{\left(A_{I-\frac{1}{2}}\right)^2}{a_{I-\frac{1}{2}}}$$

$$a_I = a_{I+1} + a_{I-1}$$

$$c = \left[(\rho u^*)_{I-\frac{1}{2}} - (\rho u^*)_{I+\frac{1}{2}}\right] A_I$$

Applying this discretization equation to pressure grid points, we have

For Node 2

$$a_{p2} p'_2 = a_{p3} p'_3 + a_{p1} p'_1 + \left[(\rho u_1^*) - (\rho u_2^*)\right] \frac{A_1 + A_2}{2}$$

where

$$a_{p3} = a_{pI+1} = \rho_{I+\frac{1}{2}} \frac{\left(A_{I+\frac{1}{2}}\right)^2}{a_{I+\frac{1}{2}}} = \frac{(A_2)^2}{u_2} = \frac{(4)^2}{25} = 0.64$$

$$a_{p1} = a_{pI-1} = \rho_{I-\frac{1}{2}} \frac{\left(A_{I-\frac{1}{2}}\right)^2}{a_{I-\frac{1}{2}}} = \frac{(A_1)^2}{u_1} = \frac{(5)^2}{20} = 1.25$$

$$a_{p2} = a_{p3} + a_{p1} = 0.64 + 1.25 = 1.89$$

$$1.89 p'_2 = 0.64 p'_3 + 1.25 p'_1 [(1 \times 25) - (1 \times 33.88)] \frac{5+4}{2}$$

$$1.89 p'_2 = 0.64 p'_3 + 0 - 39.96 \qquad \text{(E.8.1.1)}$$

For Node 3

$$a_{p3} p'_3 = a_{p4} p'_4 + a_{p2} p'_2 + c$$

where

$$a_{p4} = a_{pI+1} = \rho_{I+\frac{1}{2}} \frac{\left(A_{I+\frac{1}{2}}\right)^2}{a_{I+\frac{1}{2}}} = \frac{(A_3)^2}{u_3} = \frac{(3)^2}{33.333} = 0.270$$

$$a_{p2} = \rho_{I-\frac{1}{2}} \frac{\left(A_{I-\frac{1}{2}}\right)^2}{a_{I-\frac{1}{2}}} = \frac{(A_2)^2}{u_2} = \frac{(4)^2}{25} = 0.64$$

$$a_{p3} = a_{p2} + a_{p4} = 0.270 + 0.64 = 0.91$$

$$a_{p3}p'_3 = a_{p4}p'_4 + a_{p2}p'_2 + \left[\left(\rho u_2^*\right) - \left(\rho u_3^*\right)\right]\frac{A_2 + A_3}{2}$$

$$0.91p'_3 = 0.270 \times 0 + 0.64p'_2 + [(1 \times 33.888) - (1 \times 44.174)]\frac{4+3}{2}$$

$$0.91p'_3 = 0.64p'_2 - 36.001 \tag{E.8.1.2}$$

Solving Equations E.8.1.1 and E.8.1.2, we get

$$P'_2 = -45.332\,90 \text{ kPa} \quad \text{and} \quad P'_3 = -71.443\,97 \text{ kPa}$$

Calculate corrected pressure from Equation 8.53 or Equation 8.64 with under-relaxation

$$p_2 = p_2^* + 0.5p'_2 = 250 + 0.5(-45.332)$$
$$p_2 = 227.334 \text{ kPa}$$

and

$$p_3 = p_3^* + 0.5p'_3 = 200 + 0.5(-71.44397)$$
$$p_3 = 164.278 \text{ kPa}$$

Calculate velocity components u, v, and w from their starred and correction components using Equation 8.54 or Equation 8.63 with under-relaxation

$$u = u^* + \alpha_V u'$$

or

$$u_i = u_i^* + \alpha_V\left(p'_I - p'_{I+1}\right)\frac{\Delta y \Delta z}{a_i}$$

Node 1

$$u_1 = u_1^* + \alpha_V\left(p'_1 - p'_2\right)\frac{A_1}{a_1}$$
$$= 25 + 0.5 \times (0 - (-45.3329))\frac{5}{10}$$
$$u_1 = 36.3332 \text{ m/s}$$

Node 2

$$u_2 = u_2^* + \alpha_V(p_2' - p_3')\frac{A_2}{a_2}$$

$$= 33.888 + 0.5 \times (-45.3329 + 71.44397)\frac{4}{22.5}$$

$$u_2 = 40.238 \text{ m/s}$$

Node 3

$$u_3 = u_3^* + \alpha_V(p_3' - p_4')\frac{A_3}{a_3}$$

$$= 44.174 + 0.5(-71.44397 - 0)\frac{3}{29.166}$$

$$u_3 = 40.4999 \text{ m/s}$$

Reset the corrected pressure p as the new guessed pressure p^*

$$p_2^* = p_2 = 227.334 \text{ kPa}, \qquad p_3^* = p_3 = 164.278 \text{ kPa}$$

This concludes the first iteration.

Iteration 2 u-component velocities

Node 1

$$a_1 u_1^* = a_2 u_2^* + a_0 u_0^* + (p_1^* - p_2^*)A_1$$

$$a_2 = \left\langle 0,\ -\ (\rho u)_{i+\frac{1}{2}}\right\rangle = \left\langle 0,\ -\ (\rho u)_{1+\frac{1}{2}}\right\rangle = 0$$

$$a_0 = \left\langle 0,\ (\rho u)_{i-\frac{1}{2}}\right\rangle = \left\langle 0,\ (\rho u)_{\frac{1}{2}}\right\rangle = \frac{0 + u_1}{2} = \frac{36.332}{2} = 18.166$$

$$a_1 = a_2 + a_0 = 0 + 18.166 = 18.166$$

$$18.166 u_1^* = 0 u_2^* + 18.166 \times 0 + (300 - 227.334) \times 5$$

$$u_1^* = 20.00 \text{ m/s}$$

Node 2

$$a_2 u_2^* = a_3 u_3^* + a_1 u_1^* + (p_2^* - p_3^*)A_2$$

$$a_i = a_{i+1} + a_{i-1}$$

$$a_3 = \left\langle 0,\ -\ (\rho u)_{i+\frac{1}{2}}\right\rangle = \left\langle 0,\ -\ (\rho u)_{2+\frac{1}{2}}\right\rangle = 0$$

$$a_1 = \left\langle 0,\ (\rho u)_{1.+\frac{1}{2}}\right\rangle = \frac{u_1 + u_2}{2} = \frac{36.3332 + 40.238}{2} = 38.2856$$

$$a_2 = a_3 + a_1 = 38.2856$$

$$38.2856 u_2^* = 0 u_3^* + 38.2856 \times 20.0 + (227.334 - 164.278) \times 4$$

$$= 765.712 + 252.224$$

$$u_2^* = 26.5879 \text{ m/s}$$

Node 3

$$a_3u_3^* = a_2u_2^* + a_4u_4^* + (p_3^* - p_4^*)A_3$$

$$a_4 = \left\langle 0,\ -(\rho u)_{3.+\frac{1}{2}}\right\rangle = 0$$

$$a_2 = \left\langle 0,\ (\rho u)_{3-\frac{1}{2}}\right\rangle = \frac{u_2 + u_3}{2} = \frac{40.238 + 40.4999}{2} = 40.368\,95$$

$$a_3 = 0 + 40.368\,5 = 40.368\,95$$

$$40.368\,95u_3^* = 40.368\,95 \times 26.5879 + 0u_4^* + (164.278 - 100) \times 3$$

$$= 1073.3256 + 192.834$$

$$u_3^* = 31.3647\,\mathrm{m/s}$$

Solve for pressure correction Equation 8.61 for p'

$$a_Ip_I' = a_{I+1}p_{I+1}' + a_{I-1}p_{I-1}' + c$$

where

$$a_{I+1} = \rho_{I+\frac{1}{2}}\frac{\left(A_{I+\frac{1}{2}}\right)^2}{a_{I+\frac{1}{2}}}$$

$$a_{I-1} = \rho_{I-\frac{1}{2}}\frac{\left(A_{I-\frac{1}{2}}\right)^2}{a_{I-\frac{1}{2}}}$$

$$a_I = a_{I+1} + a_{I-1}$$

$$c = \left[(\rho u^*)_{I-\frac{1}{2}} - (\rho u^*)_{I+\frac{1}{2}}\right]A_I$$

Applying this discretization equation to pressure grid points, we have

For Node 2

$$a_{p2}p_2' = a_{p3}p_3' + a_{p1}p_1' + [(\rho u_1^*) - (\rho u_2^*)]\frac{A_1 + A_2}{2}$$

$$a_{p3} = a_{pI+1} = \rho_{I+\frac{1}{2}}\frac{\left(A_{I+\frac{1}{2}}\right)^2}{a_{I+\frac{1}{2}}} = \frac{(A_2)^2}{u_2} = \frac{(4)^2}{40.238} = 0.3976$$

$$a_{p1} = a_{pI-1} = \rho_{I-\frac{1}{2}}\frac{\left(A_{I-\frac{1}{2}}\right)^2}{a_{I-\frac{1}{2}}} = \frac{(A_1)^2}{u_1} = \frac{(5)^2}{36.3332} = 0.6880$$

$$a_{p2} = a_{p3} + a_{p1} = 0.3976 + 0.6880 = 1.0856$$

$$1.0856p_2' = 0.3976p_3' + 0.6880p_1' + [(1 \times 20) - (1 \times 26.5879)]\frac{5+4}{2}$$

$$1.0856p_2' = 0.3976p_3' + 0 - 29.645 \qquad \text{(E.8.1.3)}$$

For Node 3

$$a_{p3}p'_3 = a_{p4}p'_4 + a_{p2}p'_2 + c$$

where

$$a_{p4} = a_{pI+1} = \rho_{I+\frac{1}{2}}\frac{\left(A_{I+\frac{1}{2}}\right)^2}{a_{I+\frac{1}{2}}} = \frac{(A_3)^2}{u_3} = \frac{(3)^2}{40.238} = 0.223\,66$$

$$a_{p2} = \rho_{I-\frac{1}{2}}\frac{\left(A_{I-\frac{1}{2}}\right)^2}{a_{I-\frac{1}{2}}} = \frac{(A_2)^2}{u_2} = \frac{(4)^2}{40.238} = 0.3976$$

$$a_{p3} = a_{p2} + a_{p4} = 0.223\,66 + 0.3976 = 0.621\,26$$

$$a_{p3}p'_3 = a_{p4}p'_4 + a_{p2}p'_2 + \left[(\rho u_2^*) - (\rho u_3^*)\right]\frac{A_2 + A_3}{2}$$

$$0.621\,26p'_3 = 0.223\,66 \times 0 + 0.3976p'_2 + [(1 \times 26.5879) - (1 \times 31.3647)]\frac{4+3}{2}$$

$$0.621\,26p'_3 = 0.3976p'_2 - 16.7188 \tag{E.8.1.4}$$

Solving Equations E.8.1.3 and E.8.1.4, we get

$$P'_2 = -48.539 \text{ kPa} \quad \text{and} \quad P'_3 = -57.975 \text{ kPa}$$

Calculate corrected pressure from Equation 8.50 or Equation 8.61 with under-relaxation.

$$p_2 = p_2^* + 0.5p'_2 = 227.334 + 0.5(-48.539)$$
$$p_2 = 203.06 \text{ kPa}$$

and

$$p_3 = p_3^* + 0.5p'_3 = 164.278 + 0.5(-57.975)$$
$$p_3 = 135.290 \text{ kPa}$$

Calculate velocity components u, v, and w from their starred and correction components using Equation 8.54 or 8.60 with under-relaxation

$$u = u^* + \alpha_V u'$$

or

$$u_i = u_i^* + \alpha_V\left(p'_I - p'_{I+1}\right)\frac{\Delta y\,\Delta z}{a_i}$$

Node 1

$$u_1 = u_1^* + \alpha_V\left(p_1' - p_2'\right)\frac{A_1}{a_1} = 18.166$$

$$= 20 + 0.5 \times (0 - (-48.539))\frac{5}{18.166}$$

$$u_1 = 26.6799 \text{ m/s}$$

Node 2

$$u_2 = u_2^* + \alpha_V\left(p_2' - p_3'\right)\frac{A_2}{a_2} \qquad u_2^* = 26.5879 P_3' = -57.975$$

$$= 26.5879 + 0.5 \times (-48.539 + 57.975)\frac{4}{38.2856}$$

$$u_2 = 27.0808 \text{ m/s}$$

Node 3

$$u_3 = u_3^* + \alpha_V\left(p_3' - p_4'\right)\frac{A_3}{a_3} a_3 = 0 + 40.36895 = 40.368\,95$$

$$= 31.3647 + 0.5(-57.975 - 0)\frac{3}{40.368\,95}$$

$$u_3 = 29.210 \text{ m/s}$$

Reset the corrected pressure p as the new guessed pressure p^*

$$p_2^* = p_2 = 203.06 \text{ kPa}, \qquad p_3^* = p_3 = 135.290 \text{ kPa}$$

A summary of the two iteration results is given in the table below. The procedure can be repeated until convergence.

Iteration Number	u_1	u_2	u_3	p_2	p_3
0	20	25	33.333	250	200
1	36.3332	40.238	40.4999	227.334	164.278
2	26.6799	27.0808	29.210	203.06	135.290

SIMPLER Algorithm. The **SIMPLE** algorithm has been used very effectively for many fluid flow and heat transfer problems. However, it showed some difficulties in terms of instability, slow rate of convergence, and poor prediction of the pressure fields. This is primarily due to the simplification made in the derivation of velocity correction Equation 8.53 by dropping the velocity contributions from the neighboring nodes, i.e., $\sum a_{\mathrm{nb}} u'_{\mathrm{nb}}$ term, and using only the pressure correction equation to correct the velocity field and pressure field. Additional iterations are usually needed to obtain the correct pressure field even if the correct velocity field is achieved. In order to overcome these difficulties, some modification is made to the **SIMPLE** algorithm and a new algorithm named **SIMPLER (SIMPLE Revised)** is developed (Patankar, 1980, 1988b).

In this algorithm the solution procedure starts with a guessed velocity field, and a separate pressure equation is derived for the calculation of the pressure field in addition to the pressure correction equation. Let us discuss here the development of this pressure equation and the basic steps of this algorithm.

The Pressure Equation The discretization equation for the velocity fields given by Equations 8.46b, 8.48b, and 8.50b are rewritten as

$$u_i = \frac{\sum a_{nb} u_{nb} + d}{a_i} + (p_{I-1} - p_I)\frac{\Delta y\, \Delta z}{a_i} \tag{8.65a}$$

$$v_j = \frac{\sum a_{nb} v_{nb} + d}{a_j} + (p_{J-1} - p_J)\frac{\Delta x\, \Delta z}{a_j} \tag{8.65b}$$

$$w_k = \frac{\sum a_{nb} w_{nb} + d}{a_k} + (p_{K-1} - p_K)\frac{\Delta x\, \Delta y}{a_k} \tag{8.65c}$$

The first term on the right-hand side of Equation 8.65 is defined as a new pseudovelocity field in the following manner

$$\hat{u}_i = \frac{\sum a_{nb} u_{nb} + d}{a_i} \tag{8.66a}$$

$$\hat{v}_j = \frac{\sum a_{nb} v_{nb} + d}{a_j} \tag{8.66b}$$

$$\hat{w}_k = \frac{\sum a_{nb} w_{nb} + d}{a_k} \tag{8.66c}$$

Substituting Equation 8.66 into Equation 8.65, we get

$$u_i = \hat{u}_i + (p_{I-1} - p_I)\frac{\Delta y\, \Delta z}{a_i} \tag{8.67a}$$

$$v_j = \hat{v}_j + (p_{J-1} - p_J)\frac{\Delta x\, \Delta z}{a_j} \tag{8.67b}$$

$$w_k = \hat{w}_k + (p_{K-1} - p_K)\frac{\Delta x\, \Delta y}{a_k} \tag{8.67c}$$

In the next step, the discretization equation for the pressure field can be derived in the same manner as for the pressure correction equation discussed earlier. The only difference is that the pressure equation is derived by substituting the estimated velocity components given by Equation 8.67 into the integral form of the continuity equation. In order to demonstrate this, let us start from integral form of the continuity equation given by Equation 8.59

$$\rho_I|_t^{t+\Delta t}\,\Delta x\,\Delta y\,\Delta z + \left\{(\rho u)_i - (\rho u)_{i-1}\right\}\Delta y\,\Delta z\,\Delta t + \left\{(\rho v)_j - (\rho v)_{j-1}\right\}\Delta x\,\Delta z\,\Delta t + \left\{(\rho w)_k - (\rho w)_{k-1}\right\}\Delta x\,\Delta y\,\Delta t \tag{8.59}$$

Using the ***implicit time integration scheme*** and substituting the expressions for the velocity field given by Equation 8.67, we get

$$\begin{aligned}
&\frac{\left(\rho_I-\rho_I^0\right)\Delta x\Delta y\Delta z}{\Delta t}\\
&+\left[\left\{(\rho\hat{u})_i+\frac{\rho_i(P_I-P_{I+1})\Delta y\Delta z}{a_i}\right\}-\left\{(\rho\hat{u})_{i-1}+\frac{\rho_{i-1}(P_{I-1}-P_I)\Delta y\Delta z}{a_{i-1}}\right\}\right]\Delta y\Delta z\\
&+\left[\left\{(\rho\hat{v})_j+\frac{\rho_j(P_J-P_{J+1})\Delta x\Delta z}{a_j}\right\}-\left\{(\rho\hat{v})_{j-1}+\frac{\rho_{j-1}(P_{J-1}-P_J)\Delta x\Delta z}{a_{j-1}}\right\}\right]\Delta x\Delta z\\
&+\left[\left\{(\rho\hat{w})_k+\frac{\rho_k(P_K-P_{K+1})\Delta x\Delta y}{a_k}\right\}-\left\{(\rho\hat{w})_{k-1}+\frac{\rho_{k-1}(P_{K-1}-P_K)\Delta x\Delta y}{a_{k-1}}\right\}\right]\Delta x\Delta y=0
\end{aligned} \tag{8.68}$$

Rearranging equation 8.68, we get the ***pressure equation*** as

$$a_I p_I = a_{I+1}p_{I+1} + a_{I-1}p_{I-1} + a_{J+1}p_{J+1} + a_{J-1}p_{J-1} + a_{K+1}p_{K+1} + a_{K-1}p_{K-1} + d \tag{8.69}$$

where

$$a_{I+1} = \rho_i\frac{(\Delta y\,\Delta z)^2}{a_i} \tag{8.70a}$$

$$a_{I-1} = \rho_{i-1}\frac{(\Delta y\,\Delta z)^2}{a_{i-1}} \tag{8.70b}$$

$$a_{J+1} = \rho_j\frac{(\Delta z\,\Delta x)^2}{a_j} \tag{8.70c}$$

$$a_{J-1} = \rho_{j-1}\frac{(\Delta z\,\Delta x)^2}{a_{j-1}} \tag{8.70d}$$

$$a_{K+1} = \rho_k \frac{(\Delta x\,\Delta y)^2}{a_k} \tag{8.70e}$$

$$a_{K-1} = \rho_{k-1} \frac{(\Delta x\,\Delta y)^2}{a_{k-1}} \tag{8.70f}$$

$$a_I = a_{I+1} + a_{I-1} + a_{J+1} + a_{J-1} + a_{K+1} + a_{K-1} \tag{8.70g}$$

$$\begin{aligned} d = {} & \frac{(\rho_I^0 - \rho_I)\Delta x\,\Delta y\,\Delta z}{\Delta t} + \left[(\rho\hat{u})_{i-1} - (\rho\hat{u})_i\right]\Delta y\,\Delta z \\ & + \left[(\rho\hat{v})_{j-1} - (\rho\hat{v})_j\right]\Delta z\,\Delta x + \left[(\rho\hat{w})_{k-1} - (\rho\hat{w})_k\right]\Delta x\,\Delta y \end{aligned} \tag{8.70h}$$

It can be noted that the pressure equation given by Equations 8.69 and 8.70 is identical to that given for the pressure correction equations given by Equations 8.61 and 8.62, with the exception of the mass source term (d), which is calculated in terms of the velocity components $\hat{u}$, $\hat{v}$, and $\hat{w}$ instead of the starred velocity components. Also, as we approach a correct velocity field through iteration, the pressure field is also converged to a correct value.

The algorithm is given as follows.

1. Guess the velocity field u, v, and w.
2. Solve Equation 8.66 for $\hat{u}$, $\hat{v}$, and $\hat{v}$ using the guess velocity field for the surrounding nodes.
3. Solve the pressure Equation 8.69 for the pressure field, p.
4. Treat this pressure field as the pressure field p^* and solve discretization Equations 8.52 for u^*, v^*, and w^*.
5. Calculate the mass source term $\boldsymbol{d}$ given by Equation 8.62h and solve for pressure correction Equation 8.61 for p'.
6. Calculate correct velocity components u, v, and w from their starred components and pressure correction using Equation 8.57.
7. Solve the discretization equations for other scalar quantities such as turbulence quantities, temperature and mass concentration.
8. Repeat steps 2 to 7 with the new velocity estimates until the solution converges within a specified tolerance limit.

The **SIMPLER** algorithm is capable of achieving the correct velocity and pressure fields at the same time without requiring additional iterations for the pressure fields. However, it involves additional computational steps compared to that in the **SIMPLE** algorithm.

Example 8.2: Solution using SIMPLER: Consider a developing flow and heat transfer in a rectangular channel shown in the figure.

The flow enters the channel with uniform axial velocity w_0, pressure $\bar{p}_0$, and temperature T_0. The top surface is maintained at constant temperature T_0 and the other three surfaces are adiabatic.

The procedure for solving a three-dimensional developing fluid flow and heat transfer in a channel was given by Patankar and Spalding (1972).

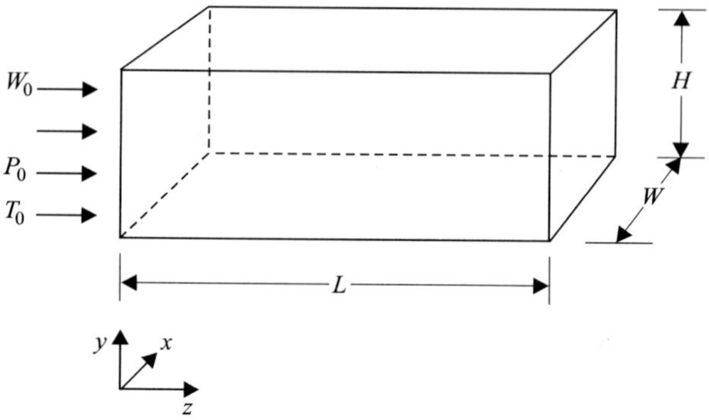

The three-dimensional elliptic nature of the differential equations are transformed into parabolic equations by making the following assumptions: (a) assume that the flow is predominant in a positive flow direction (z direction) and there is no reverse flow in that direction, (b) assume negligible diffusion of heat and momentum in the flow direction (i.e., negligible shear stress and heat diffusion flux on the xy plane). Because of this parabolic nature of the problem, the equations can be solved with marching integration from the upstream side to the downstream side. With these assumptions, associated governing equations and boundary conditions for the parabolic three-dimensional developing flow and heat transfer are given as follows.

Continuity

$$\frac{\partial u}{\partial x} + \frac{\partial v}{\partial y} + \frac{\partial w}{\partial z} = 0 \qquad \text{(E.8.2.1a)}$$

x momentum

$$u\frac{\partial u}{\partial x} + v\frac{\partial u}{\partial y} + w\frac{\partial u}{\partial z} = -\frac{1}{\rho}\frac{\partial p}{\partial x} + \nu\left(\frac{\partial^2 u}{\partial x^2} + \frac{\partial^2 u}{\partial y^2}\right) \qquad \text{(E.8.2.1b)}$$

y momentum

$$u\frac{\partial v}{\partial x} + v\frac{\partial v}{\partial y} + w\frac{\partial v}{\partial z} = -\frac{1}{\rho}\frac{\partial p}{\partial y} + \nu\left(\frac{\partial^2 v}{\partial x^2} + \frac{\partial^2 v}{\partial y^2}\right) \qquad \text{(E.8.2.1c)}$$

z momentum

$$u\frac{\partial w}{\partial x} + v\frac{\partial w}{\partial y} + w\frac{\partial w}{\partial z} = -\frac{1}{\rho}\frac{\partial \bar{p}}{\partial z} + \nu\left(\frac{\partial^2 w}{\partial x^2} + \frac{\partial^2 w}{\partial y^2}\right) \qquad \text{(E.8.2.1d)}$$

energy

$$u\frac{\partial T}{\partial x}+v\frac{\partial T}{\partial y}+w\frac{\partial T}{\partial z}=\frac{k}{\rho c_p}\left(\frac{\partial^2 T}{\partial x^2}+\frac{\partial^2 T}{\partial y^2}\right) \tag{E.8.2.1e}$$

Another important modification is made by decoupling the axial and lateral pressure gradients terms. The pressure $\bar{p}$ in the z-momentum equation represents an area average pressure at any cross-section (xy plane). The pressure gradient $d\bar{p}/dz$ is assumed to be known or calculated first, and then we proceed to solve x and y momentum equations and lateral pressure gradients $\partial p/\partial x$ and $\partial p/\partial y$.

Boundary conditions

Inlet

Uniform velocity and temperature at inlet, i.e., at $z=0$, $w=w_0$, $T=T_0$.

Wall

No-slip velocity on all walls, i.e., $u=0$, $v=0$, and $w=0$ on all walls. Constant surface temperature on top wall, i.e., at $y=0$, $T=T_w$ and rest of the walls are adiabatic.

Assume dimensionless variables as

$$u=\frac{u}{w_0},\quad v=\frac{v}{w_0},\quad w=\frac{w}{w_0},\quad p=\frac{p}{\rho w_0^2},\quad \theta=\frac{T-T_w}{T_b-T_w} \tag{E.8.2.2}$$

$$x=\frac{x}{H},\quad y=\frac{y}{H},\quad z=\frac{z}{H}$$

Equations E.8.1 to Equation E.8.5 can be transformed into dimensionless equations as

$$\frac{\partial u}{\partial x}+\frac{\partial v}{\partial y}+\frac{\partial w}{\partial z}=0 \tag{E.8.2.3a}$$

$$u\frac{\partial u}{\partial x}+v\frac{\partial u}{\partial y}+w\frac{\partial u}{\partial z}=-\frac{\partial p}{\partial x}+\frac{1}{\mathrm{Re}_H}\left(\frac{\partial^2 u}{\partial x^2}+\frac{\partial^2 u}{\partial y^2}\right) \tag{E.8.2.3b}$$

$$u\frac{\partial v}{\partial x}+v\frac{\partial v}{\partial y}+w\frac{\partial v}{\partial z}=-\frac{\partial p}{\partial y}+\frac{1}{\mathrm{Re}_H}\left(\frac{\partial^2 v}{\partial x^2}+\frac{\partial^2 v}{\partial y^2}\right) \tag{E.8.2.3c}$$

$$u\frac{\partial w}{\partial x}+v\frac{\partial w}{\partial y}+w\frac{\partial w}{\partial z}=\frac{\partial p}{\partial z}+\frac{1}{\mathrm{Re}_H}\left(\frac{\partial^2 w}{\partial x^2}+\frac{\partial^2 w}{\partial y^2}\right) \tag{E.8.2.3d}$$

$$u\frac{\partial \theta}{\partial x}+v\frac{\partial \theta}{\partial y}+w\frac{\partial \theta}{\partial z}+w\theta\lambda=\frac{1}{\mathrm{Re}_H\,\mathrm{Pr}}\left(\frac{\partial^2 \theta}{\partial x^2}+\frac{\partial^2 \theta}{\partial y^2}\right) \tag{E.8.2.3e}$$

where $\mathrm{Re}_H=\dfrac{w_0 H}{v}$, $\mathrm{Pr}=\dfrac{y}{\alpha}$, and $\alpha=\dfrac{k}{\rho_{\mathrm{cp}}}$

Derive the discretization equations for u, v, w, and p and present the sequence of steps to be used for calculating velocity and temperature field using the **SIMPLER** algorithm.

Solution. Let us first consider the discretization of Equation E.8.2.3b by rearranging the equation as

$$\left(u\frac{\partial u}{\partial x} - \Gamma\frac{\partial^2 u}{\partial x^2}\right) + \left(v\frac{\partial u}{\partial y} - \Gamma\frac{\partial^2 u}{\partial y^2}\right) = -\frac{\partial \bar{p}}{\partial x} - w\frac{\partial u}{\partial z} \tag{E.8.2.4}$$

Equation E.8.2.4 can be written in a general form as

$$\frac{\partial J_x}{\partial x} + \frac{\partial J_y}{\partial y} = S \tag{E.8.2.5}$$

where

$$J_x = u^2 - \Gamma\frac{\partial u}{\partial x}, \quad J_y = vu - \Gamma\frac{\partial u}{\partial y}$$

$$S = -\frac{dp}{dx} - w\frac{\partial u}{\partial z}, \quad \Gamma = \frac{1}{\mathrm{Re}_H}$$

Using Equation 8.46a we get the discretization equation for Equation E.8.2.3b as

$$\begin{aligned} ua_i u_i &= ua_{i+1}u_{i+1} + ua_{i-1}u_{i-1} + ua_{j+1}u_{j+1} + ua_{j-1}u_{j-1} \\ &\quad - \left(\frac{p_{I+1,J} - p_{,I,J}}{\rho\,\Delta x}\right)\Delta y\,\Delta x - w_i^u\frac{u_i - u_i^u}{\Delta z}\Delta x\,\Delta y \end{aligned}$$

or

$$\begin{aligned} \left(ua_i u_i + \frac{w_i^u}{\Delta z}\Delta x\,\Delta y\right)u_i &= ua_{i+1}u_{i+1} + ua_{i-1}u_{i-1} + ua_{j+1}u_{j+1} + ua_{j-1}u_{j-1} \\ &\quad - \left(\frac{p_{I+1,J} - p_{,I,J}}{\rho}\right)\Delta y + w_i^u\frac{u_i^u}{\Delta z}\Delta x\,\Delta y \end{aligned} \tag{E.8.2.6a}$$

or

$$ua_i u_i = \sum ua_{nb}u_{nb}^{m+1} + d \tag{E.8.2.6b}$$

where

$$ua_i = ua_{i+1} + ua_{i-1} + ua_{j+1} + ua_{j-1} + \frac{w_i^u}{\Delta z}\Delta x\,\Delta y$$

$$ua_{i+1} = \frac{\Gamma}{\Delta x}A\left|P_{i+\frac{1}{2}}\right| + \left\langle 0,\ -\ (\rho u)_{i+\frac{1}{2}}\right\rangle, \qquad ua_{i-1} = \frac{\Gamma}{\Delta x}A\left|P_{i-\frac{1}{2}}\right| + \left\langle 0,\ (\rho u)_{i-\frac{1}{2}}\right\rangle$$

$$ua_{j+1} = \frac{\Gamma}{\Delta y}A\left|P_{j+\frac{1}{2}}\right| + \left\langle 0,\ -\ (\rho v)_{j+\frac{1}{2}}\right\rangle, \qquad ua_{j-1} = \frac{\Gamma}{\Delta y_{i-\frac{1}{2}}}A\left|P_{j-\frac{1}{2}}\right| + \left\langle 0,\ (\rho v)_{j-\frac{1}{2}}\right\rangle$$

$$d = -\left(\frac{p_{I+1,J} - p_{,I,J}}{\rho}\right)\Delta y + w_i^u\frac{u_i^u}{\Delta z}\Delta x\,\Delta y$$

$$P_i = \frac{(\rho u)}{\Gamma/\Delta x}, \qquad P_j = \frac{(\rho v)}{\Gamma/\Delta y}$$

In the above discretization equation, values u_i^{u} and w_i^{u} with a superscript u indicate values at the upstream position as shown in the grid shown in the figure.

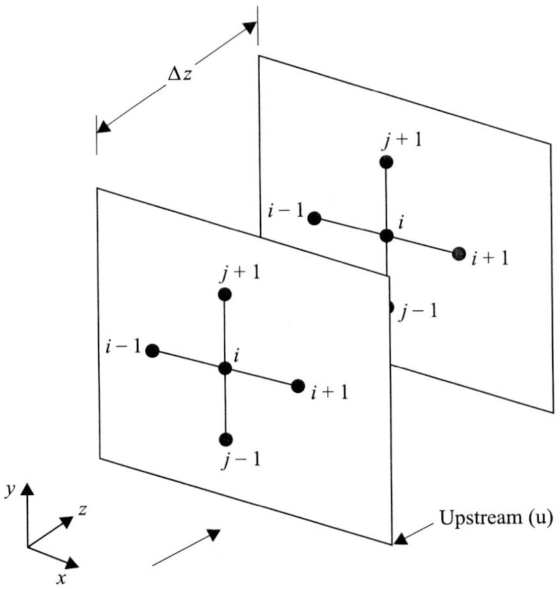

Similarly, Equation E.8.2.3c for the y-momentum equation can be discretized as

$$\left(u\frac{\partial v}{\partial x} - \Gamma\frac{\partial^2 v}{\partial x^2}\right) + \left(v\frac{\partial v}{\partial y} - \Gamma\frac{\partial^2 v}{\partial y^2}\right) = -\frac{1}{\rho}\frac{\partial p}{\partial y} - w\frac{\partial v}{\partial z}$$

Equation E.8.2.5 can be written in a general form as

$$\frac{\partial J_x}{\partial x} + \frac{\partial J_y}{\partial y} = S \tag{E.8.2.7}$$

where

$$J_x = uv - \Gamma \frac{\partial v}{\partial x}, \qquad J_y = v^2 - \Gamma \frac{\partial v}{\partial y}$$

$$S = -\frac{1}{\rho}\frac{dp}{dy} - w\frac{\partial v}{\partial z}, \qquad \Gamma = \frac{1}{\mathrm{Re}_H}$$

The discretization equation for Equation E.8.2.7 is written as

$$\begin{aligned} va_i v_i &= va_{i+1} v_{i+1} + va_{i-1} v_{i-1} + va_{j+1} v_{j+1} + va_{j-1} v_{j-1} \\ &\quad - \left(\frac{p_{I,J+1} - p_{,I,J}}{\rho}\right) \Delta y\, \Delta x - w_i^u \frac{v_i - v_i^u}{\Delta z} \Delta x\, \Delta y \end{aligned}$$

or

$$\begin{aligned} \left(va_i + \frac{v_i^u}{\Delta z} \Delta x\, \Delta y\right) v_i &= va_{i+1} v_{i+1} + va_{i-1} v_{i-1} + va_{j+1} v_{j+1} + va_{j-1} v_{j-1} \\ &\quad - \left(\frac{p_{I,J+1} - p_{,I,J}}{\rho}\right) \Delta x - w_i^u \frac{v_i^u}{\Delta z} \Delta x\, \Delta y \end{aligned} \tag{E.8.2.8a}$$

or

$$va_i v_i = \sum va_{nb} v_{nb} + c \tag{E.8.2.8b}$$

where

$$va_i = va_{i+1} + va_{i-1} + va_{j+1} + va_{j-1} + \frac{w_i^u}{\Delta z} \Delta x\, \Delta y$$

$$va_{i+1} = \frac{\Gamma}{\Delta x} A\left|P_{i+\frac{1}{2}}\right| + \left\langle 0,\ -(\rho u)_{i+\frac{1}{2}}\right\rangle, \qquad va_{i-1} = \frac{\nu}{\Delta x_{i-\frac{1}{2}}} A\left|P_{i-\frac{1}{2}}\right| + \left\langle 0,\ (\rho u)_{i-\frac{1}{2}}\right\rangle$$

$$va_{j+1} = \frac{\nu}{\Delta y_{i+\frac{1}{2}}} A\left|P_{j+\frac{1}{2}}\right| + \left\langle 0,\ -(\rho v)_{j+\frac{1}{2}}\right\rangle, \qquad va_{j-1} = \frac{\nu}{\Delta y_{i-\frac{1}{2}}} A\left|P_{j-\frac{1}{2}}\right| + \left\langle 0,\ (\rho v)_{j-\frac{1}{2}}\right\rangle$$

$$c = -\left(\frac{p_{I,J+1} - p_{,I,J}}{\rho}\right) \Delta x - w_i^u \frac{v_i^u}{\Delta z} \Delta x\, \Delta y$$

$$P_i = \frac{(\rho u)}{\Gamma/\Delta x}, \qquad P_j = \frac{(\rho v)}{\Gamma/\Delta y}$$

Similarly, Equation E.8.2.3d for the z-momentum equation can be discretized as

$$\left(u\frac{\partial w}{\partial x}-\Gamma\frac{\partial^2 w}{\partial x^2}\right)+\left(v\frac{\partial w}{\partial y}-\Gamma\frac{\partial^2 w}{\partial y^2}\right)=-\frac{\partial \bar{p}}{\partial z}-w\frac{\partial w}{\partial z}$$

Equation E.8.2.5 can be written in a general form as

$$\frac{\partial J_x}{\partial x}+\frac{\partial J_y}{\partial y}=S \tag{E.8.2.9}$$

where

$$J_x=uw-\Gamma\frac{\partial w}{\partial x},\quad J_y=vw-\Gamma\frac{\partial w}{\partial y}$$

$$S=-\Delta\bar{p}-w\frac{\partial w}{\partial z},\quad \Gamma=\frac{1}{\mathrm{Re}_H}$$

The discretization equation for Equation E.8.2.9 is written as

$$\begin{aligned} wa_i w_{i,j} &= wa_{i+1}w_{i+1,j}+wa_{i-1}w_{i-1,j}+wa_{j+1}w_{i,j+1}+wa_{j-1}w_{i,j-1} \\ &\quad -(\Delta\bar{p})\Delta y\,\Delta x-w_i^u\frac{w_i-w_i^u}{\Delta z}\Delta x\,\Delta y \end{aligned}$$

or

$$\begin{aligned} \left(wa_{i,j}+\frac{w_{i,j}^u}{\Delta z}\Delta x\,\Delta y\right)w_{i,j} &= wa_{i+1,j}w_{i+1,j}+wa_{i-1,j}w_{i-1,j}+wa_{i,j+1}w_{i,j+1} \\ &\quad +wa_{i,j-1}w_{i,j-1}-\Delta\bar{p}\,\Delta x-w_i^u\frac{w_i^u}{\Delta z}\Delta x\,\Delta y \end{aligned} \tag{E.8.2.10a}$$

or

$$wa_i w_{i,j}=\sum wa_{nb}w_{nb}+d \tag{E.8.2.10b}$$

where

$$wa_i = wa_{i+1}+wa_{j-1}+wa_{j+1}+wa_{j-1}+\frac{w_{i,j}^u}{\Delta z}\Delta x\,\Delta y$$

$$wa_{i+1}=\frac{\Gamma}{\Delta x}A\left|P_{i+\frac{1}{2}}\right|+\left\langle 0,\ -(\rho u)_{i+\frac{1}{2}}\right\rangle$$

$$wa_{i-1}=\frac{\Gamma}{\Delta x}A\left|P_{i-\frac{1}{2}}\right|+\left\langle 0,\ (\rho u)_{i-\frac{1}{2}}\right\rangle$$

$$wa_{j+1}=\frac{\Gamma}{\Delta y}A\left|P_{j+\frac{1}{2}}\right|+\left\langle 0,\ -(\rho v)_{j+\frac{1}{2}}\right\rangle$$

$$wa_{j-1} = \frac{\Gamma}{\Delta y} A \left| P_{j-\frac{1}{2}} \right| + \left\langle 0, (\rho v)_{j-\frac{1}{2}} \right\rangle$$

$$d = -\left(\frac{\Delta \bar{p}}{\rho}\right) \Delta x \, \Delta y - w_i^u \frac{w_i^u}{\Delta z} \Delta x \, \Delta y$$

$$P_i = \frac{(\rho u)}{\Gamma / \Delta x}, \qquad P_j = \frac{(\rho v)}{\Gamma / \Delta y}$$

In a similar manner, the energy Equation E.8.2.3e is first transformed into the general form as

$$\left(u\frac{\partial \theta}{\partial x} - \Gamma_\theta \frac{\partial^2 \theta}{\partial x^2}\right) + \left(v\frac{\partial \theta}{\partial y} - \Gamma_\theta \frac{\partial^2 \theta}{\partial y^2}\right) = -w\theta\lambda - w\frac{\partial \theta}{\partial z}$$

or

$$\frac{\partial J_x}{\partial x} + \frac{\partial J_y}{\partial y} = S \qquad \text{(E.8.2.11)}$$

where

$$J_x = u\theta - \Gamma_\theta \frac{\partial \theta}{\partial x}, \qquad J_y = v\theta - \Gamma_\theta \frac{\partial \theta}{\partial y}$$

$$S = -w\theta\lambda - w\frac{\partial \theta}{\partial z}, \qquad \Gamma_\theta = \frac{1}{\mathrm{Re}_H \, \mathrm{Pr}}$$

Following Equation 8.50a, we get the discretization equation for the energy Equation E.8.2.11 as

$$Ta_{i,j}\theta_{i,j} = Ta_{i+1,j}\theta_{i+1,j} + Ta_{i-1,j}\theta_{i-1,j} + Ta_{i,j+1}\theta_{i,j+1} + Ta_{i,j-1}\theta_{i,j-1} - w_{i,j}\theta_{i,j}\lambda \Delta x \, \Delta y - w_{i,j}^u \frac{\theta_{i,j} - \theta_{i,j}^u}{\Delta z} \Delta x \, \Delta y$$

or

$$\left(Ta_i + \frac{w_{i,j}^u}{\Delta z} \Delta x \, \Delta y\right)\theta_{i,j} = Ta_{i+1}\theta_{i+1} + Ta_{i-1}\theta_{i-1} + Ta_{j+1}\theta_{j+1} + Ta_{j-1}\theta_{j-1} - w_{i,j}\theta_{i,j}\lambda - \frac{w_{i,j}^u \theta_{i,j}^u}{\Delta z} \Delta x \, \Delta y \qquad \text{(E.8.2.12a)}$$

or

$$Ta_{i,j}\theta_{i,j} = \sum Ta_{nb}\theta_{nb} + d \qquad \text{(E.8.2.12b)}$$

where

$$Ta_{i,j} = Ta_{i+1,j} + Ta_{i-1,j} + Ta_{i,j+1} + Ta_{i,j-1} + \frac{w_{i,j}^{u}}{\Delta z}\Delta x\,\Delta y$$

$$Ta_{i+1,j} = \frac{\Gamma_\theta}{\Delta x}A\left|P_{i+\frac{1}{2}}\right| + \left\langle 0,\ -(\rho u)_{i+\frac{1}{2}}\right\rangle$$

$$Ta_{i-1,j} = \frac{\Gamma_\theta}{\Delta x}A\left|P_{i-\frac{1}{2}}\right| + \left\langle 0,\ (\rho u)_{i-\frac{1}{2}}\right\rangle$$

$$Ta_{i,j+1} = \frac{\Gamma_\theta}{\Delta y}A\left|P_{j+\frac{1}{2}}\right| + \left\langle 0,\ -(\rho v)_{j+\frac{1}{2}}\right\rangle$$

$$Ta_{i,j-1} = \frac{\Gamma_\theta}{\Delta y}A\left|P_{j-\frac{1}{2}}\right| + \left\langle 0,\ (\rho v)_{j-\frac{1}{2}}\right\rangle$$

$$d = -w_{i,j}\theta_{i,j}\lambda - \frac{w_{i,j}^{u}\theta_{i,j}^{u}}{\Delta z}\Delta x\,\Delta y$$

Using Equation 8.61, we get the pressure correction equation as

$$pa_I p'_I = pa_{I+1}p'_{I+1} + pa_{I-1}p'_{I-1} + pa_{J+1}p'_{J+1} + pa_{J-1}p'_{J-1} + d \qquad \text{(E.8.2.13)}$$

where

$$pa_{I+1} = \rho_i \frac{(\Delta y\,\Delta z)^2}{a_i}$$

$$pa_{I-1} = \rho_{i-1} \frac{(\Delta y\,\Delta z)^2}{a_{i+1}}$$

$$pa_{J+1} = \rho_j \frac{(\Delta z\,\Delta x)^2}{a_j}$$

$$pa_{J-1} = \rho_{j-1} \frac{(\Delta z\,\Delta x)^2}{a_{j-1}}$$

$$a_I = a_{I+1} + a_{I-1} + a_{J+1} + a_{J-1}$$

$$d = \left[(\rho u^*)_{i-1} - (\rho u^*)_i\right]\Delta y\,\Delta z + \left[(\rho v^*)_{j-1} - (\rho v^*)_j\right]\Delta z\,\Delta x$$
$$+ \left[(\rho w^*)_{k-1} - (\rho w^*)_k\right]\Delta x\,\Delta y \qquad \text{(E.8.2.14)}$$

The solution algorithm for this problem is given as follows.

1. Guess the velocity field u, v, and w, and pressure field on the xy plane and $\Delta\bar{p} = \partial\bar{p}/\partial z$.

2. Solve discretization Equations E.8.2.10 for w with the guess value of $\Delta\bar{p} = \partial\bar{p}/\partial z$ at new z location incremented.
3. Check mass flow rate based on w^* and update $\Delta\bar{p} = \partial\bar{p}/\partial z$.
4. Repeat steps 2 and 3 until convergence for w, $\Delta\bar{p} = \partial\bar{p}/\partial z$ and mass flow rate, and proceed to calculate velocity and pressure field on the xy plane.
5. Solve Equations 8.66 for $\hat{u}$ and $\hat{v}$ guess velocity field for the surrounding nodes.
6. Solve the pressure equation 8.69 for the pressure field, p.
7. Solve Equations E.8.2.6 and E.8.2.8 for u^* and v^*, respectively.
8. Calculate the mass source term $\boldsymbol{d}$ given by Equation E.8.2.14 and solve the pressure correction Equation E.8.2.13 for p'.
9. Calculate corrected velocity components u and v from their starred components (u^* and v^*) and pressure correction (p') using Equation 8.57.
10. Repeat steps 5 to 9 with the new velocity estimates until solution converges within a specified tolerance limit.
11. Solve the discretization Equation E.8.2.12 for temperature.
12. Reset values of velocity components u, v, and w, to upstream velocity components u^u, v^u, and w^u.
13. Repeat steps 2 to 12 for a new z location using increment Δz.

Computer implementation Let us briefly discuss a procedure that can be used for solving the system of equations for u, v, w, and p on the xy plane.

Each set of equations could be solved either by using the Gauss–Seidel method or by a combination of the TDMA and Gauss–Seidel method. Let us show the solution algorithm using these two methods for the discretization

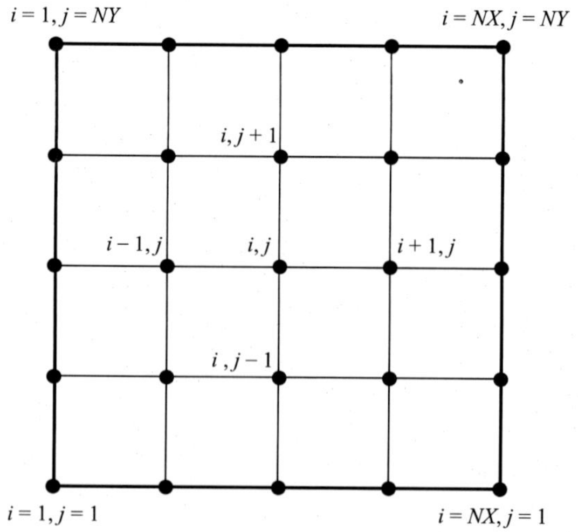

Figure 8.7 Two-dimensional grid system for computer implementation

Equation E.8.2.10 for the w-component equation given as

$$wa_{i,j}w_{i,j} = wa_{i+1,j}w_{i+1,j} + wa_{i-1,j}w_{i-1,j} + wa_{i,j+1}w_{i,j+1} + wa_{i,j-1}w_{i,j-1} - \Delta\bar{p}\,\Delta x - w_i^u \frac{w_i^u}{\Delta z}\Delta x\,\Delta y \quad \text{(E.8.2.10)}$$

Solution Algorithm Using the Gauss–Seidel Method In this method, the equation is written explicitly for the unknown

$$w_{i,j} = \left(wa_{i+1,j}w_{i+1,j} + wa_{i-1,j}w_{i-1,j} + wa_{i,j+1}w_{i,j+1} + wa_{i,j-1}w_{i,j-1} + c\right)/wa_{i,j} \quad \text{(E.8.2.14)}$$

The equation is then solved iteratively based on the values of the neighboring nodes from the previous iteration. A pseudo code for sweeping all the nodes at any iteration step can be given as

dofor I = 1, N_x

dofor J = 1, N_y

$w_{i,j} = \left(wa_{i+1,j}w_{i+1,j} + wa_{i-1,j}w_{i-1,j} + wa_{i,j+1}w_{i,j+1} + wa_{i,j-1}w_{i,j-1} + c\right)/wa_{i,j}$

enddo

enddo

Solution by Combination of TDMA and the Gauss–Seidel Methods This procedure is also referred to as the ***line-by-line method*** as suggested by Patankar (1980). The system of equations is solved by applying the ***TDMA*** algorithm along an x line or a y line and based on the values of the neighboring lines form their previous iteration. This is repeated for all lines in one direction and subsequently repeated for the other direction. For example, we can first sweep for all x lines and then all y lines. In order to apply the TDMA algorithm at any x line, we can first rewrite the discretization Equation E.8.2.10 in the form of a tridiagonal matrix format as

$$A_i w_{i,j} = B_i w_{i+1,j} + C_i w_{i-1,j} + D_i \quad \text{(E.8.2.15)}$$

where

$$A_i = wa_i, \qquad B_i = wa_{i+1,j}, \qquad C_i = wa_{i-1,j}$$

$$D_i = wa_{i,j+1}w_{i,j+1} + wa_{i,j-1}w_{i,j-1} - \Delta\bar{p}\Delta x - w_i^u \frac{w_i^u}{\Delta z}\Delta x\,\Delta y$$

Equation E.8.2.15 can now be solved using the TDMA algorithm. A pseudo code for this combination of procedure is outlined as

Dofor J = 2, $(NY - 1)$
  A(1) = 1.0
  B(1) = 0.0
  C(1) = 0.0
  D(1) = 0.0
  A(NX) = 1.0
  B(NX) = 0.0
  C(NX) = 0.0
  D(NX) = 0.0
  Dofor I = 2, $(NX - 1)$
    A(I) = wa_i
    B(I) = wa_{i+1}
    C(I) = wa_{i-1}

$$D(I) = wa_{i,j+1}w_{i,j+1} + wa_{i,j-1}w_{i,j-1} - \Delta\bar{p}\Delta x - w_i^u \frac{w_i^u}{\Delta z}\Delta x\,\Delta y$$

  Enddo
  Call tdma (NX,A,B,C,D,w)
Enddo

SIMPLEC Algorithm. The **SIMPLEC** algorithm is another variation of the **SIMPLE** algorithm, which was proposed by Doormaal and Raithby (1984) in an attempt to improve the computational cost. The **SIMPLE** algorithm requires use of optimum under-relaxation parameters for velocity and pressure in order to remove instability in the iterative solution and improve convergence rate. The **SIMPLEC** algorithm removes the need for the under-relaxation parameter α_p for pressure. In order to discuss the development of the **SIMPLEC** algorithm, let us write the x-component velocity correction Equation 8.55 in the compact form

$$a_i u_i' = \sum a_{nb} u_{nb}' + \left(p_{I-1}' - p_I'\right)\Delta y\,\Delta z \tag{8.71}$$

In order to simplify this equation, the approximation made in the **SIMPLE** algorithm is to drop the $\sum a_{nb} u_{nb}'$ term, which represents the correction contributions from the surrounding nodes. In the **SIMPLEC**, a "consistent" approximation is used by first subtracting a term $\sum a_{nb} u_i'$ from both sides of Equation 8.71 and that gives

$$\left(a_i - \sum a_{nb}\right)u_i' = \sum a_{nb}\left(u_{nb}' - u_i'\right) + \left(p_{I-1}' - p_I'\right)\Delta y\,\Delta z \tag{8.72}$$

In the next step, the term $\sum a_{nb}(u'_{nb} - u'_i)$ is neglected from the right-hand side of Equation 8.72, and a simplified form of the x component velocity correction equation is written as

$$\left(a_i - \sum a_{nb}\right)u'_i = \left(p'_{I-1} - p'_I\right)\Delta y\,\Delta z$$

or

$$u'_i = \left(p'_{I-1} - p'_I\right)\frac{\Delta y\,\Delta z}{\left(a_i - \sum a_{nb}\right)} \tag{8.73a}$$

and similarly for y- and z-component velocity correction equations

$$v'_j = \left(p'_{J-1} - p'_J\right)\frac{\Delta z\,\Delta x}{\left(a_j - \sum a_{nb}\right)} \tag{8.73b}$$

$$w'_k = \left(p'_{K-1} - p'_K\right)\frac{\Delta x\,\Delta y}{\left(a_k - \sum a_{nb}\right)} \tag{8.73c}$$

The expressions for the corrected velocity components then change to

$$u_i = u_i^* + \left(p'_{I-1} - p'_I\right)\frac{\Delta y\,\Delta z}{\left(a_i - \sum a_{nb}\right)} \tag{8.74a}$$

$$v_j = v_j^* + \left(p'_{J-1} - p'_J\right)\frac{\Delta x\,\Delta z}{\left(a_j - \sum a_{nb}\right)} \tag{8.74b}$$

$$w_k = w_k^* + \left(p'_{K-1} - p'_K\right)\frac{\Delta x\,\Delta y}{\left(a_k - \sum a_{nb}\right)} \tag{8.74c}$$

The rest of the steps are identical to those in **SIMPLE**. The **SIMPLEC** algorithm is listed below with major differences highlighted in italic.

1. Guess the pressure field p^*.
2. Solve discretization Equations 8.52 for u^*, v^*, and w^*.
3. Solve for pressure correction Equation 8.61 for p'.
4. ***Calculate the corrected pressure from Equation 8.53 without under-relaxation.***
5. ***Calculate velocity components u, v, and w from their starred and correction components using Equation 8.74 or Equation 8.63 with under-relaxation and velocity correction by Equation 8.73.***
6. Solve the discretization equations for other scalar quantities, such as turbulence quantities, temperature, and mass concentration if they influence the flow field through dependence through fluid properties such as density, viscosity, and source terms. If a particular scalar quantity does not affect the velocity field, then that quantity is calculated after a converged velocity field is achieved.

7. Reset the corrected pressure p as the new guessed pressure p^*.
8. Repeat steps 2 to 7 until the solution converges within a specified tolerance limit.

Marker-and-cell (MAC) Method. The marker-and-cell method was proposed by Harlow and Welch (1965) for the finite-difference-based solution of incompressible Navier–Stokes equations. In this method, finite difference schemes are applied to both space and time derivatives directly. The marker-cell method also uses marker particles that are convected by the fluid to the location of the free surfaces. The computational regions are divided into an Eulerian mesh set of small rectangular cells having dimensions Δx, Δy, and Δz. With respect to this set of computational cells, the velocity components are located at cell faces and pressure and other scalar values at cell centers, as shown in the Figure 8.8. The cells are numbered with indices i, j, and k, which count cell center positions in x, y, and z directions. The pressure at the center (i, j, k) of the cell is designated as $p_{i,j,k}$ while $u_{i+\frac{1}{2},j,k}$ is the u-component velocity at the center of the face between cells (i, j, k) and $(i+1, j, k)$, $v_{i,j+\frac{1}{2},k}$ is the v-component velocity at the center of the face between (i, j, k) and $(i, j+1, k)$, and $w_{i,j,k+\frac{1}{2}}$ is the w-component velocity at the center of the face between (i, j, k) and $(i, j, k+1)$. In order to number the time cycles, superscript $m+1$ is used to designate time $t = (m+1)\Delta t$. When we drop the superscript for simplicity, it is assumed that its value is m and corresponds to the time $t = m\Delta t$, i.e., it refers to values at the previous time step.

The time-dependent solution is obtained by advancing the flow field variables through small time steps, Δt.

The Finite Difference Equations Let us show the derivation of finite difference equations for viscous incompressible Navier–Stokes equations in three-dimensional Cartesian coordinates given as

$$\frac{\partial u}{\partial x} + \frac{\partial v}{\partial y} + \frac{\partial v}{\partial y} = 0 \tag{8.75a}$$

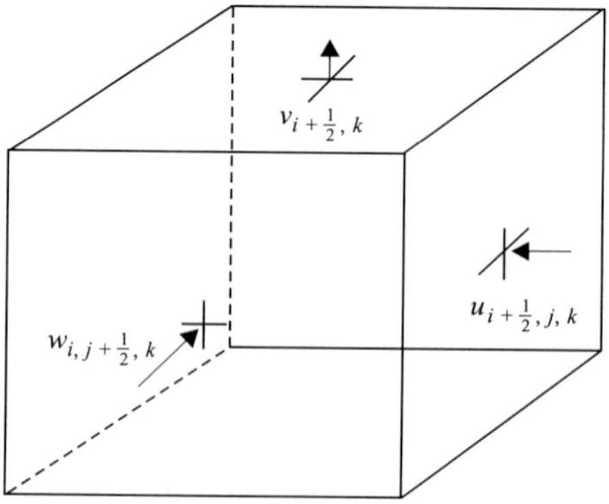

Figure 8.8 Computational cell with velocity components located at the faces

$$\frac{\partial u}{\partial t}+\frac{\partial\left(u^2\right)}{\partial x}+\frac{\partial(uv)}{\partial y}+\frac{\partial(uw)}{\partial z}=g_x-\frac{1}{\rho}\frac{\partial p}{\partial x}+\nu\left(\frac{\partial^2 u}{\partial x^2}+\frac{\partial^2 u}{\partial y^2}+\frac{\partial^2 u}{\partial z^2}\right) \tag{8.75b}$$

$$\frac{\partial v}{\partial t}+\frac{\partial(uv)}{\partial x}+\frac{\partial\left(v^2\right)}{\partial y}+\frac{\partial(vw)}{\partial z}=g_y-\frac{1}{\rho}\frac{\partial p}{\partial y}+\nu\left(\frac{\partial^2 v}{\partial x^2}+\frac{\partial^2 v}{\partial y^2}+\frac{\partial^2 v}{\partial z^2}\right) \tag{8.75c}$$

$$\frac{\partial w}{\partial t}+\frac{\partial(uw)}{\partial x}+\frac{\partial(vw)}{\partial y}+\frac{\partial\left(w^2\right)}{\partial z}=g_z-\frac{1}{\rho}\frac{\partial p}{\partial z}+\nu\left(\frac{\partial^2 w}{\partial x^2}+\frac{\partial^2 w}{\partial y^2}+\frac{\partial^2 w}{\partial z^2}\right) \tag{8.75d}$$

Substituting an explicit time approximation for the time derivatives, a forward difference scheme for the convective terms, and a central difference scheme for the diffusion terms in Equations 8.75b, 8.75c, and 8.75d, we derive the discretization equations for u, v, and w as

$$\begin{aligned}
&\frac{u^{m+1}_{i+\frac{1}{2},j,k}-u_{i+\frac{1}{2},j,k}}{\Delta t}+\frac{\left(u_{i,j,k}\right)^2-\left(u_{i+1,j,k}\right)^2}{\Delta x}\\
&+\frac{\left(u_{i+\frac{1}{2},j+\frac{1}{2},k}v_{i+\frac{1}{2}j+\frac{1}{2},k}-u_{i+\frac{1}{2},j-\frac{1}{2},k}v_{i+\frac{1}{2},j-\frac{1}{2},k}\right)}{\Delta y}\\
&+\frac{\left(u_{i+\frac{1}{2},j,k+\frac{1}{2}}w_{i+\frac{1}{2}j,k+\frac{1}{2}}-u_{i+\frac{1}{2},j,k-\frac{1}{2}}w_{i+\frac{1}{2},j,k-\frac{1}{2}}\right)}{\Delta z}=g_x-\frac{1}{\rho}\frac{p_{i,j,k}-p_{i+1,j,k}}{\Delta x}\\
&+\nu\left(\frac{u_{i+\frac{3}{2},j,k}-2u_{i+\frac{1}{2},j,k}+u_{i-\frac{1}{2},j,k}}{\Delta x^2}+\frac{u_{i+\frac{1}{2},j+1.k}-2u_{i+\frac{1}{2},j,k}+u_{i+\frac{1}{2},j-1,k}}{\Delta y^2}\right.\\
&\left.+\frac{u_{i+\frac{1}{2},j,k+1}-2u_{i+\frac{1}{2},j,k}+u_{i+\frac{1}{2},j,k-1}}{\Delta z^2}\right)
\end{aligned} \tag{8.76a}$$

$$\begin{aligned}
&\frac{v^{m+1}_{i,j+\frac{1}{2},k}-v_{i,j+\frac{1}{2},k}}{\Delta t}+\frac{\left(u_{i+\frac{1}{2},j+\frac{1}{2},k}v_{i+\frac{1}{2}j+\frac{1}{2},k}-u_{i-\frac{1}{2},j+\frac{1}{2},k}v_{i-\frac{1}{2},j-\frac{1}{2},k}\right)}{\Delta x}+\frac{\left(v_{i,j,k}\right)^2-\left(v_{i,j+1,k}\right)^2}{\Delta y}\\
&+\frac{\left(w_{i,j+\frac{1}{2},k+\frac{1}{2}}v_{i,j+\frac{1}{2},k+\frac{1}{2}}-w_{i,j+\frac{1}{2},k-\frac{1}{2}}v_{i,j+\frac{1}{2},k-\frac{1}{2}}\right)}{\Delta z}=g_y-\frac{1}{\rho}\frac{p_{i,j,k}-p_{i,j+1,k}}{\Delta y}\\
&+\nu\left(\frac{v_{i+1,j+\frac{1}{2},k}-2v_{i,j+\frac{1}{2},k}+v_{i-1,j+\frac{1}{2},k}}{\Delta x^2}+\frac{v_{i,j+\frac{3}{2}.k}-2v_{i,j+\frac{1}{2},k}+v_{i,j-\frac{1}{2},k}}{\Delta y^2}\right.\\
&\left.+\frac{v_{i,j,k+1}-2v_{i,j+\frac{1}{2},k}+v_{i,j+\frac{1}{2},k-1}}{\Delta z^2}\right)
\end{aligned} \tag{8.76b}$$

$$\frac{w^{m+1}_{i,j,k+\frac{1}{2}} - w_{i,j.k+\frac{1}{2}}}{\Delta t} + \frac{\left(u_{i+\frac{1}{2},j,k+\frac{1}{2}}w_{i+\frac{1}{2}j,k+\frac{1}{2}} - u_{i-\frac{1}{2},j,k+\frac{1}{2}}w_{i-\frac{1}{2},j,k+\frac{1}{2}}\right)}{\Delta x}$$
$$+ \frac{\left(v_{i,j+\frac{1}{2},k+\frac{1}{2}}w_{i,j+\frac{1}{2},k+\frac{1}{2}} - v_{i,j-\frac{1}{2},k+\frac{1}{2}}w_{i,j-\frac{1}{2},k+\frac{1}{2}}\right)}{\Delta y} + \frac{\left(w_{i,j,k}\right)^2 - \left(w_{i,j,k+1}\right)^2}{\Delta z}$$
$$= g_z - \frac{1}{\rho}\frac{p_{i,j,k} - p_{i,j,k+1}}{\Delta z} + \nu\left(\frac{w_{i+1,j,k+\frac{1}{2}} - 2w_{i,j,k+\frac{1}{2}} + w_{i-1,j,k+\frac{1}{2}}}{\Delta x^2}\right.$$
$$\left. + \frac{w_{i,j+1.k+\frac{1}{2}} - 2w_{i,j,k+\frac{1}{2}} + w_{i,j-1,k+\frac{1}{2}}}{\Delta y^2} + \frac{w_{i,j,k+\frac{3}{2}} - 2w_{i,j,k+\frac{1}{2}} + w_{i,j,k-\frac{1}{2}}}{\Delta z^2}\right) \tag{8.76c}$$

We can see that some of the velocity values are not centered at the points shown on the cell. Such values are expressed as an average of the adjacent nodal values. For example, we substitute expressions

$$u_{i+\frac{1}{2},j+\frac{1}{2},k} = \frac{1}{2}\left(u_{i+\frac{1}{2},j,k} + u_{i+\frac{1}{2},j+1,k}\right) \tag{8.77a}$$

$$v_{i+\frac{1}{2},j+\frac{1}{2},k} = \frac{1}{2}\left(v_{i,j+\frac{1}{2},k} + v_{i+1,j+\frac{1}{2},k}\right) \tag{8.77b}$$

$$w_{i+\frac{1}{2},j,k+\frac{1}{2}} = \frac{1}{2}\left(w_{i,j,k+\frac{1}{2}} + w_{i+1,j,k+\frac{1}{2}}\right) \tag{8.77c}$$

$$u_{i,j,k} = \frac{1}{2}\left(u_{i-\frac{1}{2},j,k} + u_{i+\frac{1}{2},j,k}\right) \tag{8.77d}$$

$$v_{i,j,k} = \frac{1}{2}\left(v_{i,j-\frac{1}{2},k} + v_{i,j+\frac{1}{2},k}\right) \tag{8.77e}$$

$$w_{i,j,k} = \frac{1}{2}\left(w_{i,j,k-\frac{1}{2}} + w_{i,j,k+\frac{1}{2}}\right) \tag{8.77f}$$

Equation 8.76 along with Equation 8.77 represents an explicit finite difference approximation of the velocity field, and can be solved with the advancement of time using small time step, Δt. The velocity field given by this explicit calculation does not necessarily result in a correct velocity field but results in a nonzero mass deficit flux D in the mass conservation equation. So, the estimated velocity field has to be corrected to ensure mass conservation in each cell. An iterative process is used for this purpose, in which the cell pressures are corrected to make all velocity field divergences become negligibly small, and so the mass deficit flux D becomes less than an assigned tolerance value ε_D. The mass deficit flux D is estimated from the mass conservation Equation 8.75a with the substitution of the estimated velocity field given by Equation 8.76

$$D = \frac{\tilde{u}^{m+1}_{i+\frac{1}{2},j,k} - \tilde{u}^{m+1}_{i-\frac{1}{2},j,k}}{\Delta x} + \frac{\tilde{v}^{m+1}_{i,j+\frac{1}{2},k} - \tilde{v}^{m+1}_{i,j-\frac{1}{2},k}}{\Delta y} + \frac{\tilde{w}^{m+1}_{i,j,k+\frac{1}{2}} - \tilde{w}^{m+1}_{i,j,k-\frac{1}{2}}}{\Delta z} \tag{8.78}$$

Note that we have changed the notation of the velocity components $\tilde{u}$, $\tilde{v}$, and $\tilde{w}$ to designate them as the estimated values. If the magnitude of the mass deficit flux D is greater than the specified tolerance limit, then the pressure correction is estimated by

$$\Delta P' = \beta D$$

where β is given by

$$\beta = \frac{\beta_0}{2\Delta t\left[\frac{1}{\Delta x^2} + \frac{1}{\Delta y^2} + \frac{1}{\Delta z^2}\right]}$$

and constant β_0 is a relaxation parameter.

The corrected pressure in each cell is then estimated as

$$p_{i,j,k} = p_{i,j,k} + \Delta p'$$

and velocity components at the faces of the cell are corrected as

$$u^{m+1}_{i+\frac{1}{2},j,k} = \tilde{u}^{m+1}_{i+\frac{1}{2},j,k} + \frac{\Delta t}{\Delta x}\Delta p'$$

$$u^{m+1}_{i-\frac{1}{2},j,k} = \tilde{u}^{m+1}_{i-\frac{1}{2},j,k} - \frac{\Delta t}{\Delta x}\Delta p'$$

$$v^{m+1}_{i,j+\frac{1}{2},k} = \tilde{v}^{m+1}_{i,j+\frac{1}{2},k} + \frac{\Delta t}{\Delta y}\Delta p'$$

$$v^{m+1}_{i,j-\frac{1}{2},k} = \tilde{v}^{m+1}_{i,j-\frac{1}{2},k} - \frac{\Delta t}{\Delta y}\Delta p'$$

$$w^{m+1}_{i,j,k+\frac{1}{2}} = \tilde{w}^{m+1}_{i,j,k+\frac{1}{2}} + \frac{\Delta t}{\Delta z}\Delta p'$$

$$w^{m+1}_{i,j,k-\frac{1}{2}} = \tilde{w}^{m+1}_{i,j,k-\frac{1}{2}} + \frac{\Delta t}{\Delta z}\Delta p'$$

This procedure is repeated for all cells to ensure that the mass deficit flux D becomes less than the tolerance value ε_D in all cells. After we reach convergence at a given time, we advance the time to the next time step.

The solution algorithm is given as follows.

1. Start with a guess velocity field. In addition coordinates of a set of marker particles are assumed to be known and this shows which region is occupied by fluid and which is empty.
2. The corresponding pressure field is computed in such a way that the rate of change of velocity divergence or mass deficit term as given by the continuity equation vanishes in all cells.
3. The two components of acceleration are calculated. The products of these with time increment per cycle then give the changes in velocity to be added to the old values.
4. The marker particles are moved according to the velocity components in their vicinities.

PROBLEMS

8.1 Consider a one-dimensional flow in a straight nozzle shown in the figure. The governing equations are given as

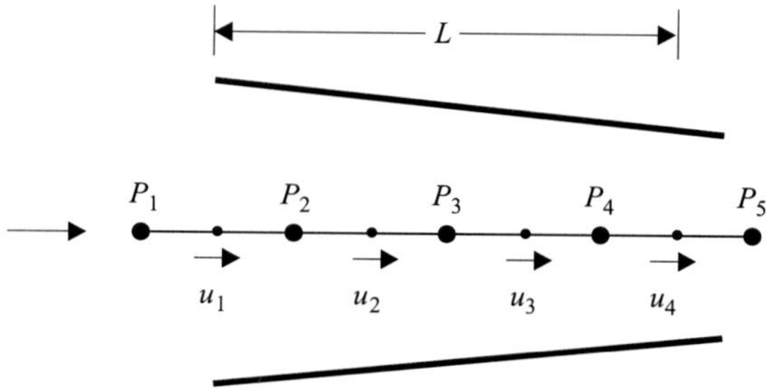

Continuity

$$\frac{\partial(uA)}{\partial x} = 0$$

x momentum

$$\frac{\partial(\rho uu)}{\partial x} = -\frac{\partial p}{\partial x}$$

where A is the cross-sectional area of the nozzle. Consider the following data: $\rho = 1.00\ \mathrm{kg/m^3}$, $A_1 = 5\ \mathrm{m^2}$, $A_2 = 4.333\ \mathrm{m^2}$, $A_3 = 3.666\ \mathrm{m^2}$, $A_4 = 3\ \mathrm{m^2}$, $P_1 = 300\ \mathrm{kPa}$, and $P_5 = 100\ \mathrm{kPa}$. Show discretization equations for velocity and pressure and use the SIMPLE algorithm to compute velocity and pressure at the nodes for two iterations. Use an initial guess for mass flow rate as $\dot{m} = \rho u A = 100\ \mathrm{kg/s}$.

8.2 Consider a fully developed heat transfer in a duct of rectangular cross section as shown in the figure.

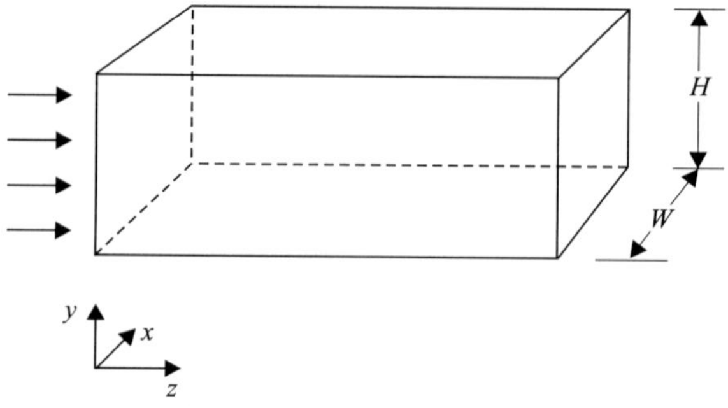

The top surface is maintained at a constant surface heat flux, q_w'', and the other three surfaces are adiabatic. The associated governing equations and boundary conditions for the fully-developed flow and heat transfer with constant surface heat flux are given as follows.

Momentum

$$\nu\left(\frac{\partial^2 w}{\partial x^2} + \frac{\partial^2 w}{\partial y^2}\right) = -\frac{1}{\rho}\frac{\partial \bar{p}}{\partial z}$$

Energy

$$\frac{k}{\rho c_p}\left(\frac{\partial^2 T}{\partial x^2} + \frac{\partial^2 T}{\partial y^2}\right) = w\frac{\partial \bar{T}}{\partial z}$$

The pressure gradient $d\bar{p}/dz$ is assumed to be known. The temperature gradient $\partial\bar{T}/\partial x$ is constant and given based on an energy balance as

$$\frac{\partial \bar{T}}{\partial x} = \frac{q_w''}{\rho c_p \bar{w}}$$

where $\bar{w}$ and $\bar{T}$ are the average velocity and average temperature given as

$$\bar{w} = \frac{1}{WH}\int_A w\,dA \quad \text{and} \quad \bar{T} = \frac{1}{WH\bar{w}}\int_A wT\,dA$$

Boundary conditions

No-slip velocity on all walls, i.e., $w = 0$ on all walls.
Constant surface heat flux on top wall, i.e., at $y = H$, $k\partial T/\partial x = q_w''$.
Rest of the walls are adiabatic, i.e., $\partial T/\partial n = 0$
Derive the discretization equations for w and T. Present the sequence of steps to be used for calculating velocity, temperature field, and convection heat transfer coefficient.

8.3 Consider the example problem 8.2 and derive all discretization equations, including the error equations for u, v, w, p, and T using the upwind scheme for the convective terms.

8.4 Consider free-convection motion and heat transfer in a two-dimensional rectangular chamber as shown. The free convection motion is created by maintaining the left wall at a high temperature T_H and right surface at a low temperature T_C. The top and bottom are assumed to be insulated. The governing equations for mass, momentum, and energy with Boussinesq assumption and for Newtonian fluid are given as

Continuity

$$\frac{\partial u}{\partial x} + \frac{\partial v}{\partial y} = 0$$

x momentum

$$u\frac{\partial u}{\partial x} + v\frac{\partial u}{\partial y} = -\frac{1}{\rho}\frac{\partial p}{\partial x} + \nu\left(\frac{\partial^2 u}{\partial x^2} + \frac{\partial^2 u}{\partial y^2}\right)$$

y momentum

$$u\frac{\partial v}{\partial x} + v\frac{\partial v}{\partial y} = -\frac{1}{\rho}\frac{\partial p}{\partial y} + \nu\left(\frac{\partial^2 v}{\partial x^2} + \frac{\partial^2 v}{\partial y^2}\right) - g[1 - \beta(T - T_0)]$$

Energy

$$u\frac{\partial T}{\partial x} + v\frac{\partial T}{\partial y} = \alpha\left(\frac{\partial^2 T}{\partial x^2} + \frac{\partial^2 T}{\partial y^2}\right)$$

Boundary conditions

Velocity No slip condition on all walls.

Temperature

1. at $x = 0$, $0 < y < H$, $T(0, y) = T_H$
2. at $x = W$, $0 < y < H$, $T(0, W) = T_C$
3. at $y = 0$, $0 < x < W$, $\left.\frac{\partial T}{\partial y}\right|_{y=0} = 0$
4. at $y = H$, $0 < x < W$, $\left.\frac{\partial T}{\partial y}\right|_{y=H} = 0$

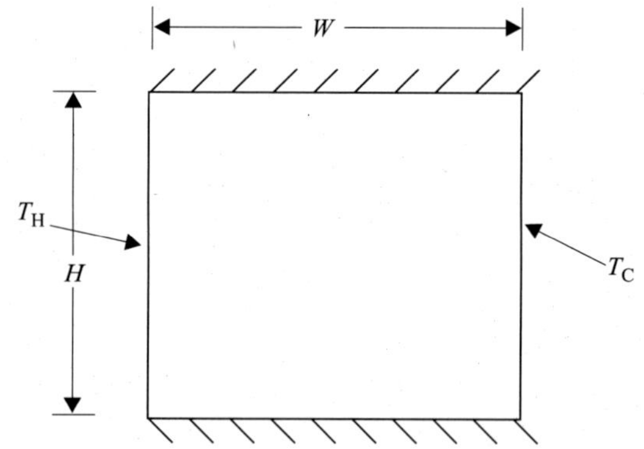

a. Derive the discretization equations for u, v, p, and T using the upwind scheme for the convective term and central difference for the diffusion terms by considering a 5 × 5 grid. Show the system of equations for each variables.
b. Derive the correction equations for the u, v, p, and T.
c. Develop a computer program implementing the SIMPLE algorithm for velocity and pressure coupling and the

line-by-line method with *tdma* solver for solving the system of equation for each variable.

8.5 Consider the developing laminar steady flow of Newtonian viscous fluid through a two-dimensional channel formed by two infinite parallel plates as shown below. The lower plate is stationary and insulated, and the upper plate is moving in the x direction with a constant speed, U_w, in the positive x direction and maintained at a constant temperature, T_W.

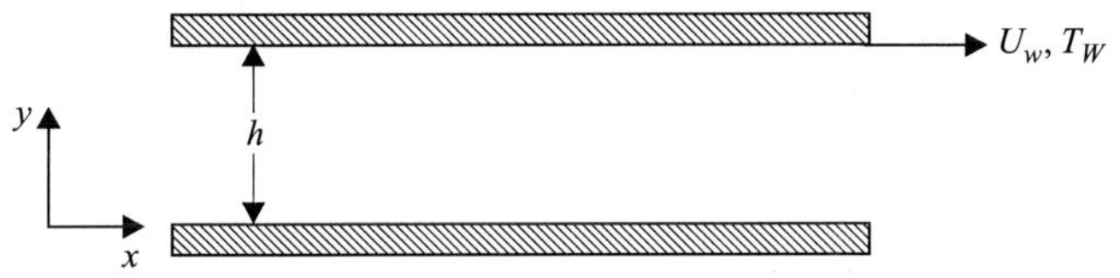

The mathematical statement of the problem is given as:

Continuity

$$\frac{\partial u}{\partial x} + \frac{\partial v}{\partial y} = 0$$

Momentum

$$\frac{\partial (u^2)}{\partial x} + \frac{\partial (uv)}{\partial y} = -\frac{1}{\rho}\frac{\partial p}{\partial x} + \nu \frac{\partial^2 u}{\partial y^2}$$

Energy

$$\frac{\partial (uT)}{\partial x} + \frac{\partial (vT)}{\partial y} = \alpha \frac{\partial^2 T}{\partial y^2}$$

Boundary conditions

Velocity

$$x = 0, \quad u = U_0, \quad v = 0$$
$$y = 0, \quad u = 0, \quad v = 0$$
$$y = h, \quad u = U_w, \quad v = 0$$

Temperature

$$x = 0, \qquad T = T_0$$
$$y = 0, \qquad \left.\frac{\partial T}{\partial y}\right|_{y=0} = 0$$
$$y = h, \qquad T = T_w$$

Derive the discretization equation for velocity, pressure and temperature using the power law scheme for the convective term for use in SIMPLER algorithm.

8.6 Consider a fully-developed laminar flow and a thermally developing flow and heat transfer, known as the thermally developing Hagen–Poiseuille flow, in a duct of rectangular cross-section as shown in the figure.

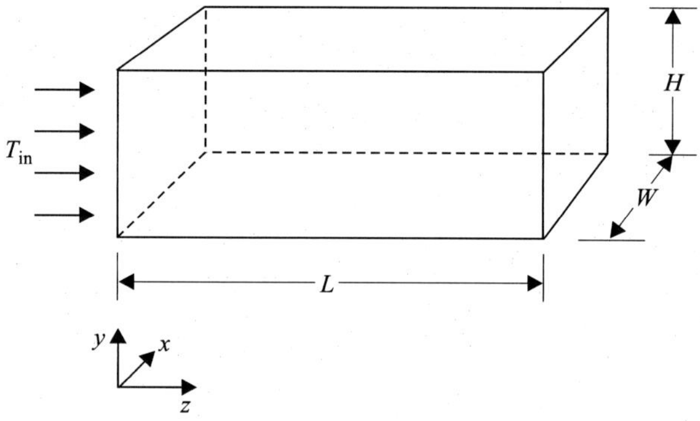

The top surface is maintained at a constant surface heat flux q_w'' and the other three surfaces are adiabatic. Assuming negligible axial conduction, the associated governing equations and boundary conditions for the fully developed flow and heat transfer with constant surface heat flux are given as follows.

Momentum

$$\nu\left(\frac{\partial^2 w}{\partial x^2} + \frac{\partial^2 w}{\partial y^2}\right) = -\frac{1}{\rho}\frac{\partial \bar{p}}{\partial z}\ .$$

Energy

$$\alpha\left(\frac{\partial^2 T}{\partial x^2} + \frac{\partial^2 T}{\partial y^2}\right) = w\frac{\partial T}{\partial z}$$

The pressure gradient $d\bar{p}/dz$ is assumed to be known for a fully-developed flow.

Boundary conditions

No-slip velocity on all walls, i.e., $w = 0$ on all walls.

Constant surface heat flux on top wall, i.e., at $y = H$, $k\partial T/\partial x = q_w''$.

Rest of the walls are adiabatic, i.e., $\partial T/\partial n = 0$.

Derive the discretization equations for w and T. Present the sequence of steps to be used for calculating velocity and temperature field and convection heat transfer coefficient.

8.7 Consider diffusion of moisture from an air stream in a two-dimensional channel to the porous adsorbing material felt lined on the bottom of the air flow channel as shown. The felt is supported at the bottom by an aluminum plate, thus making it impermeable to mass diffusion. Air enters the channel at uniform velocity u_{in} and uniform concentration C_{in}. Initial concentration distribution in the felt is uniform at C_0. Assume constant rate of moisture adsorption, m'''_{ad}.

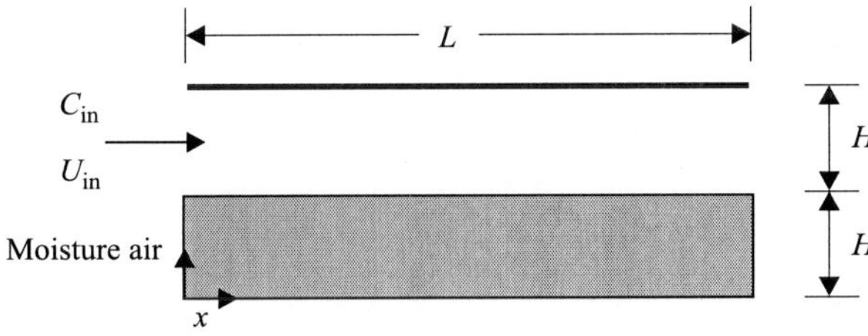

Governing equation
Channel

$$\rho H_c U_{in} \frac{\partial C_a}{\partial x} = h_D (C_a - C|_{y=H})$$

Solid

$$\frac{\partial C}{\partial t} = \frac{\partial}{\partial x}\left(D \frac{\partial C}{\partial x}\right) + \frac{\partial}{\partial y}\left(D \frac{\partial C}{\partial y}\right) + \frac{(1-\varepsilon)}{\varepsilon} \rho \dot{m}_{ad}$$

Boundary conditions

1. $y = 0 \quad \left.\frac{\partial C}{\partial x}\right|_{x=0} = 0$

2. $y = H, \quad D \left.\frac{\partial C}{\partial x}\right|_{y=H_s} = h_D (C_a - C|_{y=H_s})$

3. $x = 0, \quad \left.\frac{\partial C}{\partial x}\right|_{x=0} = 0$

4. $x = L, \quad \left.\frac{\partial C}{\partial x}\right|_{x=L} = 0$

where ε = porosity and ρ_b = material bulk density.

Apply the finite difference–control volume method using an explicit scheme and derive the system of equations for the concentration distribution in the channel and the two-dimensional transient moisture concentration distribution in the slab based on assuming a 5 × 5 grid.

Part III
FINITE ELEMENT METHOD

CHAPTER

9

BASIC STEPS

As we have discussed in previous chapters, in the ***finite difference methods*** the governing equations are approximated by a point-wise discretization scheme where derivatives are replaced by difference formulas that involve the unknown values at the nodal points. Other commonly used approximation methods are the variational methods and the finite element methods.

In the solution of a mathematical model by the ***variational method***, the governing equation is transformed into an equivalent integral form. The approximate solution to this integral form of the governing equation is obtained over the entire domain. The approximate solution is assumed to be a linear combination of an appropriately chosen function $\psi_i\ (x)$ and undetermined coefficient a_i as

$$\phi = \sum_{i=1} a_i \psi_i \tag{9.1}$$

Coefficients a_i are determined such that the integral statement equivalent to the original differential equation is satisfied. There are different kinds of variational methods, for example ***Rayleigh–Ritz***, ***collocation***, ***least-squared*** and ***Galerkin methods***. These different types of variational methods differ from each other based on the type of integral form used and selection of weighted functions. The major disadvantages of the variational method are the difficulty in selecting an approximate function that is valid over the entire solution domain satisfying the continuity and boundary conditions, and the lack of any systematic procedure for constructing them. The situation becomes even more difficult when the solution domain is geometrically complex. Due to these disadvantages, the variational methods of approximations are not as widely used as the finite difference–control volume methods or the finite element method.

The ***finite element method*** (FEM), like the finite difference method, is a numerical procedure based on discretization of the solution domain. This method overcomes the disadvantages associated with the variational method by providing a systematic procedure for discretizing the solution domain into simply shaped sub-regions, called ***finite elements***. This is followed by deriving the approximate function or solution over each of these elements, using one of the previously mentioned variational method of approximation such as ***Rayleigh–Ritz*** or ***Galerkin*** method. The total solution over the whole domain is then generated by linking together or assembling the individual element solutions satisfying the continuity at the inter-element boundaries. Hence, the finite element method can be viewed as a ***piece-wise*** or ***element-wise*** application of the variational methods. Also, a finite element method is named based on the type of variational method being used such as the Rayleigh–Ritz finite element method or the Galerkin finite element method.

The ***finite element method*** is one of the most widely used solution methods in engineering. It was originally developed for solving problems in structural mechanics. However, it became increasing popular in all branches of engineering including heat transfer and fluid mechanics. It has also become a primary analysis tool for many computer-aided design (CAD) programs. The advantages of the finite element method are quite clear. The method is easily applied to irregular-shaped objects, the medium composed of several different materials and having mixed boundary conditions.

9.1 COMPARISON BETWEEN FINITE DIFFERENCE–CONTROL VOLUME METHOD AND FINITE ELEMENT METHOD

We have discussed in Chapters 5 to 8 that in the finite difference–control volume method the solution domain is divided into a grid of discrete points, called nodes. The governing mathematical equations are then written at each node and its derivatives are expressed by the finite difference formulas, which involve unknown values at discrete grid or nodal points of the domain. This discretization procedure is referred to as ***point-wise approximation***. The system of equations resulting from all nodes including the boundary nodes is solved for the unknown values at the nodal points.

The major disadvantages of the finite difference method are: (1) difficulty in accurately representing a geometrically complex domain, (2) difficulty in imposing the boundary conditions along a non-straight boundary, and (3) inability to employ nonrectangular mesh size distribution.

In the finite element method the solution domain is subdivided into a mesh of interconnected sub-regions referred to as finite elements. A finite element model of a problem gives a ***piecewise approximation*** to the governing equations. This method has some features that accounts for its superiority over other numerical methods. These are:

1. A geometrically complex domain of the problem is represented as a collection of geometrically simple sub-domains, called ***finite elements***.
2. The approximate function is derived over each finite element by using the basic idea that any continuous function can be represented by a linear combination of interpolation functions.
3. The algebraic relations among the undetermined coefficients are obtained by satisfying the governing equations either in a ***weighted-residual integral sense*** or an equivalent ***variational form*** of the problem over each element.

One important point we should remember is that both finite difference and finite element methods are used to obtain an approximate solution of a mathematical model, which is an idealization of a physical problem subject to some assumptions. Hence the selection of the mathematical model is very important, and we should not expect any additional information regarding the physical phenomena than what is represented by the mathematical model. Also, it is necessary for us to assess the solution accuracy. If the accuracy criterion is not met, then the numerical solution has to be repeated with refined grids or mesh size distributions and/or with higher order elements until sufficient accuracy is reached. Once the mathematical model has been solved accurately and the results have been interpreted, we may decide to consider a more refined mathematical model in order to get additional insight of the physical problem. This will leads to higher-order approximations and additional grid or mesh size distributions.

9.2 BASIC STEPS IN FINITE ELEMENT METHODS

The general procedure in the finite element analysis usually involves some basic steps. Details of each step may vary from problem to problem. In this section we will briefly discuss the basic steps employed in the formulation and application of the finite element method. The details of each one of these steps will be developed and described fully in the subsequent sections and chapters. Although the particulars may vary, the implementation of the finite element method can be subdivided into the following basic steps:

1. Discretization of the Solution Domain This step involves dividing the solution domain into a finite number of sub-domains. Each sub-domain is called a ***finite element***. The collection of all elements is called the ***finite element mesh***. The shape, size, number, and orientation of the elements are selected in such a way that the solution domain is represented closely and accurately, and solution is obtained with least computational difficulties. The final selection is established through a systematic mesh refinement study until desirable accuracy is achieved. Figure 9.1 show some examples of elements employed in one, two, and three dimensions.

The points of intersection of the lines that make up the sides of the elements are referred to as ***nodes*** or ***nodal points*** and the sides themselves

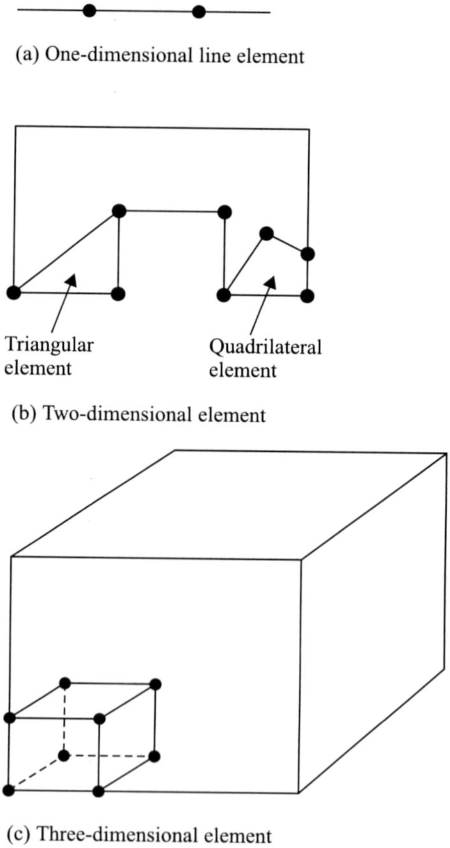

Figure 9.1 Examples of finite elements in one, two, and three dimensions

are called ***nodal lines*** or ***planes.*** This step also includes locating and numbering the nodal points, as well as specifying their coordinate values. A detail description of the type, selection, orientation of finite elements, and the numbering scheme is given Chapter 10.

2. Integral Formulation of the Governing Equation In this step, the governing differential equation is transformed into an integral form, which represents an optimum fit of the approximating function to the solution of the equation. There are several methods that are available for this purpose. The most common ones are: the ***method of weighted-residual integral*** and the ***variational formulation***. These methods specify relationships between the unknowns in the approximation function that satisfy the underlying governing differential equation in an optimal manner. The ***method of weighted-residual integral*** is based on the minimization of a residual term R, which is obtained as an approximate or a trial solution is substituted into the governing differential equation. In the ***variational formulation***, the minimization is sought for a ***variational form*** or the equivalent ***functional form I*** of the governing equation. The variational form or the functional ***I*** could be obtained either by using the principles of variational calculus or by deriving a weak form by evaluating the weighted-residual integral. Both methods, however, use ***classical variational method of approximations*** to solve the integral equations. Detailed descriptions of these methods are given in subsequent sections.

3. Selection of Approximation Function In this step, we develop equations to approximate the solutions for each element. This involves two steps. First, we must choose an appropriate function with unknown coefficients that will be used to approximate the solution. Second, we evaluate the coefficients so that the chosen function approximates the solution in an optimal fashion.

A number of mathematical functions such as polynomials or trigonometric series can be used for this purpose. Polynomials are often selected for this purpose, as they are easy to manipulate mathematically. The approximate solution is assumed to be a linear combination of appropriately selected functions $\psi_i(x)$ and undetermined coefficients a_i as expressed by Equation 9.1. For example, in one-dimensional problems, the simplest choice is a first-order polynomial or a straight line such as

$$\phi(x) = a_0 + a_1 x \tag{9.2}$$

where $\phi(x)$ is the dependent variable, a's are constants, and x is the independent variable and represents the function $\psi_1\ (x)$. Figure 9.1 shows a one-dimensional line element with length L and two nodal points i and j. The coefficients in Equation 9.2 can be found by satisfying two nodal points, i.e., $\phi_i\ (x_i)$ and $\phi_j\ (x_j)$. Substituting these two nodal points in

Equation 9.1 we have

$$\phi_i(x) = a_0 + a_1 x_i \tag{9.3a}$$

and

$$\phi_j(x) = a_0 + a_1 x_j \tag{9.3b}$$

These equations can be solved using Cramer's rule for

$$a_0 = \frac{\phi_i x_j - \phi_j x_i}{x_j - x_i} \tag{9.4a}$$

and

$$a_1 = \frac{\phi_j - \phi_i}{x_j - x_i} \tag{9.4b}$$

Substituting Equation 9.4 into Equation 9.2, we get

$$\phi = \left(\frac{x_j - x}{x_j - x}\right)\phi_i + \left(\frac{x - x_i}{x_j - x_i}\right)\phi_j \tag{9.5}$$

or

$$\phi = N_i \phi_i + N_j \phi_j \tag{9.6}$$

where

$$N_i = \frac{x_j - x}{x_j - x_i} \tag{9.7a}$$

and

$$N_j = \frac{x - x_i}{x_j - x_i} \tag{9.7b}$$

Equation 9.6 is called an ***approximation function***, which is a sum of nodal values multiplied by linear interpolation functions of x. It also provides a means to predict intermediate values between given values of ϕ at the nodal points. The interpolation functions expressed as N, given by Equation 9.7, are called ***interpolation or shape functions***. The approximation Equation 9.6 can be written in matrix form as

$$\phi = [N]\{\phi\} \tag{9.8}$$

where

$$[N] = \left[N_i, N_j\right] = \text{row vector of shape functions}$$

$$\{\phi\} = \begin{Bmatrix} \phi_i \\ \phi_j \end{Bmatrix} = \text{column vector of unknown values at the nodal points}$$

In a general form the approximation function can be written as

$$\phi = N_i\phi_i + N_2\phi_2 \cdots N_n\phi_n \tag{9.9a}$$

or

$$\phi = [N]\{\phi\} \tag{9.9b}$$

where

$$\{\phi\} = \begin{Bmatrix} \phi_1 \\ \phi_2 \\ \vdots \\ \phi_n \end{Bmatrix} = \text{a column vector of unknown nodal values}$$

$$[N] = [N_1, N_2, \ldots, N_n] = \text{a row vector of shape functions}$$

4. Formation of Element Characteristics Equation After the selection of the approximation function, we need to develop an equation which governs the behavior of the element. A direct substitution of the approximation function into the integral form of the element governing equation results in the ***element characteristics equation***, which is also referred to as element stiffness equation. Mathematically, the resulting element equation forms a system of linear algebraic equations that can be expressed in matrix form as

$$[k^e]\{\phi\} = \left\{f^e\right\} \tag{9.10}$$

where

$[k^e]$ = an element characteristics matrix or element stiffness matrix

$\{\phi\}$ = a column vector of unknown values at the nodal points of the element

$\left\{f^e\right\}$ = a column vector of element nodal forcing parameters that represents the effect of any external influences applied at the nodal points

This equation represents the optimum fit of the function to the solution of the governing differential equation. It can be mentioned here again that the procedure for obtaining the element characteristics equation is based on element-wise application of the ***classical variational method of approximation***.

5. Assembly of Element Equations The individual element equations are linked together or assembled to characterize the unified behavior of the entire system. The assembly process is governed by the concept of continuity. That is, the solutions for contiguous elements are matched so that the unknown values and/or its derivatives at their common nodes are equivalent. This will cause a continuous solution in the domain. When all the individual element equations are finally assembled, the entire system of equations, known as the global characteristic equation or global stiffness equation, can be expressed in matrix form as

$$[K]\{\Phi\} = \{F\} \tag{9.11}$$

where

K = global characteristic matrix or stiffness matrix
Φ = column vector for unknowns
F = column vector for external forces

Φ and F is the collection of all vectors ϕ and f associated with all individual elements respectively.

6. Implementation of Boundary Conditions In this step, the boundary conditions are introduced directly in the global system of equations. This may also require rearrangement of the system of equations, particularly for cases with assigned boundary conditions of the second or third kind.

7. Solution of System of Equations In this step, the appropriate solver is selected and solutions to the system of equations are obtained. In many cases the elements can be re-configured so that the resulting equations are banded, and highly efficient solution algorithms could be employed.

8. Error Estimate and Convergence Check An error analysis study is conducted in this step. Errors associated with the approximate solutions are estimated and checked for convergence. The errors in the finite element method include (1) the truncation error due to the discretization of the solution domain and error associated with selection of the approximate solution as a linear combination of interpolation functions, and (2) truncation and round-off errors associated with the numerical evaluation of integrals or selection of quadrature formulas, and solutions of systems of equations on the computer.

A mesh refinement study is conducted to improve the solution and obtain the desired accuracy or convergence. This may also, in some cases, require the selection of higher-order approximation solution.

9. Postprocessing Once the converged solution is obtained, other quantities or secondary variables of interest are calculated. These quantities are usually related to the derivative of the variables, and include quantities such as the volume flow, shear stress, and heat transfer rate.

9.3 INTEGRAL FORMULATION

We have discussed in the previous section that in a finite element method we start with an integral form of the governing equation. One of the important steps in the finite element method is to transform the governing differential equation into an equivalent integral form using one of the integration formulation techniques, namely ***variational formulation*** or ***weighted residual integration formulation***. Let us now briefly describe these two integral formulation procedures.

9.3.1 Variational Formulation

In this approach, an integral statement of the problem is obtained by deriving an equivalent ***variational form*** of the problem using the principles of ***calculus of variations***. A brief discussion of the principles of ***calculus of variations*** is given in the next sub-section. This ***variational form*** involves a ***functional, I***, such that finding a minimized or stationary or extremized value of the functional, ***I***, is equivalent to finding the solution of the original differential statement of the problem. A general expression of this functional form is given as

$$I = \int_V F\left(\tilde{x}, \phi, \frac{\partial \phi}{\partial \tilde{x}} \cdots \right) dV \tag{9.12}$$

where V is the solution domain, $\tilde{x}$ is the vector for independent variables, and $\{\phi\}$ is the unknown solution vector. The condition of minimum value of the functional with respect to the unknown solution vector is given as

$$\frac{\partial I}{\partial \{\phi\}} = \begin{Bmatrix} \dfrac{\partial I}{\partial \phi_1} \\ \dfrac{\partial I}{\partial \phi_2} \\ \vdots \\ \dfrac{\partial I}{\partial \phi_M} \end{Bmatrix} = 0 \tag{9.13}$$

where M indicates the number of unknown variables. In the case of a finite element formulation, the functional over the whole solution domain can be written as a summation of all elemental contribution, $I^{(e)}$, as

$$I = \sum_{e=1}^{n} I^{(e)} \tag{9.14}$$

Combining Equations 9.13 and 9.14 we get the condition of minimization of the functional over the entire solution domain as

$$\sum_{e=1}^{n} \frac{\partial I^{(e)}}{\partial \phi^{(e)}} = 0 \tag{9.15}$$

As we evaluate the elemental contribution $\partial I^{(e)}/\partial \phi^{(e)}$, we obtain the element characteristic equation as

$$\frac{\partial I^{(e)}}{\partial \phi^{(e)}} = \left[\kappa^{(e)}\right]\left\{\phi^{(e)}\right\} - \left\{f^{(e)}\right\} = 0 \tag{9.16}$$

Equation 9.16 is applied successively to all elements, and the resulting elemental equations are then assembled to form the global system of equations.

Principles of Calculus of Variations We have mentioned that one of the important steps in the formation of an integral statement using the variational method of approximation, as well as the finite element method, is to determine an equivalent ***variational*** or ***functional*** form using the ***calculus of variations***. Lagrange introduced an operator δ, which is similar to the differentiation operation used in 1760 (Bliss, 1946). This operator is referred to as variation, and all techniques and theory developed using this operator is referred to as the ***calculus of variation***. Application of the calculus of variation is mainly concerned with the determination of the maximum and minimum values or stationary of a certain expression known as ***functional***. A brief review of the variational form and notation, and a procedure for obtaining the equivalent variational or functional forms are given here. For a more detailed description of this method readers are suggested to read other reference books (Bliss, 1946; Hilderbrand, 1965; Ewing 1985).

The Variational Forms and Notation In order to establish the techniques, we will introduce the definition of a functional and various notation to the calculus of variation.

Functional A quantity such as

$$I = \int_{x_1}^{x_2} F\left(x, \phi(x), \frac{d\phi}{dx}, \ldots\right) dx \tag{9.17}$$

is called a functional if for any function $\phi(x)$ the quantity becomes a definite numerical value.

δ Operator For the integrand, $F(x, \phi(x), d\phi/dx, \ldots)$, we change the function $\phi(x)$ to be determined as a new function $\phi(x) + \varepsilon\eta(x)$. The change $\varepsilon\eta(x)$ in $\phi(x)$ is called the variation of ϕ and it is denoted as $\delta\phi$.

Laws of Variation Laws of variation of sums, products, ratios, powers, and other similar operations are completely analogous to the corresponding laws of differentiations. For example,

Sum: $$\delta(\Phi_1 + \Phi_2) = \delta\Phi_1 + \delta\Phi_2$$

Products: $$\delta(\Phi_1\Phi_2) = \Phi_1\delta\Phi_2 + \Phi_2\delta\Phi_1$$

Ratio: $$\delta\left(\frac{\Phi_1}{\Phi_2}\right) = \frac{\Phi_2\delta\Phi_1 - \Phi_1\delta\Phi_2}{\Phi_2^2}$$

Power: $$\delta[(\Phi_1)^n] = n\Phi_1^{n-1}\delta\Phi_1$$

Operator δ is commutative with differential and integral operators, i.e.,

$$\frac{\partial}{\partial x}(\delta\phi) = \delta\left(\frac{\partial\phi}{\partial x}\right)$$

$$\delta\int_{x_1}^{x_2} \phi(x)\, dx = \int_{x_1}^{x_2} \delta\phi(x)\, dx$$

and

$$\int_{x_1}^{x_2} \phi\delta\phi\, dx = \delta\int_{x_1}^{x_2} \frac{1}{2}\phi^2\, dx$$

Determination of a Functional or the Variational Form Let us demonstrate a general procedure for determining the functional, ***I***, or the variational form of a problem that involves governing differential equations and boundary conditions. Consider the governing differential equation of the form

$$\frac{d}{dx}\left(\Gamma_x\frac{d\phi}{dx}\right) + S_1\phi + S_0 = 0 \tag{9.18}$$

The associated boundary conditions will be prescribed as we proceed. In order to derive a corresponding variational form, we first multiply both sides of Equation 9.18 by a variation $\delta\phi$ and integrate it over the solution domain (0–L)

$$\int_0^L \frac{d}{dx}\left(\Gamma_x \frac{d\phi}{dx}\right)\delta\phi\, dx + \int_0^L S_1\phi\delta\phi\, dx + \int_0^L S_0\delta\phi\, dx = 0 \tag{9.19}$$

Using rules of variation, the second and third integrands are evaluated as

$$\int_0^L S_1\phi\delta\phi\, dx = \delta\int_0^L \frac{1}{2}S_1\phi^2\, dx \tag{9.20}$$

and

$$\int_0^L S_0\delta\phi\, dx = \delta\int_0^L S_0\phi\, dx \tag{9.21}$$

The first integral is evaluated using integration by parts and rules of variation as

$$\begin{aligned}\int_0^L \frac{d}{dx}\left(\Gamma_x \frac{d\phi}{dx}\right)\delta\phi\, dx &= \left[\Gamma_x \frac{d\phi}{dx}\delta\phi\right]\Bigg|_0^L - \int_0^L \Gamma_x \frac{d\phi}{dx}\delta\left(\frac{d\phi}{dx}\right) dx \\ &= \Gamma_x \frac{d\phi}{dx}\delta\phi\Bigg|_0^L - \delta\int_0^L \frac{1}{2}\Gamma_x\left(\frac{d\phi}{dx}\right)^2 dx\end{aligned} \tag{9.22}$$

Substituting Equations 9.20, 9.21, and 9.22 into Equation 9.19, we have

$$\delta\int_0^L \left[-\frac{1}{2}\Gamma_x\left(\frac{d\phi}{dx}\right)^2 + \frac{1}{2}S_1\phi^2 + S_0\phi\right] dx + \left[\Gamma_x \frac{d\phi}{dx}\delta\phi\right]\Bigg|_0^L = 0 \tag{9.23}$$

Let us now consider different cases involving different types of boundary conditions.

Case 1: Constant Surface Value For a constant surface value or boundary condition of the first kind, let us assume $\phi = \phi_0$ at $x = 0$ and $\phi = \phi_L$ at $x = L$. Since the values are fixed at the boundaries, $\delta\phi$ becomes zero at $x = 0$ and $x = L$, and the last integrated term $[\Gamma\,(d\phi/dx)\,\delta\phi]|_0^L$ vanishes. This reduces the Equation 9.23 to the final variational form as

$$-\delta\int_0^L\left[\frac{1}{2}\Gamma_x\left(\frac{d\phi}{dx}\right)^2-\frac{1}{2}S_1\phi^2-S_0\phi\right]dx=0 \quad (9.24)$$

and the ***functional*** is given as

$$I=\int_0^L\left[-\frac{1}{2}\Gamma_x\left(\frac{d\phi}{dx}\right)^2+\frac{1}{2}S_1\phi^2+S_0\phi\right]dx \quad (9.25)$$

with essential boundary condition $\phi(0)=\phi_0$ and $\phi(L)=\phi_L$.

Case 2: Constant Surface Flux The mathematical statement of the constant surface flux or the boundary conditions of the second kind, as depicted in Figure 9.2 is given as

$$\text{at}\,x=0,\qquad -\Gamma_x\frac{d\phi}{dx}=f''_{ls} \quad (9.26)$$

and

$$\text{at}\,x=L,\qquad \Gamma_x\frac{d\phi}{dx}=f''_{rs} \quad (9.27)$$

With the substitution of these conditions, Equation 9.23 takes the form

$$\delta\int_0^L\left[-\frac{1}{2}\Gamma_x\left(\frac{d\phi}{dx}\right)^2+\frac{1}{2}S_1\phi^2+S_0\phi\right]dx+f''_{rs}\,\delta\phi\big|_{x=L}+f''_{ls}\,\delta\phi\big|_{x=0}=0 \quad (9.28)$$

Using rules of variation we get the equivalent variational form of the problem as

$$\delta\left[\int_0^L\left\{-\frac{1}{2}\Gamma_x\left(\frac{d\phi}{dx}\right)^2+\frac{1}{2}S_1\phi^2+S_0\phi\right\}dx+f''_{rs}\phi\big|_{x=L}+f''_{ls}\phi\big|_{x=0}\right]=0 \quad (9.29a)$$

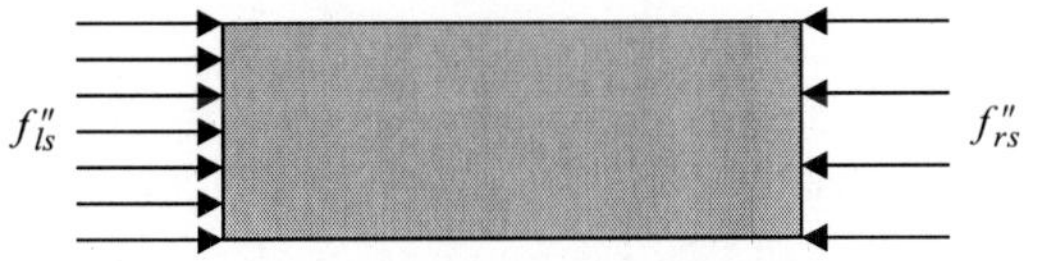

Figure 9.2 Constant surface boundary conditions on a plane slab

and the corresponding ***functional*** is given as

$$I = \int_0^L \left\{ -\frac{1}{2}\Gamma_x \left(\frac{d\phi}{dx}\right)^2 + \frac{1}{2} S_1\phi^2 + S_0\phi \right\} dx + f''_{rs}\phi\big|_{x=L} + f''_{ls}\phi\big|_{x=0} \tag{9.29b}$$

It can also be noted that for the case of zero surface flux condition, i.e.,

$$\Gamma_x \frac{d\phi}{dx}\bigg|_{x=0} = 0 \quad \text{and} \quad \Gamma_x \frac{d\phi}{dx}\bigg|_{x=L} = 0$$

and the equivalent variational form and the functional are given as

$$\delta\left[\int_0^L \left\{ -\frac{1}{2}\Gamma_x \left(\frac{d\phi}{dx}\right)^2 + \frac{1}{2} S_1\phi^2 + S_0\phi \right\} dx \right] = 0 \tag{9.30}$$

and

$$I = \int_0^L \left\{ -\frac{1}{2}\Gamma_x \left(\frac{d\phi}{dx}\right)^2 + \frac{1}{2} S_1\phi^2 + S_0\phi \right\} dx \tag{9.31}$$

Case 3: Convective Boundary Condition Let us now consider the convective boundary condition or the boundary condition of the third kind at the two boundaries shown in Figure 9.3.

The corresponding mathematical statements for the boundary conditions are

$$\text{at } x = 0, \quad \Gamma_x \frac{d\phi}{dx}\bigg|_{x=0} = h_{ls}(\phi|_{x=0} - \phi_{l\infty}) \tag{9.32}$$

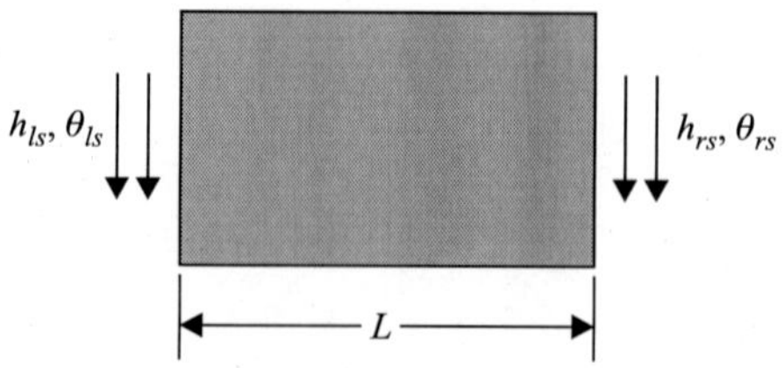

Figure 9.3 Convective boundary conditions on plane slab

and

$$\text{at } x = L, \quad -\Gamma_x \left.\frac{d\phi}{dx}\right|_{x=L} = h_{rs}(\phi|_{x=L} - \phi_{r\infty}) \tag{9.33}$$

Let us now evaluate the last integrated term in Equation 9.23 as

$$\left[\Gamma_x \frac{d\phi}{dx}\delta\phi\right]_0^L = \left[\Gamma_x \frac{d\phi}{dx}\delta\phi\right]_{x=L} - \left[\Gamma_x \frac{d\phi}{dx}\delta\phi\right]_{x=0}$$

$$= h_{rs}(\phi_{r\infty} - \phi|_{x=L})\delta\phi|_{x=L} - h_{ls}(\phi|_{x=0} - \phi_{l\infty})\delta\phi|_{x=0}$$

$$= \delta\left[h_{rs}\phi_{r\infty}\phi|_{x=L} - \frac{1}{2}h_{rs}(\phi|_{x=L})^2\right] - \delta\left[\frac{1}{2}h_{ls}(\phi|_{x=0})^2 - h_{ls}\phi_{l\infty}\phi|_{x=0}\right]$$

$$= \delta\left[h_{rs}\phi_{r\infty}\phi|_{x=L} - \frac{1}{2}h_{rs}(\phi|_{x=L})^2 - \frac{1}{2}h_{ls}(\phi|_{x=0})^2 + h_{ls}\phi_{l\infty}\phi|_{x=0}\right] \tag{9.34}$$

Now substituting Equation 9.34 into Equation 9.23, we get the variational form as

$$\delta\left[\int_0^L \left\{-\frac{1}{2}\Gamma_x\left(\frac{d\phi}{dx}\right)^2 + \frac{1}{2}S_1\phi^2 + S_0\phi\right\}\right.$$

$$\left. + \left\{h_{rs}\phi_{r\infty}\phi|_{x=L} - \frac{1}{2}h_{rs}(\phi|_{x=L})^2 - \frac{1}{2}h_{ls}(\phi|_{x=0})^2 + h_{ls}\phi_{l\infty}\phi|_{x=0}\right\}\right] = 0 \tag{9.35}$$

and the functional is given as

$$I = \int_0^L \left\{-\frac{1}{2}\Gamma_x\left(\frac{d\phi}{dx}\right)^2 + \frac{1}{2}S_1\phi^2 + S_0\phi\right\} dx$$

$$+ \left\{h_{rs}\phi_{r\infty}\phi|_{x=L} - \frac{1}{2}h_{rs}(\phi|_{x=L})^2 - \frac{1}{2}h_{ls}(\phi|_{x=0})^2 + h_{ls}\phi_{l\infty}\phi|_{x=0}\right\} \tag{9.36}$$

Example 9.1: One-dimensional Variational Formulation: Determine the variational form and functional for the following one-dimensional fin problem using the ***calculus of variation***.

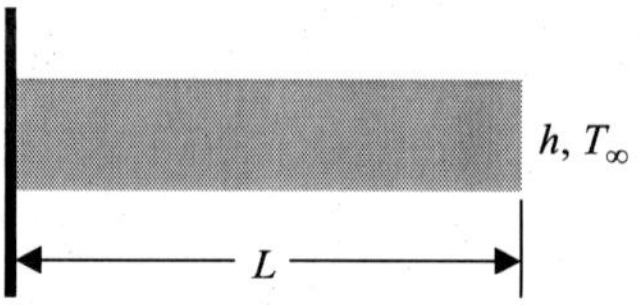

One-dimensional fin

Governing equation

$$\frac{d}{dx}\left(kA\frac{dT}{dx}\right) - h_c P(T - T_\infty) = 0 \tag{9.37}$$

Boundary conditions

1. $x = 0, \quad T = T_0$ (9.38a)

2. $x = L, \quad -k\frac{dT}{dx}\bigg|_{x=L} = h_c(T|_{x=L} - T_\infty)$ (9.38b)

Solution: In order to derive the variational form, we multiply both sides of the governing equation by a variation δT and integrate over the length of the fin

$$\int_0^L \frac{d}{dx}\left(kA\frac{dT}{dx}\right)\delta T\,dx - \int_0^L h_c PT\,\delta T\,dx + \int_0^L h_c PT_\infty\,\delta T\,dx = 0 \tag{9.39}$$

Evaluating the first term by integration-by-parts

$$\left[kA\frac{dT}{dx}\delta T\right]_0^L - \int_0^L kA\frac{dT}{dx}\frac{d}{dx}(\delta T)\,dx - \int_0^L h_c PT\,\delta T\,dx + \int_0^L h_c PT_\infty\,\delta T\,dx = 0 \tag{9.40}$$

Using rules of variational calculus

$$kA\frac{dT}{dx}\delta T\bigg|_{x=L} - kA\frac{dT}{dx}\delta T\bigg|_{x=0}$$

$$-\frac{1}{2}\delta\int_0^L kA\left(\frac{dT}{dx}\right)^2 dx - \frac{1}{2}\delta\int_0^L h_c PT^2\,dx + \delta\int_0^L h_c PT_\infty T\,dx = 0 \tag{9.41}$$

We have discussed before that for the ***boundary condition of the first kind***, i.e., of constant surface temperature value, δT must vanish at that boundary surface. So, in this problem we set $\delta T|_{x=0}=0$. At the tip of the fin we have the ***boundary condition of the third kind***, i.e., heat transfer rate is specified as convection rate and we set $-kA(dT/dx)\big|_{x=L}=hA(T|_{x=L}-T_\infty)$. Substituting these conditions directly in Equation 9.41, we get

$$h_cA(T|_{x=L}-T_\infty)\delta T|_{x=L}-\frac{1}{2}\delta\int_0^L kA\left(\frac{dT}{dx}\right)^2 dx-\frac{1}{2}\delta\int_0^L h_cPT^2\,dx+\delta\int_0^L h_cPT_\infty T\,dx=0 \tag{9.42}$$

Further using the rules of calculus of variation in the first term of the equation, we have the ***variational form*** of the problem as

$$\delta\left[-\frac{1}{2}\int_0^L kA\left(\frac{dT}{dx}\right)^2 dx-\frac{1}{2}\int_0^L h_cPT^2\,dx+\int_0^L h_cPT_\infty T\,dx+\frac{1}{2}hA(T|_{x=L})^2-h_cAT_\infty\,\delta T|_{x=L}\right]=0 \tag{9.43}$$

and the corresponding ***functional*** is given as

$$I=-\frac{1}{2}\int_0^L kA\left(\frac{dT}{dx}\right)^2 dx-\frac{1}{2}\int_0^L h_cPT^2\,dx+\int_0^L h_cPT_\infty T\,dx+\frac{1}{2}h_cA(T|_{x=L})^2-h_cAT_\infty T|_{x=L} \tag{9.44}$$

with essential condition $T(0)=T_0$.

Variational Formulation for Multi-dimensional Problems Let us consider the derivation of the variational form for a three-dimensional problem such as the one given by

$$\frac{d}{dx}\left(\Gamma_x\frac{d\phi}{dx}\right)+\frac{d}{dy}\left(\Gamma_y\frac{d\phi}{dy}\right)+\frac{d}{dz}\left(\Gamma_z\frac{d\phi}{dz}\right)+S_1\phi+S_0=0 \quad \text{over volume } \mathbf{V} \tag{9.45}$$

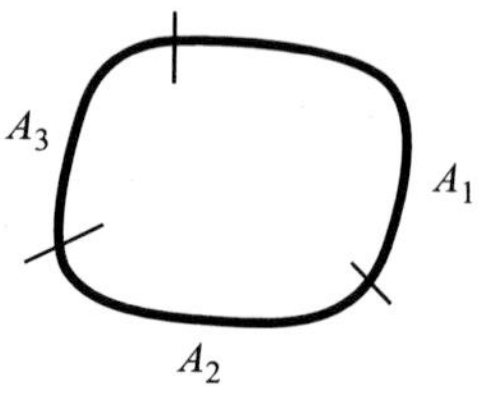

Figure 9.4 Control volume with boundary surfaces

subject to the boundary conditions

$$\phi = \phi_s \qquad \text{on } A_1 \tag{9.46a}$$

$$\Gamma_x \frac{\partial \phi}{\partial x} n_x + \Gamma_y \frac{\partial \phi}{\partial y} n_y + \Gamma_z \frac{\partial \phi}{\partial z} n_z + q_s'' = 0 \qquad \text{on } A_2 \tag{9.46b}$$

$$\Gamma_x \frac{\partial \phi}{\partial x} n_x + \Gamma_y \frac{\partial \phi}{\partial y} n_y + \Gamma_z \frac{\partial \phi}{\partial z} n_z + h_s(\phi - \phi_\infty) = 0 \qquad \text{on } A_3 \tag{9.46c}$$

where n_x, n_y, and n_z are outward direction cosines drawn normal to the surface; A_1 is the surface area on which the constant surface value ϕ_s is specified; A_2 is the surface area on which the constant surface flux ϕ_s'' is specified, and A_3 is the surface area on which the mixed or the convective boundary condition is specified.

In order to obtain the variational form of this problem, multiply the equation by $\delta\phi$

$$\iiint_V \frac{d}{dx}\left(\Gamma_x \frac{d\phi}{dx}\right) \delta\phi \, dV + \iiint_V \frac{d}{dy}\left(\Gamma_x \frac{d\phi}{dy}\right) \delta\phi \, dV + \iiint_V \frac{d}{dz}\left(\Gamma_z \frac{d\phi}{dz}\right) \delta\phi dV$$
$$+ \iiint_V S_1 \phi \, \delta\phi \, dV + \iiint_V S_0 \delta\phi \, dV = 0 \tag{9.47}$$

The first integrand term can be written using differentiation rule as

$$\iiint_V \frac{d}{dx}\left(\Gamma_x \frac{d\phi}{dx}\right) \delta\phi \, dV = \iiint_V \frac{d}{dx}\left(\Gamma_x \frac{d\phi}{dx} \delta\phi\right) dV - \iiint_V \Gamma_x \frac{d\phi}{dx} \frac{d}{dx}(\delta\phi) \, dV \tag{9.48}$$

Using the extended divergence theorem (see Appendix B) for the first term and rules of calculus of variation for the second term on the right-hand side of the Equation 9.48, we get

$$\iiint\limits_V \frac{d}{dx}\left(\Gamma_x \frac{d\phi}{dx}\right) \delta\phi\, dV = \oiint\limits_A \Gamma_x \frac{d\phi}{dx} \delta\phi n_x\, dA - \delta \iiint\limits_V \frac{1}{2}\Gamma_x \left(\frac{d\phi}{dx}\right)^2 dV \qquad (9.49a)$$

Similarly, we can express the second and third integrand terms of Equation 9.47 as

$$\iiint\limits_V \frac{d}{dy}\left(\Gamma_y \frac{d\phi}{dy}\right) \delta\phi\, dV = \oiint\limits_A \Gamma_y \frac{d\phi}{dy} \delta\phi n_y\, dA - \delta \iiint\limits_V \frac{1}{2}\Gamma_y \left(\frac{d\phi}{dy}\right)^2 dV \qquad (9.49b)$$

and

$$\iiint\limits_V \frac{d}{dz}\left(\Gamma_z \frac{d\phi}{dz}\right) \delta\phi\, dV = \oiint\limits_A \Gamma_z \frac{d\phi}{dz} \delta\phi n_z\, dA - \delta \iiint\limits_V \frac{1}{2}\Gamma_z \left(\frac{d\phi}{dz}\right)^2 dV \qquad (9.49c)$$

Substituting Equation 9.49 into Equation 9.48

$$-\delta \iiint\limits_V \left[\frac{1}{2}\left(\Gamma_x \left(\frac{d\phi}{dx}\right)^2 + \Gamma_x \left(\frac{d\phi}{dx}\right)^2 + \Gamma_x \left(\frac{d\phi}{dx}\right)^2\right) - \frac{1}{2} S_1 \phi^2 - S_0 \phi\right] dV$$

$$+ \oiint\limits_A \left(\Gamma_x \frac{d\phi}{dx} n_x + \Gamma_y \frac{d\phi}{dy} n_y + \Gamma_x \frac{d\phi}{dz} n_z\right) \delta\phi\, dA \qquad (9.50)$$

The boundary integral term can now be evaluated by considering the specified boundary conditions given by Equation 9.46 on different segments as

$$\oiint\limits_A \left(\Gamma_x \frac{d\phi}{dx} n_x + \Gamma_y \frac{d\phi}{dy} n_y + \Gamma_x \frac{d\phi}{dz} n_z\right) \delta\phi\, dA$$

$$= \oiint\limits_{A_1} \left(\Gamma_x \frac{d\phi}{dx} n_x + \Gamma_y \frac{d\phi}{dy} n_y + \Gamma_x \frac{d\phi}{dz} n_z\right) \delta\phi\, dA$$

$$+ \oiint\limits_{A_2} \left(\Gamma_x \frac{d\phi}{dx} n_x + \Gamma_y \frac{d\phi}{dy} n_y + \Gamma_x \frac{d\phi}{dz} n_z\right) \delta\phi\, dA$$

$$+ \oiint\limits_{A_3} \left(\Gamma_x \frac{d\phi}{dx} n_x + \Gamma_y \frac{d\phi}{dy} n_y + \Gamma_x \frac{d\phi}{dz} n_z\right) \delta\phi\, dA \qquad (9.51)$$

The first boundary integral on the right-hand side vanishes as $\delta\phi = 0$ due to the fact that ϕ is specified on this segment of the boundary. The remaining two boundary integrals are evaluated by directly substituting the boundary conditions given by Equations 9.46b and 9.46c. This simplifies the boundary integral as

$$\oiint_A \left(\Gamma_x \frac{d\phi}{dx} n_x + \Gamma_y \frac{d\phi}{dx} n_y + \Gamma_x \frac{d\phi}{dx} n_z \right) \delta\phi \, dA = -\oiint_{A_2} q_s'' \delta\phi \, dA - \oiint_{A_3} h_s(\phi - \phi_\infty) \delta\phi \, dA \tag{9.52}$$

Substituting Equation 9.52 into Equation 9.50, we obtain the *variational form* of the problem as

$$-\delta\Bigg[\iiint_V \left\{ \frac{1}{2} \left(\Gamma_x \left(\frac{d\phi}{dx}\right)^2 + \Gamma_x \left(\frac{d\phi}{dy}\right)^2 + \Gamma_x \left(\frac{d\phi}{dz}\right)^2 \right) - \frac{1}{2} S_1 \phi^2 - S_0 \phi \right\} dV + \oiint_{A_2} q_s'' \phi \, dA + \oiint_{A_3} h_s \left(\frac{1}{2}\phi^2 - \phi\phi_s \right) dA \Bigg] \tag{9.53}$$

and the corresponding *functional* as

$$I = \iiint_V \left\{ \frac{1}{2} \left(\Gamma_x \left(\frac{d\phi}{dx}\right)^2 + \Gamma_x \left(\frac{d\phi}{dx}\right)^2 + \Gamma_x \left(\frac{d\phi}{dx}\right)^2 \right) - \frac{1}{2} S_1 \phi^2 - S_0 \phi \right\} dV + \oiint_{A_2} q_s'' \phi \, dA + \oiint_{A_3} h \left(\frac{1}{2}\phi^2 - \phi\phi_\infty \right) dA \tag{9.54}$$

with essential condition $\phi = \phi_s$ on A_1.

Example 9.2: Two-dimensional Variational Formulation: Consider a two-dimensional steady state conduction in a rectangular plate with constant surface heat flux q_0'' in the mid section of the top surface due to an incident high-energy laser beam. The figure shows the schematic of the problem with boundary conditions specified at the surfaces.

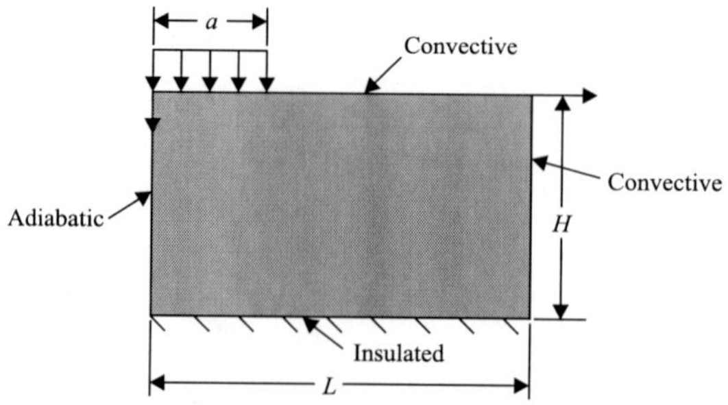

The mathematical statement of the problem is given as follows.

Governing equation

$$\frac{\partial}{\partial x}\left(k_x\frac{\partial T}{\partial x}\right)+\frac{\partial}{\partial y}\left(k_y\frac{\partial T}{\partial y}\right)=0 \tag{9.55}$$

Boundary conditions
Top surface

$$y=0,\quad 0<x\le a,\quad -k_y\frac{\partial T}{\partial y}+q_0''=h_c(T-T_\infty) \tag{9.56a}$$

$$y=0,\quad a<x\le L,\quad -k_y\frac{\partial T}{\partial y}=h_c(T-T_\infty) \tag{9.56b}$$

Bottom surface

$$y=H,\quad \frac{\partial T}{\partial y}=0 \tag{9.57}$$

Left surface

$$x=0,\quad \frac{\partial T}{\partial x}=0 \tag{9.58}$$

Right surface

$$x=L,\quad -k_x\frac{\partial T}{\partial x}=h_c(T-T_\infty) \tag{9.59}$$

Solution: In order to obtain the variational form of this problem, multiply the equation by δT and integrate over the two-dimensional solution domain A

$$\iint_A \frac{\partial}{\partial x}\left(k_x\frac{\partial T}{\partial x}\right)\delta T\,dA+\iint_A \frac{\partial}{\partial y}\left(k_y\frac{\partial T}{\partial y}\right)\delta T\,dA \tag{9.60}$$

The first integrand term can be written using the differentiation rule as

$$\iint_A \frac{\partial}{\partial x}\left(k_x\frac{dT}{dx}\right)\delta T\,dA=\iint_A \frac{\partial}{\partial x}\left(k_x\frac{\partial T}{\partial x}\delta T\right)dA-\iint_A k_x\frac{\partial T}{\partial x}\frac{\partial}{\partial x}(\delta T)\,dA \tag{9.61}$$

Using the divergence theorem (see Appendix B) for the first term and rules of calculus of variation for the second term on the right-hand side of

Equation 9.61, we get

$$\iint\limits_A \frac{\partial}{\partial x}\left(k_x \frac{dT}{dx}\right)\delta T\, dA = \oint\limits_S k_x \frac{\partial T}{\partial x}\delta T n_x\, ds - \delta \iint\limits_A \frac{1}{2}k_x\left(\frac{\partial T}{\partial x}\right)^2 dA \tag{9.62a}$$

Similarly, we can express the second term of Equation 9.60 as

$$\iint\limits_A \frac{\partial}{\partial y}\left(k_y \frac{dT}{dy}\right)\delta T\, dA = \oint\limits_S k_y \frac{\partial T}{\partial y}\delta T n_y\, ds - \delta \iint\limits_A \frac{1}{2}k_y\left(\frac{\partial T}{\partial y}\right)^2 dA \tag{9.62b}$$

Substituting Equation 9.62 into Equation 9.60

$$\delta \iint\limits_A \frac{1}{2}\left(k_x\left(\frac{\partial T}{\partial x}\right)^2 + k_y\left(\frac{\partial T}{\partial y}\right)^2\right) dA - \oint\limits_S \left(k_x \frac{\partial T}{\partial x}n_x + k_y \frac{\partial T}{\partial y}n_y\right)\delta T\, ds = 0 \tag{9.63}$$

The boundary integral term in Equation 9.63 can now be evaluated by considering the specified boundary conditions given by Equations 9.56 to 9.59 on different surfaces as

$$\begin{aligned}\oint\limits_S \left(k_x \frac{\partial T}{\partial x}n_x + k_y \frac{\partial T}{\partial y}n_y\right)\delta T\, ds = &\oint\limits_{\substack{\text{top}\\ \text{surface}}} \left(k_x \frac{\partial T}{\partial x}n_x + k_y \frac{\partial T}{\partial y}n_y\right)\delta T\, ds \\ &+ \oint\limits_{\substack{\text{right}\\ \text{surface}}} \left(k_x \frac{\partial T}{\partial x}n_x + k_y \frac{\partial T}{\partial y}n_y\right)\delta T\, ds \\ &+ \oint\limits_{\substack{\text{bottom}\\ \text{surface}}} \left(k_x \frac{\partial T}{\partial x}n_x + k_y \frac{\partial T}{\partial y}n_y\right)\delta T\, ds \\ &+ \oint\limits_{\substack{\text{left}\\ \text{surface}}} \left(k_x \frac{\partial T}{\partial x}n_x + k_y \frac{\partial T}{\partial y}n_y\right)\delta T\, ds\end{aligned} \tag{9.64}$$

For the first boundary integral on the top surface $n_x = 0$ and $n_y = -1$, and as we substitute the boundary condition given by Equation 9.56, we get

$$\oint_{\substack{\text{top}\\\text{surface}}} \left(k_x \frac{\partial T}{\partial x} n_x + k_y \frac{\partial T}{\partial y} n_y\right) \delta T\, ds$$

$$= \int_0^a \left\{h(T(x, 0) - T_\infty) - q_0''\right\} \delta T\, dx + \int_a^L h(T(x, 0) - T_\infty)\, \delta T\, dx$$

$$= -\int_0^a q_0''\, \delta T\, dx + \int_0^L \left\{h(T(x, 0) - T_\infty)\right\} \delta T\, dx$$

$$= -\delta \int_0^a q_0'' T(x, 0)\, dx + \frac{1}{2}\delta \int_0^L h[T(x, 0)]^2 dx - \delta \int_0^L h T_\infty T(x, 0)\, dx \qquad (9.65a)$$

For the second boundary integral on the right surface $n_x = 1$ and $n_y = 0$, and as we substitute the boundary condition given by Equation 9.59, we get,

$$\oint_{\substack{\text{right}\\\text{surface}}} \left(k_x \frac{\partial T}{\partial x} n_x + k_y \frac{\partial T}{\partial y} n_y\right) \delta T\, ds = -\int_0^H \left\{h(T(L, y) - T_\infty)\right\} \delta T\, dy$$

$$= -\frac{1}{2}\delta \int_0^H h[T(L, y)]^2\, dy + \delta \int_0^H h T_\infty T(L, y)\, dy \qquad (9.65b)$$

For the third boundary integral on the bottom surface $n_x = 0$ and $n_y = 1$, and as we substitute the boundary condition given by Equation 9.61, we get,

$$\oint_{\substack{\text{bottom}\\\text{surface}}} \left(k_x \frac{\partial T}{\partial x} n_x + k_y \frac{\partial T}{\partial y} n_y\right) \delta T\, ds = 0 \qquad (9.65c)$$

For the fourth boundary integral on the left surface $n_x = -1$ and $n_y = 0$, and as we substitute the boundary condition given by Equation 9.57, we get

$$\oint_{\substack{\text{left}\\\text{surface}}} \left(k_x \frac{\partial T}{\partial x} n_x + k_y \frac{\partial T}{\partial y} n_y\right) \delta T\, ds = 0 \qquad (9.65d)$$

Substituting Equation 9.65 into Equation 9.64, the boundary integral transforms into

$$\oint_S \left(k_x \frac{\partial T}{\partial x} n_x + k_y \frac{\partial T}{\partial y} n_y \right) \delta T \, ds$$

$$= -\delta \int_0^a q_0'' T(x,0) dx + \frac{1}{2} \delta \int_0^L h[T(x,0)]^2 \, dx - \delta \int_0^L h T_\infty T(x,0) \, dx$$

$$- \frac{1}{2} \delta \int_0^H h[T(L,y)]^2 dy + \delta \int_0^H h T_\infty T(L,y) \, dy \tag{9.66}$$

Substituting Equation 9.66 into Equation 9.63, we obtain the variational form of the problem as

$$\delta \left[\iint_A \frac{1}{2} \left(k_x \left(\frac{\partial T}{\partial x} \right)^2 + k_y \left(\frac{\partial T}{\partial y} \right)^2 \right) dx \, dy \right.$$

$$+ \int_0^a q_0'' T(x,0) \, dx - \frac{1}{2} \int_0^L h[T(x,0)]^2 \, dx + \int_0^L h T_\infty T(x,0) \, dx$$

$$\left. + \frac{1}{2} \int_0^H h[T(L,y)]^2 dy - \int_0^H h T_\infty T(L,y) \, dy \right] = 0 \tag{9.67}$$

and the corresponding *functional* as

$$I = \iint_A \frac{1}{2} \left(k_x \left(\frac{\partial T}{\partial x} \right)^2 + k_y \left(\frac{\partial T}{\partial y} \right)^2 \right) dx \, dy$$

$$+ \int_0^a q_0'' T(x, 0) \, dx - \frac{1}{2} \int_0^L h[T(x, 0)]^2 dx + \int_0^L h T_\infty T(x, 0) \, dx$$

$$+ \frac{1}{2} \int_0^H h[T(L, y)]^2 dy - \int_0^H h T_\infty T(L, y) \, dy = 0 \tag{9.68}$$

9.3.2 Method of Weighted-residuals

The method of weighted residuals is based on the minimization of a residual term that results as an approximate or trial solution is substituted into the governing differential equation. Let us consider the differential (in the form of an operator) equation

$$L\phi = f(\tilde{x}) \tag{9.69}$$

where L is the differential operator, ϕ is the unknown dependent variable, $\tilde{x}$ is the dependent variable, and $f(\tilde{x})$ is the source function.

As we seek an approximate solution to the above equation, let us consider an approximate or trial solution ϕ^* for ϕ as

$$\phi^* = \sum_{i=1}^{N} a_i \psi_i \tag{9.70}$$

Since the approximate solution does not satisfy the governing differential Equation 9.69 exactly, a residual term results when we substitute the approximate solution into the governing equation as

$$R(\phi, \tilde{x}) = L\phi - f(\tilde{x}) = L\left(\sum_{i=1}^{N} a_i \psi_i\right) - f(\tilde{x}) \tag{9.71}$$

We note here that the residual term becomes zero when the approximate or trial solution is exactly equal to the true solution, i.e., $\phi = \phi^*$.

In the method of weighted residuals, the objective is to select the unknown parameters a_i such that the residual $R(\phi, \tilde{x})$ is minimized along with a weighting function, W_i, as

$$\int_V W_i R(\phi, \tilde{x})\, dV = 0, \quad i = 1, 2, \ldots, n \tag{9.72}$$

where V represents the solution domain and W_i is the weighting function. The number of weighting functions equals the number of unknown coefficients in the approximate solution. There are several choices for the weighting function, and some of the most popular choices lead to different names for the ***weighted residual method*** such as the ***Galerkin method***, the ***sub-domain method***, the ***collocation method***, the ***least-square method***, and the ***Petrov–Galerkin method***. A brief description of these methods is given in Section 9.4 while describing variational methods for solving the integral form of the problem.

It can be noted here that the ***weighted-residual integral form*** given by Equation 9.72 does not involve any boundary conditions of the problem, and it can be derived for any form of differential equation.

Weak Integral Form A ***weak integral form*** can be derived from the weighted residual integral form by evaluating the integral 9.72 using integration by parts and imposing the boundary conditions of the problem. In order to satisfy a specified boundary condition of the first kind or an essential boundary condition, it is assumed that the weighting function W_i represents a variation of the variable ϕ and must vanish at the boundary. The resulting integral form is called the ***weak form*** because the order of the derivative has been reduced, and hence requires less continuity of the dependent variable. One important aspect of the weak form is that it can be obtained for both linear and nonlinear problems. But, it can only be obtained for second- or higher-order equations that contain derivatives of the order of two or higher. Further, it has been shown by Reddy (1984) that for linear problems involving even-order derivatives, the weak form will have a symmetric bilinear form. Such weak forms can be transformed into a ***variational form*** or a ***functional*** by simply substituting the weighting function as a variation of the variable $\delta\phi$. It must be remembered that for problems dealing with the first-order equation the weighted-residual integral equation 9.72 is the only choice for the variational method of approximation as well as for the finite element method.

Example 9.3: Weighted-residual Integral Formulation: The exposed surface ($x=0$) of a plane slab is subjected to radiation that causes volumetric heat generation to vary as $\dot{q} = q_0'' a\, e^{-ax}$, where a is the absorption coefficient, and q_0'' is the incident beam power intensity at the surface. The exposed surface at $x=0$ is also exposed to a fluid at temperature T_∞ and convection coefficient film h. The inner surface ($x=L$) is maintained at a constant temperature T_L.

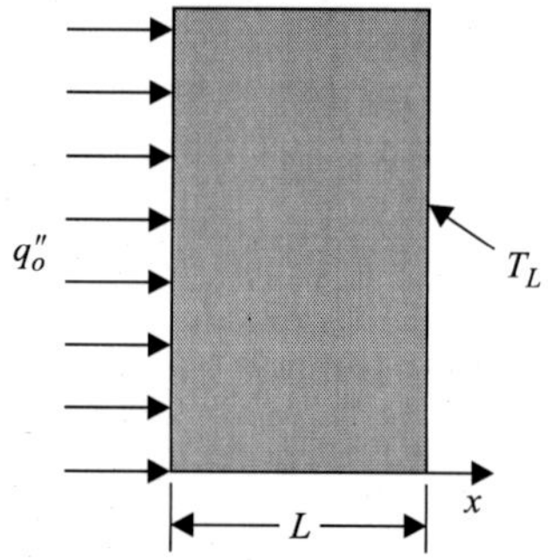

The mathematical statement of the problem is given as:

Governing equation

$$\frac{d}{dx}\left(kA\frac{dT}{dx}\right) + q_0'' a\, e^{-ax} = 0 \tag{9.73}$$

Boundary conditions

1. $x = 0,\quad k\left.\frac{dT}{dx}\right|_{x=0} = h(T|_{x=0} - T_\infty)$ (9.74a)

2. $x = L,\quad T|_{x=L} = T_L$ (9.74b)

Obtain the weighted-residual integral form, the weak form and, if possible, the variational form or the functional of the problem.

Solution: In order to derive the ***weighted-residual integral form***, we multiply the residual term of the differential equation by a weighting function W_i and integrate over the solution domain (0, L)

$$\int_0^L W_i\left[\frac{d}{dx}\left(kA\frac{dT}{dx}\right) + q_0''\,a\,\mathrm{e}^{-ax}\right]dx = 0 \tag{9.75}$$

To develop the ***weak form***, Equation 9.75 is evaluated by using the differentiation rule as

$$\int_0^L \frac{d}{dx}\left[W_i kA\frac{dT}{dx}\right]dx - \int_0^L kA\frac{dT}{dx}\frac{dW_i}{dx}dx + \int_0^L W_i q_0''a\,\mathrm{e}^{-ax}\,dx = 0$$

$$\left[W_i kA\frac{dT}{dx}\right]\Bigg|_0^L - \int_0^L kA\frac{dT}{dx}\frac{dW_i}{dx}dx + \int_0^L W_i q_0''a\,\mathrm{e}^{-ax}\,dx = 0 \tag{9.76}$$

or

$$W_i kA\frac{dT}{dx}\bigg|_{x=L} - W_i kA\frac{dT}{dx}\bigg|_{x=0} - \int_0^L kA\frac{dT}{dx}\frac{dW_i}{dx}dx + \int_0^L W_i q_0''a\,\mathrm{e}^{-ax}dx = 0 \tag{9.77}$$

We have discussed before that for the ***boundary condition of the first kind***, i.e., of constant surface temperature value, W_i must vanish at that boundary surface. So, in this problem we set $W_i|_{x=L} = 0$ at the right surface of the plane slab. At the left surface we have the ***boundary condition of the third kind***, i.e., heat transfer rate is specified by the convective cooling, and we set $kA(dT/dx)|_{x=0} = hA(T|_{x=0} - T_\infty)$. Substituting these conditions directly into Equation 9.77, we have the ***weak form*** as

$$-W_i|_{x=0}hA(T|_{x=0} - T_\infty) - \int_0^L kA\frac{dT}{dx}\frac{dW_i}{dx}dx + \int_0^L W_i q_0''a\,\mathrm{e}^{-ax}dx = 0 \tag{9.78}$$

Further, setting $W_i = \delta T$ in the equation, we have the ***variational form*** of the problem as

$$\delta\left[-\frac{1}{2}\int_0^L kA\left(\frac{dT}{dx}\right)^2 dx + \int_0^L Tq_0''a\,\mathrm{e}^{-ax}\,dx - \frac{1}{2}hA(T|_{x=0})^2 + hAT_\infty\delta T|_{x=0}\right] = 0 \tag{9.79}$$

and the corresponding ***functional*** is given as

$$I = -\frac{1}{2}\int_0^L kA\left(\frac{dT}{dx}\right)^2 dx + \int_0^L Tq_0''a\,e^{-ax}\,dx - \frac{1}{2}hA(T|_{x=0})^2 + hAT_\infty\delta T|_{x=0} \tag{9.80}$$

9.4 VARIATIONAL METHODS

We have mentioned before that the finite element method is a piecewise application of classical variational methods. In the variational methods, the governing differential problem is transformed into an equivalent integral form, and the approximate solution is assumed to be a linear combination of appropriately selected functions ψ_i and undetermined coefficients a_i given as

$$\phi = a_0\psi_0 + a_1\psi_1 + \cdots + a_n\psi_n = \sum_{i=1}^{n} a_i\psi_i \tag{9.81}$$

The coefficient a's are determined by satisfying the equivalent integral statement of the problem. Also, in the variational method, it is assumed that the approximate solution is valid over the entire solution domain.

Variational methods are classified into two primary catergories: (1) ***Rayleigh–Ritz Method*** and (2) the ***weighted residual methods***, which include the ***Galerkin method, the Collocation method***, the ***least square method***, etc. These methods differ from each other based on the selection of the equivalent integral form such as the variational form or the weighted residual integral form, the weighting function in weighted residual integral form, and the approximation functions. Let us present a brief description of some these classical variational methods.

9.4.1 The Rayleigh–Ritz Variational Method

This method is associated with the use of a functional (I) or the variational form of integral statement obtained by using the calculus of variation. The functions $\psi_0(x), \psi_1(x), \ldots, \psi_n(x)$ in the approximation solution, Equation 9.81, are selected in such a way that the specified boundary conditions are satisfied for any values of a. For example, in problems with constant surface value boundary conditions, the function $\psi_0(x)$ is selected such that it becomes equal to the constant surface values at the boundaries, and the rest of the functions, $\psi_i(x)$, become zero at the boundaries. As we substitute the approximation solution in the functional I, it becomes a function of the unknowns a_i. The necessary conditions for the functional to be stationary are given by

$$\frac{\partial I}{\partial a_i} = 0 \quad \text{for } i = 0, 1, 2, \ldots, n \tag{9.82}$$

and this leads to a system of n equations, which is solved for the n unknown a's.

It is quite evident that the accuracy of the solution depends on the selection of the approximation function, $\psi_i(x)$. The most commonly used functions are the ***polynomials***, which can be used with successively increasing degree and which satisfy the boundary values. However, other special functions such as sine, cosine, Bessel, etc. may also be used and in some cases may exhibit computational advantages. Also, the accuracy of the solution depends on the number of approximation functions retained in the approximation solution. A general procedure involves a sequence of approximations, starting with the first approximation as

$$\phi = a_0^{(1)}\psi_0 + a_1^{(1)}\psi_1 \tag{9.83a}$$

where the superscript indicates that the a's are obtained in the first stage of approximation. Additional functions are then added successively in the subsequent stages as

$$\phi = a_0^{(2)}\psi_0 + a_1^{(2)}\psi_1 + a_2^{(2)}\psi_2 \tag{9.83b}$$

$$\vdots$$

$$\phi = a_0^{(k)}\psi_0 + a_1^{(k)}\psi_1 + a_2^{(k)}\psi_2 + \cdots + a_n^{(k)}\psi_n \tag{9.83k}$$

This successive approximation process is continued until convergence, i.e., the deviations in unknown coefficients a's between two successive approximation stages are less than a specified tolerance value as

$$\left| \frac{a_i^k - a_i^{k-1}}{a_i^k} \right| \leq \varepsilon_s \tag{9.84}$$

It can be realized that with these successive stages of approximation, the functional I is monotonically converged to the stationary value.

Example 9.4: Rayleigh–Ritz Variational Method: The exposed surface ($x = L$) of a plane slab is subjected to a constant surface heat flux q''_L. The inner surface ($x = 0$) is maintained at a constant temperature T_0.

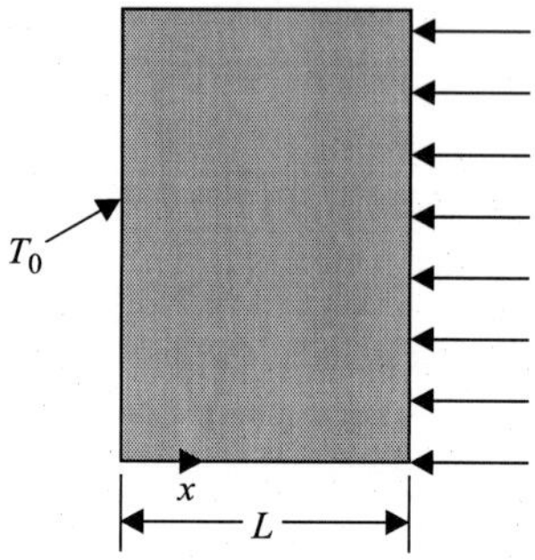

The mathematical statement of the problem is given as:

Governing equation

$$\frac{d}{dx}\left(k\frac{dT}{dx}\right) = 0 \qquad (9.85)$$

Boundary conditions

3. $x = 0, \qquad T|_{x=0} = T_0 \qquad (9.86a)$

4. $x = L, \qquad k\frac{dT}{dx}\bigg|_{x=L} = q''_L \qquad (9.86b)$

Obtain the functional form of the differential problem. Use the Rayleigh–Ritz method to obtain a solution to the equivalent variational problem.

Solution:

Step 1: Obtain the Variation Form and the Functional In order to derive the variational form, we multiply both sides of the governing equation by a variation δT and integrate over the length of the slab

$$\int_0^L \frac{d}{dx}\left(k\frac{dT}{dx}\right)\delta T\,dx = 0 \qquad (9.87)$$

Evaluating the first term by integration-by-parts

$$\left[k\frac{dT}{dx}\delta T\right]\bigg|_0^L - \int_0^L k\frac{dT}{dx}\frac{d}{dx}(\delta T)\,dx = 0 \qquad (9.88)$$

Using rules of variational calculus

$$k\frac{dT}{dx}\delta T\bigg|_{x=L} - k\frac{dT}{dx}\delta T\bigg|_{x=0} - \frac{1}{2}\delta\int_0^L k\left(\frac{dT}{dx}\right)^2 dx = 0 \tag{9.89}$$

Since at $x=0$ we have the ***boundary condition of the first kind***, i.e., of constant surface temperature value, δT must vanish at that boundary surface. So, in this problem we set $\delta T|_{x=0} = 0$. At $x=L$, we have the ***boundary condition of the second kind***, i.e., constant surface heat flux and we set $k(dT/dx)\big|_{x=0} = q''_0$. Substituting these conditions directly in Equation 9.89, we get

$$q''_L\delta T|_{x=L} - \frac{1}{2}\delta\int_0^L k\left(\frac{dT}{dx}\right)^2 dx = 0 \tag{9.90}$$

Further, using the rules of the calculus of variation in the first term of the equation, we have the ***variational form*** of the problem as

$$\delta\left[-\frac{1}{2}\int_0^L k\left(\frac{dT}{dx}\right)^2 dx + q''_L T|_{x=L}\right] = 0 \tag{9.91}$$

and the corresponding ***functional*** is given as

$$I = -\frac{1}{2}\int_0^L k\left(\frac{dT}{dx}\right)^2 dx + q''_L T|_{x=L} \tag{9.92}$$

with the essential boundary condition $T|_{x=0} = T_0$.

Step 2: Obtain Approximate Solution using the Rayleigh–Ritz Method Let us assume the approximate solution of the form

$$T = \sum_{i=0}^{2} a_i\psi_i = a_0 + a_1 x + a_2 x^2 \tag{9.93}$$

where $\psi_0(x) = 1$, $\psi_1(x) = x$, and $\psi_2(x) = x^2$.

Substituting the approximate solution Equation 9.93 into Equation 9.94, we obtain

$$I = -\frac{1}{2}\int_0^L k\left[a_1^2 + 4a_1a_2x + 4a_2^2x^2\right]dx + q''_L(a_0 + a_1L + a_2L) \tag{9.94a}$$

or

$$I = -\tfrac{1}{2}k\left(a_1^2 L + 2a_1 a_2 L^2 + \tfrac{4}{3}a_2^2 L^3\right) + q_L''(a_0 + a_1 L + a_2 L^2) \tag{9.94b}$$

Using the condition of stationary functional I, i.e., $\partial I/\partial a_i = 0$, we get

$$\text{for } i = 0 \quad \frac{\partial I}{\partial a_0} = 0, \quad q_L'' = 0 \tag{9.95a}$$

$$\text{for } i = 1 \quad \frac{\partial I}{\partial a_1} = 0, \quad a_1 + a_2 L = \frac{q_L''}{k} \tag{9.95b}$$

$$\text{for } i = 2 \quad \frac{\partial I}{\partial a_2} = 0, \quad a_1 + \frac{4}{3}a_2 L = \frac{q_L''}{k} \tag{9.95c}$$

Equations 9.95a, 9.95b and 9.95c form a system of equations for a_i, and we can write this in matrix form as

$$\begin{bmatrix} 0 & 0 & 0 \\ 0 & 1 & L \\ 0 & 1 & \frac{4}{3}L \end{bmatrix} \begin{Bmatrix} a_0 \\ a_1 \\ a_2 \end{Bmatrix} = \begin{Bmatrix} q_L'' \\ q_L''/k \\ q_L''/k \end{Bmatrix} \tag{9.96}$$

The solution of this system gives

$$a_1 = \frac{q_L''}{k} \quad \text{and} \quad a_2 = 0 \tag{9.97}$$

We now impose the essential boundary condition, i.e., $T|_{x=0} = T_0$, on the approximate solution

$$a_0 = T_0 \tag{9.98}$$

The approximate solution is then

$$T = T_0 + \frac{q_L''}{k}x \tag{9.99}$$

9.4.2 Weighted-residual Variational Methods

The weighted residual method is used along with the weighted residual integral form of a problem given by Equation 9.72. The approximate solution is of the same form as in the case of the Rayleigh–Ritz, i.e., Equation 9.81.

The weighting function $W_i(\tilde{x})$ could be selected from a set of independent functions that may be different from the approximate functions Ψ_i. Another important criterion of this method is that the selected approximate functions Ψ_i need to satisfy all specified ***essential*** as well as ***natural*** boundary conditions. This is due to the fact that the weighted residual form of the integral Equation 9.72 does not include any boundary conditions. In order to satisfy this requirement, the selected approximate functions Ψ_i should have order higher than that used in the Rayleigh–Ritz method. In other words, the functions must have nonzero derivatives up to the order of the governing differential equation.

The ***weighted-residual variational methods*** can be classified into different kinds depending on the selection of the weighting functions. Some of the popular weighted residual variational methods are the ***Galerkin method***, the c***ollocation method***, the ***sub-domain method***, the ***least square method***, and the ***Petrov–Galerkin method***. A brief description of these methods is given in the following section.

Galerkin's Method In this method the weighting function, $W_i(\tilde{x})$, is assumed to be the same as the approximation function, $\psi_i(\tilde{x})$, used in the approximate solution given by Equation 9.81, and the integral statement is given as

$$\int_V \psi_i(\tilde{x})R(\phi, \tilde{x})\,dv, \quad i = 1, 2, \ldots, n \tag{9.100}$$

Since the number of weighting functions is the same as the number of unknown coefficients, a_i, in the approximate solution, evaluation of the integral Equation 9.100 leads to a system of equations involving unknown coefficients $a_0, a_1, \ldots, a_n$.

Collocation Method In this method the weighting function is assumed to be a ***direct delta function*** of the form

$$W_i(\tilde{x}) = \delta(\tilde{x} - \tilde{x}_i) \tag{9.101}$$

where $\tilde{x}_i$ represents a set of selected points, the number of which is equal to the number of unknown coefficients. So the weighting function is $W_i = 1$ at the point $x = x_i$ and zero everywhere else in the solution domain. The equivalent integral statement is given by

$$\int_V \delta(\tilde{x} - \tilde{x}_i)R(\phi, \tilde{x}_i)\,dv \tag{9.102}$$

Such a selection of weighting functions causes the residual integral to vanish everywhere else in the solution domain, except at the selected points, and this leads to a system of equations given as

$$R(\phi, \tilde{x}_i) = 0, \quad i = 1, 2, \ldots, n \tag{9.103}$$

Subdomain Method In this method the weighting function is assumed to be unity over a specific sub-region, ΔV_i, and subsequently the integral of the residual vanishes over that sub-region or the integration interval. The number of sub-regions or integration intervals equals the number of unknown coefficients. The equivalent integral statement becomes

$$\int_{\Delta V_i} R(\phi, \tilde{x})\, dV, \quad i = 1, 2, \ldots, n \tag{9.104}$$

Least Square Method In this method the weighting function is assumed to be the same as the ***residual term***, i.e., $W_i(\tilde{x}) = R(\phi, \tilde{x})$, and an error term is evaluated as

$$E = \int_V [R(\phi, \tilde{x})]^2 \tag{9.105}$$

The unknown coefficients a_i are determined by minimizing the error with respect to the unknown coefficients as

$$\frac{\partial E}{\partial a_i} = 0, \quad i = 1, 2, \ldots, n \tag{9.106}$$

Equation (9.106) leads to a system of equations, and solutions which give the unknown coefficients a_i.

The Petrov–Galerkin Method In this method the weighting functions are selected as independent functions, which are different from the approximate functions used in the approximate solution given by Equation 9.81, i.e., $W_i(\tilde{x}) = f_i(\tilde{x}) \neq \psi_i(\tilde{x})$. The equivalent integral form is written as

$$\int_V f_i(\tilde{x}) R(\phi, \tilde{x})\, dV = 0, \quad i = 1, 2, \ldots, n \tag{9.107}$$

This equation represents a system of equations, the solution of which give the unknown coefficients a_i.

Example 9.5: Weighted-Residual Variational Method: Consider the uniform heat generation $\dot{q}$ in a plane slab with exposed surfaces $(x = 0)$ and $(x = L)$ maintained at a constant temperature $T_0 = 0$.

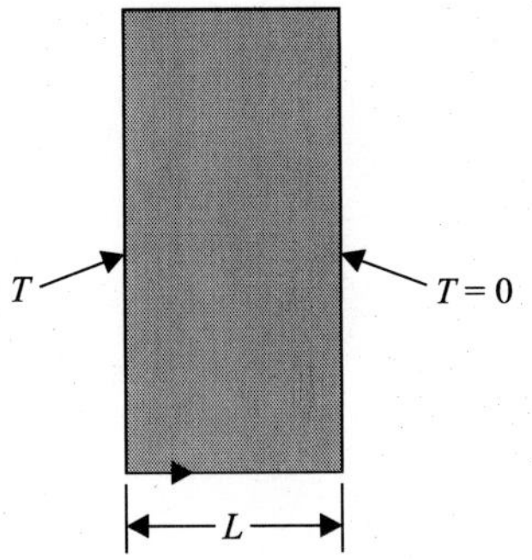

The mathematical statement of the problem is given as:

Governing equation

$$\frac{d^2T}{dx^2} + \frac{\dot{q}}{k} = 0 \tag{9.108}$$

Boundary conditions

5. $x = 0, \qquad T|_{x=0} = 0$ (9.109a)

6. $x = L, \qquad T|_{x=L}, = 0$ (9.109b)

Determine the solution using Galerkin's weighted residual method.

Solution: Let us assume an approximate solution for temperature of the form

$$T^* = a_1 \sin \frac{\pi x}{L} \tag{9.110}$$

with $\psi_1(x) = \sin(\pi x/L)$. Note that the function is selected in such a way that it satisfies the two boundary conditions. In Galerkin's method we selected the weighting function $W_1(x) = \sin(\pi x/L)$. With the substitution of the weighting function the residual Equation 9.100 is evaluated as

$$\int_0^L W_1\left(\frac{d^2T^*}{dx^2} + \frac{\dot{q}}{k}\right) dx = 0 \tag{9.111}$$

or

$$\int_0^L \sin\frac{\pi x}{L}\left(-a_1\frac{\pi^2}{L^2}\sin\frac{\pi x}{L} + \frac{\dot{q}}{k}\right) dx = 0$$

or

$$-ka_1\frac{\pi^2}{L^2}\int_0^L \sin^2\frac{\pi x}{L}dx + \frac{\dot{q}}{k}\int_0^L \sin\frac{\pi x}{L}dx = 0$$

$$ka_1\frac{\pi^2}{L^2}\frac{1}{2}\int_0^L \left(1 - \cos\frac{2\pi x}{L}\right)dx + \frac{\dot{q}}{k}\int_0^L \sin\frac{\pi x}{L}dx = 0$$

$$a_1\frac{\pi^2}{2L} - \frac{2\dot{q}L}{\pi k} = 0$$

Solving

$$a_1 = \frac{4L^2}{\pi^3 k}\dot{q} \tag{9.112}$$

the approximate temperature solution is

$$T = \frac{4\dot{q}L^2}{\pi^3 k}\sin\frac{\pi x}{L}$$

PROBLEMS

9.1 Determine the variational form and functional for the following one-dimensional conduction heat transfer problem using the calculus of variation.

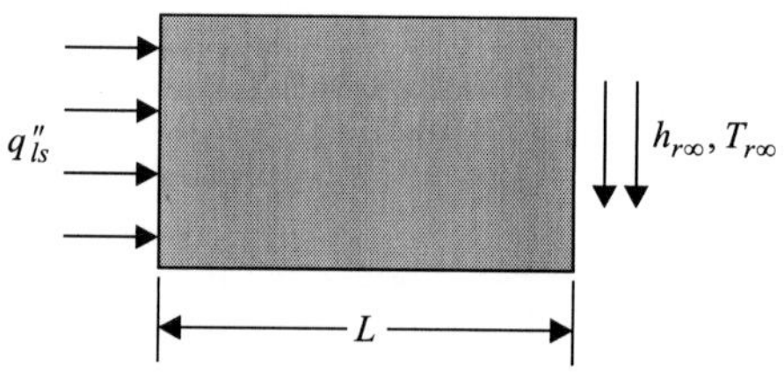

Governing equation

$$\frac{d}{dx}\left(kA\frac{dT}{dx}\right) - hP(T - T_\infty) = 0$$

Boundary conditions

1. $x = 0, \qquad -k\left.\frac{dT}{dx}\right|_{x=0} = q''_s$

2. $x = L, \qquad -k\left.\frac{dT}{dx}\right|_{x=L} = h(T|_{x=L} - T_\infty)$

9.2 Determine the variational form and functional for the following one-dimensional fin problem using the calculus of variation.

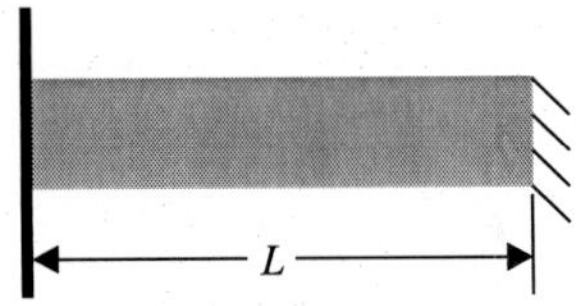

Governing equation

One-dimensional Fin $\frac{d}{dx}\left(kA\frac{dT}{dx}\right) - hP(T - T_\infty) = 0$

Boundary conditions

3. $x = 0, \quad T = T_0$
4. $x = L, \quad \left.\frac{dT}{dx}\right|_{x=L} = 0$

9.3 Consider a two-dimensional steady state conduction in a rectangular plate with a cutting groove created by a constant surface heat flux q_0'' in the mid section of the top surface due to an incident high-energy laser beam. The figure shows the schematic of the problem with boundary conditions specified at the surfaces.

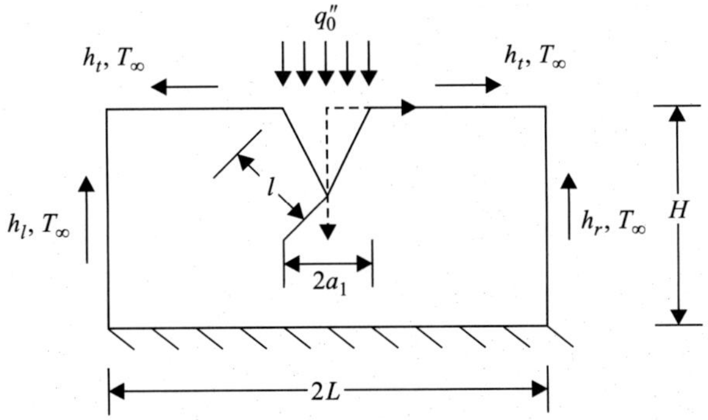

Use the calculus of variation to determine the variational form and functional considering the following mathematical statement of the problem:

Governing equation

$$\frac{\partial}{\partial x}\left(k_x\frac{\partial T}{\partial x}\right) + \frac{\partial}{\partial y}\left(k_y\frac{\partial T}{\partial y}\right) = 0$$

Boundary conditions

Top surface

$$y = 0, \quad -L < x \le -a, \quad k_y \frac{\partial T}{\partial y} = h(T - T_\infty)$$

$$y = 0, \quad a < x \le L, \quad k_y \frac{\partial T}{\partial y} = h(T - T_\infty)$$

On the groove surface

$$-k_n \frac{\partial T}{\partial n} = q_0''$$

Bottom surface $\quad y = H, \quad \frac{\partial T}{\partial y} = 0$

Left surface $\quad x = 0, \quad k_x \frac{\partial T}{\partial x} = h(T - T_\infty)$

Right surface $\quad x = L, \quad -k_x \frac{\partial T}{\partial x} = h(T - T_\infty)$

9.4 Dermine the weak form of Problem 9.3 using Galerkin's method.

9.5 Consider the problem of cooling an electronic chip that is mounted on a substrate. Assume uniform volumetric heat generation in the chip. The top and bottom surfaces are convectively cooled. The two side surfaces are assumed to be at constant temperature T_0. The conductivities of the substrate and chip materials are K_s and K_c respectively.

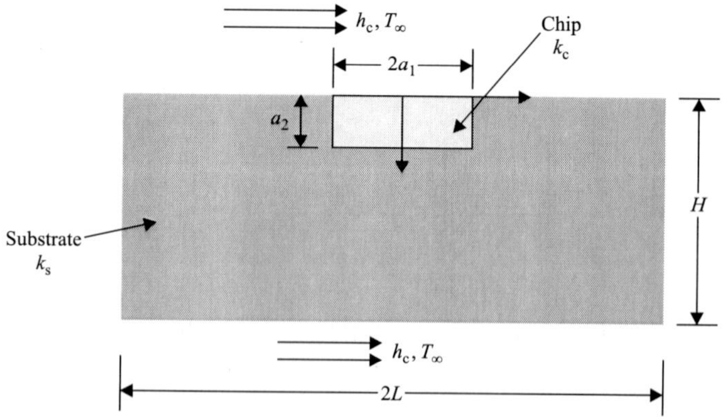

Use the ***calculus of variation*** to determine the variational form and functional by considering the following mathematical statement of the problem.

Governing equation

$$\frac{\partial}{\partial x}\left(k_x \frac{\partial T}{\partial x}\right) + \frac{\partial}{\partial y}\left(k_y \frac{\partial T}{\partial y}\right) + \dot{q} = 0$$

where $\dot{q} = 0$ in the substrate.

Boundary conditions

Top surface

$$y = 0, \quad -L < x < -a_1, \quad a_1 < x < L$$

$$k_s \frac{\partial T}{\partial y} = h(T - T_\infty)$$

$$y = 0, \quad -a_1 < x < a_1$$

$$k_c \frac{\partial T}{\partial y} = h(T - T_\infty)$$

Bottom surface $\quad y = H, \quad -k_s \frac{\partial T}{\partial x} = h(T - T_\infty)$

Left surface $\quad x = 0, \quad T = T_0$

Right surface $\quad x = L, \quad T = T_0$

9.6 Determine the integral or weak form of the three-dimensional Navier–Stokes equations for viscous incompressible flows given by the following governing equations and boundary conditions using Galerkin's method.

$$\frac{\partial u}{\partial x} + \frac{\partial v}{\partial y} + \frac{\partial w}{\partial y} = 0$$

$$\rho\left(\frac{\partial u}{\partial t} + u\frac{\partial u}{\partial x} + v\frac{\partial u}{\partial y} + w\frac{\partial u}{\partial z}\right) = b_x - \frac{\partial p}{\partial x} + \mu\left(\frac{\partial^2 u}{\partial x^2} + \frac{\partial^2 u}{\partial y^2} + \frac{\partial^2 u}{\partial z^2}\right)$$

$$\rho\left(\frac{\partial v}{\partial t} + u\frac{\partial v}{\partial x} + v\frac{\partial v}{\partial y} + w\frac{\partial v}{\partial z}\right) = b_y - \frac{\partial p}{\partial y} + \mu\left(\frac{\partial^2 v}{\partial x^2} + \frac{\partial^2 v}{\partial y^2} + \frac{\partial^2 v}{\partial z^2}\right)$$

$$\rho\left(\frac{\partial w}{\partial t} + u\frac{\partial w}{\partial x} + v\frac{\partial w}{\partial y} + w\frac{\partial w}{\partial z}\right) = b_z - \frac{\partial p}{\partial z} + \mu\left(\frac{\partial^2 w}{\partial x^2} + \frac{\partial^2 w}{\partial y^2} + \frac{\partial^2 w}{\partial z^2}\right)$$

over volume V.

With boundary conditions given in a general form as

1. $u = u_s$, $v = v_s$ and $w = w_s$ on the boundary A_1

2. $$\mu\left(\frac{\partial u}{\partial x}\hat{n}_x + \frac{\partial u}{\partial y}\hat{n}_y + \frac{\partial u}{\partial z}\hat{n}_z\right) - p\hat{n}_x = f_{sx}$$

$$\mu\left(\frac{\partial v}{\partial x}\hat{n}_x + \frac{\partial v}{\partial y}\hat{n}_y + \frac{\partial v}{\partial z}\hat{n}_z\right) - p\hat{n}_y = f_{sy}$$

$$\mu\left(\frac{\partial w}{\partial x}\hat{n}_x + \frac{\partial w}{\partial y}\hat{n}_y + \frac{\partial w}{\partial z}\hat{n}_z\right) - p\hat{n}_z = f_{sz}$$

on the boundary A_2.

9.7 Consider the mathematical statement of a problem given as:

Governing equation

$$\frac{d}{dx}\left(\Gamma\frac{d\phi}{dx}\right) + S_0\frac{x}{L} = 0$$

Boundary conditions

1. $x = 0,\qquad \phi|_{x=0} = 0$
2. $x = L,\qquad \phi|_{x=L} = \phi_L$

(a) Obtain the functional form using the variational method.
(b) Obtain an approximate solution using the Rayleigh–Ritz method, assuming the approximate solution of the form

$$\phi = \psi_0 + a_1\psi_1$$

where $\psi_0(x) = (\phi_L/L)x$ and $\psi_1(x) = x(x - L)$ that satisfies the boundary conditions.

9.8 Obtain an approximate solution to Problem 9.7 using the weighted-residual method, assuming the approximate solution of the form

$$\phi = a_1 \sin\frac{\pi x}{L}$$

Evaluate a_1 using (i) the Galerkin method, (ii) the collocation method, (iii) the sub-domain method, and (iv) the least square method.

9.9 Solve the problem outlined in Example Problem 9.5 using (i) the collocation method, (ii) the sub-domain method, and (iii) the least square method.

CHAPTER

10

ELEMENT SHAPE FUNCTIONS

We have mentioned in the previous chapter that one of the important steps in the finite element method is the selection of a simple function that is used to approximate the solution over each discretized sub-region or finite element. These simple approximate functions are also referred to as ***shape functions*** or ***interpolation functions***. They are mostly polynomial as they are easier to manipulate in mathematical operations and computer implementation compared to other functions such as trigonometric functions. Also, the accuracy of the results can be improved by increasing the order of the polynomial function.

In the following sections, we present the development of some of the common one-, two- and three-dimensional shape functions for the finite elements.

10.1 ONE-DIMENSIONAL ELEMENT

The general form of the polynomial function in one dimension is given as

$$\phi(x) = a_0 + a_1 x + a_2 x^2 + \cdots + a_n x^n \qquad (10.1)$$

where n is the order of the polynomial. Some examples of these functions are depicted in Figure 10.1. The simplest and most widely used function is

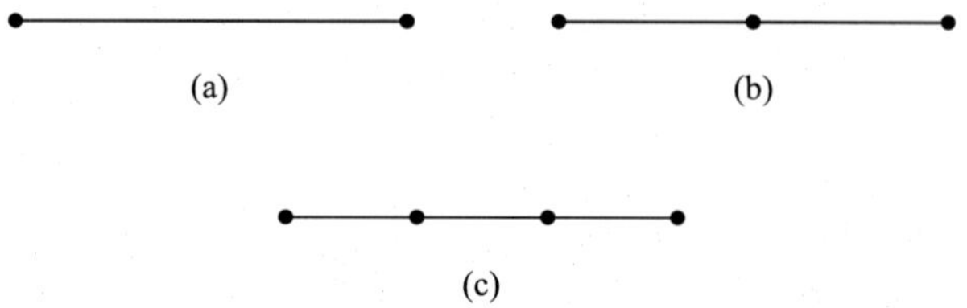

Figure 10.1 Examples of polynomial interpolation functions: (a) linear element, (b) quadratic element, and (c) cubic element

the polynomial function of order one (Figure 10.1a), which is known as a linear element involving two nodes and is given as

$$\phi(x) = a_0 + a_1 x \tag{10.2}$$

A quadratic element (Figure 10.1b) consists of three nodes and it is given by a second-order polynomial as

$$\phi(x) = a_0 + a_1 x + a_2 x^2 \tag{10.3}$$

Similarly, for the cubic element the polynomial is of the order three and expressed as

$$\phi(x) = a_0 + a_1 x + a_2 x^2 + a_3 x^3 \tag{10.4}$$

The higher the order of the polynomial, the higher the accuracy in the approximation solution as demonstrated in Figure 10.2.

We have mentioned that linear elements are generally preferred due to their simplicity. Accuracy of the results is improved further by using an increased number of elements, i.e., using finer mesh size distributions. On the other hand, one can also choose higher-order elements to achieve similar accuracies with enhanced convergence rate and using coarser mesh size. However, this will bring in additional mathematical complexity and operations involving bigger bandwidth matrices. The choice of linear or higher-order elements is eventually justified through computational savings.

10.1.1 One-dimensional Linear Element

A one-dimensional linear element is a line of length L_e and consists of two nodes i and j, one at each end as shown in Figure 10.3. The unknown values at these nodes are ϕ_i and ϕ_j.

The approximating function over the element is given as

$$\phi(x) = a_0 + a_1 x \tag{10.5a}$$

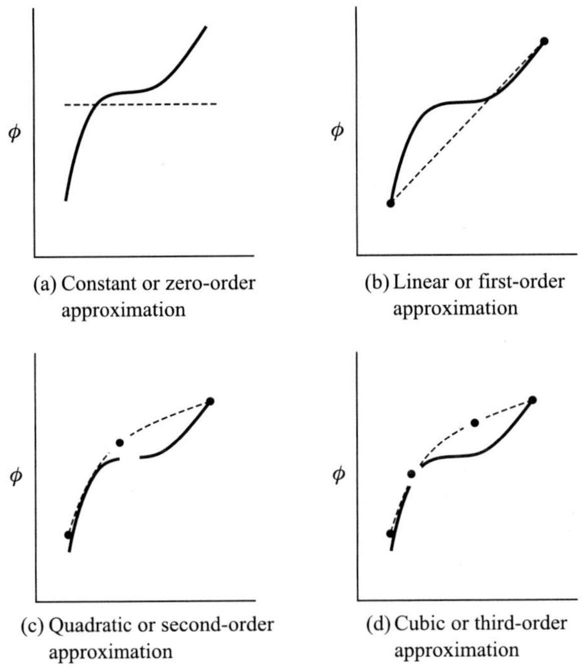

Figure 10.2 Polynomials of different order of approximation

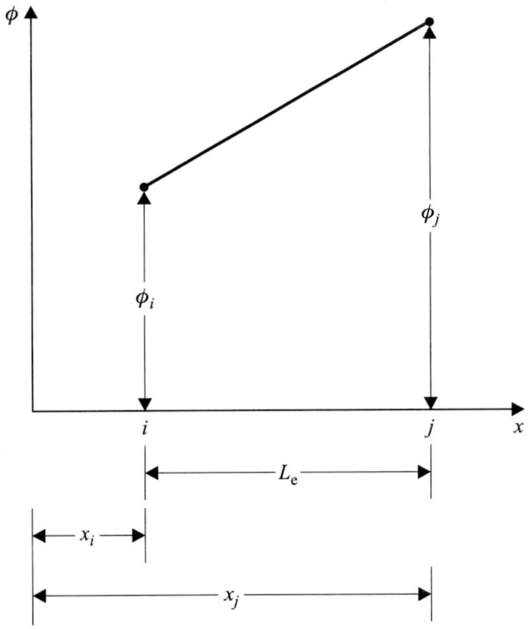

Figure 10.3 Linear line element

The two coefficient values a_0 and a_1 are determined by using two nodal values, i.e.,

$$1.\quad x = x_i, \qquad \phi = \phi_i \tag{10.5b}$$

and

$$2.\quad x = x_j, \qquad \phi = \phi_j \tag{10.5c}$$

Substituting these two conditions, we have

$$\phi_i = a_0 + a_1 x_i$$

$$\phi_j = a_0 + a_1 x_j$$

and in matrix form as

$$\begin{bmatrix} 1 & x_i \\ 1 & x_j \end{bmatrix} \begin{Bmatrix} a_0 \\ a_1 \end{Bmatrix} = \begin{Bmatrix} \phi_i \\ \phi_j \end{Bmatrix} \tag{10.6}$$

Solving Equation 10.6 using Cramer's rule

$$a_0 = \frac{\begin{vmatrix} \phi_i & x_i \\ \phi_j & x_j \end{vmatrix}}{\begin{vmatrix} 1 & x_i \\ 1 & x_j \end{vmatrix}} = \frac{\phi_i x_j - \phi_j x_i}{x_j - x_i} \tag{10.7a}$$

and

$$a_0 = \frac{\begin{vmatrix} 1 & \phi_i \\ 1 & \phi_j \end{vmatrix}}{\begin{vmatrix} 1 & x_i \\ 1 & x_j \end{vmatrix}} = \frac{\phi_j - \phi_i}{x_j - x_i} \tag{10.7b}$$

Substituting Equation 10.7 into Equation 10.5a, we get the approximating function as

$$\phi = \frac{\phi_i x_j - \phi_j x_i}{x_j - x_i} + \frac{\phi_j - \phi_i}{x_j - x_i} x \tag{10.8}$$

Rearranging

$$\phi = \frac{x_j - x}{x_j - x_i} \phi_i + \frac{x - x_i}{x_j - x_i} \phi_j \tag{10.9a}$$

or

$$\phi = N_i \phi_i + N_j \phi_j = [N]\{\phi\} \tag{10.9b}$$

where N_i and N_j are shape functions, which are defined as

$$N_i = \frac{x_j - x}{x_j - x_i}, \qquad N_j = \frac{x - x_i}{x_j - x_i} \tag{10.10}$$

and

$[N] = \begin{bmatrix} N_i & N_j \end{bmatrix}$ = row vector of shape functions

$\{\phi\} = \begin{Bmatrix} \phi_i \\ \phi_j \end{Bmatrix}$ = column vector of unknown nodal values

It can be noted that the shape functions given by Equation 10.10 are of the same polynomial type as the initial approximation Equation 10.5a. The shape functions also possess two additional properties, given as follows.

1. Each shape function has a value of one at the same node and zero at the other node, i.e.,

$$N_i(x_i) = 1, \qquad N_i(x_j) = 0 \tag{10.11a}$$

and

$$N_j(x_i) = 0, \qquad N_j(x_j) = 1 \tag{10.11b}$$

Variations of these shape functions are depicted in Figure 10.4.

2. The sum of the derivatives of the shape functions with respect to x is zero, i.e.,

$$\frac{dN_i}{dx} + \frac{dN_j}{dx} = 0 \tag{10.12a}$$

or

$$\sum_{I=i,j} \frac{dN_I}{dx} = 0 \tag{10.12b}$$

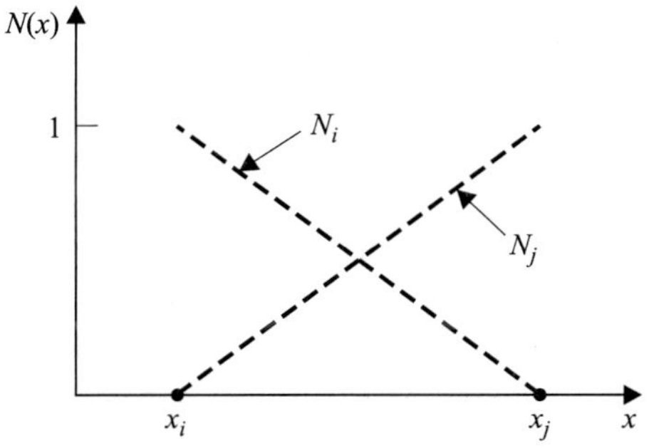

Figure 10.4 Variation of shape function for linear line element

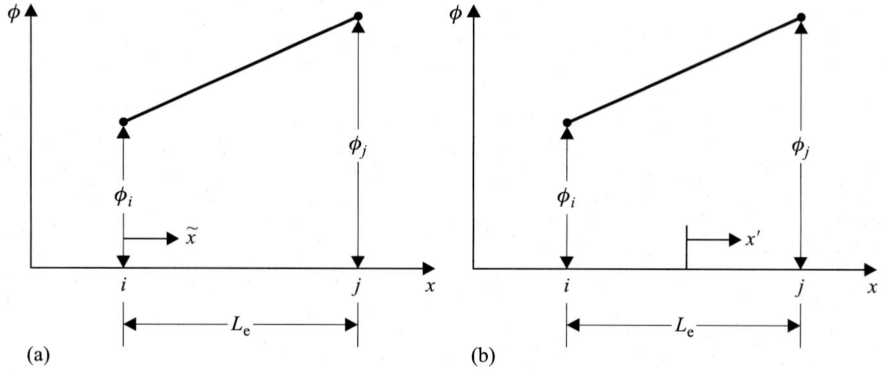

Figure 10.5 Linear line element in local coordinate system

The approximation function can be written in a general form for any type of element as

$$\phi = \sum_{I=i,j...} N_I \phi_I, \qquad I = i, j, \ldots, N_n \tag{10.13}$$

where N_n are the number of nodes in the element.

Shape Function in Local Coordinate System. So far we have considered shape functions in a global coordinate system. In many situations, it is more convenient to deal with shape functions using the local coordinate systems, particularly in evaluating integrals involving shape functions. Let us consider two such local coordinate systems for a one-dimensional line element. In the first one, the origin is located at node i or the left-hand corner node of the element as shown in Figure 10.5a.

The shape functions for the linear line element in this local coordinate system are obtained by simply substituting x by $x = x_i + \tilde{x}$ in Equation 10.10 as

$$N_i(\tilde{x}) = 1 - \frac{\tilde{x}}{L_e}, \qquad N_j(\tilde{x}) = \frac{\tilde{x}}{L_e} \tag{10.14}$$

where the coordinate $\tilde{x}$ varies from 0 to L_e.

The second local coordinate system is defined based on its origin located at the center of the element, as shown in Figure 10.5b. The shape functions based on this local coordinate system are obtained by substituting $x = x_i + (L_e/2) + x'$ and expressed as

$$N_i(x') = \left(\frac{1}{2} - \frac{x'}{L_e}\right), \quad N_j(x') = \left(\frac{1}{2} + \frac{x'}{L_e}\right) \tag{10.15}$$

where the coordinate x' varies from $-L_e/2$ to $L_e/2$.

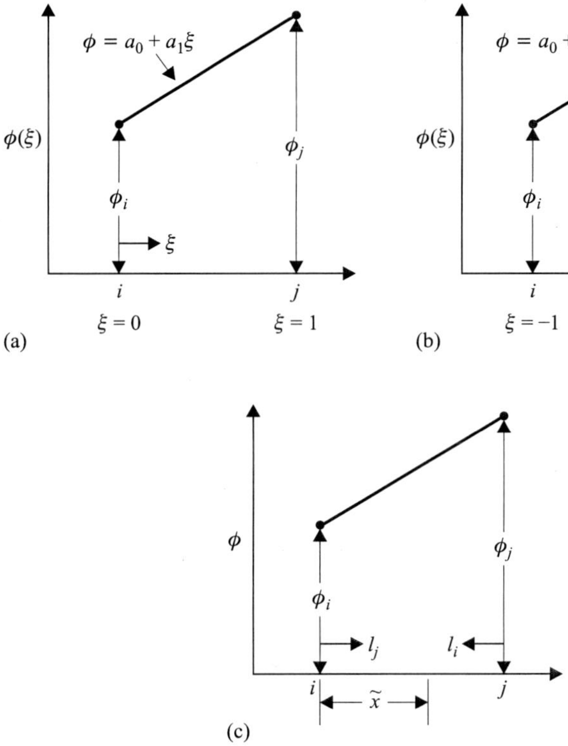

Figure 10.6 Shape function for a linear line element in a natural coordinate system

Shape Function in Natural Coordinate System. Use of the natural coordinate system is primarily preferred for cases where the integrals involving shape function can easily be evaluated analytically using the integral formulas or by numerically using Gauss–Legendre quadrature formulas. A natural coordinate system is a local coordinate system in which coordinate variables are expressed in a dimensionless form such as $\xi = x/L_e$ or $\xi = x/(L_e/2)$. In such natural coordinate systems, the coordinate variable ξ varies from 0 to 1, as in Figure 10.6a, or -1 to 1, as shown in Figure 10.6b.

Let us consider the derivation of the shape function for the natural co-ordinate system with the origin at the center of the element as in Figure 10.6b. The approximating function is assumed as

$$\phi(\xi) = a_0 + a_1\xi \tag{10.16}$$

The two coefficient values a_0 and a_1 are determined by using two nodal values, i.e.,

1. $\xi = -1, \quad \phi = \phi_i$ (10.17a)

and

$$2.\quad \xi = 1, \quad \phi = \phi_j \tag{10.17b}$$

Substituting these two conditions, we have

$$\phi_i = a_0 + a_1(-1) \tag{10.18a}$$

$$\phi_j = a_0 + a_1(1) \tag{10.18b}$$

Solving, we get

$$a_0 = \frac{\phi_i + \phi_j}{2} \quad \text{and} \quad a_1 = \frac{\phi_j - \phi_i}{2} \tag{10.19}$$

Substituting Equation 10.19 into Equation 10.16, we get

$$\phi(\xi) = \frac{\phi_i + \phi_j}{2} + \frac{\phi_j - \phi_i}{2}\xi \tag{10.20}$$

Rearranging

$$\phi(\xi) = \tfrac{1}{2}(1 - \xi)\phi_i + \tfrac{1}{2}(1 + \xi)\phi_j \tag{10.21}$$

or

$$\phi(\xi) = N_i\phi_i + N_j\phi_j \tag{10.22}$$

where the shape functions in the natural coordinate system are given as

$$N_i(\xi) = \tfrac{1}{2}(1 - \xi) \quad \text{and} \quad N_j(\xi) = \tfrac{1}{2}(1 + \xi) \tag{10.23}$$

The natural coordinate system and shape functions are used along with the coordinate transformation relation

$$x = N_i(\xi)x_i + N_j(\xi)x_j = \sum_{I=i,j} N_I x_I \tag{10.24a}$$

Additionally, we need to transform some integral expressions usually encountered in finite element formulation into natural coordinates using the following relations

$$\frac{d}{dx} = \frac{d}{d\xi}\frac{d\xi}{dx} = \frac{d}{d\xi}\frac{2}{L_e} \tag{10.24b}$$

$$dx = \frac{L_e}{2}d\xi \tag{10.24c}$$

Line Coordinates as a Natural Coordinate. Another useful natural coordinate system is given in terms of two length ratios as

$$l_i = \frac{L_e - \tilde{x}}{L_e} \quad \text{and} \quad l_j = \frac{\tilde{x}}{L_e} \tag{10.25}$$

where $\tilde{x}$ is the distance of point on the line element as shown in Figure 10.6c. Note that the two natural coordinates l_i and l_j are not independent, but related by the equation

$$l_i + l_j = 1 \tag{10.26}$$

It can be noted that these natural coordinates are identical to the shape functions given by Equation 10.14 for linear line elements in a local coordinate system, i.e.,

$$l_i = N_i(\tilde{x}) \quad \text{and} \quad l_j = N_j(\tilde{x}) \tag{10.27}$$

The main reason for expressing the shape function in terms of natural co-ordinates is the simplicity in evaluating some integrals involving shape functions through use of integral formulas, or use Gauss–Legendre quadrature as discussed in Chapter 4. One such common integral formula is given as

$$I = \int_0^1 l_i^a l_j^b \, dl_j = \frac{a!b!}{(a+b+1)} \tag{10.28}$$

For example, the integral expression of the form

$$I = \int_0^{L_e} N_i(\tilde{x}) N_j(\tilde{x}) \, d\tilde{x} \tag{10.29}$$

can be evaluated using Equations 10.25, 10.27 and 10.28 as

$$I = \int_0^{L_e} N_i(\tilde{x}) N_j(\tilde{x}) \, d\tilde{x} = \int_0^1 l_i^1 l_j^1 L_e \, dl_j \tag{10.30}$$

$$= \frac{1!1!}{(1+1+1)} L_e = \frac{L_e}{3}$$

10.1.2 One-dimensional Quadratic Line Element

A quadratic line element has three nodes: one at each end and one at the mid point as shown in Figure 10.7.

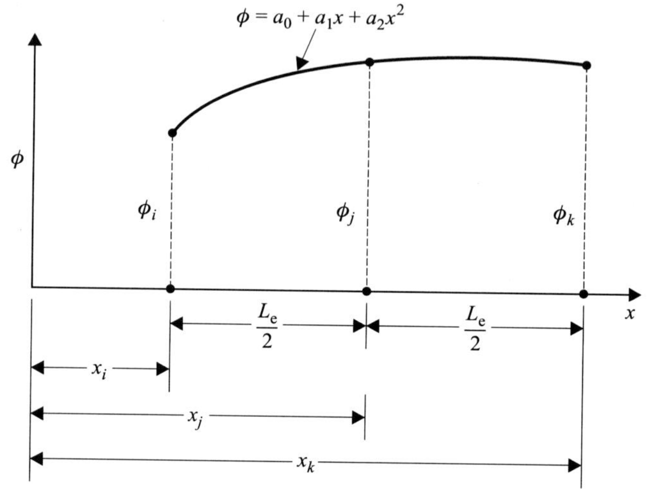

Figure 10.7 Quadratic line element

The quadratic approximation equation of this element is given as

$$\phi = a_0 + a_1 x + a_2 x^2 \tag{10.31}$$

The three coefficients a_0, a_1, and a_2 are determined by substituting the three nodal values as

$$\phi_i = a_0 + a_1 x_i + a_2 x_i^2 \tag{10.32a}$$

$$\phi_j = a_0 + a_1 x_j + a_2 x_j^2 \tag{10.32b}$$

$$\phi_k = a_0 + a_1 x_k + a_2 x_k^2 \tag{10.32c}$$

The system of equation is written in matrix form as

$$\begin{bmatrix} 1 & x_i & x_i^2 \\ 1 & x_j & x_j^2 \\ 1 & x_k & x_k^2 \end{bmatrix} \begin{Bmatrix} a_0 \\ a_1 \\ a_2 \end{Bmatrix} = \begin{Bmatrix} \phi_i \\ \phi_j \\ \phi_k \end{Bmatrix} \tag{10.33}$$

Equation 10.33 is solved by the Cramer rule as

$$a_0 = \frac{\begin{bmatrix} \phi_i & x_i & x_i^2 \\ \phi_j & x_j & x_j^2 \\ \phi_k & x_k & x_k^2 \end{bmatrix}}{\begin{bmatrix} 1 & x_i & x_i^2 \\ 1 & x_j & x_j^2 \\ 1 & x_k & x_k^2 \end{bmatrix}} = \frac{1}{D}\begin{bmatrix} \phi_i(x_j x_k^2 - x_k x_j^2) + \phi_j(x_k x_i^2 - x_i x_k^2) \\ + \phi_k(x_i x_j^2 - x_j x_i^2) \end{bmatrix} \tag{10.34a}$$

$$a_1 = \frac{\begin{bmatrix} 1 & \phi_i & x_i^2 \\ 1 & \phi_j & x_j^2 \\ 1 & \phi_k & x_k^2 \end{bmatrix}}{\begin{bmatrix} 1 & x_i & x_i^2 \\ 1 & x_j & x_j^2 \\ 1 & x_k & x_k^2 \end{bmatrix}} = \frac{1}{D}\left[\phi_i(x_j^2 - x_k^2) + \phi_j(x_k^2 - x_i^2) + \phi_k(x_i^2 - x_j^2)\right] \tag{10.34b}$$

$$a_2 = \frac{\begin{bmatrix} 1 & x_i & \phi_i \\ 1 & x_j & \phi_j \\ 1 & x_k & \phi_k \end{bmatrix}}{\begin{bmatrix} 1 & x_i & x_i^2 \\ 1 & x_j & x_j^2 \\ 1 & x_k & x_k^2 \end{bmatrix}} = \frac{1}{D}\left[\phi_i(x_k - x_j) + \phi_j(x_i - x_k) + \phi_k(x_j - x_i)\right] \tag{10.34c}$$

where

$$D = (x_j x_k^2 - x_k x_j^2) + (x_k x_i^2 - x_i x_k^2) + (x_i x_j^2 - x_j x_i^2)$$

Substituting Equation 10.34 into Equation 10.31 and rearranging, we get

$$\phi = N_i\phi_i + N_j\phi_j + N_k\phi_k \tag{10.35a}$$

or

$$\phi = \sum_{I=i,\, j,\, k} N_I\phi_I, \tag{10.35b}$$

The shape functions for the quadratic line element are written in a compact form as

$$N_I(x) = \frac{1}{D}\left(\alpha_I + \beta_I x + \gamma_I x^2\right), \qquad I = i,\, j \text{ and } k \tag{10.36}$$

where

$$\alpha_I = (x_j x_k^2 - x_k x_j^2) + (x_k x_i^2 - x_i x_k^2) + (x_i x_j^2 - x_j x_i^2) \tag{10.37a}$$

$$\beta_I = (x_j^2 - x_k^2) + (x_k^2 - x_i^2) + (x_i^2 - x_j^2) \tag{10.37b}$$

$$\gamma_I = (x_k - x_j) + (x_i - x_k) + (x_j - x_i) \tag{10.37c}$$

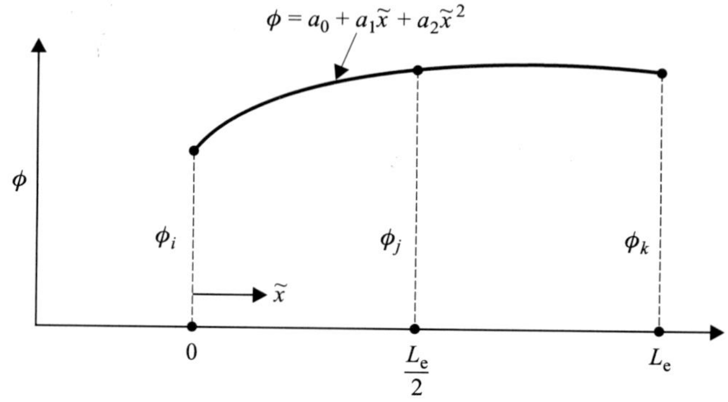

Figure 10.8 Quadratic line element in a local coordinate system

Quadratic Line Element in Local Coordinate. The quadratic shape function can be expressed in the local coordinate system with the origin located at the left-hand node as shown in Figure 10.8.

Substituting the coordinate points into the approximating function given by the quadratic Equation 10.31, in the local coordinate ($\tilde{x}$), we get

$$\phi_i = a_0 + a_1 \cdot 0 + a_2 \cdot 0 \tag{10.38a}$$

$$\phi_j = a_0 + a_1 \frac{L_e}{2} + a_2 \left(\frac{L_e}{2}\right)^2 \tag{10.38b}$$

$$\phi_k = a_0 + a_1 L_e + a_2 (L_e)^2 \tag{10.38c}$$

The system of equation is written in matrix form as

$$\begin{bmatrix} 1 & 0 & 0 \\ 1 & L_e/2 & L_e^2/4 \\ 1 & L_2 & L_e^2 \end{bmatrix} \begin{Bmatrix} a_0 \\ a_1 \\ a_2 \end{Bmatrix} = \begin{Bmatrix} \phi_i \\ \phi_j \\ \phi_k \end{Bmatrix} \tag{10.39}$$

Solution of this system gives

$$a_0 = \phi_i \tag{10.40a}$$

$$a_1 = \frac{1}{L_e}(-3\phi_i + 4\phi_j - \phi_k) \tag{10.40b}$$

$$a_2 = \frac{2}{L_e^2}(\phi_i - 2\phi_j + \phi_k) \tag{10.40c}$$

Substituting Equation 10.40 into Equation 10.31 and rearranging, we get

$$\phi(\tilde{x}) = \left(1 - \frac{\tilde{x}}{L_e}\right)\left(1 - 2\frac{\tilde{x}}{L_e}\right)\phi_i + 4\frac{\tilde{x}}{L_e}\left(1 - \frac{\tilde{x}}{L_e}\right)\phi_j + \frac{\tilde{x}}{L_e}\left(2\frac{\tilde{x}}{L_e} - 1\right)\phi_j \quad (10.41a)$$

or

$$\phi = N_i\phi_i + N_j\phi_j + N_k\phi_k \quad (10.41b)$$

where the shape functions are given as

$$N_i = \left(1 - \frac{\tilde{x}}{L_e}\right)\left(1 - 2\frac{\tilde{x}}{L_e}\right) \quad (10.42a)$$

$$N_j = 4\frac{\tilde{x}}{L_e}\left(1 - \frac{\tilde{x}}{L_e}\right) \quad (10.42b)$$

$$N_k = \frac{\tilde{x}}{L_e}\left(2\frac{\tilde{x}}{L_e} - 1\right) \quad (10.42c)$$

Quadratic Element in Natural Coordinate System. The approximation function for the quadratic line element in a natural coordinate system with the origin located at the mid point, as shown in Figure 10.9, is assumed as

$$\phi(\xi) = a_0 + a_1\xi + a_2\xi^2 \quad (10.43)$$

The coefficients are determined by substituting the coordinate points, and the approximating function is again obtained as

$$\phi = N_i\phi_i + N_j\phi_j + N_k\phi_k \quad (10.44)$$

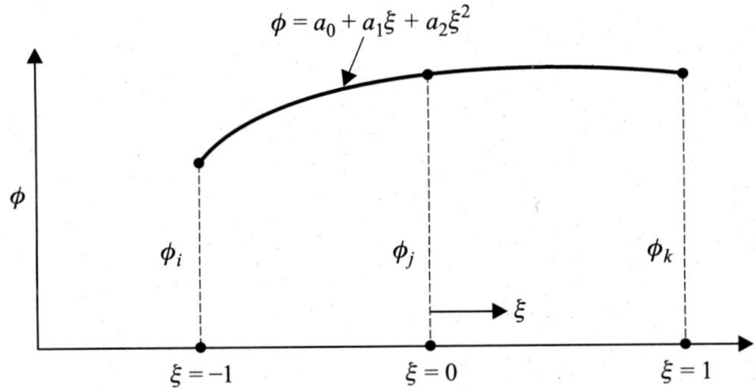

Figure 10.9 One-dimensional quadratic element in a natural coordinate system

with shape functions in natural coordinates given as

$$N_i(\xi) = \frac{\xi}{2}(\xi - 1) \tag{10.45a}$$

$$N_j(\xi) = -(\xi + 1)(\xi - 1) \tag{10.45b}$$

$$N_k(\xi) = \frac{\xi}{2}(\xi + 1) \tag{10.45c}$$

The transformation relationship between the local coordinate and natural coordinate systems is given by

$$x = \sum_{I=i,j,k} N_I x_I = N_i(\xi)x_i + N_j(\xi)x_j + N_k(\xi)x_k \tag{10.46a}$$

Additionally, we need the following relations for transforming integral relations into natural coordinate systems

$$\frac{d}{dx} = \frac{1}{dx/d\xi}\frac{d}{d\xi} \tag{10.46b}$$

where

$$\frac{dx}{d\xi} = \frac{dN_i}{d\xi}x_i + \frac{dN_j}{d\xi}x_j + \frac{dN_k}{d\xi}x_k \tag{10.46c}$$

10.1.3 One-dimensional Cubic Element

The one-dimensional cubic element in a local coordinate with the origin located at the left-hand corner node is considered as shown in Figure 10.10,

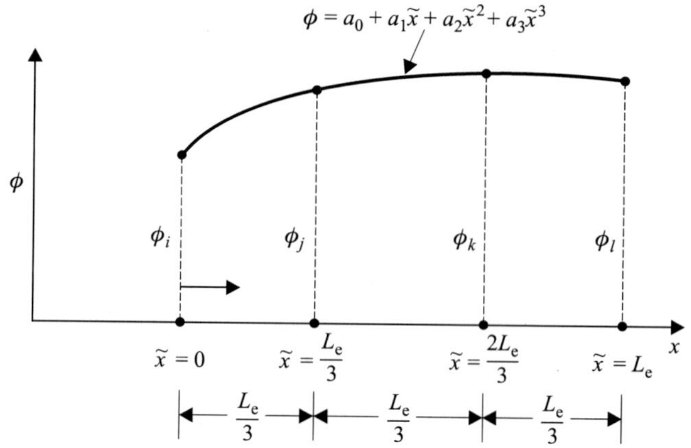

Figure 10.10 One-dimensional quadratic element in a local coordinate system

and the approximating function given as

$$\phi = a_0 + a_1\tilde{x} + a_2\tilde{x}^2 + a_3\tilde{x}^3 \tag{10.47}$$

The coefficients are derived using the nodal values at the four nodal points shown in the figure, and the approximation function derived as

$$\phi(\tilde{x}) = N_i\phi_i + N_j\phi_j + N_k\phi_k + N_l\phi_l = \sum_{i=i,j,k,l} N_I\phi_I \tag{10.48}$$

where the shape functions are given as

$$N_i(\tilde{x}) = \left(1 - \frac{3\tilde{x}}{L_e}\right)\left(1 - \frac{3\tilde{x}}{2L_e}\right)\left(1 - \frac{\tilde{x}}{L_e}\right) \tag{10.49a}$$

$$N_j(\tilde{x}) = \frac{9\tilde{x}}{L_e}\left(1 - \frac{3\tilde{x}}{2L_e}\right)\left(1 - \frac{\tilde{x}}{L_e}\right) \tag{10.49b}$$

$$N_k(\tilde{x}) = -\frac{9\tilde{x}}{2L_e}\left(1 - \frac{3\tilde{x}}{2L_e}\right)\left(1 - \frac{\tilde{x}}{L_e}\right) \tag{10.49c}$$

$$N_l(\tilde{x}) = \frac{\tilde{x}}{L}\left(1 - \frac{3\tilde{x}}{L_e}\right)\left(1 - \frac{3\tilde{x}}{2L_e}\right) \tag{10.49d}$$

One-dimensional Cubic Element in Natural Coordinate. The one-dimensional approximation function for a one-dimensional cubic element, as shown in Figure 10.11, is assumed as

$$\phi = a_0 + a_1\xi + a_2\xi^2 + a_3\xi^3 \tag{10.50}$$

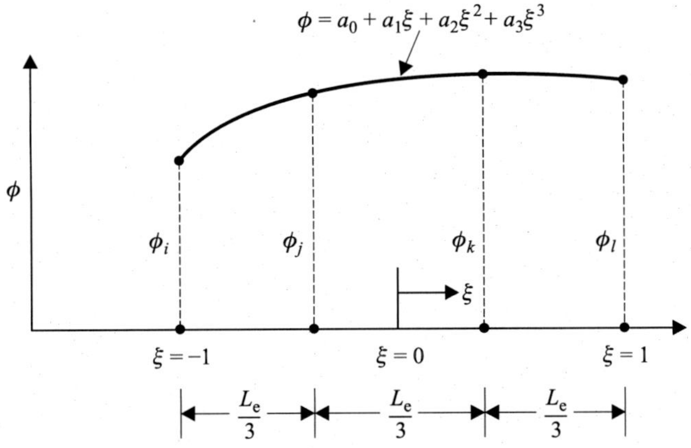

Figure 10.11 One-dimensional cubic elements in a natural coordinate

Substituting the coordinates of the four nodal points *i*, *j*, *k*, and *l*, we obtain the approximating function as

$$\xi = N_i\phi_i + N_j\phi_j + N_k\phi_k + N_l\phi_l = \sum_{i=i,j,k,l} N_I\phi_I \tag{10.51}$$

where the shape functions for the one-dimensional cubic element in a natural coordinate system are given as

$$N_i(\xi) = -\tfrac{9}{16}(1-\xi)\left(\tfrac{1}{3}+\xi\right)\left(\tfrac{1}{3}-\xi\right) \tag{10.52a}$$

$$N_j(\xi) = \tfrac{27}{16}(1+\xi)(1-\xi)\left(\tfrac{1}{3}-\xi\right) \tag{10.52b}$$

$$N_k(\xi) = \tfrac{27}{16}(1+\xi)(1-\xi)\left(\tfrac{1}{3}+\xi\right) \tag{10.52c}$$

$$N_l(\xi) = -\tfrac{9}{16}(1+\xi)\left(\tfrac{1}{3}+\xi\right)\left(\tfrac{1}{3}-\xi\right) \tag{10.52d}$$

Example 10.1: One-dimensional Element: The coordinates of the two nodes of a one-dimensional linear element are $x_i = 0.03$ m and $x_j = 0.04$ m in a media with thermal conductivity $k = 12\,\mathrm{W/m^\circ C}$. Based on the finite element analysis, the nodal temperature values are found to be $T_i = 85^\circ$C and $T_j = 80^\circ$C. Determine the temperature at a distance $x = 0.032$ m and the heat flux within the element.

Solution: The temperature distribution within a linear one-dimensional element is given by Equation 10.9 as

$$T = N_iT_i + N_jT_j$$

where the shape functions N_i and N_j are given as

$$N_i = \frac{x_j - x}{x_j - x_i}, \qquad N_j = \frac{x - x_i}{x_j - x_i}$$

Using the coordinate values, we get

$$N_i = \frac{x_j - x}{x_j - x_i} = \frac{0.04 - x}{0.04 - 0.03} = 100(0.04 - x)$$

and

$$N_j = \frac{x - x_i}{x_j - x_i} = \frac{x - 0.03}{0.04 - 0.03} = 100(x - 0.03)$$

with the substitution of the shape function expressions, the temperature distribution within the element is given as

$$T = 100(0.04 - x)T_i + 100(x - 0.03)T_j$$

Now, the temperature at a distance $x = 0.032$ within the element is given as

$$\begin{aligned} T &= 100 \times (0.04 - 0.032) \times 85.0 + 100 \times (0.032 - 0.03) \times 80.0 \\ &= 84°\text{C} \end{aligned}$$

The heat flux rate within the element is

$$\begin{aligned} q'' &= -k\frac{dT}{dx} = -k\left[\left(-\frac{1}{x_j - x_i}\right)T_i + \left(\frac{1}{x_j - x_i}\right)T_j\right] \\ &= -k\left(\frac{T_j - T_i}{x_j - x_i}\right) = -12 \times \left(\frac{80.0 - 85.0}{0.01}\right) \\ q'' &= 6000\,\text{W/m}^2 \end{aligned}$$

10.2 TWO-DIMENSIONAL ELEMENT

Two-dimensional elements are used to discretize a two-dimensional region. The most common polygonal shapes that are being used for two-dimensional elements are ***triangular*** and ***quadrilateral*** shapes or a variation of these elements as shown Figure 10.12.

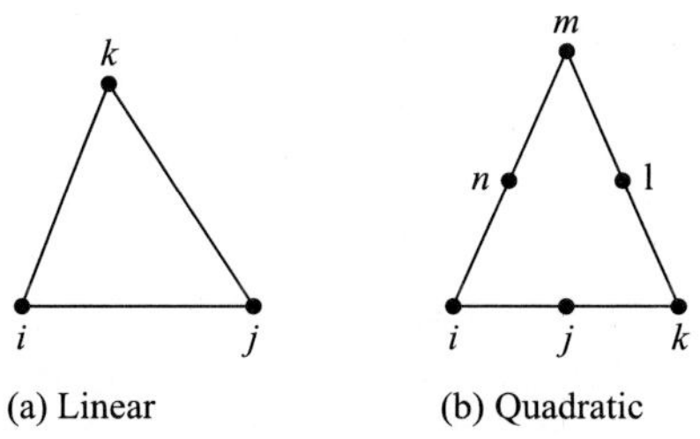

Figure 10.12 Triangular elements: (a) linear and (b) quadratic

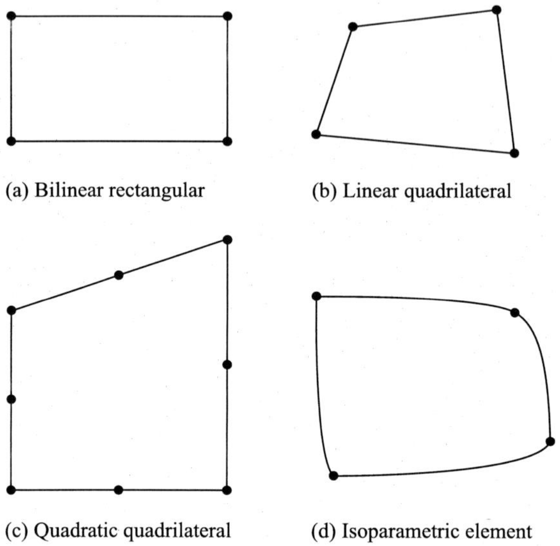

Figure 10.13 Rectangular and quadrilateral elements

The triangular elements (Figure 10.12) can take up any orientation and satisfy the continuity requirements involving adjacent elements, and gives closer approximation of irregular surfaces. On the other hand a bilinear rectangular element (Figure 10.13a) can only take up an orientation that is parallel to an x–y coordinate system, and typically used when the solution domain is rectangular in shape. However, the four-sided quadrilateral element (Figure 10.13b), other than the rectangular element, can have nonorthogonal sides, i.e., each side may have different slopes, which gives it the capability of having a closer fit of the irregular shaped solution domain. In fact, the quadrilateral element is becoming increasingly popular in discretizing the irregular solution domain.

The elements could be linear or cubic depending on the order of the polynomial used to represent the approximate function. For example, a polynomial of the form $\phi = a_0 + a_1 x + a_2 y$ is linear in x and y, and its coefficients a_0, a_1, and a_2 are expressed in terms of nodal values at the three vertices of the triangle (Figure 10.12a). A quadratic polynomial $\phi = a_0 + a_1 x + a_2 y + a_3 xy + a_4 x^2 + a_5 y^2$ with six terms can be used to form a geometrical shape or element with six nodes, such as in a triangular element with nodes at three vertices and nodes at mid-points of the three sides of the triangle (Figure 10.12b). A polynomial of the form $\phi = a_0 + a_1 x + a_2 y + a_3 xy$ has linear terms in x and y, and a bilinear term in x and y, and it requires a geometrical shape with four nodes such as the linear rectangular, which is a special case of the quadrilateral element with nodes at the vertices.

The procedure for deriving the shape functions or the interpolation functions for two-dimensional elements is the same as those given for one-dimensional elements presented in the previous sections. In this section, a brief

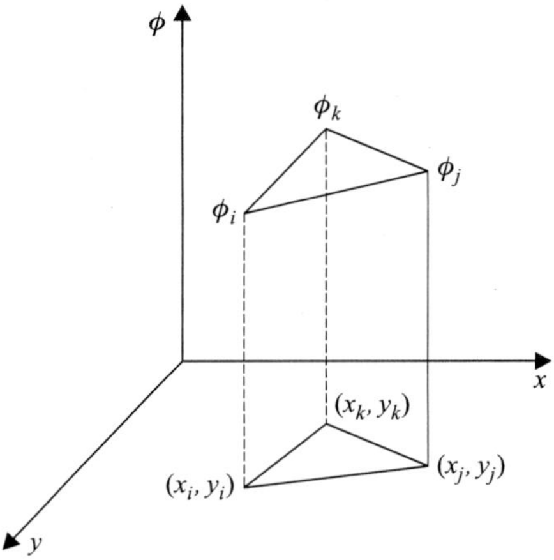

Figure 10.14 Linear triangular elements

discussion on the derivation of some of the two-dimensional elements in local and natural coordinate systems is given.

10.2.1 Linear Triangular Element

Let us consider a linear triangular element with three nodal values ϕ_i, ϕ_j, and ϕ_k with nodal coordinate (x_i, y_i), (x_j, y_j), and (x_k, y_k) as shown in Figure 10.14.

We have mentioned that the appropriate approximation is the first-order polynomial, i.e., by

$$\phi = a_0 + a_1 x + a_2 y \tag{10.53}$$

Substituting the nodal values, we get

$$\phi_i = a_0 + a_1 x_i + a_2 y_i \tag{10.54a}$$

$$\phi_j = a_0 + a_1 x_j + a_2 y_j \tag{10.54b}$$

$$\phi_k = a_0 + a_1 x_k + a_2 y_k \tag{10.54c}$$

The system of Equation 10.54 is written in a matrix notation as

$$\begin{bmatrix} 1 & x_i & y_i \\ 1 & x_j & y_j \\ 1 & x_k & y_k \end{bmatrix} \begin{Bmatrix} a_0 \\ a_1 \\ a_2 \end{Bmatrix} = \begin{Bmatrix} \phi_i \\ \phi_j \\ \phi_k \end{Bmatrix} \tag{10.55}$$

Solving by the Cramer rule

$$a_0 = \frac{\begin{vmatrix} \phi_i & x_i & y_i \\ \phi_j & x_j & y_j \\ \phi_k & x_k & y_k \end{vmatrix}}{2A} \tag{10.56}$$

or

$$a_0 = \frac{1}{2A}\left[\phi_i(x_j y_k - x_k y_j) + \phi_j(x_k y_i - x_i y_k) + \phi_k(x_i y_j - x_j y_i)\right] \tag{10.57a}$$

Similarly

$$a_1 = \frac{1}{2A}\left[\phi_i(y_j - y_k) + \phi_j(y_k - y_i) + \phi_k(y_i - y_j)\right] \tag{10.57b}$$

and

$$a_2 = \frac{1}{2A}\left[\phi_i(x_k - x_j) + \phi_j(x_i - x_k) + \phi_k(x_j - x_i)\right] \tag{10.57c}$$

where A is the area of the triangle and is given by the determinant

$$A = \frac{1}{2}\begin{vmatrix} 1 & x_i & y_i \\ 1 & x_j & y_j \\ 1 & x_k & y_k \end{vmatrix} \tag{10.58}$$

Substituting the coefficients into Equation 10.53 and rearranging, we get

$$\phi = N_i\phi_i + N_j\phi_j + N_k\phi_k \tag{10.59}$$

or

$$\phi = [N]\{\phi^{(e)}\} \tag{10.60}$$

where

$$N_i = \frac{1}{2A}(\alpha_i + \beta_i x + \gamma_i y) \tag{10.61a}$$

$$N_j = \frac{1}{2A}(\alpha_j + \beta_j x + \gamma_j y) \tag{10.61b}$$

$$N_k = \frac{1}{2A}(\alpha_k + \beta_k x + \gamma_k y) \tag{10.61c}$$

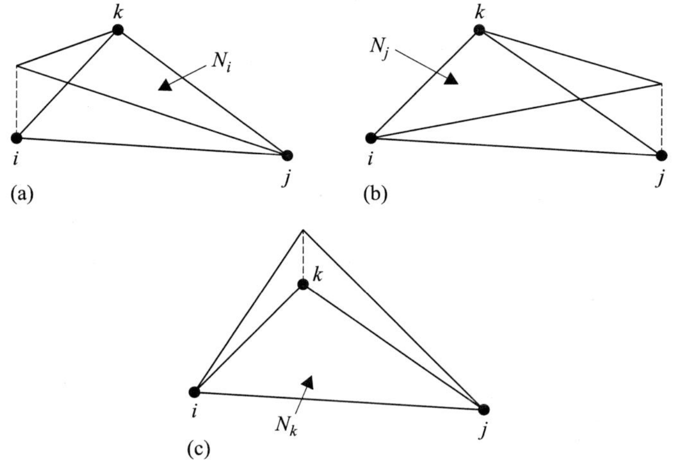

Figure 10.15 Linear triangular shape functions

and

$$\alpha_i = x_j y_k - x_k y_j, \quad \beta_i = y_j - y_k, \quad \gamma_i = x_k - x_j \tag{10.62a}$$

$$\alpha_j = x_k y_i - x_i y_k, \quad \beta_j = y_k - y_i, \quad \gamma_j = x_i - x_k \tag{10.62b}$$

$$\alpha_k = x_i y_j - x_j y_i, \quad \beta_k = y_i - y_j, \quad \gamma_k = x_j - x_i \tag{10.62c}$$

The properties of a two-dimensional triangular element are similar to those for one-dimensional elements. Each shape function $N_I, I = i, j, k$ has a value of ***one*** at its own node and is ***zero*** at the other nodes as depicted in Figure 10.14. Also, the shape function varies linearly (for the linear element) along two adjacent sides of the node, and is zero along the side opposite to the node. For example, the shape function N_i has a value of one at the node i and has a value of zero at nodes j and k. It varies linearly along ij and ik, and zero along jk.

Other relevant properties are

$$\sum_{I=i,j,k} N_I = 1, \quad \sum_{I=i,j,k} \frac{\partial N_I}{\partial x} = 0 \quad \text{and} \quad \sum_{I=i,j,k} \frac{\partial N_j}{\partial y} = 0 \tag{10.63}$$

Linear Triangular Element in Natural Coordinate. It was mentioned before that a natural coordinate system simplifies evaluation of integral expressions involving shape functions through use of integral formulas or numerical quadratures such as the Gauss–Legendre formula. A natural coordinate system

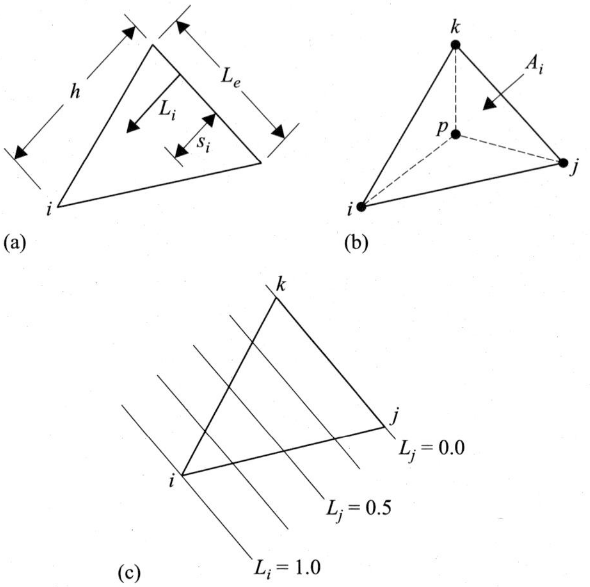

Figure 10.16 Natural or area coordinates for a linear triangular element

for a two-dimensional triangular element is known as area coordinate system. In this coordinate system three coordinate variables L_i, L_j, and L_k are defined in dimensionless form as

$$L_i = \frac{s_i}{h}, \quad L_j = \frac{s_j}{h}, \quad L_k = \frac{s_k}{h} \tag{10.64}$$

where s is the perpendicular distance of a point P from the side of the triangle and h is the altitude of the same side as shown in Figure 10.16a.

The lines of constant L_i run parallel to the opposite side of the triangle, i.e., side jk. Similarly, the lines of constant L_j and L_k run parallel to ik and ij, respectively. Also, the line coordinates, L, are called area coordinates as these are equal to the ratio of the partial area to the total area of the triangle as given by

$$L_i = \frac{A_i}{A}, \quad L_j = \frac{A_j}{A}, \quad L_k = \frac{A_k}{A} \tag{10.65}$$

where A_i, A_j, and A_k are the partial areas of the triangular element area. These partial areas are defined by joining a point P in the triangle with three vertices or nodal points i, j, and k, as shown in Figure 10.16b.

Since the sum of A_i, A_j, and A_k is equal to A, the sum of three line coordinates is equal to one, i.e.,

$$L_i + L_j + L_k = 1 \tag{10.66}$$

Considering the coordinate of the point P as (x, y) and noting that the area of the triangle A is given by Equation 10.58, we can write

$$A_i = \frac{1}{2}\begin{vmatrix} 1 & x & y \\ 1 & x_j & y_j \\ 1 & x_k & y_k \end{vmatrix} \tag{10.67a}$$

or

$$A_i = \frac{1}{2}\left[\left(x_j y_k - x_k y_j\right) + \left(y_j - y_k\right)x + \left(x_k - x_j\right)y\right] \tag{10.67b}$$

Now, substituting Equation 10.67b into the defining equation of L_i, i.e., Equation 10.65 and comparing with Equation 10.60a we get

$$L_i = N_i \tag{10.68a}$$

Similarly, we can show that

$$L_j = N_j \tag{10.68b}$$

and

$$L_k = N_k \tag{10.68c}$$

So, we can express the approximating function in terms of line coordinate s as

$$\phi = L_i\phi_i + L_j\phi_j + L_k\phi_k \tag{10.69}$$

Also, the relationship between the Cartesian and line coordinates is given by

$$x = L_i x_i + L_j x_j + L_k x_k \tag{10.70a}$$

and

$$y = L_i y_i + L_j y_j + L_k y_k \tag{10.70b}$$

We have mentioned before that the main reason for expressing the two-dimensional shape function in terms of the area coordinate is the simplicity in evaluating some integrals involving shape functions through use of integral formulas. One such common integral formula is given as

$$I = \int_A L_i^a L_j^b L_k^c \, dA = \frac{a!b!c!}{(a+b+c+2)!} 2A \tag{10.71}$$

For example the integral expression of the form

$$I = \int_A N_i N_j \, dA \tag{10.72}$$

can be evaluated using Equation 10.68 and 10.71 as

$$I = \int_A N_i N_j \, dA = \int_A L_i^1 L_j^1 L_k^0 \, dA \tag{10.73}$$

$$= \frac{1!1!0!}{(1+1+0+2)!} 2A = \frac{A}{12} \tag{10.74}$$

We will also see in later chapters that the finite element formulations require evaluation of integrals over each element, and these integrals may involve derivatives of the shape functions. In order to evaluate these integrals in natural coordinate systems, it is also required to express the derivatives in natural coordinates. We can express the derivatives in the natural coordinate system assuming that L_i and L_j are independent, L_k is given by Equation 10.66, and using the chain rule as

$$\frac{\partial N_i}{\partial L_i} = \frac{\partial N_i}{\partial x}\frac{\partial x}{\partial L_i} + \frac{\partial N_i}{\partial y}\frac{\partial y}{\partial L_i} \tag{10.75a}$$

$$\frac{\partial N_i}{\partial L_j} = \frac{\partial N_i}{\partial x}\frac{\partial x}{\partial L_j} + \frac{\partial N_i}{\partial y}\frac{\partial y}{\partial L_j} \tag{10.75b}$$

Writing this set in matrix form, we get

$$\begin{Bmatrix} \dfrac{\partial N_i}{\partial L_i} \\ \dfrac{\partial N_i}{\partial L_j} \end{Bmatrix} = \begin{bmatrix} \dfrac{\partial x}{\partial L_i} & \dfrac{\partial y}{\partial L_i} \\ \dfrac{\partial x}{\partial L_j} & \dfrac{\partial y}{\partial L_j} \end{bmatrix} \begin{Bmatrix} \dfrac{\partial N_i}{\partial x} \\ \dfrac{\partial N_i}{\partial y} \end{Bmatrix} = J \begin{Bmatrix} \dfrac{\partial N_i}{\partial x} \\ \dfrac{\partial N_i}{\partial y} \end{Bmatrix} \tag{10.75c}$$

where J is the ***Jacobian matrix***, which is defined here as

$$J = \begin{bmatrix} \dfrac{\partial x}{\partial L_i} & \dfrac{\partial y}{\partial L_i} \\ \dfrac{\partial x}{\partial L_j} & \dfrac{\partial y}{\partial L_j} \end{bmatrix} \tag{10.76}$$

Equation 10.75c can be inverted to express the derivatives in x and y coordinates as

$$\begin{Bmatrix} \dfrac{\partial N_i}{\partial x} \\ \dfrac{\partial N_i}{\partial y} \end{Bmatrix} = J^{-1} \begin{Bmatrix} \dfrac{\partial N_i}{\partial L_i} \\ \dfrac{\partial N_i}{\partial L_j} \end{Bmatrix} \tag{10.77}$$

Equation 10.77 can further be simplified by using the defining equation for the inverse of a matrix (see Appendix A) as

$$\begin{Bmatrix} \dfrac{\partial N_i}{\partial x} \\ \dfrac{\partial N_i}{\partial y} \end{Bmatrix} = \frac{1}{|J|} \begin{bmatrix} \dfrac{\partial y}{\partial L_j} & -\dfrac{\partial y}{\partial L_i} \\ -\dfrac{\partial x}{\partial L_j} & \dfrac{\partial x}{\partial L_i} \end{bmatrix} \begin{Bmatrix} \dfrac{\partial N_i}{\partial L_i} \\ \dfrac{\partial N_i}{\partial L_j} \end{Bmatrix} \tag{10.78}$$

or

$$\begin{Bmatrix} \dfrac{\partial N_i}{\partial x} \\ \dfrac{\partial N_i}{\partial y} \end{Bmatrix} = \frac{1}{|J|} \begin{Bmatrix} \dfrac{\partial y}{\partial L_j}\dfrac{\partial N_i}{\partial L_i} - \dfrac{\partial y}{\partial L_i}\dfrac{\partial N_i}{\partial L_j} \\ -\dfrac{\partial x}{\partial L_j}\dfrac{\partial N_i}{\partial L_i} + \dfrac{\partial x}{\partial L_i}\dfrac{\partial N_i}{\partial L_j} \end{Bmatrix} \tag{10.79}$$

where $|J|$ is the determinant of the Jacobian matrix and it is given as

$$|J| = \frac{\partial x}{\partial L_i}\frac{\partial y}{\partial L_j} - \frac{\partial x}{\partial L_j}\frac{\partial y}{\partial L_i} \tag{10.80}$$

Another important use of the line or area coordinate system is in evaluating an integral along one side of the element (Figure 10.17).

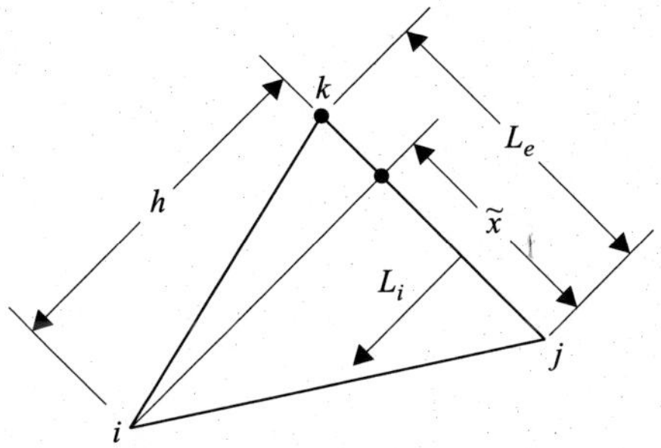

Figure 10.17 Integral along one side of an element

Consider a point p on the side j–k of the triangular element. For this point, the line coordinates are defined as

$$L_j = \frac{A_j}{A} = \frac{\frac{1}{2}h(L_e - \tilde{x})}{\frac{1}{2}hL_e} = 1 - \frac{\tilde{x}}{L_e}, \qquad L_k = \frac{\tilde{x}}{L_e} \tag{10.81}$$

Comparing these with the shape functions for one-dimensional natural coordinates, we can write

$$L_i = 0, \quad L_j = l_i \quad \text{and} \quad L_k = l_j \quad \text{for the side } j-k \tag{10.82a}$$

$$L_i = l_j, \quad L_j = 0 \quad \text{and} \quad L_k = l_i \quad \text{for the side } k-i \tag{10.82b}$$

$$L_i = l_i, \quad L_i = l_j \quad \text{and} \quad L_k = 0 \quad \text{for the side } i-j \tag{10.82c}$$

10.2.2 Quadratic Triangular Element

A quadratic triangular element has six nodes with three nodes at the three vertices and three at mid-points of the sides of the triangle as shown in Figure 10.18.

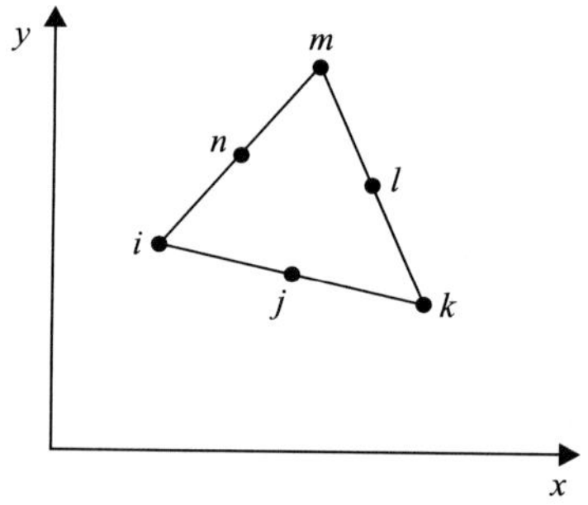

Figure 10.18 Quadratic triangular element

The shape function for the quadratic triangular element in the area coordinate is expressed as

$$N_i = L_i(2L_i - 1), \quad N_j = L_j(2L_j - 1), \quad N_k = L_k(2L_k - 1)$$
$$N_l = 4L_jL_k, \quad N_m = 4L_iL_k, \quad N_n = 4L_iL_j \tag{10.83}$$

10.2.3 Two-dimensional Quadrilateral Elements

We have mentioned before that quadrilateral elements are four-sided polygons. The sides of the element may be straight and orthogonal as in a bilinear rectangular element (Figure 10.13a) or each side may have different slopes

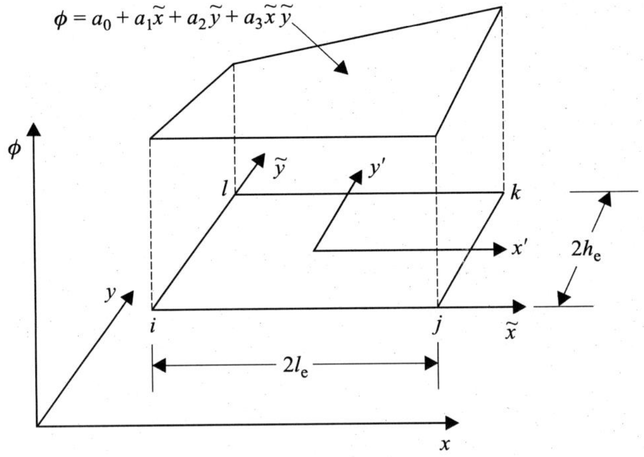

Figure 10.19 Bilinear rectangular element

(Figure 10.13b) or arbitrarily oriented curved sides as in isoparametric elements (Figure 10.13d).

Bilinear Rectangular. The bilinear rectangular element is a special class of quadrilateral element, in which the sides of the element are parallel to the coordinate axes. Let us consider a bilinear rectangular element of length $2l_e$ and height $2h_e$, as shown in Figure 10.19, in a local coordinate system $(\tilde{x}, \tilde{y})$ with the origin located at the left-hand corner node.

The approximation function is assumed as

$$\phi(\tilde{x}, \tilde{y}) = a_0 + a_1\tilde{x} + a_2\tilde{y} + a_3\tilde{x}\tilde{y} \tag{10.84}$$

The coefficients a_0, a_1, a_2, and a_3 are obtained by using the nodal values and coordinates

$$\phi_i = a_0 \tag{10.85a}$$

$$\phi_j = a_0 + 2l_e a_1 \tag{10.85b}$$

$$\phi_k = a_0 + (2l_e)a_1 + (2h_e)a_2 + (4l_e h_e)a_3 \tag{10.85c}$$

$$\phi_l = a_0 + (2h_e)a_2 \tag{10.85d}$$

Solving the set of equations, we obtain the coefficients as

$$\begin{aligned} a_0 &= \phi_i, \quad a_1 = \frac{1}{2l_e}(\phi_j - \phi_i) \\ a_2 &= \frac{1}{2h_e}(\phi_l - \phi_i) \\ a_3 &= \frac{1}{4l_e h_e}(\phi_i - \phi_j + \phi_k - \phi_l) \end{aligned} \tag{10.86}$$

Substituting Equation 10.86 into Equation 10.84, we get the approximation function for the bilinear rectangular element in the local coordinate system as

$$\phi(\tilde{x}, \tilde{y}) = N_i\phi_i + N_j\phi_j + N_k\phi_k + N_i\phi_l = \sum_{I=i,j,k,l} N_I\phi_I \tag{10.87}$$

with shape functions given as

$$\begin{aligned} N_i &= \left(1 - \frac{\tilde{x}}{2l_e}\right)\left(1 - \frac{\tilde{y}}{2h_e}\right), \quad N_j = \frac{\tilde{x}}{2l_e}\left(1 - \frac{\tilde{y}}{2h_e}\right) \\ N_k &= \frac{\tilde{x}\tilde{y}}{4l_e h_e}, \quad N_l = \frac{\tilde{y}}{2h_e}\left(1 - \frac{\tilde{x}}{2l_e}\right) \end{aligned} \tag{10.88}$$

The shape functions for bilinear rectangular element have properties similar to the triangular elements. Each shape function varies linearly along the adjacent sides of its node and it is zero along the other two sides opposite to its node. For example, the shape function N_i varies linearly along the adjacent sides *ij* and *il*, and zero along *jk* and *kl*.

Rectangular Element in Alternate Local Coordinate System The shape functions given by Equation 10.88 can be transformed for a local coordinate system (x', y') with its origin located at the center of the element as

$$N_i = \frac{1}{4}\left(1 - \frac{x'}{l_e}\right)\left(1 - \frac{y'}{h_e}\right) \tag{10.89a}$$

$$N_j = \frac{1}{4}\left(1 + \frac{x'}{l_e}\right)\left(1 - \frac{y'}{h_e}\right) \tag{10.89b}$$

$$N_k = \frac{1}{4}\left(1 + \frac{x'}{l_e}\right)\left(1 + \frac{y'}{h_e}\right) \tag{10.89c}$$

$$N_l = \frac{1}{4}\left(1 - \frac{x'}{l_e}\right)\left(1 + \frac{y'}{h_e}\right) \tag{10.89d}$$

Quadrilateral Element in Natural Coordinate System In the natural coordinate system for the two-dimensional quadrilateral element, the coordinates are expressed in a dimensionless form as $\xi = x'/l_e$ and $\eta = y'/h_e$, and the range of coordinate variables are given as $-1 \leq \xi \leq 1$ and $-1 \leq \eta \leq 1$, respectively, as shown in Figure 10.20.

With the definition of these natural coordinates, the sides of the quadrilateral element are no longer required to be parallel to the coordinate axes as in the case of a rectangular element. The shape functions for the quadrilateral element in the natural coordinate system can simply be obtained

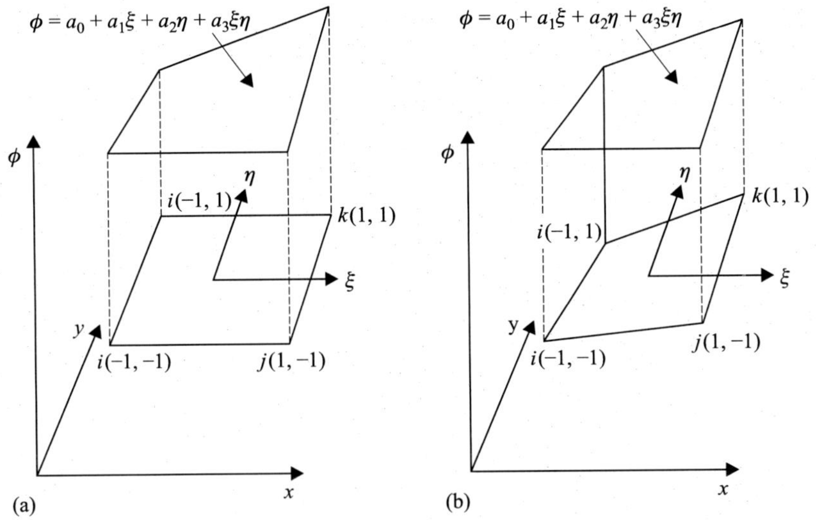

Figure 10.20 Quadrilateral element in a natural coordinate system

by substituting the dimensionless coordinate variable in the approximation function of bilinear form (Equation 10.84) in natural coordinates, and expressed as

$$N_i = \tfrac{1}{4}(1 - \xi)(1 - \eta) \tag{10.90a}$$

$$N_j = \tfrac{1}{4}(1 + \xi)(1 - \eta) \tag{10.90b}$$

$$N_k = \tfrac{1}{4}(1 + \xi)(1 + \eta) \tag{10.90c}$$

$$N_l = \tfrac{1}{4}(1 - \xi)(1 + \eta) \tag{10.90d}$$

The geometry of the quadrilateral element with arbitrarily oriented slopes (Figure 10.20b) can be represented as

$$x = N_i x_i + N_j x_j + N_k x_k + N_l x_l = \sum_{I=i}^{N_p} N_I x_I \tag{10.91a}$$

and

$$y = N_i y_i + N_j y_j + N_k y_k + N_l y_l = \sum_{I=i}^{N_p} N_I x_I \tag{10.91b}$$

Note that the approximation function ϕ and the geometry (x, y) are expressed by the shape functions of the same order.

Using the procedure outlined for triangular element in Section 10.2.1, we can express the derivative of the shape functions in natural coordinates as

$$\begin{Bmatrix} \dfrac{\partial N_i}{\partial \xi} \\ \dfrac{\partial N_i}{\partial \eta} \end{Bmatrix} = \begin{bmatrix} \dfrac{\partial x}{\partial \xi} & \dfrac{\partial y}{\partial \xi} \\ \dfrac{\partial x}{\partial \eta} & \dfrac{\partial y}{\partial \eta} \end{bmatrix} \begin{Bmatrix} \dfrac{\partial N_i}{\partial x} \\ \dfrac{\partial N_i}{\partial y} \end{Bmatrix} = J \begin{Bmatrix} \dfrac{\partial N_i}{\partial x} \\ \dfrac{\partial N_i}{\partial y} \end{Bmatrix} \tag{10.92}$$

where the Jacobian matrix is defined as

$$J = \begin{bmatrix} \dfrac{\partial x}{\partial \xi} & \dfrac{\partial y}{\partial \xi} \\ \dfrac{\partial x}{\partial \eta} & \dfrac{\partial y}{\partial \eta} \end{bmatrix} \tag{10.93}$$

The expression of the derivatives in x and y coordinates are obtained by taking the inversion of Equation 10.92 as

$$\begin{Bmatrix} \dfrac{\partial N_i}{\partial x} \\ \dfrac{\partial N_i}{\partial y} \end{Bmatrix} = J^{-1} \begin{Bmatrix} \dfrac{\partial N_i}{\partial \xi} \\ \dfrac{\partial N_i}{\partial \eta} \end{Bmatrix} \tag{10.94}$$

Evaluating the inversion of the Jacobian matrix, we get

$$\begin{Bmatrix} \dfrac{\partial N_i}{\partial x} \\ \dfrac{\partial N_i}{\partial y} \end{Bmatrix} = \frac{1}{|J|} \begin{bmatrix} \dfrac{\partial y}{\partial \eta} & -\dfrac{\partial y}{\partial \xi} \\ -\dfrac{\partial x}{\partial \eta} & \dfrac{\partial x}{\partial \xi} \end{bmatrix} \begin{Bmatrix} \dfrac{\partial N_i}{\partial \xi} \\ \dfrac{\partial N_i}{\partial \eta} \end{Bmatrix} \tag{10.95}$$

or

$$\begin{Bmatrix} \dfrac{\partial N_i}{\partial x} \\ \dfrac{\partial N_i}{\partial y} \end{Bmatrix} = \frac{1}{|J|} \begin{Bmatrix} \dfrac{\partial y}{\partial \eta}\dfrac{\partial N_i}{\partial \xi} - \dfrac{\partial y}{\partial \xi}\dfrac{\partial N_i}{\partial \eta} \\ -\dfrac{\partial x}{\partial \eta}\dfrac{\partial N_i}{\partial \xi} + \dfrac{\partial x}{\partial \xi}\dfrac{\partial N_i}{\partial \eta} \end{Bmatrix} \tag{10.96}$$

where $|J|$, the determinant of the Jacobian matrix, is given as

$$|J| = \frac{\partial x}{\partial \xi}\frac{\partial y}{\partial \eta} - \frac{\partial x}{\partial \eta}\frac{\partial y}{\partial \xi} \tag{10.97}$$

Also, $|J|$ relates the differential area $dA(\xi, \eta) = d\xi\, d\eta$ and differential area $dA(x, y) = dx\, dy$ as

$$dA(x, y) = |J|\, dA(\xi, \eta)$$

The natural coordinate system helps in simplifying the integral expressions and enables us to use Gauss–Legendre quadrature formulas. For example

$$A = \iint_A dx\, dy = \int_{-1}^{1}\int_{-1}^{1} |J|\, d\xi\, d\eta$$

$$I = \iint_A N_i N_j\, dx\, dy = \int_{-1}^{1}\int_{-1}^{1} N_i N_j |J| d\xi\, d\eta$$

and

$$I = \iint_A \frac{\partial N_i}{\partial x}\frac{\partial N_j}{\partial x}\, dx\, dy = \int_{-1}^{1}\int_{-1}^{1} \frac{\partial N_i}{\partial x}\frac{\partial N_j}{\partial y} |J|\, d\xi\, d\eta$$

Note that these integrals are now in a format suitable for numerical integration by n-term Gauss–Legendre quadrature given by Equation 4.40

$$I \cong \sum_{i=1}^{n}\sum_{j=1}^{n} w_i\, w_j g(x_i', y_j')$$

The number of points n of the quadrature formula will be one less than the order of the polynomial function g.

Quadratic Rectangular Element. A higher-order quadratic element contains eight nodes, one at each corner and one at the mid-point of each side, as shown with a local coordinate system in Figure 10.21.

The approximation function for the eight-noded quadratic rectangular element is given by

$$\phi = a_0 + a_1 x' + a_2 y' + a_3 x' y' + a_4 x'^2 + a_5 y'^2 + a_6 x'^2 y' + a_7 x' y'^2 \qquad (10.98)$$

Substituting the nodal values and coordinates, and solving the resulting system of equations, we get

$$\phi = N_i\phi_i + N_j\phi_j + N_k\phi_k + N_l\phi_l + N_m\phi_m + N_n\phi_n + N_o\phi_o + N_p\phi_p = \sum_{I=i}^{Np} N_I\phi_I \qquad (10.99)$$

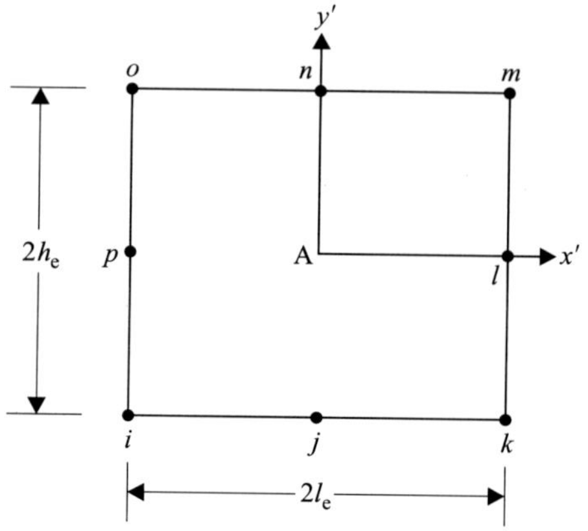

Figure 10.21 Quadratic rectangular element

where the shape functions are given as

$$N_i = -\frac{1}{4l_e^2h_e^2}(h_e - x)(l_e - y)(l_eh_e + xl_e + yh_e) \quad (10.100a)$$

$$N_j = \frac{1}{2l_eh_e^2}(h_e^2 - x^2)(l_e - y) \quad (10.100b)$$

$$N_k = \frac{1}{4l_e^2h_e^2}(h_e + x)(l_e - y)(xl_e - yh_e - l_eh_e) \quad (10.100c)$$

$$N_l = \frac{1}{2l_eh_e^2}(h_e^2 - x^2)(l_e + y) \quad (10.100d)$$

$$N_m = -\frac{1}{4l_e^2h_e^2}(h_e + x)(l_e + y)(xl_e + yh_e - l_eh_e) \quad (10.100e)$$

$$N_n = \frac{1}{2l_eh_e^2}(h_e^2 - x^2)(l_e + y) \quad (10.100f)$$

$$N_o = -\frac{1}{4l_e^2h_e^2}(h_e - x)(l_e + y)(l_eh_e + xl_e - yh_e) \quad (10.100g)$$

$$N_p = \frac{1}{2l_e^2h_e}(h_e - x)(l_e^2 - y^2) \quad (10.100h)$$

In the natural coordinate system, the approximation function is given by Equation 10.99, with shape functions given as

$$N_i = -\tfrac{1}{4}(1 - \xi)(1 - \eta)(1 + \xi + \eta) \quad (10.101a)$$

$$N_j = \tfrac{1}{2}(1 - \xi^2)(1 - \eta) \quad (10.101b)$$

$$N_k = \tfrac{1}{4}(1 + \xi)(1 - \eta)(\xi - \eta - 1) \quad (10.101c)$$

$$N_l = \tfrac{1}{2}(1 + \xi)(1 - \eta^2) \quad (10.101d)$$

$$N_m = \tfrac{1}{4}(1 + \xi)(1 + \eta)(\xi + \eta - 1) \quad (10.101e)$$

$$N_n = \tfrac{1}{2}(1 - \xi^2)(1 + \eta) \quad (10.101f)$$

$$N_o = -\tfrac{1}{4}(1 - \xi)(1 + \eta)(1 + \xi - \eta) \quad (10.101g)$$

$$N_p = \tfrac{1}{2}(1 - \xi)(1 - \eta^2) \quad (10.101h)$$

Example 10.2: Two-dimensional Quadratic Element: Derive the shape functions for the eight-noded quadratic rectangular element in the natural co-ordinate system with origin at the point $A(0, 0)$ as shown in the figure.

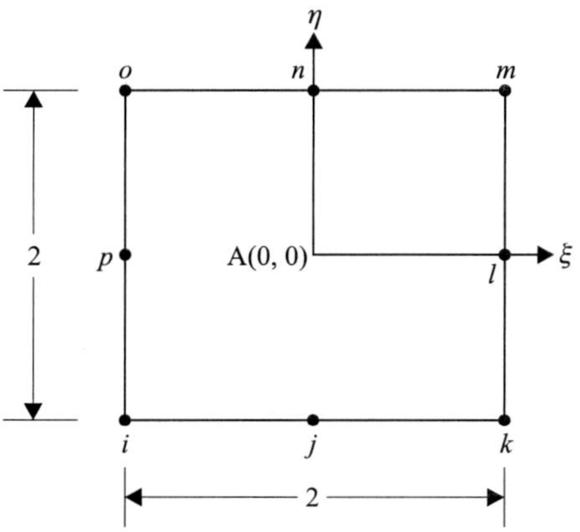

The approximate function for the eight-noded quadratic element is assumed as

$$\phi = a_0 + a_1\xi + a_2\eta + a_3\xi\eta + a_4\xi^2 + a_5\eta^2 + a_6\xi^2\eta + a_7\xi\eta^2 \quad (E.10.2.1)$$

Based on the origin located at A, the coordinates of the nodes are $i(-1, -1)$, $j(0, -1)$, $k(1, -1)$, $l(1, 0)$, $m(1, 1)$, $n(0, 1)$, $o(-1, 1)$, $p(-1, 0)$. Now, substituting the nodal values and coordinates in the approximation function given by

Equation E.10.2.1, we get

$$
\begin{aligned}
\phi_i &= a_0 - a_1 - a_2 + a_3 + a_4 + a_5 - a_6 - a_7 \\
\phi_j &= a_0 - a_2 + a_5 \\
\phi_k &= a_0 + a_1 - a_2 - a_3 + a_4 + a_5 - a_6 + a_7 \\
\phi_l &= a_0 + a_1 + a_4 \\
\phi_m &= a_0 + a_1 + a_2 + a_3 + a_4 + a_5 + a_6 + a_7 \\
\phi_n &= a_0 + a_2 + a_5 \\
\phi_o &= a_0 - a_1 + a_2 - a_3 + a_4 + a_5 + a_6 - a_7 \\
\phi_p &= a_0 - a_1 + a_4
\end{aligned}
\tag{E.10.2.2}
$$

Equation (E.10.2.2) can be expressed in matrix form as

$$
\begin{bmatrix}
1 & -1 & -1 & 1 & 1 & 1 & -1 & -1 \\
1 & -1 & 0 & 0 & 0 & 1 & 0 & 0 \\
1 & 1 & -1 & -1 & 1 & 1 & -1 & 1 \\
1 & 1 & 0 & 0 & 1 & 0 & 0 & 0 \\
1 & 1 & 1 & 1 & 1 & 1 & 1 & 1 \\
1 & 0 & 1 & 0 & 0 & 1 & 0 & 0 \\
1 & -1 & 1 & -1 & 1 & 1 & 1 & 1 \\
1 & -1 & 0 & 0 & 1 & 0 & 0 & 0
\end{bmatrix}
\begin{Bmatrix} a_0 \\ a_1 \\ a_2 \\ a_3 \\ a_4 \\ a_5 \\ a_6 \\ a_7 \end{Bmatrix}
=
\begin{bmatrix} \phi_i \\ \phi_j \\ \phi_k \\ \phi_l \\ \phi_m \\ \phi_n \\ \phi_o \\ \phi_p \end{bmatrix}
\tag{E.10.2.3}
$$

Solving the system for unknown coefficients a, we get

$$
\begin{aligned}
a_0 &= \tfrac{1}{2}\phi_j + \tfrac{1}{2}\phi_n + \tfrac{1}{2}\phi_p - \tfrac{1}{4}\phi_o + \tfrac{1}{2}\phi_l - \tfrac{1}{4}\phi_i - \tfrac{1}{4}\phi_m - \tfrac{1}{4}\phi_k \\
a_1 &= -\tfrac{1}{2}\phi_p + \tfrac{1}{2}\phi_l \\
a_2 &= -\tfrac{1}{2}\phi_j + \tfrac{1}{2}\phi_n \\
a_3 &= -\tfrac{1}{4}\phi_o + \tfrac{1}{4}\phi_i + \tfrac{1}{4}\phi_m - \tfrac{1}{4}\phi_k \\
a_4 &= -\tfrac{1}{2}\phi_j - \tfrac{1}{2}\phi_n + \tfrac{1}{4}\phi_o + \tfrac{1}{4}\phi_i + \tfrac{1}{4}\phi_m + \tfrac{1}{4}\phi_k \\
a_5 &= -\tfrac{1}{2}\phi_p + \tfrac{1}{4}\phi_o - \tfrac{1}{2}\phi_l + \tfrac{1}{4}\phi_i + \tfrac{1}{4}\phi_m + \tfrac{1}{4}\phi_k \\
a_6 &= \tfrac{1}{2}\phi_j - \tfrac{1}{2}\phi_n + \tfrac{1}{4}\phi_o - \tfrac{1}{4}\phi_i + \tfrac{1}{4}\phi_m - \tfrac{1}{4}\phi_k \\
a_7 &= \tfrac{1}{2}\phi_p - \tfrac{1}{4}\phi_o - \tfrac{1}{2}\phi_l - \tfrac{1}{4}\phi_i + \tfrac{1}{4}\phi_m + \tfrac{1}{4}\phi_k
\end{aligned}
\tag{E.10.2.4}
$$

Substituting a's into Equation (E.10.2.1) and rearranging, we get

$$\phi = \phi_i N_i + \phi_j N_j + \phi_k N_k + \phi_l N_l + \phi_m N_m + \phi_n N_n + \phi_o N_o + \phi_p N_p \qquad \text{(E.10.2.5)}$$

where the shape functions are expressed as

$$N_i = \tfrac{1}{4}\eta\xi + \tfrac{1}{4}\xi^2 + \tfrac{1}{4}\eta^2 - \tfrac{1}{4}\xi^2\eta - \tfrac{1}{4} - \tfrac{1}{4}\xi\eta^2 = -\frac{(1-\xi)(1-\eta)(1+\xi+\eta)}{4}$$

$$N_j = \tfrac{1}{2} - \tfrac{1}{2}\xi^2 - \tfrac{1}{2}\eta + \tfrac{1}{2}\xi^2\eta = \frac{(1-\xi^2)(1-\eta)}{2}$$

$$N_k = \tfrac{1}{4}\xi\eta^2 + \tfrac{1}{4}\eta^2 + \tfrac{1}{4}\xi^2 - \tfrac{1}{4}\xi\eta - \tfrac{1}{4} - \tfrac{1}{4}\xi^2\eta = \frac{(1+\xi)(1-\eta)(\xi-\eta-1)}{4}$$

$$N_l = -\tfrac{1}{2}\xi\eta^2 + \tfrac{1}{2}\xi - \tfrac{1}{2}\eta^2 + \tfrac{1}{2} = \frac{(1+\xi)(1-\eta^2)}{2}$$

$$N_m = -\tfrac{1}{4} + \tfrac{1}{4}\xi\eta^2 + \tfrac{1}{4}\xi\eta + \tfrac{1}{4}\eta^2 + \tfrac{1}{4}\xi^2 + \tfrac{1}{4}\xi^2\eta = \frac{(1+\xi)(1+\eta)(\xi+\eta-1)}{4}$$

$$N_n = \tfrac{1}{2} - \frac{1}{2}\xi^2 + \tfrac{1}{2}\eta - \tfrac{1}{2}\xi^2\eta = \frac{(1-\xi^2)(1+\eta)}{2}$$

$$N_o = -\tfrac{1}{4}\xi\eta^2 - \tfrac{1}{4} + \tfrac{1}{4}\xi^2 - \tfrac{1}{4}\xi\eta + \tfrac{1}{4}\eta^2 + \tfrac{1}{4}\xi^2\eta = \frac{(1-\xi)(1+\eta)(1+\xi-\eta)}{4}$$

$$N_p = -\tfrac{1}{2}\xi + \tfrac{1}{2}\xi\eta^2 + \tfrac{1}{2} - \tfrac{1}{2}\eta^2 = \frac{(1-\xi)(1-\eta^2)}{2} \qquad \text{(E.10.2.6)}$$

Isoparametric Element. The straight-sided triangular and quadrilateral elements are not quite suitable or accurate for approximating shapes with curved surfaces. This inaccuracy is reduced by using elements with arbitrarily oriented curved sides. One such category of these elements with curved surfaces that is increasingly being used in many commercial codes is called the isoparametric element. Basically, in such elements a procedure involving transformation of coordinates or mapping of straight regions into a curved region is used. The procedure is similar to that used in the case of quadrilateral elements in natural coordinate systems with the exception of getting an arbitrarily oriented curved surface rather than an arbitrarily oriented slope. Another important characteristic of the isoparametric element is that the same shape functions of the same order are used for both the approximation function ϕ and geometric variables (x, y). The main idea is to transform or map a simple geometric shape with straight edges in some natural coordinate system into distorted shapes with curved edges in the global coordinate system, as depicted for the case of a quadric (eight-nodes) quadrilateral element.

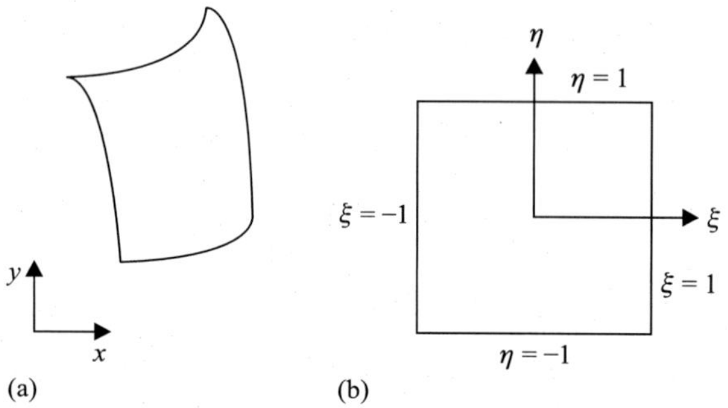

Figure 10.22 Two-dimensional isoparametric element

The quadratic variation of the approximation function is given in coordinate system as

$$\phi(\xi, \eta) = \sum_{I=i}^{N_p} N_I(\xi, \eta)\phi_I \tag{10.102}$$

The eight nodes in the natural coordinate system can be mapped into corresponding nodes in the global coordinate plane by the transformation relation

$$x = \sum_{I=i}^{N_p} N_I(\xi, \eta)x_I \tag{10.103a}$$

and

$$y = \sum_{I=i}^{N_p} N_I(\xi, \eta)y_I \tag{10.103b}$$

Equation 10.103 results in a curve-sided quadrilateral element.

A similar procedure can be used for one-dimensional and three-dimensional elements and shapes.

10.3 THREE-DIMENSIONAL ELEMENT

The most widely-used three-dimensional elements are the ***tetrahedron*** and ***hexahedron*** or ***brick*** element as shown in the Figure 10.23.

10.3.1 Three-dimensional Tetrahedron Element

The simplest tetrahedron element is the linear flat-faced tetrahedron with four nodes, one at each vertex as shown in Figure 10.23a. The approximation

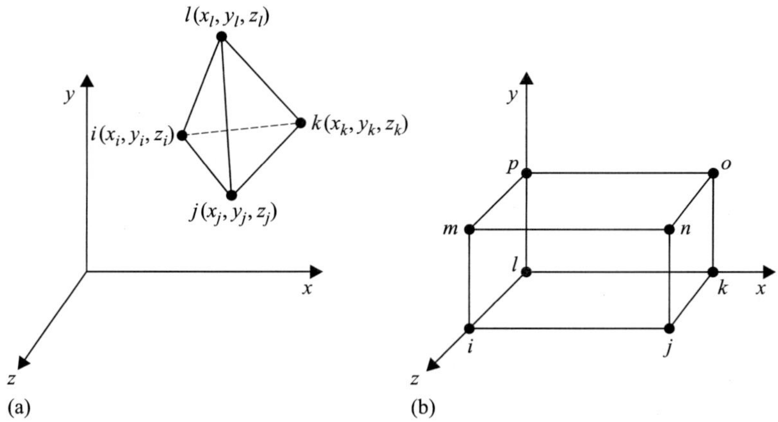

Figure 10.23 Three-dimensional elements. (a) tetrahedron element and (b) hexahedron element

function is assumed as

$$\phi(x, y, z) = a_0 + a_1 x + a_2 y + a_3 z \tag{10.104}$$

Substituting the nodal values ϕ_i, ϕ_j, ϕ_k, ϕ_l, and their nodal values in Equation 10.104, we get

$$\phi_i = a_0 + a_1 x_i + a_2 y_i + a_3 z_i \tag{10.105a}$$

$$\phi_j = a_0 + a_1 x_j + a_2 y_j + a_3 z_j \tag{10.105b}$$

$$\phi_k = a_0 + a_1 x_k + a_2 y_k + a_3 z_k \tag{10.105c}$$

$$\phi_l = a_0 + a_1 x_l + a_2 y_l + a_3 z_l \tag{10.105d}$$

Writing the equations in matrix notation we have

$$\begin{bmatrix} 1 & x_i & y_i & z_i \\ 1 & x_j & y_j & z_j \\ 1 & x_k & y_k & z_k \\ 1 & x_l & y_l & z_l \end{bmatrix} \begin{Bmatrix} a_0 \\ a_1 \\ a_2 \\ a_3 \end{Bmatrix} = \begin{Bmatrix} \phi_i \\ \phi_j \\ \phi_k \\ \phi_l \end{Bmatrix} \tag{10.106}$$

Solving the system of equations for the coefficients a as

$$a_0 = \frac{\begin{vmatrix} \phi_i & x_i & y_i & z_i \\ \phi_j & x_j & y_j & z_j \\ \phi_k & x_k & y_k & z_k \\ \phi_l & x_l & y_l & z_l \end{vmatrix}}{\begin{vmatrix} 1 & x_i & y_i & z_i \\ 1 & x_j & y_j & z_j \\ 1 & x_k & y_k & z_k \\ 1 & x_l & y_l & z_l \end{vmatrix}} = \frac{1}{6V}\left(\alpha_i\phi_i + \alpha_j\phi_j + \alpha_k\phi_k + \alpha_l\phi_l\right) \tag{10.107a}$$

where

$$\alpha_i = \begin{vmatrix} x_j & y_j & z_j \\ x_k & y_k & z_k \\ x_l & y_l & z_l \end{vmatrix}, \quad \alpha_j = -\begin{vmatrix} x_i & y_i & z_i \\ x_k & y_k & z_k \\ x_l & y_l & z_l \end{vmatrix},$$
$$\alpha_k = \begin{vmatrix} x_i & y_i & z_i \\ x_j & y_j & z_j \\ x_l & y_l & z_l \end{vmatrix}, \quad \alpha_l = -\begin{vmatrix} x_i & y_i & z_i \\ x_j & y_j & z_j \\ x_k & y_k & z_k \end{vmatrix} \tag{10.107b}$$

$$a_1 \frac{\begin{vmatrix} 1 & \phi_i & y_i & z_i \\ 1 & \phi_j & y_j & z_j \\ 1 & \phi_k & y_k & z_k \\ 1 & \phi_l & y_l & z_l \end{vmatrix}}{\begin{vmatrix} 1 & x_i & y_i & z_i \\ 1 & x_j & y_j & z_j \\ 1 & x_k & y_k & z_k \\ 1 & x_l & y_l & z_l \end{vmatrix}} = \frac{1}{6V}\left(\beta_i\phi_i + \beta_j\phi_j + \beta_k\phi_k + \beta_l\phi_l\right) \tag{10.108a}$$

where

$$\beta_i = \begin{vmatrix} 1 & y_j & z_j \\ 1 & y_k & z_k \\ 1 & y_l & z_l \end{vmatrix}, \quad \beta_j = -\begin{vmatrix} 1 & y_i & z_i \\ 1 & y_k & z_k \\ 1 & y_l & z_l \end{vmatrix},$$
$$\beta_k = \begin{vmatrix} 1 & y_i & z_i \\ 1 & y_j & z_j \\ 1 & y_l & z_l \end{vmatrix}, \quad \beta_l = -\begin{vmatrix} 1 & y_i & z_i \\ 1 & y_j & z_j \\ 1 & y_k & z_k \end{vmatrix} \tag{10.108b}$$

$$a_2 = \frac{\begin{vmatrix} 1 & x_i & \phi_i & z_i \\ 1 & x_j & \phi_j & z_j \\ 1 & x_k & \phi_k & z_k \\ 1 & x_l & \phi_l & z_l \end{vmatrix}}{\begin{vmatrix} 1 & x_i & y_i & z_i \\ 1 & x_j & y_j & z_j \\ 1 & x_k & y_k & z_k \\ 1 & x_l & y_l & z_l \end{vmatrix}} = \frac{1}{6V}\left(\gamma_i\phi_i + \gamma_j\phi_j + \gamma_k\phi_k + \gamma_l\phi_l\right) \tag{10.109a}$$

where

$$\gamma_i = \begin{vmatrix} 1 & x_j & z_j \\ 1 & x_k & z_k \\ 1 & x_i & z_i \end{vmatrix}, \quad \gamma_j = -\begin{vmatrix} 1 & x_i & z_i \\ 1 & x_k & z_k \\ 1 & x_i & z_i \end{vmatrix},$$

$$\gamma_k = \begin{vmatrix} 1 & x_i & z_i \\ 1 & x_j & z_j \\ 1 & x_i & z_i \end{vmatrix}, \quad \gamma_l = -\begin{vmatrix} 1 & x_i & z_i \\ 1 & x_j & z_j \\ 1 & x_k & z_k \end{vmatrix} \tag{10.109b}$$

$$a_3 = \frac{\begin{vmatrix} 1 & x_i & y_i & \phi_i \\ 1 & x_j & y_j & \phi_j \\ 1 & x_k & y_k & \phi_k \\ 1 & x_l & y_l & \phi_l \end{vmatrix}}{\begin{vmatrix} 1 & x_i & y_i & z_i \\ 1 & x_j & y_j & z_j \\ 1 & x_k & y_k & z_k \\ 1 & x_l & y_l & z_l \end{vmatrix}} = \frac{1}{6V}\left(\delta_i\phi_i + \delta_j\phi_j + \delta_k\phi_k + \delta_l\phi_l\right) \tag{10.110a}$$

where

$$\delta_i = \begin{vmatrix} 1 & x_j & y_j \\ 1 & x_k & y_k \\ 1 & x_l & y_l \end{vmatrix}, \quad \delta_j = -\begin{vmatrix} 1 & x_i & y_i \\ 1 & x_k & y_k \\ 1 & x_l & y_l \end{vmatrix},$$

$$\delta_k = \begin{vmatrix} 1 & x_i & y_i \\ 1 & x_j & y_j \\ 1 & x_l & y_l \end{vmatrix}, \quad \delta_l = -\begin{vmatrix} 1 & x_i & y_i \\ 1 & x_j & y_j \\ 1 & x_k & y_k \end{vmatrix} \tag{10.110b}$$

and V is the volume of the tetrahedron given by

$$V = \frac{1}{6}\begin{vmatrix} 1 & x_i & y_i & z_i \\ 1 & x_j & y_j & z_j \\ 1 & x_k & y_k & z_k \\ 1 & x_l & y_l & z_l \end{vmatrix} \tag{10.111}$$

Substituting Equation 10.107 to 10.110 into Equation 10.104, we obtain the approximation function as

$$\phi(x, y, z) = N_i\phi_i + N_j\phi_j + N_k\phi_k + N_l\phi_l = [N]^{\mathrm{T}}\{\phi\} \tag{10.112}$$

where

$$[N] = \begin{bmatrix} N_i(x, y, z) \\ N_j(x, y, z) \\ N_k(x, y, z) \\ N_l(x, y, z) \end{bmatrix} \tag{10.113}$$

$$N_i(x, y, z) = \frac{1}{6V}(\alpha_i + \beta_i x + \gamma_i y + \delta_i z) \tag{10.114a}$$

$$N_j(x, y, z) = \frac{1}{6V}(\alpha_j + \beta_j x + \gamma_j y + \delta_j z) \tag{10.114b}$$

$$N_k(x, y, z) = \frac{1}{6V}(\alpha_k + \beta_k x + \gamma_k y + \delta_k z) \tag{10.114c}$$

$$N_l(x, y, z) = \frac{1}{6V}(\alpha_l + \beta_l x + \gamma_l y + \delta_l z) \tag{10.114d}$$

Volume Coordinate System. One of the natural coordinate systems for the three-dimensional tetrahedron element is the volume coordinate system, which is developed in the same way as in one-dimensional line coordinate systems and two-dimensional area coordinate systems. In this case, the volume coordinates are the distance ratios L_i, L_j, L_k, and L_l, which are defined by considering a point p normal to each side, as shown in Figure 10.24, and taking the ratio of volumes as

$$L_i = \frac{V_i}{V}, \quad L_j = \frac{V_j}{V}, \quad L_k = \frac{V_k}{V}, \quad \text{and} \quad L_l = \frac{V_l}{V} \tag{10.115}$$

where the volume $V_I(I = i, j, k, l)$ is the volume of the tetrahedron formed by the point p and the vertices other than the vertex I.

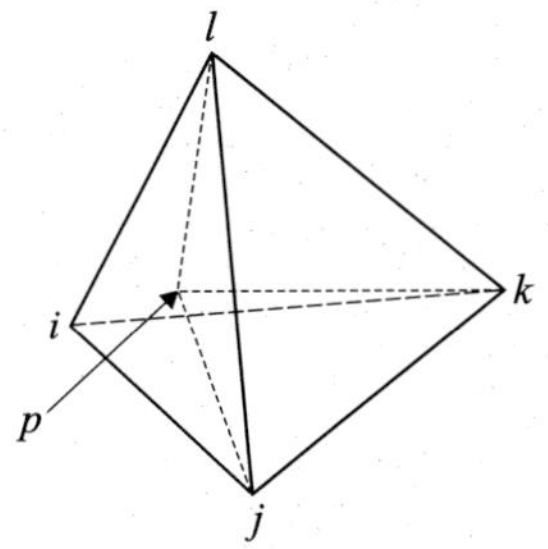

Figure 10.24 Volume coordinate system for the tetrahedron element

Based on this coordinate system, the shape functions are defined as

$$N_i = L_i, \quad N_j = L_j, \quad N_k = L_k \tag{10.116}$$

Additionally, the volume coordinates are related by

$$L_i + L_j + L_k + L_l = 1 \tag{10.117}$$

and the relationship between the volume or natural coordinates and Cartesian coordinates is given by

$$x = L_i x_i + L_j x_j + L_k x_k + L_l x_l \tag{10.118a}$$

$$y = L_i y_i + L_j y_j + L_k y_k + L_l y_l \tag{10.118b}$$

$$z = L_i z_i + L_j z_j + L_k z_k + L_l z_l \tag{10.118c}$$

The integral formula that can be used along with the three-dimensional integrals in a natural coordinate system is

$$I = \int_V L_i^a L_j^b L_k^c L_l^d \, dV = \frac{a!b!c!d!}{(a+b+c+d+3)!} 6V \tag{10.119}$$

For example the integral expression of the form

$$I = \int_V N_i N_j N_k \, dV \tag{10.120}$$

can be evaluated using Equations 10.116 and 10.119 as

$$\begin{aligned} I &= \int_V N_i N_j N_k \, dV = \int_V L_i^1 L_j^1 L_k^1 L_l^0 \, dV \\ &= \frac{1!1!1!0!}{(1+1+1+0+3)!} 6V = \frac{V}{120} \end{aligned} \tag{10.121}$$

For a more detailed discussion of higher-order tetrahedral elements, refer to the books by Zienkiewicz and Taylor (1986) and Reddy and Gartling (1994).

10.3.2 Three-dimensional Hexahedron Element

A simple and more accurate three-dimensional element, which is easier to manipulate is the ***hexahedron*** or ***brick element***. The simplest of the hexahedron elements is the trilinear eight-noded hexahedron element shown in Figure 10.23b. The approximation function is assumed to be

$$\phi(x, y, z) = a_0 + a_1x + a_2y + a_3z + a_4xy + a_5xz + a_6yz + a_7xyz \tag{10.122}$$

The a coefficients are determined by substituting the nodal values and nodal coordinates in Equation 10.122 and by solving the system of equation

$$\begin{bmatrix} 1 & x_i & y_i & z_i & x_iy_i & x_iz_i & y_iz_i & x_iy_iz_i \\ 1 & x_j & y_j & z_j & x_iy_i & x_iz_i & y_iz_i & x_iy_iz_i \\ 1 & x_k & y_k & z_k & x_ky_k & x_kz_k & y_kz_k & x_ky_kz_k \\ 1 & x_l & y_l & z_l & x_ly_l & x_lz_l & y_lz_l & x_ly_lz_l \\ 1 & x_m & y_m & z_m & x_my_m & x_mz_m & y_mz_m & x_my_mz_m \\ 1 & x_n & y_n & z_n & x_ny_n & x_nz_n & y_nz_n & x_ny_nz_n \\ 1 & x_o & y_o & z_o & x_oy_o & x_oz_o & y_oz_o & x_oy_oz_o \\ 1 & x_p & y_p & z_p & x_py_p & x_pz_p & y_pz_p & x_py_pz_p \end{bmatrix} \begin{Bmatrix} a_0 \\ a_1 \\ a_2 \\ a_3 \\ a_4 \\ a_5 \\ a_6 \\ a_7 \end{Bmatrix} = \begin{Bmatrix} \phi_i \\ \phi_j \\ \phi_k \\ \phi_l \\ \phi_m \\ \phi_n \\ \phi_o \\ \phi_p \end{Bmatrix} \tag{10.123a}$$

or

$$X\{a\} = \{\phi\} \tag{10.123b}$$

Since the matrix dimension is large, the use of Cramer's rule leads to very tedious algebraic manipulation. So, the a coefficients are determined more conveniently by using the inverse of the coefficient matrix as

$$\{a\} = X^{-1}\{\phi\} \tag{10.124}$$

Substituting Equation 10.124 into Equation 10.122, we obtain the approximation function as

$$\phi = [1 \quad x \quad y \quad z \quad xy \quad xz \quad yz \quad xyz]X^{-1}\{\phi\} \tag{10.125}$$

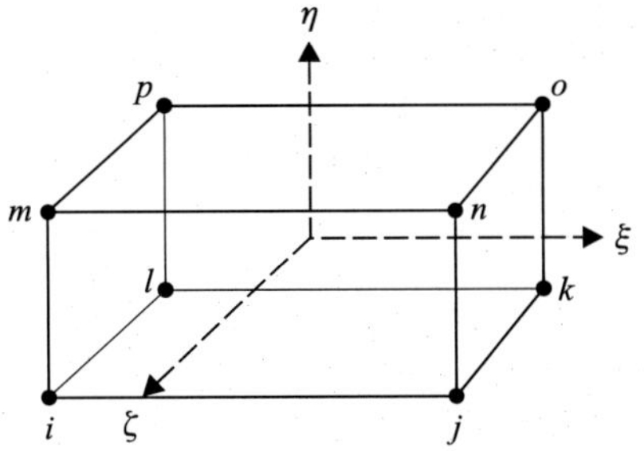

Figure 10.25 Hexahedron element in a natural coordinate

The shape functions for the hexahedron element can be deduced from Equation 10.125 as

$$[N] = [1 \quad x \quad y \quad z \quad xy \quad xz \quad yz \quad xyz]X^{-1} \tag{10.126}$$

Natural Coordinate System for Hexahedron Element. The natural coordinate system for the three-dimensional hexahedron element is defined in terms of the dimensionless form of the local coordinate system with the origin located at the center of the hexahedron (Figure 10.25) as

$$\xi = \frac{x}{l_e}, \quad \eta = \frac{y}{h_e}, \quad \text{and} \quad \zeta = \frac{z}{w_e} \tag{10.127}$$

With this definition of the natural coordinate system, the shape functions are defined as

$$N_i = \tfrac{1}{8}(1-\xi)(1-\eta)(1-\zeta), \quad N_j = \tfrac{1}{8}(1+\xi)(1-\eta)(1-\zeta) \tag{10.128a}$$

$$N_k = \tfrac{1}{8}(1+\xi)(1+\eta)(1-\zeta), \quad N_l = \tfrac{1}{8}(1-\xi)(1+\eta)(1-\zeta) \tag{10.128b}$$

$$N_m = \tfrac{1}{8}(1-\xi)(1-\eta)(1+\zeta), \quad N_n = \tfrac{1}{8}(1+\xi)(1-\eta)(1+\zeta) \tag{10.128c}$$

$$N_o = \tfrac{1}{8}(1+\xi)(1+\eta)(1+\zeta), \quad N_p = \tfrac{1}{8}(1-\xi)(1+\eta)(1+\zeta) \tag{10.128d}$$

The coordinate transformation relationship between the Cartesian and the natural coordinate systems

$$x = \sum x_I N_I(\xi, \eta, \zeta), \quad y = \sum y_I N_I(\xi, \eta, \zeta), \quad \text{and} \quad z = \sum z_I N_I(\xi, \eta, \zeta) \tag{10.129}$$

Additionally, by using the procedure outlined in this section, we can express the derivative of a shape function in the natural coordinate system by following the procedure outlined for two-dimensional elements as

$$\begin{Bmatrix} \dfrac{\partial N_I}{\partial x} \\ \dfrac{\partial N_I}{\partial y} \\ \dfrac{\partial N_I}{\partial z} \end{Bmatrix} = J^{-1} \begin{Bmatrix} \dfrac{\partial N_I}{\partial \xi} \\ \dfrac{\partial N_I}{\partial \eta} \\ \dfrac{\partial N_I}{\partial \zeta} \end{Bmatrix} \tag{10.130}$$

where the ***Jacobian*** matrix J is given as

$$J = \begin{bmatrix} \dfrac{\partial x}{\partial \xi} & \dfrac{\partial y}{\partial \xi} & \dfrac{\partial z}{\partial \xi} \\ \dfrac{\partial x}{\partial \eta} & \dfrac{\partial y}{\partial \eta} & \dfrac{\partial z}{\partial \eta} \\ \dfrac{\partial x}{\partial \zeta} & \dfrac{\partial y}{\partial \psi} & \dfrac{\partial z}{\partial \zeta} \end{bmatrix} \tag{10.131}$$

The differential volume can be transformed into a natural coordinate system using

$$dx\,dy\,dz = |J|d\xi\,d\eta\,d\zeta \tag{10.132}$$

For more discussions on higher-order ***hexahedron*** or ***brick elements***, refer to books by Zienkiewicz and Taylor (1986) and Reddy and Gartling (1994).

PROBLEMS

10.1 Evaluate the following integrals for a linear line element:

(a) $\displaystyle\int_{x_i}^{x_j} N_j\,dx$, (b) $\displaystyle\int_{x_i}^{x_j} \frac{dN_i}{dx}\frac{dN_j}{dx}\,dx$, and (c) $\displaystyle\int_{x_i}^{x_j} N_i^2\,dx$

10.2 Evaluate the following integrals using the quadratic line element with shape functions given by Equation 10.42

(a) $\displaystyle\int_{x_i}^{x_j} N_iN_j\,dx$ and (b) $\displaystyle\int_{x_i}^{x_j} \frac{dN_i}{dx}\frac{dN_j}{dx}\,dx$

10.3 Consider a local coordinate system for a triangular element as shown. The origin is at the node i and the x-axis passes though the node j. Use the coordinates of the nodes as (x_i, y_i), (x_j, y_j), and (x_k, y_k).

Consider the first-order polynomial of the form $\phi = a_0 + a_1 x + a_2 y$, and determine the shape functions for the linear triangular element.

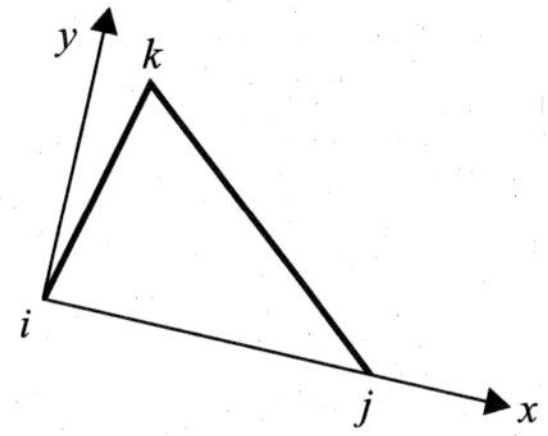

10.4 Consider a linear triangular element with global coordinates of the three nodes given as $(0.125, 0)$, $(0.25, 0)$, and $(0.25, 0.15)$. Determine the area of the triangle and the expressions for the shape functions. If the nodal values for this element are estimated as $\phi_i = 0.002$, $\phi_j = 0.0025$, and $\phi_j = 0.0031$, then determine the value of ϕ at the point (0.22, 0.125).

10.5 Evaluate the following integrals using the natural coordinate system for the linear triangular element: (a) $\int_A N_k^2\, dA$ and (b) $\int_A N_j N_k\, dA$.

10.6 Use the local coordinate system in a bilinear rectangular element with shape functions given by Equation 10.88 to evaluate the integral

$$I = \int_A N_i N_j\, dA$$

10.7 Consider a quadrilateral element shown in the figure with nodal coordinates given as (1, 1), (4, −1), (3.7, 2.8), and (0.8, 2.5).

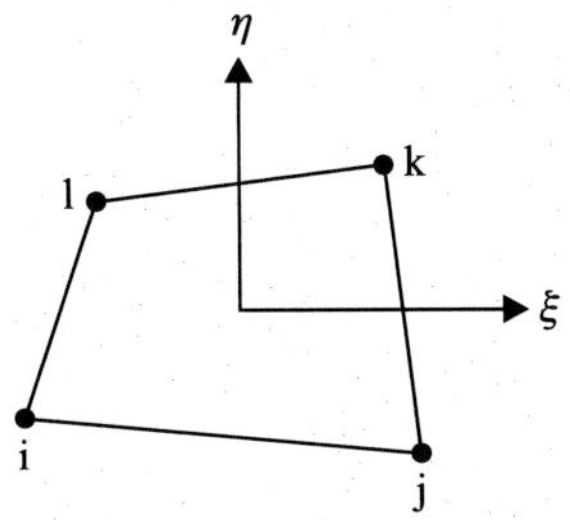

Use the natural coordinate system with shape function given by Equation 10.90 to determine the Jacobian matrix J, the inverse of the Jacobian matrix, the transformation coordinates given by Equation 10.91, and $\partial N_i/\partial x$ and $\partial N_i/\partial y$.

10.8 Use the natural coordinate system relations derived in Problem 10.7 to evaluate the following integral using Gauss–Legendre quadrature formula

$$I = \int_A \frac{\partial N_i}{\partial x}\frac{\partial N_j}{\partial x}\, dA = \int_{-1}^{1}\int_{-1}^{1} \frac{\partial N_i}{\partial x}\frac{\partial N_j}{\partial x}\, |J|\, d\xi\, d\eta$$

10.9 Evaluate the following integral numerically using the quadratic line element given by Equation 10.45 and three-point Gauss–Legendre quadrature

$$I = \int_{-1}^{1} [N]^{\mathrm{T}}[N]\, d\xi$$

CHAPTER

11

ONE-DIMENSIONAL STEADY STATE PROBLEMS

In this chapter, we will consider formulation and application of the finite element method using a general one-dimensional steady state equation containing diffusion, source and surface convection terms, and considering different types of boundary conditions. This will be followed by the presentation of a number of classical example problems. As we have mentioned in Chapter 10, there are usually two popular approaches for the development and formulation of an integral form of the governing equation and, subsequently, the formation of the element equation in the finite element method. These are (1) the ***Galerkin-based weighted residual approach*** and (2) the ***variational approach***. So, we will describe and discuss both of these approaches for forming the integral or weak form while describing the finite element formulation.

11.1 FINITE ELEMENT FORMULATION USING GALERKIN METHOD

Let us consider a general one-dimensional steady state transport process in a one-dimensional media shown in Figure 11.1.

The general mathematical statement of this transport process is given as

$$\frac{d}{dx}\left(\Gamma_x \frac{d\phi}{dx}\right) + S = 0 \tag{11.1}$$

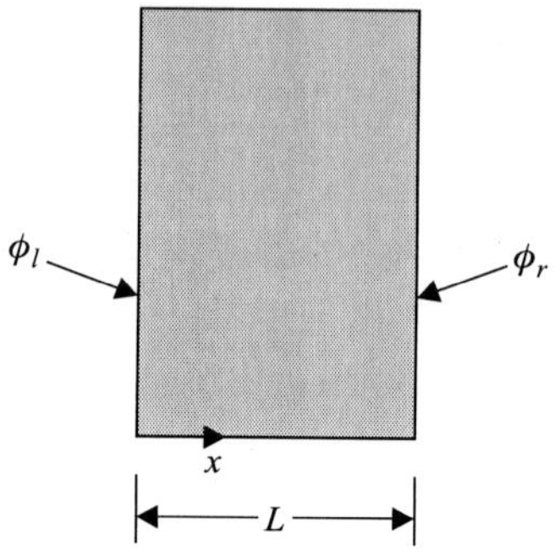

Figure 11.1 Steady state transport process in a plane slab

Boundary conditions

1. $x = 0, \quad \phi(0) = \phi_l$ (11.2a)

2. $x = L, \quad \phi(L) = \phi_r$ (11.2b)

This equation contains a diffusion term and a source term, and the boundary conditions are of the first kind, i.e., of constant surface value. Let us demonstrate the application of the finite element method to this problem through detailed descriptions of several steps that constitute the finite element method.

Step 1: Discretization of the Solution Domain. The solution domain is discretized into a mesh size distribution using one-dimensional line elements as shown in Figure 11.2.

Let us consider the elements as linear two-point line elements with a typical element shown in Figure 11.3. The approximate solution function $\phi \approx \phi^*$ over this element is assumed as

$$\phi^* = N_i\phi_i + N_j\phi_j = [N]\{\phi^{(e)}\} \tag{11.3a}$$

where N_i and N_j are the interpolation or shape functions given as

$$N_i = \frac{x_j - x}{x_j - x_i}, \qquad N_j = \frac{x - x_i}{x_j - x_i} \tag{11.3b}$$

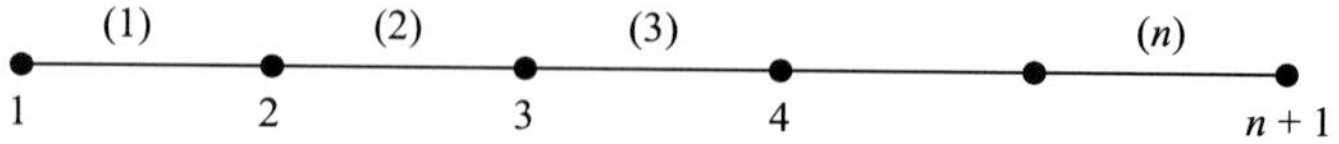

Figure 11.2 Mesh size distributions

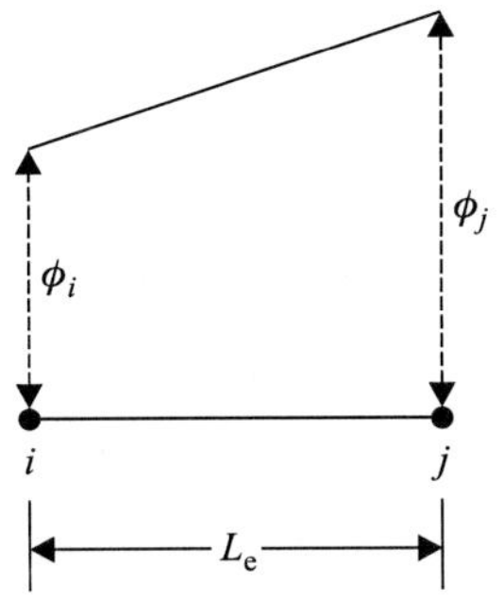

Figure 11.3 A typical two-point line element

Step 2: Formation of Integral Statement of the Problem. Let us demonstrate the formation of the integral statement of the problem using the ***Galerkin-based weighted residual method***. As we substitute the approximate solution given by Equation 11.3 into the governing differential equation, Equation 11.1, a residual term will be obtained as

$$R^{(e)} = \frac{d}{dx}\left(\Gamma_x \frac{d\phi^*}{dx}\right) + S \tag{11.4}$$

The method of weighted residual involves finding a minimum for the residual over the solution domain as

$$\int_{x_i}^{x_j} W_I R^{(e)}\, dx \tag{11.5}$$

where

W_I = weighting factors
I = index for number of weighting factors or number of nodal points in an element such as $i, j, \ldots, N_p$
N_p = number of weighting factors or number of nodal points in the element

We have mentioned before that in the ***Galerkin-based weighted residual method***, the ***interpolation*** or ***shape functions*** are used as the weighting factors. So, in the case of a two-point line element, the weighting factors are N_i and N_j. Using these weighting factors, Equation 11.5 leads to

for $I = i$

$$R_i^{(e)} = \int_{x_i}^{x_j} N_i(x)\left(\frac{d}{dx}\left(\Gamma_x \frac{d\phi^*}{dx}\right) + S\right) dx = 0 \tag{11.6a}$$

and

for $I = j$

$$R_j^{(e)} = \int_{x_i}^{x_j} N_j(x)\left(\frac{d}{dx}\left(\Gamma_x \frac{d\phi^*}{dx}\right) + S\right) dx = 0 \tag{11.6b}$$

Equations 11.6a and 11.6b can also be written in a compact form as

$$\{R^{(e)}\} = \int_{x_i}^{x_j} [N]^{\mathrm{T}}\left(\frac{d}{dx}\left(\Gamma_x \frac{d\phi^*}{dx}\right) + S\right) dx = 0 \tag{11.6c}$$

Let us now evaluate the first terms of Equations 11.6a and 11.6b using integration-by-parts as

$$\int_{x_i}^{x_j} N_i(x)\left(\frac{d}{dx}\left(\Gamma_x \frac{d\phi^*}{dx}\right)\right) dx = \left(N_i(x)\Gamma_x \frac{d\phi^*}{dx}\right)\Bigg|_{x_i}^{x_j} - \int_{x_i}^{x_j} \Gamma_x \frac{dN_i}{dx}\frac{d\phi^*}{dx} dx$$

$$= N_i(x_j)\left(\Gamma_x \frac{d\phi^*}{dx}\right)\Bigg|_{x_j} - N_i(x_i)\left(\Gamma_x \frac{d\phi^*}{dx}\right)\Bigg|_{x_i} - \int_{x_i}^{x_j} \Gamma_x \frac{dN_i}{dx}\frac{d\phi^*}{dx} \tag{11.7a}$$

and

$$\int_{x_i}^{x_j} N_j(x)\left(\frac{d}{dx}\left(\Gamma_x \frac{d\phi^*}{dx}\right)\right) dx = \left(N_j(x)\Gamma_x \frac{d\phi^*}{dx}\right)\Bigg|_{x_i}^{x_j} - \int_{x_i}^{x_j} \Gamma_x \frac{dN_j}{dx}\frac{d\phi^*}{dx} dx$$

$$= N_j(x_j)\left(\Gamma_x \frac{d\phi^*}{dx}\right)\Bigg|_{x_j} - N_j(x_i)\left(\Gamma_x \frac{d\phi^*}{dx}\right)\Bigg|_{x_i} - \int_{x_i}^{x_j} \Gamma_x \frac{dN_j}{dx}\frac{d\phi^*}{dx} \tag{11.7b}$$

Note that while evaluating integrals by integration-by-parts, we should select the two terms of the integration in such a way that the resulting integral will be easier to evaluate than the original one.

If we now use the properties of the interpolation or shape functions, i.e., for a two-points line element, $N_i(x_i) = 1$, $N_i(x_j) = 0$, and $N_j(x_i) = 0$, $N_j(x_j) = 1$, then Equations 11.7a and 11.7b transform into

$$\int_{x_i}^{x_j} N_i(x)\left(\frac{d}{dx}\left(\Gamma_x \frac{d\phi^*}{dx}\right)\right) dx == -\left(\Gamma_x \frac{d\phi^*}{dx}\right)\Bigg|_{x_i} - \int_{x_i}^{x_j} \Gamma_x \frac{dN_i}{dx}\frac{d\phi^*}{dx} dx \tag{11.8a}$$

and

$$\int_{x_i}^{x_j} N_j(x)\left(\frac{d}{dx}\left(\Gamma_x\frac{d\phi^*}{dx}\right)\right)dx == \left(\Gamma_x\frac{d\phi^*}{dx}\right)\bigg|_{x_j} - \int_{x_i}^{x_j} \Gamma_x\frac{dN_j}{dx}\frac{d\phi^*}{dx}\,dx \tag{11.8b}$$

Finally, we substitute Equations 11.8a and 11.8b into Equations 11.6a and 11.6b, respectively, to obtain the integral statement of the governing equation as

$$R_i^{(e)} = -\left(\Gamma_x\frac{d\phi^*}{dx}\right)\bigg|_{x_i} - \int_{x_i}^{x_j} \Gamma_x\frac{dN_i}{dx}\frac{d\phi^*}{dx}\,dx + \int_{x_i}^{x_j} SN_i(x)\,dx = 0 \tag{11.9a}$$

and

$$R_j^{(e)} = \left(\Gamma_x\frac{d\phi^*}{dx}\right)\bigg|_{x_j} - \int_{x_i}^{x_j} \Gamma_x\frac{dN_j}{dx}\frac{d\phi^*}{dx}\,dx + \int_{x_i}^{x_j} SN_j(x)\,dx = 0 \tag{11.9b}$$

or in vector form as

$$\left\{R^{(e)}\right\} = \left\{\begin{matrix} -\Gamma_x\left.\dfrac{d\phi}{dx}\right|_{x_i} \\ \Gamma_x\left.\dfrac{d\phi}{dx}\right|_{x_j} \end{matrix}\right\} - \int_{x_i}^{x_j} \Gamma_x\frac{d[N]^{\mathrm{T}}}{dx}\frac{d\phi^*}{dx}\,dx + \int_{x_i}^{x_j} S[N]^{\mathrm{T}}\,dx = 0 \tag{11.9c}$$

Equation (11.9) represents the ***integral statement*** of the governing differential Equation 11.1. This is also known as the ***weak formulation*** as mathematical manipulation is used to lower the highest-order derivative term from a second-order to a first-order derivative.

Step 3: Selection of Approximation Function. In this step we select an approximate solution for each element (see Equations 9.1 and 9.8) and the associated shape functions.

Step 4: Formation of Element Characteristics Equation. In this step the element characteristics equation is obtained by evaluating the integrals by substituting the approximate solution for the chosen element type. Substituting $\phi^* = [N]\left\{\phi^{(e)}\right\}$ in Equation (11.9c), we get

$$\left\{R^{(e)}\right\} = \left\{\begin{matrix} -\Gamma_x\left.\dfrac{d\phi}{dx}\right|_{x_i} \\ \Gamma_x\left.\dfrac{d\phi}{dx}\right|_{x_j} \end{matrix}\right\} - \left(\int_{x_i}^{x_j} \Gamma_x\frac{d[N]^{\mathrm{T}}}{dx}\frac{d[N]}{dx}\,dx\right)\left\{\phi^{(e)}\right\} + \int_{x_i}^{x_j} S[N]^{\mathrm{T}}\,dx = 0 \tag{11.10a}$$

or

$$\left[\kappa_\Gamma^{(e)}\right]\{\phi^{(e)}\} = \{I^{(e)}\} + \{f^{(e)}\} \tag{11.10b}$$

where

$$\{\phi^{(e)}\} = \begin{Bmatrix} \phi_i \\ \phi_j \end{Bmatrix} = \text{vector of element nodal values} \tag{11.11a}$$

$$\left[\kappa_\Gamma^{(e)}\right] = \left(\int_{x_i}^{x_j} \Gamma_x \frac{d[N]^{\mathrm{T}}}{dx}\frac{d[N]}{dx}\,dx\right) = \text{element characteristic or stiffness matrix} \tag{11.11b}$$

$$\{f^{(e)}\} = \int_{x_i}^{x_j} S[N]^{\mathrm{T}}\,dx = \text{element source or force vector} \tag{11.11c}$$

$$\{I^{(e)}\} = \begin{Bmatrix} -\Gamma_x \left.\dfrac{d\phi}{dx}\right|_{x_i} \\ \Gamma_x \left.\dfrac{d\phi}{dx}\right|_{x_j} \end{Bmatrix} = \text{vector for element contribution to inter-element conditions} \tag{11.11d}$$

Case: Linear Line Element. From the approximate solution, Equation 11.3, selected for the linear line element we can evaluate the derivatives

$$\frac{dN_i}{dx} = -\frac{1}{x_j - x_i} = -\frac{1}{L_e}, \qquad \frac{dN_j}{dx} = \frac{1}{x_j - x_i} = \frac{1}{L_e} \tag{11.12a}$$

and

$$\frac{d\phi^*}{dx} = \frac{1}{x_j - x_i}(-\phi_i) + \frac{1}{x_j - x_i}(\phi_j) = \frac{1}{L_e}(-\phi_i + \phi_j) \tag{11.12b}$$

Using Equation (11.12), we can evaluate the first integral term in Equation 11.9 as

for $I = i$

$$\int_{x_i}^{x_j} \Gamma_x \frac{dN_i}{dx}\frac{d\phi^*}{dx}\,dx = \int_{x_i}^{x_j} \Gamma_x\left(-\frac{1}{L_e}\right)\frac{\phi_j - \phi_i}{L_e}\,dx = \frac{\Gamma_x}{L_e}(\phi_i - \phi_j) \tag{11.13a}$$

for $I = j$

$$\int_{x_i}^{x_j} \Gamma_x \frac{dN_j}{dx}\frac{d\phi^*}{dx}\,dx = \int_{x_i}^{x_j} \Gamma_x\left(\frac{1}{L_e}\right)\frac{-\phi_i + \phi_j}{L_e}\,dx = \frac{\Gamma_x}{L_e}(-\phi_i + \phi_j) \tag{11.13b}$$

Equations 11.13a and 11.13b can also be written in a matrix form as

$$\int_{x_i}^{x_j} \Gamma_x \frac{d[N]^{\mathrm{T}}}{dx}\frac{d\phi^*}{dx}dx = \left(\int_{x_i}^{x_j} \Gamma_x \frac{d[N]^{\mathrm{T}}}{dx}\frac{d[N]}{dx}dx\right)\{\phi^{(e)}\} = \frac{\Gamma_x}{L_e}\begin{bmatrix} 1 & -1 \\ -1 & 1 \end{bmatrix}\begin{Bmatrix} \phi_i \\ \phi_j \end{Bmatrix} \tag{11.13c}$$

While evaluating the integrals, we assume a step-wise constant value for the transport property, Γ. For the second ***integral term***, let us assume a step-wise constant value $S = \bar{S}$ for the source in the media for this demonstration, and evaluate the integrals as

$$\int_{x_i}^{x_j} SN_i(x)\,dx = \int_{x_i}^{x_j} \bar{S}\frac{x_j - x}{x_j - x_i}dx = \frac{\bar{S}}{x_j - x_i}\left(x_j x\Big|_{x_i}^{x_j} - \frac{x^2}{2}\Big|_{x_i}^{x_j}\right)$$

$$= \frac{\bar{S}}{x_j - x_i}\left(x_j(x_j - x_i) - \frac{x_j^2 - x_i^2}{2}\right)$$

or

$$\int_{x_i}^{x_j} SN_i(x)\,dx = \frac{\bar{S}L_e}{2} \tag{11.14a}$$

and similarly

$$\int_{x_i}^{x_j} SN_j(x)dx = \int_{x_i}^{x_j} \bar{S}\frac{x - x_i}{x_j - x_i}dx = \frac{\bar{S}L_e}{2} \tag{11.14b}$$

Equations 11.14a and 11.14b can be written in vector form as

$$\int_{x_i}^{x_j} S[N]^{\mathrm{T}}\,dx = \frac{\bar{S}L_e}{2}\begin{Bmatrix} 1 \\ 1 \end{Bmatrix} \tag{11.14c}$$

Substituting Equations 11.13 and 11.14 into Equation 11.9, we obtain the element characteristic equations as

$$R_i^{(e)} = \left(\Gamma_x \frac{d\phi^*}{dx}\right)\Bigg|_{x_i} + \frac{\Gamma_x}{L_e}(\phi_i - \phi_j) - \frac{\bar{S}L_e}{2} = 0 \tag{11.15a}$$

and

$$R_j^{(e)} = -\left(\Gamma_x \frac{d\phi^*}{dx}\right)\Bigg|_{x_j} + \frac{\Gamma_x}{L_e}(-\phi_i + \phi_j) - \frac{\bar{S}L_e}{2} = 0 \tag{11.15b}$$

Equations 11.14a and 11.14b can be written in matrix form as

$$\begin{Bmatrix} R_i^{(e)} \\ R_j^{(e)} \end{Bmatrix} = \begin{Bmatrix} \Gamma_x \dfrac{d\phi}{dx}\bigg|_{x_i} \\ -\Gamma_x \dfrac{d\phi}{dx}\bigg|_{x_j} \end{Bmatrix} + \frac{\Gamma_x}{L_e}\begin{bmatrix} 1 & -1 \\ -1 & 1 \end{bmatrix}\begin{Bmatrix} \phi_i \\ \phi_j \end{Bmatrix} - \frac{\bar{S}L_e}{2}\begin{Bmatrix} 1 \\ 1 \end{Bmatrix} = 0 \qquad (11.16a)$$

or

$$\left[\kappa_\Gamma^{(e)}\right]\{\phi^{(e)}\} = \{I^{(e)}\} + \{f^{(e)}\} \qquad (11.16b)$$

where

$$\{\phi^{(e)}\} = \begin{Bmatrix} \phi_i \\ \phi_j \end{Bmatrix} = \text{vector of element nodal values} \qquad (11.17a)$$

$$\left[\kappa_\Gamma^{(e)}\right] = \frac{\Gamma_x}{L_e}\begin{bmatrix} 1 & -1 \\ -1 & 1 \end{bmatrix} = \begin{bmatrix} k_{11}^{(e)} & k_{12}^{(e)} \\ k_{21}^{(e)} & k_{22}^{(e)} \end{bmatrix}$$

$$= \text{element characteristic or stiffness matrix} \qquad (11.17b)$$

$$\{f^{(e)}\} = \begin{Bmatrix} \dfrac{\bar{S}L_e}{2} \\ \dfrac{\bar{S}L_e}{2} \end{Bmatrix} = \begin{Bmatrix} f_i^{(e)} \\ f_j^{(e)} \end{Bmatrix} = \text{element source or force vector} \qquad (11.17c)$$

$$\{I^{(e)}\} = \begin{Bmatrix} -\Gamma_x \dfrac{d\phi}{dx}\bigg|_{x_i} \\ \Gamma_x \dfrac{d\phi}{dx}\bigg|_{x_j} \end{Bmatrix} = \begin{Bmatrix} I_i^{(e)} \\ I_j^{(e)} \end{Bmatrix}$$

$$= \text{vector for element contribution to inter-element conditions} \qquad (11.17d)$$

Equation 11.16 is a system of equations with a number of equations equal to the number nodal points in the element, and it is referred to as the ***element characteristic equation***. The vector for ***inter-element contributions*** represents the boundary conditions and inter-element continuity conditions. We can now apply Equations 11.16 and 11.17 repeatedly to all elements to obtain the corresponding element equations.

In a similar manner, we can derive the element characteristics equation using quadratic and cubic line elements.

Step 5: Assembly of Element Equations to Form the Global System. In this step, the element equations are assembled to form a global system of equations,

Table 11.1 Node Numbering or Connectivity Scheme

	Node Numbers	
Element Number (e)	Local System	Global System
1	$i=1$	$i=1$
	$j=2$	$j=2$
2	$i=1$	$i=2$
	$j=2$	$j=3$
3	$i=1$	$i=3$
	$j=2$	$j=4$
⋮		
$n-1$	$i=1$	$i=n-1$
	$j=2$	$j=n$
n	$i=1$	$i=n$
	$j=2$	$j=n+1$

which consist of all nodal values in the mesh as unknowns. However, a systematic procedure is required for this assembly process, particularly for efficient computer implementations. For this purpose, a numbering or connectivity scheme is used, as described in Table 11.1.

Such a connectivity scheme may look trivial, particularly for a one-dimensional problem. However, it is an essential feature in the finite element formulations while solving multi-dimensional problems. Let us show this assembly process by considering the element equations one at a time in terms of global coordinates or a global numbering scheme, and then substituting them in the global system following the connectivity rules outlined in Table 11.1.

Element Equations

Element 1

$$\begin{bmatrix} k_{11}^{(1)} & k_{12}^{(1)} \\ k_{21}^{(1)} & k_{22}^{(1)} \end{bmatrix} \begin{Bmatrix} \phi_1^{(1)} \\ \phi_2^{(1)} \end{Bmatrix} = \begin{Bmatrix} I_1^{(1)} \\ I_2^{(1)} \end{Bmatrix} + \begin{Bmatrix} f_1^{(1)} \\ f_2^{(1)} \end{Bmatrix} \tag{11.18a}$$

Element 2

$$\begin{bmatrix} k_{11}^{(2)} & k_{12}^{(2)} \\ k_{21}^{(2)} & k_{22}^{(2)} \end{bmatrix} \begin{Bmatrix} \phi_2^{(2)} \\ \phi_3^{(2)} \end{Bmatrix} = \begin{Bmatrix} I_2^{(2)} \\ I_3^{(2)} \end{Bmatrix} + \begin{Bmatrix} f_2^{(2)} \\ f_3^{(2)} \end{Bmatrix} \tag{11.18b}$$

Element 3

$$\begin{bmatrix} k_{11}^{(3)} & k_{12}^{(3)} \\ k_{21}^{(3)} & k_{22}^{(3)} \end{bmatrix} \begin{Bmatrix} \phi_3^{(3)} \\ \phi_4^{(3)} \end{Bmatrix} = \begin{Bmatrix} I_3^{(3)} \\ I_4^{(3)} \end{Bmatrix} + \begin{Bmatrix} f_3^{(3)} \\ f_4^{(3)} \end{Bmatrix} \tag{11.18c}$$

⋮

Element (n−1)

$$\begin{bmatrix} k_{11}^{(n-1)} & k_{12}^{(n-1)} \\ k_{21}^{(n-1)} & k_{22}^{(n-1)} \end{bmatrix} \begin{Bmatrix} \phi_{n-1}^{(n-1)} \\ \phi_{n}^{(n-1)} \end{Bmatrix} = \begin{Bmatrix} I_{n-1}^{(n-1)} \\ I_{n}^{(n-1)} \end{Bmatrix} + \begin{Bmatrix} f_{n-1}^{(n-1)} \\ f_{n}^{(n-1)} \end{Bmatrix} \qquad (11.18n-1)$$

Element n

$$\begin{bmatrix} k_{11}^{(n)} & k_{12}^{(n)} \\ k_{21}^{(n)} & k_{22}^{(n)} \end{bmatrix} \begin{Bmatrix} \phi_{n}^{(n)} \\ \phi_{n+1}^{(n)} \end{Bmatrix} = \begin{Bmatrix} I_{n}^{(n)} \\ I_{n+1}^{(n)} \end{Bmatrix} + \begin{Bmatrix} f_{n}^{(n)} \\ f_{n+1}^{(n)} \end{Bmatrix} \qquad (11.18n)$$

Let us now form the global system by including element equations one at a time.

The inclusion of the ***first*** element equation, Equation 11.18a, gives the global system as

$$\begin{bmatrix} k_{11}^{(1)} & k_{12}^{(1)} & 0 & 0 & \cdots & 0 & 0 \\ k_{21}^{(1)} & k_{22}^{(1)} & 0 & 0 & \cdots & 0 & 0 \\ 0 & 0 & 0 & 0 & \cdots & 0 & 0 \\ 0 & 0 & 0 & 0 & \cdots & 0 & 0 \\ \vdots & & & & & & \\ 0 & 0 & 0 & 0 & \cdots & 0 & 0 \\ 0 & 0 & 0 & 0 & \cdots & 0 & 0 \end{bmatrix} \begin{Bmatrix} \phi_1 \\ \phi_2 \\ \phi_3 \\ \\ \vdots \\ \phi_n \\ \phi_{n+1} \end{Bmatrix} = \begin{Bmatrix} I_1^{(1)} \\ I_2^{(1)} \\ 0 \\ 0 \\ \vdots \\ 0 \\ 0 \end{Bmatrix} + \begin{Bmatrix} f_1^{(1)} \\ f_2^{(1)} \\ 0 \\ 0 \\ \vdots \\ 0 \\ 0 \end{Bmatrix}$$

Inclusion of the ***second*** element equation, Equation 11.18b, gives

$$\begin{bmatrix} k_{11}^{(1)} & k_{12}^{(1)} & 0 & 0 & \cdots & 0 & 0 \\ k_{21}^{(1)} & k_{22}^{(1)}+k_{11}^{(2)} & k_{12}^{(2)} & 0 & \cdots & 0 & 0 \\ 0 & k_{21}^{(2)} & k_{22}^{(2)} & 0 & \cdots & 0 & 0 \\ 0 & 0 & 0 & 0 & \cdots & 0 & 0 \\ \vdots & & & & & & \\ 0 & 0 & 0 & 0 & \cdots & 0 & 0 \\ 0 & 0 & 0 & 0 & \cdots & 0 & 0 \end{bmatrix} \begin{Bmatrix} \phi_1 \\ \phi_2 \\ \phi_3 \\ \\ \vdots \\ \phi_n \\ \phi_{n+1} \end{Bmatrix} = \begin{Bmatrix} I_1^{(1)} \\ I_2^{(1)}+I_1^{(2)} \\ I_2^{(2)} \\ 0 \\ \vdots \\ 0 \\ 0 \end{Bmatrix} + \begin{Bmatrix} f_1^{(1)} \\ f_2^{(1)}+f_1^{(2)} \\ f_2^{(2)} \\ 0 \\ \vdots \\ 0 \\ 0 \end{Bmatrix}$$

Inclusion of the ***third*** element equation, Equation 11.18c, gives

$$
\begin{bmatrix}
k_{11}^{(1)} & k_{12}^{(1)} & 0 & 0 & \cdots & 0 & 0 \\
k_{21}^{(1)} & k_{22}^{(1)} + k_{11}^{(2)} & k_{12}^{(2)} & 0 & \cdots & 0 & 0 \\
0 & k_{21}^{(2)} & k_{22}^{(2)} + k_{11}^{(3)} & k_{12}^{(3)} & \cdots & 0 & 0 \\
0 & 0 & k_{21}^{(3)} & k_{22}^{(3)} & \cdots & 0 & 0 \\
\vdots & & & & & & \\
0 & 0 & 0 & 0 & \cdots & 0 & 0 \\
0 & 0 & 0 & 0 & \cdots & 0 & 0
\end{bmatrix}
\begin{Bmatrix} \phi_1 \\ \phi_2 \\ \phi_3 \\ \\ \vdots \\ \phi_n \\ \phi_{n+1} \end{Bmatrix}
=
\begin{Bmatrix} I_1^{(1)} \\ I_2^{(1)} + I_1^{(2)} \\ I_2^{(2)} + I_1^{(3)} \\ I_2^{(3)} \\ \\ 0 \\ 0 \end{Bmatrix}
+
\begin{Bmatrix} f_1^{(1)} \\ f_2^{(1)} + f_1^{(2)} \\ f_2^{(2)} + f_2^{(3)} \\ I_2^{(3)} \\ \\ 0 \\ 0 \end{Bmatrix}
$$

This process of including the element equations is continued, and with the inclusion of the nth element equation, Equation 11.18n gives the final ***global system*** as

$$
\begin{bmatrix}
k_{11}^{(1)} & k_{12}^{(1)} & 0 & 0 & \cdots & 0 & 0 \\
k_{21}^{(1)} & k_{22}^{(1)} + k_{11}^{(2)} & k_{12}^{(2)} & 0 & \cdots & 0 & 0 \\
0 & k_{21}^{(2)} & k_{22}^{(2)} + k_{11}^{(3)} & k_{12}^{(3)} & \cdots & 0 & 0 \\
0 & 0 & k_{21}^{(3)} & k_{22}^{(3)} & \cdots & 0 & 0 \\
\vdots & & & & & & \\
0 & 0 & 0 & 0 & \cdots & k_{22}^{(n-1)} + k_{11}^{(n)} & k_{12}^{(n)} \\
0 & 0 & 0 & 0 & \cdots & k_{21}^{(n)} & k_{22}^{(n)}
\end{bmatrix}
\begin{Bmatrix} \phi_1 \\ \phi_2 \\ \phi_3 \\ \\ \vdots \\ \phi_n \\ \phi_{n+1} \end{Bmatrix}
=
\begin{Bmatrix} I_1^{(1)} \\ I_2^{(1)} + I_1^{(2)} \\ I_2^{(2)} + I_1^{(3)} \\ I_2^{(3)} \\ \vdots \\ I_2^{(n-1)} + I_1^{(n)} \\ I_2^{(n)} \end{Bmatrix}
+
\begin{Bmatrix} f_1^{(1)} \\ f_2^{(1)} + f_1^{(2)} \\ f_2^{(2)} + f_2^{(3)} \\ I_2^{(3)} \\ \vdots \\ f_2^{(n-1)} + f_1^{(n)} \\ f_2^{(n)} \end{Bmatrix}
\tag{11.19}
$$

Let us now introduce the expressions for the element matrix, force vector, and the inter-element vector as given by Equations 11.17b to 11.17d into Equation 11.19 to obtain the global system as

$$
\begin{bmatrix}
\frac{\Gamma_x}{L_e} & -\frac{\Gamma_x}{L_e} & & & & & \\
-\frac{\Gamma_x}{L_e} & \frac{\Gamma_x}{L_e} + \frac{\Gamma_x}{L_e} & -\frac{\Gamma_x}{L_e} & & & & \\
& -\frac{\Gamma_x}{L_e} & \frac{\Gamma_x}{L_e} + \frac{\Gamma_x}{L_e} & -\frac{\Gamma_x}{L_e} & & & \\
& & -\frac{\Gamma_x}{L_e} & \frac{\Gamma_x}{L_e} + \frac{\Gamma_x}{L_e} & -\frac{\Gamma_x}{L_e} & & \\
& & \vdots & \ddots & \ddots & \ddots & \\
& & & & -\frac{\Gamma_x}{L_e} & \frac{\Gamma_x}{L_e} + \frac{\Gamma_x}{L_e} & -\frac{\Gamma_x}{L_e} \\
& & & & & -\frac{\Gamma_x}{L_e} & \frac{\Gamma_x}{L_e}
\end{bmatrix}
\begin{Bmatrix} \phi_1 \\ \phi_2 \\ \phi_3 \\ \vdots \\ \vdots \\ \phi_{n-1} \\ \phi_n \end{Bmatrix}
$$

$$
= \begin{Bmatrix} -\Gamma_x \left.\frac{d\phi}{dx}\right|_{x_1} \\ \Gamma_x \left.\frac{d\phi}{dx}\right|_{x_2} - \Gamma_x \left.\frac{d\phi}{dx}\right|_{x_2} \\ \Gamma_x \left.\frac{d\phi}{dx}\right|_{x_3} - \Gamma_x \left.\frac{d\phi}{dx}\right|_{x_3} \\ \vdots \\ \vdots \\ \Gamma_x \left.\frac{d\phi}{dx}\right|_{x_n} - \Gamma_x \left.\frac{d\phi}{dx}\right|_{x_n} \\ +\Gamma_x \left.\frac{d\phi}{dx}\right|_{x_{n+1}} \end{Bmatrix} \begin{Bmatrix} \frac{\bar{S}L_e}{2} \\ \frac{\bar{S}L_e}{2} + \frac{\bar{S}L_e}{2} \\ \frac{\bar{S}L_e}{2} + \frac{\bar{S}L_e}{2} \\ \vdots \\ \vdots \\ \frac{\bar{S}L_e}{2} + \frac{\bar{S}L_e}{2} \\ \frac{\bar{S}L_e}{2} \end{Bmatrix} \tag{11.20}
$$

Simplifying and canceling terms due to inter-element continuity conditions, we get the final ***global system*** as

$$
\begin{bmatrix} \frac{\Gamma_x}{L_e} & -\frac{\Gamma_x}{L_e} & & & & \\ -\frac{\Gamma_x}{L_e} & \frac{2\Gamma_x}{L_e} & -\frac{\Gamma_x}{L_e} & & & \\ & -\frac{\Gamma_x}{L_e} & \frac{2\Gamma_x}{L_e} & -\frac{\Gamma_x}{L_e} & & \\ & & -\frac{\Gamma_x}{L_e} & \frac{2\Gamma_x}{L_e} & -\frac{\Gamma_x}{L_e} & \\ & & \vdots & \ddots & \ddots & \ddots \\ & & & -\frac{\Gamma_x}{L_e} & \frac{2\Gamma_x}{L_e} & -\frac{\Gamma_x}{L_e} \\ & & & & -\frac{\Gamma_x}{L_e} & \frac{\Gamma_x}{L_e} \end{bmatrix} \begin{Bmatrix} \phi_1 \\ \phi_2 \\ \phi_3 \\ \vdots \\ \vdots \\ \phi_{n-1} \\ \phi_n \end{Bmatrix} = \begin{Bmatrix} -\Gamma_x \left.\frac{d\phi}{dx}\right|_{x_1} \\ 0 \\ 0 \\ \vdots \\ \vdots \\ 0 \\ \Gamma_x \left.\frac{d\phi}{dx}\right|_{x_{n+1}} \end{Bmatrix} + \begin{Bmatrix} \frac{\bar{S}L_e}{2} \\ \bar{S}L_e \\ \bar{S}L_e \\ \vdots \\ \vdots \\ \bar{S}L_e \\ \frac{\bar{S}L_e}{2} \end{Bmatrix} \tag{11.21}
$$

It can be noted that in the assembled global system, the inter-element conditions are already implemented by satisfying the continuous flux quantity at all inter-element nodes. Again, the global system is a set of algebraic equations with a number of unknowns the same as the number of unknown nodal values in the selected mesh distribution system. The boundary conditions now can be directly introduced in the global system before solving the system of equations.

The global system can be written in a compact form as

$$
[K]\{\Phi\} = \{F\} \tag{11.22}
$$

where

$$[K] = \begin{bmatrix} \frac{\Gamma_x}{L_e} & -\frac{\Gamma_x}{L_e} & & & & & \\ -\frac{\Gamma_x}{L_e} & \frac{2\Gamma_x}{L_e} & -\frac{\Gamma_x}{L_e} & & & & \\ & -\frac{\Gamma_x}{L_e} & \frac{2\Gamma_x}{L_e} & -\frac{\Gamma_x}{L_e} & & & \\ & & -\frac{\Gamma_x}{L_e} & \frac{2\Gamma_x}{L_e} & -\frac{\Gamma_x}{L_e} & & \\ & & \vdots & \ddots & \ddots & \ddots & \\ & & & & -\frac{\Gamma_x}{L_e} & \frac{2\Gamma_x}{L_e} & -\frac{\Gamma_x}{L_e} \\ & & & & & -\frac{\Gamma_x}{L_e} & \frac{\Gamma_x}{L_e} \end{bmatrix}$$

$$= \text{global characteristics matrix} \tag{11.23a}$$

$$\{\Phi\} = \begin{Bmatrix} \phi_1 \\ \phi_2 \\ \phi_3 \\ \vdots \\ \vdots \\ \phi_{n-1} \\ \phi_n \end{Bmatrix} = \text{vector for unknown nodal values in the solution domain} \tag{11.23b}$$

$$\{F\} = \begin{Bmatrix} -\Gamma_x \left.\frac{d\phi}{dx}\right|_{x_1} \\ 0 \\ 0 \\ \vdots \\ \vdots \\ 0 \\ \Gamma_x \left.\frac{d\phi}{dx}\right|_{x_{n+1}} \end{Bmatrix} + \begin{Bmatrix} \frac{\bar{S}L_e}{2} \\ \bar{S}L_e \\ \bar{S}L_e \\ \vdots \\ \vdots \\ \bar{S}L_e \\ \bar{S}L_e \\ \frac{\bar{S}L_e}{2} \end{Bmatrix} = \text{source or force vector} \tag{11.23c}$$

Step 6: Implementation of Boundary Conditions. In this step, the boundary conditions are introduced directly in the global system of equations. For boundary conditions of the ***first kind***, i.e., of constant surface value as defined by Equation 11.2, we simply assign the given values to the boundary nodes as $\phi_1 = \phi_l$ and $\phi_{n+1} = \phi_r$, and rearranged the system so that the natural boundary conditions $d\phi/dx\big|_{x_1}$ and $d\phi/dx\big|_{x_{n+1}}$ appear as one of the unknowns as

$$\begin{bmatrix} -\Gamma_x & -\frac{\Gamma_x}{L_e} & & & & & \\ 0 & \frac{2\Gamma_x}{L_e} & -\frac{\Gamma_x}{L_e} & & & & \\ & -\frac{\Gamma_x}{L_e} & \frac{2\Gamma_x}{L_e} & -\frac{\Gamma_x}{L_e} & & & \\ & & -\frac{\Gamma_x}{L_e} & \frac{2\Gamma_x}{L_e} & -\frac{\Gamma_x}{L_e} & & \\ & & \vdots & \ddots & \ddots & \ddots & \\ & & & & -\frac{\Gamma_x}{L_e} & \frac{2\Gamma_x}{L_e} & 0 \\ & & & & & -\frac{\Gamma_x}{L_e} & -\Gamma_x \end{bmatrix} \begin{Bmatrix} \left.\frac{d\phi}{dx}\right|_{x_1} \\ \phi_2 \\ \phi_3 \\ \vdots \\ \vdots \\ \phi_n \\ \left.\frac{d\phi}{dx}\right|_{x_{n+1}} \end{Bmatrix} = \begin{Bmatrix} -\frac{\Gamma_x\phi_l}{L_e} + \frac{\bar{S}L_e}{2} \\ \frac{\Gamma_x\phi_l}{L_e} + \bar{S}L_e \\ \bar{S}L_e \\ \vdots \\ \vdots \\ \frac{\Gamma_x\phi_r}{L_e} + \bar{S}L_e \\ -\frac{\Gamma_x\phi_r}{L_e} + \frac{\bar{S}L_e}{2} \end{Bmatrix} \tag{11.24}$$

The global system can be written in a compact form as

$$[K]\{\Phi\} = \{F\} \tag{11.25}$$

where

$$[K] = \begin{bmatrix} -\Gamma_x & -\frac{\Gamma_x}{L_e} & 0 & 0 & 0 & 0 & 0 \\ 0 & \frac{2\Gamma_x}{L_e} & -\frac{\Gamma_x}{L_e} & 0 & 0 & 0 & 0 \\ 0 & -\frac{\Gamma_x}{L_e} & \frac{2\Gamma_x}{L_e} & -\frac{\Gamma_x}{L_e} & 0 & 0 & 0 \\ 0 & 0 & -\frac{\Gamma_x}{L_e} & \frac{2\Gamma_x}{L_e} & -\frac{\Gamma_x}{L_e} & 0 & 0 \\ \vdots & & & \ddots & \ddots & \ddots & \\ 0 & 0 & 0 & 0 & -\frac{\Gamma_x}{L_e} & \frac{2\Gamma_x}{L_e} & 0 \\ 0 & 0 & 0 & 0 & & -\frac{\Gamma_x}{L_e} & -\Gamma_x \end{bmatrix} \tag{11.26a}$$

= global characteristics matrix

$$\{\Phi\} = \begin{Bmatrix} \left.\dfrac{d\phi}{dx}\right|_{x_1} \\ \phi_2 \\ \phi_3 \\ \vdots \\ \vdots \\ \phi_n \\ \left.\dfrac{d\phi}{dx}\right|_{x_{n=1}} \end{Bmatrix} = \text{vector for unknown nodal values in the solution domain} \tag{11.26b}$$

$$\{F\} = \begin{Bmatrix} -\dfrac{\Gamma_x \phi_l}{L_e} + \dfrac{\bar{S}L_e}{2} \\ \dfrac{\Gamma_x \phi_l}{L_e} + \bar{S}\,L_e \\ \bar{S}L_e \\ \vdots \\ \vdots \\ \dfrac{\Gamma_x \phi_r}{L_e} + \bar{S}\,L_e \\ -\dfrac{\Gamma_x \phi_r}{L_e} + \dfrac{\bar{S}L_e}{2} \end{Bmatrix} \tag{11.26c}$$

Step 7: Solution of the Global System of Equations. The global system of equations is now solved using the appropriate solver as discussed in Chapter 3.

Example 11.1: One-dimensional Steady State Conduction: Consider a one-dimensional steady state conduction without heat generation in a plane slab as shown. The boundary surfaces at $x = 0$ and $x = L$ are maintained at constant temperatures T_l and T_r, respectively. Determine the temperature distribution in the slab and heat transfer rates at the surfaces. Use $k = 1.2\,\text{W/cm}\cdot\text{C}$, $L = 10$ cm, $T_l = 30°\text{C}$, $T_r = 100°\text{C}$.

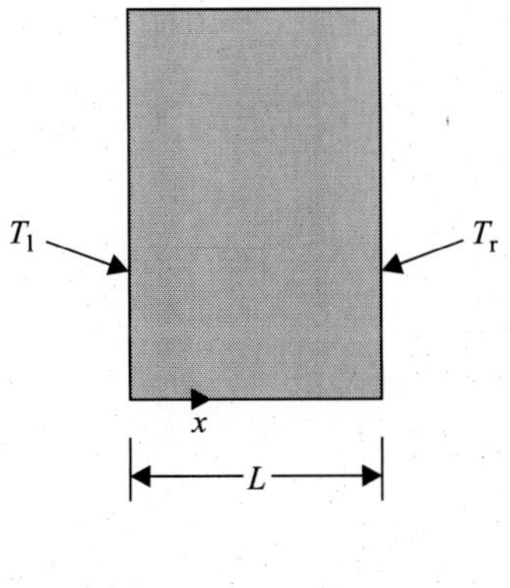

The mathematical statement of the problem is

Governing equation

$$\frac{d}{dx}\left(k\frac{dT}{dx}\right) = 0 \qquad \text{(E.11.1.1)}$$

Boundary condition

1. $x = 0, \quad T = T_l$ (E.11.1.2a)

2. $x = L, \quad T = T_r$ (E.11.1.2b)

Solution: In the ***first step***, let us discretize the domain using five uniform line elements as shown in the figure.

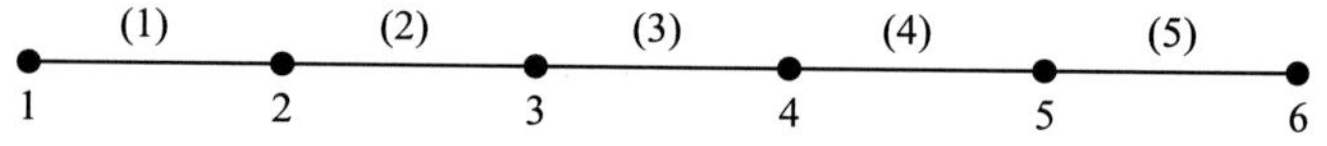

With this mesh size distribution, we have $L_e = L = 2.0\,\text{cm}$ and number of elements $N=5$ with node $i=1$ and $i=6$ located at $x=0$ and $x=10$.

In the ***second step***, we write the weak form of the governing equation following Equation 11.9, which is obtained by Galerkin's weighted residual method as

$$\left\{\begin{matrix} k\left.\dfrac{dT}{dx}\right|_{x_i} \\ -k\left.\dfrac{dT}{dx}\right|_{x_j} \end{matrix}\right\} + \int_{x}^{x_j} \Gamma_x \frac{d[N]^{\mathrm{T}}}{dx}\frac{dT^*}{dx}\,dx = 0 \qquad \text{(E.11.1.3)}$$

In the ***third step***, we assume that the temperature distribution in each element be represented by the approximate function

$$T^* = N_i T_i + N_j T_j \qquad \text{(E.11.1.4)}$$

and shape function given by linear line element (Equation 10.10)

In the ***fourth step***, the element characteristics equation is obtained following Equations 11.16 and 11.17

$$\left[\kappa_k^{(e)}\right]\left\{\phi^{(e)}\right\} = \left\{I^{(e)}\right\} \qquad \text{(E.11.1.5a)}$$

where

$$\{T^{(e)}\} = \begin{Bmatrix} T_i \\ T_j \end{Bmatrix}, \quad \left[K_k^{(e)}\right] = \frac{k}{L}\begin{bmatrix} 1 & -1 \\ -1 & 1 \end{bmatrix}, \quad \{I^{(e)}\} = \begin{Bmatrix} -k\left.\dfrac{dT}{dx}\right|_{x_i} \\ k\left.\dfrac{dT}{dx}\right|_{x_j} \end{Bmatrix} \quad \text{(E.11.1.5b)}$$

Since the material is homogeneous and the mesh is uniform, the element characteristics equation is identical for all elements.

In the ***fifth step***, we assemble all element equations and obtain the global system following Equation 11.21

$$\begin{bmatrix} 1 & -1 & & & & \\ -1 & 2 & -1 & & & \\ & -1 & 2 & -1 & & \\ & & -1 & 2 & -1 & \\ & & & -1 & 2 & -1 \\ & & & & -1 & 1 \end{bmatrix} \begin{Bmatrix} T_1 \\ T_2 \\ T_3 \\ T_4 \\ T_5 \\ T_6 \end{Bmatrix} = \begin{Bmatrix} -\left.\dfrac{dT}{dx}\right|_{x_1} \cdot L \\ 0 \\ 0 \\ 0 \\ 0 \\ \left.\dfrac{dT}{dx}\right|_{x_6} \cdot L \end{Bmatrix} \quad \text{(E.11.1.6)}$$

In the ***sixth step***, we incorporate the boundary conditions in the global system. Since the boundary temperatures are known in this problem, we substitute $T_1 = T_l = 30°C$ and $T_6 = T_r = 100°C$ and rearrange the global system as

$$\begin{bmatrix} -L & -1 & 0 & 0 & 0 & 0 \\ 0 & 2 & -1 & 0 & 0 & 0 \\ 0 & -1 & 2 & -1 & 0 & 0 \\ 0 & 0 & -1 & 2 & -1 & 0 \\ 0 & 0 & 0 & -1 & 2 & 0 \\ 0 & 0 & 0 & 0 & -1 & -L \end{bmatrix} \begin{Bmatrix} \left.\dfrac{dT}{dx}\right|_{x_1} \\ T_2 \\ T_3 \\ T_4 \\ T_5 \\ \left.\dfrac{dT}{dx}\right|_{x_6} \end{Bmatrix} = \begin{Bmatrix} -T_l \\ T_l \\ 0 \\ 0 \\ T_r \\ -T_r \end{Bmatrix} \quad \text{(E.11.1.7)}$$

In the ***seventh step***, we substitute $L = 2\,\text{cm}$, $T_l = 30°C$, $T_r = 100°C$, and choose an appropriate solver for the global system of equations. Using the

Gaussian elimination program, we can get the numerical solution as

$$\left.\frac{dT}{dx}\right|_{x_1} = -7.0^{\circ}\text{C/cm}$$

$$T_2 = 44^{\circ}\text{C}$$

$$T_3 = 58^{\circ}\text{C}$$

$$T_4 = 72^{\circ}\text{C}$$

$$T_5 = 86^{\circ}\text{C}$$

$$\left.\frac{dT}{dx}\right|_{x_6} = -7.0^{\circ}\text{C/cm}$$

The heat transfer rate at $x = 0$ is estimated as

$$q''_{x=0} = -k\left.\frac{dT}{dx}\right|_{x_1} = 1.2 \times (-7.0)$$

$$q''_{x=0} = -8.40\,\text{W/cm}^2$$

The heat transfer rate at $x = L$ is estimated as

$$q''_{x=L} = -k\left.\frac{dT}{dx}\right| = 1.2 \times (-7.0)$$

$$q''_{x=L} = -8.40\,\text{W/cm}^2$$

11.2 FINITE ELEMENT FORMULATION USING VARIATIONAL APPROACH

Most of the steps outlined for the ***Galerkin-based finite element formulation*** are also applicable to finite element formulation based on the variational approach with the exception of Steps 2 and 4, in which an integral statement of the problem and the element characteristic equations are obtained. So, while discussing the finite element formulation based on the variational approach, we will focus on the formulation of the integral statement and the element characteristic equation in Steps 2 and 4 only.

In this approach, the governing differential equation is transformed into an integral form by evaluating an equivalent ***variational form*** of the problem. This variational form primarily involves a ***functional, I***, such that finding a minimized or extremized value of the functional (I) is equivalent to finding the solution of the original differential equation. In the next step, the functional is extremized with respect to the unknown solution vector, $\{\Phi\}$, which constitutes all unknown nodal values of the mesh, i.e.

$$\frac{\partial I}{\partial \Phi} = 0 \tag{11.27}$$

where the functional I over the whole solution domain can be written as the summation of elemental contribution $I^{(e)}$ as

$$I = \sum_{e=1}^{n} I^{(e)} \tag{11.28}$$

Combining Equations 11.27 and 11.28, we get

$$\sum_{e=1}^{n} \frac{\partial I^{(e)}}{\partial \phi^{(e)}} = 0 \tag{11.29}$$

As we evaluate the elemental contribution, $\partial I^{(e)}/\partial \phi^{(e)}$, we obtain the element characteristic equation as

$$\frac{\partial I^{(e)}}{\partial \phi^{(e)}} = \left[\kappa_\Gamma^{(e)}\right]\{\phi^{(e)}\} - \{f^{(e)}\} \tag{11.30}$$

Equation 11.30 is applied successively to all elements, and the resulting element characteristic equations are then assembled through Equation 11.29 using the global connectivity system (Table 11.1) to obtain the global system of equations

$$[K]\{\Phi\} = \{F\} \tag{11.31}$$

A brief description of the variational calculus as well as the procedure for obtaining the functional (I) of a differential problem is given in Chapter 9. Let us now consider the one-dimensional steady state transport equation given by Equation 11.1 and constant surface boundary conditions given by Equation 11.2, and demonstrate the formulations of the integral form and subsequently the element characteristic equation using the variational formulation. We will consider the mesh size distribution given in Figure 11.2, and approximate solution and linear line element given by Equation 11.3.

Step 2: Formation of the Integral Statement or Variational Form. Multiply both sides of Equation 11.1 by a variation, $\delta\phi$, and integrate over an element to get

$$\int_{x_i}^{x_j} \left(\frac{d}{dx}\left(\Gamma_x \frac{d\phi}{dx}\right) + S\right) \delta\phi \, dx = 0$$

or

$$\int_{x_i}^{x_j} \frac{d}{dx}\left(\Gamma_x \frac{d\phi}{dx}\right)\delta\phi\, dx + \int_{x_i}^{x_j} S\,\delta\phi\, dx = 0 \tag{11.32}$$

Evaluating the first integral term using integration-by-parts, we get

$$\left(\Gamma_x \frac{d\phi}{dx}\delta\phi\right)\Bigg|_{x_i}^{x_j} - \int_{x_i}^{x_j} \Gamma_x \frac{d\phi}{dx}\delta\left(\frac{d\phi}{dx}\right) dx + \int_{x_i}^{x_j} S\,\delta\phi\, dx = 0$$

or

$$\left(\Gamma_x \frac{d\phi}{dx}\delta\phi\right)_{x=x_j} - \left(\Gamma_x \frac{d\phi}{dx}\delta\phi\right)_{x_i} - \int_{x_i}^{x_j} \Gamma_x \frac{1}{2}\delta\left(\frac{d\phi}{dx}\right)^2 dx + \int_{x_i}^{x_j} S\,\delta\phi\, dx = 0 \tag{11.33}$$

It can be noted that the first two integrated terms represents flux conditions at the boundaries. These terms vanish if no variation of ϕ is permitted at the boundary, i.e., $\delta\phi = 0$ or flux is zero at the boundary such as in adiabatic or symmetric boundaries. We further reformulate the rest of the terms in Equation 11.33 by bringing the δ operator outside the integrals as

$$\delta \int_0^L \left(-\frac{1}{2}\Gamma_x\left(\frac{d\phi}{dx}\right)^2 + S\phi\right) dx = 0 \tag{11.34}$$

Now, Equation 11.34 represents the variation of a functional, i.e. $\delta I = 0$. So the functional, I, corresponding to the governing differential problem, is

$$I = \int_{x_i}^{x_j} \left(\frac{1}{2}\Gamma_x\left(\frac{d\phi}{dx}\right)^2 - S\phi\right) dx \tag{11.35}$$

Minimization of the functional leads to the solution, which is also equivalent to the solution of the governing differential problem.

However, for any other types of boundary conditions such as mixed boundary condition or constant flux conditions, the integrated terms do not vanish. Such non-zero-integrated terms may also arise for any interior nodes that are located at the interface of two elements. In such situations, the integrated terms are reformulated before forming the functional as

$$\left(\Gamma_x \frac{d\phi}{dx}\bigg|_{x_j} \delta\phi(x_j)\right) - \left(\Gamma_x \frac{d\phi}{dx}\bigg|_{x_i} \delta\phi(x_i)\right) - \int_{x_i}^{x_j} \Gamma_x \frac{1}{2}\delta\left(\frac{d\phi}{dx}\right)^2 dx + \int_{x_i}^{x_j} S\,\delta\phi\, dx = 0$$

or

$$\delta\left\{\left(\Gamma_x \frac{d\phi}{dx}\bigg|_{x_j} \phi(x_j)\right) - \left(\Gamma_x \frac{d\phi}{dx}\bigg|_{x_i} \phi(x_i)\right) - \int_{x_i}^{x_j} \Gamma_x \frac{1}{2}\left(\frac{d\phi}{dx}\right)^2 dx + \int_{x_i}^{x_j} S\phi\, dx\right\} = 0$$

The corresponding functional is given as

$$I = -\left(\Gamma_x \frac{d\phi}{dx}\bigg|_{x_j} \phi(x_j)\right) + \left(\Gamma_x \frac{d\phi}{dx}\bigg|_{x_i} \phi(x_i)\right) + \int_{x_i}^{x_j} \Gamma_x \frac{1}{2}\left(\frac{d\phi}{dx}\right)^2 dx - \int_{x_i}^{x_j} S\phi\, dx$$

Substituting $\phi(x_j) = \phi_j$ and $\phi(x_i) = \phi_i$, we get

$$I = -\left(\Gamma_x \frac{d\phi}{dx}\bigg|_{x_j} \phi_j\right) + \left(\Gamma_x \frac{d\phi}{dx}\bigg|_{x_i} \phi_i\right) + \int_{x_i}^{x_j} \Gamma_x \frac{1}{2}\left(\frac{d\phi}{dx}\right)^2 dx - \int_{x_i}^{x_j} S\phi\, dx$$

Step 4: Formation of the Element Characteristic Equation. In this step, the element characteristic equation is obtained by evaluating the elemental functional quantity, $I^{(e)}$, given as

$$I^{(e)} = -\left(\Gamma_x \frac{d\phi}{dx}\bigg|_{x_j} \phi_j\right) + \left(\Gamma_x \frac{d\phi}{dx}\bigg|_{x_i} \phi_i\right) + \int_{x_i}^{x_j} \Gamma_x \frac{1}{2}\left(\frac{d\phi}{dx}\right)^2 dx - \int_{x_i}^{x_j} s\phi\, dx \quad (11.36)$$

Let us first evaluate the derivative from Equation 11.3 as follows.

From Equation 11.3a

$$\frac{d\phi}{dx} = \frac{dN_i}{dx}\phi_i + \frac{dN_j}{dx}\phi_j \quad (11.37)$$

Substituting Equations 11.3a and 11.37 into Equation 11.36, we get

$$I^{(e)} = -\left(\Gamma_x \frac{d\phi}{dx}\bigg|_{x_j} \phi_j\right) + \left(\Gamma_x \frac{d\phi}{dx}\bigg|_{x_i} \phi_i\right)$$
$$+ \frac{1}{2}\Gamma_x \int_{x_i}^{x_j} \left[\left(\frac{dN_i}{dx}\right)^2 \phi_i^2 + \left(\frac{dN_j}{dx}\right)^2 \phi_j^2 + 2\frac{dN_i}{dx}\frac{dN_i}{dx}\phi_i\phi_j\right] dx - \int_{x_i}^{x_j} S[N_i\phi_i + N_j\phi_j] dx$$
$$(11.38)$$

In order to minimize the functional, we evaluate the derivative given by Equation 11.38 as follows.

For $I = i$

$$\frac{dI^{(e)}}{d\phi_i} = \Gamma_x \frac{d\phi}{dx}\bigg|_{x_i} + \frac{\Gamma_{\bar{x}}}{2} \int_{x_i}^{x_j} \left[\left(\frac{dN_i}{dx}\right)^2 2\phi_i + 2\frac{dN_i}{dx}\frac{dN_j}{dx}\phi_j \right] dx - \bar{S} \int_{x_i}^{x_i} N_i \, dx = 0 \tag{11.39a}$$

For $I = j$

$$\frac{dI^{(e)}}{d\phi_i} = -\Gamma_x \frac{d\phi}{dx}\bigg|_{x_j} + \frac{\Gamma_{\bar{x}}}{2} \int_{x_i}^{x_j} \left[2\frac{dN_i}{dx}\frac{dN_i}{dx}\phi_j + \left(\frac{dN_j}{dx}\right)^2 2\phi_j \right] dx - \bar{S} \int_{x_i}^{x_i} N_j \, dx = 0 \tag{11.39b}$$

Note that we have assumed step-wise constant values for the transport property, $\Gamma_x = \Gamma_{\bar{x}}$, and for the source term as $S = \bar{S}$. Equation 11.39 is the ***element characteristic equation***, which can also be written in a matrix form as

$$\left\{ \frac{dI^{(e)}}{d\phi^{(e)}} \right\} = \left\{ \begin{array}{c} \Gamma_x \dfrac{d\phi}{dx}\bigg|_{x_i} \\ -\Gamma_x \dfrac{d\phi}{dx}\bigg|_{x_j} \end{array} \right\} + \Gamma_{\bar{x}} \begin{bmatrix} \int_{x_i}^{x_j} \left(\frac{dN_i}{dx}\right)^2 dx & \int_{x_i}^{x_j} \frac{dN_i}{dx}\frac{dN_j}{dx} dx \\ \int_{x_i}^{x_j} \frac{dN_i}{dx}\frac{dN_j}{dx} dx & \int_{x_i}^{x_j} \left(\frac{dN_j}{dx}\right)^2 dx \end{bmatrix} \left\{ \begin{array}{c} \phi_i \\ \phi_j \end{array} \right\} - \bar{S} \left\{ \begin{array}{c} \int_{x_i}^{x_j} N_i dx \\ \int_{x_i}^{x_j} N_j dx \end{array} \right\} = 0 \tag{11.40}$$

or

$$\left[\kappa_\Gamma^{(e)}\right]\{\phi^{(e)}\} = \{I^{(e)}\} + \{f^{(e)}\} \tag{11.41}$$

where

$$\left[\kappa_\Gamma^{(x)}\right] = \Gamma_{\bar{x}} \begin{bmatrix} \int_{x_i}^{x_j} \left(\frac{dN_i}{dx}\right)^2 dx & \int_{x_i}^{x_j} \frac{dN_i}{dx}\frac{dN_j}{dx} dx \\ \int_{x_i}^{x_j} \frac{dN_i}{dx}\frac{dN_j}{dx} dx & \int_{x_i}^{x_j} \left(\frac{dN_j}{dx}\right)^2 dx \end{bmatrix} = \text{element characteristic matrix} \tag{11.42}$$

$$\{f^{(e)}\} = \bar{S}\begin{Bmatrix} \int_{x_i}^{x_j} N_i\,dx \\ \int_{x_i}^{x_j} N_j\,dx \end{Bmatrix} = \text{element source vector} \tag{11.43}$$

$$\{I^{(e)}\} = \begin{Bmatrix} -\Gamma_x \left.\dfrac{d\phi}{dx}\right|_{x_i} \\ \Gamma_x \left.\dfrac{d\phi}{dx}\right|_{x_j} \end{Bmatrix} = \begin{Bmatrix} I_i^{(e)} \\ I_j^{(e)} \end{Bmatrix}$$

$$\{\phi^{(e)}\} = \begin{Bmatrix} \phi_i \\ \phi_j \end{Bmatrix} = \text{vector for element nodal values} \tag{11.44}$$

Let us now evaluate the elements of the characteristic matrix as follows.

From Equation 11.3b

$$\frac{dN_i}{dx} = -\frac{1}{L_e}, \quad \frac{dN_j}{dx} = \frac{1}{L_e} \tag{11.45}$$

Using these we have

$$\int_{x_i}^{x_i} \left(\frac{dN_i}{dx}\right)^2 dx = \int_{x_i}^{x_j} \frac{1}{L_e^2}\,dx = \frac{1}{L_e} \tag{11.46a}$$

$$\int_{x_i}^{x_i} \left(\frac{dN_j}{dx}\right)^2 dx = \int_{x_i}^{x_j} \frac{1}{L_e^2}\,dx = \frac{1}{L_e} \tag{11.46b}$$

$$\int_{x_i}^{x_i} \left(\frac{dN_i}{dx}\right)\left(\frac{dN_j}{dx}\right) dx = \int_{x_i}^{x_j} -\frac{1}{L_e^2}\,dx = -\frac{1}{L_e} \tag{11.46c}$$

The elements of the source vector are evaluated in Section 11.1, and we use

$$\int_{x_i}^{x_j} N_i\,dx = \frac{L_e}{2} \tag{11.47a}$$

$$\int_{x_i}^{x_j} N_j\,dx = \frac{L_e}{2} \tag{11.47b}$$

Substituting Equations 11.46 and 11.47 into Equations 11.42 and 11.43, we compute the element characteristic matrix and source vectors as

$$\left[\kappa_{\Gamma}^{(e)}\right] = \frac{\Gamma_{\bar{x}}}{L_e}\begin{bmatrix} 1 & -1 \\ -1 & 1 \end{bmatrix} \tag{11.48a}$$

and

$$\left\{f^{(e)}\right\} = \begin{Bmatrix} \dfrac{\bar{S}L_e}{2} \\ \dfrac{\bar{S}L_e}{2} \end{Bmatrix} \tag{11.48b}$$

Equations 11.41, 11.44, and 11.48 are now applied successively to all elements in the mesh and assembled through Equation 11.29 and using the global connectivity system given in Table 11.1 to form the global system of equations given by Equations 11.22 and 11.23.

Example 11.2: One-dimensional Steady State Mass Transfer in Adsorbing Desiccant Material: Consider a one-dimensional steady state diffusion of moisture in an adsorbing desiccant felt with top surface maintained at constant moisture concentration of C_0, by subjecting it to a moist air stream. The desiccant felt is supported at the bottom by an aluminum plate. Assume a constant rate of moisture adsorption, m'''_{ad}, and determine the moisture concentration profile in the felt using variational-based finite element formulation.

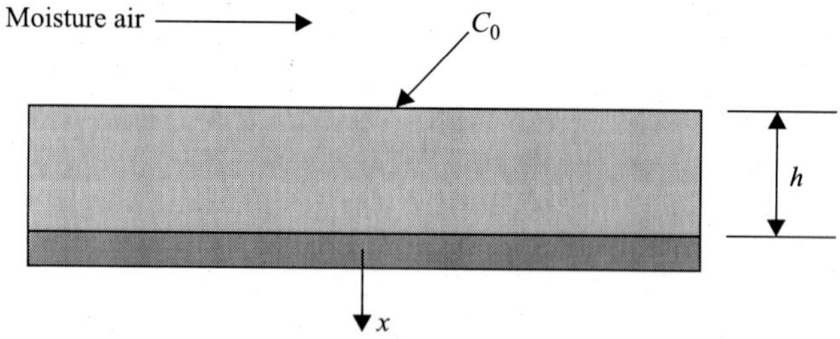

The mathematical statement of the problem is given as:

Governing equation

$$\frac{d}{dx}\left(D\frac{dC}{dx}\right) - m'''_{ad} = 0 \tag{E.11.2.1}$$

Boundary conditions

1. $x = 0, \quad C = C_0$ (E.11.2.2)

2. $x = h, \quad \dfrac{dC}{dx} = 0$ due to nonpermeable metal plate at the bottom (E.11.2.3)

Solution: Let us discretize the solution domain into a mesh size distribution using a one-dimensional line element as shown in the figure.

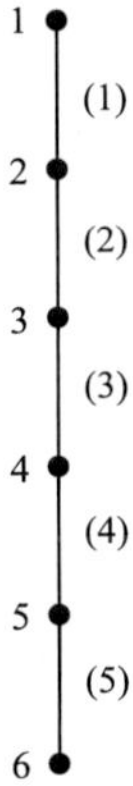

Application of the procedure outlined in Section 11.2 leads to the element characteristic equations as

$$\left[\kappa_D^{(e)}\right]\{C^{(e)}\} = \{I^{(e)}\} + \{f^{(e)}\}$$

where the element characteristic matrix and source vectors are given as

$$\left[\kappa_D^{(e)}\right] = \frac{D}{L_e}\begin{bmatrix} 1 & -1 \\ -1 & 1 \end{bmatrix}$$

$$\{I^{(e)}\} = \begin{Bmatrix} -D\left.\dfrac{dC}{dx}\right|_{x_i} \\[2ex] D\left.\dfrac{dC}{dx}\right|_{x_j} \end{Bmatrix}$$

$$\{f^{(e)}\} = \begin{Bmatrix} -\dfrac{m'''_{\text{ad}}L_e}{2} \\[2ex] -\dfrac{m'''_{\text{ad}}L_e}{2} \end{Bmatrix}$$

Assembling all element equations and applying inter-element continuity conditions, we get the final ***global system*** as

$$\frac{D}{L_e}\begin{bmatrix} 1 & -1 & 0 & 0 & 0 & 0 \\ -1 & 2 & -1 & 0 & 0 & 0 \\ 0 & -1 & 2 & -1 & 0 & 0 \\ 0 & 0 & -1 & 2 & -1 & 0 \\ 0 & 0 & 0 & -1 & 2 & -1 \\ 0 & 0 & 0 & 0 & -1 & 1 \end{bmatrix}\begin{Bmatrix} C_1 \\ C_2 \\ C_3 \\ C_4 \\ C_5 \\ C_6 \end{Bmatrix} = \begin{Bmatrix} -D\left.\dfrac{dC}{dx}\right|_{x=0} - \dfrac{m'''_{\text{ad}}L_e}{2} \\ -m'''_{\text{ad}}L_e \\ -m'''_{\text{ad}}L_e \\ -m'''_{\text{ad}}L_e \\ -m'''_{\text{ad}}L_e \\ D\left.\dfrac{dC}{dx}\right|_{x=h} - \dfrac{m'''_{\text{ad}}L_e}{2} \end{Bmatrix}$$

With the application of the boundary conditions, i.e., $C_1 = C_0$ and $dC/dx\big|_{x=h} = 0$, the global system reduces to

$$\frac{D}{L_e}\begin{bmatrix} L_e & -1 & 0 & 0 & 0 & 0 \\ 0 & 2 & -1 & 0 & 0 & 0 \\ 0 & -1 & 2 & -1 & 0 & 0 \\ 0 & 0 & -1 & 2 & -1 & 0 \\ 0 & 0 & 0 & -1 & 2 & -1 \\ 0 & 0 & 0 & 0 & -1 & 1 \end{bmatrix}\begin{Bmatrix} \left.\dfrac{dC}{dx}\right|_{x=0} \\ C_2 \\ C_3 \\ C_4 \\ C_5 \\ C_6 \end{Bmatrix} = \begin{Bmatrix} -\dfrac{DC_0}{L_e} - \dfrac{m'''_{\text{ad}}L_e}{2} \\ \dfrac{DC_0}{L_e} - m'''_{\text{ad}}L_e \\ -m'''_{\text{ad}}L_e \\ -m'''_{\text{ad}}L_e \\ -m'''_{\text{ad}}L_e \\ -\dfrac{m'''_{\text{ad}}L_e}{2} \end{Bmatrix}$$

11.3 BOUNDARY CONDITIONS

In the previous section, we have considered problems with boundary conditions of the first kind, i.e., of constant surface value. For such problems, we simply assign the given boundary values to the unknown variable at the boundary nodes and solve the global system for rest of the unknown nodal values along with the natural boundary conditions, such as the flux quantities at the boundary nodes.

11.3.1 Boundary Condition of the Second Kind or Constant Surface Flux

Let us consider the one-dimensional problem with constant surface flux boundary conditions as shown in Figure 11.4. The mathematical statement of the problem is as follows.

Governing equation

$$\frac{d}{dx}\left(\Gamma_x \frac{d\phi}{dx}\right) + \bar{S} = 0 \tag{11.49}$$

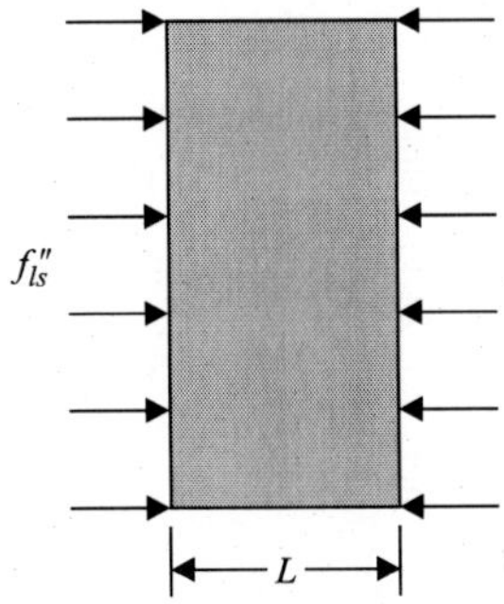

Figure 11.4 Constant surface flux boundary condition

Boundary conditions

1. $x = 0, \qquad -\Gamma_x \left.\dfrac{d\phi}{dx}\right|_{x=0} = f''_{ls}$ (11.50a)

2. $x = L, \qquad \Gamma_x \left.\dfrac{d\phi}{dx}\right|_{x=L} = f''_{rs}$ (11.50b)

The Galerkin-based finite element formulation will lead to the same element characteristic equation as in Section 11.1, i.e.

$$[k^{(e)}]\{\phi^{(e)}\} = \{I^{(e)}\} + \{f^{(e)}\} \tag{11.16}$$

where the inter-element vectors for Element 1 and Element 2 are modified based on using the boundary conditions given by Equations 11.50a and 11.50b as follows.

Element 1

$$\{I^{(1)}\} = \begin{Bmatrix} \left.\left(-\Gamma_x \dfrac{d\phi}{dx}\right)^{(1)}\right|_{x_i} \\ \left.\left(\Gamma_x \dfrac{d\phi}{dx}\right)^{(1)}\right|_{x_j} \end{Bmatrix} = \begin{Bmatrix} f''_{ls} \\ \left.\left(\Gamma_x \dfrac{d\phi}{dx}\right)^{(1)}\right|_{x_j} \end{Bmatrix} \tag{11.51}$$

and

Element n

$$\{I^{(n)}\} = \begin{Bmatrix} \left.\left(-\Gamma_x \dfrac{d\phi}{dx}\right)^{(n)}\right|_{x_i} \\ \left.\left(\Gamma_x \dfrac{d\phi}{dx}\right)^{(n)}\right|_{x_j} \end{Bmatrix} = \begin{Bmatrix} \left.\left(-\Gamma_x \dfrac{d\phi}{dx}\right)^{(n)}\right|_{x_j} \\ f''_{rs} \end{Bmatrix} \tag{11.52}$$

The expressions for the rest of the inter-element vectors, as well as the force vectors and the characteristic equations, will remain the same as that given in Equations 11.18b to Equation 11.18n−1. The assembly of all element equations will lead to the global system as equations will lead to the global system as

$$\begin{bmatrix} \frac{\Gamma_x}{L_e} & -\frac{\Gamma_x}{L_e} & & & & & \\ -\frac{\Gamma_x}{L_e} & \frac{2\Gamma_x}{L_e} & -\frac{\Gamma_x}{L_e} & & & & \\ & -\frac{\Gamma_x}{L_e} & \frac{2\Gamma_x}{L_e} & -\frac{\Gamma_x}{L_e} & & & \\ & & -\frac{\Gamma_x}{L_e} & \frac{2\Gamma_x}{L_e} & & & \\ & & & \vdots & & & \\ & & & & & \frac{2\Gamma_x}{L_e} & -\frac{\Gamma_x}{L_e} \\ & & & & & -\frac{\Gamma_x}{L_e} & \frac{\Gamma_x}{L_e} \end{bmatrix} \begin{Bmatrix} \phi_1 \\ \phi_2 \\ \phi_3 \\ \vdots \\ \vdots \\ \phi_{n-1} \\ \phi_n \end{Bmatrix} = \begin{Bmatrix} f''_{ls} \\ 0 \\ 0 \\ \vdots \\ \vdots \\ 0 \\ f'' \end{Bmatrix} + \begin{Bmatrix} \frac{\bar{S}L_e}{2} \\ \bar{S}L_e \\ \bar{S}L_e \\ \vdots \\ \vdots \\ \bar{S}L_e \\ \frac{\bar{S}L_e}{2} \end{Bmatrix} \tag{11.53}$$

The global system can be written in a compact form as

$$[K]\{\Phi\} = \{F\} \tag{11.54}$$

where

$$[K] = \begin{bmatrix} \frac{\Gamma_x}{L_e} & -\frac{\Gamma_x}{L_e} & & & & & \\ -\frac{\Gamma_x}{L_e} & \frac{2\Gamma_x}{L_e} & -\frac{\Gamma_x}{L_e} & & & & \\ & -\frac{\Gamma_x}{L_e} & \frac{2\Gamma_x}{L_e} & -\frac{\Gamma_x}{L_e} & & & \\ & & -\frac{\Gamma_x}{L_e} & \frac{2\Gamma_x}{L_e} & & & \\ & & & \vdots & & & \\ & & & & & \frac{2\Gamma_x}{L_e} & -\frac{\Gamma_x}{L_e} \\ & & & & & -\frac{\Gamma_x}{L_e} & \frac{\Gamma_x}{L_e} \end{bmatrix} = \text{global characteristics matrix} \tag{11.55a}$$

$$\{\Phi\} = \begin{Bmatrix} \phi_1 \\ \phi_2 \\ \phi_3 \\ \vdots \\ \vdots \\ \phi_n \\ \phi_{n+1} \end{Bmatrix} = \text{vector for unknown nodal values in the solution domain} \tag{11.55b}$$

$$\{F\} = \begin{Bmatrix} f''_{ls} \\ \\ \\ \\ \\ \\ f''_{rs} \end{Bmatrix} + \begin{Bmatrix} \frac{\bar{S}L_e}{2} \\ \bar{S}L_e \\ \bar{S}L_e \\ \vdots \\ \vdots \\ \bar{S}L_e \\ \frac{\bar{S}L_e}{2} \end{Bmatrix} = \begin{Bmatrix} f''_{ls} + \frac{\bar{S}L_e}{2} \\ \bar{S}L_e \\ \bar{S}L_e \\ \vdots \\ \vdots \\ \bar{S}L_e \\ f''_{rs} + \frac{\bar{S}L_e}{2} \end{Bmatrix} = \text{Source or force vector} \tag{11.55c}$$

Special case For the case of a zero surface flux condition or at a symmetric boundary as shown on the right surface of the plane slab in Figure 11.5, the mathematical statement of the problem is given as

Governing equation

$$\frac{d}{dx}\left(\Gamma_x \frac{d\phi}{dx}\right) + \bar{S} = 0 \tag{11.56}$$

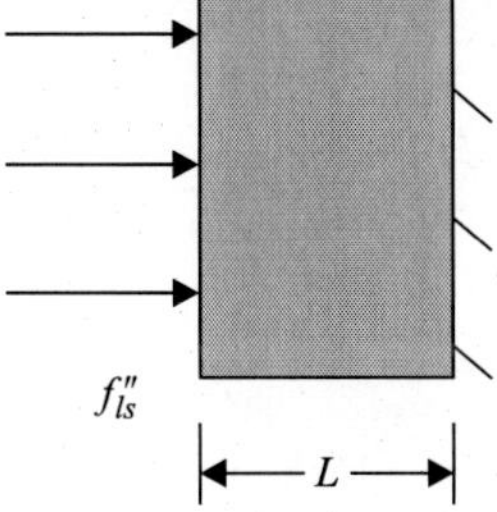

Figure 11.5 Symmetric boundary condition

Boundary conditions

1. $x = 0,\qquad -\Gamma_x \left.\frac{d\phi}{dx}\right|_{x=0} = f''_{ls}$ (11.57a)

2. $x = L,\qquad \left.\frac{d\phi}{dx}\right|_{x=L} = 0$ (11.57b)

The global system for this problem is the same as that given by Equations 11.22 and 11.23a, except for the force vector, which is modified as

$$\{F\} = \begin{Bmatrix} f''_{ls} + \frac{\bar{S}L_e}{2} \\ \bar{S}L_e \\ \bar{S}L_e \\ \vdots \\ \vdots \\ \bar{S}L_e \\ \bar{S}L_e \\ \frac{\bar{S}L_e}{2} \end{Bmatrix} \tag{11.58}$$

Example 11.3: Motion of Falling Liquid Film down an Inclined Surface: Consider the flow of a water film in steady and laminar motion, falling down an inclined surface with slope $\theta = 45^\circ$ from the horizontal. The thickness, d, of the film is constant along the length. Assume the flow as fully-developed and at zero-pressure gradient, i.e., $dP/dx = 0$.

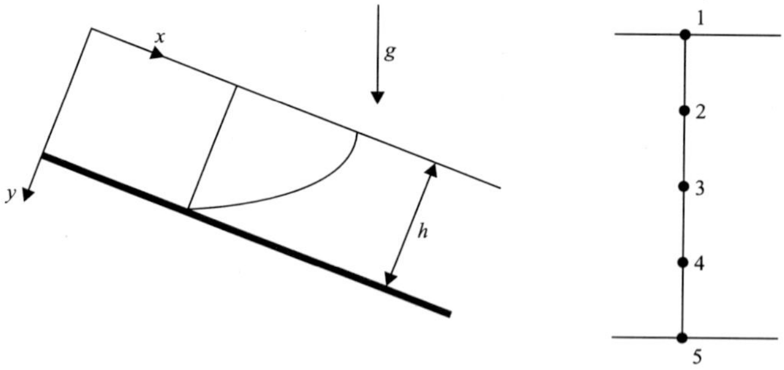

The mathematical statement of the problem can be derived from mass and momentum equations of fluid flow and it is given as follows.

Governing equation

$$\frac{d}{dx}\left(\mu \frac{du}{dx}\right) + \rho g \cos\theta = 0 \tag{E.11.3.1}$$

Boundary conditions

1. $y = 0, \quad \dfrac{du}{dx} = 0,$ zero shear stress at the free surface of the film (E.11.3.2)

2. $y = h, \quad u = 0,$ no-slip condition at the plate (E.11.3.3)

Use the following data for the calculation, $h = 10.0\,\text{mm}$, $\rho = 1000\,\text{kg/m}^3$, $\mu = 0.0005\,\text{kg/m s}$, $g = 9.8\,\text{m/s}^2$, and determine (a) the velocity distribution in the film, (b) the shear stress at the plate, and (c) validate the results by comparing the results with that given by the exact solution

$$u = \frac{\rho g}{\mu}\cos\theta\left(D(h-y) - \frac{(h-y)^2}{2}\right) \tag{E.11.3.4}$$

Solution: In the first step, let us discretize the solution domain using four uniform line elements as shown in the figure. With this mesh size distribution, we have $L_e = h/N_x = 0.01\,\text{m}/4 = 0.0025\,\text{mm}$ and the number of nodal points is $N = 5$. The nodal points $i = 1$ and $i = 6$ are located at $y = 0$ and $y = 0.01\,\text{m}$, respectively. Following Galerkin's finite element formulation and using two-point line element, we obtain the element characteristics equation, as in Equation 11.21, i.e.

$$\frac{\mu}{L_e}\begin{bmatrix} 1 & -1 & 0 & 0 & 0 \\ -1 & 2 & -1 & 0 & 0 \\ 0 & -1 & 2 & -1 & 0 \\ 0 & 0 & -1 & 2 & -1 \\ 0 & 0 & 0 & -1 & 1 \end{bmatrix}\begin{Bmatrix} u_1 \\ u_2 \\ u_3 \\ u_4 \\ u_5 \end{Bmatrix} = \begin{Bmatrix} -\mu\dfrac{du}{dy}\Big|_{y=0} + \dfrac{\rho g\cos\theta L_e}{2} \\ \rho g\cos\theta L_e \\ \rho g\cos\theta L_e \\ \rho g\cos\theta L_e \\ -\mu\dfrac{du}{dy}\Big|_{y=0} + \rho g\cos\theta\dfrac{L_e}{2} \end{Bmatrix} \tag{E.11.3.5}$$

Applying boundary conditions given by Equations E.11.3.2 and E.11.3.3, and rearranging, we get

$$\frac{\mu}{L_e}\begin{bmatrix} 1 & -1 & 0 & 0 & 0 \\ -1 & 2 & -1 & 0 & 0 \\ 0 & -1 & 2 & -1 & 0 \\ 0 & 0 & -1 & 2 & 0 \\ 0 & 0 & 0 & -1 & L_e/\mu \end{bmatrix}\begin{Bmatrix} u_1 \\ u_2 \\ u_3 \\ u_4 \\ \mu\dfrac{du}{dy}\Big|_{y=h} \end{Bmatrix} = \begin{Bmatrix} \dfrac{\rho g\cos\theta L_e}{2} \\ \rho g\cos\theta L_e \\ \rho g\cos\theta L_e \\ \rho g\cos\theta L_e \\ \dfrac{\rho g\cos\theta L_e}{2} \end{Bmatrix} \tag{E.11.3.6}$$

Substituting $\mu/L_e = 0.2$ and $\rho g \cos\theta L_e = 1000 \times 9.8 \times \cos 45 \times 0.0025 = 17.3242$, the system of equations becomes

$$\begin{bmatrix} 0.2 & -0.2 & 0 & 0 & 0 \\ -0.2 & 0.4 & -0.2 & 0 & 0 \\ 0 & -0.2 & 0.4 & -0.2 & 0 \\ 0 & 0 & -0.2 & 0.4 & 0 \\ 0 & 0 & 0 & -0.2 & 1 \end{bmatrix} \begin{Bmatrix} u_1 \\ u_2 \\ u_3 \\ u_4 \\ \mu \left.\dfrac{du}{dy}\right|_{y=h} \end{Bmatrix} = \begin{Bmatrix} 8.6621 \\ 17.3242 \\ 17.3242 \\ 17.3242 \\ 8.6621 \end{Bmatrix} \qquad \text{(E.11.3.7)}$$

The solution to the system of equations is given as

$$\begin{Bmatrix} u_1 \\ u_2 \\ u_3 \\ u_4 \\ \mu \left.\dfrac{du}{dy}\right|_{y=h} \end{Bmatrix} = \begin{Bmatrix} 692.9680 \\ 649.6575 \\ 519.1725 \\ 303.1735 \\ -69.2968 \end{Bmatrix}$$

The wall shear stress at the wall is

$$\tau_w = \mu \left.\frac{du}{dy}\right|_{y=h} = 69.2968\ \text{N/m}^2$$

Comparison of the finite element solution with the analytical solution is given in the table.

Nodal point	Analytical solution (m/s)	Finite element solution (m/s)
u_1	692.9646	692.9680
u_2	649.6543	549.6575
u_3	519.723	519.7260
u_4	303.1720	303.1725
u_5	0	0

11.3.2 Mixed Boundary Conditions

Let us consider the same one-dimensional problem with the boundary condition of the third kind or the mixed boundary condition at the left and right boundaries of the plane slab as shown in Figure 11.6. The mathematical statement of the problem is as follows.

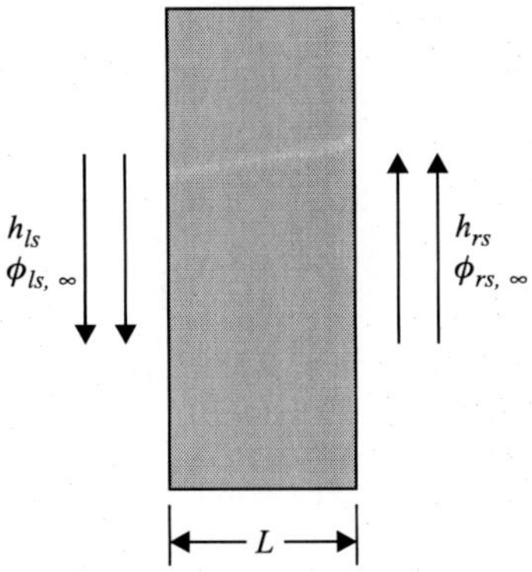

Figure 11.6 Convective boundary conditions

Governing equation

$$\frac{d}{dx}\left(\Gamma_x \frac{d\phi}{dx}\right) + \bar{S} = 0 \tag{11.59}$$

Boundary conditions

1. $x = 0, \quad \Gamma_x \left.\frac{d\phi}{dx}\right|_{x=0} = h_{ls}(\phi - \phi_{ls,\infty})$ (11.60a)

2. $x = L, \quad -\Gamma_x \left.\frac{d\phi}{dx}\right|_{x=L} = h_{rs}(\phi - \phi_{rs,\infty})$ (11.60b)

The ***Galerkin-based finite element formulation*** will lead to the same element characteristic equation as in Section 11.1, i.e.

$$\left[\kappa_\Gamma^{(e)}\right]\{\phi^{(e)}\} = -\{I^{(e)}\} + \{f^{(e)}\} \tag{11.61}$$

where the inter-element vectors for Element 1 and Element n need to be modified to implement the mixed boundary conditions. In order to do that, these inter-element vectors are split into two vectors as follows.

Element 1

$$\{I^{(1)}\} = \left\{ \begin{matrix} \left.\left(\Gamma_x \frac{d\phi}{dx}\right)^{(1)}\right|_{x_i} \\ \left.\left(-\Gamma_x \frac{d\phi}{dx}\right)^{(1)}\right|_{x_j} \end{matrix} \right\} = \left\{ \begin{matrix} \left.\left(\Gamma_x \frac{d\phi}{dx}\right)^{(1)}\right|_{x_i} \\ 0 \end{matrix} \right\} + \left\{ \begin{matrix} 0 \\ \left.\left(-\Gamma_x \frac{d\phi}{dx}\right)^{(1)}\right|_{x_j} \end{matrix} \right\} \tag{11.62a}$$

or

$$\{I^{(1)}\} = \left\{I^{(1)}_{b,ls}\right\} + \left\{I^{(1)}_{j}\right\} \tag{11.62b}$$

where $\{I^{(1)}_{b,ls}\}$ and $\{I^{(1)}_{j}\}$ are associated with the boundary conditions at node i located on the left surface and inter-element requirement at node j, respectively.

Element n

$$\{I^{(n)}\} = \left\{ \begin{array}{c} \left(\Gamma_x \dfrac{d\phi}{dx}\right)^{(n)}\Bigg|_{x_i} \\ \left(-\Gamma_x \dfrac{d\phi}{dx}\right)^{(n)}\Bigg|_{x_j} \end{array} \right\} = \left\{ \begin{array}{c} \left(\Gamma_x \dfrac{d\phi}{dx}\right)^{(n)}\Bigg|_{x_i} \\ 0 \end{array} \right\} + \left\{ \begin{array}{c} 0 \\ \left(-\Gamma_x \dfrac{d\phi}{dx}\right)^{(n)}\Bigg|_{x_j} \end{array} \right\} \tag{11.63a}$$

or

$$\{I^{(n)}\} = \left\{I^{(n)}_{i}\right\} + \left\{I^{(n)}_{b,rs}\right\} \tag{11.63b}$$

where $\{I^{(n)}_{i}\}$ and $\{I^{(n)}_{b,rs}\}$ are the components of inter-element vector associated with the inter-element requirement at node i and the boundary conditions at node j located on the right surface, respectively.

In the next step, we implement the mixed boundary conditions (Equation 11.60) directly into the inter-element vectors for boundary nodes as follows.

Element 1

$$\left\{I^{(1)}_{b,ls}\right\} = \left\{ \begin{array}{c} h_{ls}(\phi_i - \phi_{ls,\infty}) \\ 0 \end{array} \right\} = \left\{ \begin{array}{c} h_{ls}\phi_i \\ 0 \end{array} \right\} - \left\{ \begin{array}{c} h_{ls}\phi_{ls,\infty}) \\ 0 \end{array} \right\} \tag{11.64a}$$

or

$$\left\{I^{(1)}_{b,ls}\right\} = \begin{bmatrix} h_{ls} & 0 \\ 0 & 0 \end{bmatrix} \left\{ \begin{array}{c} \phi_i \\ \phi_j \end{array} \right\} - \left\{ \begin{array}{c} h_{ls}\phi_{ls,\infty}) \\ 0 \end{array} \right\} \tag{11.64b}$$

or

$$\left\{I^{(1)}_{b,ls}\right\} = \left[K^{(1)}_{b,ls}\right]\{\phi^{(1)}\} - \left\{f^{(1)}_{b,ls}\right\} \tag{11.64c}$$

where

$$\left[\kappa_{b,ls}^{(1)}\right] = \begin{bmatrix} h_{ls} & 0 \\ 0 & 0 \end{bmatrix} = \text{component of element characteristics matrix due to the mixed boundary condition at the left boundary surface} \tag{11.65a}$$

$$\left\{f_{b,ls}^{n)}\right\} = \begin{Bmatrix} h_{ls}\phi_{ls,\infty} \\ 0 \end{Bmatrix} = \text{component of element source vector due to the mixed boundary conditions at the left boundary surface} \tag{11.65b}$$

Similarly for

Element n

$$\left\{I_{b,rs}^{(n)}\right\} = \begin{Bmatrix} 0 \\ h_{rs}(\phi_j - \phi_{rs,\infty}) \end{Bmatrix} = \begin{Bmatrix} 0 \\ h_{rs}\phi_i \end{Bmatrix} - \begin{Bmatrix} 0 \\ h_{rs}\phi_{rs,\infty} \end{Bmatrix} \tag{11.66a}$$

or

$$\left\{I_{b,rs}^{(n)}\right\} = \begin{bmatrix} 0 & 0 \\ 0 & h_{rs} \end{bmatrix} \begin{Bmatrix} \phi_i \\ \phi_j \end{Bmatrix} - \begin{Bmatrix} 0 \\ h_{rs}\phi_{rs,\infty} \end{Bmatrix} \tag{11.66b}$$

or

$$\left\{I_{b,rs}^{(n)}\right\} = \left[\kappa_{b,rs}^{(n)}\right]\left\{\phi^{(n)}\right\} - \left\{f_{b,rs}^{(n)}\right\} \tag{11.66c}$$

where

$$\left[\kappa_{b,rs}^{(n)}\right] = \begin{bmatrix} 0 & 0 \\ 0 & h_{rs} \end{bmatrix} = \text{component of element characteristics matrix due to the mixed boundary condition at the right boundary surface} \tag{11.67a}$$

$$\left\{f_{b,rs}^{n)}\right\} = \begin{Bmatrix} 0 \\ h_{rs}\phi_{rs,\infty} \end{Bmatrix} = \text{component of element source vector due to the mixed boundary condition at the right boundary surface} \tag{11.67b}$$

We now substitute Equations 11.64 to 11.67 into Equations 11.62 and 11.63 respectively to obtain the inter-element vectors as follows.

Element 1

$$\left\{I^{(1)}\right\} = \left[\kappa_{b,ls}^{(1)}\right]\left\{\phi^{(1)}\right\} - \left\{f_{b,ls}^{(1)}\right\} + \left\{I_j^{(1)}\right\} \tag{11.68a}$$

and

Element n

$$\left\{I^{(n)}\right\} = \left\{I_i^{(n)}\right\} + \left[\kappa_{b,rs}^{(n)}\right]\left\{\phi^{(n)}\right\} - \left\{f_{b,rs}^{(n)}\right\} \tag{11.68b}$$

Finally, we substitute Equation 11.68 into Equation 11.61 to obtain the ***element characteristic equations*** for ***element 1 and n*** located at the boundaries as follows.

Element 1

$$\left(\left[\kappa_\Gamma^{(1)}\right] + \left[\kappa_{b,ls}^{(1)}\right]\right)\left\{\phi^{(1)}\right\} = -\left\{I_j^{(1)}\right\} + \left\{f_s^{(1)}\right\} + \left\{f_{b,ls}^{(1)}\right\} \tag{11.69a}$$

and

Element n

$$\left(\left[\kappa_\Gamma^{(n)}\right] + \left[\kappa_{b,rs}^{(n)}\right]\right)\left\{\phi^{(n)}\right\} = -\left\{I_i^{(n)}\right\} + \left\{f_s^{(n)}\right\} + \left\{f_{b,rs}^{(n)}\right\} \tag{11.69b}$$

It can be noticed that all elements which have a boundary node will have additional components for the element characteristic matrix and the source vector. The characteristic equations for rest of the elements will remain the same as that given by Equations 11.18b to 11.18n.

The assembly of all element equations will lead to the global system as

$$\begin{bmatrix} \frac{\Gamma_x}{L_e} + h_{ls} & -\frac{\Gamma_x}{L_e} & & & & \\ -\frac{\Gamma_x}{L_e} & \frac{\Gamma_x}{L_e} + \frac{\Gamma_x}{L_e} & -\frac{\Gamma_x}{L_e} & & & \\ & -\frac{\Gamma_x}{L_e} & \frac{\Gamma_x}{L_e} + \frac{\Gamma_x}{L_e} & -\frac{\Gamma_x}{L_e} & & \\ & & -\frac{\Gamma_x}{L_e} & \frac{\Gamma_x}{L_e} + \frac{\Gamma_x}{L_e} & & \\ & & & \vdots & & \\ & & & & \frac{\Gamma_x}{L_e} + \frac{\Gamma_x}{L_e} & -\frac{\Gamma_x}{L_e} \\ & & & & -\frac{\Gamma_x}{L_e} & \frac{\Gamma_x}{L_e} + h_{rs} \end{bmatrix} \begin{Bmatrix} \phi_1 \\ \phi_2 \\ \phi_3 \\ \vdots \\ \vdots \\ \phi_{n-1} \\ \phi_n \end{Bmatrix}$$

$$
= \begin{Bmatrix} 0 \\ \Gamma_x \left.\frac{d\phi}{dx}\right|_{x_2} - \Gamma_x \left.\frac{d\phi}{dx}\right|_{x_2} \\ \Gamma_x \left.\frac{d\phi}{dx}\right|_{x_3} - \Gamma_x \left.\frac{d\phi}{dx}\right|_{x_3} \\ \vdots \\ \vdots \\ \Gamma_x \left.\frac{d\phi}{dx}\right|_{x_n} - \Gamma_x \left.\frac{d\phi}{dx}\right|_{x_n} \\ 0 \end{Bmatrix} + \begin{Bmatrix} \frac{\bar{S}L_e}{2} \\ \frac{\bar{S}L_e}{2} + \frac{\bar{S}L_e}{2} \\ \frac{\bar{S}L_e}{2} + \frac{\bar{S}L_e}{2} \\ \vdots \\ \vdots \\ \frac{\bar{S}L_e}{2} + \frac{\bar{S}L_e}{2} \\ \frac{\bar{S}L_e}{2} \end{Bmatrix} + \begin{Bmatrix} h_{ls}\phi_{b,\,ls} \\ 0 \\ 0 \\ \vdots \\ \vdots \\ 0 \\ h_{rs}\phi_{b,\,rs} \end{Bmatrix} \tag{11.70}
$$

Simplifying and canceling terms due to inter-element continuity conditions, we get the final ***global system*** as

$$
\begin{bmatrix} \frac{\Gamma_x}{L_e} + h_{ls} & -\frac{\Gamma_x}{L_e} & & & & \\ -\frac{\Gamma_x}{L_e} & \frac{2\Gamma_x}{L_e} & -\frac{\Gamma_x}{L_e} & & & \\ & -\frac{\Gamma_x}{L_e} & \frac{2\Gamma_x}{L_e} & -\frac{\Gamma_x}{L_e} & & \\ & & -\frac{\Gamma_x}{L_e} & \frac{2\Gamma_x}{L_e} & & \\ \vdots & & & & & \\ & & & & \frac{2\Gamma_x}{L_e} & -\frac{\Gamma_x}{L_e} \\ & & & & -\frac{\Gamma_x}{L_e} & \frac{\Gamma_x}{L_e} + h_{rs} \end{bmatrix} \begin{Bmatrix} \phi_1 \\ \phi_2 \\ \phi_3 \\ \vdots \\ \vdots \\ \phi_{n-1} \\ \phi_n \end{Bmatrix} = \begin{Bmatrix} \frac{\bar{S}L_e}{2} + h_{ls}\phi_{b,\,ls} \\ \bar{S}L_e \\ \bar{S}L_e \\ \vdots \\ \vdots \\ \bar{S}L_e \\ \frac{\bar{S}L_e}{2} + h_{rs}\phi_{b,\,rs} \end{Bmatrix} \tag{11.71}
$$

The global system can be written in a compact form as

$$
[K]\{\Phi\} = \{F\} \tag{11.72}
$$

where

$$[K] = \begin{bmatrix} \frac{\Gamma_x}{L_e} + h_{ls} & -\frac{\Gamma_x}{L_e} & & & & & \\ -\frac{\Gamma_x}{L_e} & \frac{2\Gamma_x}{L_e} & -\frac{\Gamma_x}{L_e} & & & & \\ & -\frac{\Gamma_x}{L_e} & \frac{2\Gamma_x}{L_e} & -\frac{\Gamma_x}{L_e} & & & \\ & & -\frac{\Gamma_x}{L_e} & \frac{2\Gamma_x}{L_e} & & & \\ & & & \vdots & & & \\ & & & & & \frac{2\Gamma_x}{L_e} & -\frac{\Gamma_x}{L_e} \\ & & & & & -\frac{\Gamma_x}{L_e} & \frac{\Gamma_x}{L_e} + h_{rs} \end{bmatrix}$$

$$= \text{global characteristics matrix} \tag{11.73}$$

$$\{\Phi\} = \begin{Bmatrix} \phi_1 \\ \phi_2 \\ \phi_3 \\ \vdots \\ \vdots \\ \phi_n \\ \phi_{n+1} \end{Bmatrix} = \text{vector for unknown nodal values in the solution domain} \tag{11.74}$$

$$\{F\} = \begin{Bmatrix} \frac{\bar{S}L_e}{2} + h_{ls}\phi_{b,ls} \\ \bar{S}L_e \\ \bar{S}L_e \\ \vdots \\ \vdots \\ \bar{S}L_e \\ \frac{\bar{S}L_e}{2} + h_{rs}\phi_{b,rs} \end{Bmatrix} = \text{source or force vector} \tag{11.75}$$

The contribution of $[\kappa_b^{(e)}]$ to global characteristics matrix $[K]$ and $\{f_b^{(e)}\}$ to the global force vector occurs for boundary nodes with mixed boundary conditions or with nonzero convection coefficient, h.

Example 11.4: One-dimensional Steady State Conduction in Composite Wall: Consider a building wall made of four different homogeneous material layers. The thicknesses of these layers are $l_A = 15\,\text{cm}$, $l_B = 20\,\text{cm}$, $l_C = 10\,\text{cm}$, and $l_D = 5\,\text{cm}$. The thermal conductivities·of the corresponding layers are $k_A = 1.3\,\text{W/m}\,°\text{C}$, $k_B = 0.043\,\text{W/m}\,°\text{C}$, $k_C = 0.079\,\text{W/m°C}$ and $k_D = 0.48\,\text{W/m}\,°\text{C}$ respectively. The inside and outside convection heat transfer coefficients are $h_i = 10/\text{m}^2°\text{C}$ and $h_o = 40/\text{m}^2°\text{C}$. The inside temperature is $T_i = 25°\text{C}$, and the outside temperature is $T_o = 10°\text{C}$. Determine the temperatures at the internal and the external surfaces as well as temperatures at the interfaces of different materials using Galerkin-based finite element formulation and the mesh size distribution shown.

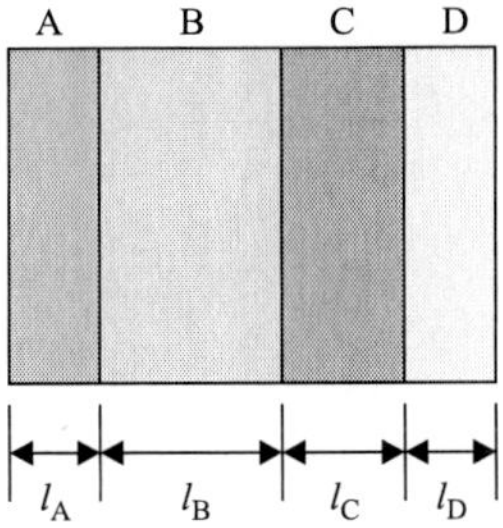

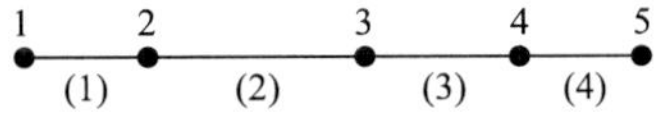

The mathematical statement of the problem is given as follows.

Governing equation

$$\frac{d}{dx}\left(kA\frac{dT_I}{dx}\right) = 0 \tag{E.11.4.1}$$

where subscript I is the index for different material layers.

Boundary conditions

1. $x = 0, \quad k\dfrac{dT}{dx} = h_{\text{in}}(T - T_{\text{in}})$ (E.11.4.2)

2. $x = L, \quad -k\dfrac{dT}{dx} = h_{\text{out}}(T - T_{\text{out}})$ (E.11.4.3)

Solution: For simplicity, let us discretize the solution domain with a non-uniform mesh size distribution with one element in each layer of material as shown in the figure. Based on the formulation presented in Section 11.3.2, we select the element characteristic equations as follows.

Element 1 Material A with convective boundary on left side. From Equations 11.61, 11.62, 11.64, 11.65, 11.68a, and 11.69a

$$\left(\left[\kappa_A^{(1)}\right] + \left[\kappa_{b,ls}^{(1)}\right]\right)\left\{T^{(1)}\right\} = -\left\{I_j^{(1)}\right\} + \left\{f_{b,ls}^{(1)}\right\}$$

where

$$\left[\kappa_{\mathrm{A}}^{(1)}\right]=\frac{k_{\mathrm{A}}}{l_{\mathrm{A}}}\begin{bmatrix}1 & -1\\ -1 & 1\end{bmatrix},\quad \left[\kappa_{b,ls}^{(1)}\right]=\begin{bmatrix}h_{\mathrm{In}} & 0\\ 0 & 0\end{bmatrix},\quad \left\{f_{b,ls}^{\,n)}\right\}=\left\{\begin{matrix}h_{\mathrm{in}}T_{\mathrm{in}}\\ 0\end{matrix}\right\}$$

$$\left\{I_j^{(1)}\right\}=\left\{\begin{matrix}0\\ \left(-k_{\mathrm{A}}\dfrac{dT}{dx}\right)^{(1)}\Bigg|_{x_j}\end{matrix}\right\},\quad \left\{T^{(1)}\right\}=\left\{\begin{matrix}T_1\\ T_2\end{matrix}\right\}$$

Substituting, we get the characteristic equation for Element 1 as

$$\begin{bmatrix}\dfrac{k_{\mathrm{A}}}{l_{\mathrm{A}}}+h_{\mathrm{in}} & -\dfrac{k_{\mathrm{A}}}{l_{\mathrm{A}}}\\ -\dfrac{k_{\mathrm{A}}}{l_{\mathrm{A}}} & \dfrac{k_{\mathrm{A}}}{l_{\mathrm{A}}}\end{bmatrix}\left\{\begin{matrix}T_1^{(1)}\\ T_2^{(1)}\end{matrix}\right\}=\left\{\begin{matrix}0\\ \left(k_{\mathrm{A}}\dfrac{dT}{dx}\right)^{(1)}\Bigg|_{x_j}\end{matrix}\right\}+\left\{\begin{matrix}h_{\mathrm{in}}T_{\mathrm{in}}\\ 0\end{matrix}\right\} \tag{E.11.4.4a}$$

We can use Eqations 11.16 and 11.17 for internal elements 2 and 3 as follows:

Element 2 Material B

$$\left[\kappa_{\mathrm{B}}^{(2)}\right]\left\{T^{(2)}\right\}=\left\{I^{(2)}\right\}$$

where

$$\left[\kappa_{\mathrm{B}}^{(2)}\right]=\frac{k_{\mathrm{B}}}{l_{\mathrm{B}}}\begin{bmatrix}1 & -1\\ -1 & 1\end{bmatrix},\quad \left\{I^{(2)}\right\}=\left\{\begin{matrix}-k_{\mathrm{B}}\dfrac{dT}{dx}\Bigg|_{x_i}\\ k_{\mathrm{B}}\dfrac{dT}{dx}\Bigg|_{x_j}\end{matrix}\right\}=\left\{\begin{matrix}I_i^{(2)}\\ I_j^{(2)}\end{matrix}\right\},\quad \left\{\phi^{(2)}\right\}=\left\{\begin{matrix}T_2\\ T_3\end{matrix}\right\}$$

Substituting, we get the characteristic equation for Element 2 as

$$\begin{bmatrix}\dfrac{k_{\mathrm{B}}}{l_{\mathrm{B}}} & -\dfrac{k_{\mathrm{B}}}{l_{\mathrm{B}}}\\ -\dfrac{k_{\mathrm{B}}}{l_{\mathrm{B}}} & \dfrac{k_{\mathrm{B}}}{l_{\mathrm{B}}}\end{bmatrix}\left\{\begin{matrix}T_2^{(2)}\\ T_3^{(2)}\end{matrix}\right\}=\left\{\begin{matrix}-k_{\mathrm{B}}\dfrac{dT}{dx}\Bigg|_{x_i}\\ k_{\mathrm{B}}\dfrac{dT}{dx}\Bigg|_{x_j}\end{matrix}\right\} \tag{E.11.4.4b}$$

Element 3 Material C

$$\left[\kappa_{\mathrm{C}}^{(3)}\right]\left\{T^{(3)}\right\}=\left\{I^{(3)}\right\}$$

where

$$\left[\kappa_{\mathrm{C}}^{(3)}\right]=\frac{k_{\mathrm{C}}}{l_{\mathrm{C}}}\begin{bmatrix}1 & -1\\ -1 & 1\end{bmatrix},\quad \left\{I^{(3)}\right\}=\left\{\begin{matrix}-k_{\mathrm{C}}\dfrac{dT}{dx}\Bigg|_{x_i}\\ k_{\mathrm{C}}\dfrac{dT}{dx}\Bigg|_{x_j}\end{matrix}\right\}=\left\{\begin{matrix}I_i^{(3)}\\ I_j^{(3)}\end{matrix}\right\},\quad \left\{\phi^{(3)}\right\}=\left\{\begin{matrix}T_3\\ T_4\end{matrix}\right\}$$

Substituting, we get the characteristic equation for Element 3 as

$$\begin{bmatrix} \dfrac{k_C}{l_C} & -\dfrac{k_C}{l_C} \\ -\dfrac{k_C}{l_C} & \dfrac{k_C}{l_C} \end{bmatrix} \begin{Bmatrix} T_3^{(3)} \\ T_4^{(3)} \end{Bmatrix} = \begin{Bmatrix} -k_C \left.\dfrac{dT}{dx}\right|_{x_i} \\ k_C \left.\dfrac{dT}{dx}\right|_{x_j} \end{Bmatrix} \qquad \text{(E.11.4.4c)}$$

Element 4 Material D with convective boundary on right side. From Equations 11.61, 11.63, 11.66, 11.67, 11.68b and 11.69b

$$\left(\left[\kappa_D^{(4)}\right] + \left[\kappa_{b,ls}^{(4)}\right]\right)\left\{T^{(4)}\right\} = -\left\{I_i^{(4)}\right\} + \left\{f_{b,rs}^{(4)}\right\}$$

where

$$\left[\kappa_A^{(4)}\right] = \frac{k_D}{l_D}\begin{bmatrix} 1 & -1 \\ -1 & 1 \end{bmatrix}, \quad \left[\kappa_{b,rs}^{(4)}\right] = \begin{bmatrix} 0 & 0 \\ 0 & h_{out} \end{bmatrix}, \quad \left\{f_{b,rs}^{(4)}\right\} = \begin{Bmatrix} 0 \\ h_{out}T_{out} \end{Bmatrix}$$

$$\left\{I_i^{(\$)}\right\} = \begin{Bmatrix} \left.\left(K_D \dfrac{dT}{dx}\right)^{(4)}\right|_{x_i} \\ 0 \end{Bmatrix}, \quad \left\{T^{(4)}\right\} = \begin{Bmatrix} T_4 \\ T_5 \end{Bmatrix}$$

Substituting, we get the characteristic equation for Element 4 as

$$\begin{bmatrix} \dfrac{k_D}{l_D} & -\dfrac{k_D}{l_D} \\ -\dfrac{k_D}{l_D} & \dfrac{k_D}{l_D} + h_{out} \end{bmatrix} \begin{Bmatrix} T_4^{(4)} \\ T_5^{(4)} \end{Bmatrix} = \begin{Bmatrix} -\left.\left(K_D \dfrac{dT}{dx}\right)^{(4)}\right|_{x_i} \\ 0 \end{Bmatrix} + \begin{Bmatrix} 0 \\ h_{out}T_{out} \end{Bmatrix}$$

(E.11.4.4d)

Assembly of all elements with the application of continuous heat flux condition at the interelement leads to the global system as

$$\begin{bmatrix} \dfrac{k_A}{l_A} + h_{in} & -\dfrac{k_A}{l_A} & & & \\ -\dfrac{k_A}{l_A} & \dfrac{k_A}{l_A} + \dfrac{k_B}{l_B} & -\dfrac{k_B}{l_B} & & \\ & -\dfrac{k_B}{l_B} & \dfrac{k_B}{l_B} + \dfrac{k_C}{l_C} & -\dfrac{k_C}{l_C} & \\ & & -\dfrac{k_C}{l_C} & \dfrac{k_C}{l_C} + \dfrac{k_D}{l_D} & -\dfrac{k_D}{l_D} \\ & & & -\dfrac{k_D}{l_D} & \dfrac{k_D}{l_D} + h_{out} \end{bmatrix} \begin{Bmatrix} T_1 \\ T_2 \\ T_3 \\ T_4 \\ T_5 \end{Bmatrix} = \begin{Bmatrix} h_{in}T_{in} \\ 0 \\ 0 \\ 0 \\ h_{out}T_{out} \end{Bmatrix}$$

(E.11.4.5)

The numerical parameters for each elements are:

Element A

$$\frac{k_A}{l_A} = \frac{1.3}{0.15} = 8.666\,\mathrm{W/m^2\,^\circ C}, \quad h_{in} = 10\,\mathrm{W/m^2\,^\circ C}, \quad h_{in}T_{in} = 10 \times 25 = 250\,\mathrm{W/m^2}$$

Element B

$$\frac{k_B}{l_B} = \frac{0.043}{0.2} = 0.215\,\mathrm{W/m^{2\circ}C}$$

Element C

$$\frac{k_C}{l_C} = \frac{0.079}{0.1} = 0.79\,\mathrm{W/m^{2\circ}C}$$

Element D

$$\frac{k_D}{l_D} = \frac{0.48}{0.05} = 9.6\,\mathrm{W/m^2\,^\circ C}, \quad h_{out} = 40\,\mathrm{W/m^2\,^\circ C}, \quad h_{out}T_{out} = 40 \times 10 = 400\,\mathrm{W/m^2}$$

Substituting all numerical parameters into Equation E.11.4.5, we get the global system as

$$\begin{bmatrix} 18.666 & -8.666 & & & \\ -8.666 & 8.881 & -0.215 & & \\ & -0.215 & 1.005 & -0.79 & \\ & & -0.79 & 10.39 & -9.6 \\ & & & -9.6 & 49.6 \end{bmatrix} \begin{Bmatrix} T_1 \\ T_2 \\ T_3 \\ T_4 \\ T_5 \end{Bmatrix} = \begin{Bmatrix} 250 \\ 0 \\ 0 \\ 0 \\ 400 \end{Bmatrix} \quad \text{(E.11.4.6)}$$

Solution to the global system is

$$\begin{Bmatrix} T_1 \\ T_2 \\ T_3 \\ T_4 \\ T_5 \end{Bmatrix} = \begin{Bmatrix} 24.76 \\ 24.48 \\ 13.34 \\ 10.3094 \\ 10.06 \end{Bmatrix} \quad \text{(E.11.4.7)}$$

11.4 VARIABLE SOURCE TERM

Let us now consider the case where the source term is assumed to be a linear function of the dependent variable as

$$S(\phi) = S_0 + S_1\phi \tag{11.76}$$

With this variable heat source, the general form of the governing equation can be written as

$$\frac{d}{dx}\left(\Gamma_x \frac{d\phi}{dx}\right) + S_1\phi + S_0 = 0 \tag{11.77}$$

Boundary conditions

3. $x = 0, \quad T(0) = T_L$ (11.78a)
4. $x = L \quad T(0) = T_R$ (11.78b)

Let us demonstrate the application of the finite element method to this problem using the Galerkin-based finite element method as outlined in Section 11.1. Based on this procedure, we get the weak form of the problem as

$$-\left(\Gamma_x \frac{d\phi^*}{dx}\right)\bigg|_{x_i} - \int_{x_i}^{x_j} \Gamma_x \frac{dN_i}{dx}\frac{d\phi^*}{dx}\,dx + \int_{x_i}^{x_j} S_0 N_i(x)\,dx + \int_{x_i}^{x_j} S_1 N_i(x)\phi^*\,dx = 0 \tag{11.79a}$$

and

$$\left(\Gamma_x \frac{d\phi^*}{dx}\right)\bigg|_{x_j} - \int_{x_i}^{x_j} \Gamma_x \frac{dN_j}{dx}\frac{d\phi^*}{dx}\,dx + \int_{x_i}^{x_j} S_0 N_j(x)\,dx + \int_{x_i}^{x_j} S_1 N_j(x)\phi^*\,dx = 0 \tag{11.79b}$$

In the next step the element characteristics equation is obtained by evaluating the integrals and by substituting the approximate solution for the chosen element type. The first three terms of Equation (11.79) are evaluated in Section 11.1, and leads to the element characteristic equations as

$$\left\{I^{(e)}\right\} + \left[\kappa_\Gamma^{(e)}\right]\left\{\phi^{(e)}\right\} - \left(f^{(e)}\right) = 0 \tag{11.80}$$

Combining Equation 11.79 and Equation 11.80, we get

$$\{I^{(e)}\} + \left[\kappa_\Gamma^{(e)}\right]\{\phi^{(e)}\} - (f^{(e)}) + \int_{x_i}^{x_j} S_1[N]^{\mathrm{T}}\phi^* \, dx \tag{11.81}$$

The fourth integral term will have additional contribution to the element equation. Let us evaluate the term as

$$\int_{x_i}^{x_j} S_1 N_i\phi^* \, dx = \int_{x_i}^{x_j} S_1 N_i(N_i\phi_i + N_j\phi_j)\, dx = \int_{x_i}^{x_j} S_1 N_i[N]\begin{Bmatrix} \phi_i \\ \phi j \end{Bmatrix} dx \tag{11.82a}$$

$$= \int_{x_i}^{x_j} S_1\left[N_i^2\phi_i + N_iN_j\phi_j\right] dx \tag{11.82b}$$

and

$$\int_{x_i}^{x_j} S_1 N_j\phi^* \, dx = \int_{x_i}^{x_j} S_1 N_j(N_i\phi_i + N_j\phi_j)\, dx = \int_{x_i}^{x_j} S_1 N_j[N]\begin{Bmatrix} \phi_i \\ \phi_j \end{Bmatrix} dx \tag{11.83a}$$

$$= \int_{x_i} S_1\left[N_jN_i\phi_i + N_j^2\right] dx \tag{11.83b}$$

Equations 11.82 and 11.83 can also be written in a matrix form as

$$\int_{x_i}^{x_j} S_1[N]^{\mathrm{T}}\phi^* \, dx = \left(\int_{x_i}^{x_j} S_1\begin{bmatrix} N_i^2 & N_iN_j \\ N_jN_i & N_j^2 \end{bmatrix}^{(e)} dx\right)\begin{Bmatrix} \phi_i \\ \phi_j \end{Bmatrix} \tag{11.84a}$$

or

$$\int_{x_i}^{x_j} S_1[N]^{\mathrm{T}}[N]\{\phi^{(e)}\}\, dx = \left[\kappa_S^{(e)}\right]\begin{Bmatrix} \phi_i \\ \phi_j \end{Bmatrix} \tag{11.84b}$$

where

$$\left[\kappa_S^{(e)}\right] = \int_{x_i}^{x_j} S_1\begin{bmatrix} N_i^2 & N_iN_j \\ N_jN_i & N_j^2 \end{bmatrix}^{(e)} dx \tag{11.85}$$

The integrals appearing as the elements of the matrix $[k_S^{(e)}]$ can be evaluated either by numerical ***quadrature formulas***, as discussed in Chapter 4, or analytically, for some cases, as discussed in *Chapter 10*. As we have demonstrated in Chapter 10, for two-point line elements these integrals can be evaluated analytically using the natural coordinate systems and expresses as

$$\int_{x_i}^{x_j} N_i^2\,dx = \frac{L_e}{3},\quad \int_{x_i}^{x_j} N_iN_j\,dx = \frac{L_e}{6},\quad \int_{x_i}^{x_j} N_j^2\,dx = \frac{L_e}{3},\quad \int_{x_i}^{x_j} N_jN_i\,dx = \frac{L_e}{6} \tag{11.86}$$

Substituting Equation 11.86 into Equations 11.82 and 11.83 we have

$$\int_{x_i}^{x_j} S_1 N_i \phi^*\,dx = \frac{S_1 L_e}{6}(2\phi_i + \phi_j) \tag{11.87a}$$

and

$$\int_{x_i}^{x_j} S_1 N_j \phi^*\,dx = \frac{S_1 L_e}{3}(\phi_i + 2\phi_j) \tag{11.87b}$$

or in matrix form as

$$\left(\int_{x_i}^{x_j} S_1[N]^{\mathrm{T}}[N]\,dx\right)\{\phi^{(e)}\} = \left[k_S^{(e)}\right]\{\phi^{(e)}\} = \frac{S_1 L_e}{6}\begin{bmatrix} 2 & 1 \\ 1 & 2 \end{bmatrix}^{(e)}\begin{Bmatrix} \phi_i \\ \phi_j \end{Bmatrix} \tag{11.87c}$$

Substituting Equation 11.87 into Equation 11.81, we obtain the element equation for the problem as

$$\left(\left[\kappa_\Gamma^{(e)}\right] + \left[\kappa_S^{(e)}\right]\right)\{\phi^{(e)}\} = -\{I^{(e)}\} + (f^{(e)}) \tag{11.88}$$

where

$$\left[\kappa_\Gamma^{(e)}\right] = \frac{\Gamma_x}{L_e}\begin{bmatrix} 1 & -1 \\ -1 & 1 \end{bmatrix} = \text{component of element characteristic matrix due to the diffusion term} \tag{11.89a}$$

$$\left[\kappa_S^{(e)}\right] = \frac{S_1 L_e}{6}\begin{bmatrix} 2 & 1 \\ 1 & 2 \end{bmatrix} = \text{component of element characteristic matrix due to the source term} \tag{11.89b}$$

$$\{f^{(e)}\} = \begin{Bmatrix} \dfrac{S_0 L_e}{2} \\ \dfrac{S_0 L_e}{2} \end{Bmatrix} = \begin{Bmatrix} f_i^{(e)} \\ f_j^{(e)} \end{Bmatrix} = \text{element source or force vector} \tag{11.89c}$$

$$\{I^{(e)}\} = \begin{Bmatrix} -\Gamma_x \dfrac{d\phi}{dx}\Big|_{x_i} \\ \Gamma_x \dfrac{d\phi}{dx}\Big|_{x_j} \end{Bmatrix} = \begin{Bmatrix} I_i^{(e)} \\ I_j^{(e)} \end{Bmatrix} \tag{11.89d}$$

Example 11.5: One-dimensional Steady State Conduction in a Circular Pin Fin: An aluminum pin fin having a diameter $D = 1$ cm and length $L = 6$ cm is exposed to the surrounding fluid with temperature $T_\infty = 20°$C and convection heat transfer coefficient, $h = 8.0\,\text{W/m}^2\,°\text{C}$. The fin base temperature is $T_0 = 100°$C. Determine the global system of equation for temperature distribution and fin heat loss. Assume thermal conductivity of the material as $k = 200\,\text{W/m}\,°\text{C}$.

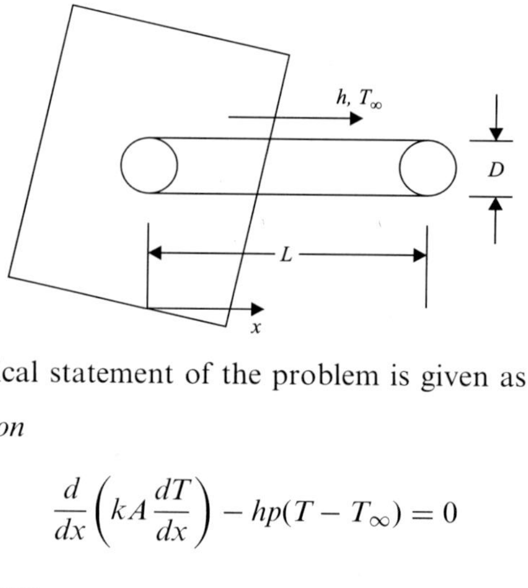

The mathematical statement of the problem is given as follows.

Governing equation

$$\frac{d}{dx}\left(kA\frac{dT}{dx}\right) - hp(T - T_\infty) = 0 \tag{E.11.6.1}$$

Boundary conditions

1. $x = 0, \quad T = T_0$ (E.11.6.2)
2. $x = L, \quad -k\dfrac{dT}{dx} = h(T - T_\infty)$ (E.11.6.3)

Solution: Let us rearrange the governing Equation E.11.6.1 as

$$\frac{d}{dx}\left(kA\frac{dT}{dx}\right) - hpT + hpT_\infty = 0$$

or

$$\frac{d}{dx}\left(kA\frac{dT}{dx}\right) + S_1T + S_0 = 0$$

where

$$S_0 = hpT_\infty = 8.0 \times \pi \times 0.01m \times 20 = 5.024\,\mathrm{W/m},$$

$$S_1 = -hp = -8.0 \times \pi \times 0.01 = -0.2512\,\mathrm{W/m\,^\circ C},$$

$$\Gamma_x = kA = 200 \times \frac{\pi \times (0.01)^2}{4} = 0.0157\,\mathrm{W\,m/^\circ C}, \quad L_e = 0.06\,m/4 = 0.015\,\mathrm{m}$$

$$\frac{kA}{L_e} = \frac{0.0157}{0.015} = 1.0466, \quad \frac{S_1 L_e}{3} = \frac{-0.2512 \times 0.015}{3} = -0.001\,256$$

$$\frac{S_1 L_e}{6} = \frac{-0.2512 \times 0.015}{6} = -0.000\,628, \quad \frac{S_0 L_e}{2} = \frac{5.024 \times 0.015}{2} = 0.037\,68$$

$$S_0 L_e = 5.024 \times 0.015 = 0.075\,36$$

Let us discretize the solution domain by four equal size line element as shown below

Following Equation 11.88, we can write the element characteristic matrix for elements 1 to 3 as

$$\left(\left[\kappa_\Gamma^{(e)}\right] + \left[\kappa_S^{(e)}\right]\right)\{T^{(e)}\} = -\{I^{(e)}\} + (f^{(e)})$$

$$\left(\frac{kA}{L_e}\begin{bmatrix} 1 & -1 \\ -1 & 1 \end{bmatrix} + \frac{S_1 L_e}{6}\begin{bmatrix} 2 & 1 \\ 1 & 2 \end{bmatrix}\right)\begin{Bmatrix} \phi_i \\ \phi_j \end{Bmatrix} = -\begin{Bmatrix} -k\left.\dfrac{d\phi}{dx}\right|_{x_i} \\ k\left.\dfrac{d\phi}{dx}\right|_{x_j} \end{Bmatrix} + \begin{Bmatrix} \dfrac{S_0 L_e}{2} \\ \dfrac{S_0 L_e}{2} \end{Bmatrix}$$

$$\begin{bmatrix} \dfrac{kA}{L_e} + \dfrac{S_1 L_e}{3} & -\dfrac{kA}{L_e} + \dfrac{S_1 L_e}{6} \\ -\dfrac{kA}{L_e} + \dfrac{S_1 L_e}{6} & \dfrac{kA}{L_e} + \dfrac{S_1 L_e}{3} \end{bmatrix}\begin{Bmatrix} \phi_i \\ \phi_j \end{Bmatrix} = -\begin{Bmatrix} -k\left.\dfrac{d\phi}{dx}\right|_{x_i} \\ k\left.\dfrac{d\phi}{dx}\right|_{x_j} \end{Bmatrix} + \begin{Bmatrix} \dfrac{S_0 L_e}{2} \\ \dfrac{S_0 L_e}{2} \end{Bmatrix}$$

Applying this equation successively to Elements 1 to 3, we get

Element 1

$$\begin{bmatrix} \dfrac{kA}{L_e} + \dfrac{S_1 L_e}{3} & -\dfrac{kA}{L_e} + \dfrac{S_1 L_e}{6} \\ -\dfrac{kA}{L_e} + \dfrac{S_1 L_e}{6} & \dfrac{kA}{L_e} + \dfrac{S_1 L_e}{3} \end{bmatrix}\begin{Bmatrix} T_1^{(1)} \\ T_2^{(1)} \end{Bmatrix} = -\begin{Bmatrix} -k\left.\dfrac{d\phi}{dx}\right|_{x_1} \\ k\left.\dfrac{d\phi}{dx}\right|_{x_2} \end{Bmatrix} + \begin{Bmatrix} \dfrac{S_0 L_e}{2} \\ \dfrac{S_0 L_e}{2} \end{Bmatrix}$$

Element 2

$$\begin{bmatrix} \dfrac{kA}{L_e}+\dfrac{S_1L_e}{3} & -\dfrac{kA}{L_e}+\dfrac{S_1L_e}{6} \\ -\dfrac{kA}{L_e}+\dfrac{S_1L_e}{6} & \dfrac{kA}{L_e}+\dfrac{S_1L_e}{3} \end{bmatrix} \begin{Bmatrix} T_2^{(2)} \\ T_3^{(2)} \end{Bmatrix} = -\begin{Bmatrix} -k\left.\dfrac{d\phi}{dx}\right|_{x_2} \\ k\left.\dfrac{d\phi}{dx}\right|_{x_3} \end{Bmatrix} + \begin{Bmatrix} \dfrac{S_0L_e}{2} \\ \dfrac{S_0L_e}{2} \end{Bmatrix}$$

Element 3

$$\begin{bmatrix} \dfrac{kA}{L_e}+\dfrac{S_1L_e}{3} & -\dfrac{kA}{L_e}+\dfrac{S_1L_e}{6} \\ -\dfrac{kA}{L_e}+\dfrac{S_1L_e}{6} & \dfrac{kA}{L_e}+\dfrac{S_1L_e}{3} \end{bmatrix} \begin{Bmatrix} T_3^{(3)} \\ T_4^{(3)} \end{Bmatrix} = -\begin{Bmatrix} -k\left.\dfrac{d\phi}{dx}\right|_{x_3} \\ k\left.\dfrac{d\phi}{dx}\right|_{x_4} \end{Bmatrix} + \begin{Bmatrix} \dfrac{S_0L_e}{2} \\ \dfrac{S_0L_e}{2} \end{Bmatrix}$$

For ***Element 4*** with convective boundary on the right side, the element characteristic equation is

$$\left(\left[\kappa_\Gamma^{(5)}\right]+\left[\kappa_S^{(5)}\right]+\left[\kappa_b^{(5)}\right]\right)\{T^{(5)}\} = -\left\{I_i^{(4)}\right\}+\left(f^{(e)}+f_{b,rs}^{(4)}\right)$$

$$\begin{bmatrix} \dfrac{kA}{L_e}+\dfrac{S_1L_e}{3} & -\dfrac{kA}{L_e}+\dfrac{S_1L_e}{6} \\ -\dfrac{kA}{L_e}+\dfrac{S_1L_e}{6} & \dfrac{kA}{L_e}+\dfrac{S_1L_e}{3}+h \end{bmatrix} \begin{Bmatrix} T_4^{(4)} \\ T_5^{(4)} \end{Bmatrix} = -\begin{Bmatrix} -k_x\left.\dfrac{d\phi}{dx}\right|_{x_4} \\ 0 \end{Bmatrix} + \begin{Bmatrix} \dfrac{S_0L_e}{2} \\ \dfrac{S_0L_e}{2}+hT_\infty \end{Bmatrix}$$

Assembly of all element equations leads to the global system as

$$\begin{bmatrix} \dfrac{kA}{L_e}+\dfrac{S_1L_e}{3} & -\dfrac{kA}{L_e}+\dfrac{S_1L_e}{6} & & & \\ -\dfrac{kA}{L_e}+\dfrac{S_1L_e}{6} & \dfrac{2kA}{L_e}+\dfrac{2S_1L_e}{3} & -\dfrac{kA}{L_e}+\dfrac{S_1L_e}{6} & & \\ & -\dfrac{kA}{L_e}+\dfrac{S_1L_e}{6} & \dfrac{2kA}{L_e}+\dfrac{2S_1L_e}{3} & -\dfrac{kA}{L_e}+\dfrac{S_1L_e}{6} & \\ & & -\dfrac{kA}{L_e}+\dfrac{S_1L_e}{6} & \dfrac{2kA}{L_e}+\dfrac{2S_1L_e}{3} & -\dfrac{kA}{L_e}+\dfrac{S_1L_e}{6} \\ & & & -\dfrac{kA}{L_e}+\dfrac{S_1L_e}{6} & \dfrac{2kA}{L_e}+\dfrac{2S_1L_e}{3}+h \end{bmatrix}$$

$$\times \begin{Bmatrix} T_1 \\ T_2 \\ T_3 \\ T_4 \\ T_5 \end{Bmatrix} = \begin{Bmatrix} k\left.\frac{\partial T}{\partial x}\right|_{x_1} + \frac{S_0 L_e}{2} \\ S_0 L_e \\ S_0 L_e \\ S_0 L_e \\ \frac{S_0 L_e}{2} + hT_\infty \end{Bmatrix}$$

With the application of the boundary condition at the base of the fin, i.e., $T_1 = T_0$, and rearranging, we get the global system as

$$\begin{bmatrix} k & -\frac{kA}{L_e} + \frac{S_1 L_e}{6} & & & \\ 0 & \frac{2kA}{L_e} + \frac{2S_1 L_e}{3} & -\frac{kA}{L_e} + \frac{S_1 L_e}{6} & & \\ & -\frac{kA}{L_e} + \frac{S_1 L_e}{6} & \frac{2kA}{L_e} + \frac{2S_1 L_e}{3} & -\frac{kA}{L_e} + \frac{S_1 L_e}{6} & \\ & & -\frac{kA}{L_e} + \frac{S_1 L_e}{6} & \frac{2kA}{L_e} + \frac{2S_1 L_e}{3} & -\frac{kA}{L_e} + \frac{S_1 L_e}{6} \\ & & & -\frac{kA}{L_e} + \frac{S_1 L_e}{6} & \frac{2kA}{L_e} + \frac{2S_1 L_e}{3} + h \end{bmatrix} \begin{Bmatrix} \left.\frac{dT}{dx}\right|_{x_1} \\ T_2 \\ T_3 \\ T_4 \\ T_5 \end{Bmatrix}$$

$$= \begin{Bmatrix} -\left(\frac{kA}{L_e} + \frac{S_1 L_e}{3}\right) T_0 + \frac{S_0 L_e}{2} \\ \left(\frac{kA}{L_e} - \frac{S_1 L_e}{3}\right) T_0 + S_0 L_e \\ S_0 L_e \\ S_0 L_e \\ \frac{S_0 L_e}{2} + hT_\infty \end{Bmatrix}$$

Substituting the numerical data, we get

$$\begin{bmatrix} 200 & -1.04728 & & & \\ 0 & 2.090688 & -1.04728 & & \\ & -1.04728 & 2.090688 & -1.04728 & \\ & & -1.04728 & 2.090688 & -1.04728 \\ & & & -1.04728 & 10.090688 \end{bmatrix} \begin{Bmatrix} \left.\frac{dT}{dx}\right|_{x_1} \\ T_2 \\ T_3 \\ T_4 \\ T_5 \end{Bmatrix} = \begin{Bmatrix} -104.4967 \\ 104.86096 \\ 0.07536 \\ 0.07536 \\ 160.0376 \end{Bmatrix}$$

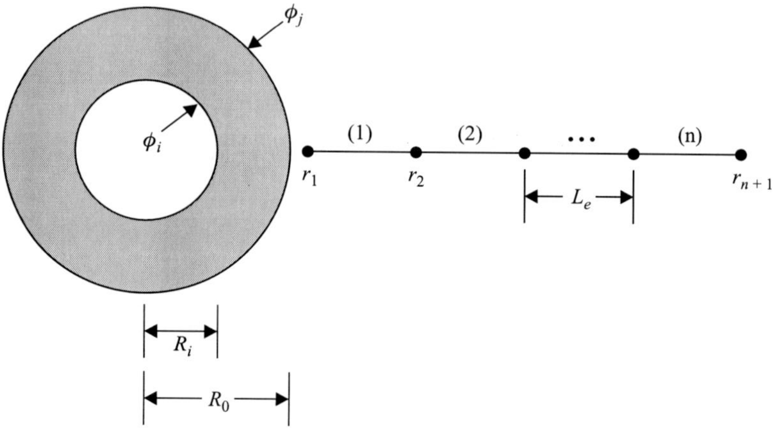

Figure 11.7 Axisymmetric problem

11.5 AXISYMMETRIC PROBLEMS

There are many problems in radial coordinate systems such as in pipes and spheres that have symmetry about the axis of rotation. This symmetry arises not only because of the geometry but also due to the boundary conditions around the circumference or the circular directions. In this section, we will consider formulation and application of the finite element method using a one-dimensional steady state equation for such an axisymmetric problem. Let us consider a one-dimensional steady state diffusion along with a volumetric generation process in a long circular hollow rod as shown. The boundary conditions are assumed to be of the first kind, i.e., constant surface values ϕ_i and ϕ_j shown in Figure 11.7.

The mathematical statement of this transport process is given as

$$\frac{1}{r}\frac{d}{dr}\left(\Gamma_r r\frac{d\phi}{dr}\right) + S = 0 \tag{11.90}$$

Boundary conditions

5. $r = R_i, \quad \phi = \phi_i$ (11.91a)

6. $r = R_0, \quad \phi = \phi_0$ (11.91b)

Let us demonstrate the application of finite element method to this problem through a detailed description of several steps that constitute the finite element method.

Using the ***Galerkin-based weighted residual method*** we obtain the integral statement of the problem as

$$\{R^{(e)}\} = \int_{r_i}^{r_j} [N]^{\mathrm{T}} \left(\frac{d}{dr} \left(\Gamma_r r \frac{d\phi^*}{dr} \right) + Sr \right) dr = 0 \tag{11.92}$$

Evaluating the first term of the equation by integration-by-parts and using the properties of the shape functions for a linear line element, i.e., $N_i(r_i) = 1$, $N_i(r_j) = 0$ and $N_j(r_i) = 0$, $N_j(r_j) = 1$, we can transform Equation 11.92 into

$$\begin{Bmatrix} -\Gamma_r r \left.\dfrac{d\phi}{dx}\right|_{x_i} \\ \Gamma_r r \left.\dfrac{d\phi}{dx}\right|_{x_j} \end{Bmatrix} - \int_{r_i}^{r_j} \Gamma_r r \frac{d[N]^{\mathrm{T}}}{dx} \frac{d\phi^*}{dx} dr + \int_{r_i}^{r_j} Sr[N]^{\mathrm{T}} \, dr = 0 \tag{11.93}$$

In the next step the element characteristics equation is obtained by evaluating the integrals by substituting the approximate solution for linear line element and following the procedure outlined in Section 11.2. The only difference is in the treatment of the term $\Gamma_r r$ in evaluating the second integral. This is assumed to be the product of a step-wise constant value for the transport property, Γ_r, in the media and a mean value for r as

$$\Gamma_r r = \Gamma_r \bar{r} \tag{11.94}$$

where mean value of r is given as

$$\bar{r} = \frac{r_i + r_j}{2} \tag{11.95}$$

Following this procedure, we can evaluate the first integrals term in Equation 11.93 as

$$\int_{r_i}^{r_j} \Gamma_r r \frac{d[N]^{\mathrm{T}}}{dr} \frac{d\phi^*}{dr} dr = \frac{\Gamma_r \bar{r}}{L_e} \begin{bmatrix} 1 & -1 \\ -1 & 1 \end{bmatrix} \begin{Bmatrix} \phi_i \\ \phi_j \end{Bmatrix} \tag{11.96}$$

For the second ***integral term***, it is more convenient to evaluate the integral using a local co-ordinate system. Let us use the shape function expression given by Equation 11.3 for the line element in a local coordinate system and substitute $r = r_i + r$ to change from the global coordinate system to the local coordinate system. Also, let us assume a step-wise constant value $S = \bar{S}$ for the source in the media for this demonstration, and evaluate the integrals as follows.

For $I=i$

$$\int_{r_i}^{r_j} SrN_i(r)\,dr = \int_0^{L_e} \bar{S}\left(1-\frac{r}{L_e}\right)(r_i+r)\,dr = \bar{S}\int_0^{L_e}\left(r_i\left(1-\frac{r}{L_e}\right)+r\left(1-\frac{r}{L_e}\right)\right)dr$$

$$= \bar{S}\left(r_i r\Big|_0^{L_e} - \frac{r_i}{L_e}\frac{r^2}{2}\Big|_0^{L_e} + \frac{r^2}{2}\Big|_0^{L_e} - \frac{r^3}{3L_e}\Big|_0^{L_e}\right)$$

or

$$\int_{r_i}^{r_j} SrN_i(r)\,dx = \frac{\bar{S}L_e}{6}(3r_i+L_e) \tag{11.97a}$$

for $I=j$,

$$\int_{r_i}^{r_j} SrN_j(r)\,dr = \int_0^{L_e} \bar{S}\frac{r}{L_e}(r_i+r)\,dr = \frac{\bar{S}}{L_e}\left(r_i\frac{r^2}{2}\Big|_0^{L_e} + \frac{r^3}{3}\Big|_0^{L_e}\right)$$

$$= \frac{\bar{S}}{L_e}\left(r_i\frac{L_e^2}{2}+\frac{L_e^3}{3}\right)$$

or

$$\int_{r_i}^{r_j} SrN_i(r)\,dx = \frac{\bar{S}L_e}{6}(3r_i+2L_e) \tag{11.97b}$$

Equations 12.99a and 12.99b can be written in vector form as

$$\int_{x_i}^{x_j} Sr[N]^{\mathrm{T}}dx = \frac{\bar{S}L_e}{6}\begin{Bmatrix} 3r_i+L_e \\ 3r_i+2L_e \end{Bmatrix} \tag{11.97c}$$

Substituting Equations 11.96 and 11.97 into Equation 11.93, we obtain the element characteristics equations as

$$\begin{Bmatrix} \Gamma_r r\dfrac{d\phi}{dr}\Big|_{r_i} \\ -\Gamma_r r\dfrac{d\phi}{dr}\Big|_{r_j} \end{Bmatrix} + \frac{\Gamma_r\bar{r}^{(e)}}{L_e}\begin{bmatrix} 1 & -1 \\ -1 & 1 \end{bmatrix}\begin{Bmatrix} \phi_i \\ \phi_j \end{Bmatrix} - \frac{\bar{S}L_e}{6}\begin{Bmatrix} 3r_i+L_e \\ 3r_i+2L_e \end{Bmatrix} = 0 \tag{11.98a}$$

or

$$\left[\kappa_\Gamma^{(e)}\right]\{\phi^{(e)}\} = \{I^{(e)}\} + \{f^{(e)}\} = 0 \tag{11.98b}$$

where

$$\{\phi^{(e)}\} = \begin{Bmatrix} \phi_i \\ \phi_j \end{Bmatrix} = \text{vector of element nodal values} \tag{11.99}$$

$$\left[\kappa_\Gamma^{(e)}\right] = \frac{\Gamma_r \bar{r}}{L_e}\begin{bmatrix} 1 & -1 \\ -1 & 1 \end{bmatrix} = \text{element characteristic or stiffness matrix} \tag{11.100}$$

$$\{f^{(e)}\} = \frac{\bar{S}L_e}{6}\begin{Bmatrix} 3r_i + L_e \\ 3r_i + 2L_e \end{Bmatrix} = \begin{Bmatrix} f_i^{(e)} \\ f_j^{(e)} \end{Bmatrix} = \text{element source or force vector} \tag{11.101}$$

$$\{I^{(e)}\} = \begin{Bmatrix} -\Gamma_r r \left.\dfrac{d\phi}{dr}\right|_{r_i} \\ \Gamma_r r \left.\dfrac{d\phi}{dr}\right|_{r_j} \end{Bmatrix} = \begin{Bmatrix} I_i^{(e)} \\ I_j^{(e)} \end{Bmatrix}$$

$$= \text{vector for element contribution to inter-element conditions} \tag{11.102}$$

We can now apply Equation 11.98 repeatedly to all elements to obtain corresponding element equations, which are then assembled to form a global system of equations as

$$\begin{bmatrix} \frac{\Gamma_r \bar{r}^{(1)}}{L_e} & -\frac{\Gamma_r \bar{r}^{(1)}}{L_e} & & & & & \\ -\frac{\Gamma_r \bar{r}^{(1)}}{L_e} & \frac{\Gamma_r \bar{r}^{(1)}}{L_e} + \frac{\Gamma_r \bar{r}^{(2)}}{L_e} & -\frac{\Gamma_r \bar{r}^{(2)}}{L_e} & & & & \\ & -\frac{\Gamma_r \bar{r}^{(2)}}{L_e} & \frac{\Gamma_r \bar{r}^{(2)}}{L_e} + \frac{\Gamma_r \bar{r}^{(3)}}{L_e} & -\frac{\Gamma_r \bar{r}^{(3)}}{L_e} & & & \\ & & -\frac{\Gamma_r \bar{r}^{(3)}}{L_e} & \frac{\Gamma_r \bar{r}^{(3)}}{L_e} + \frac{\Gamma_r \bar{r}^{(4)}}{L_e} & & & \\ & & & \vdots & & & \\ & & & & & \frac{\Gamma_r \bar{r}^{(n-1)}}{L_e} + \frac{\Gamma_r \bar{r}^{(n)}}{L_e} & -\frac{\Gamma_r \bar{r}^{(n)}}{L_e} \\ & & & & & -\frac{\Gamma_r \bar{r}^{(n)}}{L_e} & \frac{\Gamma_r \bar{r}^{(n)}}{L_e} \end{bmatrix} \begin{Bmatrix} \phi_1 \\ \phi_2 \\ \phi_3 \\ \vdots \\ \vdots \\ \phi_{n-1} \\ \phi_n \end{Bmatrix}$$

$$= \begin{Bmatrix} +\Gamma_r \bar{r}^{(1)} \left. \dfrac{d\phi}{dr} \right|_{R_i} \\ 0 \\ 0 \\ \vdots \\ \vdots \\ 0 \\ -\Gamma_r \bar{r}^{(n)} \left. \dfrac{d\phi}{dx} \right|_{R_o} \end{Bmatrix} + \frac{\bar{S}}{3} \begin{Bmatrix} 3r_1 + L_1 \\ 3r_1 + 2L_1 + 3r_2 + L_2 \\ 3r_2 + 2L_2 + 3r_3 + L_3 \\ \vdots \\ \vdots \\ 3r_{n-1} + 2L_{n-1} + 3r_n + L_n \\ 3r_n + 2L_n \end{Bmatrix} \tag{11.103}$$

The global system can be written in a compact form as

$$[K]\{\Phi\} = \{F\} \tag{11.104}$$

where

$$[K] = \begin{bmatrix} \frac{\Gamma_r \bar{r}^{(1)}}{L_e} & -\frac{\Gamma_r \bar{r}^{(1)}}{L_e} & 0 & 0 & 0 & 0 & 0 \\ -\frac{\Gamma_r \bar{r}^{(1)}}{L_e} & \frac{\Gamma_r \bar{r}^{(1)}}{L_e} + \frac{\Gamma_r \bar{r}^{(2)}}{L_e} & -\frac{\Gamma_r \bar{r}^{(2)}}{L_e} & 0 & 0 & 0 & 0 \\ & -\frac{\Gamma_r \bar{r}^{(2)}}{L_e} & \frac{\Gamma_r \bar{r}^{(2)}}{L_e} + \frac{\Gamma_r \bar{r}^{(3)}}{L_e} & -\frac{\Gamma_r \bar{r}^{(3)}}{L_e} & 0 & 0 & 0 \\ & & -\frac{\Gamma_r \bar{r}^{(3)}}{L_e} & \frac{\Gamma_r \bar{r}^{(3)}}{L_e} + \frac{\Gamma_r \bar{r}^{(4)}}{L_e} & & & \\ \vdots & & & & & & \\ 0 & 0 & 0 & 0 & & \frac{\Gamma_r \bar{r}^{(n-1)}}{L_e} + \frac{\Gamma_r \bar{r}^{(n)}}{L_e} & -\frac{\Gamma_r \bar{r}^{(n)}}{L_e} \\ 0 & 0 & 0 & 0 & 0. & -\frac{\Gamma_r \bar{r}^{(n)}}{L_e} & \frac{\Gamma_r \bar{r}^{(n)}}{L_e} \end{bmatrix} \tag{11.105}$$

$$\{\Phi\} = \begin{Bmatrix} \phi_1 \\ \phi_2 \\ \phi_3 \\ \vdots \\ \vdots \\ \phi_{n-1} \\ \phi_n \end{Bmatrix} \tag{11.106}$$

$$\{F\} = \begin{Bmatrix} +\Gamma_r \bar{r}^{(1)} \left.\frac{d\phi}{dr}\right|_{R_i} \\ 0 \\ 0 \\ \vdots \\ \vdots \\ 0 \\ -\Gamma_r \bar{r}^{(n)} \left.\frac{d\phi}{dx}\right|_{R_o} \end{Bmatrix} + \frac{\bar{S}L_e}{6} \begin{Bmatrix} 3r_1 + L_1 \\ 3r_1 + 2L_1 + 3r_2 + L_2 \\ 3r_2 + 2L_2 + 3r_3 + L_3 \\ \vdots \\ \vdots \\ 3r_{n-1} + 2L_{n-1} + 3r_n + L_n \\ 3r_n + 2L_n \end{Bmatrix} \quad (11.107)$$

In the case of a uniform mesh size distribution, i.e., for $L_1 = L_2 = \cdots = L_n = L_e$, we can set $r_1 = R_i$, $r_2 = R_i + L_e$, $r_3 = R_i + 2L_e, \ldots r_n = R_i + (n-1)L_e$, and $r_{n+1} = R_o$, and Equation 11.107 can be simplified as

$$\{F\} = \begin{Bmatrix} +\Gamma_r \bar{r}^{(1)} \left.\frac{d\phi}{dr}\right|_{R_i} \\ 0 \\ 0 \\ \vdots \\ \vdots \\ 0 \\ -\Gamma_r \bar{r}^{(n)} \left.\frac{d\phi}{dx}\right|_{R_o} \end{Bmatrix} + \frac{\bar{S}L_e}{6} \begin{Bmatrix} 3R_i + L_e \\ 6R_i + 6L_e \\ 6R_i + 12L_e \\ \vdots \\ \vdots \\ 6R_i + 6(n-2)L_e \\ 6R_i + 6(n-1)L_e \end{Bmatrix} \quad (11.108)$$

In the next step, we apply the boundary conditions directly into the global systems. For a boundary condition of the first kind or of constant surface values as defined by Equation 11.91, we directly assign the given values to the boundary nodes as $\phi_1 = \phi_i$ and $\phi_{n+1} = \phi_o$, and rearrange the global system as

$$\begin{bmatrix} -\Gamma_r \bar{r}^{(1)} & -\frac{\Gamma_r \bar{r}^{(1)}}{L_e} & 0 & 0 & 0 & 0 & 0 \\ 0 & \frac{\Gamma_r \bar{r}^{(1)}}{L_e} + \frac{\Gamma_r \bar{r}^{(2)}}{L_e} & -\frac{\Gamma_r \bar{r}^{(2)}}{L_e} & 0 & 0 & 0 & 0 \\ 0 & -\frac{\Gamma_r \bar{r}^{(2)}}{L_e} & \frac{\Gamma_r \bar{r}^{(2)}}{L_e} + \frac{\Gamma_r \bar{r}^{(3)}}{L_e} & -\frac{\Gamma_r \bar{r}^{(3)}}{L_e} & 0 & 0 & 0 \\ 0 & 0 & -\frac{\Gamma_r \bar{r}^{(3)}}{L_e} & \frac{\Gamma_r \bar{r}^{(3)}}{L_e} + \frac{\Gamma_r \bar{r}^{(4)}}{L_e} & 0 & 0 & 0 \\ & & & \vdots & & & \\ & & & & & \frac{\Gamma_r \bar{r}^{(n-1)}}{L_e} + \frac{\Gamma_r \bar{r}^{(n)}}{L_e} & 0 \\ & & & & & -\frac{\Gamma_r \bar{r}^{(n)}}{L_e} & \Gamma_r \bar{r}^{(n)} \end{bmatrix} \begin{Bmatrix} \left.\frac{d\phi}{dr}\right|_{R_i} \\ \phi_2 \\ \phi_3 \\ \vdots \\ \vdots \\ \phi_n \\ \left.\frac{d\phi}{dr}\right|_{R_o} \end{Bmatrix}$$

$$
= \begin{Bmatrix}
-\dfrac{\Gamma_r \bar{r}^{(1)}}{L_e}\phi_i + \dfrac{\bar{S}L_e}{6}(3R_i + L_e) \\
\dfrac{\Gamma_r \bar{r}^{(1)}}{L_e}\phi_i + \dfrac{\bar{S}L_e}{6}(6R_i + 6L_e) \\
\dfrac{\bar{S}L_e}{6}(6R_i + 12L_e) \\
\vdots \\
\vdots \\
\dfrac{\Gamma_r \bar{r}^{(n)}}{L_e}\phi_o + \dfrac{\bar{S}L_e}{6}(6R_i + 6(n-2)L_e) \\
-\dfrac{\Gamma_r \bar{r}^{(n)}}{L_e}\phi_o + \dfrac{\bar{S}L_e}{6}6R_i + 6(n-1)L_e
\end{Bmatrix} \tag{11.109}
$$

PROBLEMS

11.1 The exposed surface ($x=0$) of a plane wall of thermal conductivity k is subjected to thermal radiation that causes volumetric heat generation in the media to vary as

$$\dot{q}(x) = \dot{q}_0\left(1 - \frac{x}{L}\right)$$

where $\dot{q}_0$ (W/m^3) is a constant. The boundary at $x=L$ is convectively cooled, while the exposed surface is maintained at a constant temperature T_0.

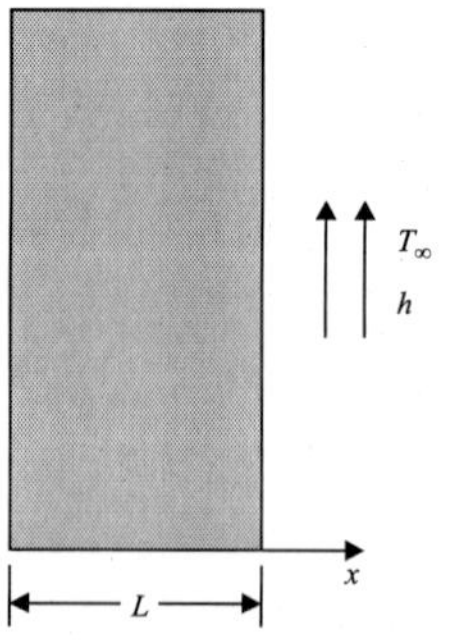

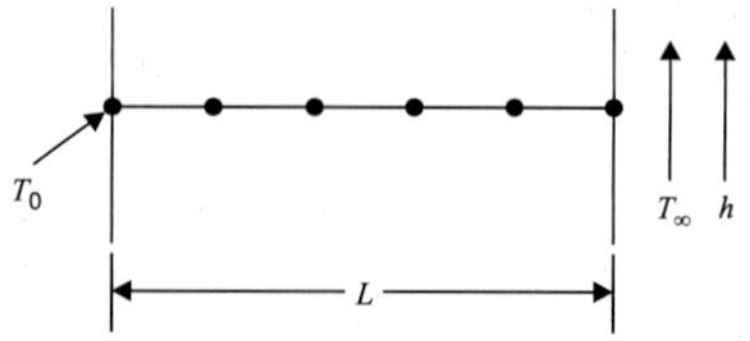

Governing equation

$$\frac{d}{dx}\left(k\frac{dT}{dx}\right) + \dot{q} = 0$$

Boundary conditions

1. $x = 0, \qquad T(x, 0) = T_0$

2. $x = L, \qquad --k\frac{\partial T}{\partial x} = h(T - T_\infty)$

Consider mesh size distribution shown with five elements and determine the temperature distribution in the slab. Use the following data for computation

$$\dot{q} = 1.0 \times 10^5\,\text{W/m}^3, \quad L = 12\,\text{cm}, T_0 = 30°\text{C},$$

$$h = 50\,\text{W/m}^{2}°\text{C} \quad T_\infty = 25°\text{C}$$

11.2 Consider a straight fin of uniform cross-sectional area, A, length $L = 2.0$ cm, width $Z = 1.0$ cm and a thickness $t = 1.4$ mm. The fin thermal conductivity is $k = 60$ W/m°C and it is exposed to a convection environment at $T_\infty = 20$°C and $h = 500$ W/m^2 °C. The base of the fin is at a constant temperature $T_0 = 150$°C and the tip of the fin is assumed to be convective.

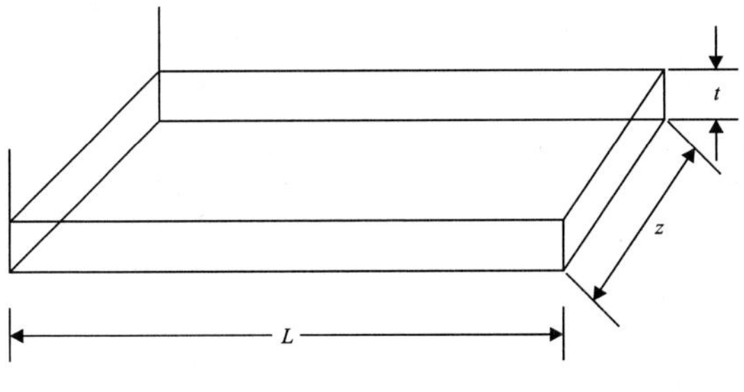

The mathematical statement of the problem is as follows.

Governing equation

$$\frac{d}{dx}\left(k\frac{dT}{dx}\right) + \frac{hP}{A}(T_\infty - T) = 0$$

Boundary conditions

1. $x = 0, \qquad T = T_0$
2. $x = L, \qquad -k\dfrac{dT}{dx} = h(T - T_\infty)$

(a) Use five linear line elements and determine the element characteristic equation for all elements.
(b) Assemble all element equations and derive the global system of equations.
(c) Determine the temperature distribution in the fin and fin heat loss.

11.3 A composite wall made of four homogeneous material layers is shown in the figure.

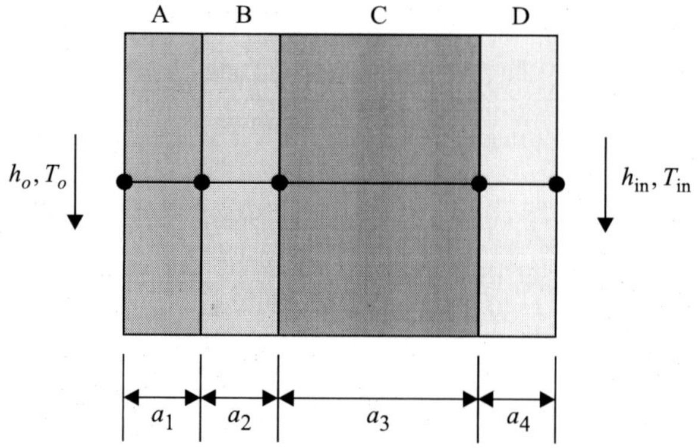

Calculate the nodal temperature values within the composite wall and evaluate the heat flow per unit width based on assuming indoor and outdoor conditions as $T_{in} = 25°C$, $h_{in} = 10\,W/m^2\,K$ and $T_0 = -20°C$, $h_o = 50\,W/m^2\,K$. Consider four linear line elements and use the data $a_1 = 2\,cm$, $a_2 = 1\,cm$, $a_3 = 5\,cm$, $a_4 = 1\,cm$, $k_A = 0.72\,W/m\,K$, $k_C = 0.026\,W/m\,K$ and $k_D = 0.22\,W/m\,K$.

11.4 Consider the electrical heat generation in a chip that is encapsulated with an aluminum cover plate of thickness $a_{al} = 2.5\,mm$ and thermal conductivity $k_{al} = 230\,W/m\,K$. The chip is mounted on a board of thickness $a_b = 4\,mm$ and thermal conductivity $k_b = 0.8\,W/m\,K$. The aluminum cover plate and the chip is cooled on the left by liquid with $h_l = 1200\,W/m^2\,K$ and $T_l = 20°C$. The board on the right is cooled by air with $h_r = 20\,W/m^2\,K$ and $T_r = 20°C$.

Considering a thin chip and assuming a uniform electric heat generation of $\dot{q}'' = 5000\,W/m^2$ at the interface of the aluminum and board materials, determine the temperature distribution and heat transfer rates at two exterior surfaces using five linear elements.

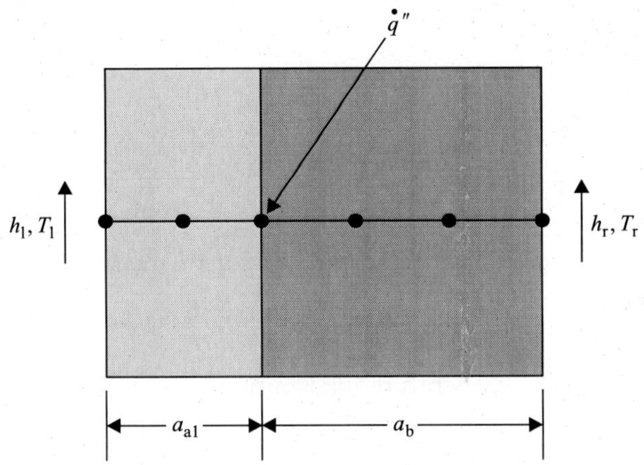

11.5 A semi-transparent wall is irradiated at the surface at $x=0$ by an incident radiation such that the absorbed radiation in the media results in an internal heat generation given as

$$\dot{q} = q_0'' a e^{-ax}$$

where a is the coefficient of absorption of the material and q_0'' is the incident radiation flux. The wall at $x=0$ is also exposed to a convective environment of h_0 and T_0, and the surface at $x=L$ is maintained at constant temperature T_L. Derive discretization equations for interior nodes and for the convective boundary node for the non uniform mesh system shown. Assuming the length of the elements as $l_{e1}, l_{e2}, \ldots, l_{e6}$, derive the system of equations for the solution of nodal temperatures and heat transfer rates at the inner and outer surfaces.

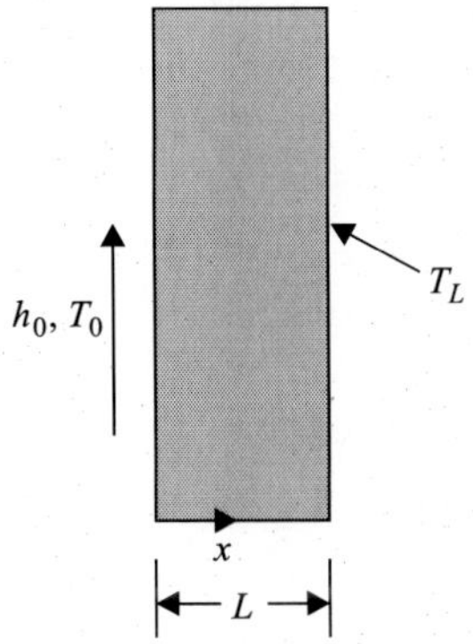

11.6 Consider the fully developed laminar steady flow of Newtonian viscous fluid through a two-dimensional channel formed by two infinite parallel plates as shown below. The lower plate is moving in the negative x-direction with a constant speed U_{w1} and the upper plate is moving in the positive x-direction with a constant speed, U_{w2} in the positive x-direction.

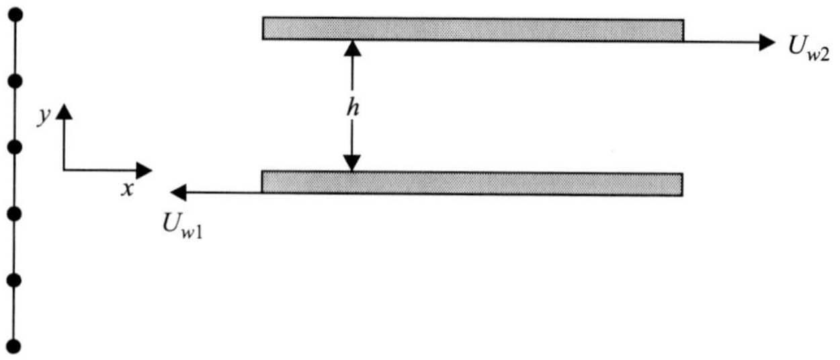

Determine (a) the velocity distribution, (b) volume flow rate, and (c) the shear stresses at the plates and use the following data for the calculation. $h = 3.00\,\text{mm}$, $\mu = 0.02\,\text{kg/m s}$, $\partial P/\partial x = \bar{S} = -1000\,\text{Pa/m}$, $U_{w1} = 10\,\text{mm/s}$, and $U_{w2} = 20\,\text{mm/s}$.

11.7 Consider a one-dimensional steady state diffusion of hydrogen in an anode electrode made of a porous diffusion layer with a thin catalyst layer placed at the right side as shown in the figure. The catalyst layer provides a heterogeneous chemical reaction for the consumption of hydrogen at a rate m''_{H_2}, defined as the consumption per unit area of the catalyst. Assume the reaction rate as given by a ***first-order reaction***, i.e., $m''_{H_2} = -k''_1 C(L)$, where k''_1 is the reaction rate constant. The gas flow channel maintains a constant hydrogen concentration of C_0 at the left surface of the electrode.

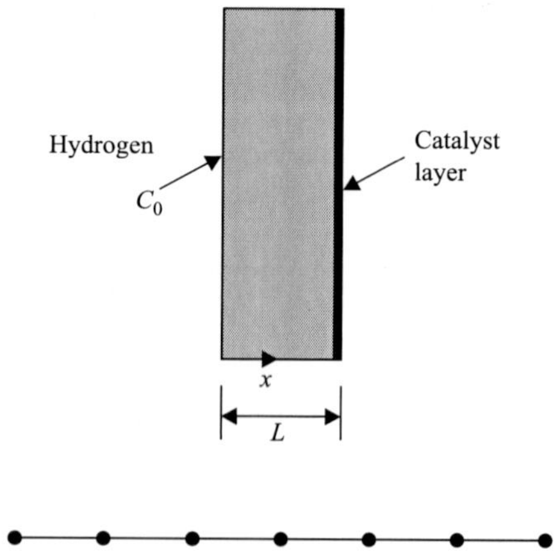

The mathematical statement of the problem is given as follows.

Governing equation

$$\frac{d}{dx}\left(D\frac{dC}{dx}\right) = 0$$

Boundary conditions

1. $x = 0, \qquad C(0) = C_0$
2. $x = L, \qquad -D\frac{dC}{dx}\Big|_{x=L} = -k_1''C(L)$

Use a uniform mesh size distribution shown to derive the system of equations for the nodal hydrogen concentration.

11.8 Derive the element characteristic equation starting from Equation 11.10 and using the quadratic line element defined by Equation 10.42.

11.9 Consider the one-dimensional steady state transport process in a circular rod with an exponentially decaying source term.

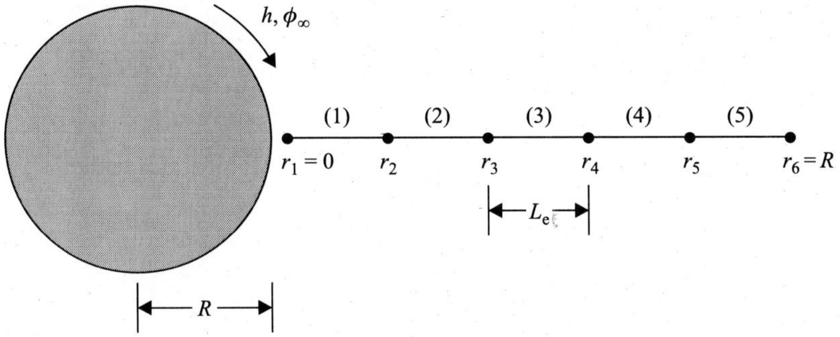

The mathematical statement of the problem is given as follows:

Governing equation

$$\frac{1}{r}\frac{d}{dr}\left(\Gamma_r r\frac{d\phi}{dr}\right) + aI_0\,\mathrm{e}^{-ar^2} = 0$$

Boundary conditions

1. $r = 0, \qquad \frac{d\phi}{dr} = 0$
2. $r = R, \qquad -\Gamma_r\frac{d\phi}{dr} = h(\phi - \phi_\infty)$

Derive the element characteristic equations for all elements. Assemble all elements to obtain the global system of equations.

11.10 Develop a computer code for the solution of a one-dimensional diffuse equation in a plane slab of thickness L. The code generates the element characteristic equation and forms the global system of equations based on the linear line element. Solve the system of equations using the Gauss-elimination solver. Use the code for the solution of the problem given in Problem 11.1.

CHAPTER

12

MULTI-DIMENSIONAL STEADY STATE PROBLEMS

In this chapter we will consider finite element procedures for solving two-dimensional and three-dimensional problems involving diffusion and source terms. Even though the basic steps and procedures used for one-dimensional problems are applicable to multi-dimensional problems, the solution of two- and three-dimensional problems pose considerably more difficulties in mesh generation, formulation of element characteristic equations, and in the assembly of element equations. Galerkin's weighted residual integral formulation is used in presenting this procedure.

12.1 TWO-DIMENSIONAL STEADY STATE DIFFUSION EQUATION

In this section we will consider the formulation and application of a finite element method for solving two-dimensional steady state problems. Let us consider a two-dimensional steady state transport process in a two-dimensional domain as shown Figure 12.1.

A general mathematical statement of the problem can be assumed as

$$\frac{\partial}{\partial x}\left(\Gamma_x \frac{\partial \phi}{\partial x}\right) + \frac{\partial}{\partial y}\left(\Gamma_y \frac{\partial \phi}{\partial y}\right) + S_1\phi + S_0 = 0 \tag{12.1}$$

The partial differential Equation 12.1 is similar to Equation 11.1 with the exception of an additional diffusion term in the y-direction.

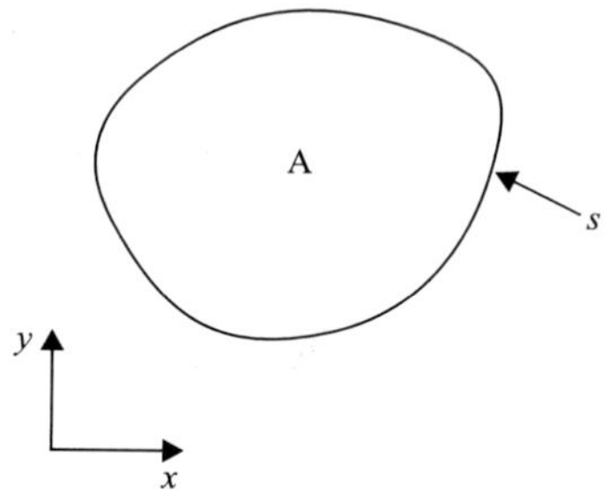

Figure 12.1 Two-dimensional domain

Step 1: Mesh Generation or Discretization of the Solution Domain We have mentioned before that the ***domain discretization*** or ***mesh generation*** involves representing the solution region into a set of elements, referred to as the mesh. The solution domain could be discretized into a mesh using two-dimensional elements such as triangular or rectangular or any other two-dimensional elements discussed in Chapter 10. A mesh of rectangular elements is created easily by constructing all elements in a horizontal row of uniform height and all elements in a vertical column of uniform length. Rectangular elements are well suited for regions consisting of straight edges. However, a combination of rectangular and triangular elements is used in irregular geometries with triangular elements being used to approximate the irregular or curved edges. For the irregular or curved boundaries, the sides of the element should closely approximate the boundary in order to reduce the error. For example, a large number of linear straight-sided triangular elements could be used to approximate the irregular boundaries with good accuracy. A better approximation could be achieved by using higher-order or nonlinear curved-sided triangular elements. So, the choice of the element type, number of elements, and element density depends on the geometry of the domain, the problem under investigation, and accuracy needed. The final selection is established by carrying out a comprehensive numerical experimentation.

Some general practices that are normally being used while creating a mesh are as follows.

1. For simplicity linear elements are always the initial elements of choice. The main advantage is that the linear shape functions and their derivatives are simpler to evaluate analytically. However, it requires a large number of elements to keep the error down to an acceptable level. Another disadvantage is that the derivative of the unknown function within an element is constant and this results in a greater discontinuity at inter-element boundaries, and causes instability and/or slower rate convergence. Higher-order or nonlinear curved-sided elements approximate curved or complex boundaries more accurately and with higher accuracy and rate of convergence, but at the expense of additional complexities in the computation process.

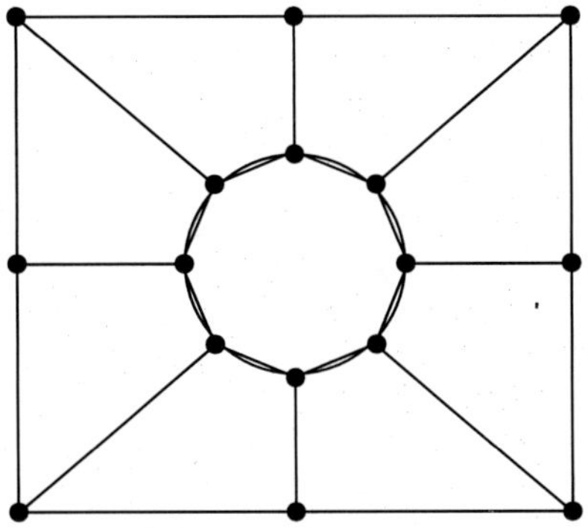

(a)

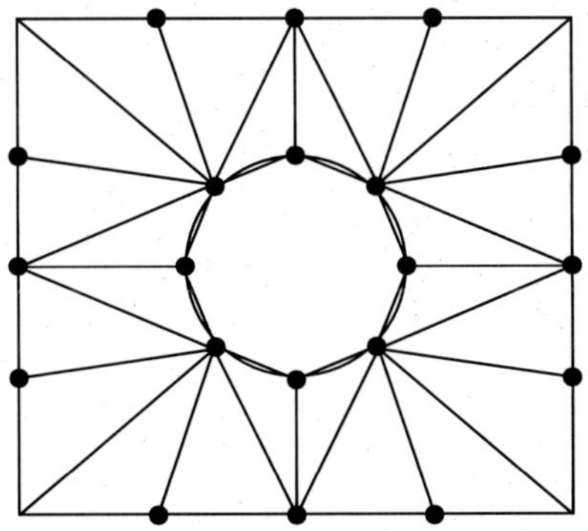

(b)

Figure 12.2 Mesh generation using quadrilateral and triangular geometries

2. While creating a mesh, a solution domain is first divided into a number of sub-regions using simpler geometries such as quadrilaterals or triangles and then progressively divided into smaller sub-domains. A mesh of triangular elements is generated by first dividing the region into a number of quadrilateral elements and then dividing these into triangular regions by joining the nodes on opposite sides or inserting a diagonal, and placing nodes at intersection points. In order to improve the accuracy, it is desirable to join the opposite nodes to insert the shortest diagonals. For example, a square plate with a hole at the center is first divided into sub-regions using quadrilateral elements as shown in Figure 12.2a.

   This is followed by dividing the quadrilateral elements into triangular elements as shown in Figure 12.2b. A more refined mesh near the circular hole can be created by dividing the triangular elements into a number of smaller triangles. A triangular region is divided into smaller triangular regions by locating the same number of nodes on each side and then joining the nodes in an orderly manner, depicted in Figure 12.3.
3. A more refined mesh distribution is used in regions that involve large gradients or variation of dependent variables as shown in Figure 12.4.

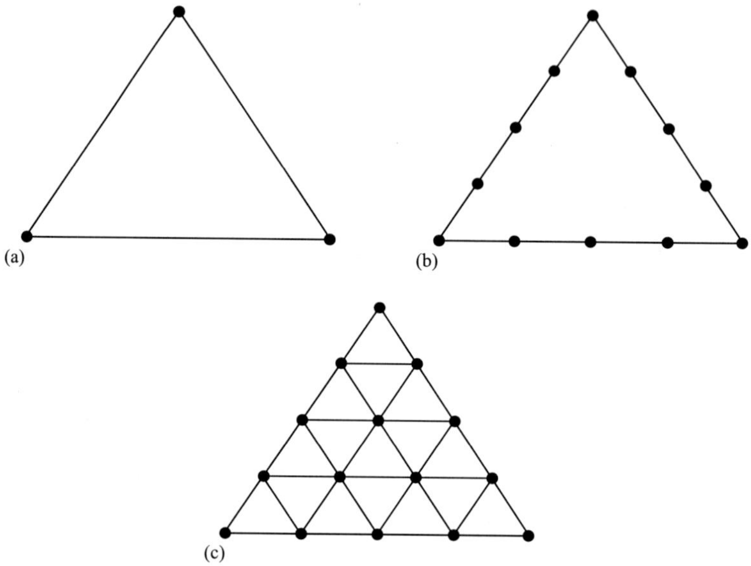

Figure 12.3 Division of a triangular region into smaller triangular elements

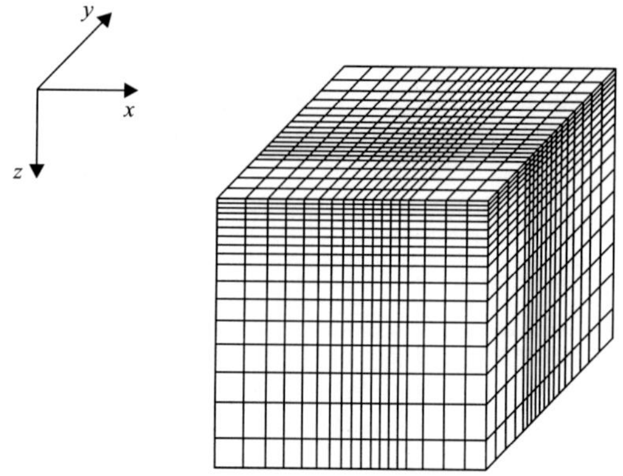

Figure 12.4 Refined mesh distributions

4. For better numerical stability and convergence, mesh refinement should vary gradually from a region with finer mesh distribution to a region with coarse mesh distribution.
5. Elements must be compatible and complete. In order to ensure the continuity in dependent variables and continuity in the flux quantities at the interface of two elements, the elements at the interface must have the same number of nodes and have the same location at the common face or edge, and be defined by the same interpolation function. An example of compatible and noncompatible adjacent elements in a mesh is shown in Figure 12.5.

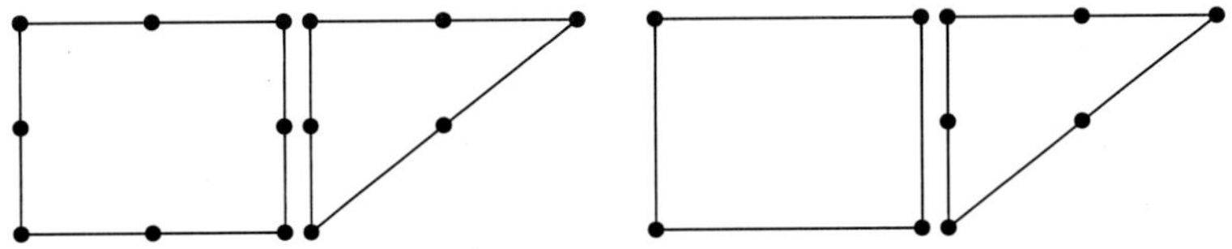

(a) Compatible (b) Non-compatible

Figure 12.5 Nodes at the interface of two elements

6. Transition elements that connect lower-order elements to higher-order elements are used in regions away from the region with a steeper gradient of dependent variables.

Step 2: Element and Node Numbering The elements as well as the nodes in a mesh can be numbered sequentially in any arbitrary manner. However, the node numbering scheme will have a significant effect on the type of global matrix system formed after the assembly of all element characteristic matrix equations. We have mentioned before that a banded system is always preferred as it only requires storing nonzero matrix elements in the band around the principal diagonal, and hence requires less computer storage and computational time. Additionally, a more efficient solver for the solution of the system of linear algebraic equations could be selected as discussed in Chapter 3. So, a preferred numbering system should be the one that leads to a banded global system with a band width (***bw***) as small as possible, preferably a band width of three as in a tridiagonal global matrix system. A random numbering system may increase the difference between node numbers in an element, and thus would increase the band width. We have noted for one-dimensional problems that when the nodes are numbered sequentially from left to right, the maximum difference between node numbers in an element was only one, and assembly of all element characteristic matrices led to a tridiagonal global system. A random numbering system would increase the bandwidth of the global characteristic matrix. Let us consider two different numbering schemes in a two-dimensional region as shown in Figure 12.6.

Comparing these two numbering schemes, we note that there is a considerable difference in node numbers in the second scheme (Figure 12.5b) compared to the first scheme (Figure 12.5a). For example, the maximum difference between node numbers in any element was three as compared to twelve in the second scheme (Figure 12.5b). An assembly of all element characteristic matrices will lead to a much bigger band width in the global system for the second scheme.

Step 3: Selection of Approximate Solution Function The approximate solution for the element is given by the general expression of the form

$$\phi^* = \sum_{I=1}^{n_p} N_I \phi_I = [N]\{\phi^{(e)}\} \tag{12.2}$$

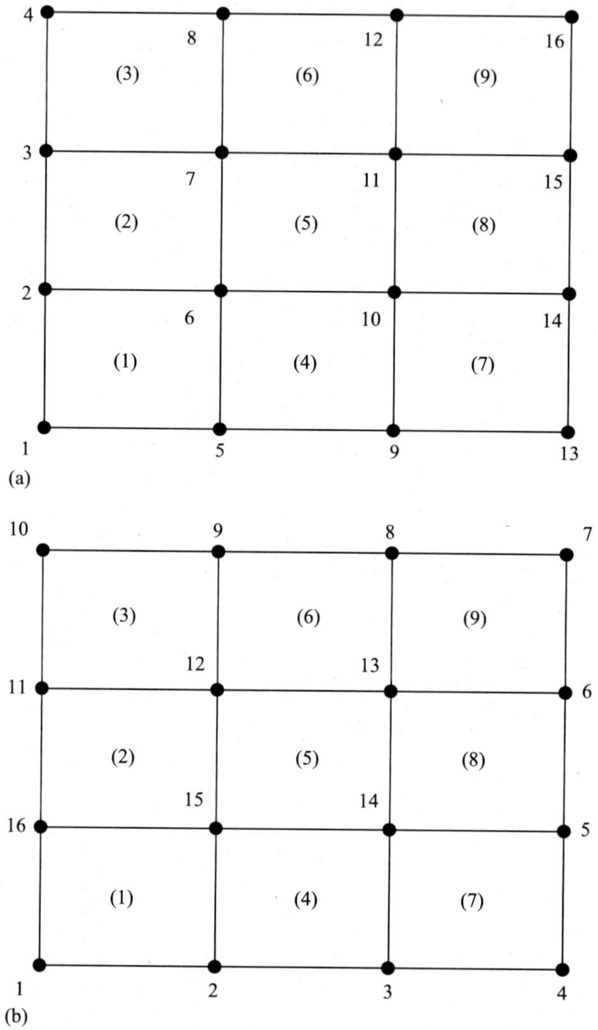

Figure 12.6 Node numbering scheme: (a) preferable; (b) not preferable

where N_I is the shape function, ϕ_I is the nodal value, and n_p is the number of nodes in an element.

Step 4: Formulation of an Integral Statement Using Galerkin's Approach The integral form of the equation over an element is obtained by applying ***Galerkin's weighted residual method*** as

$$\{R^{(e)}\} = -\int_A [N]^T \left[\frac{\partial}{\partial x}\left(\Gamma_x \frac{\partial \phi^*}{\partial x}\right) + \frac{\partial}{\partial y}\left(\Gamma_y \frac{\partial \phi^*}{\partial y}\right) + S_1\phi^* + S_0 \right] dA = 0 \quad (12.3)$$

The first two integral terms are evaluated as

$$\int_A [N]^{\mathrm{T}} \frac{\partial}{\partial x}\left(\Gamma_x \frac{\partial \phi^*}{\partial x}\right) dA = \int_A \frac{\partial}{\partial x}\left([N]^{\mathrm{T}} \Gamma_x \frac{\partial \phi^*}{\partial x}\right) dA - \int_A \Gamma_x \frac{\partial \phi^*}{\partial x} \frac{\partial [N]^{\mathrm{T}}}{\partial x} dA \tag{12.4a}$$

and

$$\int_A [N]^{\mathrm{T}} \frac{\partial}{\partial y}\left(\Gamma_y \frac{\partial \phi^*}{\partial y}\right) dA = \int_A \frac{\partial}{\partial y}\left([N]^{\mathrm{T}} \Gamma_y \frac{\partial \phi^*}{\partial y}\right) dA - \int_A \Gamma_y \frac{\partial \phi^*}{\partial y} \frac{\partial [N]^{\mathrm{T}}}{\partial y} dA \tag{12.4b}$$

The first area integral term on the right-hand side of Equation 12.4, can be transformed into a line integral using Green's theorem as

$$\int_A \frac{\partial}{\partial x}\left([N]^{\mathrm{T}} \Gamma_x \frac{\partial \phi^*}{\partial x}\right) dA = \oint_s [N]^{\mathrm{T}} \Gamma_x \frac{\partial \phi^*}{\partial x} \hat{n}_x ds \tag{12.5a}$$

and

$$\int_A \frac{\partial}{\partial y}\left([N]^{\mathrm{T}} \Gamma_y \frac{\partial \phi^*}{\partial y}\right) dA = \oint_s [N]^{\mathrm{T}} \Gamma_y \frac{\partial \phi^*}{\partial y} \hat{n}_y ds \tag{12.5b}$$

Substituting Equation 12.5 into Equation 12.4, we get

$$\int_A [N]^{\mathrm{T}} \frac{\partial}{\partial x}\left(\Gamma_x \frac{\partial \phi^*}{\partial x}\right) dA = \oint_s [N]^{\mathrm{T}} \Gamma_x \frac{\partial \phi^*}{\partial x} \hat{n}_x ds - \int_A \frac{\partial [N]^{\mathrm{T}}}{\partial x} \Gamma_x \frac{\partial \phi^*}{\partial x} dA \tag{12.6a}$$

and

$$\int_A [N]^{\mathrm{T}} \frac{\partial}{\partial y}\left(\Gamma_y \frac{\partial \phi^*}{\partial y}\right) dA = \oint_s [N]^{\mathrm{T}} \Gamma_y \frac{\partial \phi^*}{\partial y} \hat{n}_y ds - \int_A \frac{\partial [N]^{\mathrm{T}}}{\partial y} \Gamma_y \frac{\partial \phi^*}{\partial y} dA \tag{12.6b}$$

Substituting Equation 12.6 into the integral form of the elemental given by Equation 12.3, we have

$$\begin{aligned} \{R^{(e)}\} = &- \int_s [N]^{\mathrm{T}} \left(\Gamma_x \frac{\partial \phi^*}{\partial x} \hat{n}_x + \Gamma_y \frac{\partial \phi^*}{\partial y} \hat{n}_y\right) ds \\ &+ \int_A \left(\frac{\partial [N]^{\mathrm{T}}}{\partial x} \Gamma_x \frac{\partial \phi^*}{\partial x} + \frac{\partial [N]^{\mathrm{T}}}{\partial y} \Gamma_y \frac{\partial \phi^*}{\partial y}\right) dA + \int_A S_1 [N]^{\mathrm{T}} \phi^* dA \\ &- \int_A S_0 [N]^{\mathrm{T}} dA = 0 \end{aligned} \tag{12.7}$$

Step 5: Formation of Element Characteristics Equation In order to obtain the element characteristics equation, substitute the approximate solution function given by Equation 12.2 into the integral

$$-\int_{S}[N]^{\mathrm{T}}\left(\Gamma_x\frac{\partial\phi^*}{\partial x}\hat{n}_x+\Gamma_y\frac{\partial\phi^*}{\partial y}\hat{n}_y\right)ds$$

$$+\left\{\int_{A}\left(\frac{\partial[N]^{\mathrm{T}}}{\partial x}\Gamma_x\frac{\partial[N]}{\partial x}+\frac{\partial[N]^{\mathrm{T}}}{\partial y}\Gamma_y\frac{\partial[N]}{\partial y}\right)dA\right\}\{\phi^{(e)}\}+\left\{\int S_1[N]^{\mathrm{T}}[N]dA\right\}\{\phi^{(e)}\}$$

$$-\int_{A}S_0[N]^{\mathrm{T}}dA=0 \tag{12.8}$$

The element characteristics equation can be written in matrix notation as

$$\{I^{(e)}\}+[k^{(e)}]\{\phi^{(e)}\}-\left\{f_S^{(e)}\right\}=0 \tag{12.9}$$

where

$\{I^{(e)}\}$ = column vector for the element contribution to the inter-element requirement

$$=-\int_{S}[N]^{\mathrm{T}}\left(\Gamma_x\frac{\partial\phi^*}{\partial x}\hat{n}_x+\Gamma_y\frac{\partial\phi^*}{\partial y}\hat{n}_y\right)ds. \tag{12.10}$$

$\left\{f_S^{(e)}\right\}$ = element source vector due to the volumetric source term

$$=\int_{A}S_0[N]^{\mathrm{T}}dA \tag{12.11}$$

$\{\phi^{(e)}\}$ = column vector of unknown nodal values.

$\left[k^{(e)}\right]$ = element characteristics or stiffness matrix

$$=\int_{A}\left(\frac{\partial[N]^{\mathrm{T}}}{\partial x}\Gamma_x\frac{\partial[N]}{\partial x}+\frac{\partial[N]^{\mathrm{T}}}{\partial y}\Gamma_y\frac{\partial[N]}{\partial y}\right)dA+\int_{A}S_1[N]^{\mathrm{T}}[N]dA$$

$$=\left[k_\Gamma^{(e)}\right]+\left[k_S^{(e)}\right] \tag{12.12}$$

Note that the element stiffness matrix has two components: $[k_\Gamma^{(e)}]$ and $[k_S^{(e)}]$. The second term $[k_S^{(e)}]$ is contributed by the linearized source term of the governing Equation 12.1 given as

$$\left[k_S^{(e)}\right]=\int_{A}S_1[N]^{\mathrm{T}}[N]\,dA \tag{12.13}$$

The first term $[k_\Gamma^{(e)}]$ is contributed by the diffusion terms, which is expressed in a compact form as

$$\left[k_\Gamma^{(e)}\right] = \int_A \left(\frac{\partial [N]^T}{\partial x} \Gamma_x \frac{\partial [N]}{\partial x} + \frac{\partial [N]^T}{\partial y} \Gamma_y \frac{\partial [N]}{\partial y} \right) dA$$

$$= \int_A \{D\}^T [\Gamma] \{D\} dA \qquad (12.14)$$

where

$$[\Gamma] = \text{transport property matrix} = \begin{bmatrix} \Gamma_x & 0 \\ 0 & \Gamma_y \end{bmatrix} \qquad (12.15)$$

$$\{D\} = \text{column vector with derivatives of shape functions} = \begin{Bmatrix} \dfrac{\partial [N]}{\partial x} \\ \dfrac{\partial [N]}{\partial y} \end{Bmatrix} \qquad (12.16)$$

$$\{D\}^T = \text{row vector with derivatives of shape functions} = \begin{Bmatrix} \dfrac{\partial [N]}{\partial x} & \dfrac{\partial [N]}{\partial y} \end{Bmatrix} \qquad (12.17)$$

Implementation of Boundary Conditions Let us divide the boundary surface s into different surface sections s_1, s_2, and s_3 as shown in Figure 12.7 and consider different types of boundary conditions at different sections of the boundary surface as follows.

On the boundary surface s_1 Boundary condition of the first kind

$$\phi = \phi_s = \text{constant surface temperature} \qquad (12.18a)$$

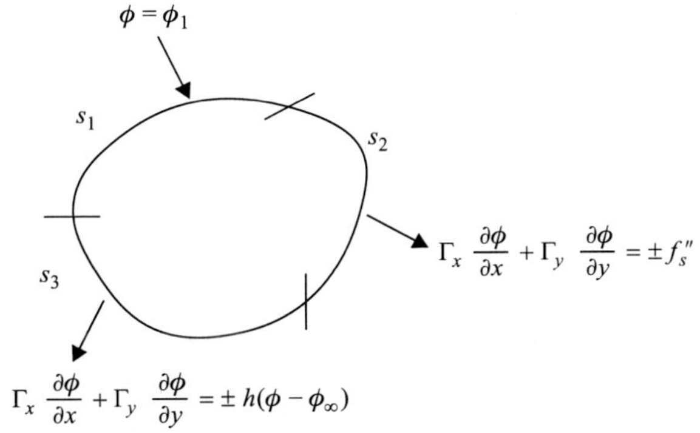

Figure 12.7 Different types of boundary conditions

On the boundary surface s_2 Boundary condition of the second kind

$$\Gamma_x \frac{\partial \phi}{\partial x}\hat{n}_x + \Gamma_y \frac{\partial \phi}{\partial y}\hat{n}_y = \pm f_s'' = \text{ constant surface flux} \tag{12.18b}$$

Note that for the special case with adiabatic conditions or for lines of symmetry, this boundary condition reduces to

$$\Gamma_x \frac{\partial \phi}{\partial x}\hat{n}_x + \Gamma_y \frac{\partial \phi}{\partial y}\hat{n}_y = 0 \tag{12.18c}$$

On the boundary surface s_3 Boundary condition of the third kind

$$\Gamma_x \frac{\partial \phi}{\partial x}\hat{n}_x + \Gamma_y \frac{\partial \phi}{\partial y}\hat{n}_y = \pm h(\phi - \phi_s) = \text{mixed conditions} \tag{12.18d}$$

In general, Equations 12.18b to 12.18d can be expressed in a general form as

$$\Gamma_x \frac{\partial \phi}{\partial x}\hat{n}_x + \Gamma_y \frac{\partial \phi}{\partial y}\hat{n}_y = \pm h(\phi - \phi_s) \pm f_s'' \tag{12.18e}$$

The implementation of the derivative boundary conditions into the finite element formulation takes place through the inter-element vector $\{I^{(e)}\}$ as given by Equation 12.10

$$\{I^{(e)}\} = -\int_s [N]^{\mathrm{T}} \left(\Gamma_x \frac{\partial \phi^*}{\partial x}\hat{n}_x + \Gamma_y \frac{\partial \phi^*}{\partial y}\hat{n}_y \right) ds \tag{12.19}$$

Note that this integral is the sum of all sides of an element including the sides that are subjected to the boundary conditions. In order to show the specific contribution from the boundary conditions, let us decompose the integral into two parts as

$$\{I^{(e)}\} = \left\{I_b^{(e)}\right\} + \left\{I_i^{(e)}\right\} \tag{12.20}$$

where $\left\{I_i^{(e)}\right\}$ is the integral contributions from all the other sides other than the boundary surface. This primarily constitutes the inter-element requirements. The component $\left\{I_b^{(e)}\right\}$ is the integral contribution from sides that represent the boundary, and it is expressed as

$$\left\{I_b^{(e)}\right\} = -\int_{s_b} [N]^{\mathrm{T}} \left(\Gamma_x \frac{\partial \phi^*}{\partial x}\hat{n}_x + \Gamma_y \frac{\partial \phi^*}{\partial y}\hat{n}_y \right) ds \tag{12.21}$$

Substituting the boundary condition given by Equation 12.18e, which represents the boundary condition in a general form, we get

$$\left\{I_b^{(e)}\right\} = -\int_{s_b} [N]^{\mathrm{T}} \{\pm h(\phi^* - \phi_\infty) \pm f_s''\}\, ds \qquad (12.22a)$$

or

$$\left\{I_b^{(e)}\right\} = \pm \int_{s_b} h[N]^{\mathrm{T}} \phi^* ds \pm \int_{s_b} [N]^{\mathrm{T}} \{(h\phi_\infty + f_s'')\}\, ds \qquad (12.22b)$$

Now, substituting the approximate solution given by (12.2), we get

$$\left\{I_b^{(e)}\right\} = \pm \left(\int_{s_b} h[N]^{\mathrm{T}}[N]\, ds \right) \phi^{(e)} \pm \int_{s_b} [N]^{\mathrm{T}} \{(h\phi_\infty + f_s'')\}\, ds \qquad (12.23a)$$

or

$$\left\{I_b^{(e)}\right\} = \left[k_b^{(e)}\right]\{\phi^{(e)}\} - \left\{f_{bs}^{(e)}\right\} \qquad (12.23b)$$

where

$$\left[k_b^{(e)}\right] = \pm \int_{s_b} h[N]^{\mathrm{T}}[N]\, ds$$

$$= \text{element stiffness matrix component due to boundary conditions} \qquad (12.24)$$

$$\left\{f_{bs}^{(e)}\right\} = \pm \int_{s_b} [N]^{\mathrm{T}} \{(h\phi_\infty + f_s'')\} ds$$

$$= \text{element source vector component due to boundary condition} \qquad (12.25)$$

Finally, we can substitute Equations 12.20 and 12.23 into Equation 12.9 to obtain element characteristic equation

$$\left\{I_i^{(e)}\right\} + \left(k_\Gamma^{(e)} + k_S^{(e)} + k_b^{(e)}\right)\{\phi^{(e)}\} - \left\{f_S^{(e)} + f_{bs}^{(e)}\right\} = 0 \qquad (12.26)$$

or

$$\left\{I_i^{(e)}\right\} + k^{(e)}\{\phi^{(e)}\} - f'' = 0 \qquad (12.27)$$

where

$$k^{(e)} = \left(k_\Gamma^{(e)} + k_S^{(e)} + k_b^{(e)}\right) = \text{element characteristics matrix} \qquad (12.28)$$

and

$$f^{(e)} = \left\{ f_S^{(e)} + f_{bs}^{(e)} \right\} = \text{element source vector} \tag{12.29}$$

Let us demonstrate the formation of this element characteristics equation for different types of elements in the following sections.

Case: Element Characteristics Equation Using Linear Triangular Element Let us consider a linear three-node triangular element as shown in Figure 10.14 and with the approximate solution and the corresponding shape functions given by Equations 10.59 and 10.61, respectively. The vectors with derivatives of shape functions are derived by substituting the shape functions given as

$$\{D\} = \begin{bmatrix} \frac{\partial N_i}{\partial x} & \frac{\partial N_j}{\partial x} & \frac{\partial N_k}{\partial x} \\ \frac{\partial N_i}{\partial y} & \frac{\partial N_j}{\partial y} & \frac{\partial N_k}{\partial y} \end{bmatrix} = \frac{1}{2A} \begin{bmatrix} \beta_i & \beta_j & \beta_k \\ \gamma_i & \gamma_j & \gamma_k \end{bmatrix} \tag{12.30a}$$

and

$$\{D\}^T = \begin{bmatrix} \frac{\partial N_i}{\partial x} & \frac{\partial N_i}{\partial y} \\ \frac{\partial N_j}{\partial x} & \frac{\partial N_j}{\partial y} \\ \frac{\partial N_k}{\partial x} & \frac{\partial N_k}{\partial y} \end{bmatrix} = \frac{1}{2A} \begin{bmatrix} \beta_i & \gamma_i \\ \beta_j & \gamma_j \\ \beta_k & \gamma_k \end{bmatrix} \tag{12.30b}$$

The component of element stiffness matrix due to the diffusion term, $[k_\Gamma^{(e)}]$, can now be computed from Equation 12.14 as

$$\left[k_\Gamma^{(e)}\right] = \int_A \{D\}^{\mathrm{T}}[\Gamma]\{D\}dA = \{D\}^{\mathrm{T}}[\Gamma]\{D\} \int_A dA$$

$$= \frac{1}{2A} \begin{bmatrix} \beta_i & \gamma_i \\ \beta_j & \gamma_j \\ \beta_k & \gamma_k \end{bmatrix} \begin{bmatrix} \Gamma_x & 0 \\ 0 & \Gamma_y \end{bmatrix} \frac{1}{2A} \begin{bmatrix} \beta_i & \beta_j & \beta_k \\ \gamma_i & \gamma_j & \gamma_k \end{bmatrix} A$$

$$= \frac{1}{4A^2} \begin{bmatrix} \beta_i & \gamma_i \\ \beta_j & \gamma_j \\ \beta_k & \gamma_k \end{bmatrix} \begin{bmatrix} \Gamma_x\beta_i & \Gamma_x\beta_j & \Gamma_x\beta_k \\ \Gamma_y\gamma_i & \Gamma_y\gamma_j & \Gamma_y\gamma_k \end{bmatrix} A$$

$$\left[k_\Gamma^{(e)}\right] = \frac{1}{4A}\begin{bmatrix} \Gamma_x\beta_i^2 + \Gamma_y\gamma_i^2 & \Gamma_x\beta_i\beta_j + \Gamma_y\gamma_i\gamma_j & \Gamma_x\beta_i\beta_k + \Gamma_y\gamma_i\gamma_k \\ \Gamma_x\beta_j\beta_i + \Gamma_y\gamma_j\gamma_i & \Gamma_x\beta_j^2 + \Gamma_y\gamma_j^2 & \Gamma_x\beta_j\beta_k + \Gamma_y\gamma_j\gamma_k \\ \Gamma_x\beta_k\beta_i + \Gamma_y\gamma_k\gamma_i & \Gamma_x\beta_k\beta_j + \Gamma_y\gamma_k\gamma_j & \Gamma_x\beta_k^2 + \Gamma_y\gamma_k^2 \end{bmatrix} \quad (12.31a)$$

or

$$\left[k_\Gamma^{(e)}\right] = \frac{\Gamma_x}{4A}\begin{bmatrix} \beta_i^2 & \beta_i\beta_j & \beta_i\beta_k \\ \beta_j\beta_i & \beta_j^2 & \beta_j\beta_k \\ \beta_k\beta_i & \beta_k\beta_j & \beta_k^2 \end{bmatrix} + \frac{\Gamma_y}{4A}\begin{bmatrix} \gamma_i^2 & \gamma_i\gamma_j & \gamma_i\gamma_k \\ \gamma_j\gamma_i & \gamma_j^2 & \gamma_j\gamma_k \\ \gamma_k\gamma_i & \gamma_k\gamma_j & \gamma_k^2 \end{bmatrix} \quad (12.31b)$$

The component of element stiffness matrix due to the linearized source term, $\left[k_S^{(e)}\right]$, can now be computed as

$$\left[k_S^{(e)}\right] = \int_A S_1[N]^{\mathrm{T}}[N]dA$$

$$= S_1\int_A \begin{Bmatrix} N_i \\ N_j \\ N_k \end{Bmatrix}\{\, N_i \quad N_j \quad N_k \,\}dA$$

$$\left[k_\beta^{(e)}\right] = S_1\int_A \begin{bmatrix} N_i^2 & N_iN_j & N_iN_k \\ N_jN_i & N_j^2 & N_jN_k \\ N_kN_i & N_kN_j & N_k^2 \end{bmatrix} dA \quad (12.32)$$

This integral can either be evaluated by a direct integral formula or by Gauss–Legendre numerical formulas. As direct integral formulas are applicable here, let us continue evaluating this integral analytically through the use of a natural coordinate system. From the expression of a natural co-ordinate system (Equation 10.68) for the triangular element

$$N_i = L_i,\ N_j = L_j \quad \text{and} \quad N_k = L_k \quad (10.68)$$

Substituting Equation 10.68 into Equation 12.32, we get

$$\left[k_S^{(e)}\right] = S_1\int_A \begin{bmatrix} L_i^2 & L_iL_j & L_iL_k \\ L_jL_i & L_j^2 & L_jL_k \\ L_kL_i & L_kL_j & L_k^2 \end{bmatrix} \quad (12.33)$$

Each of the element integrals can now be evaluated using the integral formula given by Equation 10.71. For example the integral elements in the first row are evaluated as follows.

First integral element (I_{11})

$$I_{11} = \int_A L_i^2 \, dA = \int L_i^2 L_j^0 L_k^0 \, dA \tag{12.34a}$$

$$= \frac{2!0!0!}{(2+0+0+2)!} 2A$$

$$I_{11} = \frac{A}{6}$$

Second integral element (I_{12})

$$\int_A L_i L_j dA = \int L_i^1 L_j^1 L_k^0 \, dA \tag{12.34b}$$

$$= \frac{1!1!0!}{(1+1+0+2)!} 2A$$

$$= \frac{A}{12}$$

Third integral element (I_{13})

$$\int_A L_i L_k dA = \int L_i^1 L_j^0 L_k^1 \, dA \tag{12.34c}$$

$$= \frac{1!1!0!}{(1+1+0+2)!} 2A$$

$$= \frac{A}{12}$$

Similarly, the integral elements in other rows can be evaluated and the integral expression Equation 12.33 becomes

$$\left[k_S^{(e)}\right] = \frac{S_1 A}{12} \begin{bmatrix} 2 & 1 & 1 \\ 1 & 2 & 1 \\ 1 & 1 & 2 \end{bmatrix} \tag{12.34d}$$

Substituting Equations 12.31 and 12.34 into Equation 12.28, we have the expression for the element characteristic equation for an interior element as

$$\left[k^{(e)}\right] = \frac{\Gamma_x}{4A} \begin{bmatrix} \beta_i^2 & \beta_i\beta_j & \beta_i\beta_k \\ \beta_j\beta_i & \beta_j^2 & \beta_j\beta_k \\ \beta_k\beta_i & \beta_k\beta_j & \beta_k^2 \end{bmatrix} + \frac{\Gamma_y}{4A} \begin{bmatrix} \gamma_i^2 & \gamma_i\gamma_j & \gamma_i\gamma_k \\ \gamma_j\gamma_i & \gamma_j^2 & \gamma_j\gamma_k \\ \gamma_k\gamma_i & \gamma_k\gamma_j & \gamma_k^2 \end{bmatrix}$$
$$+ \frac{S_1 A}{12} \begin{bmatrix} 2 & 1 & 1 \\ 1 & 2 & 1 \\ 1 & 1 & 2 \end{bmatrix} \tag{12.35}$$

The component of the element characteristic matrix due to convective boundary conditions as given by Equation 12.24 is computed as

$$\left[k_b^{(e)}\right] = \pm \int_{S_b} h[N]^{\mathrm{T}}[N]\, ds$$

$$= \pm h \int_{S_b} \begin{Bmatrix} N_i \\ N_j \\ N_k \end{Bmatrix} \{ N_i \quad N_j \quad N_k \}\, ds$$

$$= \pm h \int_{S_b} \begin{bmatrix} N_i^2 & N_iN_j & N_iN_k \\ N_jN_i & N_j^2 & N_jN_k \\ N_kN_i & N_kN_j & N_k^2 \end{bmatrix} ds \tag{12.36}$$

The integral is further evaluated by considering only the sides that are subjected to convection boundary conditions. For example, if convection is specified over side i–j of the element, then $N_k = 0$ and Equation 12.36 becomes

$$\left[k_b^{(e)}\right] = \pm h \int_{S_b} \begin{bmatrix} N_i^2 & N_iN_j & 0 \\ N_jN_i & N_j^2 & 0 \\ 0 & 0 & 0 \end{bmatrix} ds \tag{12.37}$$

Again, using the natural coordinate system Equation 10.68 and integral formula Equation 10.71, Equation 12.37 is transformed into

$$\left[k_b^{(e)}\right] = \pm \frac{hL_{ij}}{6} \begin{bmatrix} 2 & 1 & 0 \\ 1 & 2 & 0 \\ 0 & 0 & 0 \end{bmatrix} \tag{12.38a}$$

Similarly, for the convection condition specified on sides j–k and k–i, we get

$$\left[k_b^{(e)}\right] = \pm h \int_{S_b} \begin{bmatrix} 0 & 0 & 0 \\ 0 & N_j^2 & N_jN_k \\ 0 & N_kN_j & N_k^2 \end{bmatrix} ds = \pm \frac{hL_{jk}}{6} \begin{bmatrix} 0 & 0 & 0 \\ 0 & 2 & 1 \\ 0 & 1 & 2 \end{bmatrix} \tag{12.38b}$$

and

$$\left[k_b^{(e)}\right] = \pm h \int_{S_b} \begin{bmatrix} N_i^2 & 0 & N_iN_k \\ 0 & 0 & 0 \\ N_kN_i & 0 & N_k^2 \end{bmatrix} ds = \pm \frac{hL_{ki}}{6} \begin{bmatrix} 2 & 0 & 1 \\ 0 & 0 & 0 \\ 1 & 0 & 2 \end{bmatrix} \tag{12.38c}$$

The element characteristic matrix for the element with external boundary specified at the edge *i–j* is obtained by substituting Equations 12.31, 12.34d, and 12.38a into Equation 12.28

$$[k^{(e)}] = \frac{\Gamma_x}{4A}\begin{bmatrix} \beta_i^2 & \beta_i\beta_j & \beta_i\beta_k \\ \beta_j\beta_i & \beta_j^2 & \beta_j\beta_k \\ \beta_k\beta_i & \beta_k\beta_j & \beta_k^2 \end{bmatrix} + \frac{\Gamma_y}{4A}\begin{bmatrix} \gamma_i^2 & \gamma_i\gamma_j & \gamma_i\gamma_k \\ \gamma_j\gamma_i & \gamma_j^2 & \gamma_j\gamma_k \\ \gamma_k\gamma_i & \gamma_k\gamma_j & \gamma_k^2 \end{bmatrix}$$

$$+ \frac{S_1 A}{12}\begin{bmatrix} 2 & 1 & 1 \\ 1 & 2 & 1 \\ 1 & 1 & 2 \end{bmatrix} \pm \frac{hL_{ij}}{6}\begin{bmatrix} 2 & 1 & 0 \\ 1 & 2 & 0 \\ 0 & 0 & 0 \end{bmatrix} \tag{12.39}$$

Let us now evaluate the force vectors given by Equations 12.11 and 12.25. The element force vector due to a constant source term is computed using the natural coordinate system for a triangular element as

$$\left\{f_S^{(e)}\right\} = \int_A S_0[N]^{\mathrm{T}}\, dA \tag{12.11}$$

$$= S_0 \int_A \begin{bmatrix} N_i \\ N_j \\ N_k \end{bmatrix} dA$$

$$= S_0 \int_A \begin{bmatrix} L_i \\ L_j \\ L_k \end{bmatrix} dA \tag{12.40}$$

Computing the integral elements using an integration formula Equation 10.71, we get

$$\left\{f_S^{(e)}\right\} = \frac{S_0 A}{3}\begin{Bmatrix} 1 \\ 1 \\ 1 \end{Bmatrix} \tag{12.41}$$

Similarly, the element force vector component due to the boundary condition is computed from Equation 12.25 as

$$\left\{f_{bs}^{(e)}\right\} = \pm \int_{s_b} [N]^{\mathrm{T}}\{(h\phi_\infty + f_s'')\}\, ds \tag{12.25}$$

$$= (h\phi_\infty + f_s'') \int_{S_b} \begin{Bmatrix} N_i \\ N_j \\ N_k \end{Bmatrix} ds \tag{12.42}$$

For boundary condition specified on the i–j side of the element, $N_k = 0$, and we use the natural coordinate system for a triangle and use the integral formula Equation 10.28, to get

$$\left\{f_{bs}^{(e)}\right\} = \pm\left(h\phi_\infty + f_s''\right) \int_{S_b} \begin{Bmatrix} N_i \\ N_j \\ 0 \end{Bmatrix} ds = \pm\left(h\phi_\infty + f_s''\right) \int_{S_b} \begin{Bmatrix} L_i \\ L_j \\ 0 \end{Bmatrix} ds$$

$$\left\{f_{bs}^{(e)}\right\} = \pm \frac{\left(h\phi_\infty + f_s''\right) L_{ij}}{2} \begin{Bmatrix} 1 \\ 1 \\ 0 \end{Bmatrix} \tag{12.43a}$$

Similarly, for boundary conditions specified on the sides j–k, $N_i = 0$ and Equation 12.42 transforms into

$$\left\{f_{bs}^{(e)}\right\} = \left(h\phi_\infty + f_s''\right) \int_{S_b} \begin{Bmatrix} 0 \\ N_j \\ N_k \end{Bmatrix} ds = \pm \frac{\left(h\phi_\infty + f_s''\right) L_{jk}}{2} \begin{Bmatrix} 0 \\ 1 \\ 1 \end{Bmatrix} \tag{12.43b}$$

and for boundary conditions specified on the side k–i, $N_j = 0$, we get

$$\left\{f_{bs}^{(e)}\right\} = \left(h\phi_\infty + f_s''\right) \int_{S_b} \begin{Bmatrix} N_i \\ 0 \\ N_k \end{Bmatrix} ds = \pm \frac{\left(h\phi_\infty + f_s''\right) L_{ki}}{2} \begin{Bmatrix} 1 \\ 0 \\ 1 \end{Bmatrix} \tag{12.43c}$$

Example 12.1: Two-dimensional Problem Using Triangular Element: Determine the element characteristic equation for a triangular element in the finite element mesh used to solve the temperature distribution and heat transfer rate for a two-dimensional fin shown in the figure. The convective conditions are $h = 20.0\ \text{W/m}^2\,\text{K}$ and $T_\infty = 25°\text{C}$. The fin is made of aluminum with thermal conductivity, $k = 202\ \text{W/m}\cdot°\text{C}$.

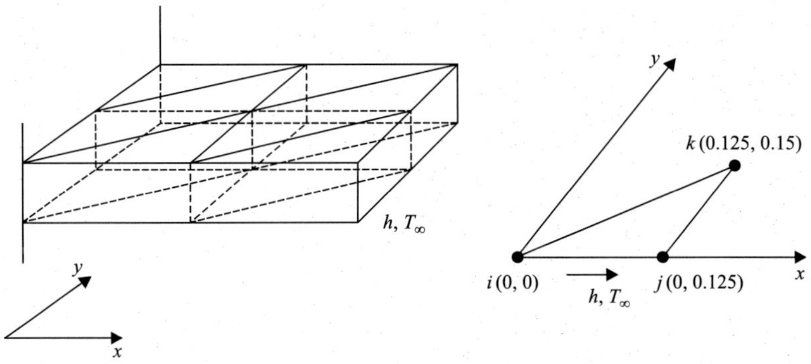

The length and width of the fin are $L = 25\,\text{cm}$ and $W = 30\,\text{cm}$ respectively. The thickness of the fin is $t = 5\,\text{mm}$. A typical triangular element in the mesh is shown in the figure. The coordinates of the nodes are: $i(0, 0)$, $j(0.125)$ and $k(0.125, 0.15)$. Determine the element characteristic equation for this element.

Solution: The governing heat equation for a two-dimensional fin is

$$\frac{\partial}{\partial x}\left(k_x A_x \frac{\partial T}{\partial x}\right) + \frac{\partial}{\partial y}\left(k_y A_y \frac{\partial T}{\partial y}\right) - h\,dA_s(T - T_\infty) = 0 \qquad \text{(E.12.1.1)}$$

Substituting $A_x = t \cdot dy$, $A_y = t \cdot dx$ and $dA_s = 2\,dx\,dy$, Equation E.12.1.1 reduces to

$$k_x \frac{\partial^2 T}{\partial x^2} + k_y \frac{\partial^2 T}{\partial y^2} - \frac{2h}{t} T + \frac{2hT_\infty}{t} = 0$$

This is of the same form as the general Equation 12.1 with

$$\Gamma_x = k_x = k = 202\,\text{W/m} \cdot {}^\circ\text{C}, \quad \Gamma_y = k_y = k = 202\,\text{W/m} \cdot {}^\circ\text{C}$$

$$S_1 = -\frac{2h}{t} = -\frac{2 \times 20}{0.005} = -8000\,\text{W/m} \cdot {}^\circ\text{C}$$

and

$$S_0 = \frac{2hT_\infty}{t} = \frac{2 \times 20 \times 25}{0.005} = 2 \times 10^5\,\text{W/m}^3$$

The element characteristic matrix is evaluated based on Equation 12.28

$$k^{(e)} = \left(k_\Gamma^{(e)} + k_S^{(e)} + k_b^{(e)}\right)$$

$$\left[k_\Gamma^{(e)}\right] = \frac{\Gamma_x}{4A}\begin{bmatrix} \beta_i^2 & \beta_i\beta_j & \beta_i\beta_k \\ \beta_j\beta_i & \beta_j^2 & \beta_j\beta_k \\ \beta_k\beta_i & \beta_k\beta_j & \beta_k^2 \end{bmatrix} + \frac{\Gamma_y}{4A}\begin{bmatrix} \gamma_i^2 & \gamma_i\gamma_j & \gamma_i\gamma_k \\ \gamma_j\gamma_i & \gamma_j^2 & \gamma_j\gamma_k \\ \gamma_k\gamma_i & \gamma_k\gamma_j & \gamma_k^2 \end{bmatrix} \qquad \text{(12.31b)}$$

$$\left[k_S^{(e)}\right] = \frac{S_1 A}{12}\begin{bmatrix} 2 & 1 & 1 \\ 1 & 1 & 1 \\ 1 & 1 & 2 \end{bmatrix}, \quad k_b^{(e)} = \int_{s_b} h[N]^{\text{T}}[N]\,ds$$

The element source vector is given by Equation 12.29 as

$$f^{(e)} = \left\{f_S^{(e)} + f_{bs}^{(e)}\right\}$$

The coordinates of the three nodes of the triangular element is

$$x_i = 0, \quad x_j = 0.125, \quad x_k = 0.125$$

$$y_i = 0, \quad y_j = 0, \quad y_k = 0.15$$

Let us first determine all the coefficients and shape functions for the triangular element based on these coordinate points and using Equation 10.62 as follows:

$$\alpha_i = x_j y_k - x_k y_j = 0.125(0.15) - 0.125(0.0) = 0.01875$$

$$\alpha_j = x_k y_i - x_i y_k = 0.125(0.0) - 0.0(0.15) = 0.0$$

$$\alpha_k = x_i y_j - x_j y_i = 0.0(0.0) - 0.125(0) = 0.0$$

$$\beta_i = y_j - y_k = 0.0 - 0.15 = -0.15$$

$$\beta_j = y_k - y_i = 0.15 - 0.0 = 0.15$$

$$\beta_k = y_i - y_j = 0.0 - 0.0 = 0.0$$

$$\gamma_i = x_k - x_j = 0.125 - 0.125 = 0.0$$

$$\gamma_j = x_i - x_k = 0.0 - 0.125 = -0.125$$

$$\gamma_k = x_j - x_i = 0.125 - 0.0 = 0.125$$

$$A = \frac{1}{2}\begin{vmatrix} 1 & x_i & y_i \\ 1 & x_j & y_j \\ 1 & x_k & y_k \end{vmatrix} = \frac{1}{2}\begin{vmatrix} 1 & 0 & 0 \\ 1 & 0.125 & 0 \\ 1 & 0.125 & 0.15 \end{vmatrix} = 0.009\,375$$

The component of element characteristics due to the diffusion terms is estimated as

$$\left[k_\Gamma^{(e)}\right] = \frac{\Gamma_x}{4A}\begin{bmatrix} \beta_i^2 & \beta_i\beta_j & \beta_i\beta_k \\ \beta_j\beta_i & \beta_j^2 & \beta_j\beta_k \\ \beta_k\beta_i & \beta_k\beta_j & \beta_k^2 \end{bmatrix} + \frac{\Gamma_y}{4A}\begin{bmatrix} \gamma_i^2 & \gamma_i\gamma_j & \gamma_i\gamma_k \\ \gamma_j\gamma_i & \gamma_j^2 & \gamma_j\gamma_k \\ \gamma_k\gamma_i & \gamma_k\gamma_j & \gamma_k^2 \end{bmatrix}$$

$$= \frac{202}{4 \times 0.009\,375}\begin{bmatrix} (-0.15)^2 & (-0.15)(0.15) & (-0.15)(0.0) \\ (0.15)(-0.15) & (0.15)^2 & (0.15)(0.0) \\ (0.0)(-0.15) & (0.0)(0.15) & (0.0)^2 \end{bmatrix}$$

$$+ \frac{202}{4 \times 0.009\,375}\begin{bmatrix} (0.0)^2 & (0.0)(-0.125) & (0.0)(0.125) \\ (-0.125)(0.0) & (-0.125)^2 & (-0.125)(0.125) \\ (0.125)(0.0) & (0.125)(-0.125) & (0.125)^2 \end{bmatrix}$$

$$= 5386.6666\begin{bmatrix} 0.0225 & -0.0225 & 0.0 \\ -0.0225 & 0.0225 & 0.0 \\ 0.0 & 0.0 & 0.0 \end{bmatrix}$$

$$+ 5386.6666\begin{bmatrix} 0.0 & 0.0 & 0.0 \\ 0.0 & 0.015\,625 & -0.015\,625 \\ 0.0 & -0.015\,625 & 0.015\,625 \end{bmatrix}$$

$$=121.1999\begin{bmatrix} 1 & -1 & 0 \\ -1 & 1 & 0 \\ 0 & 0 & 0 \end{bmatrix} + 84.1666\begin{bmatrix} 0 & 0 & 0 \\ 0 & 1 & -1 \\ 0 & -1 & 1 \end{bmatrix}$$

$$= \begin{bmatrix} 121.1999 & -121.1999 & 0.0 \\ -121.1999 & 121.1999 & 0.0 \\ 0.0 & 0.0 & 0.0 \end{bmatrix} + \begin{bmatrix} 0.0 & 0.0 & 0.0 \\ 0.0 & 84.1666 & -84.1666 \\ 0.0 & -84.1666 & 84.1666 \end{bmatrix}$$

$$\left[k_\Gamma^{(e)}\right] = \begin{bmatrix} 121.1999 & -121.1999 & 0.0 \\ -121.1999 & 205.3665 & -84.1666 \\ 0.0 & -84.1666 & 84.1666 \end{bmatrix} \tag{E.12.1.2}$$

The component of element characteristic matrix due to the linearized source term is given by Equation 12.34d.

$$\left[k_S^{(e)}\right] = \frac{S_1 A}{12}\begin{bmatrix} 2 & 1 & 1 \\ 1 & 2 & 1 \\ 1 & 1 & 2 \end{bmatrix}$$

$$= \frac{-8000 \times 0.018\,75}{12}\begin{bmatrix} 2 & 1 & 1 \\ 1 & 2 & 1 \\ 1 & 1 & 2 \end{bmatrix} = -12.5\begin{bmatrix} 2 & 1 & 1 \\ 1 & 2 & 1 \\ 1 & 1 & 2 \end{bmatrix}$$

$$\left[k_S^{(e)}\right] = -\begin{bmatrix} 25 & 12.5 & 12.5 \\ 12.5 & 25 & 12.5 \\ 12.5 & 12.5 & 25 \end{bmatrix} \tag{E.12.1.3}$$

The component of the element characteristic matrix due to boundary conditions as given by Equation 12.24 is computed as follows:

$$\left[k_b^{(e)}\right] = \pm \int_{S_b} h[N]^{\mathrm{T}}[N]\, ds$$

$$= \pm \int_{S_b} h \begin{Bmatrix} N_i \\ N_j \\ N_k \end{Bmatrix} \{ N_i \quad N_j \quad N_k \}\, ds$$

$$= \pm h \int_{S_b} \begin{bmatrix} N_i^2 & N_i N_j & N_i N_k \\ N_j N_i & N_j^2 & N_j N_k \\ N_k N_i & N_k N_j & N_k^2 \end{bmatrix} ds$$

Since convection boundary conditions are specified over the side i–j of the element, then $N_k = 0$ and Equation 12.36 is transformed into

$$\left[k_b^{(e)}\right] = \frac{hL_{ij}}{6}\begin{bmatrix} 2 & 1 & 0 \\ 1 & 2 & 0 \\ 0 & 0 & 0 \end{bmatrix} = \frac{20 \times 0.125}{6}\begin{bmatrix} 2 & 1 & 0 \\ 1 & 2 & 0 \\ 0 & 0 & 0 \end{bmatrix}$$

$$= \begin{bmatrix} 0.8333 & 0.4166 & 0 \\ 0.4166 & 0.8333 & 0 \\ 0 & 0 & 0 \end{bmatrix} \quad \text{(E.12.1.4)}$$

Combining Equations E.12.1.1, E.12.1.2, and Equation E.12.1.3, we get the element characteristics equation as

$$k^{(e)} = \left(k_\Gamma^{(e)} + k_S^{(e)} + k_b^{(e)}\right)$$

$$= \begin{bmatrix} 121.1999 & -121.1999 & 0.0 \\ -121.1999 & 205.3665 & -84.1666 \\ 0.0 & -84.1666 & 84.1666 \end{bmatrix} - \begin{bmatrix} 25 & 12.5 & 12.5 \\ 12.5 & 25 & 12.5 \\ 12.5 & 12.5 & 25 \end{bmatrix}$$

$$+ \begin{bmatrix} 0.8333 & 0.4166 & 0 \\ 0.4166 & 0.8333 & 0 \\ 0 & 0 & 0 \end{bmatrix}$$

$$k^{(e)} = \begin{bmatrix} 96.1999 & -133.6999 & -12.5 \\ -133.6999 & 180.3665 & -96.6666 \\ -12.5 & -96.6666 & 59.1666 \end{bmatrix} + \begin{bmatrix} 0.8333 & 0.4166 & 0 \\ 0.4166 & 0.8333 & 0 \\ 0 & 0 & 0 \end{bmatrix}$$

$$k^{(e)} = \begin{bmatrix} 97.0332 & -133.2833 & -12.5 \\ -133.2833 & 181.1998 & -96.6666 \\ -12.5 & -96.6666 & 59.1666 \end{bmatrix} \quad \text{(E.12.1.5)}$$

The element force vector due to a constant source term is computed by Equation 12.41 as

$$\left\{f_S^{(e)}\right\} = \frac{S_o A}{3}\begin{Bmatrix} 1 \\ 1 \\ 1 \end{Bmatrix} = \frac{2 \times 10^5 \times 0.009\,375}{3}\begin{Bmatrix} 1 \\ 1 \\ 1 \end{Bmatrix}$$

$$= \begin{Bmatrix} 625 \\ 625 \\ 625 \end{Bmatrix} \quad \text{(E.12.1.6)}$$

The element source vector component due to the boundary condition is computed from Equation 12.42 with a convective boundary condition and $f_s'' = 0$ as

$$\left\{f_{bs}^{(e)}\right\} = (hT_\infty)\int_{S_b}\begin{Bmatrix} N_i \\ N_j \\ N_k \end{Bmatrix} ds$$

Since the convective boundary condition is specified on the i–j side of the element, $N_k = 0$, and with the use of the natural co-ordinate system and integral formula, we get from Equation (12.43a),

$$\left\{f_{bs}^{(e)}\right\} = \frac{(hT_\infty)L_{ij}}{2}\begin{Bmatrix} 1 \\ 1 \\ 0 \end{Bmatrix} = \frac{(25 \times 20)0.125}{2}\begin{Bmatrix} 1 \\ 1 \\ 0 \end{Bmatrix}$$

$$\left\{f_{bs}^{(e)}\right\} = \frac{(20 \times 25)0.125}{2}\begin{Bmatrix} 1 \\ 1 \\ 0 \end{Bmatrix}$$

$$\left\{f_{bs}^{(e)}\right\} = \begin{Bmatrix} 31.25 \\ 31.25 \\ 0 \end{Bmatrix} \tag{E.12.1.6}$$

Combining Equations E.12.1.5 and E.12.1.6, we get the element source vector as

$$f^{(e)} = \left\{f_S^{(e)} + f_{bs}^{(e)}\right\}$$

$$= \begin{Bmatrix} 625 \\ 625 \\ 625 \end{Bmatrix} + \begin{Bmatrix} 31.25 \\ 31.25 \\ 0 \end{Bmatrix}$$

$$f^{(e)} = \begin{Bmatrix} 656.5 \\ 656.5 \\ 625 \end{Bmatrix} \tag{E.12.1.7}$$

The element characteristic matrix is given by Equation 12.27

$$k^{(e)}\{\phi^{(e)}\} = -\left\{I_i^{(e)}\right\} + f^{(e)}$$

$$\begin{bmatrix} 97.0332 & -133.2833 & -12.5 \\ -133.2833 & 181.1998 & -96.6666 \\ -12.5 & -96.6666 & 59.1666 \end{bmatrix} \begin{Bmatrix} \phi_i^{(e)} \\ \phi_j^{(e)} \\ \phi_k^{(e)} \end{Bmatrix}$$

$$= -\begin{Bmatrix} I_i^{(e)} \\ I_j^{(e)} \\ I_k^{(e)} \end{Bmatrix} + \begin{Bmatrix} 656.5 \\ 656.5 \\ 625 \end{Bmatrix} \tag{E.12.1.8}$$

Case: Element Characteristics Equation Using Bilinear Rectangular Element. For a bilinear rectangular element (Figure 10.19), the approximate solution and the corresponding shape functions in the local coordinate system are given by Equations 10.87 and 10.88, respectively

$$\phi(\tilde{x}, \tilde{y}) = N_i\phi_i + N_j\phi_j + N_k\phi_k + N_i\phi_l = \sum_{i=i,j,k,l} N_I\phi_I = [N]\{\phi^{(e)}\} \tag{10.87}$$

with shape functions given as

$$N_i = \left(1 - \frac{\tilde{x}}{2l_e}\right)\left(1 - \frac{\tilde{y}}{2h_e}\right), \quad N_j = \frac{\tilde{x}}{2l_e}\left(1 - \frac{\tilde{y}}{2h_e}\right)$$

$$N_k = \frac{\tilde{x}\tilde{y}}{4l_e h_e}, \quad N_l = \frac{\tilde{y}}{2h_e}\left(1 - \frac{\tilde{x}}{2l_e}\right) \tag{10.88}$$

The vectors with derivatives of shape functions are derived by substituting the shape functions given by Equations 10.88 for the bilinear rectangular element as

$$\{D\} = \begin{bmatrix} \frac{\partial N_i}{\partial x} & \frac{\partial N_j}{\partial x} & \frac{\partial N_k}{\partial x} & \frac{\partial N_l}{\partial x} \\ \frac{\partial N_i}{\partial y} & \frac{\partial N_j}{\partial y} & \frac{\partial N_k}{\partial y} & \frac{\partial N_l}{\partial y} \end{bmatrix}$$

$$= \frac{1}{4l_e h_e} \begin{bmatrix} -(2h_e - \tilde{y}) & (2h_e - \tilde{y}) & \tilde{y} & -\tilde{y} \\ -(2l_e - \tilde{x}) & -\tilde{x} & \tilde{x} & (2l_e - \tilde{x}) \end{bmatrix} \tag{12.44}$$

$$\{D^{\mathrm{T}}\} = \begin{bmatrix} \frac{\partial N_i}{\partial x} & \frac{\partial N_i}{\partial y} \\ \frac{\partial N_j}{\partial x} & \frac{\partial N_j}{\partial y} \\ \frac{\partial N_k}{\partial x} & \frac{\partial N_k}{\partial y} \\ \frac{\partial N_l}{\partial x} & \frac{\partial N_l}{\partial y} \end{bmatrix} = \begin{bmatrix} -(2h_e - \tilde{y}) & -(2l_e - \tilde{x}) \\ (2h_e - \tilde{y}) & \tilde{x} \\ \tilde{y} & \tilde{x} \\ -\tilde{y} & (2l_e - \tilde{x}) \end{bmatrix} \tag{12.45}$$

The component of element characteristic matrix due to the diffusion term $[k_\Gamma^{(e)}]$ can now be computed from Equation 12.14 as

$$\left[k_\Gamma^{(e)}\right] = \int_A \{D\}^{\mathrm{T}}[\Gamma]\{D\} dA = \{D\}^{\mathrm{T}}[\Gamma]\{D\} \int_A dA$$

$$= \int_A \begin{bmatrix} \frac{\partial N_i}{\partial x} & \frac{\partial N_i}{\partial y} \\ \frac{\partial N_j}{\partial x} & \frac{\partial N_j}{\partial y} \\ \frac{\partial N_k}{\partial x} & \frac{\partial N_k}{\partial y} \\ \frac{\partial N_l}{\partial x} & \frac{\partial N_l}{\partial y} \end{bmatrix} \begin{bmatrix} \Gamma_x & 0 \\ 0 & \Gamma_y \end{bmatrix} \begin{bmatrix} \frac{\partial N_i}{\partial x} & \frac{\partial N_j}{\partial x} & \frac{\partial N_k}{\partial x} & \frac{\partial N_l}{\partial x} \\ \frac{\partial N_i}{\partial y} & \frac{\partial N_i}{\partial y} & \frac{\partial N_i}{\partial y} & \frac{\partial N_i}{\partial y} \end{bmatrix}$$

$$= \int_A \begin{bmatrix} \frac{\partial N_i}{\partial \tilde{x}} & \frac{\partial N_i}{\partial \tilde{y}} \\ \frac{\partial N_j}{\partial \tilde{x}} & \frac{\partial N_j}{\partial \tilde{y}} \\ \frac{\partial N_k}{\partial \tilde{x}} & \frac{\partial N_k}{\partial \tilde{y}} \\ \frac{\partial N_l}{\partial \tilde{x}} & \frac{\partial N_l}{\partial \tilde{y}} \end{bmatrix} \begin{bmatrix} \Gamma_x & 0 \\ 0 & \Gamma_y \end{bmatrix} \begin{bmatrix} \frac{\partial N_i}{\partial \tilde{x}} & \frac{\partial N_j}{\partial \tilde{x}}. & \frac{\partial N_k}{\partial \tilde{x}} & \frac{\partial N_l}{\partial \tilde{x}} \\ \frac{\partial N_i}{\partial \tilde{y}} & \frac{\partial N_i}{\partial \tilde{y}} & \frac{\partial N_i}{\partial \tilde{y}} & \frac{\partial N_i}{\partial \tilde{y}} \end{bmatrix}$$

$$= \int_0^{2l_e} \int_0^{2h_e} \frac{1}{4l_e h_e} \begin{bmatrix} -(2h_e - \tilde{y}) & -(2l_e - \tilde{x}) \\ (2h_e - \tilde{y}) & -\tilde{x} \\ \tilde{y} & \tilde{x} \\ -\tilde{y} & (2l_e - \tilde{x}) \end{bmatrix} \begin{bmatrix} \Gamma_x & \\ & \Gamma_y \end{bmatrix}$$

$$\times \frac{1}{4l_e h_e} \begin{bmatrix} -(2h_e - \tilde{y}) & (2h_e - \tilde{y}) & \tilde{y} & -\tilde{y} \\ -(2l_e - \tilde{x}) & -\tilde{x} & \tilde{x} & (2l_e - \tilde{x}) \end{bmatrix} d\tilde{x}\, d\tilde{y}$$

$$= \int_0^{2l_e}\int_0^{2h_e} \frac{1}{16l_e^2h_e^2}\begin{bmatrix} -(2h_e-\tilde{y}) & -(2l_e-\tilde{x}) \\ (2h_e-\tilde{y}) & -\tilde{x} \\ \tilde{y} & \tilde{x} \\ -\tilde{y} & (2l_e-\tilde{x}) \end{bmatrix}$$

$$\times \begin{bmatrix} -\Gamma_x(2h_e-\tilde{y}) & \Gamma_y(2h_e-\tilde{y}) & \Gamma_x\tilde{y} & -\Gamma_y\tilde{y} \\ -\Gamma_y(2l_e-\tilde{x}) & -\Gamma_y\tilde{x} & \Gamma_y\tilde{x} & \Gamma_y(2l_e-\tilde{x}) \end{bmatrix} d\tilde{x}\, d\tilde{y}$$

$$= \int_0^{2l_e}\int_0^{2h_e} \frac{1}{16l_e^2h_e^2}\begin{bmatrix} -(2h_e-\tilde{y}) & -(2l_e-\tilde{x}) \\ (2h_e-\tilde{y}) & -\tilde{x} \\ \tilde{y} & \tilde{x} \\ -\tilde{y} & (2l_e-\tilde{x}) \end{bmatrix}$$

$$\times \begin{bmatrix} -\Gamma_x(2h_e-\tilde{y}) & \Gamma_y(2h_e-\tilde{y}) & \Gamma_x\tilde{y} & -\Gamma_y\tilde{y} \\ -\Gamma_y(2l_e-\tilde{x}) & -\Gamma_y\tilde{x} & \Gamma_y\tilde{x} & \Gamma_y(2l_e-\tilde{x}) \end{bmatrix} d\tilde{x}\, d\tilde{y}$$

$$= \int_0^{2l_e}\int_0^{2h_e} \frac{1}{16l_e^2h_e^2}\left[\begin{matrix} \Gamma_x(2h_e-\tilde{y})^2+\Gamma_e(2l_e-\tilde{x})^2 & -\Gamma_x(2h_e-\tilde{y})^2+\Gamma_e(2l_e-\tilde{x})\tilde{x} \\ -\Gamma_x(2h_e-\tilde{y})^2+\Gamma_e(2l_e-\tilde{x})^2\tilde{x} & \Gamma_x(2h_e-\tilde{y})^2+\Gamma_e\tilde{x}^2 \\ -\Gamma_x(2h_e-\tilde{y})\tilde{y}-\Gamma_e(2l_e-\tilde{x})\tilde{x} & \Gamma_x(2h_e-\tilde{y})\tilde{y}-\Gamma_e\tilde{x}^2 \\ \Gamma_x(2h_e-\tilde{y})\tilde{y}-\Gamma_e(2l_e-\tilde{x})^2 & \Gamma_x(2h_e-\tilde{y})\tilde{y}+\Gamma_e\tilde{x}\tilde{y} \end{matrix}\right.$$

$$\left.\begin{matrix} -\Gamma_x(2h_e-\tilde{y})\tilde{y}-\Gamma_e(2l_e-\tilde{x})\tilde{x} & \Gamma_x(2h_e-\tilde{y})\tilde{y}-\Gamma_e(2l_e-\tilde{x})^2 \\ \Gamma_x(2h_e-\tilde{y})\tilde{y}-\Gamma_e\tilde{x}^2 & -\Gamma_x(2h_e-\tilde{y})\tilde{y}-\Gamma_e(2l_e-\tilde{x})\tilde{x} \\ \Gamma_x\tilde{y}^2-\Gamma_e\tilde{x}^2 & -\Gamma_x\tilde{y}^2+\Gamma_e(2l_e-\tilde{x})\tilde{x} \\ -\Gamma_x\tilde{y}^2+\Gamma_e(2l_e-\tilde{x})\tilde{x} & \Gamma_x\tilde{y}^2+\Gamma_e(2l_e-\tilde{x})^2 \end{matrix}\right] d\tilde{x}\, d\tilde{y} \tag{12.46}$$

Each integral element of the equation can be integrated one at a time. For example, the first integral element is evaluated as

$$I_{11} = \int_0^{2l_e}\int_0^{2h_e} \frac{1}{16l_e^2h_e^2}\left(\Gamma_x(2h_e-\tilde{y})^2+\Gamma_e(2l_e-\tilde{x})^2\right) d\tilde{x}\, d\tilde{y}$$

$$= \int_0^{2l_e}\int_0^{2h_e} \frac{1}{16l_e^2h_e^2}\Gamma_x(2h_e-\tilde{y})^2\, d\tilde{x}\, d\tilde{y} + \int_0^{2l_e}\int_0^{2h_e} \frac{1}{16l_e^2h_e^2}\Gamma_e(2l_e-\tilde{x})^2\, d\tilde{x}\, d\tilde{y}$$

$$I_{11} = \frac{\Gamma_x h_e}{3l_e} + \frac{\Gamma_y l_e}{3h_e} \tag{12.47a}$$

In a similar manner, all other elements are evaluated and the expression for the element characteristic matrix due to the diffusion terms is given as

$$\left[k_\Gamma^{(e)}\right] = \frac{\Gamma_x h_e}{6l_e}\begin{bmatrix} 2 & -2 & -1 & 1 \\ -2 & 2 & 1 & -1 \\ -1 & 1 & 2 & -2 \\ 1 & -1 & -2 & 2 \end{bmatrix} + \frac{\Gamma_y l_e}{6h_e}\begin{bmatrix} 2 & 1 & -1 & -2 \\ 1 & 2 & -2 & -1 \\ -1 & -2 & 2 & 1 \\ -2 & -1 & 1 & 2 \end{bmatrix} \tag{12.47b}$$

The component of the element stiffness matrix due to the linearized source term, $[k_S^{(e)}]$, can now be computed as

$$\left[k_S^{(e)}\right] = \int_A S_1 [N]^{\mathrm{T}}[N]\, dA$$

$$= \int_A S_1 \begin{Bmatrix} N_i \\ N_j \\ N_k \\ N_l \end{Bmatrix} \{ N_i \quad N_j \quad N_k \quad N_l \}\, dA$$

$$\left[k_S^{(e)}\right] = \int_A S_1 \begin{bmatrix} N_i^2 & N_i N_j & N_i N_k & N_i N_l \\ N_j N_i & N_j^2 & N_j N_k & N_j N_l \\ N_k N_i & N_k N_j & N_k^2 & N_k N_l \\ N_l N_i & N_l N_j & N_l N_k & N_l^2 \end{bmatrix} dA \tag{12.48}$$

Each element integral can be evaluated after substituting the shape function expressions given by Equation 10.88. For cases with constant coefficient β, the integrals can be evaluated directly as

$$I_{33} = \int_A S_1 N_k^2\, dA = S_1 \int_0^{2l_e}\int_0^{2h_e} N_k^2\, d\tilde{x}\, d\tilde{y}$$

$$= S_1 \int_0^{2l_e}\int_0^{2h_e} \left(\frac{\tilde{x}\tilde{y}}{4l_e h_e}\right)^2 d\tilde{x}\, d\tilde{y} = S_1 \int_0^{2l_e} \frac{\tilde{x}^2}{16 l_e^2 h_e^2} \frac{\tilde{y}^3}{3}\bigg|_0^{2h_e} d\tilde{x}$$

$$= S_1 \int_0^{2l_e} \frac{\tilde{x}^2 h_e}{6l_e^2}\, d\tilde{x} = -S_1 \frac{h_e}{18 l_e^2} \tilde{x}^3 \Big|_0^{2l_e} = S_1 \frac{4 l_e h_e}{9}$$

$$I_{33} = \frac{S_1 A}{9} \tag{12.49a}$$

Similarly, the rest of the integral elements are evaluated, and the element stiffness matrix due to the convection term is expressed as

$$\left[k_\beta^{(e)}\right] = \frac{S_1 A}{36}\begin{bmatrix} 4 & 2 & 1 & 2 \\ 2 & 4 & 2 & 1 \\ 1 & 2 & 4 & 2 \\ 2 & 1 & 2 & 4 \end{bmatrix} \tag{12.49b}$$

The element stiffness matrix component due to boundary conditions as given by Equation 12.24 is computed as

$$\left[k_b^{(e)}\right] = \pm \int_{S_b} h[N]^{\mathrm{T}}[N]\, ds$$

$$= \pm \int_{S_b} h \begin{Bmatrix} N_i \\ N_j \\ N_k \\ N_l \end{Bmatrix} \{ N_i \quad N_j \quad N_k \quad N_l \}\, ds$$

$$\left[k_b^{(e)}\right] = \pm \int_{S_b} h \begin{bmatrix} N_i^2 & N_iN_j & N_iN_k & N_iN_l \\ N_jN_i & N_j^2 & N_jN_k & N_jN_l \\ N_kN_i & N_kN_j & N_k^2 & N_kN_l \\ N_lN_i & N_lN_j & N_lN_k & N_l^2 \end{bmatrix} ds \tag{12.50}$$

The integral is further evaluated by considering only the sides that are subjected to convection boundary conditions. For example, if convection is specified over the side i–j of the element, then $N_k = 0$ and $N_l = 0$, and Equation 12.50 becomes

$$\left[k_b^{(e)}\right] = \pm \int_{S_b} h \begin{bmatrix} N_i^2 & N_iN_j & 0 & 0 \\ N_jN_i & N_j^2 & 0 & 0 \\ 0 & 0 & 0 & 0 \\ 0 & 0 & 0 & 0 \end{bmatrix} ds \tag{12.51}$$

Again using the natural coordinate system Equation 10.27 and integral formula Equation 10.28, the integral elements can be evaluated. For example, the first integral element is evaluated as

$$\int_{S_b} hN_i^2\, ds = \frac{2l_e}{3} = \frac{L_{ij}}{3} \tag{12.52}$$

Using these integral elements, Equation 12.37 is transformed into

$$\left[k_b^{(e)}\right] = \pm\frac{hL_{ij}}{6}\begin{bmatrix} 2 & 1 & 0 & 0 \\ 1 & 2 & 0 & 0 \\ 0 & 0 & 0 & 0 \\ 0 & 0 & 0 & 0 \end{bmatrix} \quad \text{for convective boundary on the side } i-j \tag{12.53a}$$

Similarly:

for convective boundary on the side j–k

$$\left[k_b^{(e)}\right] = \pm\int_{S_b} h\begin{bmatrix} 0 & 0 & 0 & 0 \\ 0 & N_j^2 & N_jN_k & 0 \\ 0 & N_kN_j & N_k^2 & 0 \\ 0 & 0 & 0 & 0 \end{bmatrix} ds = \pm\frac{hL_{jk}}{6}\begin{bmatrix} 0 & 0 & 0 & 0 \\ 0 & 2 & 1 & 0 \\ 0 & 1 & 2 & 0 \\ 0 & 0 & 0 & 0 \end{bmatrix} \tag{12.53b}$$

for convective boundary on the side k–l

$$\left[k_b^{(e)}\right] = \pm\int_{S_b} h\begin{bmatrix} 0 & 0 & 0 & 0 \\ 0 & 0 & 0 & 0 \\ 0 & 0 & N_k^2 & N_kN_l \\ 0 & 0 & N_lN_k & N_l^2 \end{bmatrix} ds = \pm\frac{hL_{kl}}{6}\begin{bmatrix} 0 & 0 & 0 & 0 \\ 0 & 0 & 0 & 0 \\ 0 & 0 & 2 & 1 \\ 0 & 0 & 1 & 2 \end{bmatrix} \tag{12.53c}$$

for convective boundary on the side l–i

$$\left[k_b^{(e)}\right] = \pm\int_{S_b} h\begin{bmatrix} N_i^2 & 0 & 0 & N_iN_l \\ 0 & 0 & 0 & 0 \\ 0 & 0 & 0 & 0 \\ N_lN_i & 0 & 0 & N_l^2 \end{bmatrix} ds = \pm\frac{hL_{li}}{6}\begin{bmatrix} 2 & 0 & 0 & 1 \\ 0 & 0 & 0 & 0 \\ 0 & 0 & 0 & 0 \\ 1 & 0 & 0 & 2 \end{bmatrix} \tag{12.53d}$$

The element force vector due to a constant source term is computed using the natural coordinate system for a triangular element as

$$\left\{f_S^{(e)}\right\} = \int_A S_0[N]^{\mathrm{T}}\, dA \tag{12.11}$$

$$= \int_A S_0\begin{Bmatrix} N_i \\ N_j \\ N_k \\ N_l \end{Bmatrix} dA \tag{12.54}$$

$$\left\{f_S^{(e)}\right\} = \int_0^{2l_e}\int_0^{2h_e} S_0 \begin{Bmatrix} 1 - \dfrac{\tilde{x}}{2l_e} - \dfrac{\tilde{y}}{2h_e} + \dfrac{\tilde{x}\tilde{y}}{4l_eh_e} \\ \dfrac{\tilde{x}}{2l_e} - \dfrac{\tilde{x}\tilde{y}}{4l_eh_e} \\ \dfrac{\tilde{x}\tilde{y}}{4l_eh_e} \\ \dfrac{\tilde{y}}{2h_e} - \dfrac{\tilde{x}\tilde{y}}{4l_eh_e} \end{Bmatrix} d\tilde{x}\, d\tilde{y} \tag{12.55}$$

Each element integral in the force vector can be evaluated analytically for cases with a source term. For example the third element is evaluated by integral rules as

$$\begin{aligned} f_{S_3}^{(e)} &= S_0 \int_0^{2l_e}\int_0^{h_e} \frac{\tilde{x}\tilde{y}}{4l_eh_e}\, d\tilde{x}\, d\tilde{y} \\ &= S_0 \int_0^{2l_e} \frac{\tilde{x}}{4l_eh_e}\frac{\tilde{y}^2}{2}\Big|_0^{2h_e} = S_0 \frac{h_e}{2l_e}\frac{\tilde{x}^2}{2}\Big|_0^{2l_e} \\ &= S_0 l_e h_e = \frac{S_0 4 l_e h_e}{4} \end{aligned}$$

$$f_{S_3}^{(e)} = \frac{S_0 A}{4} \tag{12.56}$$

Evaluating all elements in a similar manner, the element force vector due to the source term is expressed as

$$f_S^{(e)} = \frac{S_0 A}{4}\begin{Bmatrix} 1 \\ 1 \\ 1 \\ 1 \end{Bmatrix} \tag{12.57}$$

Similarly, the element force vector component due to the boundary condition is computed as

$$\left\{f_{bs}^{(e)}\right\} = \pm \int_{S_b} [N]^{\mathrm{T}}\left\{\left(h\phi_\infty + f_s''\right)\right\} ds \tag{12.25}$$

$$= \left(h\phi_\infty + f_s''\right) \int_{S_b} \begin{Bmatrix} N_i \\ N_j \\ N_k \\ N_l \end{Bmatrix} ds \tag{12.58}$$

where shape functions can be defined by Equation 10.89 for a local co-ordinate system.

For the boundary condition specified on the i–j side of the element, $N_k = 0$, $N_l = 0$ and $y = -he$. The shape functions are $N_i = \frac{1}{2}\left(1 - \frac{x'}{le}\right)$, $N_j = \frac{1}{2}\left(1 + \frac{x'}{le}\right)$ $N_k = 0$, $N_l = 0$. Substituting these into Equation 12.58, we get

$$\left\{f_{bs}^{(e)}\right\} = \pm\left(h\phi_\infty + f_s''\right)\int_{S_b}\begin{Bmatrix} N_i \\ N_j \\ 0 \\ 0 \end{Bmatrix} ds = \left(\frac{h\phi_\infty + f_s''}{2le}\right)\int_{-le}^{le}\begin{Bmatrix} le - x' \\ le + x' \\ 0 \\ 0 \end{Bmatrix} dx$$

$$\left\{f_{bs}^{(e)}\right\} = \pm\frac{\left(h\phi_\infty + f_s''\right)L_{ij}}{2}\begin{Bmatrix} 1 \\ 1 \\ 0 \\ 0 \end{Bmatrix} \tag{12.59a}$$

Similarly, for boundary conditions specified on the sides j–k, $N_i = 0$, and $N_l = 0$, and Equation 12.26 transforms into

$$\left\{f_{bs}^{(e)}\right\} = \left(h\phi_\infty + f_s''\right)\int_{S_b}\begin{Bmatrix} 0 \\ N_j \\ N_k \\ 0 \end{Bmatrix} ds = \pm\frac{\left(h\phi_\infty + f_s''\right)L_{jk}}{2}\begin{Bmatrix} 0 \\ 1 \\ 1 \\ 0 \end{Bmatrix} \tag{12.59b}$$

For boundary conditions specified on the side k–l, $N_i = 0$, and $N_j = 0$, and we get

$$\left\{f_{bs}^{(e)}\right\} = \left(h\phi_\infty + f_s''\right)\int_{S_b}\begin{Bmatrix} 0 \\ 0 \\ N_k \\ N_l \end{Bmatrix} ds = \pm\frac{\left(h\phi_\infty + f_s''\right)L_{kl}}{2}\begin{Bmatrix} 0 \\ 0 \\ 1 \\ 1 \end{Bmatrix} \tag{12.59c}$$

For boundary conditions specified on the side l–i, $N_j = 0$, and $N_k = 0$, and we get

$$\left\{f_{bs}^{(e)}\right\} = \left(h\phi_\infty + f_s''\right)\int_{S_b}\begin{Bmatrix} N_i \\ 0 \\ 0 \\ 1 \end{Bmatrix} ds = \pm\frac{\left(h\phi_\infty + f_s''\right)L_{li}}{2}\begin{Bmatrix} 1 \\ 0 \\ 0 \\ 1 \end{Bmatrix} \tag{12.59d}$$

Note that for this bilinear rectangular element, $L_{ij} = L_{kl} = 2le$ and $L_{jk} = L_{li} = 2he$.

Example 12.2: Two-dimensional Problem Using Bilinear Rectangular Element: Determine the element characteristic equation for the bilinear rectangular element (Element 5) in the finite element mesh used to solve the steady state temperature distribution and heat transfer rate in a plane carbon steel rectangular bar irradiated by constant heat flux at the top surface as shown in the figure. All surfaces of the workpiece are convectively cooled with $h = 50.0\,\text{W/m}^2\,\text{K}$ and $T_\infty = 20°\text{C}$. The workpiece is made of plane carbon steel with thermal conductivity, $k = 43\,\text{W/m°C}$. The constant heat flux at the top surface of Element 5 is $q''_s = 5 \times 10^5\,\text{W/m}^2$. The length and height of the element are $l_e = 0.05\,\text{m}$ and $h_e = 0.05\,\text{m}$ respectively. The coordinates of the nodes are: $i(0.0, 0.0)$, $j(0.05, 0.0)$, $k(0.05, 0.05)$, and $l(0.0, 0.05)$.

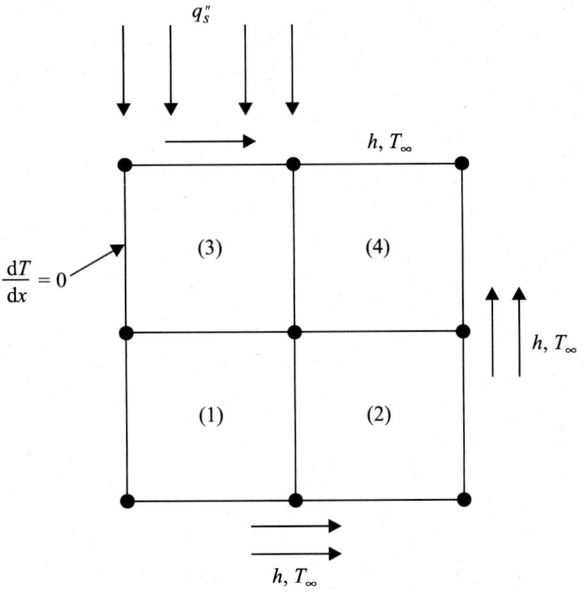

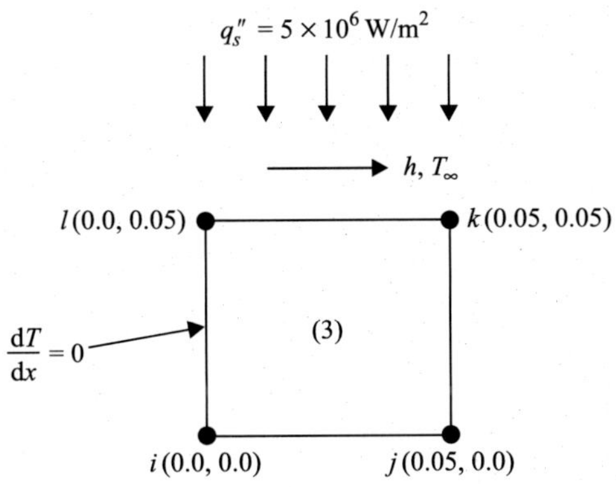

Solution: The governing heat equation for a two-dimensional region is

$$\frac{\partial}{\partial x}\left(k_x \frac{\partial T}{\partial x}\right) + \frac{\partial}{\partial y}\left(k_y \frac{\partial T}{\partial y}\right) = 0 \qquad \text{(E.12.2.1)}$$

Comparing this equation with the general equation (Equation 12.1), we get

$$\Gamma_x = k_x = k = 43\,\text{W/m°C}, \quad \Gamma_y = k_y = k = 43\,\text{W/m°C}, \quad S_1 = 0, \quad S_0 = 0$$

The boundary condition on the top surface is

$$k\frac{\partial T}{\partial y} = h(T - T_\infty) - f_s'' \qquad \text{(E.12.2.2)}$$

The rest of the sides of the element are internal.

The element characteristic matrix is evaluated based on Equation 12.28

$$k^{(e)} = \left(k_\Gamma^{(e)} + k_S^{(e)} + k_b^{(e)}\right)$$

where the element characteristic matrix component due the diffusion terms is given for a bilinear rectangular element given by Equation 12.47b as

$$\left[k_\Gamma^{(e)}\right] = \frac{kh_e}{6l_e}\begin{bmatrix} 2 & -2 & -1 & 1 \\ -2 & 2 & 1 & -1 \\ -1 & 1 & 2 & -2 \\ 1 & -1 & -2 & 2 \end{bmatrix} + \frac{kl_e}{6h_e}\begin{bmatrix} 2 & 1 & -1 & -2 \\ 1 & 2 & -2 & -1 \\ -1 & -2 & 2 & 1 \\ -2 & -1 & 1 & 2 \end{bmatrix} \qquad \text{(E.12.2.3)}$$

For the square element with $l_e = h_e$, the equation reduces to

$$\left[k_\Gamma^{(e)}\right] = \frac{43}{6}\begin{bmatrix} 4 & -1 & -2 & -1 \\ -1 & 4 & -1 & -2 \\ -2 & -1 & 4 & -1 \\ -1 & -2 & -1 & 4 \end{bmatrix} = 7.166\begin{bmatrix} 4 & -1 & -2 & -1 \\ -1 & 4 & -1 & -2 \\ -2 & -1 & 4 & -1 \\ -1 & -2 & -1 & 4 \end{bmatrix}$$

$$\left[k_\Gamma^{(e)}\right] = \begin{bmatrix} 28.666 & -7.166 & -14.333 & -7.166 \\ -7.166 & 28.666 & -7.166 & -14.333 \\ -14.333 & -7.166 & 28.666 & -7.166 \\ -7.166 & -14.333 & -7.166 & 28.666 \end{bmatrix} \qquad \text{(E.12.2.4)}$$

Since $S_1 = 0$, the component of the characteristic matrix due to the convection term is

$$\left[k_S^{(e)}\right] = 0 \qquad \text{(E.12.2.5)}$$

The component of the element characteristic matrix due to boundary conditions is given by Equation 13.24 and is computed as

$$\left[k_b^{(e)}\right] = \pm \int_{S_b} h[N]^{\mathrm{T}}[N]\, ds$$

$$\left[k_b^{(e)}\right] = \pm \int_{S_b} h \begin{bmatrix} N_i^2 & N_iN_j & N_iN_k & N_iN_l \\ N_jN_i & N_j^2 & N_jN_k & N_jN_l \\ N_kN_i & N_kN_j & N_k^2 & N_kN_l \\ N_lN_i & N_lN_j & N_lN_k & N_l^2 \end{bmatrix} ds$$

The integral is further evaluated by considering only the sides that are subjected to convection boundary conditions. Since the side k–l of the element is subjected to a convective condition, the matrix is evaluated as

$$\left[k_b^{(e)}\right] = \pm \int_{S_b} h \begin{bmatrix} 0 & 0 & 0 & 0 \\ 0 & 0 & 0 & 0 \\ 0 & 0 & N_k^2 & N_kN_l \\ 0 & 0 & N_lN_k & N_l^2 \end{bmatrix} ds = \frac{hL_{kl}}{6} \begin{bmatrix} 0 & 0 & 0 & 0 \\ 0 & 0 & 0 & 0 \\ 0 & 0 & 2 & 1 \\ 0 & 0 & 1 & 2 \end{bmatrix}$$

$$= \frac{50 \times 0.05}{6} \begin{bmatrix} 0 & 0 & 0 & 0 \\ 0 & 0 & 0 & 0 \\ 0 & 0 & 2 & 1 \\ 0 & 0 & 1 & 2 \end{bmatrix}$$

$$= \begin{bmatrix} 0 & 0 & 0 & 0 \\ 0 & 0 & 0 & 0 \\ 0 & 0 & 0.8333 & 0.4166 \\ 0 & 0 & 0.4166 & 0.8333 \end{bmatrix} \qquad \text{(E.12.2.6)}$$

The element characteristic matrix is now obtained by combining Equations E.12.2.4, E.12.2.5, and E.12.2.6

$$k^{(e)} = \left(k_\Gamma^{(e)} + k_S^{(e)} + k_b^{(e)}\right)$$

$$= \begin{bmatrix} 28.666 & -7.166 & -14.333 & -7.166 \\ -7.166 & 28.666 & -7.166 & -14.333 \\ -14.333 & -7.166 & 28.666 & -7.166 \\ -7.166 & -14.333 & -7.166 & 28.666 \end{bmatrix} + \begin{bmatrix} 0 & 0 & 0 & 0 \\ 0 & 0 & 0 & 0 \\ 0 & 0 & 0.8333 & 0.4166 \\ 0 & 0 & 0.4166 & 0.8333 \end{bmatrix}$$

$$k^{(e)} = \begin{bmatrix} 28.666 & -7.166 & -14.333 & -7.166 \\ -7.166 & 28.666 & -7.166 & -14.333 \\ -14.333 & -7.166 & 29.4993 & -6.75 \\ -7.166 & -14.333 & -6.75 & 29.4993 \end{bmatrix} \tag{E.12.2.7}$$

The element source vector is given as

$$f^{(e)} = \left\{f_S^{(e)} + f_{bs}^{(e)}\right\} \tag{E.12.2.8}$$

where $\left\{f_S^{(e)}\right\}$ is the element source vector due to the volumetric source term, which for the case of no volumetric heat generation is given as

$$\left\{f_S^{(e)}\right\} = \int_A s[N]^T \, dA = 0 \tag{E.12.2.9}$$

For the bilinear rectangular element with the mixed boundary condition specified on the side k–l, $N_i = 0$, and $N_j = 0$, and we get the source vector due to the boundary condition for the element as

$$\left\{f_{bs}^{(e)}\right\} = (hT_\infty + q_s'') \int_{S_b} \begin{Bmatrix} 0 \\ 0 \\ N_k \\ N_l \end{Bmatrix} ds = \frac{(h\phi_\infty + q_s'')L_{kl}}{2} \begin{Bmatrix} 0 \\ 0 \\ 1 \\ 1 \end{Bmatrix}$$

$$\left\{f_{bs}^{(e)}\right\} = \frac{(hT_\infty + q_s'')2l_e}{2} \begin{Bmatrix} 0 \\ 0 \\ 1 \\ 1 \end{Bmatrix} \tag{E.12.2.10}$$

Substituting Equations E.12.2.9 and E.12.2.10 into Equation E.12.2.8, we get the element source vector as

$$f^{(e)} = \frac{(hT_\infty + q_s'')2l_e}{2}\begin{Bmatrix}0\\0\\1\\1\end{Bmatrix} = \frac{(50\times 20 + 5\times 10^5)2\times 0.05}{2}\begin{Bmatrix}0\\0\\1\\1\end{Bmatrix}$$

$$f^{(e)} = \begin{Bmatrix}0\\0\\50+0.25\times 10^5\\50+0.25\times 10^5\end{Bmatrix} \tag{E.12.2.11}$$

The element characteristic matrix for the bilinear rectangular element is

$$k^{(e)}\{\phi^{(e)}\} = -\left\{I_i^{(e)}\right\} + f^{(e)}$$

$$\begin{bmatrix}28.666 & -7.166 & -14.333 & -7.166\\ -7.166 & 28.666 & -7.166 & -14.333\\ -14.333 & -7.166 & 29.4993 & -6.75\\ -7.166 & -14.333 & -6.75 & 29.4993\end{bmatrix}\begin{Bmatrix}\phi_i\\ \phi_j\\ \phi_k\\ \phi_l\end{Bmatrix}$$

$$= \begin{Bmatrix}I_i\\ I_j\\ I_k\\ I_l\end{Bmatrix} + \begin{Bmatrix}0\\0\\50+0.25\times 10^5\\50+0.25\times 10^5\end{Bmatrix} \tag{E.12.2.12}$$

Step 6: Assembly of Element Equations and Formation of Global System The assembly of all element characteristic equations to form the global system is done by using a procedure known as the direct stiffness method. Let us demonstrate the procedure by considering two triangular elements with both local and global numbers shown in Figure 12.8.

A connectivity array between the local and global numbering of elements needs to be constructed as shown in Table 12.1.

Let us consider a general expression for these elements as follows:

Element 1

$$\begin{matrix} & 1 & 3 & 4 & & & & \\ & \begin{bmatrix}k_{ii}^{(1)} & k_{ij}^{(1)} & k_{ik}^{(1)}\\ k_{ji}^{(1)} & k_{jj}^{(1)} & k_{jk}^{(1)}\\ k_{ki}^{(1)} & k_{kj}^{(1)} & k_{kk}^{(1)}\end{bmatrix} & & & \begin{Bmatrix}\phi_i\\ \phi_j\\ \phi_k\end{Bmatrix} = \begin{Bmatrix}f_i^{(1)}\\ f_j^{(1)}\\ f_k^{(1)}\end{Bmatrix} & \begin{matrix}1\\3\\4\end{matrix}\end{matrix} \tag{12.60a}$$

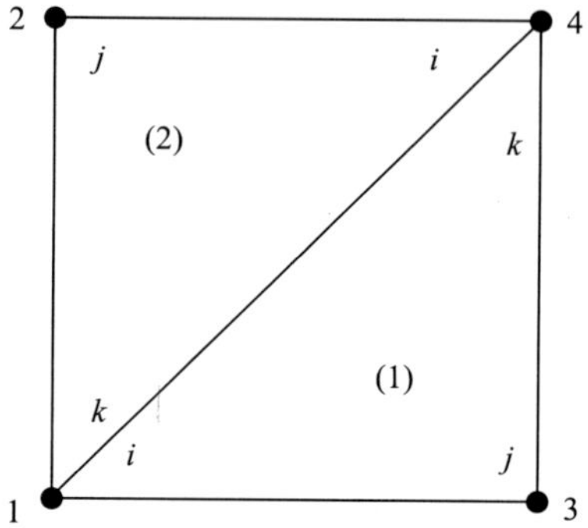

Figure 12.8 Assembly of elements

Table 12.1 Connectivity array

Element Number(e)	Global Node Numbers		
	Local Node (i)	Local Node (j)	Local Node (k)
1	1	3	4
2	4	2	1

Element 2

$$\begin{array}{c} \begin{array}{ccc} 4 & 2 & 1 \end{array} \\ \begin{bmatrix} k_{ii}^{(2)} & k_{ij}^{(2)} & k_{ik}^{(2)} \\ k_{ji}^{(2)} & k_{jj}^{(2)} & k_{jk}^{(2)} \\ k_{ki}^{(2)} & k_{kj}^{(2)} & k_{kk}^{(2)} \end{bmatrix} \end{array} \begin{Bmatrix} \phi_i \\ \phi_j \\ \phi_k \end{Bmatrix} = \begin{Bmatrix} f_i^{(1)} \\ f_j^{(1)} \\ f_k^{(1)} \end{Bmatrix} \begin{array}{c} 4 \\ 2 \\ 1 \end{array} \tag{12.60b}$$

Note that the global node numbers are written over the columns and along rows of the element characteristic matrix $[k^{(e)}]$ and along rows of the force vector $\{f^{(e)}\}$.

Let us start with the initial global characteristic equation as

$$\begin{array}{c} \begin{array}{cccc} 1 & 2 & 3 & 4 \end{array} \\ \begin{bmatrix} 0 & 0 & 0 & 0 \\ 0 & 0 & 0 & 0 \\ 0 & 0 & 0 & 0 \\ 0 & 0 & 0 & 0 \end{bmatrix} \end{array} \begin{Bmatrix} \phi_1 \\ \phi_2 \\ \phi_3 \\ \phi_4 \end{Bmatrix} = \begin{Bmatrix} 0 \\ 0 \\ 0 \\ 0 \end{Bmatrix} \begin{array}{c} 1 \\ 2 \\ 3 \\ 4 \end{array} \tag{12.61}$$

In order to form the global system, add the coefficients of the element characteristic matrix and force vector one at a time. For example, the coefficients of the element characteristics equation for element 1 can be added as

$$\text{add } k_{ii}^{(1)} \text{ to } K_{11},\ \text{add } k_{ij}^{(1)} \text{ to } K_{13},\ \text{and add } k_{ik}^{(1)} \text{ to } K_{14}$$
$$\text{add } k_{ji}^{(1)} \text{ to } K_{31},\ \text{add } k_{ij}^{(1)} \text{ to } K_{33},\ \text{and add } k_{jk}^{(1)} \text{ to } K_{34}$$
$$\text{add } k_{ki}^{(1)} \text{ to } K_{41},\ \text{add } k_{ij}^{(1)} \text{ to } K_{43},\ \text{and add } k_{kk}^{(1)} \text{ to } K_{44}$$
$$f_i^{(1)} \text{ to } F_1,\ f_j^{(1)} \text{ to } F_3,\ \text{and } f_k^{(1)} \text{ to } F_4$$

With these additions of Element 1, the global system becomes

$$\begin{matrix} 1 & 2 & 3 & 4 \end{matrix}$$
$$\begin{bmatrix} k_{ii}^{(1)} & 0 & k_{ij}^{(1)} & k_{ik}^{(1)} \\ 0 & 0 & 0 & 0 \\ k_{ji}^{(1)} & 0 & k_{jj}^{(1)} & k_{jk}^{(1)} \\ k_{ki}^{(1)} & 0 & k_{kj}^{(1)} & k_{kk}^{(1)} \end{bmatrix} \begin{Bmatrix} \phi_1 \\ \phi_2 \\ \phi_3 \\ \phi_4 \end{Bmatrix} = \begin{Bmatrix} f_i^{(1)} \\ 0 \\ f_j^{(1)} \\ f_k^{(1)} \end{Bmatrix} \begin{matrix} 1 \\ 2 \\ 3 \\ 4 \end{matrix} \qquad (12.62)$$

In a similar manner, coefficients of the element characteristic matrix and force vector of Element 2 are added to the coefficient positions of the global system and the global system takes the form

$$\begin{matrix} 1 & 2 & 3 & 4 \end{matrix}$$
$$\begin{bmatrix} k_{ii}^{(1)} + k_{kk}^{(2)} & k_{kj}^{(2)} & k_{ij}^{(1)} & k_{ik}^{(1)} + k_{ki}^{(2)} \\ k_{jk}^{(2)} & k_{jj}^{(2)} & 0 & k_{ji}^{(2)} \\ k_{ji}^{(1)} & 0 & k_{jj}^{(1)} & k_{jk}^{(1)} \\ k_{ki}^{(1)} + k_{ik}^{(2)} & k_{ij}^{(2)} & k_{kj}^{(1)} & k_{kk}^{(1)} + k_{ii}^{(2)} \end{bmatrix} \begin{Bmatrix} \phi_1 \\ \phi_2 \\ \phi_3 \\ \phi_4 \end{Bmatrix} = \begin{Bmatrix} f_i^{(1)} + f_k^{(1)} \\ f_j^{(2)} \\ f_j^{(1)} \\ f_k^{(1)} + f_i^{(2)} \end{Bmatrix} \begin{matrix} 1 \\ 2 \\ 3 \\ 4 \end{matrix} \qquad (12.63)$$

A pseudo code for the formation of the global system by assembling all the element characteristic equations using the connectivity array is given in Table 12.2.

Table 12.2 Pseudo Code for the Assembly of Element Characteristic Equations

```
do for i 1 = ne                          : ne = number of element
  do for j = 1, np                       : np = number of node in the element
    i1 = cr (j)                          : cr = connectivity array
    F(i1) = F (i1) + f(j)                : F,f = global and element force vectors
    do for k = 1, np
      j1 = cr (k)
      K(i1,j1) = K(i1,j1) + k(j,k)
    end do
  end do
end do
```

Example 12.3: Consider a square rod with uniform volumetric heat generation $\dot{q} = 3.0 \times 10^4\ \text{W/m}^3$. The rod is convectively cooled at the surface with $h = 200\ \frac{W}{m^2\,{}^\circ C}$ and $T_\infty = 10°\text{C}$

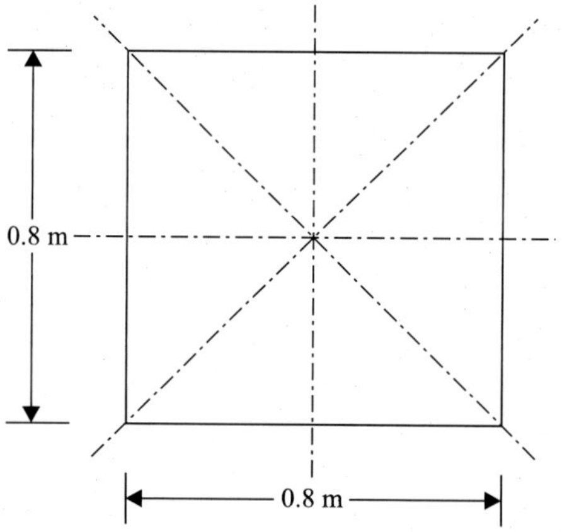

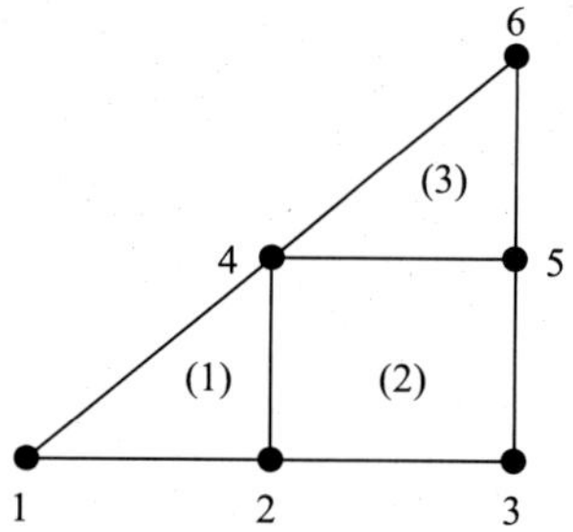

Due to the symmetry of the problem, consider only one-eighth of the region with the mesh shown in the figure. Element numbers 1 and 3 are linear triangular and element 2 is bilinear rectangular

The appropriate governing equation is

$$\frac{\partial}{\partial x}\left(k_x \frac{\partial T}{\partial x}\right) + \frac{\partial}{\partial y}\left(k_y \frac{\partial \phi}{\partial y}\right) + \dot{q} = 0 \qquad \text{(E.12.3.1)}$$

Galerkin's finite element formulation leads to

$$-\int_S [N]^{\mathrm{T}}\left(k_x \frac{\partial T^*}{\partial x}\hat{n}_x + k_y \frac{\partial T^*}{\partial y}\hat{n}_y\right) ds$$

$$+\left\{\int_A \left(\frac{\partial [N]^{\mathrm{T}}}{\partial x} k_x \frac{\partial [N]}{\partial x} + \frac{\partial [N]^{\mathrm{T}}}{\partial y} k_y \frac{\partial [N]}{\partial y}\right) dA\right\}\{T^{(e)}\}$$

$$-\int_A \dot{q}[N]^{\mathrm{T}}\, dA = 0 \qquad (12.8)$$

The element characteristics equation can be written in matrix notation as

$$\left\{I_i^{(e)}\right\} + k^{(e)}\{\phi^{(e)}\} - f'' = 0 \qquad (12.27)$$

where

$$k^{(e)} = \left(k_\Gamma^{(e)} + k_b^{(e)}\right) = \text{element characteristics matrix} \qquad (12.28)$$

and

$$\left[k_{\Gamma}^{(e)}\right] = \int_{A} \left(\frac{\partial [N]^{\mathrm{T}}}{\partial x} \Gamma_x \frac{\partial [N]}{\partial x} + \frac{\partial [N]^{\mathrm{T}}}{\partial y} \Gamma_y \frac{\partial [N]}{\partial y} \right) dA$$

$$= \int_{A} \{D\}^{\mathrm{T}} [\Gamma] \{D\} dA \tag{12.14}$$

For the ***triangular elements*** (1) and (3)

$$\left[k_{\Gamma}^{(e)}\right] = \frac{\Gamma_x}{4A} \begin{bmatrix} \beta_i^2 & \beta_i\beta_j & \beta_i\beta_k \\ \beta_j\beta_i & \beta_j^2 & \beta_j\beta_k \\ \beta_k\beta_i & \beta_k\beta_j & \beta_k^2 \end{bmatrix} + \frac{\Gamma_y}{4A} \begin{bmatrix} \gamma_i^2 & \gamma_i\gamma_j & \gamma_i\gamma_k \\ \gamma_j\gamma_i & \gamma_j^2 & \gamma_j\gamma_k \\ \gamma_k\gamma_i & \gamma_k\gamma_j & \gamma_k^2 \end{bmatrix} \tag{12.31b}$$

For the elements (1) and (3), the co-ordinates of the nodes are (0.0, 0.0), (0.2, 0.0) and (0.2, 0.2). The coefficients, α, β, and γ are calculated as

$$\alpha_i = x_j y_k - x_k y_j = 0.2(0.2) - 0.2(0.0) = 0.04$$

$$\alpha_j = x_k y_i - x_i y_k = 0.2(0.0) - 0.0(0.2) = 0.0$$

$$\alpha_k = x_i y_j - x_j y_i = 0.0(0.0) - 0.2(0) = 0.0$$

$$\beta_i = y_j - y_k = 0.0 - 0.2 = -0.2$$

$$\beta_j = y_k - y_i = 0.2 - 0.0 = 0.2$$

$$\beta_k = y_i - y_j = 0.0 - 0.0 = 0.0$$

$$\gamma_i = x_k - x_j = 0.2 - 0.2 = 0.0$$

$$\gamma_j = x_i - x_k = 0.0 - 0.2 = -0.2$$

$$\gamma_k = x_j - x_i = 0.2 - 0.0 = 0.2$$

$$A = \frac{1}{2} \begin{vmatrix} 1 & x_i & y_i \\ 1 & x_j & y_j \\ 1 & x_k & y_k \end{vmatrix} \frac{1}{2} \begin{vmatrix} 1 & x_i & y_i \\ 1 & x_j & y_j \\ 1 & x_k & y_k \end{vmatrix} = 0.02 \tag{12.3.2}$$

Substituting the coefficients, we get

$$\left[k_\Gamma^{(1)}\right] = \left[k_\Gamma^{(3)}\right] = \frac{k\beta_i^2}{4 \times A}\begin{bmatrix} 1 & -1 & 0 \\ -1 & 1 & 0 \\ 0 & 0 & 0 \end{bmatrix} + \frac{k\gamma_j^2}{4A}\begin{bmatrix} 0 & 0 & 0 \\ 0 & 1 & -1 \\ 0 & -1 & 1 \end{bmatrix}$$

$$= \frac{2.0 \times (-0.2)^2}{4 \times 0.02}\begin{bmatrix} 1 & -1 & 0 \\ -1 & 1 & 0 \\ 0 & 0 & 0 \end{bmatrix} + \frac{2.0 \times (-0.2)^2}{4 \times 0.02}\begin{bmatrix} 0 & 0 & 0 \\ 0 & 1 & -1 \\ 0 & -1 & 1 \end{bmatrix}$$

$$= \frac{2.0 \times (-0.2)^2}{4 \times 0.02}\begin{bmatrix} 1 & -1 & 0 \\ -1 & 2 & -1 \\ 0 & -1 & 1 \end{bmatrix}$$

$$\left[k_\Gamma^{(1)}\right] = \left[k_\Gamma^{(3)}\right] = 1.0\begin{bmatrix} 1 & -1 & 0 \\ -1 & 2 & -1 \\ 0 & -1 & 1 \end{bmatrix} \qquad \text{(E.12.3.3)}$$

For the ***rectangular element (2)***

$$\left[k_\Gamma^{(2)}\right] = \frac{k_x h_e}{6 l_e}\begin{bmatrix} 2 & -2 & -1 & 1 \\ -2 & 2 & 1 & -1 \\ -1 & 1 & 2 & -2 \\ 1 & -1 & -2 & 2 \end{bmatrix} + \frac{k_y l_e}{6 h_e}\begin{bmatrix} 2 & 1 & -1 & -2 \\ 1 & 2 & -2 & -1 \\ -1 & -2 & 2 & 1 \\ -2 & -1 & 1 & 2 \end{bmatrix}$$

$$= \frac{2.0 \times 0.2}{6 \times 0.2}\begin{bmatrix} 2 & -2 & -1 & 1 \\ -2 & 2 & 1 & -1 \\ -1 & 1 & 2 & -2 \\ 1 & -1 & -2 & 2 \end{bmatrix} + \frac{2.0 \times 0.2}{6 \times 0.2}\begin{bmatrix} 2 & 1 & -1 & -2 \\ 1 & 2 & -2 & -1 \\ -1 & -2 & 2 & 1 \\ -2 & -1 & 1 & 2 \end{bmatrix}$$

$$\left[k_\Gamma^{(2)}\right] = \frac{1}{3}\begin{bmatrix} 4 & -1 & -2 & -1 \\ -1 & 4 & -1 & -2 \\ -2 & -1 & 4 & -1 \\ -1 & -2 & -1 & 4 \end{bmatrix} \qquad \text{(E.12.3.4)}$$

The component of element characteristic matrix due to convective boundary conditions as given by Equation 12.24 for the triangular element is computed as

$$\left[k_b^{(e)}\right] = \pm \int_{s_b} h[N]^{\mathrm{T}}[N]\,ds$$

$$= \pm h \int_{S_b} \begin{bmatrix} N_i^2 & N_iN_j & N_iN_k \\ N_jN_i & N_j^2 & N_jN_k \\ N_kN_i & N_kN_j & N_k^2 \end{bmatrix} ds \tag{E.12.3.5}$$

For the ***Element 1***, there are no convective boundary conditions on the sides. So, the element characteristic matrix due to the convective boundary is

$$\left[k_b^{(1)}\right] = 0 \tag{E.12.3.6a}$$

For ***Element 3***, the convective boundary conditions is specified on the side j–k. So, the element characteristic matrix due to the convective boundary is given by Equation 12.38b as

$$\left[k_b^{(3)}\right] = h \int_{S_b} \begin{bmatrix} 0 & 0 & 0 \\ 0 & N_j^2 & N_jN_k \\ 0 & N_kN_j & N_k^2 \end{bmatrix} ds = \frac{hL_{jk}}{6} \begin{bmatrix} 0 & 0 & 0 \\ 0 & 2 & 1 \\ 0 & 1 & 2 \end{bmatrix}$$

$$= \frac{200 \times 0.2}{6} \begin{bmatrix} 0 & 0 & 0 \\ 0 & 2 & 1 \\ 0 & 1 & 2 \end{bmatrix}$$

$$= \frac{20}{3} \begin{bmatrix} 0 & 0 & 0 \\ 0 & 2 & 1 \\ 0 & 1 & 2 \end{bmatrix} \tag{E.12.3.6b}$$

The component of the element characteristic matrix due to convective boundary conditions for the bilinear rectangular element is given by Equation 12.50 as

$$\left[k_b^{(e)}\right] = \int_{s_b} h[N]^{\mathrm{T}}[N]\,ds$$

$$= \int_{S_b} h \begin{bmatrix} N_i^2 & N_iN_j & N_iN_k & N_iN_l \\ N_jN_i & N_j^2 & N_jN_k & N_jN_l \\ N_kN_i & N_kN_j & N_k^2 & N_kN_l \\ N_lN_i & N_lN_j & N_lN_k & N_l^2 \end{bmatrix} ds$$

Since the convective boundary condition is specified on the side j–k of ***Element 2***, Equation 12.50 reduces to Equation 12.53b and given as

$$\left[k_b^{(2)}\right] = \int_{S_b} h \begin{bmatrix} 0 & 0 & 0 & 0 \\ 0 & N_j^2 & N_jN_k & 0 \\ 0 & N_kN_j & N_k^2 & 0 \\ 0 & 0 & 0 & 0 \end{bmatrix} ds = \frac{hL_{jk}}{6}\begin{bmatrix} 0 & 0 & 0 & 0 \\ 0 & 2 & 1 & 0 \\ 0 & 1 & 2 & 0 \\ 0 & 0 & 0 & 0 \end{bmatrix}$$

$$= \frac{200 \times 0.2}{6}\begin{bmatrix} 0 & 0 & 0 & 0 \\ 0 & 2 & 1 & 0 \\ 0 & 1 & 2 & 0 \\ 0 & 0 & 0 & 0 \end{bmatrix}$$

$$\left[k_b^{(2)}\right] = \frac{20}{3}\begin{bmatrix} 0 & 0 & 0 & 0 \\ 0 & 2 & 1 & 0 \\ 0 & 1 & 2 & 0 \\ 0 & 0 & 0 & 0 \end{bmatrix} \qquad \text{(E.12.3.6c)}$$

The element source vector is given as

$$f^{(e)} = \left\{f_S^{(e)} + f_{bs}^{(e)}\right\} = \text{element source vector}$$

where the force vector for the constant volumetric heat source is

$$\left\{f_S^{(e)}\right\} = \int_A \dot{q}[N]^{\mathrm{T}}\, dA \qquad (12.11)$$

For triangular ***Elements 1*** and ***3***, Equation 12.11 reduces to

$$\left\{f_S^{(e)}\right\} = \dot{q}\int_A \begin{bmatrix} N_i \\ N_j \\ N_k \end{bmatrix} dA$$

$$\left\{f_S^{(1)}\right\} = \left\{f_S^{(3)}\right\} = \frac{\dot{q}A}{3}\begin{Bmatrix} 1 \\ 1 \\ 1 \end{Bmatrix} = \frac{3.0 \times 10^4 \times 0.02}{3}\begin{Bmatrix} 1 \\ 1 \\ 1 \end{Bmatrix}$$

$$= 0.02 \times 10^4 \begin{Bmatrix} 1 \\ 1 \\ 1 \end{Bmatrix} \qquad \text{(E.12.3.7a)}$$

For the bilinear rectangular ***Element 2***, Equation 12.11 reduces to

$$f_S^{(2)} = \int_A \dot{q} \begin{Bmatrix} N_i \\ N_j \\ N_k \\ N_l \end{Bmatrix} dA$$

$$f_S^{(2)} = \frac{\dot{q}A}{4} \begin{Bmatrix} 1 \\ 1 \\ 1 \\ 1 \end{Bmatrix} = \frac{3.0 \times 10^4}{4} \begin{Bmatrix} 1 \\ 1 \\ 1 \\ 1 \end{Bmatrix} = 0.75 \times 10^4 \begin{Bmatrix} 1 \\ 1 \\ 1 \\ 1 \end{Bmatrix} \quad \text{(E.12.3.7b)}$$

The element force vector component due to the convective boundary condition is computed by Equation 12.25 as

$$\left\{ f_{bs}^{(e)} \right\} = \pm \int_{s_b} [N]^{\mathrm{T}} \{(h\phi_\infty)\} \, ds$$

For triangular ***Element 1***, there is no convective boundary. So, the equation gives

$$\left\{ f_{bs}^{(1)} \right\} = 0 \quad \text{(E.12.3.8a)}$$

For triangular ***Element 3***, the convective boundary condition is specified on the side *j*–*k*. So, the source vector due to convective boundary for ***Element 3*** reduces to

$$\left\{ f_{bs}^{(3)} \right\} = (hT_\infty) \int_{S_b} \begin{Bmatrix} N_i \\ N_j \\ N_k \end{Bmatrix} ds$$

$$= (h\phi_\infty) \int_{S_b} \begin{Bmatrix} 0 \\ N_j \\ N_k \end{Bmatrix} ds = \frac{(hT_\infty)L_{jk}}{2} \begin{Bmatrix} 0 \\ 1 \\ 1 \end{Bmatrix}$$

$$= \frac{(200 \times 10)0.2}{2} \begin{Bmatrix} 0 \\ 1 \\ 1 \end{Bmatrix}$$

$$\left\{ f_{bs}^{(3)} \right\} = 200 \begin{Bmatrix} 0 \\ 1 \\ 1 \end{Bmatrix} \quad \text{(E.12.3.8b)}$$

For the bilinear rectangular ***Element 2***, the convective boundary condition is specified on the side *j–k*. So, the source vector due to the convective boundary for Element 2 reduces to

$$\left\{f_{bs}^{(2)}\right\} = (hT_\infty)\int_{S_b}\begin{Bmatrix} N_i \\ N_j \\ N_k \\ N_l \end{Bmatrix} ds$$

$$\left\{f_{bs}^{(2)}\right\} = (hT_\infty)\int_{S_b}\begin{Bmatrix} 0 \\ N_j \\ N_k \\ 0 \end{Bmatrix} ds = \frac{(h\phi_\infty)L_{jk}}{2}\begin{Bmatrix} 0 \\ 1 \\ 1 \\ 0 \end{Bmatrix}$$

$$= \frac{(200\times 10)0.2}{2}\begin{Bmatrix} 0 \\ 1 \\ 1 \\ 0 \end{Bmatrix}$$

$$\left\{f_{bs}^{(2)}\right\} = 200\begin{Bmatrix} 0 \\ 1 \\ 1 \\ 0 \end{Bmatrix} \tag{E.12.3.8c}$$

The element characteristic equations for the elements are as follows.

Element 1

$$k^{(1)}\{\phi^{(1)}\} = -\left\{I_i^{(1)}\right\} + f^{(1)}$$

$$\left\{\left[k_\Gamma^{(1)}\right] + \left[k_b^{(1)}\right]\right\}\{\phi^{(1)}\} = -\left\{I_i^{(1)}\right\} + \left\{f_S^{(1)} + f_{bs}^{(1)}\right\}$$

$$1.0\begin{bmatrix} 1 & -1 & 0 \\ -1 & 2 & -1 \\ 0 & -1 & 1 \end{bmatrix}\begin{Bmatrix} \phi_i^{(1)} \\ \phi_j^{(1)} \\ \phi_k^{(1)} \end{Bmatrix} = -\begin{Bmatrix} I_i^{(1)} \\ I_j^{(1)} \\ I_k^{(1)} \end{Bmatrix} + 0.02\times 10^4\begin{Bmatrix} 1 \\ 1 \\ 1 \end{Bmatrix}$$

$$\frac{1}{3}\begin{bmatrix} 3 & -3 & 0 \\ -3 & 6 & -3 \\ 0 & -3 & 3 \end{bmatrix}\begin{Bmatrix} \phi_i^{(1)} \\ \phi_j^{(1)} \\ \phi_k^{(1)} \end{Bmatrix} = -\begin{Bmatrix} I_i^{(1)} \\ I_j^{(1)} \\ I_k^{(1)} \end{Bmatrix} + 0.02\times 10^4\begin{Bmatrix} 1 \\ 1 \\ 1 \end{Bmatrix} \tag{E.12.3.9a}$$

Element 2

$$k^{(2)}\{\phi^{(2)}\} = -\left\{I_i^{(2)}\right\} + f^{(2)}$$

$$\left\{\left[k_\Gamma^{(2)}\right] + \left[k_b^{(2)}\right]\right\}\{\phi^{(2)}\} = -\left\{I_i^{(2)}\right\} + \left\{f_S^{(2)} + f_{bs}^{(2)}\right\}$$

$$\left\{\frac{1}{3}\begin{bmatrix} 4 & -1 & -2 & -1 \\ -1 & 4 & -1 & -2 \\ -2 & -1 & 4 & -1 \\ -1 & -2 & -1 & 4 \end{bmatrix} + \frac{20}{3}\begin{bmatrix} 0 & 0 & 0 & 0 \\ 0 & 2 & 1 & 0 \\ 0 & 1 & 2 & 0 \\ 0 & 0 & 0 & 0 \end{bmatrix}\right\}\begin{Bmatrix} \phi_i^{(2)} \\ \phi_j^{(2)} \\ \phi_k^{(2)} \\ \phi_k^{(2)} \end{Bmatrix}$$

$$= -\left\{I_i^{(2)}\right\} + 0.75 \times 10^4\begin{Bmatrix} 1 \\ 1 \\ 1 \\ 1 \end{Bmatrix} + 200\begin{Bmatrix} 0 \\ 1 \\ 1 \\ 0 \end{Bmatrix}$$

or

$$\frac{1}{3}\begin{bmatrix} 4 & -1 & -2 & -1 \\ -1 & 44 & 19 & -2 \\ -2 & 19 & 44 & -1 \\ -1 & -2 & -1 & 4 \end{bmatrix}\begin{Bmatrix} \phi_i^{(2)} \\ \phi_j^{(2)} \\ \phi_k^{(2)} \\ \phi_k^{(2)} \end{Bmatrix}$$

$$= -\begin{Bmatrix} I_i^{(2)} \\ I_j^{(2)} \\ I_k^{(2)} \\ I_l^{(2)} \end{Bmatrix} + 0.75 \times 10^4\begin{Bmatrix} 1 \\ 1 \\ 1 \\ 1 \end{Bmatrix} + 200\begin{Bmatrix} 0 \\ 1 \\ 1 \\ 0 \end{Bmatrix} \qquad \text{(E.12.3.9b)}$$

Element 3

$$k^{(3)}\{\phi^{(3)}\} = -\left\{I_i^{(3)}\right\} + f^{(3)}$$

$$\left[k_\Gamma^{(3)}\right] + \left[k_b^{(3)}\right]\{\phi^{(3)}\} = -\left\{I_i^{(3)}\right\} + \left\{f_S^{(3)} + f_{bs}^{(3)}\right\}$$

$$\left\{\frac{1}{3}\begin{bmatrix} 3 & -3 & 0 \\ -3 & 6 & -3 \\ 0 & -3 & 3 \end{bmatrix} + \frac{20}{3}\begin{bmatrix} 0 & 0 & 0 \\ 0 & 2 & 1 \\ 0 & 1 & 2 \end{bmatrix}\right\}\begin{Bmatrix} \phi_i^{(3)} \\ \phi_j^{(3)} \\ \phi_k^{(3)} \end{Bmatrix}$$

$$= -\begin{Bmatrix} I_i^{(3)} \\ I_j^{(3)} \\ I_k^{(3)} \end{Bmatrix} + 0.02 \times 10^4 \begin{Bmatrix} 1 \\ 1 \\ 1 \end{Bmatrix} + 200 \begin{Bmatrix} 0 \\ 1 \\ 1 \end{Bmatrix}$$

$$\frac{1}{3}\begin{bmatrix} 3 & -3 & 0 \\ -3 & 46 & -17 \\ 0 & -17 & 43 \end{bmatrix}\begin{Bmatrix} \phi_i^{(3)} \\ \phi_j^{(3)} \\ \phi_k^{(3)} \end{Bmatrix}$$

$$= -\begin{Bmatrix} I_i^{(3)} \\ I_j^{(3)} \\ I_k^{(3)} \end{Bmatrix} + 0.02 \times 10^4 \begin{Bmatrix} 1 \\ 1 \\ 1 \end{Bmatrix} + 200 \begin{Bmatrix} 0 \\ 1 \\ 1 \end{Bmatrix} \tag{E.12.3.9c}$$

In the next step all element equations are assembled to form the global system. In order to assemble the systems in a systematic manner, a connectivity array is defined for the mesh in the following manner.

Element Number (e)	i	j	k	l
1	1	2	4	
2	2	3	5	4
3	4	5	6	

In order to demonstrate the formation of the global system let us start with an initial global system for the mesh as

$$\begin{bmatrix} 0 & 0 & 0 & 0 & 0 & 0 \\ 0 & 0 & 0 & 0 & 0 & 0 \\ 0 & 0 & 0 & 0 & 0 & 0 \\ 0 & 0 & 0 & 0 & 0 & 0 \\ 0 & 0 & 0 & 0 & 0 & 0 \\ 0 & 0 & 0 & 0 & 0 & 0 \end{bmatrix}\begin{Bmatrix} T_1 \\ T_2 \\ T_3 \\ T_4 \\ T_5 \\ T_6 \end{Bmatrix} = \begin{Bmatrix} 0 \\ 0 \\ 0 \\ 0 \\ 0 \\ 0 \end{Bmatrix} + \begin{Bmatrix} 0 \\ 0 \\ 0 \\ 0 \\ 0 \\ 0 \end{Bmatrix}$$

Addition of Element 1: Associated connectivity array $-\{1,2,4\}$

$$\frac{1}{3}\begin{bmatrix} 3 & -3 & 0 & 0 & 0 & 0 \\ -3 & 6 & 0 & -3 & 0 & 0 \\ 0 & 0 & 0 & 0 & 0 & 0 \\ 0 & -3 & 0 & 3 & 0 & 0 \\ 0 & 0 & 0 & 0 & 0 & 0 \\ 0 & 0 & 0 & 0 & 0 & 0 \end{bmatrix}\begin{Bmatrix} T_1 \\ T_2 \\ T_3 \\ T_4 \\ T_5 \\ T_6 \end{Bmatrix} = \begin{Bmatrix} I_1^{(1)} \\ I_2^{(1)} \\ \\ I_4^{(1)} \\ 0 \\ 0 \end{Bmatrix} + \begin{Bmatrix} 0.02\times 10^4 \\ 0.02\times 10^4 \\ 0 \\ 0.02\times 10^4 \\ 0 \\ 0 \end{Bmatrix}$$

Addition of Element 2: Associated connectivity array $-\{2,3,5,4\}$

$$\frac{1}{3}\begin{bmatrix} 3 & -3 & 0 & 0 & 0 & 0 \\ -3 & 6+4 & 0-1 & -3-1 & 0-2 & 0 \\ 0 & 0-1 & 0+44 & 0-2 & 0+19 & 0 \\ 0 & -3-1 & 0-2 & 3+4 & 0-1 & 0 \\ 0 & 0-2 & 0+19 & 0-1 & 0+44 & 0 \\ 0 & 0 & 0 & 0 & 0 & 0 \end{bmatrix}\begin{Bmatrix} T_1 \\ T_2 \\ T_3 \\ T_4 \\ T_5 \\ T_6 \end{Bmatrix}$$

$$= \begin{Bmatrix} I_1^{(1)} \\ I_2^{(1)} + I_2^{(2)} \\ I_3^{(2)} \\ I_4^{(1)} + I_4^{(2)} \\ I_5^{(2)} \\ 0 \end{Bmatrix} + \begin{Bmatrix} 0.02\times 10^4 \\ (0.02+0.75)\times 10^4 \\ 0+0.75\times 10^4 \\ (0.02+0.75)\times 10^4 \\ 0+0.75\times 10^4 \\ 0 \end{Bmatrix} + 200\begin{Bmatrix} 0 \\ 0 \\ 1 \\ 0 \\ 1 \\ 0 \end{Bmatrix}$$

Addition of Element 3: Associated connectivity array $-\{4,5,6\}$

$$\frac{1}{3}\begin{bmatrix} 3 & -3 & 0 & 0 & 0 & 0 \\ -3 & 6+4 & 0-1 & -3-1 & 0-2 & 0 \\ 0 & 0-1 & 0+44 & 0-2 & 0+19 & 0 \\ 0 & -3-1 & 0-2 & 3+4+3 & 0-1-3 & 0 \\ 0 & 0-2 & 0+19 & 0-1-3 & 0+44+46 & -17 \\ 0 & 0 & 0 & 0 & -17 & 43 \end{bmatrix}\begin{Bmatrix} T_1 \\ T_2 \\ T_3 \\ T_4 \\ T_5 \\ T_6 \end{Bmatrix}$$

$$= \begin{Bmatrix} I_1^{(1)} \\ I_2^{(1)} + I_2^{(2)} \\ I_3^{(1)} + I_3^{(2)} \\ I_4^{(2)} + I_4^{(3)} \\ I_5^{(2)} + I_5^{(3)} \\ I_6^{(3)} \end{Bmatrix} + \begin{Bmatrix} 0.02\times 10^4 \\ (0.02+0.75)\times 10^4 \\ 0+0.75\times 10^4 \\ (0.02+0.75+0.02)\times 10^4 \\ 0+(0.75+0.02)\times 10^4 \\ 0.02\times 10^4 \end{Bmatrix} + 200\begin{Bmatrix} 0 \\ 0 \\ 1 \\ 0 \\ 1+1 \\ 0+1 \end{Bmatrix}$$

Because of the continuity condition, the inter-element condition vanishes and the global system becomes

$$\frac{1}{3}\begin{bmatrix} 3 & -3 & 0 & 0 & 0 & 0 \\ -3 & 10 & -1 & -4 & -2 & 0 \\ 0 & -1 & 44 & -2 & 19 & 0 \\ 0 & -4 & -2 & 10 & -4 & 0 \\ 0 & -2 & 19 & -4 & 90 & -17 \\ 0 & 0 & 0 & 0 & -17 & 43 \end{bmatrix}\begin{Bmatrix} T_1 \\ T_2 \\ T_3 \\ T_4 \\ T_5 \\ T_6 \end{Bmatrix} = \begin{Bmatrix} 0.02 \times 10^4 \\ 0.77 \times 10^4 \\ 0.75 \times 10^4 \\ 0.79 \times 10^4 \\ 0.77 \times 10^4 \\ 0.02 \times 10^4 \end{Bmatrix} + 200\begin{Bmatrix} 0 \\ 0 \\ 1 \\ 0 \\ 2 \\ 1 \end{Bmatrix}$$

12.2 TWO-DIMENSIONAL AXISYMMETRIC PROBLEMS

We have discussed before that many higher-dimensional problems can be approximated as two-dimensional problems due to symmetry in geometry and boundary conditions. Problems dealing with cylindrical and spherical geometries are analyzed using cylindrical (r, θ, z) and spherical coordinates (r, θ, ϕ). Many of these may also exhibit symmetry in geometry and boundary conditions around the axis of rotation or in the θ directions. In such situations, a three-dimension problem such as in (r, θ, z) for cylindrical geometries or in (r, θ, ϕ) for spheres can be approximated as a two-dimensional axisymmetric problem, i.e., a two-dimensional (r, z) problem for cylindrical geometries and a two dimensional (r, ϕ) problem for spherical geometries. Let us demonstrate the finite element formulation of a two-dimensional axisymmetric problem for the cylindrical rod shown in Figure 12.9.

It can be seen that the problem can be solved in the r–z plane, which is identical to the x–y plane discussed in previous sections. However, as we have seen in one-dimensional axisymmetric problems, even though the finite element

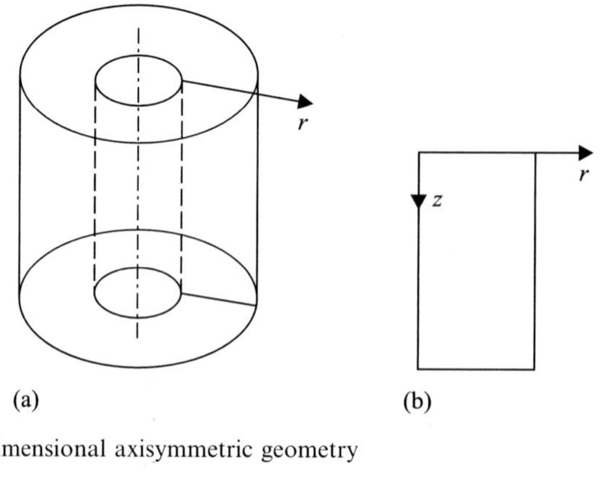

Figure 12.9 Two-dimensional axisymmetric geometry

formulation is similar to that in a Cartesian coordinate there is some difference due to the variable surface area in the radial direction.

Let us start with the three-dimensional steady state diffusion equation in cylindrical coordinate of the form

$$\frac{1}{r}\frac{\partial}{\partial r}\left(\Gamma_r r\frac{\partial\phi}{\partial r}\right)+\frac{1}{r^2}\frac{\Gamma\partial}{\partial\theta}\left(\Gamma_\theta\frac{\partial\phi}{\partial\theta}\right)+\frac{\partial}{\partial z}\left(\Gamma_z\frac{\partial\phi}{\partial z}\right)+S=0 \tag{12.64}$$

For a two-dimensional axisymmetric problem with symmetry around the axis of rotation or θ direction, Equation 12.64 reduces to

$$\frac{1}{r}\frac{\partial}{\partial r}\left(\Gamma_r r\frac{\partial\phi}{\partial r}\right)+\frac{\partial}{\partial z}\left(\Gamma_z\frac{\partial\phi}{\partial z}\right)+S=0$$

Let us assume the associate general boundary conditions as follows.

On the part of boundary surface s_1

Boundary condition is of the first kind

$$\phi=\phi_s=\text{constant surface temperature} \tag{12.65a}$$

On the part of boundary surface s_2

Boundary condition of the second kind

$$\Gamma_r\frac{\partial\phi}{\partial r}\hat{n}_r+\Gamma_z\frac{\partial\phi}{\partial z}\hat{n}_z=\pm f_s''=\text{constant surface flux} \tag{12.65b}$$

Again note that for the special case with adiabatic conditions or for the line of symmetry, this boundary condition reduces to

$$\Gamma_r\frac{\partial\phi}{\partial r}\hat{n}_r+\Gamma_z\frac{\partial\phi}{\partial z}\hat{n}_z=0 \tag{12.65c}$$

On the boundary surface s_3

Boundary condition of the third kind

$$\Gamma_r\frac{\partial\phi}{\partial r}\hat{n}_r+\Gamma_z\frac{\partial\phi}{\partial z}\hat{n}_z=\pm h(\phi-\phi_s)\pm f_s'' \tag{12.65d}$$

All these boundary conditions are assumed to be independent of the angular directions θ.

Selection of Approximate Solution Function The approximate solution function for a typical element is assumed to be

$$\phi^*=\sum_{I=1}^{n_p}N_I\phi_I=[N]\{\phi^{(e)}\} \tag{12.2}$$

where ϕ_I represents the nodal values, n_p is the number of nodes in an element, and N_I is the shape function. Discussion of a shape function for an element in an axisymmetric problem was given in a subsequent section.

Formulation of an Integral Statement The integral form of the equation over an element is obtained by applying ***Galerkin's weighted residual method*** as

$$\{R^{(e)}\} = -\int_V [N]^{\mathrm{T}}\left[\frac{1}{r}\frac{\partial}{\partial r}\left(\Gamma_r r\frac{\partial\phi^*}{\partial r}\right) + \frac{\partial}{\partial z}\left(\Gamma_z\frac{\partial\phi^*}{\partial z}\right) + S\right] dV = 0 \qquad (12.66)$$

The first two integral terms are evaluated as

$$\int_V [N]^{\mathrm{T}}\frac{1}{r}\frac{\partial}{\partial r}\left(\Gamma_r r\frac{\partial\phi^*}{\partial r}\right)dV = \int_V \frac{1}{r}\frac{\partial}{\partial r}\left([N]^{\mathrm{T}}\Gamma_r r\frac{\partial\phi^*}{\partial r}\right)dV - \int_V \Gamma_r\frac{\partial\phi^*}{\partial r}\frac{\partial[N]^{\mathrm{T}}}{\partial r}dV \qquad (12.67a)$$

and

$$\int_V [N]^{\mathrm{T}}\frac{\partial}{\partial z}\left(\Gamma_z\frac{\partial\phi^*}{\partial z}\right)dV = \int_V \frac{\partial}{\partial z}\left([N]^{\mathrm{T}}\Gamma_z\frac{\partial\phi^*}{\partial z}\right)dV - \int_V \Gamma z\frac{\partial\phi^*}{\partial z}\frac{\partial[N]^{\mathrm{T}}}{\partial z}dV \qquad (12.67b)$$

Substituting Equations 12.67 into Equation 12.66, we have

$$\int_V\left[\Gamma_r\frac{\partial\phi^*}{\partial r}\frac{\partial[N]^{\mathrm{T}}}{\partial r} + \Gamma_z\frac{\partial\phi^*}{\partial z}\frac{\partial[N]^{\mathrm{T}}}{\partial z} - N^{\mathrm{T}}S\right]dV$$

$$-\int_V\left[\frac{1}{r}\frac{\partial}{\partial r}\left([N]^{\mathrm{T}}\Gamma_r r\frac{\partial\phi^*}{\partial r}\right) + \frac{\partial}{\partial z}\left([N]^{\mathrm{T}}\Gamma_z\frac{\partial\phi^*}{\partial z}\right)\right]dV = 0 \qquad (12.68)$$

Using the Gauss theorem, the second volume integral term on the right-hand side of Equation 12.68 can be transformed into a surface integral as

$$\int_V\left[\frac{1}{r}\frac{\partial}{\partial r}\left([N]^{\mathrm{T}}\Gamma_r r\frac{\partial\phi^*}{\partial r}\right) + \frac{\partial}{\partial z}\left([N]^{\mathrm{T}}\Gamma_z\frac{\partial\phi^*}{\partial z}\right)\right]dV$$

$$= \oint_s\left[[N]^{\mathrm{T}}\Gamma_r\frac{\partial\phi^*}{\partial r}\hat{n}_r + [N]^T\Gamma_z\frac{\partial\phi^*}{\partial z}\hat{n}_z\right]ds = 0 \qquad (12.69)$$

Substituting Equation 12.69 into Equation 12.68

$$\int_V\left[\Gamma_r\frac{\partial\phi^*}{\partial r}\frac{\partial[N]^{\mathrm{T}}}{\partial r} + \Gamma_z\frac{\partial\phi^*}{\partial z}\frac{\partial[N]^{\mathrm{T}}}{\partial z} - N^{\mathrm{T}}S\right]dV$$

$$-\oint_s[N]^{\mathrm{T}}\left[\Gamma_r\frac{\partial\phi^*}{\partial r}\hat{n}_r + \Gamma_z\frac{\partial\phi^*}{\partial z}\hat{n}_z\right]ds = 0 \qquad (12.70)$$

Formation of Element Characteristics Equation Formation of the element characteristics equation is done by substituting the approximate solution function given by Equation 12.2 into the integral form

$$\left\{\int_V \left(\frac{\partial[N]^{\mathrm{T}}}{\partial r}\Gamma_r\frac{\partial[N]}{\partial r}+\frac{\partial[N]^{\mathrm{T}}}{\partial z}\Gamma_z\frac{\partial[N]}{\partial z}\right)dV\right\}\{\phi^{(e)}\}$$

$$-\int_S [N]^{\mathrm{T}}\left(\Gamma_r\frac{\partial\phi^*}{\partial r}\hat{n}_r+\Gamma_z\frac{\partial\phi^*}{\partial z}\hat{n}_z\right)ds-\int_V S[N]^{\mathrm{T}}dV=0 \tag{12.71}$$

The element characteristics equation can be written in matrix notation as

$$\{I^{(e)}\}+[k^{(e)}]\{\phi^{(e)}\}-\left\{f_S^{(e)}\right\}=0 \tag{12.72}$$

where

$\{I^{(e)}\}$ = column vector for the element contribution to the inter-element requirement

$$=-\int_S [N]^{\mathrm{T}}\left(\Gamma_r\frac{\partial\phi^*}{\partial r}\hat{n}_r+\Gamma_z\frac{\partial\phi^*}{\partial z}\hat{n}_z\right)ds \tag{12.73}$$

$\left\{f_S^{(e)}\right\}$ = element source vector due to the volumetric source term

$$=\int_V S[N]^{\mathrm{T}}dV \tag{12.74}$$

$\{\phi^{(e)}\}$ = column vector of unknown nodal values

$[k^{(e)}]$ = element characteristics or stiffness matrix

$$=\int_V \left(\frac{\partial[N]^{\mathrm{T}}}{\partial r}\Gamma_r\frac{\partial[N]}{\partial r}+\frac{\partial[N]^{\mathrm{T}}}{\partial z}\Gamma_z\frac{\partial[N]}{\partial z}\right)dV$$

$$=\left[k_\Gamma^{(e)}\right]$$

$$=\int_V \{D\}^{\mathrm{T}}[\Gamma]\{D\}dV \tag{12.75}$$

where

$$[\Gamma] = \text{transport property matrix} = \begin{bmatrix} \Gamma_r & 0 \\ 0 & \Gamma_z \end{bmatrix} \quad (12.76)$$

$$\{D\} = \text{column vector with derivatives of shape functions}$$

$$= \begin{Bmatrix} \partial[N]/\partial r \\ \partial[N]/\partial z \end{Bmatrix} \quad (12.77)$$

$$\{D\}^{\mathrm{T}} = \text{row vector with derivatives of shape functions}$$

$$= \begin{Bmatrix} \dfrac{\partial[N]}{\partial r} & \dfrac{\partial[N]}{\partial z} \end{Bmatrix} \quad (12.78)$$

Implementation of Boundary Conditions The implementation of the boundary conditions into the finite element formulation takes place through the inter-element vector $\{I^{(e)}\}$ as given by Equation 12.10

$$\{I^{(e)}\} = -\int_{s} [N]^{\mathrm{T}}\left(\Gamma_r \frac{\partial \phi^*}{\partial r}\hat{n}_r + \Gamma_z \frac{\partial \phi^*}{\partial z}\hat{n}_z\right) ds \quad (12.79)$$

We have indicated before that this integral is the sum of all sides of an element including the side or sides that are subjected to the boundary conditions. In order to show the specific contribution from the boundary conditions, let us decompose the integral into two parts as

$$\{I^{(e)}\} = \left\{I_b^{(e)}\right\} + \left\{I_i^{(e)}\right\} \quad (12.80)$$

where $\left\{I_i^{(e)}\right\}$ is the integral contribution from all the other sides other than the boundary surface and represents the inter-element requirements. The component $\left\{I_b^{(e)}\right\}$ is the integral contribution from sides that represent the boundary, and it is expressed as follows:

$$\left\{I_b^{(e)}\right\} - \int_{s_b} [N]^{\mathrm{T}}\left(\Gamma_r \frac{\partial \phi^*}{\partial r}\hat{n}_r + \Gamma_z \frac{\partial \phi^*}{\partial z}\hat{n}_z\right) ds \quad (12.81)$$

Substituting the boundary condition given by Equation 12.65d, we get

$$\left\{I_b^{(e)}\right\} = -\int_{s_b} [N]^{\mathrm{T}}\{\pm h(\phi^* - \phi_\infty) \pm f_s''\} ds \quad (12.82a)$$

or

$$\left\{I_b^{(e)}\right\} = \pm\int_{s_b} h[N]^{\mathrm{T}}\phi^* ds \pm \int_{s_b} [N]^{\mathrm{T}}\{(h\phi_\infty + f_s'')\}ds \tag{12.82b}$$

Substituting the approximate solution given by Equation 12.2, we get

$$\left\{I_b^{(e)}\right\} = \pm\left(\int_{s_b} h[N]^{\mathrm{T}}[N]\,ds\right)\phi^{(e)} \pm \int_{s_b} [N]^{\mathrm{T}}\{(h\phi_\infty + f_s'')\}ds \tag{12.83a}$$

or

$$\left\{I_b^{(e)}\right\} = \left[k_b^{(e)}\right]\{\phi^{(e)}\} - \left\{f_{bs}^{(e)}\right\} \tag{12.83b}$$

where

$$\left[k_b^{(e)}\right] = \pm\int_{s_b} h[N]^{\mathrm{T}}[N]\,ds$$

$$= \text{element stiffness matrix component due to boundary conditions} \tag{12.84}$$

$$\left\{f_{bs}^{(e)}\right\} = \pm\int_{s_b} [N]^{\mathrm{T}}\{(h\phi_\infty + f_s'')\}ds$$

$$= \text{element source vector component due to boundary condition} \tag{12.85}$$

Finally, we can substitute equations 12.80 and 12.84 into 12.72.

$$\left\{I_i^{(e)}\right\} + \left(k_\Gamma^{(e)} + k_b^{(e)}\right)\{\phi^{(e)}\} - \left\{f_S^{(e)} + f_{bs}^{(e)}\right\} = 0 \tag{12.86}$$

or

$$\left\{I_i^{(e)}\right\} + k^{(e)}\{\phi^{(e)}\} - f'' = 0 \tag{12.87}$$

where

$$k^{(e)} = \left(k_\Gamma^{(e)} + k_b^{(e)}\right) = \text{element characteristics matrix} \tag{12.88}$$

and

$$f^{(e)} = \left\{f_S^{(e)} + f_{bs}^{(e)}\right\} = \text{element source vector} \tag{12.89}$$

Axisymmetric Element An axisymmetric element in r–z coordinates is the same as a two-dimensional element in x–z coordinates. Figure 12.10 show a linear triangular axisymmetric element with three nodal values ϕ_i, ϕ_j, and ϕ_k with nodal coordinate (r_i, z_i) (r_j, z_j) (r_k, z_k).

The approximation function and the associated shape function for linear triangular elements can be deduced from Equations 10.59 and 10.60 as

$$\phi = N_i\phi_i + N_j\phi_j + N_k\phi_k = [N]\{\phi^{(e)}\} \tag{12.90}$$

where

$$N_i = \frac{1}{2A}(\alpha_i + \beta_i r + \gamma_i z) \tag{12.91a}$$

$$N_j = \frac{1}{2A}(\alpha_j + \beta_j r + \gamma_j z) \tag{12.91b}$$

$$N_k = \frac{1}{2A}(\alpha_k + \beta_k r + \gamma_k z) \tag{12.91c}$$

and

$$\alpha_i = r_j z_k - r_k z_j, \quad \beta_i = z_j - z_k, \quad \gamma_i = r_k - r_j \tag{12.92a}$$

$$\alpha_j = r_k z_i - r_i z_k, \quad \beta_j = z_k - z_i, \quad \gamma_j = r_i - r_k \tag{12.92b}$$

$$\alpha_k = r_i z_j - r_j z_i, \quad \beta_k = z_i - z_j, \quad \gamma_k = r_j - r_i \tag{12.92c}$$

A is the area of the triangle and is given by the determinant

$$A = \frac{1}{2}\begin{vmatrix} 1 & r_i & z_i \\ 1 & r_j & z_j \\ 1 & r_k & z_k \end{vmatrix} \tag{12.93}$$

In a similar manner, other two-dimensional axisymmetric elements in the r–z plane can be deduced from two-dimensional elements in the x–y plane.

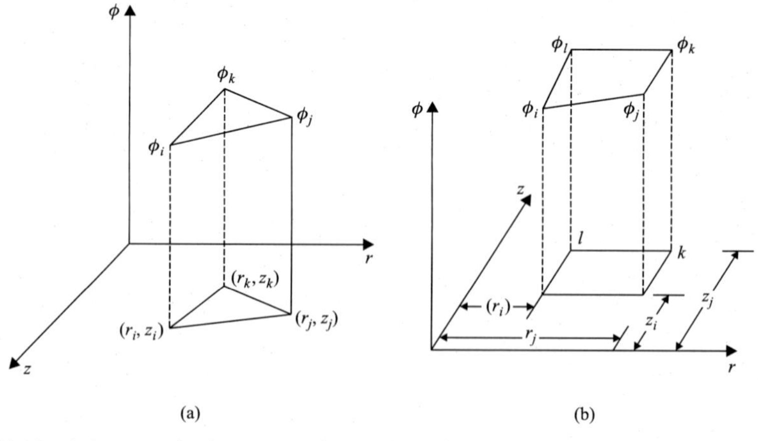

Figure 12.10 Axisymmetric element (a) Linear triangular (b) bilinear rectangular

For example, the rectangular element in the r–z plane can be deduced from Equation 10.88 by sustituting $\tilde{x} = r - r_i$ and $\tilde{y} = z - z_i$. The approximation and the shape function are summarized as

$$\phi^* = N_i\phi_i + N_j\phi_j + N_k\phi_k + N_l\phi_l$$

where

$$N_i = \frac{1}{4l_e h_e}(r_j - r)(z_l - z), \quad N_j = \frac{1}{4l_e h_e}(r - r_i)(z_l - z)$$
$$N_k = \frac{1}{4l_e h_e}(r - r_i)(z - z_i), \quad N_l = \frac{1}{4l_e h_e}(r_j - r)(z - z_i)$$

Case: Element Characteristic Equation for Axisymmetric Triangular Element The element characteristic matrix is evaluated as follows

$$\left[k_\Gamma^{(e)}\right] = \int_V \{D\}^{\mathrm{T}}[\Gamma]\{D\}dV = \{D\}^{\mathrm{T}}[\Gamma]\{D\}\int_V dV$$

$$= \frac{1}{2A}\begin{bmatrix} \beta_i & \gamma_i \\ \beta_j & \gamma_j \\ \beta_k & \gamma_k \end{bmatrix}\begin{bmatrix} \Gamma_r & 0 \\ 0 & \Gamma_z \end{bmatrix}\frac{1}{2A}\begin{bmatrix} \beta_i & \beta_j & \beta_k \\ \gamma_i & \gamma_j & \gamma_k \end{bmatrix}V$$

$$\left[k_\Gamma^{(e)}\right] = \frac{1}{4A^2}\begin{bmatrix} \Gamma_r\beta_i^2 + \Gamma_z\gamma_i^2 & \Gamma_r\beta_i\beta_j + \Gamma_z\gamma_i\gamma_j & \Gamma_r\beta_i\beta_k + \Gamma_z\gamma_i\gamma_k \\ \Gamma_r\beta_j\beta_i + \Gamma_z\gamma_j\gamma_i & \Gamma_r\beta_j^2 + \Gamma_z\gamma_j^2 & \Gamma_r\beta_j\beta_k + \Gamma_z\gamma_j\gamma_k \\ \Gamma_r\beta_k\beta_i + \Gamma_z\gamma_k\gamma_i & \Gamma_r\beta_k\beta_j + \Gamma_z\gamma_k\gamma_j & \Gamma_r\beta_k^2 + \Gamma_y\gamma_k^2 \end{bmatrix}V \quad (12.94)$$

The volume V is the volume formed by the rotation of a two-dimensional axisymetric element with area A about the line of symmetry as shown Figure 12.11.

The volume can approximated based on an average radius as

$$V = 2\pi\bar{r}A \quad (12.95)$$

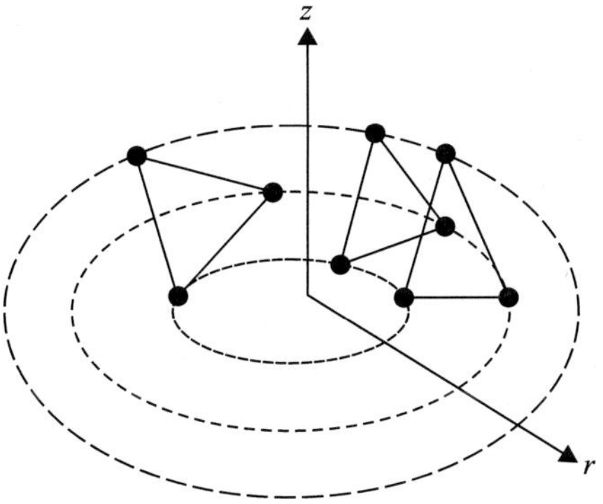

Figure 12.11 Volume generated by the rotation of an axisymmetric element

where

$$\bar{r} = \text{average radial distance of the element} = \frac{r_i + r_j + r_k}{3}$$

Substituting Equation 12.95 into Equation 12.94, we have

$$\left[k_{\Gamma}^{(e)}\right] = \frac{1}{4A^2}\begin{bmatrix} \Gamma_r\beta_i^2 + \Gamma_z\gamma_i^2 & \Gamma_r\beta_i\beta_j + \Gamma_z\gamma_i\gamma_j & \Gamma_r\beta_i\beta_k + \Gamma_z\gamma_i\gamma_k \\ \Gamma_r\beta_j\beta_i + \Gamma_z\gamma_j\gamma_i & \Gamma_r\beta_j^2 + \Gamma_z\gamma_j^2 & \Gamma_r\beta_j\beta_k + \Gamma_z\gamma_j\gamma_k \\ \Gamma_r\beta_k\beta_i + \Gamma_z\gamma_k\gamma_i & \Gamma_r\beta_k\beta_j + \Gamma_z\gamma_k\gamma_j & \Gamma_r\beta_k^2 + \Gamma_y\gamma_k^2 \end{bmatrix} 2\pi\bar{r}A \quad (12.96a)$$

$$\left[k_{\Gamma}^{(e)}\right] = \frac{2\pi\bar{r}\Gamma_r}{4A}\begin{bmatrix} \beta_i^2 & \beta_i\beta_j & \beta_i\beta_k \\ \beta_j\beta_i & \beta_j^2 & \beta_j\beta_k \\ \beta_k\beta_i & \beta_k\beta_j & \beta_k^2 \end{bmatrix} + \frac{2\pi\bar{r}\Gamma_z}{4A}\begin{bmatrix} \gamma_i^2 & \gamma_i\gamma_j & \gamma_i\gamma_k \\ \gamma_j\gamma_i & \gamma_j^2 & \gamma_j\gamma_k \\ \gamma_k\gamma_i & \gamma_k\gamma_j & \gamma_k^2 \end{bmatrix} \quad (12.96b)$$

The element source vector due to the volumetric source term is given and is evaluated as

$$\begin{aligned} \left\{f_S^{(e)}\right\} &= \int_V S[N]^{\mathrm{T}} dV \\ &= \int_A S[N]^{\mathrm{T}} 2\pi r\, dA \\ &= 2\pi S \int_A \begin{bmatrix} N_i r \\ N_j r \\ N_k r \end{bmatrix} dA \end{aligned} \quad (12.97)$$

Equation 12.97 can be evaluated numerically for cases where S is not constant. For a triangular element that involves simple algebraic shape functions and a constant source term, we can evaluate the equation analytically using area coordinates and integral formulas as discussed in Chapter 10. Replacing the shape functions by the area coordinate, i.e., $N_i = L_i$, $N_j = L_j$, and $N_k = L_k$, and writing the radial distance inside an element as

$$r = N_i r_i + N_j r_j + N_k r_k \quad (12.98)$$

or

$$r = L_i r_i + L_j r_j + L_k r_k \quad (12.99)$$

the integral Equation 12.97 is expressed as

$$\left\{f_S^{(e)}\right\} = 2\pi S \int_A \begin{Bmatrix} L_i(L_i r_i + L_j r_j + L_k) \\ L_j(L_i r_i + L_j r_j + L_k) \\ L_k(L_i r_i + L_j r_j + L_k) \end{Bmatrix} dA \quad (12.100)$$

We can now evaluate each element of the integral using the integral formula given by Equation 12.97. For example the first row of elements of the integral is evaluated as

$$\int_A L_i(L_i r_i + L_j r_j + L_k r_k)dA = \int_A L_i^2 r_i dA + \int_A L_i L_j r_i dA + \int_A L_i L_k r_k dA$$

$$= r_i \int_A \left(L_i^2 L_j^0 L_k^0\right) dA + r_i \int_A \left(L_i^1 L_j^1 L_k^0\right) dA + r_i \int_A \left(L_i^1 L_j^0 L_k^1\right) dA$$

$$= r_i \frac{2!0!0!}{(2+0+0+2)} 2A + r_j \frac{1!1!0!}{(1+1+0+2)} 2A + r_k \frac{1!0!1!}{(1+0+1+2)} 2A$$

$$= \frac{2Ar_i}{12} + \frac{Ar_i}{12} + \frac{Ar_k}{12} \tag{12.101a}$$

Similarly, the second and third row of the integral is evaluated as

$$\int_A L_j(L_i r_i + L_j r_j + L_k r_k)dA = \frac{Ar_i}{12} + \frac{2Ar_i}{12} + \frac{Ar_k}{12} \tag{12.101b}$$

and

$$\int_A L_k(L_i r_i + L_j r_j + L_k r_k)dA = \frac{Ar_i}{12} + \frac{Ar_i}{12} + \frac{2Ar_k}{12} \tag{12.101c}$$

Substituting Equation 12.101 into Equation 12.100, the element source vector due to the volumetric source term is expressed as

$$\left\{f_S^{(e)}\right\} = \frac{\pi SA}{6} \begin{Bmatrix} (2r_i + r_j + r_k) \\ (r_i + 2r_j + r_k) \\ (r_i + r_j + 2r_k) \end{Bmatrix} \tag{12.102}$$

Note that the source is distributed among the three nodes based on radial distances.

12.3 THREE-DIMENSIONAL PROBLEMS

In this section we will consider the formulation and application of the finite element method for solving three-dimensional steady state problems. The three-dimensional formulation is similar to those discussed for two- and one-dimensional problems. Let us consider a three-dimensional steady state transport process in a three-dimensional domain as shown in Figure 12.12.

A general mathematical statement of the problem can be assumed to be

$$\frac{\partial}{\partial x}\left(\Gamma_x \frac{\partial \phi}{\partial x}\right) + \frac{\partial}{\partial y}\left(\Gamma_y \frac{\partial \phi}{\partial y}\right) + \frac{\partial}{\partial z}\left(\Gamma_z \frac{\partial \phi}{\partial z}\right) + S = 0 \tag{12.103}$$

subject to the associated general boundary conditions.

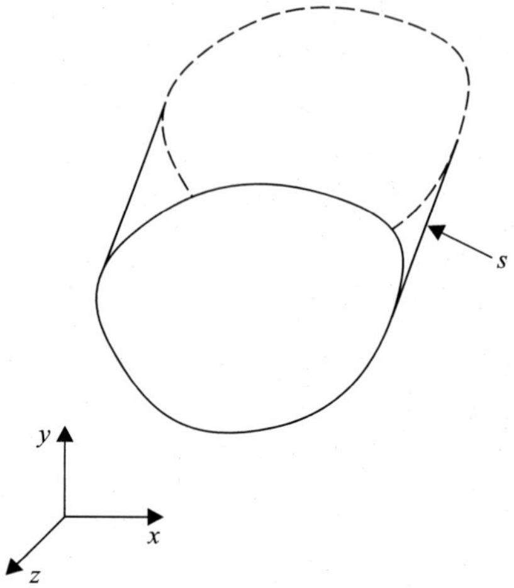

Figure 12.12 Three-dimensional domain

On the boundary surface A_b
Boundary condition of the third kind

$$\Gamma_x \frac{\partial \phi}{\partial x} n_x + \Gamma_y \frac{\partial \phi}{\partial y} n_y + \Gamma_z \frac{\partial \phi}{\partial z} n_z = \pm h(\phi - \phi_s) \pm f_s'' \tag{12.104}$$

Approximation Function The approximate solution for the element is given by a similar expression as for the case of a two-dimensional problem, i.e.

$$\phi^* = \sum_{I=1}^{n_p} N_I \phi_I = [N]\{\phi^{(e)}\} \tag{12.2}$$

where N_I are the shape functions of the three-dimensional element, ϕ_I are the nodal values, and n_p are the number of nodes in the three-dimensional element.

Integral Formulation The integral form of the equation over an element is obtained by applying ***Galerkin's weighted residual method*** as

$$\{R^{(e)}\} = -\int_V [N]^{\mathrm{T}} \left[\frac{\partial}{\partial x}\left(\Gamma_x \frac{\partial \phi^*}{\partial x}\right) + \frac{\partial}{\partial y}\left(\Gamma_y \frac{\partial \phi^*}{\partial y}\right) + \frac{\partial}{\partial z}\left(\Gamma_z \frac{\partial \phi^*}{\partial z}\right) + S \right] dV = 0 \tag{12.105}$$

The first three integral terms are evaluated as

$$\int_V [N]^{\mathrm{T}} \frac{\partial}{\partial x}\left(\Gamma_x \frac{\partial \phi^*}{\partial x}\right) dV = \int_V \frac{\partial}{\partial x}\left([N]^{\mathrm{T}} \Gamma_x \frac{\partial \phi^*}{\partial x}\right) dV - \int_V \Gamma_x \frac{\partial \phi^*}{\partial x} \frac{\partial [N]^{\mathrm{T}}}{\partial x} dV \tag{12.106a}$$

$$\int_V [N]^{\mathrm{T}} \frac{\partial}{\partial y}\left(\Gamma_y \frac{\partial \phi^*}{\partial y}\right) dV = \int_V \frac{\partial}{\partial y}\left([N]^{\mathrm{T}} \Gamma_y \frac{\partial \phi^*}{\partial y}\right) dV - \int_V \Gamma_y \frac{\partial \phi^*}{\partial y} \frac{\partial [N]^{\mathrm{T}}}{\partial y} dV \tag{12.106b}$$

and

$$\int_V [N]^{\mathrm{T}} \frac{\partial}{\partial z}\left(\Gamma_z \frac{\partial \phi^*}{\partial z}\right) dV = \int_V \frac{\partial}{\partial z}\left([N]^{\mathrm{T}} \Gamma_z \frac{\partial \phi^*}{\partial z}\right) dV - \int_V \Gamma_z \frac{\partial \phi^*}{\partial z} \frac{\partial [N]^{\mathrm{T}}}{\partial z} dV \quad (12.106c)$$

The first volume integral term on the right-hand side of Equation 12.106 can be transformed into an area integral using Gauss's divergence theorem as

$$\int_V \frac{\partial}{\partial x}\left([N]^{\mathrm{T}} \Gamma_x \frac{\partial \phi^*}{\partial x}\right) dV = \oint_s [N]^{\mathrm{T}} \Gamma_x \frac{\partial \phi^*}{\partial x} \hat{n}_x \, dA \quad (12.107a)$$

$$\int_V \frac{\partial}{\partial y}\left([N]^{\mathrm{T}} \Gamma_y \frac{\partial \phi^*}{\partial y}\right) dV = \oint_A [N]^{\mathrm{T}} \Gamma_y \frac{\partial \phi^*}{\partial y} \hat{n}_y \, dA \quad (12.107b)$$

and

$$\int_V \frac{\partial}{\partial z}\left([N]^{\mathrm{T}} \Gamma_z \frac{\partial \phi^*}{\partial z}\right) dV = \oint_A [N]^{\mathrm{T}} \Gamma_y \frac{\partial \phi^*}{\partial y} \hat{n}_z \, dA \quad (12.107c)$$

Substituting Equation 12.107 into Equation 12.106, we get

$$\int_V [N]^{\mathrm{T}} \frac{\partial}{\partial x}\left(\Gamma_x \frac{\partial \phi^*}{\partial x}\right) dV = \oint_A [N]^{\mathrm{T}} \Gamma_x \frac{\partial \phi^*}{\partial x} n_x \, dA - \int_V \frac{\partial [N]^{\mathrm{T}}}{\partial x} \Gamma_x \frac{\partial \phi^*}{\partial x} dV \quad (12.108a)$$

$$\int_V [N]^{\mathrm{T}} \frac{\partial}{\partial y}\left(\Gamma_y \frac{\partial \phi^*}{\partial y}\right) dV = \oint_A [N]^{\mathrm{T}} \Gamma_y \frac{\partial \phi^*}{\partial y} n_y \, dA - \int_V \frac{\partial [N]^{\mathrm{T}}}{\partial y} \Gamma_y \frac{\partial \phi^*}{\partial y} dV \quad (12.108b)$$

and

$$\int_V [N]^{\mathrm{T}} \frac{\partial}{\partial z}\left(\Gamma_z \frac{\partial \phi^*}{\partial z}\right) dV = \oint_A [N]^{\mathrm{T}} \Gamma_z \frac{\partial \phi^*}{\partial z} n_z \, dA \int_V \frac{\partial [N]^{\mathrm{T}}}{\partial z} \Gamma_z \frac{\partial \phi^*}{\partial z} dV \quad (12.108c)$$

Substituting Equation 12.108 into the integral form of the equation over an element given by Equation 12.105, we have

$$\begin{aligned} \{R^{(e)}\} = & -\int_A [N]^{\mathrm{T}} \left(\Gamma_x \frac{\partial \phi^*}{\partial x} n_x + \Gamma_y \frac{\partial \phi^*}{\partial y} n_y + \Gamma_z \frac{\partial \phi^*}{\partial z} n_z\right) dA \\ & + \int_V \left(\frac{\partial [N]^{\mathrm{T}}}{\partial x} \Gamma_x \frac{\partial \phi^*}{\partial x} + \frac{\partial [N]^{\mathrm{T}}}{\partial y} \Gamma_y \frac{\partial \phi^*}{\partial y} + \frac{\partial [N]^{\mathrm{T}}}{\partial z} \Gamma_z \frac{\partial \phi^*}{\partial z}\right) dV \\ & - \int_V S[N]^{\mathrm{T}} \, dV = 0 \end{aligned} \quad (12.109)$$

Formation of Element Characteristics Equation In order to obtain the element characteristics equation, substitute the approximate solution function given by Equation 12.2 into the integral Equation 12.109

$$-\int_A [N]^{\mathrm{T}}\left(\Gamma_x \frac{\partial \phi^*}{\partial x} n_x + \Gamma_y \frac{\partial \phi^*}{\partial y} n_y + \Gamma_z \frac{\partial \phi^*}{\partial z} n_z\right) dA$$
$$+\left\{\int_V \left(\frac{\partial [N]^{\mathrm{T}}}{\partial x}\Gamma_x \frac{\partial [N]}{\partial x} + \frac{\partial [N]^{\mathrm{T}}}{\partial y}\Gamma_y \frac{\partial [N]}{\partial y} + \frac{\partial [N]^{\mathrm{T}}}{\partial z}\Gamma_z \frac{\partial [N]}{\partial z}\right) dV\right\}\{\phi^{(e)}\}$$
$$-\int_V S[N]^{\mathrm{T}}\, dV = 0 \tag{12.110}$$

The element characteristics equation can be written in matrix notation as

$$\{I^{(e)}\} + [k^{(e)}]\{\phi^{(e)}\} - \left\{f_S^{(e)}\right\} = 0 \tag{12.111}$$

where

$\{I^{(e)}\}$ = column vector for the element contribution to the inter-element requirement

$$= -\int_A [N]^{\mathrm{T}}\left(\Gamma_x \frac{\partial \phi^*}{\partial x} n_x + \Gamma_y \frac{\partial \phi^*}{\partial y} n_y + \Gamma_z \frac{\partial \phi^*}{\partial z} n_z\right) dA \tag{12.112}$$

$\left\{f_S^{(e)}\right\}$ = element source vector due to the volumetric source term

$$= \int_V S[N]^{\mathrm{T}}\, dV \tag{12.113}$$

$\{\phi^{(e)}\}$ = column vector of unknown nodal values

$[k^{(e)}]$ = element characteristics matrix

$$= \int_V \left(\frac{\partial [N]^{\mathrm{T}}}{\partial x}\Gamma_x \frac{\partial [N]}{\partial x} + \frac{\partial [N]^{\mathrm{T}}}{\partial y}\Gamma_y \frac{\partial [N]}{\partial y} + \frac{\partial [N]^{\mathrm{T}}}{\partial z}\Gamma_z \frac{\partial [N]}{\partial z}\right) dV$$
$$= \left[k_\Gamma^{(e)}\right] = \left[k_{\Gamma x}^{(e)}\right] + \left[k_{\Gamma y}^{(e)}\right] + \left[k_{\Gamma z}^{(e)}\right] \tag{12.114}$$

The first term $\left[k_\Gamma^{(e)}\right]$ is contributed by the three diffusion terms, which is expressed in compact form as

$$\left[k_\Gamma^{(e)}\right] = \int_V \left(\frac{\partial [N]^{\mathrm{T}}}{\partial x}\Gamma_x \frac{\partial [N]}{\partial x} + \frac{\partial [N]^{\mathrm{T}}}{\partial y}\Gamma_y \frac{\partial [N]}{\partial y} + \frac{\partial [N]^{\mathrm{T}}}{\partial z}\Gamma_z \frac{\partial [N]}{\partial z}\right) dV$$
$$= \int_V \{D\}^{\mathrm{T}}[\Gamma]\{D\}\, dV \tag{12.115}$$

where

$$[\Gamma] = \text{transport property matrix} = \begin{bmatrix} \Gamma_x & 0 & 0 \\ 0 & \Gamma_y & 0 \\ 0 & 0 & \Gamma_z \end{bmatrix} \quad (12.116)$$

$$\{D\} = \text{column vector with derivatives of shape functions}$$

$$= \begin{Bmatrix} \partial[N]/\partial x \\ \partial[N]/\partial y \\ \partial[N]/\partial z \end{Bmatrix} \quad (12.116)$$

$$\{D\}^{\mathrm{T}} = \text{row vector with derivatives of shape functions}$$

$$= \begin{Bmatrix} \dfrac{\partial[N]}{\partial x} & \dfrac{\partial[N]}{\partial y} & \dfrac{\partial[N]}{\partial z} \end{Bmatrix} \quad (12.117)$$

The implementation of the derivative boundary conditions into the finite element formulation takes place in the same manner as in the case of one- and two-dimensional problems through the inter-element vector $\{I^{(e)}\}$ as given by Equation 12.10

$$\{I^{(e)}\} = -\int_A [N]^{\mathrm{T}}\left(\Gamma_x \frac{\partial\phi^*}{\partial x} n_x + \Gamma_y \frac{\partial\phi^*}{\partial y} n_y + \Gamma_y \frac{\partial\phi^*}{\partial z} n_z\right) dA \quad (12.118)$$

In order to show the specific contribution from the boundary conditions, let us decompose the integral into two parts as

$$\{I^{(e)}\} = \left\{I_b^{(e)}\right\} + \left\{I_i^{(e)}\right\} \quad (12.119)$$

where $\left\{I_i^{(e)}\right\}$ is the integral contribution from all the other sides other than the boundary surface, constituting the inter-element requirements. The component $\left\{I_b^{(e)}\right\}$ is the integral contribution from the surfaces that represent the external boundary surfaces, and it is expressed as

$$\left\{I_b^{(e)}\right\} = -\int_{A_b} [N]^{\mathrm{T}}\left(\Gamma_x \frac{\partial\phi^*}{\partial x} n_x + \Gamma_y \frac{\partial\phi^*}{\partial y} n_y + \Gamma_z \frac{\partial\phi^*}{\partial z} n_z\right) dA \quad (12.120)$$

Substituting the boundary condition given by Equation 12.18e, we get

$$\left\{I_b^{(e)}\right\} = -\int_{A_b} [N]^{\mathrm{T}}\{\pm h(\phi^* - \phi_\infty) \pm f_s''\}\, dA \quad (12.121a)$$

or

$$\left\{I_b^{(e)}\right\} = \pm \int_{A_b} h[N]^{\mathrm{T}} \phi^* dA \pm \int_{s_b} [N]^{\mathrm{T}}\{(h\phi_\infty + f_s'')\}\, dA \tag{12.121b}$$

Now, substituting the approximate solution given by Equation 12.2, we get

$$\left\{I_b^{(e)}\right\} = \pm\left(\int_{A_b} h[N]^{\mathrm{T}}[N]\, ds\right)\phi^{(e)} \pm \int_{A_b} [N]^{\mathrm{T}}\{(h\phi_\infty + f_s'')\}\, dA \tag{12.122a}$$

or

$$\left\{I_b^{(e)}\right\} = \left[k_b^{(e)}\right]\{\phi^{(e)}\} - \left\{f_{bs}^{(e)}\right\} \tag{12.122b}$$

where

$$\left[k_b^{(e)}\right] = \pm \int_{A_b} h[N]^{\mathrm{T}}[N] dA$$

$$= \text{element stiffness matrix component due to boundary conditions} \tag{12.123a}$$

$$\left\{f_{bs}^{(e)}\right\} = \pm \int_{s_b} [N]^{\mathrm{T}}\{(h\phi_\infty + f_s'')\}\, dA$$

$$= \text{element source vector component due to boundary condition} \tag{12.123b}$$

Finally, we can substitute Equations 12.119 and 12.122b into 12.111 to obtain

$$\left\{I_i^{(e)}\right\} + \left(k_\Gamma^{(e)} + k_b^{(e)}\right)\{\phi^{(e)}\} - \left\{f_S^{(e)} + f_{bs}^{(e)}\right\} = 0 \tag{12.124}$$

or

$$\left\{I_i^{(e)}\right\} + k^{(e)}\{\phi^{(e)}\} - f'' = 0 \tag{12.125}$$

where

$$k^{(e)} = \left(k_\Gamma^{(e)} + k_b^{(e)}\right) = \text{element characteristics matrix} \tag{12.126}$$

and

$$f^{(e)} = \left\{f_S^{(e)} + f_{bs}^{(e)}\right\} = \text{element source vector} \tag{12.127}$$

Case: Element Characteristics Equation Using Linear Tetrahedron Element: Let us consider a linear four-node tetrahedron element as shown in Figure 10.22a and with the approximate solution and the corresponding shape functions given by Equations 10.112 and 10.114, respectively.

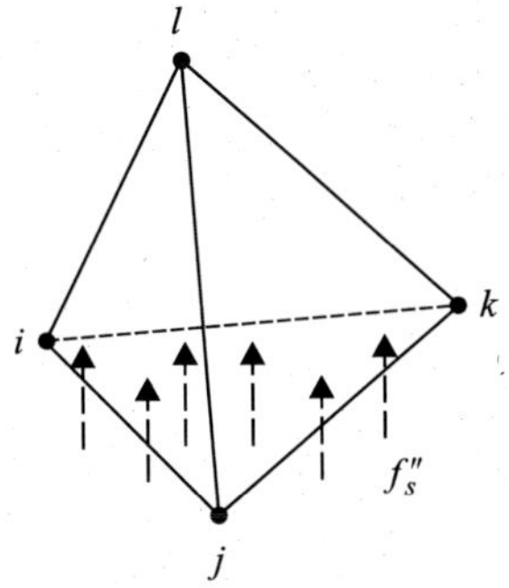

Figure 12.13 Tetrahedron element with mixed boundary condition at the area surface *i–j–k*

The vectors with derivatives of shape functions are derived by substituting the shape functions given as

$$\{D\} = \begin{Bmatrix} \dfrac{\partial[N]}{\partial x} \\ \dfrac{\partial[N]}{\partial y} \\ \dfrac{\partial[N]}{\partial z} \end{Bmatrix}$$

$$\{D\} = \begin{bmatrix} \dfrac{\partial N_i}{\partial x} & \dfrac{\partial N_j}{\partial x} & \dfrac{\partial N_k}{\partial x} & \dfrac{\partial N_l}{\partial x} \\ \dfrac{\partial N_i}{\partial y} & \dfrac{\partial N_j}{\partial y} & \dfrac{\partial N_k}{\partial y} & \dfrac{\partial N_l}{\partial y} \\ \dfrac{\partial N_i}{\partial z} & \dfrac{\partial N_j}{\partial z} & \dfrac{\partial N_k}{\partial z} & \dfrac{\partial N_l}{\partial z} \end{bmatrix} \tag{12.128}$$

Evaluating the derivatives from Equation 10.114, we can get

$$\{D\} = \frac{1}{6V} \begin{bmatrix} \beta_i & \beta_j & \beta_k & \beta_l \\ \gamma_i & \gamma_j & \gamma_k & \gamma_l \\ \delta_i & \delta_j & \delta_k & \delta_l \end{bmatrix}$$

Similarly, we get

$$\{D\}^{\mathrm{T}} = \left\{ \frac{\partial[N]}{\partial x} \quad \frac{\partial[N]}{\partial y} \quad \frac{\partial[N]}{\partial z} \right\}$$

$$\{D\}^{\mathrm{T}} = \begin{bmatrix} \dfrac{\partial N_i}{\partial x} & \dfrac{\partial N_i}{\partial y} & \dfrac{\partial N_i}{\partial z} \\ \dfrac{\partial N_j}{\partial x} & \dfrac{\partial N_j}{\partial y} & \dfrac{\partial N_j}{\partial z} \\ \dfrac{\partial N_k}{\partial x} & \dfrac{\partial N_k}{\partial y} & \dfrac{\partial N_k}{\partial z} \\ \dfrac{\partial N_l}{\partial x} & \dfrac{\partial N_l}{\partial y} & \dfrac{\partial N_l}{\partial z} \end{bmatrix} = \frac{1}{6V} \begin{bmatrix} \beta_i & \gamma_i & \delta_i \\ \beta_j & \gamma_j & \delta_j \\ \beta_k & \gamma_k & \delta_k \\ \beta_l & \gamma_l & \delta_l \end{bmatrix} \tag{12.129}$$

The component of the element characteristic matrix due to the diffusion term, $\left[k_\Gamma^{(e)}\right]$ can now be computed as

$$\left[k_\Gamma^{(e)}\right] = \int_V \{D\}^{\mathrm{T}}[\Gamma]\{D\}\, dV = \{D\}^{\mathrm{T}}[\Gamma]\{D\} \int_V dV$$

$$= \frac{1}{6V} \begin{bmatrix} \beta_i & \gamma_i & \delta_i \\ \beta_j & \gamma_j & \delta_j \\ \beta_k & \gamma_k & \delta_k \\ \beta_l & \gamma_l & \delta_l \end{bmatrix} \begin{bmatrix} \Gamma_x & 0 & 0 \\ 0 & \Gamma_y & 0 \\ 0 & 0 & \Gamma_z \end{bmatrix} \frac{1}{6V} \begin{bmatrix} \beta_i & \beta_j & \beta_k & \beta_l \\ \gamma_i & \gamma_j & \gamma_k & \gamma_l \\ \delta_i & \delta_j & \delta_k & \delta_l \end{bmatrix} V$$

$$= \frac{1}{36V^2} \begin{bmatrix} \beta_i & \gamma_i & \delta_i \\ \beta_j & \gamma_j & \delta_j \\ \beta_k & \gamma_k & \delta_k \\ \beta_l & \gamma_l & \delta_l \end{bmatrix} \begin{bmatrix} \Gamma_x\beta_i & \Gamma_x\beta_j & \Gamma_x\beta_k & \Gamma_x\beta_l \\ \Gamma_y\gamma_i & \Gamma_y\gamma_j & \Gamma_y\gamma_k & \Gamma_z\gamma_l \\ \Gamma_z\delta_i & \Gamma_z\delta_j & \Gamma_z\delta_k & \Gamma_z\delta_l \end{bmatrix}$$

$$= \frac{1}{36V^2} \begin{bmatrix} \Gamma_x\beta_i^2 + \Gamma_y\gamma_i^2 + \Gamma_z\delta_i^2 & \Gamma_x\beta_i\beta_j + \Gamma_y\gamma_i\gamma_j + \Gamma_z\delta_i\delta_j & \Gamma_x\beta_i\beta_k + \Gamma_y\gamma_i\gamma_k + \Gamma_z\delta_i\delta_k & \Gamma_x\beta_i\beta_l + \Gamma_y\gamma_i\gamma_l + \Gamma_z\delta_i\delta_l \\ \Gamma_x\beta_i\beta_j + \Gamma_y\gamma_i\gamma_j + \Gamma_z\delta_i\delta_j & \Gamma_x\beta_j^2 + \Gamma_y\gamma_j^2 + \Gamma_z\delta_j^2 & \Gamma_x\beta_j\beta_k + \Gamma_y\gamma_j\gamma_k + \Gamma_z\delta_j\delta_k & \Gamma_x\beta_j\beta_l + \Gamma_y\gamma_j\gamma_l + \Gamma_z\delta_j\delta_l \\ \Gamma_x\beta_i\beta_k + \Gamma_y\gamma_i\gamma_k + \Gamma_z\delta_i\delta_k & \Gamma_x\beta_j\beta_k + \Gamma_y\gamma_j\gamma_k + \Gamma_z\delta_j\delta_k & \Gamma_x\beta_k^2 + \Gamma_y\gamma_k^2 + \Gamma_z\delta_k^2 & \Gamma_x\beta_l\beta_k + \Gamma_y\gamma_l\gamma_k + \Gamma_z\delta_l\delta_k \\ \Gamma_x\beta_i\beta_l + \Gamma_y\gamma_i\gamma_l + \Gamma_z\delta_i\delta_l & \Gamma_x\beta_j\beta_l + \Gamma_y\gamma_j\gamma_l + \Gamma_z\delta_j\delta_l & \Gamma_x\beta_k\beta_l + \Gamma_y\gamma_k\gamma_l + \Gamma_z\delta_k\delta_l & \Gamma_x\beta_l^2 + \Gamma_y\gamma_l^2 + \Gamma_z\delta_l^2 \end{bmatrix} \tag{12.130a}$$

or

$$\left[k_{\Gamma}^{(e)}\right] = \frac{\Gamma_x}{36V^2}\begin{bmatrix} \beta_i^2 & \beta_i\beta_j & \beta_i\beta_k & \beta_i\beta_l \\ \beta_i\beta_j & \beta_j^2 & \beta_j\beta_k & \beta_j\beta_l \\ \beta_i\beta_k & \beta_j\beta_k & \beta_k^2 & \beta_l\beta_k \\ \beta_i\beta_l & \beta_j\beta_l & \beta_k\beta_l & \beta_l^2 \end{bmatrix}$$

$$+\frac{\Gamma_y}{36V^2}\begin{bmatrix} \gamma_i^2 & \gamma_i\gamma_j & \gamma_i\gamma_k & \gamma_i\gamma_l \\ \gamma_i\gamma_j & \gamma_j^2 & \gamma_j\gamma_k & \gamma_j\gamma_l \\ \gamma_i\gamma_k & \gamma_j\gamma_k & \gamma_k^2 & \gamma_l\gamma_k \\ \gamma_i\gamma_l & \gamma_j\gamma_l & \gamma_k\gamma_l & \gamma_l^2 \end{bmatrix}$$

$$+\frac{\Gamma_z}{36V^2}\begin{bmatrix} \delta_i^2 & \delta_i\delta_j & \delta_i\delta_k & \delta_i\delta_l \\ \delta_i\delta_j & \delta_j^2 & \delta_j\delta_k & \delta_j\delta_l \\ \delta_i\delta_k & \delta_j\delta_k & \delta_k^2 & \delta_l\delta_k \\ \delta_i\delta_l & \delta_j\delta_l & \delta_k\delta_l & \delta_l^2 \end{bmatrix} \qquad (12.130b)$$

where the βs, γs, and δs are given by Equations 10.108b, 10.109b, and 10.110b.

The component of the element characteristic matrix due to the boundary condition term $\lfloor k_b^{(e)} \rfloor$ can now be computed as

$$\left[k_b^{(e)}\right] = \pm \int_{A_b} h[N]^{\mathrm{T}}[N]\,dA$$

$$= \int_{A_b} h \left\{ \begin{matrix} N_i \\ N_j \\ N_k \\ N_l \end{matrix} \{N_i \quad N_j \quad N_k \quad N_l\} \right\} dA$$

$$= \int_{A_b} h \begin{bmatrix} N_i^2 & N_iN_j & N_iN_k & N_iN_l \\ N_jN_i & N_j^2 & N_jN_k & N_jN_l \\ N_kN_i & N_kN_j & N_k^2 & N_kN_l \\ N_lN_i & N_lN_j & N_lN_k & N_l^2 \end{bmatrix} dA \qquad (12.131)$$

The integral is further evaluated by considering only the sides that are subject to external boundary conditions. For example, if the boundary condition given by Equation 10.104 is specified over the area surface side *i–j–k* of the element, then we set $N_l = 0$ for the area integral, and Equation 12.131 becomes

$$\left[k_b^{(e)}\right] = \int_{A_{ijk}} h \begin{bmatrix} N_i^2 & N_iN_j & N_iN_k & 0 \\ N_jN_i & N_j^2 & N_jN_k & 0 \\ N_kN_i & N_kN_j & N_k^2 & 0 \\ 0 & 0 & 0 & 0 \end{bmatrix} dA \qquad (12.132)$$

Again using the natural volume coordinate system (Equation 10.116), the integral elements can be evaluated as

$$\left[k_b^{(e)}\right] = \pm \int_{A_{ijk}} h \begin{bmatrix} L_i^2 & L_iL_j & L_iL_k & 0 \\ L_jL_i & L_j^2 & L_jL_k & 0 \\ L_kL_i & L_kL_j & L_k^2 & 0 \\ 0 & 0 & 0 & 0 \end{bmatrix} \qquad (12.133)$$

$$\left[k_b^{(e)}\right] = \pm h \begin{bmatrix} \int_{A_{ijk}} L_i^2\,dA & \int_{A_{ijk}} L_iL_j\,dA & \int_{A_{ijk}} L_iL_k\,dA & 0 \\ \int_{A_{ijk}} L_jL_i\,dA & \int_{A_{ijk}} L_j^2\,dA & \int_{A_{ijk}} L_jL_k\,dA & 0 \\ \int_{A_{ijk}} L_kL_i\,dA & \int_{A_{ijk}} L_kL_j\,dA & \int_{A_{ijk}} L_k^2\,dA & 0 \\ 0 & 0 & 0 & 0 \end{bmatrix}$$

$$= \pm h \begin{bmatrix} \int_{A_{ijk}} L_i^2L_j^0L_k^0\,dA & \int_{A_{ijk}} L_i^1L_j^1L_k^0\,dA & \int_{A_{ijk}} L_i^1L_j^0L_k^1\,dA & 0 \\ \int_{A_{ijk}} L_i^1L_j^1L_k^0\,dA & \int_{A_{ijk}} L_i^0L_j^2L_k^0\,dA & \int_{A_{ijk}} L_i^0L_j^1L_k^1\,dA & 0 \\ \int_{A_{ijk}} L_i^1L_j^0L_k^1\,dA & \int_{A_{ijk}} L_i^0L_j^1L_k^1\,dA & \int_{A_{ijk}} L_i^0L_j^0L_k^2\,dA & 0 \\ 0 & 0 & 0 & 0 \end{bmatrix}$$

Using analytical integral formulas involving an area natural coordinate system, we get

$$\left[k_b^{(e)}\right] = \pm h \begin{bmatrix} \frac{2!0!0!}{(2+0+0+2)!}2A_{ijk} & \frac{1!1!0!}{(1+1+0+2)!}2A_{ijk} & \frac{1!0!1!}{(1+0+1+2)!}2A_{ijk} & 0 \\ \frac{1!1!0!}{(1+1+0+2)!}2A_{ijk} & \frac{0!2!0!}{(0+2+0+2)!}2A_{ijk} & \frac{0!1!1!}{(0+1+1+2)}2A_{ijk} & 0 \\ \frac{1!0!1!}{(1+0+1+2)!}2A_{ijk} & \frac{0!1!1!}{(0+1+1+2)}2A_{ijk} & \frac{0!0!2!}{(0+0+2+2)!}2A_{ijk} & 0 \\ 0 & 0 & 0 & 0 \end{bmatrix}$$

$$\pm h \begin{bmatrix} \frac{A_{ijk}}{6} & \frac{A_{ijk}}{12} & \frac{A_{ijk}}{12} & 0 \\ \frac{A_{ijk}}{12} & \frac{A_{ijk}}{6} & \frac{A_{ijk}}{12} & 0 \\ \frac{A_{ijk}}{12} & \frac{A_{ijk}}{12} & \frac{A_{ijk}}{6} & 0 \\ 0 & 0 & 0 & 0 \end{bmatrix}$$

$$\left[k_b^{(e)}\right] = \pm \frac{hA_{ijk}}{12} \begin{bmatrix} 2 & 1 & 1 & 0 \\ 1 & 2 & 1 & 0 \\ 1 & 1 & 2 & 0 \\ 0 & 0 & 0 & 0 \end{bmatrix} \tag{12.134a}$$

Similarly:

for boundary condition specified on the side j–k–l

$$\left[k_b^{(e)}\right] = \pm \int_{A_b} h \begin{bmatrix} 0 & 0 & 0 & 0 \\ 0 & N_j^2 & N_jN_k & N_jN_l \\ 0 & N_kN_j & N_k^2 & N_kN_l \\ 0 & N_lN_j & N_lN_k & N_l^2 \end{bmatrix} dA$$

or

$$= \pm \frac{hA_{jkl}}{12} \begin{bmatrix} 0 & 0 & 0 & 0 \\ 0 & 2 & 1 & 1 \\ 0 & 1 & 2 & 1 \\ 0 & 1 & 1 & 2 \end{bmatrix} \tag{12.134b}$$

for boundary condition specified on the side *i–j–l*

$$\left[k_b^{(e)}\right] = \int_{A_b} h \begin{bmatrix} N_i^2 & N_iN_j & 0 & N_iN_l \\ N_jN_i & N_j^2 & 0 & N_jN_l \\ 0 & 0 & 0 & 0 \\ N_lN_i & N_lN_j & 0 & N_l^2 \end{bmatrix} dA$$

or

$$\left[k_b^{(e)}\right] = \pm\frac{hA_{ijl}}{12} \begin{bmatrix} 2 & 1 & 0 & 1 \\ 1 & 2 & 0 & 1 \\ 0 & 0 & 0 & 0 \\ 1 & 1 & 0 & 2 \end{bmatrix} \tag{12.134c}$$

for boundary condition specified on the side *i–k–l*

$$\left[k_b^{(e)}\right] = \int_{A_b} h \begin{bmatrix} N_i^2 & 0 & N_iN_k & N_iN_l \\ 0 & 0 & 0 & 0 \\ N_kN_i & 0 & N_k^2 & N_kN_l \\ N_lN_i & 0 & N_lN_k & N_l^2 \end{bmatrix} dA$$

or

$$\left[k_b^{(e)}\right] = \pm\frac{hA_{ikl}}{12} \begin{bmatrix} 2 & 0 & 1 & 1 \\ 0 & 0 & 0 & 0 \\ 1 & 0 & 2 & 1 \\ 1 & 0 & 1 & 2 \end{bmatrix} \tag{12.134d}$$

The source term is evaluated as

$$\left\{f_S^{(e)}\right\} = \int_V S[N]^{\mathrm{T}}\, dV$$

$$= \int_V S \begin{Bmatrix} N_i \\ N_j \\ N_k \\ N_l \end{Bmatrix} dV \tag{12.135}$$

Using the natural volume coordinate system given by Equation 10.116 and the integral formula in the natural volume coordinate system as given by Equation 10.119, we get

$$\left\{f_S^{(e)}\right\} = \int_V S \begin{Bmatrix} N_i \\ N_j \\ N_k \\ N_l \end{Bmatrix} dV = \int_V S \begin{Bmatrix} L_i \\ L_j \\ L_k \\ L_l \end{Bmatrix} dV \tag{12.136}$$

$$= S \begin{Bmatrix} \int_V L_i^1 L_j^0 L_k^0 L_l^0 \, dV \\ \int_V L_i^0 L_j^1 L_k^0 L_l^0 \, dV \\ \int_V L_i^0 L_j^0 L_k^1 L_l^0 \, dV \\ \int_V L_i^0 L_j^0 L_k^0 L_l^1 \, dV \end{Bmatrix}$$

$$= S \begin{Bmatrix} \dfrac{1!0!0!0!}{(1+0+0+0+3)!} 6V \\ \dfrac{0!1!0!0!}{(0+1+0+0+3)!} 6V \\ \dfrac{0!0!1!0!}{(0+0+1+0+3)!} 6V \\ \dfrac{0!0!0!1!}{(0+0+0+1+3)!} 6V \end{Bmatrix}$$

$$\left\{f_S^{(e)}\right\} = \frac{SV}{4} \begin{Bmatrix} 1 \\ 1 \\ 1 \\ 1 \end{Bmatrix} \tag{12.137}$$

The element source vector component due to the boundary condition is evaluated as

$$\left\{f_{bs}^{(e)}\right\} = \pm \int_{S_b} [N]^{\mathrm{T}} \left\{ \left(h\phi_\infty + f_s'' \right) \right\} dA$$

$$= \left(h\phi_\infty + f_s'' \right) \int_{A_b} \begin{Bmatrix} N_i \\ N_j \\ N_k \\ N_l \end{Bmatrix} dA \tag{12.138}$$

For a boundary condition specified on the i–j–k side of the element, $N_l = 0$, and using the natural coordinate system and integral formula Equation 10.71, we get

$$\left\{f_{bs}^{(e)}\right\} = \left(h\phi_\infty + f_s''\right) \int_{A_b} \begin{Bmatrix} N_i \\ N_j \\ N_k \\ 0 \end{Bmatrix} dA = \left(h\phi_\infty + f_s''\right) \int_{A_b} \begin{Bmatrix} L_i \\ L_j \\ L_k \\ 0 \end{Bmatrix} dA$$

$$\left\{f_{bs}^{(e)}\right\} = \pm \frac{\left(h\phi_\infty + f_s''\right) A_{ijk}}{3} \begin{Bmatrix} 1 \\ 1 \\ 1 \\ 0 \end{Bmatrix} \tag{12.139a}$$

Similarly, for a boundary condition specified on the surface area j–k–l, we get

$$\left\{f_{bs}^{(e)}\right\} = \left(h\phi_\infty + f_s''\right) \int_{A_{jkl}} \begin{Bmatrix} 0 \\ N_j \\ N_k \\ N_l \end{Bmatrix} dA = \pm \frac{\left(h\phi_\infty + f_s''\right) A_{jkl}}{3} \begin{Bmatrix} 0 \\ 1 \\ 1 \\ 1 \end{Bmatrix} \tag{12.139b}$$

For a boundary condition specified on the surface area i–j–l, we get

$$\left\{f_{bs}^{(e)}\right\} = \left(h\phi_\infty + f_s''\right) \int_{A_{ijl}} \begin{Bmatrix} N_i \\ N_j \\ 0 \\ N_l \end{Bmatrix} dA = \pm \frac{\left(h\phi_\infty + f_s''\right) A_{ijl}}{3} \begin{Bmatrix} 1 \\ 1 \\ 0 \\ 1 \end{Bmatrix} \tag{12.139c}$$

For a boundary condition specified on the surface area i–k–l, we get

$$\left\{f_{bs}^{(e)}\right\} = \left(h\phi_\infty + f_s''\right) \int_{A_{ikl}} \begin{Bmatrix} N_i \\ 0 \\ N_k \\ N_l \end{Bmatrix} dA = \pm \frac{\left(h\phi_\infty + f_s''\right) A_{ikl}}{3} \begin{Bmatrix} 1 \\ 0 \\ 1 \\ 1 \end{Bmatrix} \tag{12.139d}$$

The element characteristic equation for the tetrahedron element with area surface i–j–k as the external boundary surface with mixed boundary condition is obtained by substituting Equations 12.130a, 12.134a, 12.137, and 12.139a into 12.124

$$\left\{ \frac{1}{36V^2} \begin{bmatrix} \Gamma_x\beta_i^2 + \Gamma_y\gamma_i^2 + \Gamma_z\delta_i^2 & \Gamma_x\beta_i\beta_j + \Gamma_y\gamma_i\gamma_j + \Gamma_z\delta_i\delta_j & \Gamma_x\beta_i\beta_k + \Gamma_y\gamma_i\gamma_k + \Gamma_z\delta_i\delta_k & \Gamma_x\beta_i\beta_l + \Gamma_y\gamma_i\gamma_l + \Gamma_z\delta_i\delta_l \\ \Gamma_x\beta_i\beta_j + \Gamma_y\gamma_i\gamma_j + \Gamma_z\delta_i\delta_j & \Gamma_x\beta_j^2 + \Gamma_y\gamma_j^2 + \Gamma_z\delta_j^2 & \Gamma_x\beta_j\beta_k + \Gamma_y\gamma_j\gamma_k + \Gamma_z\delta_j\delta_k & \Gamma_x\beta_j\beta_l + \Gamma_y\gamma_j\gamma_l + \Gamma_z\delta_j\delta_l \\ \Gamma_x\beta_i\beta_k + \Gamma_y\gamma_i\gamma_k + \Gamma_z\delta_i\delta_k & \Gamma_x\beta_j\beta_k + \Gamma_y\gamma_j\gamma_k + \Gamma_z\delta_j\delta_k & \Gamma_x\beta_k^2 + \Gamma_y\gamma_k^2 + \Gamma_z\delta_k^2 & \Gamma_x\beta_l\beta_k + \Gamma_y\gamma_l\gamma_k + \Gamma_z\delta_l\delta_k \\ \Gamma_x\beta_i\beta_l + \Gamma_y\gamma_i\gamma_l + \Gamma_z\delta_i\delta_l & \Gamma_x\beta_j\beta_l + \Gamma_y\gamma_j\gamma_l + \Gamma_z\delta_j\delta_l & \Gamma_x\beta_k\beta_l + \Gamma_y\gamma_k\gamma_l + \Gamma_z\delta_k\delta_l & \Gamma_x\beta_l^2 + \Gamma_y\gamma_l^2 + \Gamma_z\delta_l^2 \end{bmatrix} \pm \frac{hA_{ijk}}{12} \begin{bmatrix} 2 & 1 & 1 & 0 \\ 1 & 2 & 1 & 0 \\ 1 & 1 & 2 & 0 \\ 0 & 0 & 0 & 0 \end{bmatrix} \right\}$$

$$\begin{Bmatrix} \phi_i \\ \phi_j \\ \phi_k \\ \phi_l \end{Bmatrix} = \frac{SV}{4} \begin{Bmatrix} 1 \\ 1 \\ 1 \\ 1 \end{Bmatrix} \pm \frac{\left(h\phi_\infty + f_s''\right)A_{ijk}}{6} \begin{Bmatrix} 1 \\ 1 \\ 1 \\ 0 \end{Bmatrix} \tag{12.140}$$

12.4 POINT SOURCE

In many cases the source term may not be distributed over the entire domain, but applied at some discrete points. Let us consider a three-dimensional tetrahedron element with source located at point (x_0, y_0, z_0).

The source term can now be expressed as a delta source function over the element as

$$S = S_0\delta(x - x_0)\delta(y - y_0)\delta(z - z_0) \tag{12.141}$$

The element force vector for a four-noded three-dimensional element can be estimated as

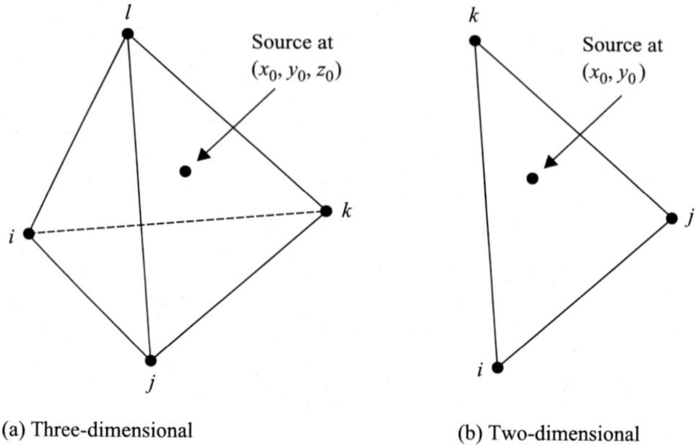

Figure 12.14 Source located at a discrete point

$$\left\{f_S^{(e)}\right\} = \int_V S[N]^{\mathrm{T}}\, dV$$

$$= S_0 \int_A \begin{bmatrix} N_i \\ N_j \\ N_k \\ N_l \end{bmatrix} \delta(x - x_0)\delta(y - y_0)\delta(z - z_0)\, dx\, dy\, dz$$

$$\left\{f_S^{(e)}\right\} = S_0 \begin{Bmatrix} N_i(x_0, y_0, z_0) \\ N_j(x_0, y_0, z_0) \\ N_k(x_0, y_0, z_0) \\ N_l(x_0, y_0, z_0) \end{Bmatrix} \tag{12.142}$$

Basically, the source term at a point is distributed among the nodes of the element with weights given by the relative values of shape functions calculated at the coordinates of the point.

In a similar manner, the element force vector for a three-noded two-dimensional element can be estimated as

$$\left\{f_S^{(e)}\right\} = \int_A S[N]^{\mathrm{T}}\, dA$$

$$= S_0 \int_A \begin{bmatrix} N_i \\ N_j \\ N_k \end{bmatrix} \delta(x - x_0)\delta(y - y_0)\, dx\, dy$$

$$\left\{f_S^{(e)}\right\} = S_0 \begin{Bmatrix} N_i(x_0, y_0) \\ N_j(x_0, y_0) \\ N_k(x_0, y_0) \end{Bmatrix}$$

PROBLEMS

12.1 Consider a two-dimensional steady state conduction in a 1.5 m by 1.5 m square slab with the top and the left surfaces maintained at a high temperature, $T_H = 600°C$ and a low temperature, $T_C = 20°C$, respectively. The right and the bottom surfaces are subjected to convection condition with $h_c = 30\ W/m^{2}°C$ and $T_\infty = 25°C$. The material conductivity is $k = 30\ W/m°C$. Apply a finite element method using six bilinear rectangular elements to calculate the two-dimensional temperature distribution in the slab and heat transfer rates at all surfaces, and show the overall energy balance.

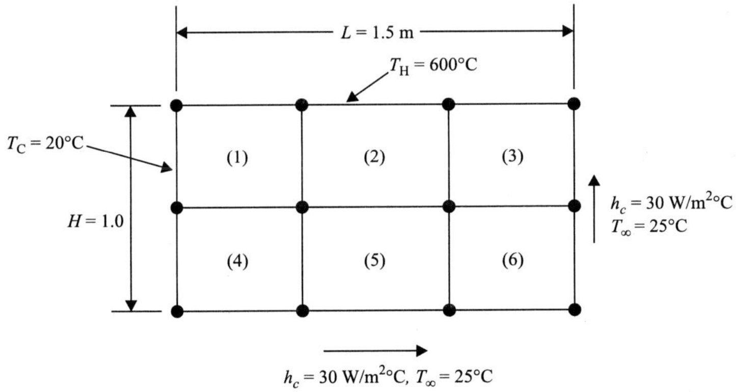

12.2 Repeat Problem 12.1 using linear triangular elements using the mesh shown in the figure.

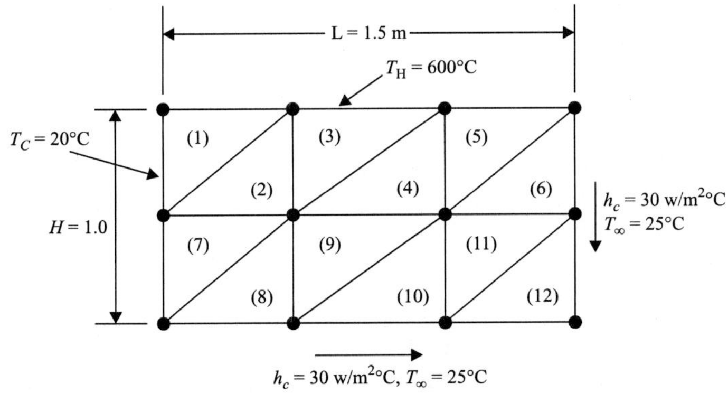

12.3 The fin shown has a base maintained at 300°C and is exposed to the convection environment shown in the figure. Consider the mesh consisting of four-node rectangular elements and write the characteristics equation for each element. Assemble them to obtain the global system of equation.

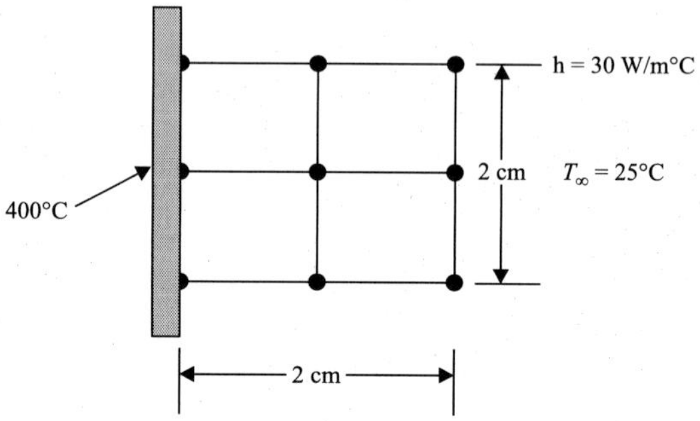

12.4 Consider a temperature rise in a metal cutting tool due to heat generation in the shear zones of a metallic workpiece and metal chips during metal cutting processes.

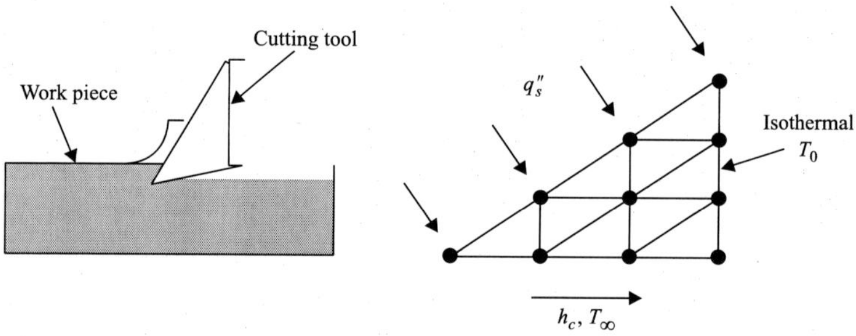

A simplified heating condition is considered in this problem by considering a constant surface heat flux at the top rake surface and forced convective cooling at the bottom surface. The further interior regions, i.e., the right surface is assumed to be isothermal or at a constant temperature T_o. Consider the mesh size distribution with linear triangular elements and derive the element characteristic equation and the global system of equations for the nodal temperatures.

12.5 Repeat Problem 12.4 with a mesh size distribution consisting of triangular and rectangular elements as shown.

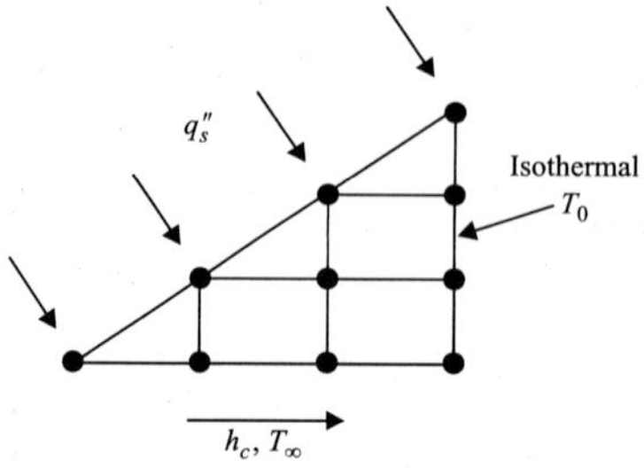

12.6 Use the finite element method to solve two-dimensional steady state conduction in a rectangular carbon steel ($k = 200\,\text{W/m°C}$) slab subjected to a constant surface heat flux irradiated by a high-energy beam at the top surface. For simplicity, assume the heat flux distribution to be a constant average value $q_0'' = 2 \times 10^8\,\text{W/m}^2$, acting over a section of the surface. The top surface is also subjected to forced convection with $h_c = 500\,\text{W/m}^2\text{°C}$. The left surface is assumed to be adiabatic as a line of symmetry. All other surfaces are assumed to be subjected to free convection with $T_\infty = 25\text{°C}$ and $h = 40\,\text{W/m}^2\text{°C}$. Derive the finite element equations for all elements and derive the global system of equations.

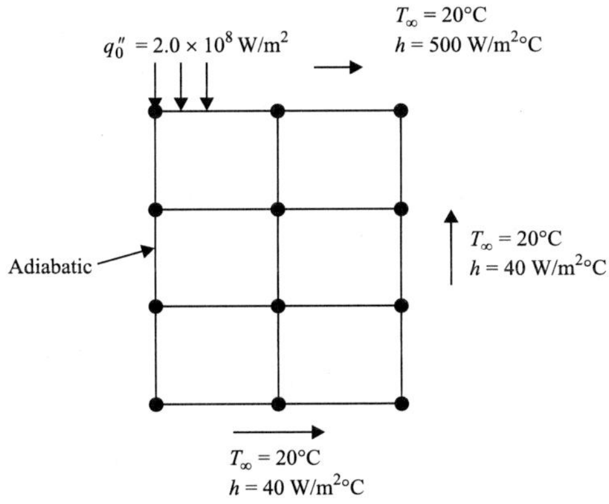

Solve the system of equations using the Gauss elimination solver and check for the overall energy balance.

12.7 Consider a three-dimensional steady state heat conduction within a tetrahedron element with a convective boundary condition shown in the figure. Derive the element characteristic equation for the element with conditions shown in the figure.

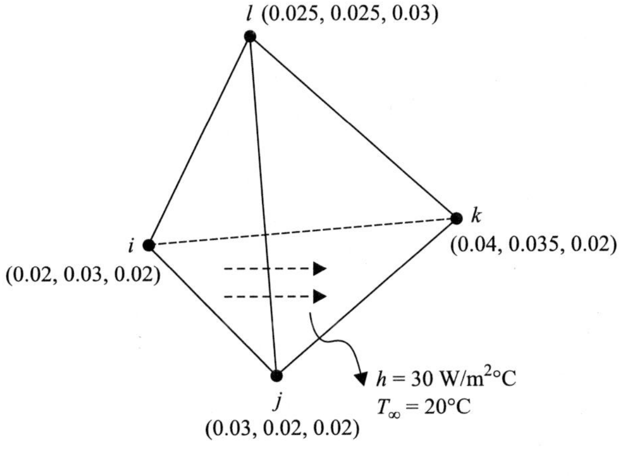

12.8 Use a bilinear rectangular element to derive the element characteristic equation for the two-dimensional axisymmetric steady state problem without heat generation.

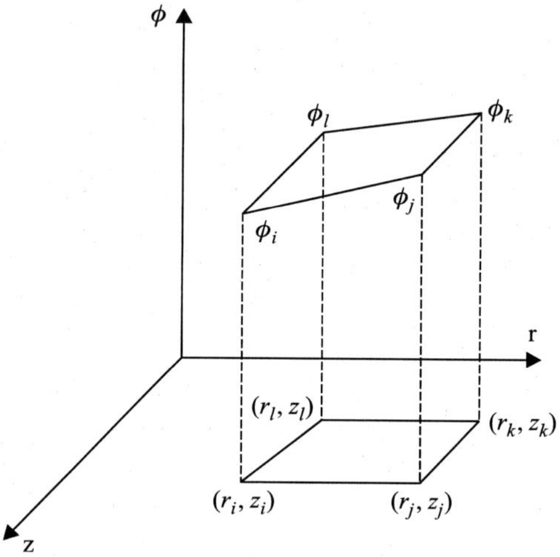

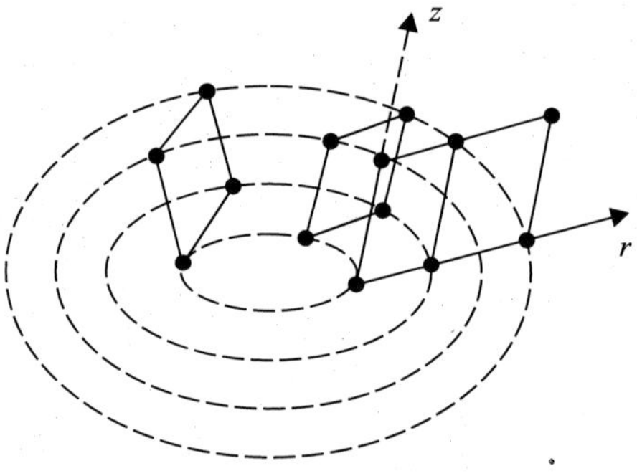

12.9 Use a trilinear eight-noded hexahedron element to derive the three-dimensional element characteristic equation starting from Equations 12.125 to 12.127.

CHAPTER

13

UNSTEADY STATE PROBLEMS

In this chapter we will consider the development of a finite element solution for unsteady state problems. The treatment of the diffusion terms, source term, and the boundary condition will remain the same as outlined for the steady state problems, and we will focus primarily on the treatment of the first-order time derivative term. We will first demonstrate the development of a finite element solution for a general unsteady state one-dimensional problem. The procedure will then be extended to two- and three-dimensional unsteady state problems.

13.1 DISCRETIZATION SCHEME

The basic steps for the finite element formulation of unsteady state problems are identical to those described in Chapters 11 and 12 with the exception of the treatment of the unsteady state term. There are two major changes in the basic steps: (1) the ***spatial discretization scheme*** in selecting the element approximation solution and (2) the ***time approximation scheme*** for the unsteady term.

In the ***spatial discretization scheme***, the approximate solution for the element is assumed to be

$$\phi^* = \sum_{I=1}^{n_p} N_I(\tilde{x})\phi_I(t) = [N(\tilde{x})]\{\phi^{(e)}(t)\} \tag{13.1}$$

where N_I is the shape function, which is assumed to be a function of the space variable only, $\phi_I(t)$ are the nodal values as a function time, and n_p is the number of nodes in an element. Basically, it is assumed that the spatial and time variations of the dependent variable are separable, and this is appropriate for linear problems and in nonlinear problems with small time steps. Next, the finite element formulation is carried out in the same manner as in the case of a steady state problem, but retaining all time dependent terms in the equation. This procedure leads to a system of ordinary equations in time for the nodal values $\phi^{(e)}(t)$. This formulation is also referred to as ***semi-discrete finite element formulation***.

In the ***next step***, the system of ordinary differential equations are solved numerically. In the numerical approach, the time derivatives are approximated by finite difference schemes such as the one discussed in Chapters 5 and 6. This procedure transforms the system of ordinary differential equations into a set of algebraic equations.

13.2 ONE-DIMENSIONAL UNSTEADY STATE PROBLEMS

Let us consider a general one-dimensional unsteady state problem of the form

$$C\frac{\partial \phi}{\partial t} = \frac{\partial}{\partial x}\left(\Gamma_x \frac{\partial \phi}{\partial x}\right) + S_1\phi + S_0 \tag{13.2}$$

13.2.1 Semi-discrete Finite Element Formulation

Galerkin's based weight integral statement of Equation 13.2 over an element is

$$R_i^{(e)} = \int_{x_i}^{x_j} [N]^{\mathrm{T}}\left(\frac{\partial}{\partial x}\left(\Gamma_x \frac{\partial \phi^*}{\partial x}\right) + S_1\phi^* + S_0 - C\frac{\partial \phi^*}{\partial t}\right) dx = 0 \tag{13.3}$$

Substituting the approximate solution function given by Equation 13.1 for one-dimensional space coordinate, we get the integral statement of the problem as

$$\{R^{(e)}\} = \left\{\begin{array}{c} -\Gamma_x \left.\dfrac{d\phi}{dx}\right|_{x_i} \\ \Gamma_x \left.\dfrac{d\phi}{dx}\right|_{x_j} \end{array}\right\} - \int_{x_i}^{x_j} \Gamma_x \frac{d[N]^{\mathrm{T}}}{dx}\frac{d\phi^*}{dx}\, dx + \int_{x_i}^{x_j} [N]^{\mathrm{T}}(S_1\phi^*)\, dx + \int_{x_i}^{x_j} S_0[N]^{\mathrm{T}}\, dx$$
$$- \int_{x_i}^{x_j} C[N]^{\mathrm{T}}\frac{\partial\{\phi^{(e)}\}}{\partial t}\, dx \tag{13.4}$$

The first four terms in Equation 13.2 were evaluated in Chapter 11 for a one-dimensional element, and given as

$$\int_{x_i}^{x_j} \Gamma_x \frac{d[N]^{\mathrm{T}}}{dx}\frac{d\phi^*}{dx}\,dx = [k_\Gamma]\{\phi^{(e)}\} \tag{13.5a}$$

where

$$[k_\Gamma] = \int_{x_i}^{x_j} \Gamma_x \frac{d[N]^{\mathrm{T}}}{dx}\frac{d[N]}{dx}\,dx = \text{element characteristic matrix due to the diffusion term}$$

$$= \frac{\Gamma_x}{L_e}\begin{bmatrix} 1 & -1 \\ -1 & 1 \end{bmatrix} \quad \text{for two-points line elements} \tag{13.5b}$$

$$\int_{x_i}^{x_j} [N]^{\mathrm{T}}(S_1\phi^*)\,dx = \left[k_S^{(e)}\right]\{\phi^{(e)}\} \tag{13.6a}$$

where

$$\left[k_S^{(e)}\right] = \int_{x_i}^{x_j} S_1[N]^{\mathrm{T}}[N]\,dx = \text{element characteristic matrix due to the source term}$$

$$= \frac{S_1 L_e}{6}\begin{bmatrix} 2 & 1 \\ 1 & 2 \end{bmatrix} \quad \text{for two-points line elements} \tag{13.6b}$$

$$\int_{x_i}^{x_j} [N]^{\mathrm{T}}(S_0)\,dx = \begin{Bmatrix} f_i^{(e)} \\ f_j^{(e)} \end{Bmatrix} = \{f^{(e)}\} = \text{element force vector}$$

$$= \begin{Bmatrix} \dfrac{S_0 L_e}{2} \\ \dfrac{S_0 L_e}{2} \end{Bmatrix} \quad \text{for two-points line elements} \tag{13.7}$$

and

$$\begin{Bmatrix} -\Gamma_x \left.\dfrac{d\phi}{dx}\right|_{x_i} \\ \Gamma_x \left.\dfrac{d\phi}{dx}\right|_{x_j} \end{Bmatrix} = \begin{Bmatrix} I_i^{(e)} \\ I_j^{(e)} \end{Bmatrix} = \{I^{(e)}\} = \text{inter-element vector} \tag{13.8}$$

The unsteady state term is evaluated as

$$\left(\int_{x_i}^{x_j} C[N]^{\mathrm{T}}[N]\right)\frac{\partial\{\phi^{(e)}\}}{\partial t}\,dx = \left[c^{(e)}\right]\{\phi^{(e)}\} \tag{13.9a}$$

where

$$\left[c^{(e)}\right] = \left(\int_{x_i}^{x_j} C[N]^{\mathrm{T}}[N]\right) dx = \text{local capacitance matrix} \tag{13.9b}$$

and

$$\left\{\dot{\phi}^{(e)}\right\} = \begin{Bmatrix} \dfrac{d\phi_i}{dt} \\ \dfrac{d\phi_j}{dt} \end{Bmatrix} \tag{13.9c}$$

The capacitance matrix $c^{(e)}$ can be evaluated by selecting an appropriate element type and associated shape functions. For example, for a ***one-dimensional linear element***, Equation 13.9b can be evaluated in the same manner as the $\left[k_S^{(e)}\right]$ given by Equation 11.87 as

$$\left[c^{(e)}\right] = \int_{x_i}^{x_j} C[N]^{\mathrm{T}}[N]\, dx = \frac{Cl_e}{6}\begin{bmatrix} 2 & 1 \\ 1 & 2 \end{bmatrix} \tag{13.10}$$

Substituting Equation 13.5 to Equation 13.9 into Equation 13.4, we get the ***element characteristic equation*** as

$$\left[c^{(e)}\right]\frac{d\left\{\phi^{(e)}\right\}}{dt} = \left\{I^{(e)}\right\} - \left[k^{(e)}\right]\left\{\phi^{(e)}\right\} + \left\{f^{(e)}\right\} \tag{13.11a}$$

or

$$\left[c^{(e)}\right]\left\{\dot{\phi}^{(e)}\right\} = \left\{I^{(e)}\right\} - \left[k^{(e)}\right]\left\{\phi^{(e)}\right\} + \left\{f^{(e)}\right\} \tag{13.11b}$$

The element characteristic Equation 13.11 represents a system of first-order ordinary differential equations.

13.2.2 Time Approximation

The system of first-order differential equations in time can be solved numerically by approximating the first-order time derivatives by finite difference formulas presented in Chapter 5. For example, for a continuous variation of the dependent variable $\phi(t)$ as shown in Figure 13.1, the time derivative at a time step $m + \frac{1}{2}$ can be estimated based on the ***central difference scheme*** using nodal values at $m+1$ and m as

$$\frac{d\{\phi\}}{dt} = \frac{\{\phi\}^{m+1} - \{\phi\}^{m}}{\Delta t} + O\left(\Delta t^2\right) \tag{13.12}$$

Let us write the element characteristic equation at the time step $m + \frac{1}{2}$ as

$$\left[c^{(e)}\right]\frac{d\left\{\phi^{(e)}\right\}}{dt}\Bigg|^{m+\frac{1}{2}} = \left\{I^{(e)}\right\}^{m+\frac{1}{2}} - \left[k^{(e)}\right]\left\{\phi^{(e)}\right\}^{m+\frac{1}{2}} + \left\{f^{(e)}\right\}^{m+\frac{1}{2}} \tag{13.13}$$

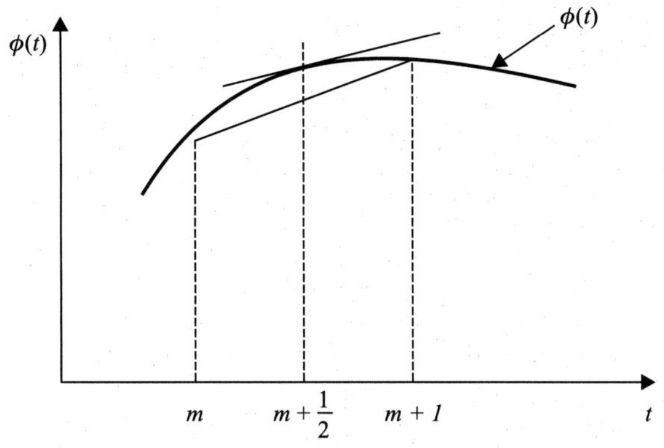

Figure 13.1 Central difference time approximations

Using the central difference scheme for the time derivative and a linear piecewise profile for ϕ, f, and I in the time interval, the element characteristic Equation 13.13 becomes

$$\left[c^{(e)}\right]\frac{\{\phi\}^{m+1}-\{\phi\}^{m}}{\Delta t}=\frac{\left\{I^{(e)}\right\}^{m+1}+\left\{I^{(e)}\right\}^{m}}{2}-\left[k^{(e)}\right]\frac{\{\phi\}^{m+1}+\{\phi\}^{m}}{2}+\frac{\left\{f^{(e)}\right\}^{m+1}+\left\{f^{(e)}\right\}^{m}}{2} \tag{13.14}$$

Rearranging

$$\left(\left[c^{(e)}\right]+\tfrac{1}{2}\Delta t\left[k^{(e)}\right]\right)\{\phi\}^{m+1}=\left(\left[c^{(e)}\right]-\tfrac{1}{2}\Delta t\left[k^{(e)}\right]\right)\{\phi\}^{m}+\frac{\Delta t}{2}\left(\left\{I^{(e)}\right\}^{m+1}+\left\{I^{(e)}\right\}^{m}\right)+\frac{\Delta t}{2}\left(\left\{f^{(e)}\right\}^{m+1}+\left\{f^{(e)}\right\}^{m}\right) \tag{13.15}$$

The element characteristic equation given by Equation 13.15 is referred to as the ***Crank–Nicolson*** or ***semi-implicit time integration scheme***, which is obtained by using the ***central difference scheme***. In a similar manner, we can derive the element characteristic equation based on the forward difference scheme and the backward difference scheme.

Forward Difference Scheme In this scheme it is assumed that the old values of$\left\{\phi^{(e)}\right\}^{m}$, $\left\{f^{(e)}\right\}^{m}$ and $\left\{I^{(e)}\right\}^{m}$ prevail throughout the time interval $(t, \Delta t)$ except at the time $t+\Delta t$ as shown in Figure 7.2. The time derivative at the time step i is approximated by the forward difference scheme in terms of nodal values at steps m and $m+1$ as shown in Figure 13.2 and by the equation

$$\frac{d\{\phi\}^{m}}{dt}=\frac{\{\phi\}^{m+1}-\{\phi\}^{m}}{\Delta t}+O(\Delta t) \tag{13.16}$$

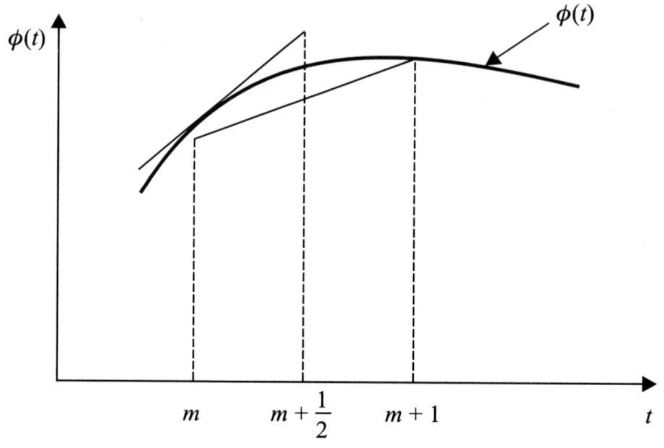

Figure 13.2 Forward difference time approximation

Let us write the element characteristic equation at the time step m as

$$\left[c^{(e)}\right]\frac{d\{\phi^{(e)}\}}{dt}\bigg|^{m} = \left\{I^{(e)}\right\}^{m} - \left[k^{(e)}\right]\left\{\phi^{(e)}\right\}^{m} + \left\{f^{(e)}\right\}^{m} \tag{13.17}$$

Now, substituting the forward difference scheme for the first time derivative given by Equation 13.16, we get

$$\left[c^{(e)}\right]\frac{\{\phi\}^{m+1} - \{\phi\}^{m}}{\Delta t} = \left\{I^{(e)}\right\}^{m} - \left[k^{(e)}\right]\left\{\phi^{(e)}\right\}^{m} + \left\{f^{(e)}\right\}^{m} \tag{13.18}$$

Rearranging

$$\left[c^{(e)}\right]\left\{\phi^{(e)}\right\}^{m+1} = \left(\left[c^{(e)}\right] - \Delta t\left[k^{(e)}\right]\right)\left\{\phi^{(e)}\right\}^{m} + \Delta t\left\{I^{(e)}\right\}^{m} + \Delta t\left\{f^{(e)}\right\}^{m} \tag{13.19}$$

The element characteristic equation given by Equation 13.16 is referred to as the ***explicit time integration scheme***, which is obtained by using the forward difference scheme.

Backward Difference Scheme In this scheme it is assumed that the new values of $\left\{\phi^{(e)}\right\}^{m+1}$, $\left\{f^{(e)}\right\}^{m+1}$, and $\left\{I^{(e)}\right\}^{m+1}$ prevail throughout the time interval $(t, \Delta t)$ except at the time t as shown in Figure 7.2. The time derivative at the time step $m+1$ is approximated by the backward difference scheme in terms of nodal values at steps m and $m+1$ as depicted in Figure 13.3 and by the equation

$$\frac{d\{\phi\}^{m+1}}{dt} = \frac{\{\phi\}^{m+1} - \{\phi\}^{m}}{\Delta t} + O(\Delta t) \tag{13.20}$$

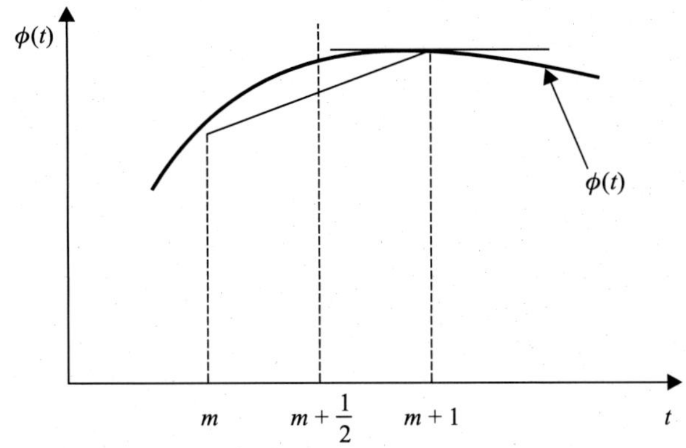

Figure 13.3 Backward difference time approximations

Let us write the element characteristic equation at the time step $m+1$ as

$$\left[c^{(e)}\right]\frac{d\{\phi^{(e)}\}}{dt}\Bigg|^{m+1} = \{I^{(e)}\}^{m+1} - \left[k^{(e)}\right]\{\phi^{(e)}\}^{m+1} + \{f^{(e)}\}^{m+1} \tag{13.21}$$

Substituting the backward difference scheme for the first time derivative given by Equation 13.20, we get

$$\left[c^{(e)}\right]\frac{\{\phi\}^{m+1} - \{\phi\}^{m}}{\Delta t} = \{I^{(e)}\}^{m+1} - \left[k^{(e)}\right]\{\phi^{(e)}\}^{m+1} + \{f^{(e)}\}^{m+1} \tag{13.22}$$

Rearranging

$$\left(\left[c^{(e)}\right] + \Delta t\left[k^{(e)}\right]\right)\{\phi^{(e)}\}^{m+1} = \left[c^{(e)}\right]\{\phi^{(e)}\}^{m} + \Delta t\{I^{(e)}\}^{m+1} + \Delta t\{f^{(e)}\}^{m+1} \tag{13.23}$$

The element characteristic equation given by Equation 13.20 is referred to as the ***implicit time integration scheme***, which is obtained by using a backward difference scheme.

Equations 3.15, 3.19, and 3.23 can be expressed in a general form

$$\begin{aligned}\left(\left[c^{(e)}\right] + \lambda\Delta t\left[k^{(e)}\right]\right)\{\phi\}^{m+1} &= \left(\left[c^{(e)}\right] - (1-\lambda)\Delta t\left[k^{(e)}\right]\right)\{\phi\}^{m} \\ &\quad + \Delta t\left(\lambda\{I^{(e)}\}^{m+1} + (1-\lambda)\{I^{(e)}\}^{m}\right) \\ &\quad + \Delta t\left(\lambda\{f^{(e)}\}^{m+1} + (1-\lambda)\{f^{(e)}\}^{m}\right)\end{aligned} \tag{13.24a}$$

Equation 13.24a can further be written in a compact form as

$$A\{\phi\}^{m+1} = B\{\phi\}^{m} + C \tag{13.24b}$$

where

$$A = \left[c^{(e)}\right] + \lambda\Delta t\left[k^{(e)}\right] \tag{13.24c}$$

$$B = \left[c^{(e)}\right] + (1-\lambda)\Delta t\left[k^{(e)}\right] \tag{13.24d}$$

$$C = \Delta t\left(\lambda\left\{I^{(e)}\right\}^{m+1} + (1-\lambda)\left\{I^{(e)}\right\}^{m}\right) - \Delta t\left(\lambda\left\{f^{(e)}\right\}^{m+1} + (1-\lambda)\left\{f^{(e)}\right\}^{m}\right) \tag{13.24e}$$

where λ is the weighting factor for the time approximation scheme. We can derive a different time approximation scheme by selecting different values for the weighting factor. This weighting factor is usually specified in the range of $0 \le \lambda \le 1$. For example

- $\lambda = 0$ leads to the ***explicit scheme*** (Equation 13.19) based on the ***forward difference scheme***
- $\lambda = 1$ leads to the fully ***implicit scheme*** (Equation 13.23) based on the ***backward difference scheme***
- $\lambda = \frac{1}{2}$ leads to the semi-implicit or **Crank–Nicolson scheme** based on the ***central difference scheme*** (Equation 13.15)

With the known initial values of $\{\phi\}^0$, Equation (13.24) is solved with marching in time. For example, with the known values of $\{\phi\}^0$ at time $t=0$ and a time step Δt, we can solve Equation 13.24 for $\{\phi\}^1$ at time $t=\Delta t$. Now, with the known value of $\{\phi\}^1$, we can proceed to determine $\{\phi\}^2$ at time $t=2\Delta t$, and so on. So, at each time step the solution is equivalent to solving a steady state problem. It can also be noted that for constant material properties and constant time step Δt, matrices A and B are the same for all time steps. For variable material properties as in a nonlinear problem and for variable time steps, the matrices A and B are reevaluated at each time step.

13.2.3 Stability Consideration

As we have discussed in Chapter 7, the advantage of a fully-implicit scheme (based on the backward difference approximation ($\lambda = 1$)) is that it is unconditionally stable. Whereas an explicit scheme ($\lambda = 0$) based on the forward difference approximation is associated with the stability problem, the temporal approximation error grows with time and the estimated solution oscillates when the selected time step Δt exceeds a certain minimum value. A convergence for such a scheme is achieved only if the time steps satisfy the stability criterion. A stability analysis for such a system shows (Segerlind, 1984) that for $\lambda < \frac{1}{2}$ the largest Δt is given by

$$\Delta t = \frac{\beta_{\min}}{1-\lambda} \tag{13.25}$$

where $\beta_{\min}$ is the smallest eigenvalue of the equation given by

$$\left|\left(\left[c^{(e)}\right] - \beta\left[k^{(e)}\right]\right)\right| = 0 \tag{13.26}$$

Example 13.1: One-dimensional Unsteady State Conduction: Consider a one-dimensional unsteady state conduction without heat generation in a plane slab as shown. Initially the slab is at a uniform temperature of $T_I = 25°\text{C}$. The boundary surface at $x = 0$ is subject to a constant surface heat flux, $f''_{ls} = 1 \times 10^5\,\text{W/m}^2$, and the boundary surface $x = L$ is maintained at constant surface temperature, $T_R = 25°\text{C}$, respectively. Determine the temperature distribution in the slab and heat transfer rates at the right surface as a function of time using implicit time approximation and time step $\Delta t = 1\,\text{s}$. Use $k = 60.5\,\text{W/m}\cdot\text{C}$, $\rho = 7854\,\text{kg/m}^3$, $c_p = 434\,\text{J/kg}\cdot\text{K}$, $L = 0.08\,\text{m}$.

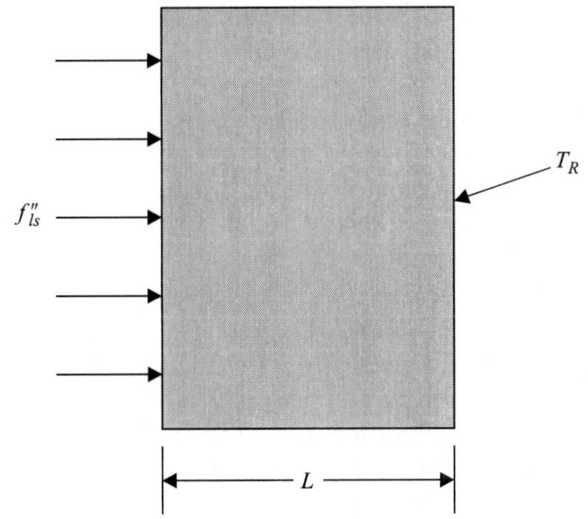

The mathematical statement of the problem is

Governing equation

$$\rho c_p \frac{\partial T}{\partial t} = \frac{d}{dx}\left(k\frac{dT}{dx}\right) \tag{E.13.1.1}$$

Boundary condition

1. $x = 0, \qquad -k\dfrac{\partial T}{\partial x} = f''_{ls}$ (E.13.1.2a)

2. $x = L, \qquad T = T_R$ (E.13.1.2b)

Solution: In the ***first step***, let us discretize the domain using four uniform line elements as shown in the figure.

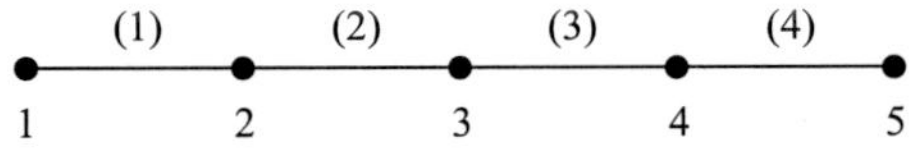

With this mesh size distribution, we have $l_e = l = 0.02\,\text{m}$ and number of grid points $N = 5$ with node $i = 1$ and $i = 5$ located at $x = 0$ and $x = 0.08\,\text{m}$.

In the ***second step***, we write the weak form of the governing equation following Equation 11.9, which is obtained by Galerkin's weighted residual method as

$$\left(\int_{x_i}^{x_j} \rho c_p [N]^{\mathrm{T}}[N]\right) \frac{\partial\{T^{(e)}\}}{\partial t}\,dx = \begin{Bmatrix} -k\left.\dfrac{dT}{dx}\right|_{x_i} \\ +k\left.\dfrac{dT}{dx}\right|_{x_j} \end{Bmatrix} - \int_{x}^{x_j} k \frac{d[N]^{\mathrm{T}}}{dx}\frac{dT^*}{dx}\,dx = 0 \tag{E.13.1.3}$$

In the ***third step***, we assume that the temperature distribution in each element be represented by the approximate function

$$T^* = N_i T_i + N_j T_j \tag{E.13.1.4}$$

In the ***fourth step***, we can write the ***element characteristic equation*** by comparing with equation Equation 13.11 as

$$\left[c^{(e)}\right]\frac{d\{T^{(e)}\}}{dt} = \{I^{(e)}\} - \left[k_k^{(e)}\right]\{T^{(e)}\} \tag{13.8a}$$

or

$$\left[c^{(e)}\right]\{\dot{T}^{(e)}\} = \{I^{(e)}\} - \left[k_k^{(e)}\right]\{T^{(e)}\} \tag{13.8b}$$

Using the fully implicit time approximation scheme as given by Equation 13.24 with $\lambda = 1$, we get

$$\left(c^{(e)} + \Delta t\left[k^{(e)}\right]\right)\{T\}^{m+1} = \left[c^{(e)}\right]\{T\}^m + \Delta t\left\{I^{(e)}\right\}^{m+1} \tag{E.13.1.5}$$

where

$$\left[c^{(e)}\right] = \frac{Cl_e}{6}\begin{bmatrix} 2 & 1 \\ 1 & 2 \end{bmatrix}, \quad C = \rho c_p \tag{E.13.1.6a}$$

$$[k_k] = \frac{k}{l_e}\begin{bmatrix} 1 & -1 \\ -1 & 1 \end{bmatrix} \tag{E.13.1.6b}$$

$$\{I^{(e)}\} = \begin{Bmatrix} -k\left.\dfrac{dT}{dx}\right|_{x_i} \\ +k\left.\dfrac{dT}{dx}\right|_{x_j} \end{Bmatrix} \tag{E.13.1.6c}$$

$$\{T^{(e)}\} = \begin{Bmatrix} T_i \\ T_j \end{Bmatrix} \tag{E.13.1.6d}$$

Based on these expressions the element characteristics Equation E.13.1.5 becomes

$$\left(\frac{Cl_e}{6}\begin{bmatrix}2 & 1\\ 1 & 2\end{bmatrix}+\frac{k\Delta t}{l_e}\begin{bmatrix}1 & -1\\ -1 & 1\end{bmatrix}\right)\begin{Bmatrix}T_i^{m+1}\\ T_j^{m+1}\end{Bmatrix}=\frac{Cl_e}{6}\begin{bmatrix}2 & 1\\ 1 & 2\end{bmatrix}\begin{Bmatrix}T_i^{m}\\ T_j^{m}\end{Bmatrix}+\Delta t\begin{Bmatrix}-k\left.\dfrac{dT}{dx}\right|_{x_i}^{m+1}\\ +k\left.\dfrac{dT}{dx}\right|_{x_j}^{m+1}\end{Bmatrix}$$

$$\left(\begin{bmatrix}\dfrac{Cl_e}{3}+\dfrac{k\Delta t}{l_e} & \dfrac{Cl_e}{6}-\dfrac{k\Delta t}{l_e}\\ \dfrac{Cl_e}{6}-\dfrac{k\Delta t}{l_e} & \dfrac{Cl_e}{3}+\dfrac{k\Delta t}{l_e}\end{bmatrix}\right)\begin{Bmatrix}T_i^{m+1}\\ T_j^{m+1}\end{Bmatrix}=\begin{bmatrix}\dfrac{Cl_e}{3} & \dfrac{Cl_e}{6}\\ \dfrac{Cl_e}{6} & \dfrac{Cl_e}{3}\end{bmatrix}\begin{Bmatrix}T_i^{m}\\ T_j^{m}\end{Bmatrix}+\Delta t\begin{Bmatrix}-k\left.\dfrac{dT}{dx}\right|_{x_i}^{m+1}\\ +k\left.\dfrac{dT}{dx}\right|_{x_j}^{m+1}\end{Bmatrix} \tag{E.13.1.7}$$

Since the material is homogeneous and the mesh is uniform, the element characteristics equation is identical for all elements.

In the ***fifth step***, we assemble all element equations and obtain the global system following Equation 11.21 through the connectivity relations. For example addition of ***Element 1*** into the global system results in

$$\begin{bmatrix}\dfrac{Cl_e}{3}+\dfrac{k\Delta t}{l_e} & \dfrac{Cl_e}{6}-\dfrac{k\Delta t}{l_e} & 0 & 0 & 0\\ \dfrac{Cl_e}{6}-\dfrac{k\Delta t}{l_e} & \dfrac{Cl_e}{3}+\dfrac{k\Delta t}{l_e} & 0 & 0 & 0\\ 0 & 0 & 0 & 0 & 0\\ 0 & 0 & 0 & 0 & 0\\ 0 & 0 & 0 & 0 & 0\end{bmatrix}\begin{Bmatrix}T_1^{m+1}\\ T_2^{m+1}\\ T_3^{m+1}\\ T_4^{m+1}\\ T_5^{m+1}\end{Bmatrix}=\begin{bmatrix}\dfrac{Cl_e}{3} & \dfrac{Cl_e}{6} & 0 & 0 & 0\\ \dfrac{Cl_e}{6} & \dfrac{Cl_e}{3} & 0 & 0 & 0\\ 0 & 0 & 0 & 0 & 0\\ 0 & 0 & 0 & 0 & 0\\ 0 & 0 & 0 & 0 & 0\end{bmatrix}\begin{Bmatrix}T_1^{m}\\ T_2^{m}\\ T_3^{m}\\ T_4^{m}\\ T_5^{m}\end{Bmatrix}+\Delta t\begin{Bmatrix}-k\left.\dfrac{dT}{dx}\right|_{x_1}^{m+1}\\ +k\left.\dfrac{dT}{dx}\right|_{x_2}^{m+1}\\ 0\\ 0\\ 0\end{Bmatrix} \tag{E.13.1.8a}$$

Similarly, addition of ***Element 2*** gives

$$\begin{bmatrix} \frac{Cl_e}{3}+\frac{k\Delta t}{l_e} & \frac{Cl_e}{6}-\frac{k\Delta t}{l_e} & 0 & 0 & 0 \\ \frac{Cl_e}{6}-\frac{k\Delta t}{l_e} & \frac{2Cl_e}{3}+\frac{2k\Delta t}{l_e} & \frac{Cl_e}{6}-\frac{k\Delta t}{l_e} & 0 & 0 \\ 0 & \frac{Cl_e}{6}-\frac{k\Delta t}{l_e} & \frac{Cl_e}{3}+\frac{k\Delta t}{l_e} & 0 & 0 \\ 0 & 0 & 0 & 0 & 0 \\ 0 & 0 & 0 & 0 & 0 \end{bmatrix} \begin{Bmatrix} T_1^{m+1} \\ T_2^{m+1} \\ T_3^{m+1} \\ T_4^{m+1} \\ T_5^{m+1} \end{Bmatrix}$$

$$= \begin{bmatrix} \frac{Cl_e}{3} & \frac{Cl_e}{6} & 0 & 0 & 0 \\ \frac{Cl_e}{6} & \frac{2Cl_e}{3} & \frac{Cl_e}{6} & 0 & 0 \\ 0 & \frac{Cl_e}{6} & \frac{Cl_e}{3} & 0 & 0 \\ 0 & 0 & 0 & 0 & 0 \\ 0 & 0 & 0 & 0 & 0 \end{bmatrix} \begin{Bmatrix} T_1^m \\ T_2^m \\ T_3^m \\ T_4^m \\ T_5^m \end{Bmatrix} + \Delta t \begin{Bmatrix} -k\left.\frac{dT}{dx}\right|_{x_1}^{m+1} \\ +k\left.\frac{dT}{dx}\right|_{x_2}^{m+1} -k\left.\frac{dT}{dx}\right|_{x_2}^{m+1} \\ +k\left.\frac{dT}{dx}\right|_{x_3}^{m+1} \\ 0 \\ 0 \end{Bmatrix} \quad \text{(E.13.1.8b)}$$

Addition of ***Element 3*** gives

$$\begin{bmatrix} \frac{Cl_e}{3}+\frac{k\Delta t}{l_e} & \frac{Cl_e}{6}-\frac{k\Delta t}{l_e} & 0 & 0 & 0 \\ \frac{Cl_e}{6}-\frac{k\Delta t}{l_e} & \frac{2Cl_e}{3}+\frac{2k\Delta t}{l_e} & \frac{Cl_e}{6}-\frac{k\Delta t}{l_e} & 0 & 0 \\ 0 & \frac{Cl_e}{6}-\frac{k\Delta t}{l_e} & \frac{2Cl_e}{3}+\frac{2k\Delta t}{l_e} & \frac{Cl_e}{6}-\frac{k\Delta t}{l_e} & 0 \\ 0 & 0 & \frac{Cl_e}{6}-\frac{k\Delta t}{l_e} & \frac{Cl_e}{3}+\frac{k\Delta t}{l_e} & 0 \\ 0 & 0 & 0 & 0 & 0 \end{bmatrix} \begin{Bmatrix} T_1^{m+1} \\ T_2^{m+1} \\ T_3^{m+1} \\ T_4^{m+1} \\ T_5^{m+1} \end{Bmatrix}$$

$$= \begin{bmatrix} \frac{Cl_e}{3} & \frac{Cl_e}{6} & 0 & 0 & 0 \\ \frac{Cl_e}{6} & \frac{2Cl_e}{3} & \frac{Cl_e}{6} & 0 & 0 \\ 0 & \frac{Cl_e}{6} & \frac{2Cl_e}{3} & \frac{Cl_e}{6} & 0 \\ 0 & 0 & \frac{Cl_e}{6} & \frac{Cl_e}{3} & 0 \\ 0 & 0 & 0 & 0 & 0 \end{bmatrix} \begin{Bmatrix} T_1^m \\ T_2^m \\ T_3^m \\ T_4^m \\ T_5^m \end{Bmatrix} + \Delta t \begin{Bmatrix} -k\left.\frac{dT}{dx}\right|_{x_1}^{m+1} \\ +k\left.\frac{dT}{dx}\right|_{x_2}^{m+1} -k\left.\frac{dT}{dx}\right|_{x_2}^{m+1} \\ +k\left.\frac{dT}{dx}\right|_{x_3}^{m+1} -k\left.\frac{dT}{dx}\right|_{x_3}^{m+1} \\ +k\left.\frac{dT}{dx}\right|_{x_3}^{m+1} \\ 0 \end{Bmatrix} \quad \text{(E.13.1.8c)}$$

Finally, addition of ***Element 4*** gives the final global system as

$$\begin{bmatrix} \frac{Cl_e}{3}+\frac{k\Delta t}{l_e} & \frac{Cl_e}{6}-\frac{k\Delta t}{l_e} & 0 & 0 & 0 \\ \frac{Cl_e}{6}-\frac{k\Delta t}{l_e} & \frac{2Cl_e}{3}+\frac{2k\Delta t}{l_e} & \frac{Cl_e}{6}-\frac{k\Delta t}{l_e} & 0 & 0 \\ 0 & \frac{Cl_e}{6}-\frac{k\Delta t}{l_e} & \frac{2Cl_e}{3}+\frac{2k\Delta t}{l_e} & \frac{Cl_e}{6}-\frac{k\Delta t}{l_e} & 0 \\ 0 & 0 & \frac{Cl_e}{6}-\frac{k\Delta t}{l_e} & \frac{2Cl_e}{3}+\frac{2k\Delta t}{l_e} & \frac{Cl_e}{6}-\frac{k\Delta t}{l_e} \\ 0 & 0 & 0 & \frac{Cl_e}{6}-\frac{k\Delta t}{l_e} & \frac{Cl_e}{3}+\frac{k\Delta t}{l_e} \end{bmatrix} \begin{Bmatrix} T_1^{m+1} \\ T_2^{m+1} \\ T_3^{m+1} \\ T_4^{m+1} \\ T_5^{m+1} \end{Bmatrix}$$

$$= \begin{bmatrix} \frac{Cl_e}{3} & \frac{Cl_e}{6} & 0 & 0 & 0 \\ \frac{Cl_e}{6} & \frac{2Cl_e}{3} & \frac{Cl_e}{6} & 0 & 0 \\ 0 & \frac{Cl_e}{6} & \frac{2Cl_e}{3} & \frac{Cl_e}{6} & 0 \\ 0 & 0 & \frac{Cl_e}{6} & \frac{2Cl_e}{3} & \frac{Cl_e}{6} \\ 0 & 0 & 0 & \frac{Cl_e}{6} & \frac{Cl_e}{3} \end{bmatrix} \begin{Bmatrix} T_1^m \\ T_2^m \\ T_3^m \\ T_4^m \\ T_5^m \end{Bmatrix} + \Delta t \begin{Bmatrix} -k\left.\frac{dT}{dx}\right|_{x_1}^{m+1} \\ +k\left.\frac{dT}{dx}\right|_{x_2}^{m+1} - k\left.\frac{dT}{dx}\right|_{x_2}^{m+1} \\ +k\left.\frac{dT}{dx}\right|_{x_3}^{m+1} - k\left.\frac{dT}{dx}\right|_{x_3}^{m+1} \\ +k\left.\frac{dT}{dx}\right|_{x_4}^{m+1} - k\left.\frac{dT}{dx}\right|_{x_4}^{m+1} \\ +k\left.\frac{dT}{dx}\right|_{x_5}^{m+1} \end{Bmatrix} \tag{E.13.1.8d}$$

In the ***sixth step***, we incorporate the inter-element continuity conditions and apply the boundary conditions at the left boundary node as $-k(\partial T/\partial x)\big|_{x_1} = f''_{ls}$ and at the right boundary as $T_5 = T_R$, we get

$$\begin{bmatrix} \frac{Cl_e}{3}+\frac{k\Delta t}{l_e} & \frac{Cl_e}{6}-\frac{k\Delta t}{l_e} & 0 & 0 & 0 \\ \frac{Cl_e}{6}-\frac{k\Delta t}{l_e} & \frac{2Cl_e}{3}+\frac{2k\Delta t}{l_e} & \frac{Cl_e}{6}-\frac{k\Delta t}{l_e} & 0 & 0 \\ 0 & \frac{Cl_e}{6}-\frac{k\Delta t}{l_e} & \frac{2Cl_e}{3}+\frac{2k\Delta t}{l_e} & \frac{Cl_e}{6}-\frac{k\Delta t}{l_e} & 0 \\ 0 & 0 & \frac{Cl_e}{6}-\frac{k\Delta t}{l_e} & \frac{2Cl_e}{3}+\frac{2k\Delta t}{l_e} & \frac{Cl_e}{6}-\frac{k\Delta t}{l_e} \\ 0 & 0 & 0 & \frac{Cl_e}{6}-\frac{k\Delta t}{l_e} & \frac{Cl_e}{3}+\frac{k\Delta t}{l_e} \end{bmatrix} \begin{Bmatrix} T_1^{m+1} \\ T_2^{m+1} \\ T_3^{m+1} \\ T_4^{m+1} \\ T_R \end{Bmatrix}$$

$$= \begin{bmatrix} \frac{Cl_e}{3} & \frac{Cl_e}{6} & 0 & 0 & 0 \\ \frac{Cl_e}{6} & \frac{2Cl_e}{3} & \frac{Cl_e}{6} & 0 & 0 \\ 0 & \frac{Cl_e}{6} & \frac{2Cl_e}{3} & \frac{Cl_e}{6} & 0 \\ 0 & 0 & \frac{Cl_e}{6} & \frac{2Cl_e}{3} & \frac{Cl_e}{6} \\ 0 & 0 & 0 & \frac{Cl_e}{6} & \frac{Cl_e}{3} \end{bmatrix} \begin{Bmatrix} T_1^m \\ T_2^m \\ T_3^m \\ T_4^m \\ T_R \end{Bmatrix} + \Delta t \begin{Bmatrix} +f''_{ls} \\ 0 \\ 0 \\ 0 \\ +k\left.\frac{dT}{dx}\right|_{x_5}^{m+1} \end{Bmatrix} \tag{E.13.1.9}$$

Since the boundary conditions at the right surface is of the first kind, we rearrange

$$\begin{bmatrix} \frac{Cl_e}{3}+\frac{k\Delta t}{l_e} & \frac{Cl_e}{6}-\frac{k\Delta t}{l_e} & 0 & 0 & 0 \\ \frac{Cl_e}{6}-\frac{k\Delta t}{l_e} & \frac{2Cl_e}{3}+\frac{2k\Delta t}{l_e} & \frac{Cl_e}{6}-\frac{k\Delta t}{l_e} & 0 & 0 \\ 0 & \frac{Cl_e}{6}-\frac{k\Delta t}{l_e} & \frac{2Cl_e}{3}+\frac{2k\Delta t}{l_e} & \frac{Cl_e}{6}-\frac{k\Delta t}{l_e} & 0 \\ 0 & 0 & \frac{Cl_e}{6}-\frac{k\Delta t}{l_e} & \frac{2Cl_e}{3}+\frac{2k\Delta t}{l_e} & \Delta t \\ 0 & 0 & 0 & \frac{Cl_e}{6}-\frac{k\Delta t}{l_e} & \Delta t \end{bmatrix} \begin{Bmatrix} T_1^{m+1} \\ T_2^{m+1} \\ T_3^{m+1} \\ T_4^{m+1} \\ -k\frac{dT}{dx}\Big|_{x_5}^{m+1} \end{Bmatrix}$$

$$= \begin{bmatrix} \frac{Cl_e}{3} & \frac{Cl_e}{6} & 0 & 0 & 0 \\ \frac{Cl_e}{6} & \frac{2Cl_e}{3} & \frac{Cl_e}{6} & 0 & 0 \\ 0 & \frac{Cl_e}{6} & \frac{2Cl_e}{3} & \frac{Cl_e}{6} & 0 \\ 0 & 0 & \frac{Cl_e}{6} & \frac{2Cl_e}{3} & 0 \\ 0 & 0 & 0 & \frac{Cl_e}{6} & 0 \end{bmatrix} \begin{Bmatrix} T_1^l \\ T_2^l \\ T_3^l \\ T_4^l \\ -k\frac{dT}{dx}\Big|_{x_5}^{m} \end{Bmatrix} + \begin{Bmatrix} +f''_{ls}\Delta t \\ 0 \\ 0 \\ \frac{k\Delta t T_R}{l_e} \\ -\frac{k\Delta t T_R}{l_e} \end{Bmatrix} \quad \text{(E.13.1.10)}$$

For the problem with the given data, we get

$C=\rho c_p=7854\times 434=3.408\,636\times 10^6\,\text{J/m}^3\cdot\text{K}$, $k=60.5\,\text{W/m}\cdot\text{K}$, $l_e=0.02\,\text{m}$

The element matrices and inter-element vector are

$$\left[\kappa_k^{(e)}\right]=\frac{k}{l_e}\begin{bmatrix}1 & -1\\ -1 & 1\end{bmatrix}=\frac{60.5}{0.02}\begin{bmatrix}1 & -1\\ -1 & 1\end{bmatrix}=3025\begin{bmatrix}1 & -1\\ -1 & 1\end{bmatrix}$$

$$\frac{k\Delta t T_R}{l_e}=7.5625\times 10^4$$

$$\left[c^{(e)}\right]=\int_{x_i}^{x_j}\rho c_p[N]^{\mathrm{T}}[N]\,dx=\frac{\rho c_p l_e}{6}\begin{bmatrix}2 & 1\\ 1 & 2\end{bmatrix}=\frac{7854\times 434\times 0.02}{6}\begin{bmatrix}2 & 1\\ 1 & 2\end{bmatrix}$$

$$=1.136\,212\times 10^4\begin{bmatrix}2 & 1\\ 1 & 2\end{bmatrix},\quad \frac{Cl_e}{6}=1.1362\times 10^4\,\text{J/m}^2\,\text{K}$$

$$\frac{Cl_e}{3}+\frac{k\Delta t}{l_e}=22\,724+3025=2.5749\times 10^4$$

$$\frac{2Cl_e}{3}+\frac{2k\Delta t}{l_e}=5.1498\times 10^4$$

$$\frac{Cl_e}{6}-\frac{k\Delta t}{l_e}=11\,362-3025=0.8337\times 10^4$$

Substituting the numerical values, we obtain the global system as

$$\begin{bmatrix} 2.5749\times10^4 & 0.8337\times10^4 & 0 & 0 & 0 \\ 0.8337\times10^4 & 5.1498\times10^4 & 0.8337\times10^4 & 0 & 0 \\ 0 & 0.8337\times10^4 & 5.1498\times10^4 & 0.8337\times10^4 & 0 \\ 0 & 0 & 0.8337\times10^4 & 5.1498\times10^4 & 0 \\ 0 & 0 & 0 & 0.8337\times10^4 & 1 \end{bmatrix} \begin{Bmatrix} T_1^{m+1} \\ T_2^{m+1} \\ T_3^{m+1} \\ T_4^{m+1} \\ -k\left.\dfrac{dT}{dx}\right|_{x_5}^{m+1} \end{Bmatrix}$$

$$= \begin{bmatrix} 2.2724\times10^4 & 1.1362\times10^4 & 0 & 0 & 0 \\ 1.1362\times10^4 & 4.5448\times10^4 & 1.1362\times10^4 & 0 & 0 \\ 0 & 1.1362\times10^4 & 4.5448\times10^4 & 1.1362\times10^4 & 0 \\ 0 & 0 & 1.1362\times10^4 & 4.5448\times10^4 & 0 \\ 0 & 0 & 0 & 1.1362\times10^4 & 0 \end{bmatrix} \begin{Bmatrix} T_1^{m} \\ T_2^{m} \\ T_3^{m} \\ T_4^{m} \\ k\left.\dfrac{dT}{dx}\right|_{x_5}^{m} \end{Bmatrix}$$

$$+ \begin{Bmatrix} +0.1\times10^6 \\ 0 \\ 0 \\ 7.5625\times10^4 \\ -7.5625\times10^4 \end{Bmatrix} \tag{E.13.1.11}$$

Considering the initial values at $t=0$ as

$$\begin{Bmatrix} T_1^0 \\ T_2^0 \\ T_3^0 \\ T_4^0 \\ k\left.\dfrac{dT}{dx}\right|_{x_5}^{0} \end{Bmatrix} = \begin{Bmatrix} 25 \\ 25 \\ 25 \\ 25 \\ 0 \end{Bmatrix}$$

we can solve the system of Equation E.13.1.11 for solution at $t = t + \Delta t = 1$ s as

$$\begin{Bmatrix} T_1^1 \\ T_2^1 \\ T1 \\ T_4^1 \\ -k\left.\dfrac{dT}{dx}\right|_{x_5}^{1} \end{Bmatrix} = \begin{Bmatrix} 29.1048 \\ 24.3170 \\ 25.1141 \\ 24.9780 \\ 183.80 \end{Bmatrix}$$

The heat transfer rate at $x = L$ is estimated as

$$\dot{q}_{x=L} = -k\frac{dT}{dx}\bigg|_{x_5} = 183.80\,\text{W/m}^2$$

The computation can be continued with increase in time. Note that the interior temperature distribution shows a little fluctuation, which is against the physical phenomena and caused by the time approximation used with larger time steps. This discrepancy may disappear after several time steps and with the use of smaller time steps.

Example 13.2: One-dimensional Unsteady State Conduction with Volume Heat Generation: Consider a one-dimensional unsteady state bio-heat equation (Ravi and Majumdar) in a plane slab of biological tissue irradiated by a laser beam as shown. The boundary surface at $x = 0$ is subjected to a convective cooling and the boundary surface $x = H$ is maintained at constant surface temperature T_b.

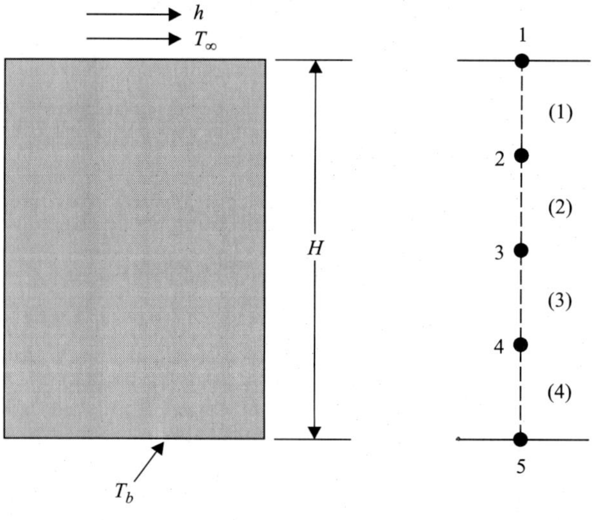

The mathematical statement of the problem is given as

$$\rho c\frac{\partial T}{\partial t} = \frac{\partial}{\partial x}\left(k\frac{\partial T}{\partial x}\right) + I_0 a\,\mathrm{e}^{-ax} - \dot{m}_{bl}c_{bl}(T - T_b) \qquad \text{(E.13.2.1)}$$

Boundary conditions

1. $x = 0, \qquad k\dfrac{\partial T}{\partial x} = h(T - T_\infty)$ (E.13.2.2a)

2. $x = L, \qquad T = T_b$ (E.13.2.2b)

Initial conditions

$$t = 0, \qquad T(x, 0) = T_b \tag{E.13.2.3}$$

Derive the element and global characteristic equation for temperature distribution of the tissue as a function of time.

Solution: Let us consider a mesh size distribution with five linear line elements as shown in the figure. With this mesh size distribution, we have $l_e = H/5$ and number of grid points $N=5$ with node $i=1$ and $i=5$ located at $x=0$ and $x=H$.

In the ***second step***, we assume that the temperature distribution in each element be represented by the approximate function

$$T^* = N_i T_i + N_j T_j \tag{E.13.2.4}$$

In the ***third step***, the integral form of the equation over an element is obtained by applying ***Galerkin's weighted residual method*** as

$$\{R^{(e)}\} = -\int_{x_i}^{x_j} [N]^{\mathrm{T}} \left[\frac{\partial}{\partial x}\left(\Gamma_x \frac{\partial T^*}{\partial x} \right) - \dot{m}_{bl} c_{bl} T^* + \dot{m}_{bl} c_{bl} T_b + I_0 a\, \mathrm{e}^{-ax} \right] dx = 0 \tag{E.13.2.5}$$

Carrying out the integrations term by term and substituting the approximate solution, we get

$$\left(\int_{x_i}^{x_j} \rho c_p [N]^{\mathrm{T}}[N] \right) \frac{\partial \{T^{(e)}\}}{\partial t}\, dx = \begin{Bmatrix} k\left.\dfrac{dT}{dx}\right|_{x_i} \\ -k\left.\dfrac{dT}{dx}\right|_{x_j} \end{Bmatrix} - \left(\int_{x_i}^{x_j} k \frac{d[N]^{\mathrm{T}}}{dx} \frac{dT^*}{dx}\, dx \right) \{T^{(e)}\}$$

$$- \left\{ \int_{x_i}^{x_j} \dot{m}_{bl} c_{bl} [N]^{\mathrm{T}}[N]\, dx \right\} \{T^{(e)}\}$$

$$+ \int_{x_i}^{x_j} \dot{m}_{bl} c_{bl} T_b [N]^{\mathrm{T}}\, dx$$

$$+ \int_{x_i}^{x_j} I_0 a\, \mathrm{e}^{-ax} [N]^{\mathrm{T}}\, dx = 0 \tag{E.13.2.6}$$

$$\frac{\rho c_p \Delta x}{6}\begin{bmatrix} 2 & 1 \\ 1 & 2 \end{bmatrix}\begin{Bmatrix} \left.\dfrac{\partial T}{\partial t}\right|_{x_i} \\ \left.\dfrac{\partial T}{\partial t}\right|_{x_j} \end{Bmatrix} = \begin{Bmatrix} \left.k\dfrac{dT}{dx}\right|_{x_i} \\ \left.-k\dfrac{dT}{dx}\right|_{x_j} \end{Bmatrix}$$

$$- \frac{k}{\Delta x}\begin{bmatrix} 1 & -1 \\ -1 & 1 \end{bmatrix}\begin{Bmatrix} T_i \\ T_j \end{Bmatrix} - \frac{\dot{m}_{bl} c_{bl} \Delta x}{6}\begin{bmatrix} 2 & 1 \\ 1 & 2 \end{bmatrix}\begin{Bmatrix} T_i \\ T_j \end{Bmatrix}$$

$$+ \frac{\dot{m}_{bl} c_{bl} \Delta x}{2}\begin{Bmatrix} 1 \\ 1 \end{Bmatrix}$$

$$+ \frac{I_0}{\Delta x}\begin{Bmatrix} \dfrac{1}{a}(\mathrm{e}^{-ax_j} - \mathrm{e}^{-ax_i}) + \Delta x\,\mathrm{e}^{-ax_i} \\ -\dfrac{1}{a}(\mathrm{e}^{-ax_j} - \mathrm{e}^{-ax_i}) - \Delta x\,\mathrm{e}^{-ax_j} \end{Bmatrix} \quad \text{(E.13.2.7)}$$

The element characteristics equation can be written in matrix notation as

$$\left[c^{(e)}\right]\frac{d\{T^{(e)}\}}{dt} = \{I^{(e)}\} - \left[k^{(e)}\right]\{T^{(e)}\} + \{f^{(e)}\} \quad \text{(E.13.2.8a)}$$

or

$$\left[c^{(e)}\right]\{\dot{T}^{(e)}\} = \{I^{(e)}\} - \left[k^{(e)}\right]\{T^{(e)}\} + \{f^{(e)}\} \quad \text{(E.13.2.8b)}$$

where

$$\left[c^{(e)}\right] = \frac{\rho c_p \Delta x}{6}\begin{bmatrix} 2 & 1 \\ 1 & 2 \end{bmatrix} = \begin{bmatrix} 2c_1 & c_1 \\ c_1 & 2c_1 \end{bmatrix} \quad \text{(E.13.2.9a)}$$

$$\left[k^{(e)}\right] = \frac{k}{\Delta x}\begin{bmatrix} 1 & -1 \\ -1 & 1 \end{bmatrix} - \frac{\dot{m}_{bl} c_{bl} \Delta x}{6}\begin{bmatrix} 2 & 1 \\ 1 & 2 \end{bmatrix}$$

$$= \begin{bmatrix} k_1 - 2m & -k_1 - m \\ -k_1 - m & k_1 - 2m \end{bmatrix} \quad \text{(E.13.2.9b)}$$

$$\{f^{(e)}\}\frac{\dot{m}_{bl} c_{bl} \Delta x}{2}\begin{Bmatrix} 1 \\ 1 \end{Bmatrix} + \frac{I_0}{\Delta x}\begin{Bmatrix} \dfrac{1}{a}(\mathrm{e}^{-ax_j} - \mathrm{e}^{-ax_i}) + \Delta x\,\mathrm{e}^{-ax_i} \\ -\dfrac{1}{a}(\mathrm{e}^{-ax_j} - \mathrm{e}^{-ax_i}) - \Delta x\,\mathrm{e}^{-ax_j} \end{Bmatrix} = \begin{Bmatrix} f_i^{(e)} \\ f_j^{(e)} \end{Bmatrix}$$

(E.13.2.9c)

$$\{I^{(e)}\} = \begin{Bmatrix} \left.k\dfrac{dT}{dx}\right|_{x_i} \\ \left.-k\dfrac{dT}{dx}\right|_{x_j} \end{Bmatrix} \quad \text{(13.2.9.d)}$$

For a fully implicit scheme, we use Equation 13.24 with $\lambda = 1$ as

$$\left([c^{(e)}] + \Delta t[k^{(e)}]\right)\{T\}^{m+1} = [c^{(e)}]\{T\}^m + \Delta t\{I^{(e)}\}^{m+1} + \Delta t\{f^{(e)}\}^{m+1} \quad \text{(E.13.2.10a)}$$

$$\begin{bmatrix} 2c_1 + \Delta t(k_1 - 2m) & c_1 - \Delta t(k_1 + m) \\ c_1 - \Delta t(k_1 + m) & 2c_1 + \Delta t(k_1 - 2m) \end{bmatrix} \begin{Bmatrix} T_i^{(e)} \\ T_j^{(e)} \end{Bmatrix}^{m+1}$$

$$= \begin{bmatrix} 2c_1 & c_1 \\ c_1 & 2c_1 \end{bmatrix} \begin{Bmatrix} T_i^{(e)} \\ T_j^{(e)} \end{Bmatrix}^{m}$$

$$+ \Delta t \begin{Bmatrix} k\left.\dfrac{dT}{dx}\right|_{x_i} \\ -k\left.\dfrac{dT}{dx}\right|_{x_j} \end{Bmatrix} + \Delta t \begin{Bmatrix} f_i^{(e)} \\ f_j^{(e)} \end{Bmatrix} \quad \text{(E.13.2.10b)}$$

$$\begin{bmatrix} 2c_1 + \Delta t(k_1 - 2m) & c_1 - \Delta t(k_1 + m) & & & & \\ c_1 - \Delta t(k_1 + m) & 4c_1 + 2\Delta t(k_1 - 2m) & c_1 - \Delta t(k_1 + m) & & & \\ & c_1 - \Delta t(k_1 + m) & 4c_1 + 2\Delta t(k_1 - 2m) & c_1 - \Delta t(k_1 + m) & & \\ & & c_1 - \Delta t(k_1 + m) & 4c_1 + 2\Delta t(k_1 - 2m) & c_1 - \Delta t(k_1 + m) & \\ & & & c_1 - \Delta t(k_1 + m) & 4c_1 + 2\Delta t(k_1 - 2m) & c_1 - \Delta t(k_1 + m) \\ & & & & c_1 - \Delta t(k_1 + m) & 2c_1 + \Delta t(k_1 - 2m) \end{bmatrix}$$

$$\times \begin{Bmatrix} T_1 \\ T_2 \\ T_3 \\ T_4 \\ T_5 \\ T_6 \end{Bmatrix}^{m+1} = \begin{bmatrix} 2c_1 & c_1 & & & & \\ c_1 & 4c_1 & c_1 & & & \\ & c_1 & 4c_1 & c_1 & & \\ & & c_1 & 4c_1 & c_1 & \\ & & & c_1 & 4c_1 & c_1 \\ & & & & c_1 & 4c_1 \end{bmatrix} \begin{Bmatrix} T_1 \\ T_2 \\ T_3 \\ T_4 \\ T_5 \\ T_6 \end{Bmatrix}^{m} + \Delta t \begin{Bmatrix} k\left.\dfrac{dT}{dx}\right|_{x_1} \\ 0 \\ 0 \\ 0 \\ 0 \\ -k\left.\dfrac{dT}{dx}\right|_{x_6} \end{Bmatrix}^{m+1} + \Delta t \begin{Bmatrix} f_1^{(1)} \\ f_2^{(1)} + f_1^{(2)} \\ f_2^{(2)} + f_1^{(3)} \\ f_2^{(3)} + f_1^{(4)} \\ f_2^{(4)} + f_1^{(5)} \\ f_2^{(5)} \end{Bmatrix}^{m+1} \quad \text{(E.13.2.11)}$$

Applying the boundary conditions, we get

$$
\begin{bmatrix}
2c_1+\Delta t(k_1-2m) & c_1-\Delta t(k_1+m) & & & & \\
c_1-\Delta t(k_1+m) & 4c_1+2\Delta t(k_1-2m) & c_1-\Delta t(k_1+m) & & & \\
 & c_1-\Delta t(k_1+m) & 4c_1+2\Delta t(k_1-2m) & c_1-\Delta t(k_1+m) & & \\
 & & c_1-\Delta t(k_1+m) & 4c_1+2\Delta t(k_1-2m) & c_1-\Delta t(k_1+m) & \\
 & & & c_1-\Delta t(k_1+m) & 4c_1+2\Delta t(k_1-2m) & c_1-\Delta t(k_1+m) \\
 & & & & c_1-\Delta t(k_1+m) & 2c_1+\Delta t(k_1-2m)
\end{bmatrix}
$$

$$
\times\begin{Bmatrix} T_1 \\ T_2 \\ T_3 \\ T_4 \\ T_5 \\ T_b \end{Bmatrix}^{m+1}
=\begin{bmatrix}
2c_1 & c_1 & & & & \\
c_1 & 4c_1 & c_1 & & & \\
 & c_1 & 4c_1 & c_1 & & \\
 & & c_1 & 4c_1 & c_1 & \\
 & & & c_1 & 4c_1 & c_1 \\
 & & & & c_1 & 4c_1
\end{bmatrix}
\begin{Bmatrix} T_1 \\ T_2 \\ T_3 \\ T_4 \\ T_5 \\ T_b \end{Bmatrix}^{m}
+\Delta t\begin{Bmatrix} h(T_1-T_\infty) \\ 0 \\ 0 \\ 0 \\ 0 \\ -k\left.\dfrac{dT}{dx}\right|_{x_6} \end{Bmatrix}^{m+1}
+\Delta t\begin{Bmatrix} f_1^{(1)} \\ f_2^{(1)}+f_1^{(2)} \\ f_2^{(2)}+f_1^{(3)} \\ f_2^{(3)}+f_1^{(4)} \\ f_2^{(4)}+f_1^{(5)} \\ f_2^{(5)} \end{Bmatrix}^{m+1}
\quad \text{(E.13.2.12)}
$$

$$
\begin{bmatrix}
2c_1+\Delta t(k_1-2m)-\Delta th & c_1-\Delta t(k_1+m) & & & & \\
c_1-\Delta t(k_1+m) & 4c_1+2\Delta t(k_1-2m) & c_1-\Delta t(k_1+m) & & & \\
 & c_1-\Delta t(k_1+m) & 4c_1+2\Delta t(k_1-2m) & c_1-\Delta t(k_1+m) & & \\
 & & c_1-\Delta t(k_1+m) & 4c_1+2\Delta t(k_1-2m) & c_1-\Delta t(k_1+m) & \\
 & & & c_1-\Delta t(k_1+m) & 4c_1+2\Delta t(k_1-2m) & 0 \\
 & & & & c_1-\Delta t(k_1+m) & \Delta t
\end{bmatrix}
$$

$$
\times\begin{Bmatrix} T_1 \\ T_2 \\ T_3 \\ T_4 \\ T_5 \\ k\left.\dfrac{dT}{dx}\right|_{x_6} \end{Bmatrix}^{m+1}
=\begin{bmatrix}
2c_1 & c_1 & & & & \\
c_1 & 4c_1 & c_1 & & & \\
 & c_1 & 4c_1 & c_1 & & \\
 & & c_1 & 4c_1 & c_1 & \\
 & & & c_1 & 4c_1 & c_1 \\
 & & & & c_1 & 4c_1
\end{bmatrix}
\begin{Bmatrix} T_1 \\ T_2 \\ T_3 \\ T_4 \\ T_5 \\ T_b \end{Bmatrix}^{m}
+\Delta t\begin{Bmatrix} f_1^{(1)}-hT_\infty \\ f_2^{(1)}+f_1^{(2)} \\ f_2^{(2)}+f_1^{(3)} \\ f_2^{(3)}+f_1^{(4)} \\ f_2^{(4)}+f_1^{(5)}-\left\{\frac{c_1}{\Delta t}-(k_1+m)\right\}T_b \\ f_2^{(5)}-\left\{2\frac{c_1}{\Delta t}+(k_1-2m)\right\}T_b \end{Bmatrix}^{m+1}
\quad \text{(E.13.2.13)}
$$

13.3 TWO-DIMENSIONAL UNSTEADY STATE DIFFUSION EQUATION

In this section we will consider the formulation and application finite element method for solving two-dimensional unsteady state problems. Let us consider a two-dimensional unsteady state transport process in a two-dimensional domain as show in Figure 13.4.

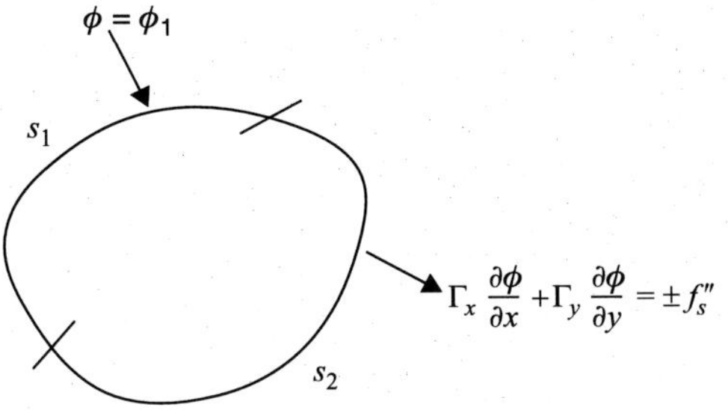

Figure 13.4 Two-dimensional domain and boundary conditions

A general mathematical statement of the problem can be assumed as

$$C\frac{\partial \phi}{\partial t} = \frac{\partial}{\partial x}\left(\Gamma_x \frac{\partial \phi}{\partial x}\right) + \frac{\partial}{\partial y}\left(\Gamma_y \frac{\partial \phi}{\partial y}\right) + S_1\phi + S_0 \tag{13.27}$$

The partial differential Equation 13.27 is similar to Equation 12.1 with the exception of a storage term on the left-hand side. The associated boundary conditions at different sections of the boundary surface are assumed as follows.

On the boundary surface s_1

Boundary condition of the first kind

$$\phi = \phi_s = \text{constant surface temperature} \tag{13.28a}$$

On the boundary surface s_2

$$\Gamma_x \frac{\partial \phi}{\partial x}\hat{n}_x + \Gamma_y \frac{\partial \phi}{\partial y}\hat{n}_y = \pm h(\phi - \phi_s) \pm f_s'' \tag{13.28b}$$

Initial condition

$$\phi(x, y, z, 0) = \phi_I \tag{13.29}$$

In the semi-discrete finite element formulation, the ***spatial discretization scheme*** assumes the approximate solution for the element as

$$\phi^* = \sum_{I=1}^{n_p} N_I(\tilde{x}, \tilde{y})\phi_I(t) = [N(\tilde{x}, \tilde{y})]\{\phi^{(e)}(t)\} \tag{13.30}$$

where N_I is a two-dimensional shape function, which is assumed to be a function of space variable only, $\phi_I(t)$ are the nodal values as a function time, and n_p are the number of nodes in an element.

Galerkin's integral procedure can be applied for obtaining the integral form of the equation over an element as

$$\{R^{(e)}\} = -\int_A [N]^{\mathrm{T}}\left[\frac{\partial}{\partial x}\left(\Gamma_x \frac{\partial \phi^*}{\partial x}\right) + \frac{\partial}{\partial y}\left(\Gamma_y \frac{\partial \phi^*}{\partial y}\right) + S_1\phi^* + S_0 - C\frac{\partial \phi^*}{\partial t}\right] dA = 0 \tag{13.31}$$

Following the procedure outlined in Chapter 12, Equation 13.31 reduces to

$$\begin{aligned}\{R^{(e)}\} = &-\int_s [N]^{\mathrm{T}}\left(\Gamma_x \frac{\partial \phi^*}{\partial x}\hat{n}_x + \Gamma_y \frac{\partial \phi^*}{\partial y}\hat{n}_y\right) ds \\ &+ \int_A \left(\frac{\partial [N]^{\mathrm{T}}}{\partial x}\Gamma_x \frac{\partial \phi^*}{\partial x} + \frac{\partial [N]^{\mathrm{T}}}{\partial y}\Gamma_y \frac{\partial \phi^*}{\partial y}\right) dA \\ &- \int_A S_1[N]^{\mathrm{T}}\phi^*\, dA - \int_A S_0[N]^{\mathrm{T}}\, dA - \int_A [N]^{\mathrm{T}}\left[C\frac{\partial \phi^*}{\partial t}\right] dA = 0\end{aligned} \tag{13.32}$$

Substituting the approximate solution function given by Equation 13.28 into the integral

$$\begin{aligned}&-\int_s [N]^{\mathrm{T}}\left(\Gamma_x \frac{\partial \phi^*}{\partial x}\cos\theta + \Gamma_y \frac{\partial \phi^*}{\partial y}\sin\theta\right) ds \\ &+ \left\{\int_A \left(\frac{\partial [N]^{\mathrm{T}}}{\partial x}\Gamma_x \frac{\partial [N]}{\partial x} + \frac{\partial [N]^{\mathrm{T}}}{\partial y}\Gamma_y \frac{\partial [N]}{\partial y}\right) dA\right\}\{\phi^{(e)}\} \\ &- \left\{\int S_1[N]^{\mathrm{T}}[N]dA\right\}\{\phi^{(e)}\} - \int_A S_0[N]^{\mathrm{T}}\, dA + \int_A C[N]^{\mathrm{T}}[N]\left[\frac{\partial \phi^{(e)}}{\partial t}\right] dA\end{aligned} \tag{13.33}$$

The element characteristics equation can be written in matrix notation as

$$\left[c^{(e)}\right]\frac{d\{\phi^{(e)}\}}{dt} = \{I^{(e)}\} + [k^{(e)}]\{\phi^{(e)}\} - \left\{f_S^{(e)}\right\} = 0 \tag{13.34}$$

where

$$\{I^{(e)}\} = -\int_s [N]^{\mathrm{T}}\left(\Gamma_x \frac{\partial \phi^*}{\partial x}\hat{n}_x + \Gamma_y \frac{\partial \phi^*}{\partial y}\hat{n}_y\right) ds \tag{13.35a}$$

$$\left\{f_S^{(e)}\right\} = \int_A S_0[N]^{\mathrm{T}} dA \tag{13.35b}$$

$$\left[k^{(e)}\right] = \int_A \left(\frac{\partial [N]^T}{\partial x} \Gamma_x \frac{\partial [N]}{\partial x} + \frac{\partial [N]^T}{\partial y} \Gamma_y \frac{\partial [N]}{\partial y} \right) dA - \int_A S_1 [N]^T [N]\, dA$$

$$= \left[k_\Gamma^{(e)}\right] - \left[k_S^{(e)}\right] \tag{13.35c}$$

$$\left[k_S^{(e)}\right] = \int_A S_1 [N]^T [N]\, dA \tag{13.35d}$$

$$\left[k_\Gamma^{(e)}\right] = \int_A \left(\frac{\partial [N]^T}{\partial x} \Gamma_x \frac{\partial [N]}{\partial x} + \frac{\partial [N]^T}{\partial y} \Gamma_y \frac{\partial [N]}{\partial y} \right) dA$$

$$= \int_A \{D\}^T [\Gamma] \{D\}\, dA \tag{13.35e}$$

$$\left[c^{(e)}\right] = \int_A C [N]^T [N]\, dA \tag{13.35f}$$

The implementation of the derivative boundary conditions into the finite element formulation takes place through the inter-element vector $\left\{I^{(e)}\right\}$ and results in the element characteristic equation

$$\left[c^{(e)}\right] \frac{d\{\phi^{(e)}\}}{dt} = \left\{I_i^{(e)}\right\} + \left(k_\Gamma^{(e)} - k_S^{(e)} - k_b^{(e)}\right)\{\phi^{(e)}\} - \left\{f_S^{(e)} + f_{bs}^{(e)}\right\} = 0 \tag{13.36a}$$

or

$$\left[c^{(e)}\right] \frac{d\{\phi^{(e)}\}}{dt} = \left\{I_i^{(e)}\right\} + k^{(e)}\{\phi^{(e)}\} - f'' \tag{13.36b}$$

where

$$k^{(e)} = \left(k_\Gamma^{(e)} - k_S^{(e)} - k_b^{(e)}\right) = \text{element characteristic matrix} \tag{13.36a}$$

$$f^{(e)} = \left\{f_S^{(e)} + f_{bs}^{(e)}\right\} = \text{element source vector} \tag{13.36b}$$

$$\left[k_b^{(e)}\right] = \pm \int_{s_b} h [N]^T [N]\, ds \tag{13.36c}$$

$$\left\{f_{bs}^{(e)}\right\} = \pm \int_{s_b} [N]^T \left\{\left(h\phi_\infty + f_s''\right)\right\} ds \tag{13.36d}$$

$$\left[c^{(e)}\right] = \int_A C [N]^T [N]\, dA = \text{element characteristic matrix} \tag{13.36e}$$

The element characteristic matrix, force vector, and capacitance matrix can now be evaluated for a particular element type. For example, we have seen the development of the element characteristic matrix and force vector for the ***linear triangular element*** and by the ***bilinear rectangular element*** in Chapter 12 for steady state problems. The capacitance matrix $\left[c^{(e)}\right]$ can also be evaluated by selecting appropriate element type and associated shape functions in the same manner as the $\left[k_S^{(e)}\right]$ given by Equation 12.32. For the ***linear triangular element***

$$\left[c^{(e)}\right] = \int_A C[N]^{\mathrm{T}}[N]\,dA$$

$$= C\int_A \begin{Bmatrix} N_i \\ N_j \\ N_k \end{Bmatrix} \{ N_i \quad N_j \quad N_k \}\,dA$$

$$\left[c^{(e)}\right] = C\int_A \begin{bmatrix} N_i^2 & N_iN_j & N_iN_k \\ N_jN_i & N_j^2 & N_jN_k \\ N_kN_i & N_kN_j & N_k^2 \end{bmatrix} dA \tag{13.37}$$

Using the natural coordinate system (Equation 10.68) for the triangular element

$$N_i = L_i, \quad N_j = L_j \quad \text{and} \quad N_k = L_k \tag{10.68}$$

we get

$$\left[c^{(e)}\right] = C\int_A \begin{bmatrix} L_i^2 & L_iL_j & L_iL_k \\ L_jL_i & L_j^2 & L_jL_k \\ L_kL_i & L_kL_j & L_k^2 \end{bmatrix} \tag{13.38}$$

Using the integral formula given by equation 10.71, we can evaluate the integral elements and the integral expression (Equation 13.38) becomes

$$\left[c^{(e)}\right] = \frac{CA}{12}\begin{bmatrix} 2 & 1 & 1 \\ 1 & 2 & 1 \\ 1 & 1 & 2 \end{bmatrix} \quad \text{for the triangular element} \tag{13.39}$$

Similarly, for the ***bilinear rectangular element*** the capacitance matrix is given as

$$\left[c^{(e)}\right] = \int_A C[N]^{\mathrm{T}}[N]\, dA$$

$$= \int_A C \begin{Bmatrix} N_i \\ N_j \\ N_k \\ N_l \end{Bmatrix} \{ N_i \quad N_j \quad N_k \quad N_l \} \, dA$$

$$\left[c^{(e)}\right] = \int_A C \begin{bmatrix} N_i^2 & N_iN_j & N_iN_k & N_iN_l \\ N_jN_i & N_j^2 & N_jN_k & N_jN_l \\ N_kN_i & N_kN_j & N_k^2 & N_kN_l \\ N_lN_i & N_lN_j & N_lN_k & N_l^2 \end{bmatrix} dA \qquad (13.40a)$$

For cases with constant capacitance value, the integrals can be evaluated directly by substituting the shape function expressions in a local coordinate system given by Equation 10.88, and the element capacitance matrix is expressed as

$$\left[C^{(e)}\right] = \frac{CA}{36} \begin{bmatrix} 4 & 2 & 1 & 2 \\ 2 & 4 & 2 & 1 \\ 1 & 2 & 4 & 2 \\ 2 & 1 & 2 & 4 \end{bmatrix} \quad \text{for the rectangular element} \qquad (13.40b)$$

In the next step, the element characteristic Equation 13.36, which represents a set of first-order differential equations, can be solved using the time approximation scheme used for a one-dimensional problem in Section 13.2.2.

13.4 THREE-DIMENSIONAL UNSTEADY STATE DIFFUSION EQUATION

Solution of three-dimensional unsteady state problems follows the same procedure as outlined for one- and two-dimensional problems. Let us consider a three-dimensional steady state transport process in a three-dimensional domain with the mathematical statement given as

$$C\frac{\partial \phi}{\partial t} = \frac{\partial}{\partial x}\left(\Gamma_x \frac{\partial \phi}{\partial x}\right) + \frac{\partial}{\partial y}\left(\Gamma_y \frac{\partial \phi}{\partial y}\right) + \frac{\partial}{\partial y}\left(\Gamma_y \frac{\partial \phi}{\partial y}\right) + S \qquad (13.41)$$

with associated general boundary conditions given as

On the boundary surface A_b

Boundary condition of the third kind

$$\Gamma_x \frac{\partial \phi}{\partial x} n_x + \Gamma_y \frac{\partial \phi}{\partial y} n_y + \Gamma_z \frac{\partial \phi}{\partial z} n_z = \pm h(\phi - \phi_s) \pm f_s'' \tag{13.42a}$$

On the remaining boundary surface: $\phi = \phi_s$ (13.42b)

and with initial condition

$$\phi(x, y, z, 0) = \phi_I \tag{13.43}$$

In the semi-discrete finite element formulation, the ***spatial discretization scheme*** assumes the approximate solution for the element as

$$\phi^* = \sum_{I=1}^{n_p} N_I(x, y, z)\phi_I(t) = [N(x, y, z)]\{\phi^{(e)}(t)\} \tag{13.44}$$

where N_I is a three-dimensional shape function, which is assumed to be a function of the space variable only, $\phi_I(t)$ are nodal values as a function time, and n_p are the number of nodes in an element.

Galerkin's integral procedure can be applied for obtaining the integral form of the equation over an element as

$$\{R^{(e)}\} = -\int_V [N]^{\mathrm{T}} \left[\frac{\partial}{\partial x}\left(\Gamma_x \frac{\partial \phi^*}{\partial x}\right) + \frac{\partial}{\partial y}\left(\Gamma_y \frac{\partial \phi^*}{\partial y}\right) + S - C\frac{\partial \phi^*}{\partial t} \right] dA = 0 \tag{13.45}$$

Evaluating the integral we, get

$$\begin{aligned}\int_V C[N]^{\mathrm{T}}[N]\,dV = &-\int_A [N]^{\mathrm{T}} \left(\Gamma_x \frac{\partial \phi^*}{\partial x} n_x + \Gamma_y \frac{\partial \phi^*}{\partial y} n_y + \Gamma_z \frac{\partial \phi^*}{\partial z} n_z\right) dA \\ &+ \int_V \left(\frac{\partial [N]^{\mathrm{T}}}{\partial x} \Gamma_x \frac{\partial \phi^*}{\partial x} + \frac{\partial [N]^{\mathrm{T}}}{\partial y} \Gamma_y \frac{\partial \phi^*}{\partial y} + \frac{\partial [N]^{\mathrm{T}}}{\partial z} \Gamma_z \frac{\partial \phi^*}{\partial z}\right) dV \\ &- \int_V S[N]^{\mathrm{T}}\,dV = 0\end{aligned} \tag{13.46}$$

In order to obtain the element characteristics equation, substitute the approximate solution function given by Equation 13.28 into the integral Equation 13.42

$$\int_V C[N]^T[N]\left[\frac{\partial\phi^{(e)}}{\partial t}\right]dV = -\int_A [N]^T\left(\Gamma_x\frac{\partial\phi^*}{\partial x}n_x + \Gamma_y\frac{\partial\phi^*}{\partial y}n_y + \Gamma_z\frac{\partial\phi^*}{\partial z}n_z\right)dA$$
$$+\left\{\int_V\left(\frac{\partial[N]^T}{\partial x}\Gamma_x\frac{\partial[N]}{\partial x} + \frac{\partial[N]^T}{\partial y}\Gamma_y\frac{\partial[N]}{\partial y} + \frac{\partial[N]^T}{\partial z}\Gamma_z\frac{\partial[N]}{\partial z}\right)dV\right\}\{\phi^{(e)}\}$$
$$-\int_V S[N]^T dV \tag{13.47}$$

The terms on the right-hand side of the Equation 13.43 have already been evaluated in Chapter 12 for three-dimensional steady state problems along with the implementation of the boundary conditions. With the addition of the new storage term on the left-hand side, the element characteristic equation takes the form as

$$\left[c^{(e)}\right]\frac{d\{\phi^{(e)}\}}{dt} = \left\{I_i^{(e)}\right\} + k^{(e)}\{\phi^{(e)}\} - f'' \tag{13.48}$$

where

$$\left[c^{(e)}\right] = \int_V C[N]^T[N]\,dV \tag{13.49a}$$

$$k^{(e)} = \left(k_\Gamma^{(e)} + k_b^{(e)}\right) \tag{13.49b}$$

$$\left[k^{(e)}\right] = \int_V\left(\frac{\partial[N]^T}{\partial x}\Gamma_x\frac{\partial[N]}{\partial x} + \frac{\partial[N]^T}{\partial y}\Gamma_y\frac{\partial[N]}{\partial y} + \frac{\partial[N]^T}{\partial z}\Gamma_z\frac{\partial[N]}{\partial z}\right)dV$$
$$= \int_V \{D\}^T[\Gamma]\{D\}\,dV \tag{13.49c}$$

$$[\Gamma] = \text{transport property matrix} = \begin{bmatrix}\Gamma_x & 0 & 0\\ 0 & \Gamma_y & 0\\ 0 & 0 & \Gamma_z\end{bmatrix} \tag{13.49d}$$

$$\{D\} = \begin{Bmatrix} \partial[N]/\partial x \\ \partial[N]/\partial y \\ \partial[N]/\partial z \end{Bmatrix} \tag{13.49e}$$

$$\left[k_b^{(e)}\right] = \pm \int_{A_b} h[N]^{\mathrm{T}}[N]\,dA \tag{13.49f}$$

$$f^{(e)} = \left\{f_S^{(e)} + f_{bs}^{(e)}\right\} \tag{13.49g}$$

$$\left\{f_S^{(e)}\right\} = \int_V S[N]^{\mathrm{T}} dV \tag{13.49h}$$

$$\left\{f_{bs}^{(e)}\right\} = \pm \int_{S_b} [N]^{\mathrm{T}}\{(h\phi_\infty + f_s'')\}\,dA \tag{13.49i}$$

The element characteristic matrix, force vector, and capacitance matrix can be now be evaluated for a particular element type. For the tetrahedron element, we have already derived the element characteristic matrix and force vector in Chapter 12 for steady state problems. The capacitance matrix $\left[c^{(e)}\right]$ can now be evaluated for the three-dimensional such as the tetrahedron element as

$$\begin{aligned}
\left[c^{(e)}\right] &= \int_V C[N]^{\mathrm{T}}[N]\,dV \\
&= \int_V C\left\{\begin{matrix} N_i \\ N_j \\ N_k \\ N_l \end{matrix}\{\,N_i \quad N_j \quad N_k \quad N_l\,\}\right\} dV \\
&= \int_V C\begin{bmatrix} N_i^2 & N_iN_j & N_iN_k & N_iN_l \\ N_jN_i & N_j^2 & N_jN_k & N_jN_l \\ N_kN_i & N_kN_j & N_k^2 & N_kN_l \\ N_lN_i & N_lN_j & N_lN_k & N_l^2 \end{bmatrix} dV
\end{aligned} \tag{13.50}$$

Using the natural volume coordinate system (Equation 10.116), the integral elements can be evaluated as

$$\left[c^{(e)}\right] = \int_V C\begin{bmatrix} L_i^2 & L_iL_j & L_iL_k & L_iL_l \\ L_jL_i & L_j^2 & L_jL_k & L_jL_l \\ L_kL_i & L_kL_j & L_k^2 & L_kL_l \\ L_lL_i & L_lL_j & L_lL_k & L_l^2 \end{bmatrix} dV \tag{13.51}$$

$$c^{(e)} = C\begin{bmatrix} \int\limits_V L_i^2 dV & \int\limits_V L_iL_j dV & \int\limits_V L_iL_k dV & \int\limits_V L_iL_l dV \\ \int\limits_V L_jL_i dV & \int\limits_V L_j^2 dV & \int\limits_V L_jL_k dV & \int\limits_V L_jL_l dV \\ \int\limits_V L_kL_i dV & \int\limits_V L_kL_j dV & \int\limits_V L_k^2 dV & \int\limits_V L_kL_l dV \\ \int\limits_V L_lL_i dV & \int\limits_V L_lL_j dV & \int\limits_V L_lL_k dV & \int\limits_V L_l^2 dV \end{bmatrix}$$

$$= C\begin{bmatrix} \int\limits_V L_i^2L_j^0L_k^0L_l^0 dV & \int\limits_V L_i^1L_j^1L_k^0L_l^0 dV & \int\limits_V L_i^1L_j^0L_k^1L_l^0 dV & \int\limits_V L_i^1L_j^0L_k^0L_l^1 dV \\ \int\limits_V L_i^1L_j^1L_k^0L_l^0 dV & \int\limits_V L_i^0L_j^2L_k^0L_l^0 dV & \int\limits_V L_i^0L_j^1L_k^1L_l^0 dV & \int\limits_V L_i^0L_j^1L_k^0L_l^1 dV \\ \int\limits_V L_i^1L_j^0L_k^1L_l^0 dV & \int\limits_V L_i^0L_j^1L_k^1L_l^0 dV & \int\limits_V L_i^0L_j^0L_k^2L_l^0 dV & \int\limits_V L_i^0L_j^0L_k^1L_l^1 dV \\ \int\limits_V L_i^1L_j^0L_k^0L_l^1 dV & \int\limits_V L_i^0L_j^1L_k^0L_l^1 dV & \int\limits_V L_i^0L_j^0L_k^1L_l^1 dV & \int\limits_V L_i^0L_j^0L_k^0L_l^2 dV \end{bmatrix} \tag{13.52}$$

Using analytical integral formulas involving a natural volume co-ordinate system given by Equation 10.119, we get

$$[c^{(e)}] = 6VC\begin{bmatrix} \frac{2!0!0!0!}{(2+0+0+0+3)!} & \frac{1!1!0!0!}{(1+1+0+0+3)!} & \frac{1!0!1!0!}{(1+0+1+0+3)!} & \frac{1!0!0!1!}{(1+0+0+1+3)!} \\ \frac{1!1!0!0!}{(1+1+0+0+3)!} & \frac{0!2!0!0!}{(0+2+0+0+3)!} & \frac{0!1!1!0!}{(0+1+1+0+3)} & \frac{0!1!0!1!}{(0+1+0+1+3)!} \\ \frac{1!0!1!0!}{(1+0+1+0+2)!} & \frac{0!1!10!!}{(0+1+1+0+2)} & \frac{0!0!2!0!}{(0+0+2+0)!} & \frac{1!0!1!}{(0+0+1+1+3)!} \\ \frac{1!0!0!1!}{(1+0+0+1+3)!} & \frac{0!1!0!1!}{(0+1+0+1+3)!} & \frac{0!0!1!1!}{(0+0+1+1+3)!} & \frac{0!0!0!2!}{(0+0+0+2+3)!} \end{bmatrix}$$

or

$$[c^{(e)}] = \frac{CV}{20}\begin{bmatrix} 2 & 1 & 1 & 1 \\ 1 & 2 & 1 & 1 \\ 1 & 1 & 2 & 1 \\ 1 & 1 & 1 & 2 \end{bmatrix} \tag{13.53}$$

PROBLEMS

13.1 Consider a one-dimensional unsteady state conduction in a plane slab [$k = 50\,\text{W/m}\,°\text{C}$, $\alpha = 2 \times 10^{-5}\,\text{m}^2/\text{s}$] of thickness $L = 4\,\text{mm}$. Initially the slab is at a temperature of 20°C. The right side of the plate is suddenly brought to a temperature of 200°C. The left side is maintained at 20°C.

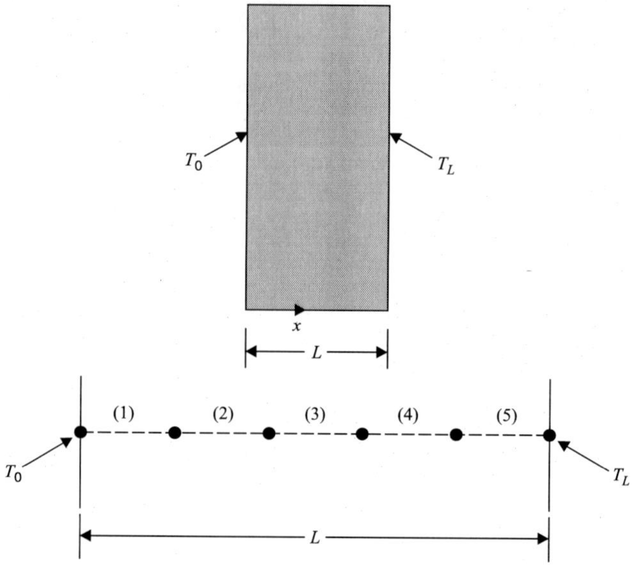

Use the finite element method and fully implicit time approximation to derive the global system of equations. Carry out the computation for five consecutive time steps using a time step $\Delta t = 0.2\,\text{s}$. Plot the heat flux at the right surface as a function of time.

13.2 Redo Problem 13.1 using the Crank–Nicolson time approximation scheme.

13.3 Derive the global system of equations in Problem 13.1 using the explicit time approximation scheme. Estimate the largest $\Delta t_{\text{largest}}$ based on Equations 13.25 and 13.26. Solve the global system of equations using time step $\Delta t < \Delta t_{\text{largest}}$ for temperature distribution and heat transfer rates. Plot the heat transfer rate at the right surface as function of time.

13.4 Derive the global system of equations in Problem 13.1 using the quadratic line element.

13.5 Consider a one-dimensional unsteady state conduction in a plane slab [$k = 50\,\text{W/m}\,°\text{C}$, $\alpha = 2 \times 10^{-5}\,\text{m}^2/\text{s}$] of thickness $L = 4\,\text{mm}$. Initially the slab is at a temperature of 20°C. The right side of the plate is suddenly brought to a convective environment with a temperature of $T_\infty = 200°\text{C}$ and $h = 80\,\text{W/m}^2\,°\text{C}$. The left side is maintained at 20°C. Consider the same mesh size distribution as in Problem 13.1 and determine the transient temperature distribution in the slab. Plot the heat transfer rate at the right surface as a function of time.

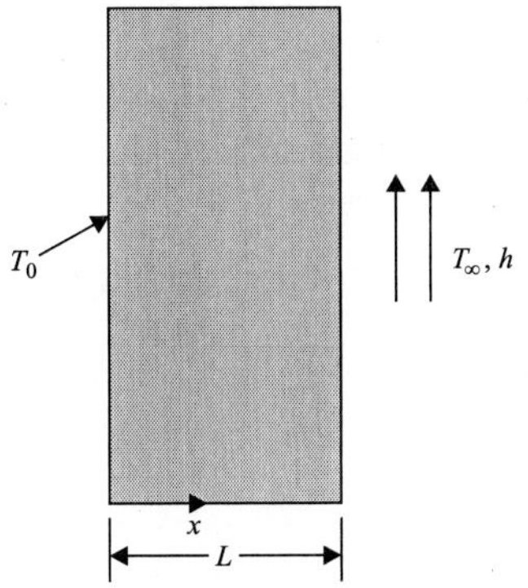

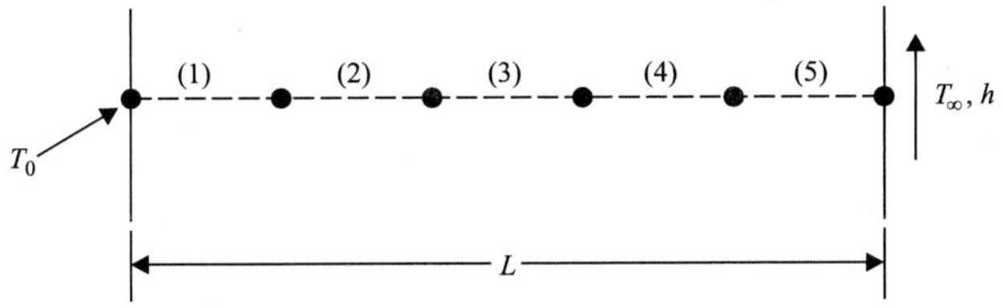

13.6 Consider a one-dimensional unsteady state heat conduction in a plane slab thickness $L = 10\,\text{cm}$. Initially the slab is at a uniform temperature of $T_I = 20°\text{C}$. At time $t > 0$, the left side of the slab is subjected to a uniform constant surface heat flux $q_s'' = 2.0 \times 10^4\,\text{W/cm}^2$ and the right surface is maintained at $T_R = 20°\text{C}$.

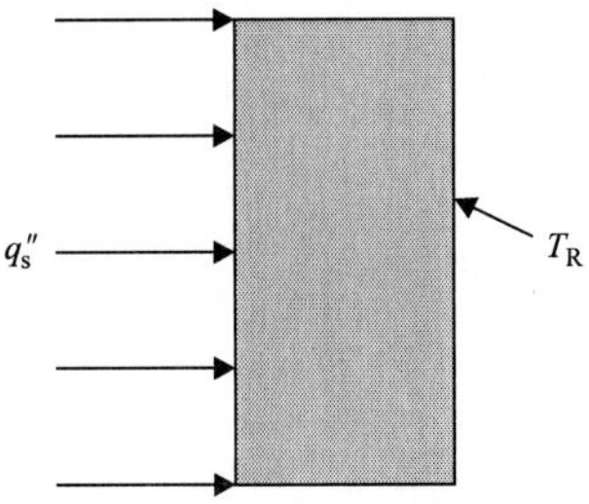

The slab is carbon steel with $\alpha = 0.835\,\text{cm}^2/\text{s}$. Use a fully implicit scheme to obtain the global system of equations with six linear line elements and a time step of $\Delta t = 0.2\,\text{s}$. Show computations for two time steps.

13.7 Develop a computer program using a fully implicit scheme and Gauss elimination solver to study the transient temperature distribution and rate of heat transfer for Problem 13.3. Present results in terms of temperature distribution and heat transfer rates at left and right surfaces with a time step of $\Delta t = 0.2\,\text{s}$ up to 5 s. Perform a mesh refinement study by using the number of elements as 10, 15, and 20.

Present results for the variation of heat transfer rate at $x = L$ with an increase in time for different numbers of elements.

13.8 Redo Problem 13.7 based on developing a computer code that uses a matrix inversion procedure with a Gauss–Jordan solver.

13.9 Let us consider a two-dimensional unsteady state problem with the bilinear rectangular mesh distribution system and boundary conditions as shown in the figure.

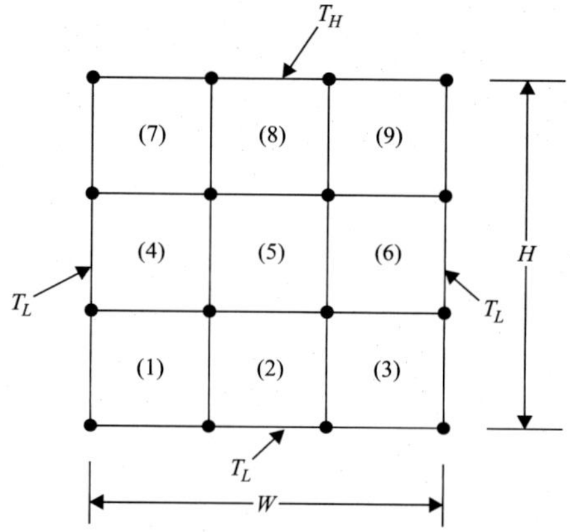

Derive the system of equations using the fully implicit scheme and using the following data. $W = 8$ cm, $H = 8$ cm, $T_I = 25°C$, $T_H = 400°C$, $T_L = 20°C$. Solve the system for two time steps with $\Delta t = 0.2$ s.

13.10 A two-dimensional 8 cm × 8 cm square slab, initially at a uniform temperature of 300°C, is suddenly exposed to a convection environment at a temperature of 20°C and $h = 40\ W/m^2 °C$ all around. Consider a two-dimensional unsteady state conduction in the slab and determine temperature distribution and rate of heat loss as a function of time.

(a) Use the 3 × 3 mesh and derive the system of equations for grid temperatures using the fully implicit scheme.

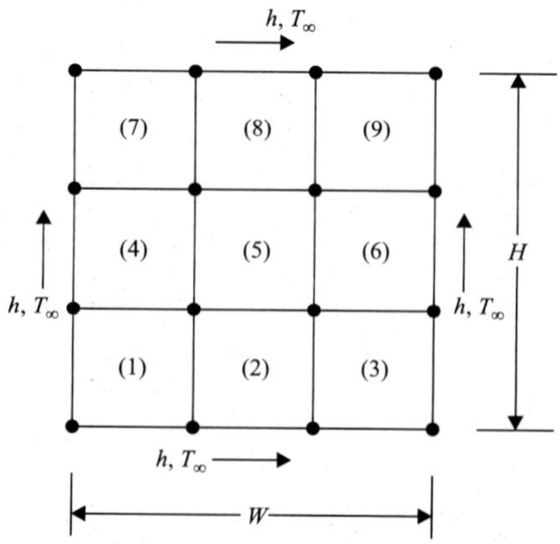

(b) Select an appropriate Δt and calculate the grid points temperature for five time increments. Material thermo-physical properties are $\rho = 7900\,\text{kg/m}^3$, $c = 477\,\text{J/kg K}$ and $k = 14.9\,\text{W/m K}$.

13.11 Repeat Problem 13.10 using the Crank–Nicolson scheme.

13.12 Develop a two-dimensional finite element program using the fully implicit time approximation scheme and bilinear rectangular elements, and using input data given for Problem 13.9. The computer program should incorporate the following features: (a) construct the element capacitance and matrix $\left[c^{(e)}\right]$, the element characteristic matrix $\left[k^{(e)}\right]$, and element force vector $\left\{f^{(e)}\right\}$based on the bilinear rectangular element; (b) assemble the global system of equations and (c) solve the global system using the Gauss elimination solver. Redo Problem 13.9 with refined grid size distributions as 4×4, 6×6, and 8×8 using this code. Present the temperature distribution along with the percent relative error for the vertical center line with refined grid size distributions at different time steps.

13.13 Consider diffusion of moisture from air stream in a two-dimensional porous adsorbing material felt lined on the side of the air flow channel as shown. All other three sides of the felt are lined with thin aluminum foil, thus making it impermeable to mass diffusion. Initial moisture concentration in the felt is $C_I = 0.001\,\text{kg}\,H_2O/\text{m}^3$. The left surface of the felt is exposed to moving air in the channel with mass transfer coefficient, $h_D = 0.02\,\text{m/s}$, and moisture concentration, $C_\infty = 0.01\,\text{kg}\,H_2O/\text{m}^3$. The material is assumed to be homogeneous with constant moisture adsorption rate of $\dot{m}_{\text{ad}} = 0.000\,04\,\text{kg}\,H_2O/\text{kg s}$. The mathematical statement of the problem is as follows.

Governing equation

$$\frac{\partial C}{\partial t} = \frac{\partial}{\partial x}\left(D\frac{\partial C}{\partial x}\right) + \frac{\partial}{\partial y}\left(D\frac{\partial C}{\partial y}\right) + \frac{(1-\varepsilon)}{\varepsilon}\rho\dot{m}_{\text{ad}}$$

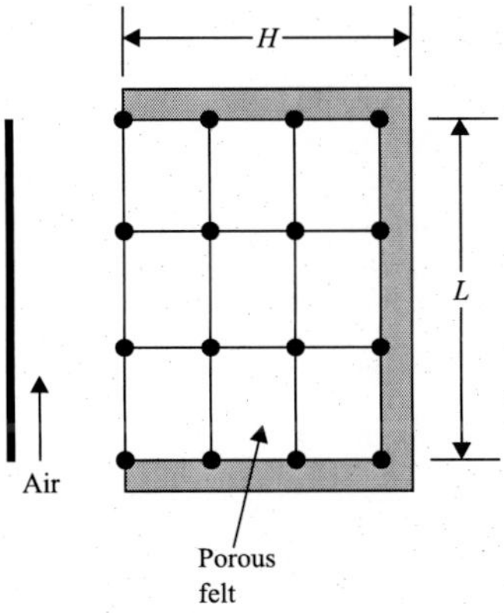

Boundary conditions

1. $y = 0, \qquad -D\frac{\partial C}{\partial x}\bigg|_{y=0} = h_D(C_\infty - C|_{y=0})$
2. $y = H, \qquad \frac{\partial C}{\partial x}\bigg|_{x=H} = 0$
3. $x = 0, \qquad \frac{\partial C}{\partial x}\bigg|_{x=0} = 0$
4. $x = L, \qquad \frac{\partial C}{\partial x}\bigg|_{x=L} = 0$

where ε = porosity and ρ_b = material bulk density.

(a) Apply the finite element method with the fully implicit time approximation scheme and derive the system of equations to calculate the two-dimensional transient moisture concentration distribution in the slab based on the grid distribution shown in the figure. Assume the diffusion coefficient as $D = 4.0 \times 10^{-6}\,\text{m}^2/\text{s}$ and use the following data for computation: $H = 4\,\text{mm}$, $L = 0.1\,\text{m}$, $\rho_b = 450\,\text{kg/m}^3$, $\varepsilon = 0.5$.

(b) Solve the system of equations using a direct solver and present the results along the four vertical lines as a function time.

13.14 Use the finite element method to solve the two-dimensional unsteady state conduction in a rectangular carbon steel ($k = 200\,\text{W/m}\,^\circ\text{C}$) slab subjected to a constant surface heat flux irradiated by a continuous high-energy laser beam at the top surface. Assume the heat flux distribution to be a constant average value, $q_0'' = 2 \times 10^8\,\text{W/m}^2$, acting over a section of the surface. The top surface is also subjected to forced convection with $h_c = 500\,\text{W/m}^2\,^\circ\text{C}$. The left surface is assumed to be adiabatic as a line of symmetry. All other surfaces are assumed to be subjected to free convection with $T_\infty = 25^\circ\text{C}$ and $h = 40\,\text{W/m}^2\,^\circ\text{C}$.

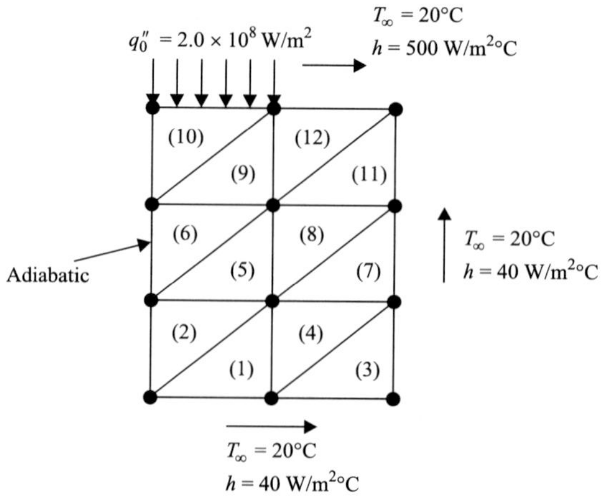

(a) Derive the element characteristic equations for all elements using bilinear rectangular elements and the fully implicit scheme. Assemble all element equations to derive the global system in matrix form.
(b) Solve the system of equations using the Gauss elimination solver and present results for temperature distribution as a function of time.

13.15 Use the bilinear rectangular element to derive the element characteristic equation for the two-dimensional axisymmetric unsteady state problem with heat generation.

13.16 Use the trilinear eight-noded hexahedron element to derive the three-dimensional element characteristic equation starting from Equations 13.48 and 13.49.

CHAPTER

14

CONVECTION PROBLEMS

We have mentioned in Chapter 8 that the convection problems involve distribution and dissipation of scalar transport quantities such as heat and mass by molecular diffusion as well as by bulk motion of the fluid. Since the convection problems are primarily influenced by the associated flow field, the solution of convection problems starts with the solution of flow field governed by the conservation laws of mass and momentum equations such as the ***Navier–Stokes equations***, followed by the solution of energy or mass concentration equations. These governing equations constitute a set of coupled nonlinear partial differential equations for three velocity components, pressure, temperature, and/or mass concentration. Major computational difficulties arise due to the presence of the nonlinear convective terms as well as for pressure term for which there is no distinct equation. In many problems, the inertial forces or the convective terms are significantly small compared to viscous and pressure force terms, as in the case of large Reynolds number laminar flow problems, and hence the nonlinear convective terms are dropped as an approximation giving rise to so-called ***Stokes equations***. Solution of the Stokes equations is computationally similar to the solution of diffusion equations, as discussed in the previous chapters. In another class of problems, simplification is achieved due to a constant velocity of the fluid body, making the problem a linear one. However, solution of this class of problem also exhibits considerable difficulties when the convective term becomes significantly greater in magnitude compared to the diffusion term in large ***Peclet number*** flows. A discussion of this class of problems was given in Chapter 8 while discussing finite difference methods for nonlinear convection problems involving Navier–Stokes equations and scalar transport equations. We will start our discussion of the finite element method for the solution of convection problems by first considering the linear convection problems.

Also, as we have discussed before, there are two major approaches for solving the flow field: (1) the ***direct solution method involving the primitive variable***, and (2) the ***Stream function/vorticity-based method***. The stream function/vorticity-based method is computationally less convenient, as discussed in Chapter 8. The direct solution method is becoming increasingly more popular and is the method of choice for most commercial finite element codes. We will concentrate our discussion of the finite element methods based on the direct solution method involving the primitive variables.

14.1 CLASSIFICATION OF FINITE ELEMENT METHODS FOR CONVECTION PROBLEMS

There are primarily two finite element formulation methods for convection problems, namely ***velocity-pressure formulation or mixed formulation*** and ***penalty finite element formulation***. The velocity-pressure or mixed formulation starts with the weak form of the governing equations retaining both velocity and pressure variables. The continuity equation leads to the formation of a pressure equation. In the penalty formulation, the pressure term is effectively eliminated by satisfying the continuity equation in a least square sense. More discussion on the penalty method is given in the books by Reddy (1994), Reddy and Gartling (1994), and Bathe (1996). In this book, we restrict our discussion to only velocity-pressure formulation or mixed formulation.

14.2 VELOCITY-PRESSURE OR MIXED FORMULATION

This is a direct formulation in terms of the primitive variables velocity and pressure given by the original mass and momentum equations. The procedure for developing the finite element model is similar to the basic steps outlined for the diffusion equation presented in the previous chapters, with the exception of the additional convective terms. In order to keep the derivation simple, we will first demonstrate the procedure by first considering one-dimensional linear convection problems. We will subsequently consider a two-dimensional steady state nonlinear flow problem. This will be followed by inclusion of the unsteady state term, and considerations of scalar transport equations for convective heat and mass transfer problems.

14.2.1 One-dimensional Convection-Diffusion Problem

Let us consider a one-dimensional steady state problem involving convection and diffusion terms as

$$\frac{d}{dx}(Cu\phi) = \frac{d}{dx}\left(\Gamma_x \frac{d\phi}{dx}\right) + S \tag{14.1}$$

In this problem, it is assumed that the value of the convective velocity u is a known constant value. Let us demonstrate the application of the finite element method to this problem through a brief description of several main steps that constitute the finite element method.

Step 1: Discretization of the Solution Domain The solution domain is discretized into a mesh size distribution using one-dimensional line elements as shown in Figure 14.1.

Step 2: Formation of Integral Statement of the Problem Galerkin's based weighted integral statement of Equation 14.1 over an element is given as

$$R_i^{(e)} = \int_{x_i}^{x_j} [N]^{\mathrm{T}} \left(\frac{\partial}{\partial x}\left(\Gamma_x \frac{\partial \phi^*}{\partial x}\right) + Cu\frac{\partial \phi}{\partial x} + S \right) dx = 0 \tag{14.2}$$

Evaluating the first integral, we get the integral weak statement of the problem as

$$\left\{ \begin{matrix} -\Gamma_x \left.\dfrac{d\phi}{dx}\right|_{x_i} \\ \Gamma_x \left.\dfrac{d\phi}{dx}\right|_{x_j} \end{matrix} \right\} - \int_{x_i}^{x_j} \Gamma_x \frac{d[N]^T}{dx}\frac{d\phi^*}{dx}\,dx + \int_{x_i}^{x_j} [N]^T Cu \frac{d\phi^*}{dx}\,dx + \int_{x_i}^{x_j} S[N]^T\,dx = 0 \tag{14.3}$$

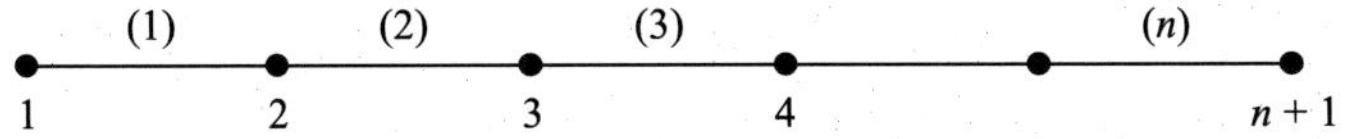

Figure 14.1 Mesh size distributions

Step 3: Let us consider the elements as linear two-point line elements with a typical element shown in Figure 14.1. The approximate solution function $\phi \approx \phi^*$ over this element is given as

$$\phi^* = \sum N_I \phi_I^{(e)} = [N]\{\phi^{(e)}\} \tag{14.4}$$

Step 4: Formation of Element Characteristics Equation Substituting the approximate solution function given by Equation 14.2 for a one-dimensional space coordinate and evaluating the integrals for a one-dimensional two-point line element, we get the ***element characteristic equation*** as

$$\begin{Bmatrix} -\Gamma_x \left.\dfrac{d\phi}{dx}\right|_{x_i} \\ \Gamma_x \left.\dfrac{d\phi}{dx}\right|_{x_j} \end{Bmatrix} - \int_{x_i}^{x_j} \Gamma_x \frac{d[N]^{\mathrm{T}}}{dx}\frac{d[N]}{dx}\{\phi^{(e)}\}dx + \int_{x_i}^{x_j} [N]^{\mathrm{T}} Cu \frac{d[N]}{dx}\{\phi^{(e)}\}dx$$

$$+ \int_{x_i}^{x_j} S[N]^{\mathrm{T}} dx = 0 \tag{14.5}$$

or

$$\{I^{(e)}\} - [k^{(e)}]\{\phi^{(e)}\} + \{f^{(e)}\} = 0 \tag{14.6}$$

where

$[k] = [k_\Gamma] + [k_I] =$ element characteristic matrix

$$= \frac{\Gamma_x}{L_e}\begin{bmatrix} 1 & -1 \\ -1 & 1 \end{bmatrix} + \frac{Cu}{2}\begin{bmatrix} -1 & 1 \\ -1 & 1 \end{bmatrix} \tag{14.7a}$$

$$[k_\Gamma] = \int_{x_i}^{x_j} \Gamma_x \frac{d[N]^{\mathrm{T}}}{dx}\frac{d[N]}{dx}dx$$

$=$ element characteristic matrix due to the diffusion term

$$= \frac{\Gamma_x}{L_e}\begin{bmatrix} 1 & -1 \\ -1 & 1 \end{bmatrix} \tag{14.7b}$$

$[k_I] = \int_{x_i}^{x_j} Cu[N]^{\mathrm{T}}\dfrac{d[N]}{dx}dx =$ element characteristic matrix due to the inertia term

$$= \frac{Cu}{2}\begin{bmatrix} -1 & 1 \\ -1 & 1 \end{bmatrix} \tag{14.7c}$$

$\{f^{(e)}\} = \int_{x_i}^{x_j} [N]^{\mathrm{T}}(S)\, dx =$ element force vector

$$= \begin{Bmatrix} \dfrac{SL_e}{2} \\ \dfrac{SL_e}{2} \end{Bmatrix} \tag{14.7d}$$

and

$$\{I^{(e)}\} = \left\{ \begin{array}{c} -\Gamma_x \left.\dfrac{d\phi}{dx}\right|_{x_i} \\ \Gamma_x \left.\dfrac{d\phi}{dx}\right|_{x_j} \end{array} \right\} = \text{inter-element vector} \tag{14.7e}$$

Step 5: Assembly of Element Equations to form the Global System Equation 14.6 is applied repeatedly to all elements to obtain corresponding element equations and are assembled through use of the connectivity scheme to form a global system of equations

$$[K]\{\Phi\} = \{F\} \tag{14.8a}$$

or

$$\begin{bmatrix} \dfrac{\Gamma_x}{L_e} - \dfrac{Cu}{2} & -\dfrac{\Gamma_x}{L_e} + \dfrac{Cu}{2} & & & & \\ -\dfrac{\Gamma_x}{L_e} - \dfrac{Cu}{2} & \dfrac{2\Gamma_x}{L_e} & -\dfrac{\Gamma_x}{L_e} + \dfrac{Cu}{2} & & & \\ & -\dfrac{\Gamma_x}{L_e} - \dfrac{Cu}{2} & \dfrac{2\Gamma_x}{L_e} & -\dfrac{\Gamma_x}{L_e} + \dfrac{Cu}{2} & & \\ & & -\dfrac{\Gamma_x}{L_e} - \dfrac{Cu}{2} & \dfrac{2\Gamma_x}{L_e} & & \\ \vdots & & & & & \\ & & & & \dfrac{2\Gamma_x}{L_e} & -\dfrac{\Gamma_x}{L_e} + \dfrac{Cu}{2} \\ & & & & -\dfrac{\Gamma_x}{L_e} - \dfrac{Cu}{2} & \dfrac{\Gamma_x}{L_e} + \dfrac{Cu}{2} \end{bmatrix}$$

$$\times \left\{ \begin{array}{c} \phi_1 \\ \phi_2 \\ \phi_3 \\ \vdots \\ \vdots \\ \phi_{n-1} \\ \phi_n \end{array} \right\} = \left\{ \begin{array}{c} -\Gamma_x \left.\dfrac{d\phi}{dx}\right|_{x_1} \\ 0 \\ 0 \\ \vdots \\ \vdots \\ 0 \\ \Gamma_x \left.\dfrac{d\phi}{dx}\right|_{x_{n+1}} \end{array} \right\} + \left\{ \begin{array}{c} \dfrac{SL_e}{2} \\ SL_e \\ SL_e \\ \vdots \\ \vdots \\ SL_e \\ \dfrac{SL_e}{2} \end{array} \right\} \tag{14.8b}$$

Step 6: Implementation of Boundary Conditions In this step, the boundary conditions are introduced directly into the global system of equations. For boundary conditions of the *first kind*, i.e., of constant surface value, we simply assign the specified values to the boundary nodes as $\phi_1 = \phi_l$ and $\phi_{n+1} = \phi_r$, and rearrange the system so that the natural boundary conditions $d\phi/dx\big|_{x_1}$ and $d\phi/dx\big|_{x_{n+1}}$ appear as one of the unknowns as

$$[K]\{\Phi\} = \{F\}$$

or

$$\begin{bmatrix} \frac{\Gamma_x}{L_e} - \frac{Cu}{2} & -\frac{\Gamma_x}{L_e} + \frac{Cu}{2} & & & & \\ -\frac{\Gamma_x}{L_e} - \frac{Cu}{2} & \frac{2\Gamma_x}{L_e} & -\frac{\Gamma_x}{L_e} + \frac{Cu}{2} & & & \\ & -\frac{\Gamma_x}{L_e} - \frac{Cu}{2} & \frac{2\Gamma_x}{L_e} & -\frac{\Gamma_x}{L_e} + \frac{Cu}{2} & & \\ & & -\frac{\Gamma_x}{L_e} - \frac{Cu}{2} & \frac{\Gamma_x}{L_e} + \frac{\Gamma_x}{L_e} & & \\ \vdots & & & & & \\ & & & & \frac{2\Gamma_x}{L_e} & -\frac{\Gamma_x}{L_e} + \frac{Cu}{2} \\ & & & & -\frac{\Gamma_x}{L_e} - \frac{Cu}{2} & \frac{\Gamma_x}{L_e} + \frac{Cu}{2} \end{bmatrix}$$

$$\times \begin{Bmatrix} \left.\frac{d\phi}{dx}\right|_{x_1} \\ \phi_2 \\ \phi_3 \\ \vdots \\ \vdots \\ \phi_n \\ \left.\frac{d\phi}{dx}\right|_{x_{n+1}} \end{Bmatrix} = \begin{Bmatrix} -\frac{\Gamma_x \phi_l}{L_e} + \frac{\bar{S}L_e}{2} \\ \frac{\Gamma_x \phi_l}{L_e} + \bar{S}L_e \\ \bar{S}L_e \\ \vdots \\ \vdots \\ \frac{\Gamma_x \phi_r}{L_e} + \bar{S}L_e \\ -\frac{\Gamma_x \phi_r}{L_e} + \frac{\bar{S}L_e}{2} \end{Bmatrix} \tag{14.9}$$

Step 7: Solution of the Global System of Equations Before deciding on an appropriate solver for solving the global system of equations, let us examine the nature of the element characteristic matrix K. It can be noted that due to the presence of the convective term, the characteristic matrix is non-symmetric. Also, as the inertia term becomes greater in magnitude compared to the diffusion term, i.e., $(Cu/2) > \Gamma_x/L_e$ or in dimensionless form, $uL_e/(\Gamma/C) > 2$, the characteristic matrix no longer remains diagonally dominant. So, the use of an iterative solver may experience instability, and may not lead to a converged solution. In the heat transfer problem, this dimensionless number is named the Peclet number ($Pe = uL_e/\alpha$), and the criterion for a diagonally dominant characteristic matrix is given as $Pe \leq 2$. Severity of this difficulty caused by the convection terms grows as the local element Peclet number becomes greater and greater.

This difficulty is also experienced in the use of a central difference scheme in ***finite difference–control volume methods*** as discussed in Chapter 8. As we have discussed in Chapter 8, this difficulty was resolved through use of alternate schemes such as the ***upwind scheme***, the ***exponential scheme***, the ***hybrid scheme*** and the ***power law scheme***. Similar concepts are also utilized in developing finite element techniques as an alternative to Galerkin finite element methods. The alternative finite element techniques that have shown improved accuracy in one-dimensional problems, and increasingly being used in two- and three-dimensional fluid flow and heat transfer, are the ***Petrov–Galerkin method*** and the ***Galerkin least squares method***. More detailed discussions of these techniques are given by Bathe (1996) and Pepper and Heinrich (1992).

14.2.2 Two-dimensional Viscous Incompressible Flow

The governing equations for two-dimensional Newtonian viscous incompressible flow over a flow domain A are given as follows.

Continuity

$$\frac{\partial u}{\partial x} + \frac{\partial v}{\partial y} = 0 \tag{14.10a}$$

x momentum

$$\rho\left(u\frac{\partial u}{\partial x} + v\frac{\partial u}{\partial y}\right) = b_x - \frac{\partial p}{\partial x} + \mu\left(\frac{\partial^2 u}{\partial x^2} + \frac{\partial^2 u}{\partial y^2}\right) \tag{14.10b}$$

y momentum

$$\rho\left(u\frac{\partial v}{\partial x} + v\frac{\partial v}{\partial y}\right) = b_y - \frac{\partial p}{\partial y} + \mu\left(\frac{\partial^2 v}{\partial x^2} + \frac{\partial^2 v}{\partial y^2}\right) \tag{14.10c}$$

Boundary conditions are given in a general form as

1. $u = u_s$, and $v = v_s$ on the boundary s_1 (14.11a)

2. $$\mu\left(\frac{\partial u}{\partial x}\hat{n}_x + \frac{\partial u}{\partial y}\hat{n}_y\right) - p\hat{n}_x = f_{sx}$$

$$\mu\left(\frac{\partial v}{\partial x}\hat{n}_x + \frac{\partial v}{\partial y}\hat{n}_y\right) - p\hat{n}_y = f_{sy} \quad \text{on the boundary } s_2 \tag{14.11b}$$

Note that the boundary surface s is divided into two sections s_1 and s_2. A boundary condition of the first kind, i.e., of constant surface value, is specified on the boundary surface s_1. A boundary condition of the second kind, i.e., of constant stress, is specified on the boundary surface s_2. Note that this constant stress boundary condition given by Equation 8.2b is a general boundary condition for the total stress that includes the normal hydrostatic pressure and the viscous boundary shear stress on the boundary.

Let us now demonstrate the finite element formulation following the basic steps outlined in Chapters 11 to 13.

Step 1: Discretization of the Solution Domain As a first step, the solution domain is discretized into a finite element mesh following the procedure outlined for the diffusion problems involving only a single variable, as in Chapter 12. However, solution of the flow field involves two different kinds of variables such as velocity, $\vec{V}$, and pressure, p. This requires the need for two different finite element meshes: one for the velocity and the other for the pressure. In general the mesh for the pressure is a subset of the primary mesh for the velocity and so only one mesh is displayed along with the nodal degrees of freedom associated with the nodes of the element.

Step 2: Formulation of an Integral or Weak Form Using Galerkin's Method In order to obtain the weak forms of the governing set of equations using the weighted residual method, we substitute the approximate solutions in the governing equations. The weighted residual method involves finding a minimum for the residuals over an element $A^{(e)}$ by multiplying the residuals of Equations 14.10a, 14.10b and 14.10c by the weighting functions W_p, W_u, and W_v, and integrating over the element as

$$\left\{R_P^{(e)}\right\} = -\int_{A^{(e)}} W_p\left\{\frac{\partial u^*}{\partial x} + \frac{\partial v^*}{\partial y}\right\} dA = 0 \tag{14.12a}$$

$$\{R_u^{(e)}\} = \int\limits_{A^{(e)}} W_u\left\{\rho\left(u^*\frac{\partial u^*}{\partial x} + v\frac{\partial u^*}{\partial y}\right) - b_x + \frac{\partial p^*}{\partial x} - \mu\left(\frac{\partial^2 u^*}{\partial x^2} + \frac{\partial^2 u^*}{\partial y^2}\right)\right\} dA = 0 \tag{14.12b}$$

$$\{R_v^{(e)}\} = \int\limits_{A^{(e)}} W_v\left\{\rho\left(u\frac{\partial v^*}{\partial x} + v\frac{\partial v^*}{\partial y}\right) - b_y + \frac{\partial p^*}{\partial y} - \mu\left(\frac{\partial^2 v^*}{\partial x^2} + \frac{\partial^2 v^*}{\partial y^2}\right)\right\} dA = 0 \tag{14.12c}$$

In the ***Galerkin finite element method*** the weighting functions are assumed to be the same as the shape functions used for the approximation functions given by Equation 14.2, i.e.,

$$W_u = [N]^{\mathrm{T}}, \quad W_v = [N]^{\mathrm{T}}, \quad \text{and} \quad W_p = [N_P]^{\mathrm{T}} \tag{14.13}$$

where $[N]$ and $\left[N_p\right]$ are the shape functions associated with velocity and pressure approximation functions, respectively.

The integral equations are transformed into a weak form by evaluating the integral as follows.

x momentum

$$\int\limits_{A^{(e)}} [N^{\mathrm{T}}]\rho\left(u^*\frac{\partial u^*}{\partial x} + v\frac{\partial u^*}{\partial y}\right) dA - \int\limits_{A^{(e)}} [N]^{\mathrm{T}} b_x\, dA + \int\limits_{A^{(e)}} [N]^{\mathrm{T}}\frac{\partial p^*}{\partial x}\, dA$$
$$- \int\limits_{A^{(e)}} [N]^{\mathrm{T}}\mu\frac{\partial^2 u^*}{\partial x^2}\, dA + \int\limits_{A^{(e)}} [N]^{\mathrm{T}}\mu\frac{\partial^2 u^*}{\partial y^2}\, dA = 0$$

or

$$\int\limits_{A^{(e)}} [N^{\mathrm{T}}]\rho\left(u^*\frac{\partial u^*}{\partial x} + v\frac{\partial u^*}{\partial y}\right) dA - \int\limits_{A^{(e)}} [N]^{\mathrm{T}} b_x\, dA$$
$$+ \int\limits_{A^{(e)}} \frac{\partial}{\partial x}([N]^{\mathrm{T}}p^*)\, dA - \int\limits_{A^{(e)}} p^*\frac{\partial [N]^{\mathrm{T}}}{\partial x}\, dA$$
$$- \int\limits_{A^{(e)}} \frac{\partial}{\partial x}\left(\mu[N]^{\mathrm{T}}\frac{\partial u^*}{\partial x}\right) dA + \int\limits_{A^{(e)}} \mu\frac{\partial u^*}{\partial x}\frac{\partial [N]^{\mathrm{T}}}{\partial x}\, dA$$
$$- \int\limits_{A^{(e)}} \frac{\partial}{\partial y}\left(\mu[N]^{\mathrm{T}}\frac{\partial u^*}{\partial y}\right) dA + \int\limits_{A^{(e)}} \mu\frac{\partial u^*}{\partial y}\frac{\partial [N]^{\mathrm{T}}}{\partial y}\, dA = 0$$

or

$$\int_{A^{(e)}} [N^{\mathrm{T}}]\rho\left(u^*\frac{\partial u^*}{\partial x}+v\frac{\partial u^*}{\partial y}\right)dA-\int_{A^{(e)}}[N]^{\mathrm{T}}b_x dA$$
$$+\oint_S [N]^{\mathrm{T}}p^*\hat{n}_x ds-\int_{A^{(e)}} p^*\frac{\partial[N]^{\mathrm{T}}}{\partial x}dA$$
$$-\oint_S \mu[N]^{\mathrm{T}}\frac{\partial u^*}{\partial x}\hat{n}_x ds+\int_{A^{(e)}}\mu\frac{\partial u^*}{\partial x}\frac{\partial[N]^{\mathrm{T}}}{\partial x}dA$$
$$-\oint_S \mu[N]^{\mathrm{T}}\frac{\partial u^*}{\partial y}\hat{n}_y ds+\int_{A^{(e)}}\mu\frac{\partial u^*}{\partial y}\frac{\partial[N]^{\mathrm{T}}}{\partial y}dA=0$$

or

$$\int_{A^{(e)}} [N^{\mathrm{T}}]\rho\left(u^*\frac{\partial u^*}{\partial x}+v\frac{\partial u^*}{\partial y}\right)dA-\int_{A^{(e)}}[N]^{\mathrm{T}}b_x\,dA-\int_{A^{(e)}}p^*\frac{\partial[N]^{\mathrm{T}}}{\partial x}dA$$
$$+\int_{A^{(e)}}\mu\frac{\partial u^*}{\partial x}\frac{\partial[N]^{\mathrm{T}}}{\partial x}dA+\int_{A^{(e)}}\mu\frac{\partial u^*}{\partial y}\frac{\partial[N]^{\mathrm{T}}}{\partial y}dA$$
$$+\oint_S [N]^{\mathrm{T}}\left(p^*\,n_x-\mu\left(\frac{\partial u^*}{\partial x}\hat{n}_x+\frac{\partial u^*}{\partial y}\hat{n}_y\right)\right)ds=0 \tag{14.14}$$

Similarly, the y momentum can be transformed in a weak form as

$$\int_{A^{(e)}} [N^{\mathrm{T}}]\rho\left(u^*\frac{\partial v^*}{\partial x}+v\frac{\partial v^*}{\partial y}\right)dA-\int_{A^{(e)}}[N]^{\mathrm{T}}b_y\,dA-\int_{A^{(e)}}p^*\frac{\partial[N]^{\mathrm{T}}}{\partial y}dA$$
$$+\int_{A^{(e)}}\mu\frac{\partial v^*}{\partial x}\frac{\partial[N]^{\mathrm{T}}}{\partial x}dA+\int_{A^{(e)}}\mu\frac{\partial v^*}{\partial y}\frac{\partial[N]^{\mathrm{T}}}{\partial y}dA$$
$$+\oint_S [N]^{\mathrm{T}}\left(p^*\,n_y-\mu\left(\frac{\partial v^*}{\partial x}\hat{n}_x+\frac{\partial v^*}{\partial y}\hat{n}_y\right)\right)ds=0 \tag{14.15}$$

Step 3: Selection of Approximate Solution Function The approximate solutions for velocity and pressure fields can be assumed as

$$u^*=\sum_{I=1}^{n_p} N_I u_I=[N]\{u^{(e)}\} \tag{14.16a}$$

$$v^* = \sum_{I=1}^{n_p} N_I v_I = [N]\{v^{(e)}\} \tag{14.16b}$$

$$p^* = \sum_{I=1}^{n_p} N_{pI} p_I = [N_p]\{p^{(e)}\} \tag{14.16c}$$

Note that two different shape functions or interpolation functions are used for velocity and pressure fields. This is due to the fact that the order of the derivative for pressure in the governing equations as well as in its weak form is one order less than that of the velocity components. In general, the order of the shape function N_p used for pressure is one order less than the order for the shape function N for the velocity components. Additionally, the pressure need not be approximated as a continuous function across inter-element boundaries. So, we could use the quadratic shape function for the velocity components and the discontinuous linear shape function for the pressure. For example, we could use the linear triangular element with pressure calculated at three nodes located at the vertices of the triangle and use the quadratic triangular element with velocity components calculated at six nodes of the triangle as shown in Figure 14.2a.

Figure 14.2 shows a quadratic rectangular element for velocity and a bilinear rectangular element for pressure. Figure 14.3 shows that the velocity

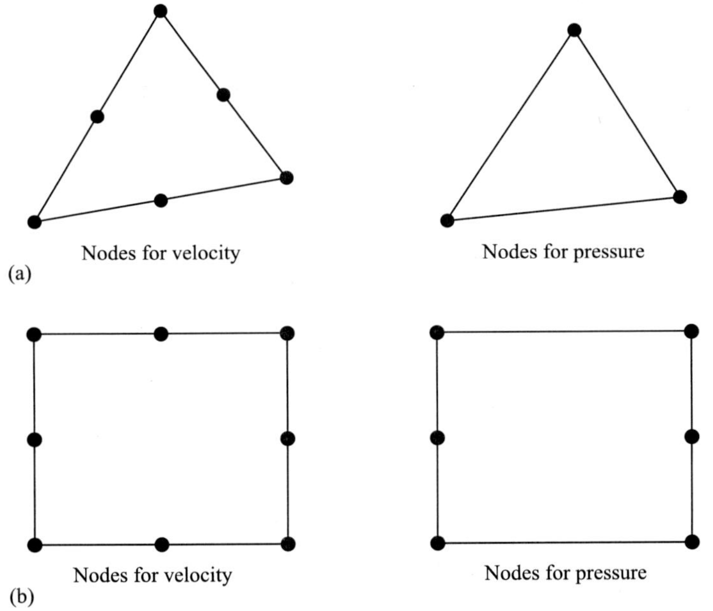

Figure 14.2 Different element types for velocity and pressure: (a) quadratic triangular element for velocity and linear triangular element for pressure, (b) quadratic rectangular element for velocity and bilinear rectangular element for pressure

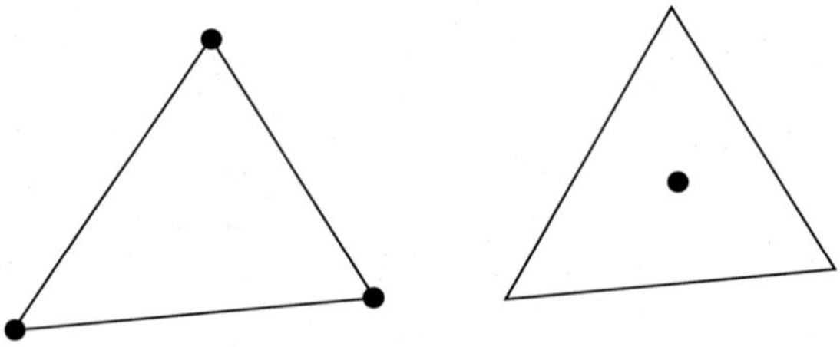

Figure 14.3 Different element types for velocity and pressure: (a) three-noded linear triangular element for velocity and (b) discontinuous constant pressure at the center of the triangle

components are calculated at three nodes in a linear triangular element, whereas pressure is calculated at a node located at the center of the triangle approximated by a constant discontinuous shape function. The pressure is assumed as constant over the entire element.

So, in essence there are two different finite element meshes, one for velocity components and the other for pressure. However, since the mesh for the pressure appears as a subset of the general mesh for the velocity, a common practice is to display the general mesh and indicate the total degrees of freedom at each node.

Step 4: Formation of Element Characteristic Equation The element characteristic equation can be derived by substituting the approximate solution functions given by Equation 14.2 in the integral form of the governing Equations 14.12a, 14.14 and 14.15. Let us demonstrate this by considering the x momentum Equation 14.14 as

$$\int_{A^{(e)}} [N^{\mathrm{T}}]\rho\left([N]\{u^{(e)}\}\frac{\partial[N]}{\partial x}\{u^{(e)}\} + [N]\{v^{(e)}\}\frac{\partial[N]}{\partial y}\{u^{(e)}\}\right)dA - \int_{A^{(e)}} [N]^{\mathrm{T}} b_x\, dA$$

$$- \int_{A^{(e)}} [N_p]\{p^{(e)}\}\frac{\partial[N]^{\mathrm{T}}}{\partial x}\, dA + \int_{A^{(e)}} \mu\frac{\partial[N]}{\partial x}\{u^{(e)}\}\frac{\partial[N]^{\mathrm{T}}}{\partial x}\, dA$$

$$+ \int_{A^{(e)}} \mu\frac{\partial[N]}{\partial y}\{u^{(e)}\}\frac{\partial[N]^{\mathrm{T}}}{\partial y}\, dA$$

$$+ \oint_{S} [N]^{\mathrm{T}}\left(p^{*}\, n_x - \mu\left(\frac{\partial u^{*}}{\partial x}\hat{n}_x + \frac{\partial u^{*}}{\partial y}\hat{n}_y\right)\right) ds = 0$$

or

$$\left\{ \int_{A^{(e)}} \rho [N^{\mathrm{T}}][N] \left(\{u^{(e)}\} \frac{\partial [N]}{\partial x} + \{v^{(e)}\} \frac{\partial [N]}{\partial y} \right) dA \right\} \{u^{(e)}\}$$
$$+ \left[\left\{ \int_{A^{(e)}} \mu \frac{\partial [N]}{\partial x} \frac{\partial [N]^{\mathrm{T}}}{\partial x} dA \right\} + \left\{ \int_{A^{(e)}} \mu \frac{\partial [N]}{\partial y} \frac{\partial [N]^{\mathrm{T}}}{\partial y} dA \right\} \right] \{u^{(e)}\}$$
$$- \left\{ \int_{A^{(e)}} [N_p] \frac{\partial [N]^{\mathrm{T}}}{\partial x} dA \right\} \{p^{(e)}\}$$
$$+ \oint_S [N]^{\mathrm{T}} \left(p^* n_x - \mu \left(\frac{\partial u^*}{\partial x} \hat{n}_x + \frac{\partial u^*}{\partial y} \hat{n}_y \right) \right) ds - \int_{A^{(e)}} [N]^{\mathrm{T}} b_x \, dA \tag{14.17a}$$

Writing the above equation in matrix notation, we get

$$\left[k_I^{(e)}(u, v) \right] \{u^{(e)}\} + \left[k_\mu^{(e)} \right] \{u^{(e)}\} - \left[k_{px}^{(e)} \right] \{p^{(e)}\} + \{I_x^{(e)}\} - \left\{ f_{bx}^{(e)} \right\} = 0 \tag{14.17b}$$

where

$$\left[k_I^{(e)}(u, v) \right] = \int_{A^{(e)}} \rho [N^{\mathrm{T}}][N] \left(\{u^{(e)}\} \frac{\partial [N]}{\partial x} + \{v^{(e)}\} \frac{\partial [N]}{\partial y} \right) dA$$

$$= \text{element characteristic matrix due to inertia forces} \tag{14.18a}$$

$$k_\mu^{(e)} = k_{xx}^{(e)} + k_{yy}^{(e)}$$

$$= \text{element characteristic matrix due to viscous forces} \tag{14.18b}$$

$$k_{xx}^{(e)} = \int_{A^{(e)}} \mu \frac{\partial [N]}{\partial x} \frac{\partial [N]^{\mathrm{T}}}{\partial x} dA \tag{14.18c}$$

$$k_{yy}^{(e)} = \int_{A^{(e)}} \mu \frac{\partial [N]}{\partial y} \frac{\partial [N]^{\mathrm{T}}}{\partial y} dA \tag{14.18d}$$

$$\left[k_{px}^{(e)}\right] = \int_{A^{(e)}} [N_p] \frac{\partial [N]^{\mathrm{T}}}{\partial x} dA$$

= element characteristic matrix due to the pressure force in the x direction (14.18e)

$\left\{f_{bx}^{(e)}\right\}$ = element source vector due to the x component body force term

$$= \int_{A^{(e)}} [N]^{\mathrm{T}} b_x \, dA \tag{14.18f}$$

$\left\{I_x^{(e)}\right\}$ = column vector for the element contribution to the inter-element requirement

$$= \oint_S [N]^{\mathrm{T}} \left(p^* n_x - \mu \left(\frac{\partial u^*}{\partial x} \hat{n}_x + \frac{\partial u^*}{\partial y} \hat{n}_y \right) \right) ds \tag{14.18g}$$

Similarly, the element characteristic equation corresponding to the y momentum is derived from Equation 14.15 as

$$\int_{A^{(e)}} [N^{\mathrm{T}}] \rho \left([N]\{u^{(e)}\} \frac{\partial [N]}{\partial x} \{v^{(e)}\} + [N]\{v^{(e)}\} \frac{\partial [N]}{\partial y} \{v^{(e)}\} \right) dA - \int_{A^{(e)}} [N]^{\mathrm{T}} b_y \, dA$$
$$- \int_{A^{(e)}} [N_p]\{p^{(e)}\} \frac{\partial [N]^{\mathrm{T}}}{\partial y} dA + \int_{A^{(e)}} \mu \frac{\partial [N]\{v^{(e)}\}}{\partial x} \frac{\partial [N]^{\mathrm{T}}}{\partial x} dA$$
$$+ \int_{A^{(e)}} \mu \frac{\partial [N]}{\partial y} \{v^{(e)}\} \frac{\partial [N]^{\mathrm{T}}}{\partial y} dA + \oint_S [N]^{\mathrm{T}} \left(p^* n_y - \mu \left(\frac{\partial v^*}{\partial x} \hat{n}_x + \frac{\partial v^*}{\partial y} \hat{n}_y \right) \right) ds = 0$$

or

$$\left\{ \int_{A^{(e)}} \rho [N^{\mathrm{T}}][N] \left(\{u^{(e)}\} \frac{\partial [N]}{\partial x} + \{v^{(e)}\} \frac{\partial [N]}{\partial y} \right) dA \right\} \{v^{(e)}\}$$
$$+ \left\{ \int_{A^{(e)}} \mu \frac{\partial [N]}{\partial x} \frac{\partial [N]^{\mathrm{T}}}{\partial x} dA + \int_{A^{(e)}} \mu \frac{\partial [N]}{\partial y} \frac{\partial [N]^{\mathrm{T}}}{\partial y} dA \right\} \{v^{(e)}\} - \left\{ \int_{A^{(e)}} [N_p] \frac{\partial [N]^{\mathrm{T}}}{\partial y} dA \right\} \{p^{(e)}\}$$
$$+ \oint_S [N]^{\mathrm{T}} \left(p^* n_y - \mu \left(\frac{\partial v^*}{\partial x} \hat{n}_x + \frac{\partial v^*}{\partial y} \hat{n}_y \right) \right) ds - \int_{A^{(e)}} [N]^{\mathrm{T}} b_y \, dA = 0 \tag{14.19a}$$

In matrix notation, we get

$$\left[k_I^{(e)}(u,v)\right]\left\{v^{(e)}\right\}+\left[k_\mu^{(e)}\right]\left\{v^{(e)}\right\}-\left[k_{py}^{(e)}\right]\left\{p^{(e)}\right\}+\left\{I_y^{(e)}\right\}-\left\{f_{by}^{(e)}\right\}=0 \quad (14.19b)$$

where

$$\left[k_I^{(e)}(u,v)\right]=\int_{A^{(e)}}\rho\left[N^{\mathrm{T}}\right][N]\left(\left\{u^{(e)}\right\}\frac{\partial[N]}{\partial x}+\left\{v^{(e)}\right\}\frac{\partial[N]}{\partial y}\right)dA \quad (14.20a)$$

$$k_\mu^{(e)}=k_{xx}^{(e)}+k_{yy}^{(e)} \quad (14.20b)$$

$$k_{xx}^{(e)}=\int_{A^{(e)}}\mu\frac{\partial[N]}{\partial x}\frac{\partial[N]^{\mathrm{T}}}{\partial x}dA \quad (14.20c)$$

$$k_{yy}^{(e)}=\int_{A^{(e)}}\mu\frac{\partial[N]}{\partial y}\frac{\partial[N]^{\mathrm{T}}}{\partial y}dA \quad (14.20d)$$

$$[k_{py}]=\int_{A^{(e)}}[N_p]\frac{\partial[N]^{\mathrm{T}}}{\partial y}dA$$

= element characteristic matrix due to the pressure force in the y direction (14.20e)

$\left\{f_{by}^{(e)}\right\}$ = element source vector due to the body force term

$$=\int_{A^{(e)}}[N]^{\mathrm{T}}b_y\,dA \quad (14.20f)$$

$\left\{I_y^{(e)}\right\}$ = column vector for the element contribution to the inter-element requirement

$$=\oint_s[N]^{\mathrm{T}}\left(p^*n_y-\mu\left(\frac{\partial u^*}{\partial x}\hat{n}_x+\frac{\partial u^*}{\partial y}\hat{n}_y\right)\right)ds \quad (14.20g)$$

In a similar manner we can reduce the integral form of the continuity equation given by Equation 14.12a to the following element characteristic equation as

$$-\int_{A^{(e)}}[N_p]^{\mathrm{T}}\left\{\frac{\partial[N]}{\partial x}\left\{u^{(e)}\right\}+\frac{\partial[N]}{\partial y}\left\{v^{(e)}\right\}\right\}dA=0$$

or

$$-\left\{\int_{A^{(e)}} [N_p]^{\mathrm{T}} \frac{\partial [N]}{\partial x} dA\right\}\{u^{(e)}\} - \left\{\int_{A^{(e)}} [N_p]^{\mathrm{T}} \frac{\partial [N]}{\partial y} dA\right\}\{v^{(e)}\} = 0 \qquad (14.21a)$$

In the matrix notation, the above equation is written as

$$-\left[k_{px}^{(e)}\right]^{\mathrm{T}}\{u^{(e)}\} - \left[k_{py}^{(e)}\right]^{\mathrm{T}}\{v^{(e)}\} = 0 \qquad (14.21b)$$

Finally, the set of element characteristic equations for u, v, and p are assembled by collecting Equations 14.17, 14.19 and 14.21 as

$$\left[k_I^{(e)}(u,v)\right]\{u^{(e)}\} + \left[k_\mu^{(e)}\right]\{u^{(e)}\} - \left[k_{px}^{(e)}\right]\{p^{(e)}\} + \{I_x^{(e)}\} - \left\{f_{bx}^{(e)}\right\} = 0 \qquad (14.22a)$$

$$\left[k_I^{(e)}(u,v)\right]\{v^{(e)}\} + \left[k_\mu^{(e)}\right]\{v^{(e)}\} - \left[k_{py}^{(e)}\right]\{p^{(e)}\} + \left\{I_y^{(e)}\right\} - \left\{f_{by}^{(e)}\right\} = 0 \qquad (14.22b)$$

$$-\left[k_{px}^{(e)}\right]^{\mathrm{T}}\{u^{(e)}\} - \left[k_{py}^{(e)}\right]^{\mathrm{T}}\{v^{(e)}\} \qquad (14.22c)$$

In the matrix form, the element characteristic equations are written as

$$\begin{bmatrix} k_I^{(e)}(u,v) & 0 & 0 \\ 0 & k_I^{(e)}(u,v) & 0 \\ 0 & 0 & 0 \end{bmatrix} \begin{Bmatrix} \{u^{(e)}\} \\ \{v^{(e)}\} \\ \{p^{(e)}\} \end{Bmatrix} + \begin{bmatrix} k_\mu^{(e)} & 0 & -k_{px}^{(e)} \\ 0 & k_\mu^{(e)} & -k_{yy}^{(e)} \\ -k_{px}^{(e)^T} & -k_{py}^{(e)^T} & 0 \end{bmatrix} \begin{Bmatrix} \{u^{(e)}\} \\ \{v^{(e)}\} \\ \{p^{(e)}\} \end{Bmatrix}$$

$$+ \begin{Bmatrix} \{I_x^{(e)}\} \\ \{I_y^{(e)}\} \\ 0 \end{Bmatrix} - \begin{Bmatrix} \{f_{bx}^{(e)}\} \\ f_{by}^{(e)} \\ 0 \end{Bmatrix} = 0 \qquad (14.23a)$$

Equation 14.23a can be further simplified as

$$\begin{bmatrix} k_I^{(e)}(u,v) + k_\mu^{(e)} & 0 & -k_{px}^{(e)} \\ 0 & k_I^{(e)}(u,v) + k_\mu^{(e)} & -k_{py}^{(e)} \\ -k_{px}^{(e)^T} & -k_{py}^{(e)^T} & 0 \end{bmatrix} \begin{Bmatrix} \{u^{(e)}\} \\ \{v^{(e)}\} \\ \{p^{(e)}\} \end{Bmatrix}$$

$$+ \begin{Bmatrix} \{I_x^{(e)}\} \\ \{I_y^{(e)}\} \\ 0 \end{Bmatrix} - \begin{Bmatrix} \{f_{bx}^{(e)}\} \\ f_{by}^{(e)} \\ 0 \end{Bmatrix} \qquad (14.23b)$$

Step 5: Implementation of the Boundary Conditions The implementation of the boundary conditions of the second kind into the finite element formulation takes place through the inter-element vector $\{I_x^{(e)}\}$given by Equation 14.18g for x-component. In order to show the specific contribution from the boundary condition, the inter-element vector can be decomposed into two parts

$$\{I_x^{(e)}\} = \{I_{sx}^{(e)}\} + \left\{I_{ix}^{(e)}\right\} \tag{14.24}$$

The vector $\{I_{ix}^{(e)}\}$ is the integral contribution from all interior sides of the element other than the boundary sides, and constitutes the inter-element requirements. The vector $\{I_{sx}^{(e)}\}$ constitutes the integral contributions from the boundary forces and it is expressed as

$$\{I_{sx}^{(e)}\} = \oint_s [N]^T\left(p^* n_x - \mu\left(\frac{\partial u^*}{\partial x}\hat{n}_x + \frac{\partial u^*}{\partial y}\hat{n}_y\right)\right) ds \tag{14.25}$$

Substituting the boundary conditions given by Equation 14.11, we get

$$\{I_{sx}^{(e)}\} = \oint_s [N]^{\mathrm{T}} f_{sx}\, ds = \left\{f_{bsx}^{(e)}\right\} \tag{14.26}$$

$= \text{element source vector due to boundary conditions}$

Substituting Equations 14.24 and 14.26 into Equation 14.22a, we get the x-momentum element characteristics equation as

$$\left[k_I^{(e)}(u,v)\right]\{u^{(e)}\} + \left[k_\mu^{(e)}\right]\{u^{(e)}\} - \left[k_{px}^{(e)}\right]\{p^{(e)}\} + \left\{I_{ix}^{(e)}\right\} - \left\{f_{bx}^{(e)} - f_{bsx}^{(e)}\right\} = 0$$

or

$$\left[k_I^{(e)}(u,v)\right]\{u^{(e)}\} + \left[k_\mu^{(e)}\right]\{u^{(e)}\} - \left[k_{px}^{(e)}\right]\{p^{(e)}\} + \left\{I_{ix}^{(e)}\right\} - \{f_x^{(e)}\} = 0 \tag{14.27}$$

In a similar manner, the boundary condition for the y-momentum equation can be implemented in Equation 14.22b, and the set of element characteristic equations for u, v, and p are assembled as

$$\left[k_I^{(e)}(u,v) + \left[k_\mu^{(e)}\right]\right]\{u^{(e)}\} - \left[k_{px}^{(e)}\right]\{p^{(e)}\} + \left\{I_{ix}^{(e)}\right\} - \{f_x^{(e)}\} = 0 \tag{14.28a}$$

$$\left[k_I^{(e)}(u,v) + \left[k_\mu^{(e)}\right]\right]\{v^{(e)}\} - \left[k_{py}^{(e)}\right]\{p^{(e)}\} + \left\{I_{iy}^{(e)}\right\} - \left\{f_y^{(e)}\right\} = 0 \tag{14.28b}$$

$$-\left[k_{px}^{(e)}\right]^{\mathrm{T}}\left\{u^{(e)}\right\}-\left[k_{py}^{(e)}\right]^{T}\left\{v^{(e)}\right\}=0 \tag{14.28c}$$

and in the matrix form as

$$\begin{bmatrix} k_I^{(e)}(u,v)+k_\mu^{(e)} & 0 & -k_{px}^{(e)} \\ 0 & k_I^{(e)}(u,v)+k_\mu^{(e)} & -k_{py}^{(e)} \\ -k_{px}^{(e)^T} & -k_{py}^{(e)^T} & 0 \end{bmatrix}\begin{Bmatrix} \left\{u^{(e)}\right\} \\ \left\{v^{(e)}\right\} \\ \left\{p^{(e)}\right\} \end{Bmatrix}$$

$$+\begin{Bmatrix} \left\{I_{ix}^{(e)}\right\} \\ \left\{I_{iy}^{(e)}\right\} \\ 0 \end{Bmatrix}-\begin{Bmatrix} \left\{f_x^{(e)}\right\} \\ f_y^{(e)} \\ 0 \end{Bmatrix}=0 \tag{14.29a}$$

and in a more compact form as

$$\left[k^{(e)}\right]\left\{U^{(e)}\right\}+\left\{I^{(e)}\right\}-\left\{F^{(e)}\right\}=0 \tag{14.29b}$$

where

$$\left[k_I^{(e)}(u,v)\right]=\int_{A^{(e)}}\rho\left[N^{\mathrm{T}}\right][N]\left(\left\{u^{(e)}\right\}\frac{\partial[N]}{\partial x}+\left\{v^{(e)}\right\}\frac{\partial[N]}{\partial y}\right)dA \tag{14.30a}$$

$$k_\mu^{(e)}=k_{xx}^{(e)}+k_{yy}^{(e)} \tag{14.30b}$$

$$k_{xx}^{(e)}=\int_{A^{(e)}}\mu\frac{\partial[N]}{\partial x}\frac{\partial[N]^{\mathrm{T}}}{\partial x}\,dA \tag{14.30c}$$

$$k_{yy}^{(e)}=\int_{A^{(e)}}\mu\frac{\partial[N]}{\partial y}\frac{\partial[N]^{\mathrm{T}}}{\partial y}\,dA \tag{14.30d}$$

$$\left[k_{px}^{(e)}\right]=\int_{A^{(e)}}[N_p]\frac{\partial[N]^{\mathrm{T}}}{\partial x}\,dA \tag{14.30e}$$

$$\left[k_{py}^{(e)}\right]=\int_{A^{(e)}}[N_p]\frac{\partial[N]^{\mathrm{T}}}{\partial y}\,dA \tag{14.30f}$$

$$\left\{f_x^{(e)}\right\}=\left\{f_{bx}^{(e)}\right\}-\left\{f_{bsx}^{(e)}\right\} \tag{14.30g}$$

$$\left\{f_{bx}^{(e)}\right\}=\int_{A^{(e)}}[N]^{\mathrm{T}}b_x\,dA \tag{14.30h}$$

$$\left\{f_{bsx}^{(e)}\right\}=\oint_s[N]^{\mathrm{T}}f_{sx}\,ds \tag{14.30i}$$

$$\left\{f_y^{(e)}\right\} = \left\{f_{by}^{(e)}\right\} - \left\{f_{bsy}^{(e)}\right\} \tag{14.30j}$$

$$\left\{f_{by}^{(e)}\right\} = \int\limits_{A^{(e)}} [N]^{\mathrm{T}} b_y \, dA \tag{14.30k}$$

$$\left\{f_{bsy}^{(e)}\right\} = \oint\limits_{s} [N]^{\mathrm{T}} f_{sy} \, ds \tag{14.30l}$$

$$\left\{U^{(e)}\right\} = \begin{Bmatrix} u^{(e)} \\ v^{(e)} \\ p^{(e)} \end{Bmatrix} \tag{14.30m}$$

$$\left\{F^{(e)}\right\} = \begin{Bmatrix} f_x^{(e)} \\ f_y^{(e)} \\ 0 \end{Bmatrix} \tag{14.30n}$$

An examination of the matrix in Equation 4.30a clearly shows that the characteristic matrix $k_I^{(e)}(u, v)$ is nonlinear, and that also makes Equation 14.29 nonlinear.

Case: Element Characteristic Equations Using Linear Triangular Element Let us consider a linear three-node triangular element for the velocity components as shown in Figure 10.14 or Figure 14.3a and with the approximate solution and the corresponding shape functions given by Equations 10.59 and 10.61, respectively. For pressure, we assume discontinuous constant pressure at the center of the triangle as shown in Figure 14.3b. The vectors with derivatives of shape functions are derived by substituting the shape functions given as

$$\frac{\partial[N]}{\partial x} = \begin{bmatrix} \frac{\partial N_i}{\partial x} & \frac{\partial N_j}{\partial x} & \frac{\partial N_k}{\partial x} \end{bmatrix} = \frac{1}{2A}\begin{bmatrix} \beta_i & \beta_j & \beta_k \end{bmatrix} \tag{14.31a}$$

and

$$\frac{\partial[N]}{\partial y} = \begin{bmatrix} \frac{\partial N_i}{\partial y} & \frac{\partial N_j}{\partial y} & \frac{\partial N_k}{\partial y} \end{bmatrix} = \frac{1}{2A}\begin{bmatrix} \gamma_i & \gamma_j & \gamma_k \end{bmatrix} \tag{14.31b}$$

Using these expressions, we can evaluate the element stiffness matrices given by Equations 14.30a and 14.30b in the following manner.

The element characteristic matrix due to inertia forces is given as

$$\left[k_I^{(e)}(u,v)\right] = \int_{A^{(e)}} \rho[N^{\mathrm{T}}][N]\left(\left\{u^{(e)}\right\}\frac{\partial[N]}{\partial x} + \left\{v^{(e)}\right\}\frac{\partial[N]}{\partial y}\right) dA$$

$$= \int_{A^{(e)}} \rho \begin{Bmatrix} N_i \\ N_j \\ N_k \end{Bmatrix} \{ N_i \quad N_j \quad N_k \} \begin{Bmatrix} u_i^{(e)} \\ u_j^{(e)} \\ u_k^{(e)} \end{Bmatrix} \left\{ \frac{\partial N_i}{\partial x} \quad \frac{\partial N_j}{\partial x} \quad \frac{\partial N_k}{\partial x} \right\} dA$$

$$+ \int_{A^{(e)}} \rho \begin{Bmatrix} N_i \\ N_j \\ N_k \end{Bmatrix} \{ N_i \quad N_j \quad N_k \} \begin{Bmatrix} v_i^{(e)} \\ v_j^{(e)} \\ v_k^{(e)} \end{Bmatrix} \left\{ \frac{\partial N_i}{\partial y} \quad \frac{\partial N_j}{\partial y} \quad \frac{\partial N_k}{\partial y} \right\} dA$$

$$= \rho \int_{A^{(e)}} \begin{bmatrix} N_i^2 & N_iN_j & N_iN_k \\ N_jN_i & N_j^2 & N_jN_k \\ N_kN_i & N_kN_j & N_k^2 \end{bmatrix} \begin{Bmatrix} u_i^{(e)} \\ u_j^{(e)} \\ u_k^{(e)} \end{Bmatrix} \frac{1}{2A} \{ \beta_i \quad \beta_j \quad \beta_k \} dA$$

$$+ \rho \int_{A^{(e)}} \begin{bmatrix} N_i^2 & N_iN_j & N_iN_k \\ N_jN_i & N_j^2 & N_jN_k \\ N_kN_i & N_kN_j & N_k^2 \end{bmatrix} \begin{Bmatrix} v_i^{(e)} \\ v_j^{(e)} \\ v_k^{(e)} \end{Bmatrix} \frac{1}{2A} \{ \gamma_i \quad \gamma_j \quad \gamma_k \} dA$$

$$= \frac{\rho}{2A} \begin{Bmatrix} u_i^{(e)} \\ u_j^{(e)} \\ u_k^{(e)} \end{Bmatrix} \{ \beta_i \quad \beta_j \quad \beta_k \} \int_{A^{(e)}} \begin{bmatrix} N_i^2 & N_iN_j & N_iN_k \\ N_jN_i & N_j^2 & N_jN_k \\ N_kN_i & N_kN_j & N_k^2 \end{bmatrix} dA$$

$$+ \frac{\rho}{2A} \begin{Bmatrix} v_i^{(e)} \\ v_j^{(e)} \\ v_k^{(e)} \end{Bmatrix} \{ \gamma_i \quad \gamma_j \quad \gamma_k \} \int_{A^{(e)}} \begin{bmatrix} N_i^2 & N_iN_j & N_iN_k \\ N_jN_i & N_j^2 & N_jN_k \\ N_kN_i & N_kN_j & N_k^2 \end{bmatrix} dA$$

$$= \frac{\rho}{24} \begin{bmatrix} 2 & 1 & 1 \\ 1 & 2 & 1 \\ 1 & 1 & 2 \end{bmatrix} \begin{bmatrix} u_i^{(e)}\beta_i & u_i^{(e)}\beta_j & u_i^{(e)}\beta_k \\ u_j^{(e)}\beta_i & u_j^{(e)}\beta_j & u_j^{(e)}\beta_k \\ u_k^{(e)}\beta_i & u_k^{(e)}\beta_j & u_k^{(e)}\beta_k \end{bmatrix}$$

$$+ \frac{\rho}{24} \begin{bmatrix} 2 & 1 & 1 \\ 1 & 2 & 1 \\ 1 & 1 & 2 \end{bmatrix} \begin{bmatrix} v_i^{(e)}\gamma_i & v_i^{(e)}\gamma_j & v_i^{(e)}\gamma_k \\ v_j^{(e)}\gamma_i & v_j^{(e)}\gamma_j & v_j^{(e)}\gamma_k \\ v_k^{(e)}\gamma_i & v_k^{(e)}\gamma_j & v_k^{(e)}\gamma_k \end{bmatrix} \tag{14.32}$$

The element characteristics matrix due to viscous forces is given as

$$k_{xx}^{(e)} = \int_{A^{(e)}} \mu \frac{\partial[N]}{\partial x}\frac{\partial[N]^{\mathrm{T}}}{\partial x}\, dA$$

$$= \int_{A^{(e)}} \mu \left\{ \frac{\partial N_i}{\partial x} \quad \frac{\partial N_j}{\partial x} \quad \frac{\partial N_k}{\partial x} \right\} \left\{ \begin{matrix} \frac{\partial N_i}{\partial x} \\ \frac{\partial N_j}{\partial x} \\ \frac{\partial N_k}{\partial x} \end{matrix} \right\} dA$$

$$= \frac{\mu}{4A^2} \{ \beta_i \quad \beta_j \quad \beta_k \} \left\{ \begin{matrix} \beta_i \\ \beta_j \\ \beta_k \end{matrix} \right\} A$$

$$k_{xx}^{(e)} = \frac{\mu}{4A} \begin{bmatrix} \beta_i^2 & \beta_i\beta_j & \beta_i\beta_k \\ \beta_j\beta_i & \beta_j^2 & \beta_j\beta_i \\ \beta_k\beta_i & \beta_k\beta_j & \beta_k^2 \end{bmatrix} \tag{14.33a}$$

Similarly

$$k_{yy}^{(e)} = \int_{A^{(e)}} \mu \frac{\partial[N]}{\partial y}\frac{\partial[N]^{\mathrm{T}}}{\partial y}\, dA = \frac{\mu}{4A} \begin{bmatrix} \gamma_i^2 & \gamma_i\gamma_j & \gamma_i\gamma_k \\ \gamma_j\gamma_i & \gamma_j^2 & \gamma_j\gamma_i \\ \gamma_k\gamma_i & \gamma_k\gamma_j & \gamma_k^2 \end{bmatrix} \tag{14.33b}$$

Combining Equations 14.33a and 14.33b, we get the element characteristics matrix as

$$\left[k_{\mu}^{(e)}\right] = \frac{\mu}{4A} \begin{bmatrix} \beta_i^2 + \gamma_i^2 & \beta_i\beta_j + \gamma_i\gamma_j & \beta_i\beta_k + \gamma_i\gamma_k \\ \beta_j\beta_i + \gamma_j\gamma_i & \beta_j^2 + \gamma_j^2 & \beta_j\beta_k + \gamma_j\gamma_k \\ \beta_k\beta_i + \gamma_k\gamma_i & \beta_k\beta_j + \gamma_k\gamma_j & \beta_k^2 + \gamma_k^2 \end{bmatrix} \tag{14.34}$$

The component of the element characteristics matrix due to the pressure force term can now be computed as

$$\left[k_{px}^{(e)}\right] = \int_{A^{(e)}} [N_p] \frac{\partial[N]^{\mathrm{T}}}{\partial x}\, dA = \int_A N_p \left\{ \begin{matrix} \frac{\partial N_i}{\partial x} \\ \frac{\partial N_j}{\partial x} \\ \frac{\partial N_k}{\partial x} \end{matrix} \right\} dA = N_p \int_A \frac{1}{2A} \left\{ \begin{matrix} \beta_i \\ \beta_j \\ \beta_k \end{matrix} \right\} dA$$

$$\left[k_{px}^{(e)}\right] = \frac{N_p}{2} \left\{ \begin{matrix} \beta_i \\ \beta_j \\ \beta_k \end{matrix} \right\} \tag{14.35a}$$

Similarly for the y-component pressure force

$$\left[k_{py}^{(e)}\right] = \int\limits_{A^{(e)}} [N_p]\frac{\partial[N]^{\mathrm{T}}}{\partial y}dA = \int\limits_{A} N_p \begin{Bmatrix} \frac{\partial N_i}{\partial y} \\ \frac{\partial N_j}{\partial y} \\ \frac{\partial N_k}{\partial y} \end{Bmatrix} dA = N_p \int\limits_{A} \frac{1}{2A}\begin{Bmatrix} \gamma_i \\ \gamma_j \\ \gamma_k \end{Bmatrix} dA$$

$$\left[k_{py}^{(e)}\right] = \frac{N_p}{2}\begin{Bmatrix} \gamma_i \\ \gamma_j \\ \gamma_k \end{Bmatrix} \tag{14.35b}$$

The element source vector due the body forces can be evaluated as

$$\begin{aligned} \left\{f_{bx}^{(e)}\right\} &= \int\limits_{A^{(e)}} [N]^{\mathrm{T}} b_x \, dA \\ &= b_x \int\limits_{A} \begin{Bmatrix} N_i \\ N_j \\ N_k \end{Bmatrix} dA = b_x \int\limits_{A} \begin{Bmatrix} L_i \\ L_j \\ L_k \end{Bmatrix} dA \end{aligned} \tag{14.36}$$

Using the natural coordinate system for a triangle and integral formula (Equation 10.71), we get

$$\left\{f_{bx}^{(e)}\right\} = \frac{b_x A}{3}\begin{Bmatrix} 1 \\ 1 \\ 1 \end{Bmatrix} \tag{14.37a}$$

Similarly for the y-component body force term

$$\left\{f_{by}^{(e)}\right\} = \int\limits_{A^{(e)}} [N]^{\mathrm{T}} b_y \, dA = \frac{b_y A}{3}\begin{Bmatrix} 1 \\ 1 \\ 1 \end{Bmatrix} \tag{14.37b}$$

The element source vector due to the x-component boundary surface forces

$$\begin{aligned} \left\{f_{bsx}^{(e)}\right\} &= \oint\limits_{s} [N]^{\mathrm{T}} f_{sx} \, ds \\ &= f_{sx} \oint\limits_{s} \begin{Bmatrix} N_i \\ N_j \\ N_k \end{Bmatrix} ds \end{aligned} \tag{14.38}$$

For a boundary condition specified on the i–j side of the element, $N_k = 0$. Using a natural coordinate system for a triangle and integral formula (Equation 10.28), we get

$$\left\{f_{bsx}^{(e)}\right\} = f_{sx} \int_{S_b} \begin{Bmatrix} N_i \\ N_j \\ 0 \end{Bmatrix} ds = f_{sx} \int_{S_b} \begin{Bmatrix} L_i \\ L_j \\ 0 \end{Bmatrix} ds$$

$$\left\{f_{bsx}^{(e)}\right\} = \frac{f_{sx} L_{ij}}{2} \begin{Bmatrix} 1 \\ 1 \\ 0 \end{Bmatrix} \tag{14.39a}$$

Similarly, for boundary conditions specified on the sides j–k, $N_i = 0$ and Equation 14.38 transforms into

$$\left\{f_{bsx}^{(e)}\right\} = f_{sx} \int_{S_b} \begin{Bmatrix} 0 \\ N_j \\ N_k \end{Bmatrix} ds = \frac{f_{sx} L_{jk}}{2} \begin{Bmatrix} 0 \\ 1 \\ 1 \end{Bmatrix} \tag{14.39b}$$

and for boundary conditions specified on the side k–i, $N_j = 0$, we get

$$\left\{f_{bsx}^{(e)}\right\} = f_{sx} \int_{S_b} \begin{Bmatrix} N_i \\ 0 \\ N_k \end{Bmatrix} ds = \frac{f_{sx} L_{ki}}{2} \begin{Bmatrix} 1 \\ 0 \\ 1 \end{Bmatrix} \tag{14.39c}$$

The element source vector due to y-component boundary surface forces is

$$\left\{f_{bsy}^{(e)}\right\} = \oint_S [N]^{\mathrm{T}} f_{sy}\, ds$$

$$= f_{sy} \oint_S \begin{Bmatrix} N_i \\ N_j \\ N_k \end{Bmatrix} ds \tag{14.40}$$

For the boundary condition specified on the i–j side of the element Equation 14.40 reduces to

$$\left\{f_{bsy}^{(e)}\right\} = \frac{f_{sy} L_{ij}}{2} \begin{Bmatrix} 1 \\ 1 \\ 0 \end{Bmatrix} \tag{14.41a}$$

For boundary conditions specified on the side j–k, Equation 14.40 reduces to

$$\left\{f_{bsy}^{(e)}\right\} = \frac{f_{sy} L_{jk}}{2} \begin{Bmatrix} 0 \\ 1 \\ 1 \end{Bmatrix} \tag{14.41b}$$

For boundary conditions specified on the side k–i, Equation 14.40 reduces to

$$\left\{f_{bsy}^{(e)}\right\} = \frac{f_{sy}L_{ki}}{2}\begin{Bmatrix}1\\0\\1\end{Bmatrix} \qquad (14.41c)$$

Substituting Equations 14.32, 14.34, 14.35a, and 14.36a into Equation 14.28a, we have the expression for the element characteristic equation for an ***interior element*** with $f_{bsx}^{(e)} = 0$ as

$$\left[k_I^{(e)}(u,v) + \left[k_\mu^{(e)}\right]\right]\{u^{(e)}\} - \left[k_{px}^{(e)}\right]\{p^{(e)}\} + \left\{I_{ix}^{(e)}\right\} - \left\{f_{bx}^{(e)}\right\} = 0$$

$$\left\{\frac{\rho}{24}\begin{bmatrix}2&1&1\\1&2&1\\1&1&2\end{bmatrix}\left\{\begin{bmatrix}u_i^{(e)}\beta_i & u_i^{(e)}\beta_j & u_i^{(e)}\beta_k\\ u_j^{(e)}\beta_i & u_j^{(e)}\beta_j & u_j^{(e)}\beta_k\\ u_k^{(e)}\beta_i & u_k^{(e)}\beta_j & u_k^{(e)}\beta_k\end{bmatrix} + \begin{bmatrix}v_i^{(e)}\gamma_i & v_i^{(e)}\gamma_j & v_i^{(e)}\gamma_k\\ v_j^{(e)}\gamma_i & v_j^{(e)}\gamma_j & v_j^{(e)}\gamma_k\\ v_k^{(e)}\gamma_i & v_k^{(e)}\gamma_j & v_k^{(e)}\gamma_k\end{bmatrix}\right\}\right\}\begin{Bmatrix}u_i^{(e)}\\u_j^{(e)}\\u_k^{(e)}\end{Bmatrix}$$
$$+\frac{\mu}{4A}\begin{bmatrix}\beta_i^2+\gamma_i^2 & \beta_i\beta_j+\gamma_i\gamma_j & \beta_i\beta_k+\gamma_i\gamma_k\\ \beta_j\beta_i+\gamma_j\gamma_i & \beta_j^2+\gamma_j^2 & \beta_j\beta_k+\gamma_j\gamma_k\\ \beta_k\beta_i+\gamma_k\gamma_i & \beta_k\beta_j+\gamma_k\gamma_j & \beta_k^2+\gamma_k^2\end{bmatrix}\begin{Bmatrix}u_i^{(e)}\\u_j^{(e)}\\u_k^{(e)}\end{Bmatrix} - \frac{N_p}{2}\begin{Bmatrix}\beta_i\\\beta_j\\\beta_k\end{Bmatrix}\{p^{(e)}\} + \left\{I_{ix}^{(e)}\right\}$$
$$-\frac{b_xA}{3}\begin{Bmatrix}1\\1\\1\end{Bmatrix} = 0 \qquad (14.42)$$

The element characteristic matrix for a ***boundary element*** with boundary surface force specified at the edge i–j is obtained by substituting Equations 14.32, 14.34, 14.35a, 14.36a, and 14.38a into Equation 14.28a as

$$\left\{\frac{\rho}{24}\begin{bmatrix}2&1&1\\1&2&1\\1&1&2\end{bmatrix}\left\{\begin{bmatrix}u_i^{(e)}\beta_i & u_i^{(e)}\beta_j & u_i^{(e)}\beta_k\\ u_j^{(e)}\beta_i & u_j^{(e)}\beta_j & u_j^{(e)}\beta_k\\ u_k^{(e)}\beta_i & u_k^{(e)}\beta_j & u_k^{(e)}\beta_k\end{bmatrix} + \begin{bmatrix}v_i^{(e)}\gamma_i & v_i^{(e)}\gamma_j & v_i^{(e)}\gamma_k\\ v_j^{(e)}\gamma_i & v_j^{(e)}\gamma_j & v_j^{(e)}\gamma_k\\ v_k^{(e)}\gamma_i & v_k^{(e)}\gamma_j & v_k^{(e)}\gamma_k\end{bmatrix}\right\}\right\}\begin{Bmatrix}u_i^{(e)}\\u_j^{(e)}\\u_k^{(e)}\end{Bmatrix}$$
$$+\frac{\mu}{4A}\begin{bmatrix}\beta_i^2+\gamma_i^2 & \beta_i\beta_j+\gamma_i\gamma_j & \beta_i\beta_k+\gamma_i\gamma_k\\ \beta_j\beta_i+\gamma_j\gamma_i & \beta_j^2+\gamma_j^2 & \beta_j\beta_k+\gamma_j\gamma_k\\ \beta_k\beta_i+\gamma_k\gamma_i & \beta_k\beta_j+\gamma_k\gamma_j & \beta_k^2+\gamma_k^2\end{bmatrix}\begin{Bmatrix}u_i^{(e)}\\u_j^{(e)}\\u_k^{(e)}\end{Bmatrix} - \frac{N_p}{2}\begin{Bmatrix}\beta_i\\\beta_j\\\beta_k\end{Bmatrix}\{p^{(e)}\} + \left\{I_{ix}^{(e)}\right\}$$
$$-\frac{b_xA}{3}\begin{Bmatrix}1\\1\\1\end{Bmatrix} + \frac{f_{sx}L_{ij}}{2}\begin{Bmatrix}1\\1\\0\end{Bmatrix} = 0 \qquad (14.43)$$

In a similar manner, we can derive the element characteristic equation for the y-component velocity v and pressure p starting from Equations 14.28b and 14.28c, respectively.

Step 6: Assembly of Elements and Formation of Global System The assembly of all element characteristics equations to form the global system is done by following the procedure and the connectivity array presented in Chapter 12. In a compact form the global system of equations is written in a compact form as

$$K\{U\} = \{F\} \tag{14.44}$$

Example 14.1: Consider a two-dimensional viscous incompressible flow induced in a rectangular cavity by the motion of an upper surface at constant velocity. Consider ***Element*** *7* for the mesh consisting of linear triangular elements for velocity components and a constant pressure at the center of the triangle as shown in the figure and derive the element characteristic equation. The governing equations over a flow domain Ω are given as

Continuity equation

$$\frac{\partial u}{\partial x} + \frac{\partial v}{vy} = 0$$

x momentum

$$\rho\left(u\frac{\partial u}{\partial x} + v\frac{\partial u}{\partial y}\right) = b_x - \frac{\partial P}{\partial x} + \mu\left(\frac{\partial^2 u}{\partial x^2} + \frac{\partial^2 u}{\partial y^2}\right)$$

y momentum

$$\rho\left(u\frac{\partial v}{\partial x} + v\frac{\partial v}{\partial y}\right) = b_y - \frac{\partial P}{\partial y} + \mu\left(\frac{\partial^2 v}{\partial x^2} + \frac{\partial^2 v}{\partial y^2}\right)$$

with boundary conditions given as

1. $x = 0, \quad u = 0; \quad v = 0$
2. $x = 1, \quad u = 0; \quad v = 0$
3. at $y = 0, \quad u = 0; \quad v = 0$
4. at $y = 1, \quad u = u_w; \quad v = 0$

Steps for the finite element formulation are:

Step 1: Discretization of the Solution Domain Solution for flow field involves two different kinds of variables such as velocity, V, and pressure, P. This requires the need for two different finite element meshes: one for the velocity and the other for the pressure.

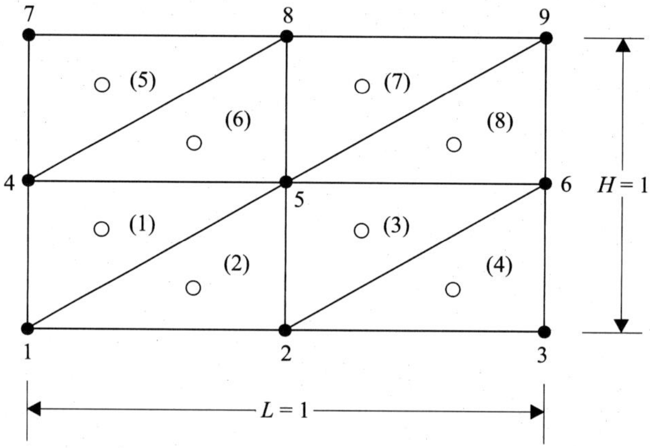

Step 2: Formulation of an Integral or Weak Form Using Galerkin's Method Weighted residual involves finding a minimum for the residuals over an element $A^{(e)}$ by multiplying the residuals of governing equations by the weighted functions W_p, W_u, and W_v, and integrating over the element as

$$\left\{R_p^{(e)}\right\} = -\int W_P\left\{\frac{\partial u^*}{\partial x} + \frac{\partial v^*}{\partial y}\right\} dA = 0$$

$$\{R_u^{(e)}\} = -\int_{A^{(e)}} W_u\left\{\rho\left(u^*\frac{\partial u^*}{\partial x} + v^*\frac{\partial u^*}{\partial y}\right) - b_x + \frac{\partial P^*}{\partial x} - \mu\left(\frac{\partial^2 u^*}{\partial x^2} + \frac{\partial^2 u^*}{\partial y^2}\right)\right\} dA = 0$$

$$\{R_v^{(e)}\} = -\int_{A^{(e)}} W_v\left\{\rho\left(u^*\frac{\partial v^*}{\partial x} + v^*\frac{\partial v^*}{\partial y}\right) - b_v + \frac{\partial P^*}{\partial y} - \mu\left(\frac{\partial^2 v^*}{\partial x^2} + \frac{\partial^2 v^*}{\partial y^2}\right)\right\} dA = 0$$

In the Galerkin finite element method the weighting functions are assumed to be the same as the shape functions used for the approximate functions

$$W_u = [N]^{\mathrm{T}}, \quad W_y = [N]^{\mathrm{T}} \quad \text{and} \quad W_p = \left[N_p\right]^{\mathrm{T}}$$

where $[N]$ and $[N_p]$ are the shape functions associated with velocity and the pressure approximation function respectively.

The integral equations are transformed into a weak form

$$\int_{A^{(e)}} [N^{\mathrm{T}}]\rho\left(u^*\frac{\partial u^*}{\partial x}+v\frac{\partial u^*}{\partial y}\right)dA-\int_{A^{(e)}} [N]^{\mathrm{T}}b_x\,dA-\int_{A^{(e)}} p^*\frac{\partial[N]^{\mathrm{T}}}{\partial x}dA+\int_{A^{(e)}} \mu\frac{\partial u^*}{\partial x}\frac{\partial[N]^{\mathrm{T}}}{\partial x}dA$$
$$+\int_{A^{(e)}} \mu\frac{\partial u^*}{\partial y}\frac{\partial[N]^{\mathrm{T}}}{\partial y}dA+\oint_{s}[N]^{\mathrm{T}}\left(p^*n_x-\mu\left(\frac{\partial u^*}{\partial x}\hat{n}_x+\frac{\partial u^*}{\partial x}\hat{n}_y\right)\right)dS=0$$

and

$$\int_{A^{(e)}} [N^{\mathrm{T}}]\rho\left(u^*\frac{\partial v^*}{\partial x}+v\frac{\partial v^*}{\partial y}\right)dA-\int_{A^{(e)}} [N]^{\mathrm{T}}b_y\,dA-\int_{A^{(e)}} p^*\frac{\partial[N]^{\mathrm{T}}}{\partial y}dA+\int_{A^{(e)}} \mu\frac{\partial v^*}{\partial x}\frac{\partial[N]^{\mathrm{T}}}{\partial x}dA$$
$$+\int_{A^{(e)}} \mu\frac{\partial v^*}{\partial y}\frac{\partial[N]^{\mathrm{T}}}{\partial y}dA+\oint_{s}[N]^{\mathrm{T}}\left(p^*n_x-\mu\left(\frac{\partial v^*}{\partial x}\hat{n}_x+\frac{\partial v^*}{\partial x}\hat{n}_y\right)\right)dS=0$$

Step 3: Selection of Approximate Solution Functions The approximate solutions for velocity and pressure fields can be assumed as follows:

$$u^* = \sum_{I=1}^{n_P} N_I u_I = [N]\{u^{(e)}\}$$
$$v^* = \sum_{I=1}^{n_P} N_I v_I = [N]\{v^{(e)}\}$$
$$p^* = \sum_{I=1}^{n_P} N_I p_I = [N_p]\{p^{(e)}\}$$

Step 4: Formation of Element Characteristic Equation The element characteristic equation can be derived by substituting the approximate solution functions in the integral form of the governing equations. The set of element characteristic equations for u, v, and p are assembled as

$$\left[k_1^{(e)}(u,v)\right]\{u^{(e)}\}+\left[k_\mu^{(e)}\right]\{u^{(e)}\}-\left[k_{px}^{(e)}\right]\{p^{(e)}\}+\{I_x^{(e)}\}-\left\{f_{bx}^{(e)}\right\}=0$$

$$\left[k_1^{(e)}(u,v)\right]\{v^{(e)}\}+\left[k_\mu^{(e)}\right]\{v^{(e)}\}-\left[k_{py}^{(e)}\right]\{p^{(e)}\}+\left\{I_y^{(e)}\right\}-\left\{f_{by}^{(e)}\right\}=0$$

$$-\left[k_{px}^{(e)}\right]\{u^{(e)}\}-\left[k_{py}^{(e)}\right]\{v^{(e)}\}=0$$

This can be written in matrix form as

$$\begin{bmatrix} k_1^{(e)}(u,v)+k_\mu^{(e)} & 0 & -k_{px}^{(e)} \\ 0 & k_1^{(e)}(u,v)+k_\mu^{(e)} & -k_{py}^{(e)} \\ -k_{px}^{(e)^T} & -k_{py}^{(e)^T} & 0 \end{bmatrix} \begin{Bmatrix} \{u^{(e)}\} \\ \{v^{(e)}\} \\ \{p^{(e)}\} \end{Bmatrix} + \begin{Bmatrix} \{I_x^{(e)}\} \\ \{I_y^{(e)}\} \\ 0 \end{Bmatrix} - \begin{Bmatrix} \{f_{bx}^{(e)}\} \\ \{f_{by}^{(e)}\} \\ 0 \end{Bmatrix} = 0$$

The matrices and vectors in the element characteristic equations are now evaluated using linear triangular element as

$$\left[k_1^{(e)}(u,v)\right] = \int_{A^{(e)}} \rho[N^{\mathrm{T}}][N]\left(\{u^{(e)}\}\frac{\partial[N]}{\partial x} + \{v^{(e)}\}\frac{\partial[N]}{\partial y}\right) dA$$

$$= \frac{\rho}{24}\begin{bmatrix} 2 & 1 & 1 \\ 1 & 2 & 1 \\ 1 & 1 & 2 \end{bmatrix} \begin{bmatrix} u_i^{(e)}\beta_i + v_i^{(e)}\gamma_i & u_i^{(e)}\beta_j + v_i^{(e)}\gamma_j & u_i^{(e)}\beta_k + v_i^{(e)}\gamma_k \\ u_j^{(e)}\beta_i + v_j^{(e)}\gamma_i & u_j^{(e)}\beta_j + v_j^{(e)}\gamma_j & u_j^{(e)}\beta_k + v_j^{(e)}\gamma_k \\ u_k^{(e)}\beta_i + v_k^{(e)}\gamma_i & u_k^{(e)}\beta_j + v_k^{(e)}\gamma_j & u_k^{(e)}\beta_k + v_k^{(e)}\gamma_k \end{bmatrix}$$

The stiffness matrix due to viscous forces is given as

$$k_{xx}^{(e)} = \int_{A^{(e)}} \mu \frac{\partial[N]}{\partial x}\frac{\partial[N]^{\mathrm{T}}}{\partial x} dA$$

$$= \frac{\mu}{4A}\begin{bmatrix} \beta_i^2 & \beta_i\beta_j & \beta_i\beta_k \\ \beta_j\beta_j & \beta_j^2 & \beta_j\beta_k \\ \beta_k\beta_j & \beta_k\beta_j & \beta_k^2 \end{bmatrix}$$

$$k_{yy}^{(e)} = \int_{A^{(e)}} \mu \frac{\partial[N]}{\partial y}\frac{\partial[N]^{\mathrm{T}}}{\partial y} dA$$

$$= \frac{\mu}{24}\begin{bmatrix} \gamma_i^2 & \gamma_i\gamma_j & \gamma_i\gamma_k \\ \gamma_j\gamma_i & \gamma_j^2 & \gamma_j\gamma_k \\ \gamma_k\gamma_i & \gamma_k\gamma_j & \gamma_k^2 \end{bmatrix}$$

The component of the element stiffness matrix due to the pressure force term can now be computed as

$$\left[k_{px}^{(e)}\right] = \int_{A^{(e)}} [N_{px}]\frac{\partial[N]^{\mathrm{T}}}{\partial x} dA = \frac{N_p}{2}\begin{Bmatrix} \beta_i \\ \beta_j \\ \beta_k \end{Bmatrix}$$

$$\left[k_{py}^{(e)}\right] = \int_{A^{(e)}} [N_{py}]\frac{\partial[N]^{\mathrm{T}}}{\partial y} dA = \frac{N_p}{2}\begin{Bmatrix} \beta_i \\ \beta_j \\ \beta_k \end{Bmatrix}$$

The element source vector due to the body force can be evaluated as

$$\left\{f_{bx}^{(e)}\right\} = \int\limits_{A^{(e)}} [N]^{\mathrm{T}} b_x \, dA = 0$$

similarly for y-component

$$\left\{f_{by}^{(e)}\right\} = \int\limits_{A^{(e)}} [N]^{\mathrm{T}} b_y \, dA = 0$$

The element source vector due to the x-component boundary surface forces is

$$\left\{f_{bsx}^{(e)}\right\} = \oint\limits_{S} [N]^{\mathrm{T}} f_{sx} \, ds = 0$$

For the boundary condition specified on the i–j side of the element, $N_k = 0$. Using the natural coordinate system for a triangle and integral formula, we get

$$\left\{f_{bsx}^{(e)}\right\} = \frac{f_{sx} L_{ij}}{2} \begin{Bmatrix} 1 \\ 1 \\ 0 \end{Bmatrix} = 0$$

The element source vector due to the y-component boundary surface forces is

$$\left\{f_{bsy}^{(e)}\right\} = \oint\limits_{S} [N]^{\mathrm{T}} f_{sy} \, ds$$

$$= f_{sy} \oint\limits_{s} \begin{Bmatrix} N_i \\ N_j \\ N_k \end{Bmatrix} ds$$

For the boundary condition specified on the i–j side of the element, $N_k = 0$. Using the natural coordinate system for a triangle and integral formula, we get

$$\left\{f_{bsy}^{(e)}\right\} = \frac{f_{sy} L_{ij}}{2} \begin{Bmatrix} 1 \\ 1 \\ 0 \end{Bmatrix} = 0$$

Model Calculation for Element 7

$$(x_i, y_i) = (0.5, 0.5)$$

$$(x_j, y_j) = (1, 1)$$

$$(x_k, y_k) = (0.5, 1)$$

$$\beta_i = (y_j - y_k) = 1 - 1 = 0$$

$$\beta_j = (y_k - y_i) = 1 - 0.5 = 0.5$$

$$\beta_i = (y_i - y_j) = 0.5 - 1 = -0.5$$

$$\gamma_i = -(x_j - x_k) = 0.5 - 1 = -0.5$$

$$\gamma_j = -(x_k - x_i) = 0.5 - 0.5 = 0$$

$$\gamma_i = -(x_i - x_j) = 1 - 0.5 = 0.5$$

$k_1^{(7)}(u, v)$

$$= \frac{\rho}{24}\begin{bmatrix} 2 & 1 & 1 \\ 1 & 2 & 1 \\ 1 & 1 & 2 \end{bmatrix}\begin{bmatrix} u_i^{(e)}\beta_i + v_i^{(e)}\gamma_i & u_i^{(e)}\beta_j + v_i^{(e)}\gamma_j & u_i^{(e)}\beta_k + v_i^{(e)}\gamma_k \\ u_j^{(e)}\beta_i + v_j^{(e)}\gamma_i & u_j^{(e)}\beta_j + v_j^{(e)}\gamma_j & u_j^{(e)}\beta_k + v_j^{(e)}\gamma_k \\ u_k^{(e)}\beta_i + v_k^{(e)}\gamma_i & u_k^{(e)}\beta_j + v_k^{(e)}\gamma_j & u_k^{(e)}\beta_k + v_k^{(e)}\gamma_k \end{bmatrix}$$

$$= \frac{\rho}{24}\begin{bmatrix} 2 & 1 & 1 \\ 1 & 2 & 1 \\ 1 & 1 & 2 \end{bmatrix}\begin{bmatrix} (u_5 * 0) + (v_5 * -0.5) & (u_5 * 0.5) + (v_5 * 0) & (u_5 * -0.5) + (v_5 * 0.5) \\ (u_9 * 0) + (v_9 * -0.5) & (u_9 * 0.5) + (v_9 * 0) & (u_9 * -0.5) + (v_9 * 0.5) \\ (u_8 * 0) + (v_8 * -0.5) & (u_8 * 0.5) + (v_8 * 0) & (u_8 * -0.5) + (v_8 * 0.5) \end{bmatrix}$$

$$= \frac{\rho}{24}\begin{bmatrix} 2 & 1 & 1 \\ 1 & 2 & 1 \\ 1 & 1 & 2 \end{bmatrix}\begin{bmatrix} -0.5v_5 & 0.5u_5 & 0.5 * (v_5 - u_5) \\ -0.5v_9 & 0.5u_9 & 0.5 * (v_9 - u_9) \\ -0.5v_8 & 0.5u_8 & 0.5 * (v_8 - u_8) \end{bmatrix}$$

$$k_\mu^{(7)} = \frac{\mu}{4A}\begin{bmatrix} \beta_i^2 + \gamma_i^2 & \beta_i\beta_j + \gamma_i\gamma_j & \beta_i\beta_k + \gamma_i\gamma_k \\ \beta_j\beta_i + \gamma_j\gamma_i & \beta_j^2 + \gamma_j^2 & \beta_j\beta_k + \gamma_j\gamma_k \\ \beta_k\beta_i + \gamma_k\gamma_i & \beta_k\beta_j + \gamma_k\gamma_j & \beta_k^2 + \gamma_k^2 \end{bmatrix}$$

$$= \frac{\mu}{4A}\begin{bmatrix} 0.25 & 0 & -0.25 \\ 0 & 0.25 & -0.25 \\ -0.25 & -0.25 & 0.5 \end{bmatrix}$$

$$k_{px}^{(7)} = \frac{N_p}{2}\begin{Bmatrix} \beta_i \\ \beta_j \\ \beta_k \end{Bmatrix} = \frac{N_p}{2}\begin{Bmatrix} 0 \\ 0.5 \\ -0.5 \end{Bmatrix}$$

$$k_{py}^{(7)} = \frac{N_p}{2}\begin{Bmatrix} \gamma_i \\ \gamma_j \\ \gamma_k \end{Bmatrix} = \frac{N_p}{2}\begin{Bmatrix} -0.5 \\ 0 \\ 0.5 \end{Bmatrix}$$

14.2.3 Unsteady Two-dimensional Viscous Incompressible Flow

Let us consider the development of the transient problem by considering the two-dimensional ***Stokes flow*** which is applicable for problems with low speed. The fluid is assumed to Newtonian. As we have discussed in Chapter 1, appropriate governing equations for Stokes flow are derived by neglecting the nonlinear inertial or convective terms, and are given as follows.

Continuity

$$\frac{\partial u}{\partial x} + \frac{\partial v}{\partial y} = 0 \tag{14.45a}$$

x momentum

$$\rho \frac{\partial u}{\partial t} = b_x - \frac{\partial p}{\partial x} + \frac{\partial}{\partial x}\left[\mu\left(\frac{\partial u}{\partial x}\right)\right] + \frac{\partial}{\partial y}\left[\mu\left(\frac{\partial u}{\partial y}\right)\right] \tag{14.45b}$$

y momentum

$$\rho \frac{\partial v}{\partial t} = b_y - \frac{\partial p}{\partial y} + \frac{\partial}{\partial x}\left[\mu\left(\frac{\partial v}{\partial x}\right)\right] + \frac{\partial}{\partial y}\left[\mu\left(\frac{\partial v}{\partial y}\right)\right] \tag{14.45c}$$

The treatment of the unsteady state term is similar to that discussed in Chapter 13. Following the procedure outlined in Section 14.2.2, we can obtain the weak forms for the equations using Galerkin's weighted residual method as

$$\left\{R_P^{(e)}\right\} = -\int_{A^{(e)}} [N_p]^{\mathrm{T}} \left\{\frac{\partial u^*}{\partial x} + \frac{\partial v^*}{\partial y}\right\} dA = 0 \tag{14.46a}$$

$$\left\{R_u^{(e)}\right\} = \int_{A^{(e)}} [N]^{\mathrm{T}} \left\{\rho \frac{\partial u^*}{\partial t} - b_x + \frac{\partial p^*}{\partial x} - \frac{\partial}{\partial x}\left(\mu \frac{\partial u^*}{\partial x}\right) - \frac{\partial}{\partial y}\left(\mu \frac{\partial u^*}{\partial y}\right)\right\} dA = 0 \tag{14.46b}$$

$$\left\{R_v^{(e)}\right\} = \int_{A^{(e)}} [N]^{\mathrm{T}} \left\{\rho \frac{\partial v^*}{\partial t} - b_y + \frac{\partial p^*}{\partial y} - \frac{\partial}{\partial x}\left(\mu \frac{\partial v^*}{\partial x}\right) - \frac{\partial}{\partial y}\left(\mu \frac{\partial v^*}{\partial y}\right)\right\} dA = 0 \tag{14.46c}$$

In the ***semi-discrete finite element formulation*** as outlined in Chapter 13, the approximate solutions for velocity and pressure fields are assumed to be

$$u^* = \sum_{I=1}^{n_p} N_I(x) u_I(t) = [N(x)]\left\{u^{(e)}(t)\right\} \tag{14.47a}$$

$$v^* = \sum_{I=1}^{n_p} N_I(x) v_I(t) = [N(x)]\left\{v^{(e)}(t)\right\} \tag{14.47b}$$

$$p^* = \sum_{I=1}^{n_p} N_{pI}(x) p_I(t) = \left[N_p(x)\right]\left\{p^{(e)}(t)\right\} \tag{14.47c}$$

where the shape functions N_I and N_{pI} are assumed to be a function of the space variable only, and the nodal values u_I, v_I, and p_I as a function of time. With the substitution of the approximation solutions and evaluation of integrals, we get the system of element characteristics equations as

$$\left[c^{(e)}\right]\frac{d\{u^{(e)}\}}{dt} + \left[k_{\mu}^{(e)}\right]\{u^{(e)}\} - \left[k_{px}^{(e)}\right]\{p^{(e)}\} + \left\{I_{ix}^{(e)}\right\} - \{f_x^{(e)}\} = 0 \tag{14.48a}$$

$$\left[c^{(e)}\right]\frac{d\{v^{(e)}\}}{dt} + \left[k_{\mu}^{(e)}\right]\{v^{(e)}\} - \left[k_{py}^{(e)}\right]\{p^{(e)}\} + \left\{I_{iy}^{(e)}\right\} - \left\{f_y^{(e)}\right\} = 0 \tag{14.48b}$$

$$-\left[k_{px}^{(e)}\right]^{\mathrm{T}}\{u^{(e)}\} - \left[k_{py}^{(e)}\right]^{\mathrm{T}}\{v^{(e)}\} = 0 \tag{14.48c}$$

and in the matrix form as

$$\begin{bmatrix} c^{(e)} & 0 & 0 \\ 0 & c^{(e)} & 0 \\ 0 & 0 & 0 \end{bmatrix} \begin{Bmatrix} \dfrac{\partial u^{(e)}}{\partial t} \\ \dfrac{\partial v^{(e)}}{\partial t} \\ \dfrac{\partial p}{\partial t} \end{Bmatrix} + \begin{bmatrix} k_{\mu}^{(e)} & 0 & -k_{px}^{(e)} \\ 0 & k_{\mu}^{(e)} & -k_{py}^{(e)} \\ -k_{px}^{(e)\mathrm{T}} & -k_{py}^{(e)\mathrm{T}} & 0 \end{bmatrix} \begin{Bmatrix} \{u^{(e)}\} \\ \{v^{(e)}\} \\ \{p^{(e)}\} \end{Bmatrix}$$

$$+ \begin{Bmatrix} \left\{I_{ix}^{(e)}\right\} \\ \left\{I_{iy}^{(e)}\right\} \\ 0 \end{Bmatrix} - \begin{Bmatrix} \{f_x^{(e)}\} \\ f_y^{(e)} \\ 0 \end{Bmatrix} = 0 \tag{14.49a}$$

and in a compact form as

$$C^{(e)}\left\{\frac{\partial U^{(e)}}{\partial t}\right\} + k^{(e)}\{U^{(e)}\} + \{I^{(e)}\} - \{F^{(e)}\} \tag{14.49b}$$

where

$$c^{(e)} = \int_A \rho[N]^{\mathrm{T}}[N]\,dA \tag{14.50a}$$

$$C^{(e)} = \begin{bmatrix} c^{(e)} & 0 & 0 \\ 0 & c^{(e)} & 0 \\ 0 & 0 & 0 \end{bmatrix} \tag{14.50b}$$

All other terms are the same as those defined by Equation 14.29.

14.2.4 Unsteady Three-dimensional Viscous Incompressible Flow

The procedure described for the two-dimensional problems can easily be extended to a three-dimensional problem. Let us consider the three-dimensional incompressible viscous flow problem given as follows.

Continuity

$$\frac{\partial u}{\partial x}+\frac{\partial v}{\partial y}+\frac{\partial w}{\partial y}=0 \tag{14.51a}$$

x momentum

$$\rho\left(\frac{\partial u}{\partial t}+u\frac{\partial u}{\partial x}+v\frac{\partial u}{\partial y}+w\frac{\partial u}{\partial z}\right)=b_x-\frac{\partial p}{\partial x}+\mu\left(\frac{\partial^2 u}{\partial x^2}+\frac{\partial^2 u}{\partial y^2}+\frac{\partial^2 u}{\partial z^2}\right) \tag{14.51b}$$

y momentum

$$\rho\left(\frac{\partial v}{\partial t}+u\frac{\partial v}{\partial x}+v\frac{\partial v}{\partial y}+w\frac{\partial v}{\partial z}\right)=b_y-\frac{\partial p}{\partial y}+\mu\left(\frac{\partial^2 v}{\partial x^2}+\frac{\partial^2 v}{\partial y^2}+\frac{\partial^2 v}{\partial z^2}\right) \tag{14.51c}$$

z momentum

$$\rho\left(\frac{\partial w}{\partial t}+u\frac{\partial w}{\partial x}+v\frac{\partial w}{\partial y}+ww\right)=b_z-\frac{\partial p}{\partial z}+\mu\left(\frac{\partial^2 w}{\partial x^2}+\frac{\partial^2 w}{\partial y^2}+\frac{\partial^2 w}{\partial z^2}\right) \tag{14.51d}$$

with boundary conditions given in a general form as

1. $u=u_s, \quad v=v_s \quad \text{and} \quad w=w_s \quad \text{on the boundary } s_1$ (14.52a)
2. $\mu\left(\frac{\partial u}{\partial x}\hat{n}_x+\frac{\partial u}{\partial y}\hat{n}_y+\frac{\partial u}{\partial z}\hat{n}_z\right)-p\hat{n}_x=f_{sx}$

$$\mu\left(\frac{\partial v}{\partial x}\hat{n}_x+\frac{\partial v}{\partial y}\hat{n}_y+\frac{\partial v}{\partial z}\hat{n}_z\right)-p\hat{n}_y=f_{sy}$$

$$\mu\left(\frac{\partial w}{\partial x}\hat{n}_x+\frac{\partial w}{\partial y}\hat{n}_y+\frac{\partial w}{\partial z}\hat{n}_z\right)-p\hat{n}_z=f_{sz} \quad \text{on the boundary } s_2 \tag{14.52b}$$

Following the procedure outlined in Sections 14.2.1, 14.2.2 and 14.2.3, the set of element characteristic equations for u, v, w, and p are assembled as

$$\left[c^{(e)}\right]\frac{d\{u^{(e)}\}}{dt}+\left[k_I^{(e)}(u,v,w)+\left[k_\mu^{(e)}\right]\right]\{u^{(e)}\}$$
$$-\left[k_{px}^{(e)}\right]\{p^{(e)}\}+\left\{I_{ix}^{(e)}\right\}-\{f_x^{(e)}\}=0 \tag{14.53a}$$

$$\left[c^{(e)}\right]\frac{d\{v^{(e)}\}}{dt}+\left[k_I^{(e)}(u,v,w)+\left[k_\mu^{(e)}\right]\right]\{v^{(e)}\}$$
$$-\left[k_{py}^{(e)}\right]\{p^{(e)}\}+\left\{I_{iy}^{(e)}\right\}-\left\{f_y^{(e)}\right\}=0 \tag{14.53b}$$

$$\left[c^{(e)}\right]\frac{d\{w^{(e)}\}}{dt}+\left[k_I^{(e)}(u,v,w)+\left[k_\mu^{(e)}\right]\right]\{w^{(e)}\}$$
$$-\left[k_{pz}^{(e)}\right]\{p^{(e)}\}+\left\{I_{iz}^{(e)}\right\}-\{f_z^{(e)}\}=0 \tag{14.53c}$$

$$-\left[k_{px}^{(e)}\right]^{\mathrm{T}}\{u^{(e)}\}-\left[k_{py}^{(e)}\right]^{\mathrm{T}}\{v^{(e)}\}-\left[k_{pz}^{(e)}\right]^{\mathrm{T}}\{w^{(e)}\}=0 \tag{14.53d}$$

and in the matrix form as

$$\begin{bmatrix} c^{(e)} & 0 & 0 & 0 \\ 0 & c^{(e)} & 0 & 0 \\ 0 & 0 & c^{(e)} & 0 \\ 0 & 0 & 0 & 0 \end{bmatrix}\begin{Bmatrix} \dfrac{\partial u}{\partial t} \\ \dfrac{\partial v}{\partial t} \\ \dfrac{\partial w}{\partial t} \\ \dfrac{\partial p}{\partial t} \end{Bmatrix}$$
$$+\begin{bmatrix} k_I^{(e)}(u,v,w)+k_\mu^{(e)} & 0 & 0 & -k_{px}^{(e)} \\ 0 & k_I^{(e)}(u,v,w)+k_\mu^{(e)} & 0 & -k_{py}^{(e)} \\ 0 & 0 & k_I^{(e)}(u,v,w)+k_\mu^{(e)} & -k_{pz}^{(e)} \\ -k_{px}^{(e)\mathrm{T}} & -k_{py}^{(e)\mathrm{T}} & -k_{pz}^{(e)\mathrm{T}} & 0 \end{bmatrix}$$
$$+\begin{Bmatrix} u^{(e)} \\ v^{(e)} \\ w^{(e)} \\ p^{(e)} \end{Bmatrix}+\begin{Bmatrix} I_{ix}^{(e)} \\ I_{iy}^{(e)} \\ I_{iz}^{(e)} \\ 0 \end{Bmatrix}-\begin{Bmatrix} f_{sx}^{(e)} \\ f_{sy}^{(e)} \\ f_{sz}^{(e)} \\ 0 \end{Bmatrix}=0 \tag{14.54a}$$

and in compact notation as

$$C^{(e)}\left\{\frac{\partial U^{(e)}}{\partial t}\right\}+k^{(e)}\{U^{(e)}\}+\{I^{(e)}\}=\{F^{(e)}\} \tag{14.54b}$$

Assembly of all element characteristic equations gives the global system as

$$C\left\{\frac{\partial U}{\partial t}\right\}+K\{U\}=\{F\} \tag{14.54c}$$

14.2.5 Convective Transport

For convection heat and mass transfer, we need to consider the solution of the flow field followed by the solution of the energy or mass concentration equations.

Let us consider the following general set of governing equations and boundary conditions for forced convection problems involving three-dimensional incompressible viscous flow as

Continuity

$$\frac{\partial u}{\partial x} + \frac{\partial v}{\partial y} + \frac{\partial w}{\partial y} = 0 \tag{14.55a}$$

x momentum

$$\rho\left(\frac{\partial u}{\partial t} + u\frac{\partial u}{\partial x} + v\frac{\partial u}{\partial y} + w\frac{\partial u}{\partial z}\right) = b_x - \frac{\partial p}{\partial x} + \mu\left(\frac{\partial^2 u}{\partial x^2} + \frac{\partial^2 u}{\partial y^2} + \frac{\partial^2 u}{\partial z^2}\right) \tag{14.55b}$$

y momentum

$$\rho\left(\frac{\partial v}{\partial t} + u\frac{\partial v}{\partial x} + v\frac{\partial v}{\partial y} + w\frac{\partial v}{\partial z}\right) = b_y - \frac{\partial p}{\partial y} + \mu\left(\frac{\partial^2 v}{\partial x^2} + \frac{\partial^2 v}{\partial y^2} + \frac{\partial^2 v}{\partial z^2}\right) \tag{14.55c}$$

z momentum

$$\rho\left(\frac{\partial w}{\partial t} + u\frac{\partial w}{\partial x} + v\frac{\partial w}{\partial y} + w\frac{\partial w}{\partial z}\right) = b_z - \frac{\partial p}{\partial z} + \mu\left(\frac{\partial^2 w}{\partial x^2} + \frac{\partial^2 w}{\partial y^2} + \frac{\partial^2 w}{\partial z^2}\right) \tag{14.55d}$$

Energy equation

$$C_\phi\left(\frac{\partial \phi}{\partial t} + u\frac{\partial \phi}{\partial x} + v\frac{\partial \phi}{\partial y} + w\frac{\partial \phi}{\partial z}\right) = \Gamma\left(\frac{\partial^2 \phi}{\partial x^2} + \frac{\partial^2 \phi}{\partial y^2} + \frac{\partial^2 \phi}{\partial z^2}\right) + S \tag{14.55e}$$

where ϕ represents a scalar transport quantity such as temperature, T, in the energy equation or concentration, c, in the mass transport equation. S represents a generation term such as the volumetric heat generation, Q''', and/or viscous dissipation term, Φ, in the energy equation, or mass source or sink term in the mass transport equation.

Boundary conditions are given in a general form as

1. $u = u_s, \quad v = v_s$ and $w = w_s$ on the boundary s_1 (14.56a)

2. $$\mu\left(\frac{\partial u}{\partial x}\hat{n}_x + \frac{\partial u}{\partial y}\hat{n}_y + \frac{\partial u}{\partial z}\hat{n}_z\right) - p\hat{n}_x = f_{sx}$$
$$\mu\left(\frac{\partial v}{\partial x}\hat{n}_x + \frac{\partial v}{\partial y}\hat{n}_y + \frac{\partial v}{\partial z}\hat{n}_z\right) - p\hat{n}_y = f_{sy}$$
$$\mu\left(\frac{\partial w}{\partial x}\hat{n}_x + \frac{\partial w}{\partial y}\hat{n}_y + \frac{\partial w}{\partial z}\hat{n}_z\right) - p\hat{n}_z = f_{sz} \quad \text{on the boundary } s_2 \tag{14.56b}$$

3. $\phi = \phi_s$ on the boundary s_1 (14.56c)

4. $$\Gamma\left(\frac{\partial \phi}{\partial x}\hat{n}_x + \frac{\partial \phi}{\partial y}\hat{n}_y + \frac{\partial \phi}{\partial z}\hat{n}_z\right) = f''_{\phi s} \tag{14.56d}$$

where $f''_{\phi s}$ represents a surface flux quantity such as the surface heat flux, q''_s, or surface mass flux, m''_s.

Following the procedure outlined for the ***velocity-pressure*** or ***mixed finite element method***, the weighted residual statement of the equations over an element $A^{(e)}$ is

$$-\int_{A^{(e)}} W_p\left\{\frac{\partial u^*}{\partial x} + \frac{\partial v^*}{\partial y} + \frac{\partial w^*}{\partial z}\right\} dA = 0 \tag{14.57a}$$

$$\int_{A^{(e)}} W_u\left\{\rho\left(\frac{\partial u^*}{\partial t} + u\frac{\partial u^*}{\partial x} + v\frac{\partial u^*}{\partial y} + w\frac{\partial u^*}{\partial z}\right) - b_x + \frac{\partial p^*}{\partial x} - \mu\left(\frac{\partial^2 u^*}{\partial x^2} + \frac{\partial^2 u^*}{\partial y^2} + \frac{\partial^2 u^*}{\partial z^2}\right)\right\} dA = 0 \tag{14.57b}$$

$$\int_{A^{(e)}} W_v\left\{\rho\left(\frac{\partial v^*}{\partial t} + u\frac{\partial v^*}{\partial x} + v\frac{\partial v^*}{\partial y} + w\frac{\partial v^*}{\partial z}\right) - b_y + \frac{\partial p^*}{\partial y} - \mu\left(\frac{\partial^2 v^*}{\partial x^2} + \frac{\partial^2 v^*}{\partial y^2} + \frac{\partial^2 v^*}{\partial z^2}\right)\right\} dA = 0 \tag{14.57c}$$

$$\int_{A^{(e)}} W_w\left\{\rho\left(\frac{\partial w^*}{\partial t} + u\frac{\partial w^*}{\partial x} + v\frac{\partial w^*}{\partial y} + w\frac{\partial w^*}{\partial z}\right) - b_z + \frac{\partial p^*}{\partial z} - \mu\left(\frac{\partial^2 w^*}{\partial x^2} + \frac{\partial^2 w^*}{\partial y^2} + \frac{\partial^2 w^*}{\partial z^2}\right)\right\} dA = 0 \tag{14.57d}$$

$$\int_{A^{(e)}} W_\phi \left\{ C\left(\frac{\partial \phi}{\partial t} + u\frac{\partial \phi^*}{\partial x} + v\frac{\partial \phi^*}{\partial y} + w\frac{\partial \phi^*}{\partial z}\right) - S \right.$$

$$\left. -\mu\left(\frac{\partial^2 \phi^*}{\partial x^2} + \frac{\partial^2 \phi^*}{\partial y^2} + \frac{\partial^2 \phi^*}{\partial z^2}\right)\right\} dA = 0 \tag{14.57e}$$

The approximate solutions for velocity, pressure, and transport quantity are assumed to be

$$u^* = \sum_{I=1}^{n_p} N_I(x) u_I(t) = [N(x)]\{u^{(e)}(t)\} \tag{14.58a}$$

$$v^* = \sum_{I=1}^{n_p} N_I(x) v_I(t) = [N(x)]\{v^{(e)}(t)\} \tag{14.58b}$$

$$p^* = \sum_{I=1}^{n_p} N_{pI}(x) p_I(t) = [N_p(x)]\{p^{(e)}(t)\} \tag{14.58c}$$

$$\phi^* = \sum_{I=1}^{n_p} N_{\phi I}(x) \phi_I(t) = [N_\phi(x)]\{\phi^{(e)}(t)\} \tag{14.58d}$$

It can be seen that three different shape functions are used for velocity, pressure, and scalar transport quantity requiring three different finite element mesh. However, since the form of the momentum equations and scalar transport equations are of the same form, we could also use the same type of shape functions for the velocity (u, v, w) and scalar transport (ϕ).

In the ***Galerkin finite element method*** the weighting functions are assumed to be the same as the shape functions used for the approximation functions given by Equation 14.58, i.e.

$$W_u = W_v = W_w = [N]^{\mathrm{T}}, \quad W_p = [N_P]^{\mathrm{T}}, \quad W_\phi = [N_\phi]^{\mathrm{T}} \tag{14.59}$$

where $[N]$, $[N_p]$, and $[N_\phi]$ are the shape functions associated with velocity, pressure, and scalar transport approximation functions respectively.

The weighted residual statements given by Equation 14.57 are transformed into weak forms by integrating the integrals. This is followed by substitution of approximation functions given by Equations 14.58 and formation of element characteristic equations of the form

$$[c^{(e)}]\frac{d\{u^{(e)}\}}{dt} + \left[k_I^{(e)}(u, v, w) + \left[k_\mu^{(e)}\right]\right]\{u^{(e)}\}$$

$$-\left[k_{px}^{(e)}\right]\{p^{(e)}\} + \left\{I_{ix}^{(e)}\right\} - \{f_x^{(e)}\} = 0 \tag{14.60a}$$

$$[c^{(e)}]\frac{d\{v^{(e)}\}}{dt} + \left[k_I^{(e)}(u, v, w) + \left[k_\mu^{(e)}\right]\right]\{v^{(e)}\}$$

$$-\left[k_{py}^{(e)}\right]\{p^{(e)}\} + \left\{I_{iy}^{(e)}\right\} - \left\{f_y^{(e)}\right\} = 0 \tag{14.60b}$$

$$\left[c^{(e)}\right]\frac{d\{w^{(e)}\}}{dt}+\left[k_I^{(e)}(u,v,w)+\left[k_\mu^{(e)}\right]\right]\{w^{(e)}\}$$
$$-\left[k_{pz}^{(e)}\right]\{p^{(e)}\}+\left\{I_{iz}^{(e)}\right\}-\{f_z^{(e)}\}=0 \tag{14.60c}$$

$$-\left[k_{px}^{(e)}\right]^{\mathrm{T}}\{u^{(e)}\}-\left[k_{py}^{(e)}\right]^{\mathrm{T}}\{v^{(e)}\}-\left[k_{pz}^{(e)}\right]^{\mathrm{T}}\{w^{(e)}\}=0 \tag{14.60d}$$

$$\left[d^{(e)}\right]\frac{d\{\phi^{(e)}\}}{dt}+\left[k_{\phi I}^{(e)}(u,v,w)+\left[k_\Gamma^{(e)}\right]\right]\{\phi^{(e)}\}+\left\{I_{i\phi}^{(e)}\right\}-\left\{f_\phi^{(e)}\right\}=0 \tag{14.60e}$$

and in a matrix form

$$\begin{bmatrix} c^{(e)} & 0 & 0 & 0 & 0 \\ 0 & c^{(e)} & 0 & 0 & 0 \\ 0 & 0 & c^{(e)} & 0 & 0 \\ 0 & 0 & 0 & 0 & 0 \\ 0 & 0 & 0 & 0 & d^{(e)} \end{bmatrix}\begin{Bmatrix} \dfrac{\partial u}{\partial t} \\ \dfrac{\partial v}{\partial t} \\ \dfrac{\partial w}{\partial t} \\ \dfrac{\partial p}{\partial t} \\ \dfrac{\partial \phi}{\partial t} \end{Bmatrix}$$
$$+\begin{bmatrix} k_I+k_\mu & 0 & 0 & 0 & -k_{px}^{\mathrm{T}} \\ 0 & k_I+k_\mu & 0 & 0 & -k_{py}^{\mathrm{T}} \\ 0 & 0 & k_I+k_\mu & 0 & -k_{pz}^{\mathrm{T}} \\ -k_{px}^{\mathrm{T}} & -k_{py}^{\mathrm{T}} & -k_{pz}^{\mathrm{T}} & 0 & 0 \\ 0 & 0 & 0 & 0 & k_{\phi l}+k_\Gamma \end{bmatrix} 0 \begin{Bmatrix} u \\ v \\ w \\ p \\ \phi \end{Bmatrix}+\begin{Bmatrix} I_{ix} \\ I_{iy} \\ I_{iz} \\ 0 \\ I_{i\phi} \end{Bmatrix}$$
$$=\begin{Bmatrix} f_x \\ f_y \\ f_z \\ 0 \\ f_\phi \end{Bmatrix} \tag{14.61a}$$

and in compact notation as

$$C^{(e)}\left\{\frac{\partial U^{(e)}}{\partial t}\right\}+k^{(e)}\{U^{(e)}\}+\{I^{(e)}\}=\{F^{(e)}\} \tag{14.61b}$$

where

$$c^{(e)} = \int_A \rho[N]^{\mathrm{T}}[N]dA, \quad d^{(e)} = \int_A C_\phi[N]^{\mathrm{T}}[N]dA$$

$$\left[k_I^{(e)}(u,v)\right] = \int_{A^{(e)}} \rho[N^{\mathrm{T}}][N]\left(\left\{u^{(e)}\right\}\frac{\partial[N]}{\partial x} + \left\{v^{(e)}\right\}\frac{\partial[N]}{\partial y}\right)dA$$

$$\left[k_{\phi I}^{(e)}(u,v)\right] = \int_{A^{(e)}} C\left[N_\phi^{\mathrm{T}}\right][N_\phi]\left(\left\{u^{(e)}\right\}\frac{\partial[N_\phi]}{\partial x} + \left\{v^{(e)}\right\}\frac{\partial[N_\phi]}{\partial y}\right)dA$$

$$k_\mu^{(e)} = \int_{A^{(e)}} \mu\frac{\partial[N]}{\partial x}\frac{\partial[N]^{\mathrm{T}}}{\partial x}dA + \int_{A^{(e)}} \mu\frac{\partial[N]}{\partial y}\frac{\partial[N]^{\mathrm{T}}}{\partial y}dA + \int_{A^{(e)}} \mu\frac{\partial[N]}{\partial z}\frac{\partial[N]^{\mathrm{T}}}{\partial z}dA$$

$$k_\Gamma^{(e)} = \int_{A^{(e)}} \Gamma\frac{\partial[N]}{\partial x}\frac{\partial[N]^{\mathrm{T}}}{\partial x}dA + \int_{A^{(e)}} \Gamma\frac{\partial[N]}{\partial y}\frac{\partial[N]^{\mathrm{T}}}{\partial y}dA + \int_{A^{(e)}} \Gamma\frac{\partial[N]}{\partial z}\frac{\partial[N]^{\mathrm{T}}}{\partial z}dA$$

$$\left[k_{px}^{(e)}\right] = \int_{A^{(e)}} [N_p]\frac{\partial[N]^{\mathrm{T}}}{\partial x}dA, \quad \left[k_{py}^{(e)}\right] = \int_{A^{(e)}} [N_p]\frac{\partial[N]^{\mathrm{T}}}{\partial y}dA$$

$$\left\{f_x^{(e)}\right\} = \int_{A^{(e)}} [N]^{\mathrm{T}}b_x\,dA - \oint_s [N]^{\mathrm{T}}f_{sx}\,ds, \quad \left\{f_y^{(e)}\right\} = \int_{A^{(e)}} [N]^{\mathrm{T}}b_y\,dA - \oint_s [N]^{\mathrm{T}}f_{sy}\,ds$$

$$\left\{f_z^{(e)}\right\} = \int_{A^{(e)}} [N]^{\mathrm{T}}b_z\,dA - \oint_s [N]^{\mathrm{T}}f_{sz}\,ds, \quad \left\{f_\phi^{(e)}\right\} = \int_{A^{(e)}} [N_\phi]^{\mathrm{T}}S\,dA - \oint_s [N_\phi]^{\mathrm{T}}f''_{\phi s}\,ds$$

$$C^{(e)} = \begin{bmatrix} c^{(e)} & 0 & 0 & 0 & 0 \\ 0 & c^{(e)} & 0 & 0 & 0 \\ 0 & 0 & c^{(e)} & 0 & 0 \\ 0 & 0 & 0 & 0 & 0 \\ 0 & 0 & 0 & 0 & d^{(e)} \end{bmatrix}, \quad \left\{U^{(e)}\right\} = \begin{Bmatrix} u^{(e)} \\ v^{(e)} \\ w^{(e)} \\ p^{(e)} \\ \phi^{(e)} \end{Bmatrix}, \quad \left\{F^{(e)}\right\} = \begin{Bmatrix} f_x \\ f_y \\ f_z \\ 0 \\ f_\phi \end{Bmatrix} \tag{14.61c}$$

Assembly of all element characteristic equations gives the global system as

$$C\left\{\frac{\partial U}{\partial t}\right\} + K\{U\} = \{F\} \tag{14.62}$$

14.3 SOLUTION METHODS

Before selecting any particular solution techniques, let us first examine the nature and type of element and global system of equations given by Equations 14.61 and 14.62. For unsteady state problems, Equation 14.62 represents a set of first-order nonlinear differential equations. The nonlinearity is caused due to the nonlinear characteristics of the matrices, $k_I^{(e)}(u, v, w)$ and $k_{\phi I}^{(e)}(u, v, w)$, and by variable material properties such as Γ, μ, ρ, and C. Also, the system is nonsymmetric because the element characteristic matrix due to convection is also nonsymmetric. For constant material properties and very low fluid velocities, the system becomes linear and symmetric. Choice of solution technique differs depending on whether the system is linear or nonlinear, has a degree of nonlinearity, and is symmetric or nonsymmetric. For unsteady problems, the nonlinear differential equations are reduced to a system of nonlinear algebraic equations by using the time approximation scheme. The general solution techniques can be categorized into two basic types: (1) a linear system of equations that is solved either by a direct solver or by iterative methods; or (2) for nonlinear problems, the system of nonlinear equations is first linearized or reduced to a system of linear algebraic equations, and then an iterative technique is used to solve the linear system. Let us begin with the discussion of these methods for steady state problems first.

14.3.1 Steady State Problems

For steady state problems, the global system of equations is given as

$$K\{U\} = \{F\} \tag{14.63}$$

and component forms as

$$\left[K_I(u, v, w) + K_\mu\right]\{u\} - \left[K_{px}\right]\{p\} - \{F_x\} = 0 \tag{14.64a}$$

$$\left[K_I(u, v, w) + K_\mu\right]\{v\} - \left[K_{py}\right]\{p\} - \left\{F_y\right\} = 0 \tag{14.64b}$$

$$\left[K_I(u, v, w) + K_\mu\right]\{w\} - \left[K_{pz}\right]\{p\} - \{F_z\} = 0 \tag{14.64c}$$

$$-\left[k_{px}\right]^{\mathrm{T}}\{u\} - \left[k_{py}\right]^{\mathrm{T}}\{v\} - \left[k_{pz}\right]^{\mathrm{T}}\{w\} = 0 \tag{14.64d}$$

The solution to this set of nonlinear equations is usually solved by methods such as the ***Picard method***, the ***Newton–Raphson method*** and the ***velocity-pressure correction method***.

Picard Method The Picard method for the solution of equations is based on the ***fixed-point iteration method*** or ***successive substitution method*** used for the solution of nonlinear algebraic equations (Chapra and Canale, 2002; Rao, 2002). The iterative algorithm for the ***Picard method*** is given as

$$K(U^i)\{U^{i+1}\} = \{F(U^i)\} \tag{14.65}$$

where U^i is the unknown vector at the previous iteration step and U^{i+1} is the unknown vector at the present iteration step. The iteration process is carried out until the convergence criteria given as $(U^{i+1} - U^i)/U^{i+1} \leq \varepsilon_s$ is satisfied. In order to enhance the convergence rate and/or eliminate any oscillation in the convergence process, a relaxation scheme is used as

$$U^{i+1} = \lambda U^i + (1 - \lambda)U^{i+1} \tag{14.66}$$

The inherent slow convergence rate in the method allows a poor initial guess vector. For example, an initial starting vector $U^0 = 0$ can always be used if no better guess can be made. However, a better initial guess will always enhance the convergence rate.

Also note that Equation 14.65 basically represents a system of linear algebraic equations, which can be solved by using methods used for a linear system of equations, i.e., either by a direct solver such as the Gauss elimination or by an iterative solver such as the Gauss–Seidel method.

Newton–Raphson Method The Newton–Raphson method is one of the most efficient solution algorithms for solving nonlinear equations (Chapra and Canale 2002). The algorithm is derived by using a truncated Taylor series expansion. In order to derive the algorithm, let us rewrite Equation 14.63 as

$$g(U) = K\{U\} - \{F\} = 0 \tag{14.67}$$

and in component form as

$$g_u(u, v, w, p) = \left[K_I(u, v, w) + K_\mu\right]\{u\} - \left[K_{px}\right]\{p\} - \{F_x\} = 0 \tag{14.68a}$$

$$g_v(u, v, w, p) = \left[K_I(u, v, w) + K_\mu\right]\{v\} - \left[K_{py}\right]\{p\} - \{F_y\} = 0 \tag{14.68b}$$

$$g_w(u, v, w, p) = \left[K_I(u, v, w) + K_\mu\right]\{w\} - \left[K_{pz}\right]\{p\} - \{F_z\} = 0 \tag{14.68c}$$

$$g_p(u, v, w, p) = -\left[k_{px}\right]^{\mathrm{T}}\{u\} - \left[k_{py}\right]^{\mathrm{T}}\{v\} - \left[k_{pz}\right]^{\mathrm{T}}\{w\} = 0 \tag{14.68d}$$

Truncated multi-variable Taylor series expansion of the function $g(U)$ is given as

$$g(U^{i+1}) = g(U^i) + \left.\frac{\partial g}{\partial U}\right|^i (U^{i+1} - U^i) + \cdots \tag{14.69}$$

and in component form as

$$g_u(u^{i+1}, v^{i+1}, w^{i+1}, p^{i+1}) = g_u(u^i, v^i, w^i) + \left.\frac{\partial g_u}{\partial u}\right|^i (u^{i+1} - u^i) + \left.\frac{\partial g_u}{\partial v}\right|^i (v^{i+1} - v^i)$$
$$+ \left.\frac{\partial g_u}{\partial w}\right|^i (w^{i+1} - w^i) + \left.\frac{\partial g_u}{\partial p}\right|^i (p^{i+1} - p^i) + \cdots \quad (14.70a)$$

$$g_v(u^{i+1}, v^{i+1}, w^{i+1}, p^{i+1}) = g_v(u^i, v^i, w^i) + \left.\frac{\partial g_v}{\partial u}\right|^i (u^{i+1} - u^i) + \left.\frac{\partial g_v}{\partial v}\right|^i (v^{i+1} - v^i)$$
$$+ \left.\frac{\partial g_v}{\partial w}\right|^i (w^{i+1} - w^i) + \left.\frac{\partial g_v}{\partial p}\right|^i (p^{i+1} - p^i) + \cdots \quad (14.70b)$$

$$g_w(u^{i+1}, v^{i+1}, w^{i+1}, p^{i+1}) = g_w(u^i, v^i, w^i) + \left.\frac{\partial g_w}{\partial u}\right|^i (u^{i+1} - u^i) + \left.\frac{\partial g_w}{\partial v}\right|^i (v^{i+1} - v^i)$$
$$+ \left.\frac{\partial g_w}{\partial w}\right|^i (w^{i+1} - w^i) + \left.\frac{\partial g_w}{\partial p}\right|^i (p^{i+1} - p^i) \cdots \quad (14.70c)$$

$$g_p(u^{i+1}, v^{i+1}, w^{i+1}, p^{i+1}) = g_p(u^i, v^i, w^i) + \left.\frac{\partial g_p}{\partial u}\right|^i (u^{i+1} - u^i) + \left.\frac{\partial g_p}{\partial v}\right|^i (v^{i+1} - v^i)$$
$$+ \left.\frac{\partial g_p}{\partial w}\right|^i (w^{i+1} - w^i) + \left.\frac{\partial g_p}{\partial p}\right|^i (p^{i+1} - p^i) + \cdots \quad (14.70d)$$

Equations 14.70 constitute a system of linear equations, which can be written in matrix form as

$$\begin{bmatrix} \frac{\partial g_u}{\partial u} & \frac{\partial g_u}{\partial v} & \frac{\partial g_u}{\partial w} & \frac{\partial g_u}{\partial p} \\ \frac{\partial g_v}{\partial u} & \frac{\partial g_v}{\partial v} & \frac{\partial g_v}{\partial w} & \frac{\partial g_v}{\partial p} \\ \frac{\partial g_w}{\partial u} & \frac{\partial g_w}{\partial v} & \frac{\partial g_w}{\partial w} & \frac{\partial g_w}{\partial p} \\ \frac{\partial g_p}{\partial u} & \frac{\partial g_p}{\partial u} & \frac{\partial g_p}{\partial u} & \frac{\partial g_p}{\partial u} \end{bmatrix} \begin{Bmatrix} (u^{i+1} - u^i) \\ (v^{i+1} - v^i) \\ (w^{i+1} - w^i) \\ (p^{i+1} - p^i) \end{Bmatrix} = \begin{Bmatrix} g_u(u^i, v^i, w^i, p^i) \\ g_v(u^i, v^i, w^i, p^i) \\ g_w(u^i, v^i, w^i, p^i) \\ g_p(u^i, v^i, w^i, p^i) \end{Bmatrix} \quad (14.71)$$

Defining the ***Jacobian matrix*** as

$$J = \begin{bmatrix} \frac{\partial g_u}{\partial u} & \frac{\partial g_u}{\partial v} & \frac{\partial g_u}{\partial w} & \frac{\partial g_u}{\partial p} \\ \frac{\partial g_v}{\partial u} & \frac{\partial g_v}{\partial v} & \frac{\partial g_v}{\partial w} & \frac{\partial g_v}{\partial p} \\ \frac{\partial g_w}{\partial u} & \frac{\partial g_w}{\partial v} & \frac{\partial g_w}{\partial w} & \frac{\partial g_w}{\partial p} \\ \frac{\partial g_p}{\partial u} & \frac{\partial g_p}{\partial u} & \frac{\partial g_p}{\partial u} & \frac{\partial g_p}{\partial u} \end{bmatrix} \tag{14.72}$$

and solving, we get the Newton–Raphson scheme as

$$U^{i+1} = U^i - J^{-1}(U^i)g(U^i) \tag{14.73}$$

and in component form as

$$u^{i+1} = u^i - J^{-1}(u^i, v^i, w^i, p^i)g_u(u^i, v^i, w^i, p^i) \tag{14.74a}$$

$$v^{i+1} = v^i - J^{-1}(u^i, v^i, w^i, p^i)g_v(u^i, v^i, w^i, p^i) \tag{14.74b}$$

$$w^{i+1} = w^i - J^{-1}(u^i, v^i, w^i, p^i)g_w(u^i, v^i, w^i, p^i) \tag{14.74c}$$

$$p^{i+1} = p^i - J^{-1}(u^i, v^i, w^i, p^i)g_p(u^i, v^i, w^i, p^i) \tag{14.74d}$$

The Jacobian for the problem under consideration can be evaluated from Equation 14.68 as

$$J = \begin{bmatrix} K_I(u^i, v^i, w^i) + K_\mu & 0 & 0 & -K_{px} \\ 0 & K_I(u^i, v^i, w^i) + K_\mu & 0 & -K_{py} \\ 0 & 0 & K_I(u^i, v^i, w^i) + K_\mu & -K_{pz} \\ -K_{px}^{\mathrm{T}} & -K_{py}^{\mathrm{T}} & -K_{pz}^{\mathrm{T}} & 0 \end{bmatrix} \tag{14.75}$$

Velocity-Pressure Correction Method The concept of the velocity-pressure correction method in finite element method originated from algorithms such as SIMPLE and SIMPLER (Patankar, 1980) that are successfully used in the finite difference solution of fluid flow and heat transfer problems, as discussed in Chapter 8. In order to describe this method let us consider the three-dimensional steady state fluid flow equations given as

$$K\{U\} = \{F\} \tag{14.63}$$

and in component form as

$$[K_I(u, v, w) + K_\mu]\{u\} - [K_{px}]\{p\} - \{F_x\} = 0 \tag{14.76a}$$

$$[K_I(u, v, w) + K_\mu]\{v\} - [K_{py}]\{p\} - \{F_y\} = 0 \tag{14.76b}$$

$$[K_I(u, v, w) + K_\mu]\{w\} - [K_{pz}]\{p\} - \{F_z\} = 0 \tag{14.76c}$$

$$-[K_{px}]^{\mathrm{T}}\{u\} - \left[K_{py}^{()}\right]^{\mathrm{T}}\{v\} - \left[K_{pz}^{()}\right]^{\mathrm{T}}\{w\} = 0 \tag{14.76d}$$

Equations 14.76a, 14.76b, and 14.76c are written explicitly for u, v, and w as

$$\{u\} = [K_I(u, v, w) + K_\mu]^{-1}([K_{px}]\{p\} - \{F_x\}) \tag{14.77a}$$

$$\{v\} = [K_I(u, v, w) + K_\mu]^{-1}([K_{py}]\{p\} - \{F_y\}) \tag{14.77b}$$

$$\{w\} = [K_I(u, v, w) + K_\mu]^{-1}([K_{pz}]\{p\} - \{F_z\}) \tag{14.77c}$$

Substituting Equation 14.77 into Equation 14.76d, we get

$$\begin{aligned} &-[K_{px}]^{\mathrm{T}}[K_I(u, v, w) + K_\mu]^{-1}([K_{px}]\{p\} - \{F_x\}) \\ &-[K_{py}]^{\mathrm{T}}[K_I(u, v, w) + K_\mu]^{-1}([K_{py}]\{p\} - \{F_y\}) \\ &-[K_{pz}]^{\mathrm{T}}[K_I(u, v, w) + K_\mu]^{-1}([K_{pz}]\{p\} - \{F_z\}) = 0 \end{aligned}$$

or

$$\begin{aligned} &\left(\left[K_{px}^{\mathrm{T}}\right][K_I(u, v, w) + K_\mu]^{-1} + \left[K_{px}^{\mathrm{T}}\right] + \left[K_{py}^{\mathrm{T}}\right][K_I(u, v, w) + K_\mu]^{-1}\left[K_{pu}^{\mathrm{T}}\right]\right. \\ &\quad \left.+\left[K_{pz}^{\mathrm{T}}\right][K_I(u, v, w) + K_\mu]^{-1}\left[K_{pz}^{\mathrm{T}}\right]\right)p = -\left[K_{px}^{\mathrm{T}}\right][K_I(u, v, w) + K_\mu]^{-1}\{F_x\} \\ &\quad - \left[K_{py}^{\mathrm{T}}\right][K_I(u, v, w) + K_\mu)]^{-1}\{F_y\} \\ &\quad - \left[K_{pz}^{\mathrm{T}}\right][K_I(u, v, w) + K_\mu)]^{-1}\{F_z\} \end{aligned} \tag{14.78}$$

Equation 14.78 represents a direct form of the pressure discretization equation. However, an approximation is often made by approximating the $[K_I + K_\mu]$ by matrix $[\hat{K}]$ and retaining only the diagonal elements. The approximate pressure discretization equation is written as

$$\begin{aligned} &\left(\left[K_{px}^{\mathrm{T}}\right]\left[\hat{K}\right]^{-1} + \left[K_{px}^{\mathrm{T}}\right] + \left[K_{py}^{\mathrm{T}}\right]\left[\hat{K}\right]^{-1}[K_{py}] + \left[K_{pz}^{\mathrm{T}}\right]\left[\hat{K}\right]^{-1}[K_{pz}]\right)p^* \\ &\quad = -\left[K_{px}^{\mathrm{T}}\right]\left[\hat{K}\right]^{-1}\{F_x\} - \left[K_{py}^{\mathrm{T}}\right]\left[\hat{K}\right]^{-1}\{F_y\} - \left[K_{pz}^{\mathrm{T}}\right]\left[\hat{K}\right]^{-1}\{F_z\} \end{aligned} \tag{14.79}$$

where the superscript * is used to indicate an approximate quantity. The matrices $\hat{K}$ and force vector F are evaluated based on velocities u^i, v^i, and w^i at previous iteration step. A relaxation scheme is also performed at this time step to establish the pressure estimate at the present iteration step as

$$p^{i+1} = \lambda_p p^i + (1 - \lambda_p)p^* \tag{14.80}$$

With the known pressure values, an estimate of the velocity components is now given by Equation 14.76 as

$$\left[K_I(u, v, w) + K_\mu\right]\{u^*\} = \left[K_{px}\right]\left\{p^{i+1}\right\} + \{F_x\} \tag{14.81a}$$

$$\left[K_I(u, v, w) + K_\mu\right]\{v^*\} - \left[K_{py}\right]\left\{p^{i+1}\right\} + \left\{F_y\right\} = 0 \tag{14.81b}$$

$$\left[K_I(u, v, w) + K_\mu\right]\{w^*\} = \left[K_{pz}\right]\left\{p^{i+1}\right\} + \{F_z\} \tag{14.81c}$$

In the next step, a discretization equation for the corrected pressure is given as

$$\begin{aligned}&\left(\left[K_{px}^{\mathrm{T}}\right]\left[\hat{K}\right]^{-1}\left[K_{px}\right] + \left[K_{px}^{\mathrm{T}}\right] + \left[K_{py}^{\mathrm{T}}\right]\left[\hat{K}\right]^{-1}\left[K_{py}\right] + \left[K_{pz}^{\mathrm{T}}\right]\left[\hat{K}\right]^{-1}\left[K_{pz}\right]\right)p \\ &\quad = -\left[K_{px}^{\mathrm{T}}\right]u^* - \left[K_{py}^{\mathrm{T}}\right]v^* - \left[K_{pz}^{\mathrm{T}}\right]w^*\end{aligned} \tag{14.82}$$

Finally, the velocity components are corrected by the equations

$$u^{i+1} = u^* + \left[\hat{K}\right]^{-1}\left[K_{px}\right]p \tag{14.83a}$$

$$v^{i+1} = v^* + \left[\hat{K}\right]^{-1}\left[K_{py}\right]p \tag{14.83b}$$

$$w^{i+1} = w^* + \left[\hat{K}\right]^{-1}\left[K_{pz}\right]p \tag{14.83c}$$

Equations 14.79 to 14.83 are used iteratively until convergence is achieved. An algorithm based on these equations and concepts is given as follows.

1. Guess the velocity field u, v, and w.
2. Solve the pressure Equation 14.79 for the pressure field, p^*.
3. Use Equation 14.80 to update the pressure estimate at the present iteration using relaxation.
4. Use this pressure estimate to solve discretization Equations 14.81 for u^*, v^*, and w^*.

5. Solve for corrected pressure using Equation 14.82.
6. Calculate velocity components u, v, and w from their starred components and corrected pressure using Equation 14.83.
7. Solve the discretization equation for other scalar quantities such as temperature or mass concentration.
8. Repeat Steps 2 to 7 with the new velocity estimates until the solution converges within a specified tolerance limit.

14.3.2 Unsteady State Problem

For the unsteady state problems, the governing system of equations is given as

$$C\left\{\frac{\partial U}{\partial t}\right\} + K\{U\} = \{F\} \tag{14.62}$$

and in component form as

$$[c]\frac{d\{u^{(e)}\}}{dt} + \left[k_I^{(e)}(u, v, w) + \left[k_\mu^{(e)}\right]\right]\{u^{(e)}\} - [K_{px}]\{p\} - \{F_x\} = 0 \tag{14.84a}$$

$$[c]\frac{d\{v^{(e)}\}}{dt} + \left[k_I^{(e)}(u, v, w) + \left[k_\mu^{(e)}\right]\right]\{v^{(e)}\} - [K_{py}]\{p\} - \{F_y\} = 0 \tag{14.84b}$$

$$[c]\frac{d\{w^{(e)}\}}{dt} + \left[k_I^{(e)}(u, v, w) + \left[k_\mu^{(e)}\right]\right]\{w^{(e)}\} - [K_{pz}]\{p\} - \{F_z\} = 0 \tag{14.84c}$$

$$-\left[k_{px}^{(e)}\right]^{\mathrm{T}}\{u^{(e)}\} - \left[k_{py}^{(e)}\right]^{\mathrm{T}}\{v^{(e)}\} - \left[k_{pz}^{(e)}\right]^{\mathrm{T}}\{w^{(e)}\} = 0 \tag{14.84d}$$

The first-order ordinary differential equations representing the system of global characteristic equations are of the same form as for the characteristic Equation 13.11 for the diffusion problem, and so the equations can be discretized using similar time approximation schemes as discussed in Chapter 13. Since the equations do not involve any time derivative for pressure, complete explicit time integration is not possible. If we use an implicit time approximation, the system of equations given by Equation 14.84 reduces to

$$\left([c^{(e)}] + \Delta t[k^{(e)}]\right)\{u^{(e)}\}^{m+1} = [c^{(e)}]\{u^{(e)}\}^m - \left[k_{px}^{(e)}\right]\{p^{(e)}\}^{m+1} + \Delta t\{I_{ix}\}^{m+1} + \Delta t\{f_x^{(e)}\}^{m+1} \tag{14.85a}$$

$$\left([c^{(e)}] + \Delta t[k^{(e)}]\right)\{v^{(e)}\}^{m+1} = [c^{(e)}]\{v^{(e)}\}^m - \left[k_{py}^{(e)}\right]\{p^{(e)}\}^{m+1} + \Delta t\{I_{iy}\}^{m+1} + \Delta t\left\{f_y^{(e)}\right\}^{m+1} \tag{14.85b}$$

$$\left([c^{(e)}] + \Delta t[k^{(e)}]\right)\{w^{(e)}\}^{m+1} = [c^{(e)}]\{w^{(e)}\}^m - \left[k_{pz}^{(e)}\right]\{p^{(e)}\}^{m+1} + \Delta t\{I_{iz}\}^{m+1} + \Delta t\{f_z^{(e)}\}^{m+1} \tag{14.85c}$$

Note that with the use of the time approximation, the equations transform into a set of nonlinear algebraic equations. The procedure discussed for solving nonlinear and linear systems of equations can now be used to solve Equation 14.84. More discussions on time approximation schemes, including an explicit time integration scheme, are given in Reddy and Gartling (1994).

PROBLEMS

14.1 Derive the element characteristic equation for all elements in the example Problem 14.1. Show the global system of equations for the mesh shown.

14.2 Derive the element characteristic equation for u, v, and pressure p using the bilinear rectangular element for velocity components and constant pressure at the center of the rectangle based on Equations 14.28 and 14.30.

14.3 Consider a fully developed flow and heat transfer in a duct of rectangular cross-section as shown in the figure.

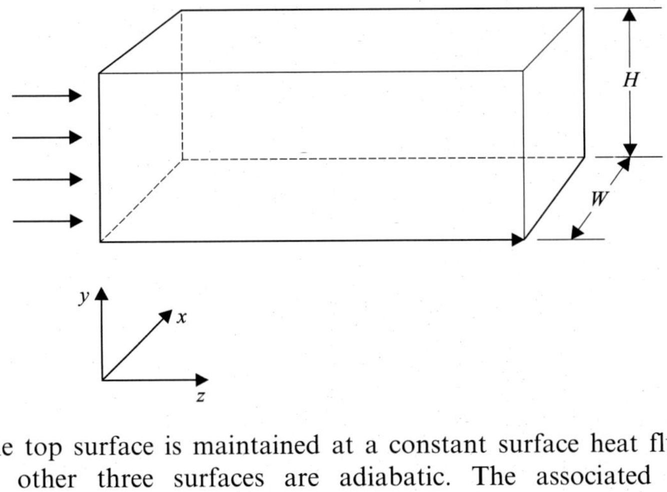

The top surface is maintained at a constant surface heat flux q''_w and all other three surfaces are adiabatic. The associated governing equations and boundary conditions for the fully developed flow and heat transfer with constant surface heat flux are given as follows.

Momentum

$$\nu\left(\frac{\partial^2 w}{\partial x^2}+\frac{\partial^2 w}{\partial y^2}\right)=-\frac{1}{\rho}\frac{\partial \bar{p}}{\partial z}$$

Energy

$$\frac{k}{\rho c_p}\left(\frac{\partial^2 T}{\partial x^2}+\frac{\partial^2 T}{\partial y^2}\right)=w\frac{\partial \bar{T}}{\partial z}$$

The pressure gradient $d\bar{p}/dz$ is assumed to be known. The temperature gradient $\partial\bar{T}/\partial x$ is constant and given based on an energy balance as

$$\frac{\partial \bar{T}}{\partial x} = \frac{q''_w}{\rho c_p \bar{w}}$$

where $\bar{w}$ and $\bar{T}$ are the average velocity and average temperature given as

$$\bar{w} = \frac{1}{WH}\int_A w\,dA \quad \text{and} \quad \bar{T} = \frac{1}{WH\bar{w}}\int_A wT\,dA$$

Boundary conditions

No-slip velocity on all walls, i.e., $w = 0$ on all walls.

Constant surface heat flux on top wall, i.e., at $y = H$, $k\partial T/\partial x = q''_w$.

Rest of the walls are adiabatic, i.e., $\partial T/\partial n = 0$.

Derive the element characteristic equations for w and T using Galerkin's weighted residual formulation and using a mesh consisting of bilinear rectangular elements. Present the sequence of steps to be used for calculating velocity and temperature field and convection heat transfer coefficient.

14.4 Consider free-convection motion and heat transfer in a two-dimensional rectangular chamber as shown. The free convection motion is created by maintaining the left wall at a high temperature T_H and right surface at a low temperature T_C. The top and bottom are assumed as insulated. The governing equations for mass, momentum, and energy with Boussinesq assumption and for Newtonian fluid are given as follows.

Continuity

$$\frac{\partial u}{\partial x} + \frac{\partial v}{\partial y} = 0$$

x momentum

$$u\frac{\partial u}{\partial x} + v\frac{\partial u}{\partial y} = -\frac{1}{\rho}\frac{\partial p}{\partial x} + \nu\left(\frac{\partial^2 u}{\partial x^2} + \frac{\partial^2 u}{\partial y^2}\right)$$

y momentum

$$u\frac{\partial v}{\partial x} + v\frac{\partial v}{\partial y} = -\frac{1}{\rho}\frac{\partial p}{\partial y} + \nu\left(\frac{\partial^2 v}{\partial x^2} + \frac{\partial^2 v}{\partial y^2}\right) - g[1 - \beta(T - T_0)]$$

Energy

$$u\frac{\partial T}{\partial x} + v\frac{\partial T}{\partial y} = \alpha\left(\frac{\partial^2 T}{\partial x^2} + \frac{\partial^2 T}{\partial y^2}\right)$$

Boundary conditions

Velocity: No slip condition on all walls

Temperature

1. at $x = 0, 0 < y < H, T(0, y) = T_H$
2. at $x = W, 0 < y < H, T(W, y) = T_C$
3. at $y = 0, 0 < x < W, \left.\frac{\partial T}{\partial y}\right|_{y=0} = 0$
4. at $y = H, 0 < x < W, \left.\frac{\partial T}{\partial y}\right|_{y=H} = 0$

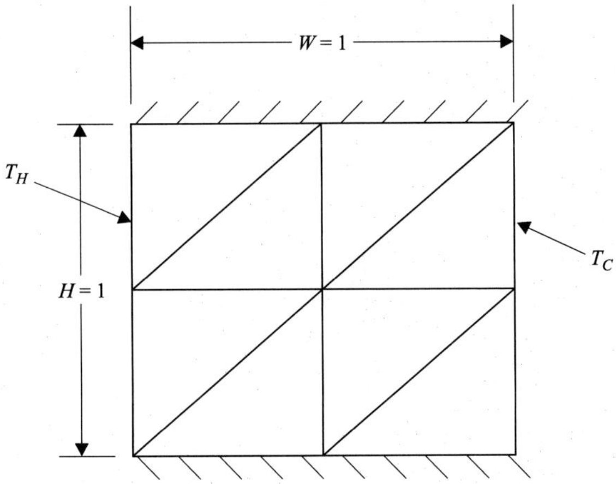

(a) Derive the element characteristic equations for u, v, p, and T using a mesh consisting of linear triangular elements as shown. Show the system of equations for each variables.
(b) Derive the correction equations for the u, v, and p.
(c) Develop a computer program implementing the velocity–pressure algorithm for velocity and pressure coupling and a direct solver for solving the system of equations for each variable.

14.5 Derive the element characteristic equation for an interior element in a two-dimensional channel assuming linear triangular elements. Consider the flow as unsteady Stokes flow with governing equations given by Equations 14.45a to 14.45c.

Appendix

A

REVIEW OF VECTORS AND MATRICES

A.1 VECTORS

A.1.1 Symbols

A vector is written in terms of its components as

$$\tilde{a} = \hat{i}a_x + \hat{j}a_y + \hat{k}a_z \tag{A.1a}$$

where $\hat{i}$, $\hat{j}$, and $\hat{k}$ are unit vectors in x, y, and z directions respectively.

The unit vector in the direction of $\tilde{a}$ is given as

$$\hat{i}_a = \frac{\tilde{a}}{|\tilde{a}|} \tag{A.1b}$$

Another notation for the unit vectors in x, y, and z directions are given as n_x, n_y, and n_z, which are simply the components of a unit vector $\hat{n}$ normal to an arbitrary surface. Physically, these components are represented by the direction cosine of the unit vector $\hat{n}$ in x, y, and z directions, and written as

$$\hat{n}_x = \cos\theta_{xn}, \quad \hat{n}_y = \cos\theta_{yn}, \quad \text{and} \quad \hat{n}_z = \cos\theta_{zn}$$

A.1.2 Vector Norm

A ***norm*** is a single-valued quantity that represents the size of vectors or/and matrices. For example, for a vector $\{x\}$, the ***Euclidean norm*** is defined as

$$\|x\|_e = \sqrt{\sum_{i=1}^{n} x_i^2} = \sqrt{((x_1)^2 + (x_2)^2 + \cdots + (x_n))^2} \tag{A.2}$$

An alternative to the *Euclidean norm* is given in terms of the ***uniform vector norm***, which is defined as

$$\|x\|_\infty = \max_{1 \le i \le n} |x_i| \tag{A.3}$$

With a *uniform vector norm*, the element with the largest absolute value is used to represent the size of the vector.

A.1.3 Dot or Scalar Product

The dot product of two vectors $\tilde{a}$ and $\tilde{b}$ is defined as

$$\tilde{a}.\tilde{b} = |\tilde{a}||\tilde{b}| \cos\theta_{ab} \tag{A.4}$$

where θ_{ab} is the included angle between two vectors $\tilde{a}$ and $\tilde{b}$. Physically, a dot product represents the product of $|\tilde{a}|$ with the component of $\tilde{b}$ in the direction of $\tilde{a}$.

Example A.1: Volume flow rate through a differential area vector $\tilde{dA}$ is given by the product of the component of velocity $\tilde{V}$ normal to an area and expressed as

$$d\forall = \left|\tilde{V}\right| \cos\theta_{vdA} \left|\tilde{dA}\right| = \tilde{V} \cdot \tilde{dA} = \hat{V} \cdot \hat{n}\, dA \tag{A.5}$$

A.1.4 Cross or Vector Product or Curl

Cross product or curl of two vectors $\tilde{a}$ and $\tilde{b}$ is a vector $\tilde{c}$ and it is written as

$$\tilde{c} = \tilde{a} \times \tilde{b} \tag{A.6}$$

Physically, its magnitude is given as

$$\tilde{c} = \tilde{a} \times \tilde{b} = |\tilde{a}||\tilde{b}| \sin \theta_{ab} \tag{A.7}$$

and its sense is given by the right-hand rule, i.e., as $\tilde{a}$ is rotated into $\tilde{b}$, the vector $\tilde{c} = \tilde{a} \times \tilde{b}$ points vertically upward and perpendicular to both $\tilde{a}$ and $\tilde{b}$.

The cross product vector $\tilde{c} = \tilde{a} \times \tilde{b}$ is evaluated as:

$$\begin{aligned} \tilde{c} = \tilde{a} \times \tilde{b} &= \begin{vmatrix} \hat{i} & \hat{j} & \hat{k} \\ a_x & a_y & a_z \\ b_x & b_y & b_z \end{vmatrix} \\ &= \hat{i}(a_y b_z - a_z b_y) - \hat{j}(a_x b_z - a_z b_x) + \hat{k}(a_x b_y - a_y b_x) \end{aligned} \tag{A.8}$$

Example A.2: Torque given by a force vector $\tilde{F}$ acting at a distance $\tilde{r}$ is given as

$$\tilde{T} = |\tilde{r}||\tilde{F}| \sin \theta_{Fr} = \tilde{r} \times \tilde{F} \tag{A.9}$$

A.1.5 Vector Operator

Del, ∇

Cartesian coordinate

$$\nabla = \hat{i}\frac{\partial}{\partial x} + \hat{j}\frac{\partial}{\partial y} + \hat{k}\frac{\partial}{\partial z} \tag{A.10a}$$

or

$$\nabla = n_x \frac{\partial}{\partial x} + n_y \frac{\partial}{\partial y} + n_z \frac{\partial}{\partial z} = \hat{n} \cdot \nabla \tag{A.10b}$$

Cylindrical coordinate

$$\nabla = \hat{i}_r \frac{\partial}{\partial r} + \hat{i}_\theta \frac{1}{r}\frac{\partial}{\partial \theta} + \hat{i}_z \frac{\partial}{\partial z} \tag{A.10c}$$

Gradient Gradient is the operation of ∇ on a differentiable scalar function such as pressure or temperature. For example, the gradient of a scalar field ϕ is written as

$$\nabla \phi = \hat{i}\frac{\partial \phi}{\partial x} + \hat{j}\frac{\partial \phi}{\partial y} + \hat{k}\frac{\partial \phi}{\partial z} \tag{A.11}$$

Divergence Divergence is the dot product of ∇ and a vector function such as velocity field. For example, the divergence of the velocity field is given as

$$\begin{aligned}\nabla.\vec{V} &= \left(\hat{i}\frac{\partial}{\partial x}+\hat{j}\frac{\partial}{\partial y}+\hat{k}\frac{\partial}{\partial z}\right)\cdot\left(\hat{i}u+\hat{j}v+\hat{k}w\right) \\ &= \left(\frac{\partial u}{\partial x}+\frac{\partial v}{\partial y}+\frac{\partial w}{\partial z}\right)\end{aligned} \tag{A.12}$$

A.1.6 Laplacian Operator

It is the dot product of two del operators.

Cartesian coordinate

$$\nabla\cdot\nabla = \nabla^2 = \frac{\partial^2}{\partial x^2}+\frac{\partial^2}{\partial y^2}+\frac{\partial^2}{\partial z^2} \tag{A.13a}$$

Laplacian of a function ϕ is given as

$$\nabla^2\phi = \frac{\partial^2\phi}{\partial x^2}+\frac{\partial^2\phi}{\partial y^2}+\frac{\partial^2\phi}{\partial z^2} \tag{A.13b}$$

Cylindrical coordinate

$$\nabla\cdot\nabla = \nabla^2 = \frac{\partial^2}{\partial r^2}+\frac{1}{r}\frac{\partial}{\partial r}+\frac{1}{r^2}\frac{\partial^2}{\partial\theta^2}+\frac{\partial^2}{\partial z^2} \tag{A.14a}$$

Laplacian of the function ϕ is

$$\nabla^2\phi = \frac{\partial^2\phi}{\partial r^2}+\frac{1}{r}\frac{\partial\phi}{\partial r}+\frac{1}{r^2}\frac{\partial^2\phi}{\partial\theta^2}+\frac{\partial^2\phi}{\partial z^2} \tag{A.14b}$$

A.2 MATRICES

A.2.1 Coefficient Matrix

A coefficient matrix is a matrix whose elements are the coefficients of the unknowns in a system of equations. For example, the coefficient matrix for the system of equations given by Equation 3.1 is

$$A = \begin{bmatrix} a_{11} & a_{12} & \cdots & \cdots & \cdots & a_{1n} \\ a_{21} & a_{22} & \cdots & \cdots & \cdots & a_{2n} \\ \vdots & & & & & \\ \vdots & & & & & \\ \vdots & & & & & \\ a_{n1} & a_{n2} & \cdots & \cdots & \cdots & a_{nn} \end{bmatrix} \tag{A.15}$$

The elements of the matrix **A** are specified a_{ij}, where index i represents row number and varies from 1, 2, . . . , n. The index j represents column number and varies from 1, 2. . . , n.

If matrix A is symmetric, then

$$a_{ij} = a_{ji} \tag{A.16}$$

A.2.2 Augmented Matrix

Augmented matrix is formed by adding the constant vector of the system to the coefficient matrix as a column of elements

$$\mathrm{A} = \begin{bmatrix} a_{11} & a_{11} & & a_{11} & c_1 \\ a_{21} & a_{21} & & a_{21} & c_2 \\ \vdots & & & & \\ \vdots & & & & \\ \vdots & & & & \\ a_{n1} & a_{n1} & & a_{n1} & c_n \end{bmatrix}$$

$$= \begin{bmatrix} a_{11} & a_{11} & & a_{11} & a_{1m} \\ a_{21} & a_{21} & & a_{21} & a_{2m} \\ \vdots & & & & \\ \vdots & & & & \\ \vdots & & & & \\ a_{n1} & a_{n2} & & a_{nn} & c_{nm} \end{bmatrix} = a_{i,j} \tag{A.17}$$

where index i represents the row number and varies from 1, 2, . . . , n. The index j represents the column number and varies from 1, 2, . . . , m, and $m = n + 1$, respectively.

A ***square matrix*** is a matrix where the number of rows is the same as the number of columns, i.e., $m = n$. A ***diagonal matrix*** is a square matrix where all elements off the main diagonal are equal to zero.

$$\begin{bmatrix} a_{11} & & & & & \\ & a_{22} & & & & \\ & & a_{33} & & & \\ & & & \ddots & & \\ & & & & \ddots & \\ & & & & & a_{nn} \end{bmatrix} \tag{A.18}$$

Note that all zero elements are left blank.

A.2.3 Lower Triangular Matrix

A matrix with all elements above the main diagonal are zero. For example

$$L = A = \begin{bmatrix} a_{11} & & & & & \\ a_{21} & a_{22} & & & & \\ a_{31} & a_{32} & a_{33} & & & \\ \vdots & & & \ddots & & \\ \vdots & & & & \ddots & \\ a_{n1} & a_{n2} & a_{n3} & \cdots & \cdots & a_{nn} \end{bmatrix} \tag{A.19}$$

A.2.4 Upper Triangular Matrix

A matrix with all elements below the main diagonal are zero.

$$U = A = \begin{bmatrix} a_{11} & a_{12} & a_{13} & \cdots & \cdots & a_{1n} \\ & a_{22} & a_{23} & \cdots & \cdots & a_{2n} \\ & & a_{33} & & & a_{3n} \\ & & & \ddots & & \vdots \\ & & & & \ddots & \vdots \\ & & & & & a_{nn} \end{bmatrix} \tag{A.20}$$

A.2.5 Banded Matrix

A matrix that has all elements equal to zero with the exception of a band centered around the main diagonal. For example

$$A = \begin{bmatrix} a_{11} & a_{12} & a_{12} & & & \\ a_{21} & a_{22} & a_{23} & a_{24} & & \\ a_{31} & a_{32} & a_{33} & a_{34} & \ddots & \\ & a_{42} & a_{43} & \ddots & \ddots & \\ & & \ddots & \ddots & \ddots & a_{n-1,n} \\ & & & a_{n,n-2} & a_{n,n-1} & a_{nn} \end{bmatrix} \tag{A.21}$$

A.2.6 Tridiagonal Matrix

A tridiagonal matrix is a banded matrix with a bandwidth of three. For example

$$A = \begin{bmatrix} a_{11} & a_{12} & & & & \\ a_{21} & a_{22} & a_{23} & & & \\ & a_{32} & a_{33} & a_{34} & & \\ & & a_{43} & \ddots & \ddots & \\ & & & \ddots & \ddots & a_{n-1,n} \\ & & & & a_{n,n-1} & a_{nn} \end{bmatrix} \tag{A.22}$$

A.2.7 Equality of Matrices

$$A = B, \quad \text{if } a_{ij} = b_{ij}, \quad i = 1, 2, \ldots, n; \quad j = 1, 2, \ldots, n$$

A.2.8 Summation of Matrices

$$A + B = C, \quad \text{if } a_{ij} + b_{ij} = c_{ij}, \quad i = 1, 2, \ldots, n; \quad j = 1, 2, \ldots, n$$

A.2.9 Difference of Matrices

$$A - B = C, \quad \text{if } a_{ij} - b_{ij} = c_{ij}, \quad i = 1, 2 \ldots n; \quad j = 1, 2, \ldots n$$

A.2.10 Matrix Multiplication

Product of two matrices $A = [a_{ij}]$ and $B = [b_{ij}]$ is written as

$$[a_{ij}][b_{ij}] = [c_{ij}] \quad \text{(A.23)}$$

or

$$AB = C$$

where the elements of the product matrix are given by

$$c_{ij} = \sum_{k=1}^{n} a_{ik} b_{kj} \quad \text{(A.24)}$$

A.2.11 Properties of Matrix Multiplication

1. Matrix multiplication is not commutative, i.e., $AB \neq BA$.
2. Product of two matrices A and B is defined if the number of columns in A is equal to number of rows in B.
3. Multiplication of a matrix A $(n \times n)$ by a column matrix B $(n \times 1)$ results in a column matrix C $(n \times 1)$.
4. Matrix multiplication is associative. For example,

$$ABC = (AB)C = A(BC)$$

5. Premultiplication AB by C means

$$CAB = D$$

6. Postmultiplication AB by C means

$$ABC = E$$

A.2.12 Identity Matrix or Unit Matrix

It is a matrix with 1s on the main diagonal and all off diagonal elements as zero. So, basically it is diagonal matrix where all elements on the main diagonal are 1s.

$$\mathrm{I} = \begin{bmatrix} 1 & & & & & \\ & 1 & & & & \\ & & 1 & & & \\ & & & \ddots & & \\ & & & & \ddots & \\ & & & & & 1 \end{bmatrix} \quad \text{(A.25)}$$

It gives matrix properties

$$IA = AI = A \tag{A.26}$$

A.2.13 Zero Matrix

A matrix with all elements as zero. For example

$$\begin{bmatrix} 0 & 0 & 0 & \cdots & \cdots & 0 \\ 0 & 0 & 0 & \cdots & \cdots & 0 \\ 0 & 0 & 0 & \cdots & \cdots & 0 \\ \vdots & & & \ddots & & \\ \vdots & & & & \ddots & \\ 0 & 0 & 0 & \cdots & \cdots & 0 \end{bmatrix} \tag{A.27}$$

A.2.14 Transpose of a Matrix

Transpose of a matrix is obtained by interchanging the row and columns. For example the transpose of the matrix A as given by Equation (A.1) is

$$A^{\mathrm{T}} = \begin{bmatrix} a_{11} & a_{21} & \cdots & \cdots & \cdots & a_{n1} \\ a_{12} & a_{22} & \cdots & \cdots & \cdots & a_{n2} \\ \vdots & & & & & \\ \vdots & & & & & \\ \vdots & & & & & \\ a_{1n} & a_{2n} & \cdots & \cdots & \cdots & a_{nn} \end{bmatrix} \tag{A.28}$$

If A is a symmetric matrix, then

$$A = A^{\mathrm{T}}$$

One important property of transpose can be demonstrated as

$$[BA]^{\mathrm{T}} = B^{\mathrm{T}} A^{\mathrm{T}} \tag{A.29}$$

A.2.15 Minor (M_{ij})

The minor of an element a_{ij} in a coefficient matrix $\boldsymbol{A}$ is defined as the determinant of the square array obtained by deleting the ith row and jth column. For example, the minor of the element a_{11} of the matrix

$$A = \begin{bmatrix} a_{11} & a_{12} & a_{13} \\ a_{21} & a_{22} & a_{23} \\ a_{31} & a_{32} & a_{33} \end{bmatrix} \tag{A.30a}$$

is

$$M_{11} = \begin{vmatrix} a_{22} & a_{23} \\ a_{32} & a_{33} \end{vmatrix} \tag{A.30b}$$

A.2.16 Co-factor (F_{ij})

The co-factor F_{ij} is defined as the result of changing the sign of the minor by $(-1)^{i+j}$ as

$$F_{ij} = (-1)^{i+j} M_{ij} \tag{A.31}$$

A.2.17 Determinant ($|A|$)

It is the sum of the products of each element by its co-factor in any row of the matrix. For example, the determinant of the matrix A is given as

$$|A| = \sum_{\substack{i=1 \\ j=1,2...n}} a_{ij} F_{ij} = \sum_{\substack{i=1 \\ j=1,2...n}} (-1)^{i+j} a_{ij} M_{ij} \tag{A.32}$$

Example A.3: Determinant of the 2 × 2 matrix

$$A = \begin{bmatrix} a_{11} & a_{12} \\ a_{21} & a_{22} \end{bmatrix}$$

is

$$|A| = \begin{vmatrix} a_{11} & a_{12} \\ a_{21} & a_{22} \end{vmatrix} = a_{11}a_{22} - a_{12}a_{21}$$

Determinant of the 3 × 3 matrix

$$A = \begin{bmatrix} a_{11} & a_{12} & a_{13} \\ a_{21} & a_{22} & a_{23} \\ a_{31} & a_{32} & a_{33} \end{bmatrix}$$

is

$$|A| = \begin{vmatrix} a_{11} & a_{12} & a_{13} \\ a_{21} & a_{22} & a_{23} \\ a_{31} & a_{32} & a_{33} \end{vmatrix}$$

Using equation (A.30)

$$|A| = a_{11}\begin{vmatrix} a_{22} & a_{23} \\ a_{32} & a_{33} \end{vmatrix} - a_{12}\begin{vmatrix} a_{21} & a_{23} \\ a_{31} & a_{33} \end{vmatrix} + a_{13}\begin{vmatrix} a_{21} & a_{22} \\ a_{31} & a_{32} \end{vmatrix}$$

$$|A| = a_{11}(a_{22}a_{33} - a_{23}a_{32}) - a_{12}(a_{21}a_{33} - a_{23}a_{31}) + a_{13}(a_{21}a_{32} - a_{22}a_{31})$$

A.2.18 Inverse of a Matrix

The inverse of a square matrix A is written as A^{-1} and it is defined by the relationship

$$A^{-1}A = AA^{-1} = I \tag{A.33}$$

The inverse of a matrix is determined by

$$A^{-1} = \frac{\text{adj}[A]}{|A|} \tag{A.34}$$

where adj[A] is defined as the adjoint of a matrix A and it is obtained by replacing each element a_{ij} of matrix by its co-factor F_{ij} and then taking a transpose of the resulting matrix. So, adjoint of a matrix is written as

$$Adj[A] = \begin{bmatrix} F_{11} & F_{12} & \cdots & \cdots & F_{1n} \\ F_{21} & F_{22} & & & F_{2n} \\ \vdots & & & & \\ \vdots & & & & \\ F_{n1} & F_{n2} & \cdots & \cdots & F_{nn} \end{bmatrix}^T = \begin{bmatrix} F_{11} & F_{21} & \cdots & \cdots & F_{n1} \\ F_{12} & F_{22} & & & F_{n2} \\ \vdots & & & & \\ \vdots & & & & \\ F_{1n} & F_{2n} & \cdots & \cdots & F_{nn} \end{bmatrix} \tag{A.35}$$

A.2.19 Orthogonal Matrix

A square matrix A is said to be orthogonal if

$$A^{\mathrm{T}} = A^{-1} \tag{A.36}$$

A.2.20 Dense Matrix

A matrix is termed as dense if it contains few zero elements and the size is relatively small.

A.2.21 Sparse Matrix

A large matrix with many zero elements is termed as sparse matrix.

A.2.22 Well-conditioned System

A system is termed as ***well-conditioned*** if a small relative change in one or more of the coefficients of a system of equations results in a small change in the solution.

A.2.23 Ill-conditioned System

A system is ill-conditioned if a small change in one or more of the coefficients results in a large change in the solution of the system.

A.2.24 Singular Matrix

A matrix is termed as singular if the determinant of the matrix $|A|$ is zero. A system that is close to singular poses considerable difficulties in resulting into a solution.

A.2.25 Quadratic Form

For a $n \times n$ matrix A and a $n \times 1$ vector, the scalar quantity $x^{\mathrm{T}}Ax$ is called a quadratic form. The quadratic form can be shown by expanding $x^{\mathrm{T}}Ax$ as:

$$
\begin{aligned}
x^{\mathrm{T}}Ax &= \left\{ x_1 \quad x_2 \quad x_3 \quad \cdots \quad x_n \right\} \begin{bmatrix} a_{11} & a_{12} & a_{13} & \cdots & a_{1n} \\ a_{21} & a_{22} & a_{23} & \cdots & a_{2n} \\ a_{31} & a_{32} & a_{33} & \cdots & a_{3n} \\ \vdots & & & & \\ a_{n1} & a_{n2} & a_{n3} & \cdots & a_{nn} \end{bmatrix} \begin{Bmatrix} x_1 \\ x_2 \\ x_3 \\ \vdots \\ x_n \end{Bmatrix} \\
&= x_1 a_{11} x_1 + x_1 a_{12} x_2 + x_1 a_{13} x_3 + \cdots + x_1 a_{1n} x_n \\
&\quad + x_2 a_{21} x_1 + x_2 a_{22} x_2 + x_2 a_{23} x_3 + \cdots + x_2 a_{2n} x_n \\
&\quad + x_3 a_{31} x_1 + x_3 a_{32} x_2 + x_3 a_{33} x_3 + \cdots + x_3 a_{3n} x_n \\
&\quad \vdots \\
&\quad + x_n a_{n1} x_1 + x_n a_{n2} x_2 + x_n a_{n3} x_3 + \cdots + x_n a_{nn} x_n
\end{aligned} \tag{A.37}
$$

A.2.26 Eigenvalues and Eigenvectors

For a nontrivial solution to exist for a homogeneous system of equations such as $Ax = c$, an ***eigenvalue problem*** is defined as

$$Ax = \lambda x \tag{A.38a}$$

and the matrix $A - \lambda I$ is singular, i.e.

$$|A - \lambda I| = 0 \tag{A.38b}$$

Equation (A.38b) is referred to as the characteristic equation and its n roots $\lambda_1, \lambda_2, \ldots, \lambda_n$ are the eigenvalues. For each eigenvalue λ_i, the corresponding eigenvector x is obtained from

$$(A - \lambda_i I)x = 0 \tag{A.39}$$

A.2.27 Positive Definite Matrix

A symmetric matrix is ***positive definite*** if all its eigenvalues are strictly positive. Alternatively, a symmetric $n \times n$ matrix A is positive definite if for a nonzero vector $n \times 1$ vector x

$$x^{\mathrm{T}} A x > 0 \tag{A.40}$$

A.2.28 Matrix Norm

In the case of a two-dimensional vector array, a ***matrix norm*** is a single-valued quantity that represents the size of a two-dimensional vector array or matrix array $a_{i,j}$. A ***Frobenius norm*** is defined as

$$\|a_{ij}\|_e = \sqrt{\sum_{i=1}^{n}\sum_{j=1}^{n}(a_{ij})^2} \tag{A.41}$$

a ***uniform matrix norm*** or ***row sum norm*** is defined as

$$\left\|a_{ij}\right\|_\infty = \max_{1 \le i \le n} \sum_{j=1}^{n} |a_{ij}| \tag{A.42}$$

A.2.29 Matrix Condition Number

The matrix condition number is used to assess the condition of a system and it is defined as

$$\mathrm{Cond}\,[A] = \|A\| \cdot \left\|A^{-1}\right\| \tag{A.43}$$

The condition number is equal to or greater than one. The greater the condition number of a system compared to one, the more ill-conditioned the system is.

Appendix

B

INTEGRAL THEOREMS

B.1 GRADIENT THEOREM

If $\forall$ is volume bounded by a surface A and ϕ is a scalar, then the gradient theorem states that

$$\iiint_{\forall} \nabla\phi d\forall = \iint_{A} \hat{n}\phi\, dA = \iint_{A} \phi \cdot \tilde{d}A \tag{B.1}$$

Similarly, for an area A bounded by a contour s

$$\iint_{A} \nabla\phi\, dA = \oint_{s} \hat{n}\phi\, ds \tag{B.2}$$

Basically, the gradient theorem changes the volume integral into a surface integral and the area integral into a line integral involving a scalar, and vice versa.

B.2 DIVERGENCE THEOREM

If $\forall$ is volume bounded by a surface A and $\tilde{\phi}$ is a vector, then, the divergence theorem states that

$$\iiint_{\forall} \nabla \cdot \tilde{\phi} d\forall = \iint_{A} \hat{n} \cdot \tilde{\phi}\, dA = \iint_{A} \tilde{\phi} \cdot \tilde{d}A \tag{B.3}$$

where dA is the differential area vector on the surface A.

Similarly, for an area A bounded by a contour s

$$\iint_A \nabla \cdot \tilde{\phi}\, dA = \oint_s \hat{n} \cdot \tilde{\phi}\, ds = \oint_s \tilde{\phi} \cdot \tilde{ds} \tag{B.4}$$

Basically, the divergence theorem changes the volume integral into a surface integral and the area integral into a line integral involving a vector, and vice versa.

BIBLIOGRAPHY

Abdel-Wahed, R. M., S. V. Patankar and E. M. Sparrow, *Letters in Heat and Mass Transfer*, **3**, 355–364, 1976.

Akai, T. J., *Applied Numerical Methods for Engineers*. John Wiley & Sons, New York, 1994.

Anderson, D. A., J. C. Tannehill and R. H. Pletcher, *Computational Fluid Mechanics and Heat Transfer*. McGraw-Hill, New York, 1984.

Angirasa, D., M. J. B. Pourquie and F. T. M. Nieuwstadt, Numerical study of transient and steady laminar buoyancy-driven flows and heat transfer in a square open cavity. *Journal of Numerical Heat Transfer, Part A*, **22**, 223–239, 1992.

Arpaci, V., *Conduction Heat Transfer*. Addison-Wesley, Menlo Park, CA, 1966.

Baker, A. J., *Finite Element Computational Fluid Mechanics*. Taylor & Francis, New York, 1991.

Baker, A. J. and D. W. Pepper, *Finite Elements 1–2–3*. McGraw-Hill, New York, 1991.

Batchelor, G. K., *Fluid Dynamics*. Cambridge University Press, New York, 1967.

Bathe, K.-J., *Finite Element Procedures in Engineering Analysis*. Prentice Hall, Englewood Cliffs, NJ, 1982.

Bathe, K.-J., *Finite Element Procedures*. Prentice Hall, Englewood Cliffs, NJ, 1996.

Bathe, K.-J. and E.L. Wilson, *Numerical Methods in Finite Element Analysis*. Prentice Hall, Englewood Cliffs, NJ, 1976.

Bayazittoglu, Y. and M. N. Ozisik, *Elements of Heat Transfer*. McGraw-Hill, New York, 1988.

Becker, M., *The Principles and Applications of Variational Methods*. MIT Press, Cambridge, MA, 1964.

Bejan, A., *Convection Heat Transfer*. John Wiley & Sons, New York, 1984.

Bejan, A., *Heat Transfer*. John Wiley & Sons, New York, 1993.

Benim, A. and W. Zinser, A segregated formulation of Navier–Stokes equations with finite elements, *Computer-Methods in Applied Mechanics and Engineering*, **57**, 223–237, 1986.

Bennett, C. O. and J. E. Myers, *Momentum, Heat, and Mass Transfer*. McGraw-Hill, New York, 1982.

Bickford. W. B., *A First Course in the Finite Element Method*. Eichard D. Irwin, Inc., Burr Ridge, IL, 1994.

Biot, M. A., *Variational Principles in Heat Transfer*.

Bird, R. B., W. E. Stewart and E. N. Lightfoot, *Transport Phenomena*, Second Edition. John Wiley & Sons, New York, 2002.

Bliss, G. A., *Calculus of Variations*. Mathematical Association of America, The Open Court Publishing Company, Lasalle, IL., 1925.

Bliss, G. A., *Lectures on the Calculus of Variations*. University of Chicago Press, Chicago, 1946.

Bolza, O., *Lectures on the Calculus of Variations*. Stechert-Hafner, Inc., New York, 1946.

Burden, R. L., J. D. Faires and A. C. Reynolds, *Numerical Analysis*. PWS Publishers, Boston, 1981.

Carslaw, H. S. and J. C. Jaeger, *Conduction of Heat in Solids*. Oxford University Press, New York, 1959.

Cengel, Y. A., Heat Transfer – A Practical Approach. McGraw-Hill, New York, 2003.

Chandrupatla, T. R. and Belegundu, A. D., *Introduction to Finite Elements in Engineering*. Prentice Hall, New York, 1997.

Chapra, S. C and R. P. Canale, *Numerical Methods for Engineers*. McGraw-Hill, New York, 2002.

Constantinides, A., *Applied Numerical Methods with Personal Computers*. McGraw-Hill, New York, 1987.

Courant, R. and Hilbert, D., *Methods of Mathematical Physics*, Vols. I and II. Interscience, New York, 1962.

Datta, B. N., *Numerical Linear Algebra and Applications*. Brooks/Cole Publishing Company, CA, 1995.

Desai, C. S., *Elementary Finite Element Method*. Prentice Hall, Inc., Englewood Cliffs, NJ, 1979.

Desai, C. S. and J. F. Abel, *Introduction to the Finite Element Method*. Van Nostrand Reinhold, New York, 1972.

Doormaal, J. P. van and G. D. Raithby, Enhancements of the SIMPLE method for predicting incompressible fluid flows. *Journal of Numerical Heat Transfer*, **7**, 147–163, 1984.

Eckert, E. R. G. and R. M. Drake, Jr., *Analysis of Heat and Mass Transfer*. McGraw-Hill, New York, 1972.

Elsgolc, E. L., Calculus of variations. Addison-Wesley, Reading, MA, 1962.

Ewing, G. M., *Calculus of Variations with Applications*. Dover Publications, New York, 1985.

Finlayson, B. A., *The Method of Weighted Residuals and Variational Principles*. Academic Press, New York, 1972.

Fletcher, A. J., Computational Techniques for Fluid Dynamics, Vol. I, 2nd ed. Springer-Verlag, 1991.

Foray, M. J., *Variational Calculus in Science and Engineering*. McGraw-Hill, New York, 1968.

Forsythe, G. E., M. A. Malcolm and C. B. Moler, *Computer Methods for Mathematical Computations*. Prentice Hall, Englewood Cliffs, NJ, 1977.

Fox, C., *An Introduction to the Calculus of Variations*. Oxford University Press, New York, 1950.

Fox, R. W. and A. T. McDonald, *Introduction to Fluid Mechanics*, Fifth Edition. John Wiley & Sons, New York, 1998.

Gallagher, R. H., *Finite Element Analysis Fundamentals*. Prentice Hall, Inc, Englewood Cliffs, NJ, 1975.

Gelfand, I. M. and S. V. Fomin, *Calculus of Variations*. Prentice Hall, Inc., Englewood Cliffs, NJ, 1963.

Gerald, C. F and P. O. Wheatley, *Applied Numerical Analysis*, Fifth Edition. Addison Wesley, New York, 1994.

Goldstein, S., Modern Developments in Fluid Dynamics. Dover, New York, 1965.

Golub, G. H. and C. F. Van Loan, *Matrix Computations*. Johns Hopkins University Press, Baltimore, MD, 1989.

Grandin, H., *Fundamentals of the Finite Element Method*. Macmillan Publishing Company, 866 Third Avenue, New York, 1986.

Harlow, F. H. and J. E. Welch, Numerical calculation of time-dependent viscous incompressible flow of fluid with free surface, *Journal of the Physics of Fluid*, **8**, 1965.

Haroutunian, V., M. S. Engelman and I. Hasbani, Segregated finite element algorithms for the numerical solution of large-scale incompressible flow problems, *International Journal for Numerical Methods in Fluids*, **17**, 323–348, 1993.

Hestenes, M. E., and E. Stiefel, Methods of conjugate gradients for solving linear systems, *J. Res. Nat. Bus Standards*, **49**, 1952.
Hilderbrand, F. B., *Methods of Applied Mathematics*. Prentice-Hall, Englewood Cliffs, NJ, 1965.
Hilderband, F. B., Introduction to Numerical Analysis, Second Edition. McGraw-Hill, New York, 1974.
Hirt, C. W and J. L. Cook, Calculating three-dimensional flows around structures and over rough terrain, *Journal of Computational Physics*, **10**, 324–340, 1972.
Holman, J. P., *Heat Transfer*, 8th Edition. McGraw-Hill, New York, 1997.
Howel, J. R and R. O. Buckius, *Fundamentals of Engineering Thermodynamics*. McGraw-Hill, New York, 1987.
Incropera, F. P. and D. P. DeWitt, *Fundamentals of Heat and Mass Transfer*, Fifth Edition. John Wiley & Sons, New York, 2002.
Jakob, M., *Heat Transfer*. John Wiley & Sons, New York, 1949.
Jaluria, Y. and K. E. Torrance, *Computational Heat Transfer*. Hemisphere, New York, 1986.
James, M. L., G.M. Smith and J. C. Wolford, *Applied Numerical Methods for Digital Computation*, Third Edition. Harper Collins Publishers, New York, 1985.
Kays, W. M. and M. E. Crawford, *Convective Heat and Mass Transfer*. McGraw-Hill, New York, 1993.
Keenan, J. H., *Thermodynamics*. The MIT Press, Cambridge, MA, 1970.
Kincaid, D. and W. Cheney, *Numerical Analysis*. Brooks/Cole Publishing, CA, 1991.
Kreith, F. K. and M. S. Bohn, *Principles of Heat Transfer*, Sixth Edition. Brooks/Cole, Pacific Grove, CA, 2001.
Lamb, H., *Hydrodynamics*. Cambridge University Press, New York, 1945.
Lanczos, C., *The Variational Principles of Mechanics*. The University of Toronto Press, Toronto, 1964.
Leipholz, H., *Direct Variational Methods and Eigenvalue Problems in Engineering*. Nordholf International Publishing, Leyden, Netherlands, 1977.
Lippmann, H., *Extremum and Variational Principles in Mechanics*. Springer-Verlag, New York, 1972.
Logan, D. L., *A First Course in the Finite Element Method*. Brooks/Cole Publishers, Pacific Grove, CA, 2002.
Majumdar, P. and P. Deb, Computational analysis of turbulent fluid flow and heat transfer over an array of heated modules using turbulence models. *Journal of Numerical Heat Transfer*, **43**, 669–692, 2003.
Mikhlin, S. G., *Variational Methods in Mathematical Physics*. Pergamon Press, New York, 1964.
Mills, A. F., *Basic Heat and Mass Transfer*, Second Edition. Prentice Hall, Englewood Cliffs, NJ, 1999.
Mills, A. F., *Mass Transfer*. Prentice Hall, New Jersey, 2001.
Minkowycz, W. J., E. M. Sparrow, G. E. Schneider and R. H. Pletcher, *Handbook of Numerical Heat Transfer*. John Wiley & Sons, New York, 1988.
Mitchell, A. R., *Computational Methods in Partial Differential Equations*. Wiley, New York, 1969.
Montgomery, S. B., and P. Wibulswas, Laminar flow and heat transfer in ducts of rectangular cross-section. *Proceedings of the International Heat Transfer Conference*, Vol. 1, pp. 104–112, 1966.
Moran, M. J. and H. N. Shapiro, *Fundamentals of Engineering Thermodynamics*. John Wiley & Sons, New York, 1993.
Morse, R. M and H. Feshbach, *Methods of Theoretical Physics*. McGraw-Hill, New York, 1953.
Myers, G. E., *Analytical Methods in Conduction Heat Transfer*. AMCHT Publications, Madison, Wisconsin, 1998.
Noye, B. J., *Numerical Solution of Differential Equations*. North-Holland, Amsterdam, 1983.
Ortega, J. M. and W. G. Poole, *An Introduction to Numerical Methods for Differential Equations*. Pitman Publishers Inc., MA, 1981.
Ozisik, M. N., *Heat Conduction*. John Wiley & Sons, New York, 1980.

Panton, R. L., *Incompressible Flow*. John Wiley & Sons, New York, 1984.

Patankar, S. V., *Numerical Heat Transfer and Fluid Flow*, Taylor & Francis, Washington D.C, 1980.

Patankar, S. V., Parabolic systems: finite difference method – I, *Handbook of Numerical Heat Transfer*, Edited by Minkowycz, W. J., E. M. Sparrow, G. E. Schneider, and R. H. Pletcher, Chapter 2, pp. 89–115. John Wiley & Sons, New York, 1988.

Patankar, S. V., Elliptic systems: finite difference method – I, *Handbook of Numerical Heat Transfer*, Edited by Minkowycz, W. J., E. M. Sparrow, G. E. Schneider, and R. H. Pletcher, Chapter 6, pp. 215–240. John Wiley & Sons, New York, 1988.

Patankar, S. V., *Computation of Conduction and Duct Flow Heat Transfer*. Innovative Research, Inc. Maple Grove 1991, New York, 1991.

Patankar, S. V. and D. B. Spalding, A calculation procedure for heat, mass and momentum transfer in three-dimensional parabolic flows. *International Journal of Heat and Mass Transfer*, **15**, 1787–1806, 1972.

Payvar, P. and Majumdar, P., Developing flow and heat transfer in a rectangular duct with a moving wall. *Journal of Numerical Heat Transfer, Part A*, **26**, 17–30, 1994.

Peaceman, D. W. and H. H., Rachford Numerical solution of parabolic and elliptic differential equations. *J. Soc. Ind. Appl. Math.*, **3**, 28–41, 1955.

Pepper, D. W. and J. C. Heinrich, *The Finite Element Method – Basic Concepts and Applications*. Hemisphere Publishing Corp, 1992.

Plybon, B. F., *An Introduction to Applied Numerical Analysis*. PWS-Kent Publishing, Boston, 1992.

Poulikakos, D., *Conduction Heat Transfer*. Prentice Hall, Englewood Cliffs, NJ, 1994.

Rao, S. S., *The Finite Element Method in Engineering*. Butterworth and Heinemann, Boston, 1999.

Rao, S. S., *Applied Numerical Methods for Engineers and Scientists*. Prentice Hall, Englewood Cliffs, NJ, 2002.

Recktenwald, G., *Numerical Methods with MatLab – Implementation and Application*. Prentice Hall, Englewood Cliffs, NJ, 2000.

Reddy, J. N., *Energy and Variational Methods in Applied Mechanics*. John Wiley & Sons, New York, 1984.

Reddy, J. N., *Applied Functional Analysis and Variational Methods in Engineering*. McGraw-Hill, New York, 1986.

Reddy, J. N, *An Introduction to the Finite Element Method*, Second Edition. McGraw-Hill, New York, 1994.

Reddy, J. N and D. K. Gartling, *The Finite Element Method in Heat Transfer and Fluid Dynamics*. CRC Press, Boca Raton, Florida, 1994.

Rektorys, K., *Variational Methods in Mathematics, Science and Engineering*. Reidel, Boston, 1977.

Roaches, P. J., *Computational Fluid Dynamics*. Hermosa Publishers, Albuquerque, NM, 1982.

Robertson, J. A. and C. T. Crowe, *Engineering Fluid Mechanics*. John Wiley & Sons, New York, 1997.

Saad, Y., *Iterative Methods for Sparse Linear Systems*. PWS Publishers, Boston, 1996.

Schilling, R. J. and S. L. Harris, *Applied Numerical Methods for Engineers Using MATLAB* and C. Brooks/Cole Publishing, CA, 1999.

Schlichting, H, *Boundary Layer Theory*. McGraw-Hill, New York, 1955.

Schlichting, H. and K. Gersten, *Boundary Layer Theory*. Springer-Verlag, New York, 2000

Schechter, R. S., *The Variational Methods in Engineering*. McGraw-Hill, New York, 1967.

Segerlind, L. J., *Applied Finite Element Analysis*. John Wiley & Sons, New York, 1984.

Sherman, F. S., *Viscous Flow*. McGraw-Hill, New York, 1990.

Sherwood, T. K., R. L. Pigford and C. R. Wilke, *Mass Transfer*. McGraw-Hill, New York, 1975.

Shih, T. M., *Numerical Heat Transfer*. Hemisphere Publishing Corp., New York, 1984.

Siegel, R., and J. R. Howell, *Thermal Radiation Heat Transfer*. Hemisphere Publishing Corp., New York, 1992.

Skeel, R. D. and J. B. Keiper, *Elementary Numerical Computing with Mathematica*. McGraw-Hill, New York, 1993.

Smith, G. D., *Numerical Solution of Partial Differential Equation.* Oxford University Press, London, 1978.

Sonntag, R. E., C. Borgnakke and G. J. Van Wylen, *Fundamentals of Thermodynamics.* John Wiley & Sons, New York, 1998.

Sparrow, E. M. and R. D. Cess, *Radiation Heat Transfer.* McGraw-Hill, New York, 1978.

Stakgold, I., *Green's Functions and Boundary Value Problems.* John Wiley & Sons, New York, 1979.

Thomas, L. C., *Heat Transfer.* Prentice-Hall, Inc, Englewood Cliffs, NJ, 1992.

Thompson, P. A., *Compressible Fluid Dynamics.* McGraw-Hill, New York, 1972.

Treybal, R.E., *Mass Transfer Operations*, Third Edition. McGraw-Hill, New York, 1980.

Tzou, D. Y., *Macro- to Microscale Heat Transfer*, Taylor & Francis, Washington, 1997.

Van Doormaal, J. P. and G. D. Rathby, Enhancements of the SIMPLE method for predicting incompressible fluid flows, *Journal of Numerical Heat Transfer*, **7**, 147–163, 1984.

Van Loan, C. F., *Introduction to Scientific Computing: A Matrix-Vector Approach using Matlab.* Prentice-Hall, Inc, Englewood Cliffs, NJ, 1997.

Weinstock, R., *Calculus of Variation with Applications to Physics and Engineering.* McGraw-Hill, New York, 1952.

White, F. M., *Heat and Mass Transfer.* Addison-Wesley, New York, 1988.

White, F. M., *Viscous Fluid Flow.* McGraw-Hill, New York, 1991.

Yuan, S. W., *Foundations of Fluid Mechanics.* Prentice-Hall, Englewood Cliffs, New Jersey, 1967.

Zienkiewicz, O. C. and R. L. Taylor, *The Finite Element Method*, Vol. 1. McGraw-Hill, New York, 1986.

Zienkiewicz, O. C. and R. L. Taylor, *The Finite Element Method*, Vol. 2. McGraw-Hill, New York, 1986.

Zienkiewicz, O. C. *The Finite Element Method in Engineering Science.* McGraw-Hill, London, 1971.

INDEX